# Advanced Transport Phenome.

The term "transport phenomena" describes the fundamental processes of momentum, energy, and mass transfer. This text provides a thorough discussion of transport phenomena, laying the foundation for understanding a wide variety of operations used by engineers.

The book is arranged in three parallel parts covering the major topics of momentum, energy, and mass transfer. Each part begins with the theory, followed by illustrations of the way the theory can be used to obtain fairly complete solutions, and concludes with the four most common types of averaging used to obtain approximate solutions. A broad range of technologically important examples, as well as numerous exercises, are provided throughout the text. Based on the author's extensive teaching experience, a suggested lecture outline is also included.

This book is intended for first-year graduate engineering students; it will be an equally useful reference for researchers in this field.

John C. Slattery is the Jack E. and Frances Brown Chair of Engineering at Texas A&M University.

# CAMBRIDGE SERIES IN CHEMICAL ENGINEERING

*Books in the Series*

E. L. Cussler, *Diffusion: Mass Transfer in Fluid Systems, second edition*

Liang-Shih Fan and Chao Zhu, *Principles of Gas–Solid Flows*

Hasan Orbey and Stanley I. Sandler, *Modeling Vapor–Liquid Equilibria: Cubic Equations of State and Their Mixing Rules*

T. Michael Duncan and Jeffrey A. Reimer, *Chemical Engineering Design and Analysis: An Introduction*

John C. Slattery, *Advanced Transport Phenomena*

A. Varma, M. Morbidelli and H. Wu, *Parametric Sensitivity in Chemical Systems*

# Advanced Transport Phenomena

**John C. Slattery**

CAMBRIDGE
UNIVERSITY PRESS

PUBLISHED BY THE PRESS SYNDICATE OF THE UNIVERSITY OF CAMBRIDGE
The Pitt Building, Trumpington Street, Cambridge, United Kingdom

CAMBRIDGE UNIVERSITY PRESS
The Edinburgh Building, Cambridge CB2 2RU, UK          http: //www.cup.cam.ac.uk
40 West 20th Street, New York, NY 10011-4211, USA     http: //www.cup.org
10 Stamford Road, Oakleigh, Melbourne 3166, Australia

First published 1999

Printed in the United States of America

Typeset in Gill Sans and Times Roman LaTeX $2_\varepsilon$ [TB]

*A catalog record for this book is available from the British Library*

*Library of Congress Cataloging-in-Publication Data*

Slattery, John Charles, 1932–
    Advanced transport phenomena / John C. Slattery.
        p.   cm. – (Cambridge series in chemical engineering)
    ISBN 0-521-63203-X (hb). – ISBN 0-521-63565-9 (pb)
    1. Transport theory.   2. Chemical engineering.   I. Title.
    II. Series.
    TP156.T7S57   1999
    660′.2842 – dc21                                        98-44872
                                                              CIP

ISBN 0 521 63203X hardback
ISBN 0 521 635659  paperback

# Contents

# Preface

*Transport phenomena* is the term popularized, if not coined, by Bird, Stewart, and Lightfoot (1960) to describe momentum, energy, and mass transfer. This book is based upon roughly thirty-five years of teaching first-year graduate students, first at Northwestern University and later at Texas A&M University. It draws upon *Momentum, Energy, and Mass Transfer in Continua* (Slattery 1981), the first edition of which was written at an early stage of my career, as well as *Interfacial Transport Phenomena* (Slattery 1990), which is intended for a much more specialized audience.

As you examine the table of contents, you will observe that, superimposed upon the three major themes of momentum, energy, and mass transfer, there are three minor themes. I begin each of these blocks of material with a discussion of the theory that we are to use: Chapters 1, 2, 5, and 8. This is followed by illustrations of the various ways in which the theory can be used to obtain fairly complete solutions: Chapters 3, 6, and 9. Finally, I conclude by examining the four most common types of averaging that are used to obtain approximate solutions: Chapters 4, 7, and 10.

This book is intended both as a textbook for first-year graduate students and as a self-study text for students and researchers interested in the field. The book is relatively long, in order to accommodate a variety of course syllabi as well as a range of self-study interests.

In spite of its length, you easily will find topics that I have left out. For example, I could have included more on the mathematical techniques that are used in solving various transport phenomena problems. Recognizing that mathematics is the language most naturally used to explain transport phenomena, I have included an appendix outlining tensor analysis. But the primary purpose of this text is to discuss transport phenomena, not mathematics.

In this context, I encourage my students to use software such as Mathematica (1996). I have found its symbolic capabilities particularly useful in doing perturbation analyses. More generally, I have employed it in developing numerical solutions of transcendental equations and nonlinear ordinary differential equations. Continuing with the concept that the primary purpose of this text is to discuss transport phenomena, I have not included as part of the text the details of my Mathematica notebooks.

In the lecture outline that follows, you will notice that I do not cover the entire book in my own lectures. Nor do I begin with Appendix A before proceeding to Chapters 1 and 2. Although it sounds sensible to begin with a thorough grounding in the fundamentals of tensor analysis, my experience is that this is relatively difficult for students. They don't understand initially why all of the mathematical apparatus is required. Instead, I begin with Chapters 1 and 2, referring back to Appendix A as the various ideas are needed. For example, it is much easier to introduce students to (second-order) tensors when they understand why this structure is required.

On a more personal note, I would like to express my gratitude to the many graduate students with whom I have discussed this material both at Northwestern University and at Texas A&M University. I have listened to your questions and comments, and I have done my best to take them into account. I am particularly grateful to a smaller group of activist students at both universities, my own students as well as the students of others, who have helped me develop new material for the book. There are too many to name here, but I have done my best to reference your papers and your suggestions in the text. You have made teaching exciting.

Special thanks are due to Professor Pierce Cantrell (Department of Electrical Engineering, Texas A&M University) who made this book possible. He suggested that I use LaTeX, built a bigger version on his faster machine, created special symbols, showed me how to introduce computer modern fonts in the figures, and installed a high-speed link between my home and my office.

Finally, I would also like to thank Christine Sanchez, who assisted me in converting my notes to LaTeX, Peter Weiss, who created many of the drawings, and Professor Nelson H. F. Beebe, who allowed me to use a prepublication version of his author index (Beebe 1998) for LaTeX.

*May 12, 1997*

## Sample Lecture Outline

What follows is a sample outline for 28 weeks of lectures (three hours per week of lectures and at least one hour per week of problem discussion) over two semesters. There is some change from year to year for several reasons. I attempt to adjust the pace of the class to the response of the students. In discussing Chapters 3, 6, and 9, I usually base my lectures on exercises rather than the examples worked out in the text. Finally, I am always looking for new material. Some instructors will wish to consider problems resulting in numerical solution of partial differential equations as software appropriate for a busy class becomes available.

In examining this outline, keep in mind that I do not cover all sections with the same intensity. In some cases, I am trying to emphasize only the concept and not the details. In other cases, my students may have seen similar material in other classes.

## First Semester

| Week | Tentative reading assignment |
| --- | --- |
| 1 | Introduction, 1.1, 1.3.1, 1.3.2, Appendix B |
| 2 | 1.3.3 through 1.3.7, A.1.1 through A.1.7, A.2.1, A.3.1, A.4.1 |
| 3 | Chapter 2 through 2.2.3, A.5.1 through A.5.4, A.6.1, A.8.1, A.11.1, A.11.2 |
| 4 | 2.2.4, 2.3.1 through 2.3.3 |
| 5 | 3.1, 3.2.1, first exam |
| 6 | 3.2.2, 3.2.3 |
| 7 | 3.2.4 |
| 8 | 3.2.5 |
| 9 | 3.3 through 3.3.1, second exam |
| 10 | 3.3.4, 3.4.1 |
| 11 | 3.4.2, 3.4.3 |
| 12 | 3.5.1, 3.5.2, third exam |
| 13 | 3.5.3 through 3.5.6 |
| 14 | 3.6.1 |

## Second semester

| Week | Tentative reading assignment |
| --- | --- |
| 1 | 5.1.1 through 5.1.3 |
| 2 | 5.2.1 through 5.4.1, 6.1, 6.2 through 6.2.3 |
| 3 | 6.3 through 6.3.1 |
| 4 | 6.3.2 |
| 5 | 6.3.3, 6.4.1 |
| 6 | 6.5.1, 6.6 through 6.6.1, 6.7.1, first exam |
| 7 | 6.7.2 through 6.7.6, 6.8.1, 6.8.2 |
| 8 | 8.2.1, 8.2.2, 8.3, 8.4, 9.2 |
| 9 | 9.3.1 |
| 10 | 9.3.2 |
| 11 | 9.3.3, 9.3.4 |
| 12 | 9.3.6, 9.3.7, second exam |
| 13 | 9.4.1, 9.4.2 |
| 14 | 9.4.3 |

# List of Notation

| | |
|---|---|
| $a_{(A)}$ | Relative activity on a mass basis of species $A$ defined by (8.4.5-5). |
| $a_{(A)}^{(m)}$ | Relative activity on a molar basis of species $A$ defined in Exercise 8.4.4-1. |
| $\hat{A}$ | Helmholtz free energy per unit mass. See, for example, Exercises 5.3.2-2 and 8.4.2-2. |
| $\mathbf{A}$ | Angular velocity tensor of starred frame with respect to unstarred frame defined by (1.2.2-8). |
| $c$ | Total molar density defined in Table 8.5.1-1. |
| $\hat{c}$ | Heat capacity per unit mass for an incompressible material. |
| $c_{(A)}$ | Molar density of species $A$, in moles per unit volume. |
| $\hat{c}_P$ | Heat capacity per unit mass at constant pressure defined in Exercises 5.3.2-4 and 8.4.2-7. |
| $\hat{c}_V$ | Heat capacity per unit mass at constant volume defined in Exercises 5.3.2-4 and 8.4.2-7. |
| $\mathbf{C}_t$ | Right relative Cauchy–Green strain tensor defined by (2.3.4-3). |
| $\tilde{C}_{(AB)}$ | Defined by (8.4.4-23) |
| $\mathbf{d}_{(A)}$ | Defined by (8.4.3-4). |
| $\mathbf{D}$ | Rate of deformation tensor defined by (2.3.2-3). |
| $\mathcal{D}_{(AB)}$ | Maxwell–Stefan diffusion coefficient introduced in (8.4.4-9). For dilute gas mixtures, this is known as the binary diffusion coefficient. |
| $\mathcal{D}_{(AB)}^0$ | Defined in Table 8.5.1-7 |
| $\tilde{D}_{(AB)}$ | Curtiss diffusion coefficient defined in (8.4.4-1) |
| $\hat{D}_{(AB)}$ | Defined by (8.4.4-18) |
| $dA$ | Indicates that a surface integration is to be performed. |
| $dV$ | Indicates that a volume integration is to be performed. |
| $\mathbf{e}$ | Thermal energy flux vector introduced in (5.2.3-5) and (8.3.5-4). See also (5.3.1-25) and (8.4.1-24). |
| $\mathbf{e}_i$ | Rectangular Cartesian basis vector introduced in Section A.1.5. |
| $\mathbf{f}$ | Sum of the external and mutual body force per unit mass. See the introduction to Section 2.1. |
| $\mathbf{f}_{(A)}$ | External or mutual force acting on species $A$. |
| $\mathbf{F}$ | Deformation gradient introduced in Exercise 2.3.2-1. |
| $\mathbf{F}_t$ | Relative deformation gradient introduced in (2.3.4-2). |

| | |
|---|---|
| $g$ | Acceleration of gravity. Also see Exercises A.4.1-4 and A.4.1-5. |
| $g_{ij}$ | Defined by (A.4.1-10). |
| $g^{ij}$ | Defined by (A.4.2-2) and (A.4.2-5). |
| $\mathbf{g}_i$ | Defined by (A.4.1-4). |
| $\hat{G}$ | Gibbs free energy per unit mass. See, for example, Exercises 5.3.2-2 and 8.4.2-2. |
| $H$ | Mean curvature. See Tables 2.4.3-1 through 2.4.3-8. |
| $\hat{H}$ | Enthalpy per unit mass. See, for example, Exercises 5.3.2-2 and 8.4.2-2. |
| $\mathbf{j}_{(A)}$ | Mass flux relative to $\mathbf{v}$ as defined in Table 8.5.1-4. |
| $\mathbf{j}_{(A)}^{\diamond}$ | Mass flux relative to $\mathbf{v}^{\diamond}$ as defined in Table 8.5.1-4. |
| $\mathbf{J}_{(A)}$ | Molar flux relative to $\mathbf{v}$ as defined in Table 8.5.1-4. |
| $\mathbf{J}_{(A)}^{\diamond}$ | Molar flux relative to $\mathbf{v}^{\diamond}$ as defined in Table 8.5.1-4. |
| $M$ | Molar-averaged molecular weight defined in Table 8.5.1-1. |
| $M_{(A)}$ | Molecular weight of species $A$. |
| $\mathbf{n}$ | Outwardly directed unit normal to a closed surface. |
| $\mathbf{n}_{(A)}$ | Mass flux of species $A$ relative to a fixed frame of reference as defined in Table 8.5.1-4. |
| $\mathbf{N}_{(A)}$ | Molar flux of species $A$ relative to a fixed frame of reference as defined in Table 8.5.1-4. |
| $N_{Br}$ | Brinkman number as in (6.4.0-5). |
| $N_{Da}$ | Damköhler number as in Exercise 9.6.2-10. |
| $N_{Fr}$ | Froude number as in (6.4.0-5). |
| $N_{Kn}$ | Knudsen number as in (10.3.2-1). |
| $N_{Pe}$ | Peclet number as in (6.4.0-5). |
| $N_{Pe,m}$ | Peclet number for mass transfer as in (9.2.0-8). |
| $N_{Pr}$ | Prandtl number as in (6.4.0-5). |
| $N_{Re}$ | Reynolds number as in (6.4.0-5). |
| $N_{Ru}$ | Ruark number as in (6.4.0-5). |
| $N_{Sc}$ | Schmidt number as in (9.2.0-8). |
| $N_{St}$ | Strouhal number as in (6.4.0-5). |
| $p$ | Mean pressure defined by (2.3.2-26). |
| $\mathbf{p}$ | Position vector introduced in Section A.1.2. |
| $P$ | Thermodynamic pressure. See (2.3.2-21), (5.3.1-30), and (8.4.1-28). |
| $\mathcal{P}$ | Modified pressure for an incompressible fluid defined by (2.4.1-5). |
| $\mathbf{q}$ | Energy flux vector introduced in (5.1.3-5). |
| $Q$ | Sum of the external and mutual energy transmission rates per unit mass. See Section 5.1.2. |
| $\mathbf{Q}$ | Time-dependent orthogonal transformation introduced in Section 1.2.1. |
| $r$ | Cylindrical coordinate introduced in Exercise A.4.1-4 and spherical coordinate introduced in Exercise A.4.1-5. |
| $r_{(A)}$ | Rate of production of species $A$ per unit volume by homogeneous chemical reactions. |
| $r_{(A)}^{(\sigma)}$ | Rate of production of species $A$ per unit area by heterogeneous chemical reactions. |
| $R$ | Gas law constant. Also used as the radius of a tube or sphere. |
| $R_{(m)}$ | Region occupied by a material body. |
| $\hat{S}$ | Entropy per unit mass. |
| $\mathbf{S}$ | Extra stress or viscous portion of the stress tensor defined by (2.3.2-27) for an incompressible fluid and by (5.3.1-27) or (8.4.1-26) for a compressible fluid. |

$S_{(m)}$      Closed surface bounding a material body.

$t$      Time.

$\mathbf{t}$      Stress vector, in force per unit area.

$T$      Temperature introduced in (5.2.3-5) and (8.3.5-4).

$\mathbf{T}$      Stress tensor introduced in Section 2.2.2.

$\mathbf{u}$      Velocity of a point on the dividing surface as introduced in Section 1.3.5.

$\mathbf{u}_{(A)}$      Velocity of species $A$ relative to $\mathbf{v}$ defined in Table 8.5.1-3.

$\mathbf{u}_{(A)}^{\diamond}$      Velocity of species $A$ relative to $\mathbf{v}^{\diamond}$ defined in Table 8.5.1-3.

$\hat{U}$      Internal energy per unit mass introduced in Section 5.1.1.

$\mathbf{v}$      Velocity defined by (1.1.0-13); also mass-averaged velocity defined in Table 8.5.1-3.

$\mathbf{v}^{\diamond}$      Molar-averaged velocity defined in Table 8.5.1-3.

$\mathbf{v}_{(A)}$      Velocity of species $A$ defined by (8.1.1-9).

$x_{(A)}$      Mole fraction defined in Table 8.5.1-1.

$X(\zeta)$      Configuration of body as introduced in Section 1.1.

$X^{-1}(z)$      Material particle whose place in $E$ is $z$.

$z$      Point in $E$ space introduced in Section A.1.1 as well as cylindrical coordinate introduced in Exercise A.4.1-4.

$z_i$      Rectangular Cartesian coordinate introduced in (A.1.5-5).

$\mathbf{z}$      Position vector for the point $z$ introduced in Section A.1.2.

$z_{\kappa i}$      Coordinate of $\mathbf{z}_\kappa$ in the reference configuration $\kappa$.

$\mathbf{z}_\kappa$      Position vector of a material particle in the reference configuration $\kappa$.

$\gamma$      Surface or interfacial tension introduced in (2.4.3-1)

$\gamma_{(A)}$      Activity coefficient for species $A$ defined by (8.4.8-9).

$\epsilon$      Defined by (8.4.3-4).

$\zeta$      Material particle introduced in Section 1.1.

$\eta$      Apparent viscosity for an incompressible generalized Newtonian fluid introduced in (2.3.3-1).

$\theta$      Cylindrical coordinate introduced in Exercise A.4.1-4 as well as spherical coordinate introduced in Exercise A.4.1-5.

$\kappa$      Bulk viscosity of a Newtonian fluid defined by (2.3.2-24).

$\kappa$      Reference configuration of body as introduced in Section 1.1.

$\kappa^{-1}(\mathbf{z}_\kappa)$      Material particle at the place $\mathbf{z}_\kappa$ as introduced in Section 1.1.

$\mu$      Shear viscosity of a Newtonian fluid introduced in (2.3.2-21) as well as chemical potential on a mass basis for a single-component material defined by (5.3.2-6).

$\mu_{(A)}$      Chemical potential for species $A$ on a mass basis defined by (8.4.2-6).

$\mu^{(m)}$      Chemical potential on a molar basis for a single-component material defined by (5.3.2-7).

$\mu_{(A)}^{(m)}$      Chemical potential for species $A$ on a molar basis defined by (8.4.2-7).

$\boldsymbol{\xi}$      Unit normal to a dividing surface. See Section 1.3.5 for sign convention. See Tables 2.4.3-1 through 2.4.3-8 for specific examples, when using interfacial tension $\gamma$.

$\rho$      Mass density, in mass per unit volume; also total mass density defined in Table 8.5.1-1.

$\rho_{(A)}$      Mass density of species $A$, in mass per unit volume.

$\Sigma$      Dividing surfaces within a material body.

$\phi$      Gravitational potential introduced in (2.4.1-3).

| | |
|---|---|
| $\varphi$ | Spherical coordinate introduced in Exercise A.4.1-5 as well as the fluidity for an incompressible generalized Newtonian fluid introduced in (2.3.3-19). |
| $\chi^{-1}(\mathbf{z}, t)$ | Material particle at $\mathbf{z}$ and $t$. |
| $\chi(\zeta, t)$ | Motion of body as introduced in Section 1.1. |
| $\chi_\kappa(\mathbf{z}_\kappa, t)$ | Defined by (1.1.0-9). |
| $\psi$ | Stream function introduced in Section 1.3.7. |
| $\omega_{(A)}$ | Mass fraction of species $A$ defined in Table 8.5.1-1. |
| $\boldsymbol{\omega}$ | Angular velocity vector of the unstarred frame with respect to the starred frame as defined by (1.2.2-11). |
| $d_{(m)}/dt$ | Material derivative defined by (1.1.0-12). |
| $\hat{\phantom{x}}$ | Indicates that the quantity is per unit mass. |
| $\check{\phantom{x}}$ | Indicates that the quantity is per unit volume. |
| $\tilde{\phantom{x}}$ | Indicates that the quantity is per unit mole. |
| $\overline{\Phi}_{(A)}$ | Indicates that the quantity is a partial mass variable for species $A$ as defined in Exercise 8.4.2-4. |
| $\overline{\Phi}_{(A)}^{(m)}$ | Indicates that the quantity is a partial molar variable for species $A$ as defined in Exercise 8.4.2-5. |
| $\llbracket A\xi \rrbracket$ | Defined by (1.3.5-5) as the jump of the quantity enclosed across an interface. |

# Advanced Transport Phenomena

# 1

# Kinematics

THIS ENTIRE CHAPTER IS INTRODUCTORY in much the same way as Appendix A is. In Appendix A, I introduce the mathematical language that I shall be using in describing physical problems. In this chapter, I indicate some of the details involved in representing from the continuum point of view the motions and deformations of real materials. This chapter is important not only for the definitions introduced, but also for the viewpoint taken in some of the developments. For example, the various forms of the transport theorem will be used repeatedly throughout the text in developing differential equations and integral balances from our basic postulates.

Perhaps the most difficult point for a beginner is to properly distinguish between the continuum model for real materials and the particulate or molecular model. We can all agree that the most factually detailed picture of real materials requires that they be represented in terms of atoms and molecules. In this picture, mass is distributed discontinuously throughout space; mass is associated with the protons, neutrons, electrons, ..., which are separated by relatively large voids. In the continuum model for materials, mass is distributed continuously through space, with the exception of surfaces of discontinuity, which represent phase interfaces or shock waves.

The continuum model is less realistic than the particulate model but far simpler. For many purposes, the detailed accuracy of the particulate model is unnecessary. To our sight and touch, mass appears to be continuously distributed throughout the water that we drink and the air that we breathe. The problem is analogous to a study of traffic patterns on an expressway. The speed and spacings of the automobiles are important, but we probably should not worry about whether the automobiles have four, six, or eight cylinders.

This is not to say that the particulate theories are of no importance. Information is lost in a continuum picture. It is only through the use of statistical mechanics that a complete a priori prediction about the behavior of the material can be made. I will say more about this in the next chapter.

## 1.1 Motion

My goal in this book is to lay the foundation for understanding a wide variety of operations employed in the chemical and petroleum industries. To be specific, consider the extrusion of

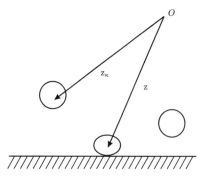

**Figure 1.1.0-1.** A rubber ball in three configurations as it strikes a wall and rebounds. The particle that was in the position $\mathbf{z}_\kappa$ in the reference configuration is in the position $\mathbf{z}$ at time $t$.

a molten polymer to produce a fiber, catalytic cracking in a fluidized reaction, the production of oil and gas from a sandstone reservoir, or the flow of a coal slurry through a pipeline. One important feature that these operations have in common is that at least some of the materials concerned are undergoing deformation and flow.

How might we describe a body of material as it deforms? Figure 1.1.0-1 shows a rubber ball in three configurations as it strikes a wall and rebounds. How should we describe the deformation of this rubber body from its original configuration as a sphere? How should velocity be defined in order to take into account that it must surely vary as a function of position within the ball as well as time as the ball reaches the wall and begins to deform? We need a mathematical description for a body that allows us to describe where its various components go as functions of time.

Let's begin by rather formally defining a body to be set, any element $\zeta$ of which is called a particle or a *material particle*. A material particle is a *primitive*, in the sense that it is not defined but its properties are described. I will give an experimentally oriented description of a material particle a little later in this section. Meanwhile, be careful not to confuse a material particle with a molecule. Molecules play no role in continuum mechanics; they are introduced in the context of the other model for real materials – statistical mechanics.

A one-to-one continuous mapping of this set of material particles onto a region of the space $E$ studied in elementary geometry exists and is called a *configuration* of the body:

$$z = X(\zeta) \tag{1.1.0-1}$$

$$\zeta = X^{-1}(z) \tag{1.1.0-2}$$

The point $z = X(\zeta)$ of $E$ is called the place occupied by the particle $\zeta$, and $\zeta = X^{-1}(z)$, the particle whose place in $E$ is $z$.

It is completely equivalent to describe the configuration of a body in terms of the position vector $\mathbf{z}$ of the point $z$ with respect to the origin $O$ (Section A.1.2):

$$\mathbf{z} = \chi(\zeta) \tag{1.1.0-3}$$

$$\zeta = \chi^{-1}(\mathbf{z}) \tag{1.1.0-4}$$

Here $\chi^{-1}$ indicates the inverse mapping of $\chi$. With an origin $O$ having been defined, it is unambiguous to refer to $\mathbf{z} = \chi(\zeta)$ as the place occupied by the particle $\zeta$ and $\zeta = \chi^{-1}(\mathbf{z})$ as the particle whose place is $\mathbf{z}$.

In what follows, I choose to refer to points in $E$ by their position vectors relative to a previously defined origin $O$.

A *motion* of a body is a one-parameter family of configurations; the real parameter $t$ is time. We write

$$\mathbf{z} = \chi(\zeta, t) \tag{1.1.0-5}$$

and

$$\zeta = \chi^{-1}(\mathbf{z}, t) \tag{1.1.0-6}$$

I have introduced the material particle as a primitive concept, without definition but with a description of its attributes. A set of material particles is defined to be a body; there is a one-to-one continuous mapping of these particles onto a region of the space $E$ in which we visualize the world about us. But clearly we need a link with what we can directly observe.

Whereas the body $B$ should not be confused with any of its spatial configurations, it is available to us for observation and study only in these configurations. We will describe a material particle by its position in a *reference configuration* $\kappa$ of the body. This reference configuration may be, but need not be, one actually occupied by the body in the course of its motion. The place of a particle in $\kappa$ will be denoted by

$$\mathbf{z}_\kappa = \kappa(\zeta) \tag{1.1.0-7}$$

The particle at the place $\mathbf{z}_\kappa$ in the configuration $\kappa$ may be expressed as

$$\zeta = \kappa^{-1}(\mathbf{z}_\kappa) \tag{1.1.0-8}$$

The motion of a body is described by

$$\begin{aligned}
\mathbf{z} &= \chi(\zeta, t) \\
&= \chi_\kappa(\mathbf{z}_\kappa, t) \\
&\equiv \chi(\kappa^{-1}(\mathbf{z}_\kappa), t)
\end{aligned} \tag{1.1.0-9}$$

Referring to Figure 1.1.0-1, we find that the particle that was in the position $\mathbf{z}_\kappa$ in the reference configuration is at time $t$ in the position $\mathbf{z}$. This expression defines a family of *deformations* from the reference configuration. The subscript $\ldots_\kappa$ is to remind you that the form of $\chi_\kappa$ depends upon the choice of reference configuration $\kappa$.

The position vector $\mathbf{z}_\kappa$ with respect to the origin $O$ may be written in terms of its rectangular Cartesian coordinates:

$$\mathbf{z}_\kappa = z_{\kappa i} \mathbf{e}_i \tag{1.1.0-10}$$

The $z_{\kappa i}$ ($i = 1, 2, 3$) are referred to as the *material coordinates* of the material particle $\zeta$. They locate the position of $\zeta$ relative to the origin $O$, when the body is in the reference configuration $\kappa$. In terms of these material coordinates, we may express (1.1.0-9) as

$$\begin{aligned}
\mathbf{z} &= \chi_\kappa(\mathbf{z}_\kappa, t) \\
&= \hat{\chi}_\kappa(z_{\kappa 1}, z_{\kappa 2}, z_{\kappa 3}, t)
\end{aligned} \tag{1.1.0-11}$$

Let $A$ be any quantity: scalar, vector, or tensor. We shall have occasion to talk about the time derivative of $A$ following the motion of a particle. We define

$$\frac{d_{(m)}A}{dt} \equiv \left(\frac{\partial A}{\partial t}\right)_\zeta$$

$$\equiv \left(\frac{\partial A}{\partial t}\right)_{z_\kappa}$$

$$\equiv \left(\frac{\partial A}{\partial t}\right)_{z_{\kappa 1}, z_{\kappa 2}, z_{\kappa 3}} \tag{1.1.0-12}$$

We refer to the operation $d_{(m)}/dt$ as the *material derivative* [or substantial derivative (Bird et al. 1960, p. 73)]. For example, the *velocity vector* $\mathbf{v}$ represents the time rate of change of position of a material particle:

$$\mathbf{v} \equiv \frac{d_{(m)}\mathbf{z}}{dt}$$

$$\equiv \left[\frac{\partial \chi(\zeta, t)}{\partial t}\right]_\zeta$$

$$\equiv \left[\frac{\partial \chi_\kappa(\mathbf{z}_\kappa, t)}{\partial t}\right]_{z_\kappa}$$

$$\equiv \left[\frac{\partial \hat{\chi}_\kappa(z_{\kappa 1}, z_{\kappa 2}, z_{\kappa 3}, t)}{\partial t}\right]_{z_{\kappa 1}, z_{\kappa 2}, z_{\kappa 3}} \tag{1.1.0-13}$$

We are involved with several derivative operations in the chapters that follow. Bird et al. (1960, p. 73) have suggested some examples that serve to illustrate the differences.

**The partial time derivative $\partial c/\partial t$** Suppose we are in a boat that is anchored securely in a river, some distance from the shore. If we look over the side of our boat and note the concentration of fish as a function of time, we observe how the fish concentration changes with time at a fixed position in space:

$$\frac{\partial c}{\partial t} \equiv \left(\frac{\partial c}{\partial t}\right)_\mathbf{z}$$

$$\equiv \left(\frac{\partial c}{\partial t}\right)_{z_1, z_2, z_3}$$

**The material derivative $d_{(m)}c/dt$** Suppose we pull up our anchor and let our boat drift along with the river current. As we look over the side of our boat, we report how the concentration of fish changes as a function of time while following the water (the material):

$$\frac{d_{(m)}c}{dt} = \frac{\partial c}{\partial t} + \nabla c \cdot \mathbf{v} \tag{1.1.0-14}$$

**The total derivative $dc/dt$** We now switch on our outboard motor and race about the river, sometimes upstream, sometimes downstream, or across the current. As we peer over the side of our boat, we measure fish concentration as a function of time while following an arbitrary path across the water:

$$\frac{dc}{dt} = \frac{\partial c}{\partial t} + \nabla c \cdot \mathbf{v}_{(b)} \tag{1.1.0-15}$$

Here $\mathbf{v}_{(b)}$ denotes the velocity of the boat.

**Exercise 1.1.0-1**   Let $A$ be any real scalar field, spatial vector field, or second-order tensor field. Show that[1]

$$\frac{d_{(m)}A}{dt} = \frac{\partial A}{\partial t} + \nabla A \cdot \mathbf{v}$$

**Exercise 1.1.0-2**   Let $\mathbf{a} = \mathbf{a}(\mathbf{z}, t)$ be some vector field that is a function of position and time.

i) Show that

$$\frac{d_{(m)}\mathbf{a}}{dt} = \left( \frac{\partial a^n}{\partial t} + a^n{}_{,i}\, v^i \right) \mathbf{g}_n$$

ii) Show that

$$\frac{d_{(m)}\mathbf{a}}{dt} = \left( \frac{\partial a_n}{\partial t} + a_{n,i}\, v^i \right) \mathbf{g}^n$$

**Exercise 1.1.0-3**   Consider the second-order tensor field $\mathbf{T} = \mathbf{T}(\mathbf{z}, t)$.

i) Show that

$$\frac{d_{(m)}\mathbf{T}}{dt} = \left( \frac{\partial T^{ij}}{\partial t} + T^{ij}{}_{,k}\, v^k \right) \mathbf{g}_i \mathbf{g}_j$$

ii) Show that

$$\frac{d_{(m)}\mathbf{T}}{dt} = \left( \frac{\partial T^i{}_j}{\partial t} + T^i{}_{j,k}\, v^k \right) \mathbf{g}_i \mathbf{g}^j$$

**Exercise 1.1.0-4**   Show that

$$\frac{d_{(m)}(\mathbf{a} \cdot \mathbf{b})}{dt} = \frac{d_{(m)}}{dt}(a^i b_i)$$

$$= \frac{d_{(m)}a^i}{dt} b_i + a^i \frac{d_{(m)}b_i}{dt}$$

---

[1]  Where I write $(\nabla A) \cdot \mathbf{v}$, some authors say instead $\mathbf{v} \cdot (\nabla A)$. When $A$ is a scalar, there is no difference. When $A$ is either a vector or second-order tensor, the change in notation is the result of a different definition for the gradient operation. See Sections A.6.1 and A.8.1.

**Exercise 1.1.0-5**

   i) Starting with the definition for the velocity vector, prove that

$$\mathbf{v} = \frac{d_{(m)}x^i}{dt}\mathbf{g}_i$$

   ii) Determine that, with respect to the cylindrical coordinate system defined in Exercise A.4.1-4,

$$\mathbf{v} = \frac{d_{(m)}r}{dt}\mathbf{g}_r + r\frac{d_{(m)}\theta}{dt}\mathbf{g}_\theta + \frac{d_{(m)}z}{dt}\mathbf{g}_z$$

   iii) Determine that, with respect to the spherical coordinate system defined in Exercise A.4.1-5,

$$\mathbf{v} = \frac{d_{(m)}r}{dt}\mathbf{g}_r + r\frac{d_{(m)}\theta}{dt}\mathbf{g}_\theta + r\sin\theta\frac{d_{(m)}\varphi}{dt}\mathbf{g}_\varphi$$

**Exercise 1.1.0-6** *Path lines*   The curve in space along which the material particle $\zeta$ travels is referred to as the *path line* for the material particle $\zeta$. The path line may be determined from the motion of the material as described in Section 1.1:

$$\mathbf{z} = \chi(\mathbf{z}_\kappa, t)$$

Here $\mathbf{z}_\kappa$ represents the position of the material particle $\zeta$ in the reference configuration $\kappa$; time $t$ is a parameter along the path line that corresponds to any given position $\mathbf{z}_\kappa$.

   The path lines may be determined conveniently from the velocity distribution, since velocity is the derivative of position with respect to time following a material particle. The parametric equations of a particle path are the solutions of the differential system

$$\frac{d\mathbf{z}}{dt} = \mathbf{v}$$

or

$$\frac{dz_i}{dt} = v_i \quad \text{for } i = 1, 2, 3$$

The required boundary conditions may be obtained by choosing the reference configuration to be a configuration that the material assumed at some time $t_0$.

   As an example, let the rectangular Cartesian components of $\mathbf{v}$ be

$$v_1 = \frac{z_1}{1+t}, \qquad v_2 = \frac{z_2}{1+2t}, \qquad v_3 = 0$$

and let the reference configuration be that which the material assumed at time $t = 0$. Prove that, in the plane $z_3 = z_{\kappa 3}$, the particle paths or the path lines have the form

$$\frac{z_2}{z_{\kappa 2}} = \left(2\frac{z_1}{z_{\kappa 1}} - 1\right)^{1/2}$$

**Exercise 1.1.0-7** *Streamlines*  The streamlines for time $t$ form that family of curves to which the velocity field is everywhere tangent at a fixed time $t$. The parametric equations for the streamlines are solutions of the differential equations

$$\frac{dz_i}{d\alpha} = v_i \quad \text{for } i = 1, 2, 3$$

Here $\alpha$ is a parameter with the units of time, and $d\mathbf{z}/d\alpha$ is tangent to the streamline [see (A.4.1-1)]. Alternatively, we may think of the streamlines as solutions of the differential system

$$\frac{d\mathbf{z}}{d\alpha} \wedge \mathbf{v} = 0$$

or

$$e_{ijk} \frac{dz_j}{d\alpha} v_k = 0 \quad \text{for } i = 1, 2, 3$$

As an example, show that, for the velocity distribution introduced in Exercise 1.1.0-6, the streamlines take the form

$$z_2 = z_{2(0)} \left( \frac{z_1}{z_{1(0)}} \right)^{(1+t)/(1+2t)}$$

for different reference points $(z_{1(0)}, z_{2(0)})$.

Experimentalists sometimes sprinkle particles over a gas–liquid phase interface and take a photograph in which the motion of the particles is not quite stopped (see Figures 3.5.1-1 and 3.5.1-3). The traces left by the particles are proportional to the velocity of the fluid at the surface (so long as we assume that very small particles move with the fluid). For a steady-state flow, such a photograph may be used to construct the particle paths. For an unsteady-state flow, it depicts the streamlines, the family of curves to which the velocity vector field is everywhere tangent.

In two-dimensional flows, the streamlines have a special significance. They are curves along which the stream function (Sections 1.3.7) is a constant. See Exercise 1.3.7-2.

**Exercise 1.1.0-8**  For the limiting case of steady-state, plane potential flow past a stationary cylinder of radius $a$ with no circulation, the physical components of velocity in cylindrical coordinates are (see Exercise 3.4.2-2)

$$v_r = V \left( 1 - \frac{a^2}{r^2} \right) \cos \theta$$

$$v_\theta = -V \left( 1 + \frac{a^2}{r^2} \right) \sin \theta$$

and

$$v_z = 0$$

Show that the family of streamlines is described by

$$\left( 1 - \frac{a^2}{r^2} \right) r \sin \theta = C$$

Plot representative members of this family as in Figure 1.1.0-2.

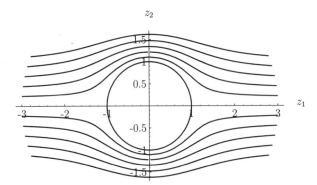

**Figure 1.1.0-2.** Streamlines for the limiting case of steady-state, plane potential flow past a stationary cylinder with no circulation corresponding to $C = 0.2, 0.4, 0.6, 0.8, 1$.

Exercise 1.1.0-9 *Streak lines*    The streak line through the point $\mathbf{z}_{(0)}$ at time $t$ represents the positions at time $t$ of the material particles that at any time $\tau \le t$ have occupied the place $\mathbf{z}_{(0)}$.

Experimentally we might visualize that smoke, dust, or dye are continuously injected into a fluid at a position $\mathbf{z}_{(0)}$ and that the resulting trails are photographed as functions of time. Each photograph shows a streak line corresponding to the position $\mathbf{z}_{(0)}$ and the time at which the photograph was taken.

We saw in Section 1.1 that the motion $\chi$ describes the position $\mathbf{z}$ at time $t$ of the material particle that occupied the position $\mathbf{z}_\kappa$ in the reference configuration:

$$\mathbf{z} = \chi(\mathbf{z}_\kappa, t)$$

In constructing a streak line, we focus our attention on those material particles that were in the place $\mathbf{z}_{(0)}$ at any time $\tau \le t$:

$$\mathbf{z}_\kappa = \chi^{-1}\left(\mathbf{z}_{(0)}, \tau\right)$$

The parametric equations of the streak line through the point $z_{(0)}$ at time $t$ are obtained by eliminating $\mathbf{z}_\kappa$ between these equations:

$$\mathbf{z} = \chi\left(\chi^{-1}\left(\mathbf{z}_{(0)}, \tau\right), t\right)$$

Time $\tau \le t$ is the parameter along the streak line.

As an example, show that, for the velocity distribution of Exercise 1.1.0-6, the streak line through $\mathbf{z}_{(0)}$ at time $t$ is specified by

$$z_1 = z_{1(0)}\left(\frac{1+t}{1+\tau}\right)$$

$$z_2 = z_{2(0)}\left(\frac{1+2t}{1+2\tau}\right)^{1/2}$$

$$z_3 = z_{3(0)}$$

A streak line corresponding to $t = 4$ is shown in Figure 1.1.0-3. This figure also presents two of the path lines from Exercise 1.1.0-6 corresponding to $\tau = 0$ and $0.5$ that contribute

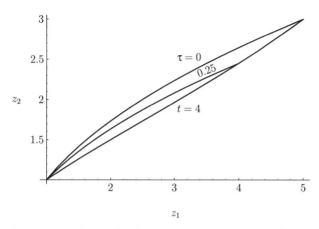

**Figure 1.1.0-3.** Starting from the top, we see two path lines from Exercise 1.1.0-6) corresponding to $\tau = 0$ and 0.25. The bottom curve is the streak line corresponding to $t = 4$.

to this streak line. The path line corresponding to $\tau = 0$ extends to the right tip of the streak line. It represents the first particle contributing to the streak line. A path line corresponding to $\tau = 4$ would merely be a point at the left tip of the streak line, the origin of the streak line $(0, 0)$. It would represent the particle currently leaving the origin.

**Exercise 1.1.0-10** Show that, for a velocity distribution that is independent of time, the path lines, streamlines, and streak lines coincide.

*Hint:* In considering the path line, take as the boundary condition

at $t = \tau : \mathbf{z} = \mathbf{z}_{(0)}$

This suggests the introduction of a new variable $\alpha \equiv t - \tau$, which denotes time measured since the particle passed through the position $\mathbf{z}_{(0)}$.

## 1.2  Frame

### 1.2.1  Changes of Frame

The Chief of the United States Weather Bureau in Milwaukee announces that a tornado was sighted in Chicago at 3 P.M. (Central Standard Time). In Chicago, Harry reports that he saw a black funnel cloud about two hours ago at approximately 800 North and 2400 West. Both men described the same event with respect to their own particular frame of reference.

The time of some occurrence may be specified only with respect to the time of some other event, the *frame of reference for time*. This might be the time at which a stopwatch was started or an electric circuit was closed. The Chief reported the time at which the tornado was sighted relative to the mean time at which the sun appeared overhead on the Greenwich meridian. Harry gave the time relative to his conversation.

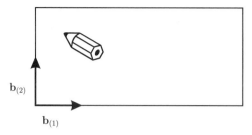

**Figure 1.2.1-1.** Pencil points away from the direction of $\mathbf{b}_{(1)}$ and toward the direction of $\mathbf{b}_{(2)}$.

A *frame of reference for position* might be the walls of a laboratory, the fixed stars, or the shell of a space capsule that is following an arbitrary trajectory. When the Chief specified Chicago, he meant the city at 41° north and 87° west measured relative to the equator and the Greenwich meridian. Harry thought in terms of eight blocks north of Madison Avenue and 24 blocks west of State Street. More generally, a frame of reference for position is a set of objects whose mutual distances remain unchanged during the period of observation and which do not all lie in the same plane.

To help you get a better physical feel for these ideas, let us consider two more examples.

Extend your right arm and take as your frame of reference for position the direction of your right arm, the direction of your eyes, and the direction of your spine. Stand out at the street with your eyes fixed straight ahead. A car passes in the direction of your right arm. If you were standing facing the street on the opposite side, the automobile would appear to pass in the opposite direction from your right arm.

Lay a pencil on your desk as shown in Figure 1.2.1-1 and take the edges of the desk that meet in the left-hand front corner as your frame of reference for position. The pencil points away from $\mathbf{b}_{(1)}$ and toward $\mathbf{b}_{(2)}$. Without moving the pencil, walk around to the left-hand side and take as your new frame of reference for position the edges of the desk that meet at the left-hand rear corner. The pencil now appears to point toward the intersection of $\mathbf{b}_{(1)}^*$ and $\mathbf{b}_{(2)}^*$ in Figure 1.2.1-2.

Since all of the objects defining a frame of reference do not lie in the same plane, we may visualize replacing them by three mutually orthogonal unit vectors. Let us view a typical point $z$ in this space with respect to two such frames of reference: the $\mathbf{b}_{(i)}$ ($i = 1, 2, 3$) in Figure 1.2.1-3 and the $\mathbf{b}_{(j)}^*$ ($j = 1, 2, 3$) in Figure 1.2.1-4.

An orthogonal transformation preserves both lengths and angles (Section A.5.2). Let $\mathbf{Q}$ be the orthogonal transformation that describes the rotation and (possibly) reflection that takes the $\mathbf{b}_{(i)}$ in Figure 1.2.1-3 into the vectors $\mathbf{Q} \cdot \mathbf{b}_{(i)}$, which are seen in Figure 1.2.1-4 with respect to the starred frame of reference for position. A reflection allows for the possibility that an observer in the new frame looks at the old frame through a mirror. Alternatively, a reflection allows for the possibility that two observers orient themselves oppositely, one choosing to work in terms of a right-handed frame of reference for position and the other in terms of a left-handed one. [For more on this point, I suggest that you read Truesdell (1966a, p. 22) as well as Truesdell and Noll (1965, pp. 24 and 47).]

The vector $\left(\mathbf{z} - \mathbf{z}_{(0)}\right)$ in Figure 1.2.1-3 becomes $\mathbf{Q} \cdot \left(\mathbf{z} - \mathbf{z}_{(0)}\right)$ when viewed in the starred frame shown in Figure 1.2.1-4. From Figure 1.2.1-4, it follows as well that

$$\mathbf{z}^* - \mathbf{z}_{(0)}^* = \mathbf{Q} \cdot \left(\mathbf{z} - \mathbf{z}_{(0)}\right) \tag{1.2.1-1}$$

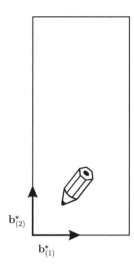

**Figure 1.2.1-2.** Pencil points toward the direction of $\mathbf{b}^*_{(1)}$ and $\mathbf{b}^*_{(2)}$.

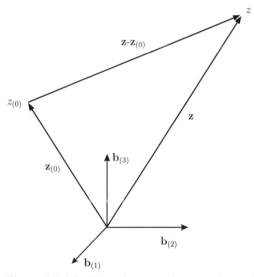

**Figure 1.2.1-3.** The points $z$ and $z_{(0)}$ are located by the position vectors $\mathbf{z}$ and $\mathbf{z}_{(0)}$ with respect to the frame of reference for position $(\mathbf{b}_{(1)}, \mathbf{b}_{(2)}, \mathbf{b}_{(3)})$.

Similarly, $(\mathbf{z}^* - \mathbf{z}^*_{(0)})$ in Figure 1.2.1-4 is seen as $\mathbf{Q}^T \cdot (\mathbf{z}^* - \mathbf{z}^*_{(0)})$ when observed with respect to the unstarred frame in Figure 1.2.1-5. Figure 1.2.1-5 also makes it clear that

$$\mathbf{z} - \mathbf{z}_{(0)} = \mathbf{Q}^T \cdot (\mathbf{z}^* - \mathbf{z}^*_{(0)}) \tag{1.2.1-2}$$

Let $\mathbf{z}$ and $t$ denote a position and time in the old frame; $\mathbf{z}^*$ and $t^*$ are the corresponding position and time in the new frame. We can extend the discussion above to conclude that the

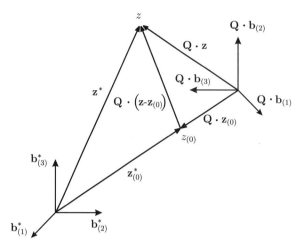

**Figure 1.2.1-4.** The points $z$ and $z_{(0)}$ are located by the position vectors $\mathbf{z}^*$ and $\mathbf{z}_{(0)}^*$ with respect to the starred frame of reference for position $\left( \mathbf{b}_{(1)}^*, \ \mathbf{b}_{(2)}^*, \ \mathbf{b}_{(3)}^* \right)$. With respect to the starred frame of reference, the unstarred frame is seen as $\left( \mathbf{Q} \cdot \mathbf{b}_{(1)}, \ \mathbf{Q} \cdot \mathbf{b}_{(2)}, \ \mathbf{Q} \cdot \mathbf{b}_{(3)} \right)$.

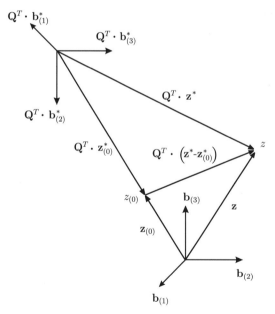

**Figure 1.2.1-5.** With respect to the unstarred frame of reference, the starred frame is seen as $\left( \mathbf{Q}^T \cdot \mathbf{b}_{(1)}^*, \ \mathbf{Q}^T \cdot \mathbf{b}_{(2)}^*, \ \mathbf{Q}^T \cdot \mathbf{b}_{(3)}^* \right)$.

most general change of frame is of the form

$$\mathbf{z}^* = \mathbf{z}^*_{(0)}(t) + \mathbf{Q}(t) \cdot \left(\mathbf{z} - \mathbf{z}_{(0)}\right) \tag{1.2.1-3}$$

$$t^* = t - a \tag{1.2.1-4}$$

where we allow the two frames discussed in Figures 1.2.1-3 and 1.2.1-4 to rotate and translate with respect to one another as functions of time. The quantity $a$ is a real number. Equivalently, we could also write

$$\mathbf{z} = \mathbf{z}_{(0)}(t) + \mathbf{Q}^T \cdot \left(\mathbf{z}^* - \mathbf{z}^*_{(0)}\right) \tag{1.2.1-5}$$

$$t = t^* + a \tag{1.2.1-6}$$

It is important to carefully distinguish between a frame of reference for position and a coordinate system. Any coordinate system whatsoever can be used to locate points in space with respect to three vectors defining a frame of reference for position and their intersection, although I recommend that admissible coordinate systems be restricted to those whose axes have a time-invariant orientation with respect to the frame. Let $(z_1, z_2, z_3)$ be a rectangular Cartesian coordinate system associated with the frame of reference $(\mathbf{b}_{(1)}, \mathbf{b}_{(2)}, \mathbf{b}_{(3)})$; similarly, let $(z^*_1, z^*_2, z^*_3)$ be a rectangular Cartesian coordinate system associated with another frame of reference $(\mathbf{b}^*_{(1)}, \mathbf{b}^*_{(2)}, \mathbf{b}^*_{(3)})$. We will say that these two coordinate systems are the *same* if the orientation of the basis fields $\mathbf{e}_i$ with respect to the vectors $\mathbf{b}_{(j)}$ is identical to the orientation of the basis fields $\mathbf{e}^*_i$ with respect to the vectors $\mathbf{b}^*_{(j)}$:

$$\mathbf{e}_i \cdot \mathbf{b}_{(j)} = \mathbf{e}^*_i \cdot \mathbf{b}^*_{(j)} \quad \text{for all } i, j = 1, 2, 3 \tag{1.2.1-7}$$

We will generally find it convenient to use the same coordinate system in discussing two different frames of reference.

Let us use the *same* rectangular Cartesian coordinate system to discuss the change of frame illustrated in Figures 1.2.1-4 and 1.2.1-5. The orthogonal tensor

$$\mathbf{Q} = Q_{ij}\mathbf{e}^*_i\mathbf{e}_j \tag{1.2.1-8}$$

describes the rotation (and possibly reflection) that transforms the basis vectors $\mathbf{e}_j$ ($j = 1, 2, 3$) into the vectors

$$\mathbf{Q} \cdot \mathbf{e}_j = Q_{ij}\mathbf{e}^*_i \tag{1.2.1-9}$$

which are vectors expressed in terms of the starred frame of reference for position. The rectangular Cartesian components of $\mathbf{Q}$ are defined by the angles between the $\mathbf{e}^*_i$ and the $\mathbf{Q} \cdot \mathbf{e}_j$:

$$Q_{ij} = \mathbf{e}^*_i \cdot \left(\mathbf{Q} \cdot \mathbf{e}_j\right) \tag{1.2.1-10}$$

The vector $\left(\mathbf{z} - \mathbf{z}_{(0)}\right)$ in Figure 1.2.1-3 becomes

$$\mathbf{Q} \cdot \left(\mathbf{z} - \mathbf{z}_{(0)}\right) = Q_{ij}\left(z_j - z_{(0)j}\right)\mathbf{e}^*_i \tag{1.2.1-11}$$

when viewed in the starred frame shown in Figure 1.2.1-4. From Figure 1.2.1-4, it follows as well that

$$z^*_i\mathbf{e}^*_i = z^*_{(0)i}\mathbf{e}^*_i + Q_{ij}\left(z_j - z_{(0)j}\right)\mathbf{e}^*_i \tag{1.2.1-12}$$

We speak of a quantity as being *frame indifferent* if it remains unchanged or invariant under all changes of frame. A *frame-indifferent scalar b* does not change its value:

$$b^* = b \tag{1.2.1-13}$$

A *frame-indifferent spatial vector* remains the same directed line element under a change of frame in the sense that if

$$\mathbf{u} = \mathbf{z}_1 - \mathbf{z}_2$$

then

$$\mathbf{u}^* = \mathbf{z}_1^* - \mathbf{z}_2^*$$

From (1.2.1-3),

$$\mathbf{u}^* = \mathbf{Q} \cdot (\mathbf{z}_1 - \mathbf{z}_2)$$
$$= \mathbf{Q} \cdot \mathbf{u} \tag{1.2.1-14}$$

A *frame-indifferent second-order tensor* is one that transforms frame-indifferent spatial vectors into frame-indifferent spatial vectors. If

$$\mathbf{u} = \mathbf{T} \cdot \mathbf{w} \tag{1.2.1-15}$$

the requirement that $\mathbf{T}$ be a frame-indifferent second-order tensor is

$$\mathbf{u}^* = \mathbf{T}^* \cdot \mathbf{w}^* \tag{1.2.1-16}$$

where

$$\mathbf{u}^* = \mathbf{Q} \cdot \mathbf{u}$$
$$\mathbf{w}^* = \mathbf{Q} \cdot \mathbf{w} \tag{1.2.1-17}$$

This means that

$$\mathbf{Q} \cdot \mathbf{u} = \mathbf{T}^* \cdot \mathbf{Q} \cdot \mathbf{w}$$
$$= \mathbf{Q} \cdot \mathbf{T} \cdot \mathbf{w} \tag{1.2.1-18}$$

which implies

$$\mathbf{T} = \mathbf{Q}^T \cdot \mathbf{T}^* \cdot \mathbf{Q} \tag{1.2.1-19}$$

or

$$\mathbf{T}^* = \mathbf{Q} \cdot \mathbf{T} \cdot \mathbf{Q}^T \tag{1.2.1-20}$$

The importance of changes of frame will become apparent in Section 2.3.1, where the principle of frame indifference is introduced. This principle will be used repeatedly in discussing representations for material behavior and in preparing empirical data correlations.

The material in this section is drawn from Truesdell and Toupin (1960, p. 437), Truesdell and Noll (1965, p. 41), and Truesdell (1966a, p. 22).

**Exercise 1.2.1-1**  Let $T$ be a frame-indifferent scalar field. Starting with the definition of the gradient of a scalar field in Section A.3.1, show that the gradient of $T$ is frame indifferent:

$$\nabla T^* \equiv (\nabla T)^* = \mathbf{Q} \cdot \nabla T$$

**Exercise 1.2.1-2**  In order that $\epsilon$ (defined in Exercise A.7.2-11) be a frame-indifferent third-order tensor field, prove that

$$\epsilon^* = (\det \mathbf{Q}) e_{ijk} \mathbf{e}_i \mathbf{e}_j \mathbf{e}_k$$

### 1.2.2  Equivalent Motions

In Section 1.1, I described the motion of a material with respect to some frame of reference by

$$\mathbf{z} = \chi(\mathbf{z}_\kappa, t) \tag{1.2.2-1}$$

where we understand that the form of this relation depends upon the choice of reference configuration $\kappa$. According to our discussion in Section 1.2.1, the same motion with respect to some new frame of reference is represented by

$$\begin{aligned}
\mathbf{z}^* &= \chi^*(\mathbf{z}_\kappa^*, t^*) \\
&= \mathbf{z}_0^*(t) + \mathbf{Q}(t) \cdot [\chi(\mathbf{z}_\kappa, t) - \mathbf{z}_0]
\end{aligned} \tag{1.2.2-2}$$

We will say that any two motions $\chi$ and $\chi^*$ related by an equation of the form of (1.2.2-2) are *equivalent motions*.

Let us write (1.2.2-2) in an abbreviated form:

$$\mathbf{z}^* = \mathbf{z}_0^* + \mathbf{Q} \cdot (\mathbf{z} - \mathbf{z}_0) \tag{1.2.2-3}$$

The material derivative of this equation gives

$$\mathbf{v}^* = \frac{d\mathbf{z}_0^*}{dt} + \frac{d\mathbf{Q}}{dt} \cdot (\mathbf{z} - \mathbf{z}_0) + \mathbf{Q} \cdot \mathbf{v} \tag{1.2.2-4}$$

or

$$\mathbf{v}^* - \mathbf{Q} \cdot \mathbf{v} = \frac{d\mathbf{z}_0^*}{dt} + \frac{d\mathbf{Q}}{dt} \cdot (\mathbf{z} - \mathbf{z}_0) \tag{1.2.2-5}$$

In view of (1.2.2-3), we may write

$$\begin{aligned}
\mathbf{z} - \mathbf{z}_0 &= \mathbf{Q}^T \cdot \mathbf{Q} \cdot (\mathbf{z} - \mathbf{z}_0) \\
&= \mathbf{Q}^T \cdot (\mathbf{z}^* - \mathbf{z}_0^*)
\end{aligned} \tag{1.2.2-6}$$

This allows us to express (1.2.2-5) as

$$\begin{aligned}
\mathbf{v}^* - \mathbf{Q} \cdot \mathbf{v} &= \frac{d\mathbf{z}_0^*}{dt} + \left(\frac{d\mathbf{Q}}{dt} \cdot \mathbf{Q}^T\right) \cdot (\mathbf{z}^* - \mathbf{z}_0^*) \\
&= \frac{d\mathbf{z}_0^*}{dt} + \mathbf{A} \cdot (\mathbf{z}^* - \mathbf{z}_0^*)
\end{aligned} \tag{1.2.2-7}$$

where

$$\mathbf{A} \equiv \frac{d\mathbf{Q}}{dt} \cdot \mathbf{Q}^T \tag{1.2.2-8}$$

We refer to the second-order tensor $\mathbf{A}$ as the *angular velocity tensor of the starred frame with respect to the unstarred frame* (Truesdell 1966a, p. 24).

Since $\mathbf{Q}$ is an orthogonal tensor,

$$\mathbf{Q} \cdot \mathbf{Q}^T = \mathbf{I}^* \tag{1.2.2-9}$$

Taking the material derivative of this equation, we have

$$\mathbf{A} = \frac{d\mathbf{Q}}{dt} \cdot \mathbf{Q}^T = -\mathbf{Q} \cdot \frac{d\mathbf{Q}^T}{dt}$$

$$= -\mathbf{Q} \cdot \left(\frac{d\mathbf{Q}}{dt}\right)^T$$

$$= -\mathbf{A}^T \tag{1.2.2-10}$$

In this way we see that the angular velocity tensor is skew symmetric.

The *angular velocity vector of the unstarred frame with respect to the starred frame* $\boldsymbol{\omega}$ is defined as

$$\boldsymbol{\omega} \equiv \frac{1}{2}\text{tr}(\boldsymbol{\epsilon}^* \cdot \mathbf{A}) \tag{1.2.2-11}$$

The third-order tensor $\boldsymbol{\epsilon}$ is introduced in Exercises A.7.2-11 and A.7.2-12 (see also Exercise 1.2.1-2), where tr denotes the trace operation defined in Section A.7.3. Let us consider the following spatial vector in rectangular Cartesian coordinates:

$$\boldsymbol{\omega} \wedge \left(\mathbf{z}^* - \mathbf{z}_0^*\right) = \text{tr}\left(\boldsymbol{\epsilon}^* \cdot \left[\left(\mathbf{z}^* - \mathbf{z}_0^*\right)\boldsymbol{\omega}\right]\right)$$

$$= \text{tr}\left(\boldsymbol{\epsilon}^* \cdot \left\{\left[\mathbf{z}^* - \mathbf{z}_0^*\right]\left[\frac{1}{2}\text{tr}(\boldsymbol{\epsilon}^* \cdot \mathbf{A})\right]\right\}\right)$$

$$= e_{ijk}\left(z_k^* - z_{0k}^*\right)\left(\frac{1}{2}e_{jmn}A_{nm}\right)\mathbf{e}_i^*$$

$$= \frac{1}{2}\left(z_k^* - z_{0k}^*\right)\left(A_{ij} - A_{ki}\right)\mathbf{e}_i^*$$

$$= \left(z_k^* - z_{0k}^*\right)A_{ik}\mathbf{e}_i^*$$

$$= \mathbf{A} \cdot \left(\mathbf{z}^* - \mathbf{z}_0^*\right) \tag{1.2.2-12}$$

We may consequently write (1.2.2-7) in terms of the angular velocity of the unstarred frame with respect to the starred frame (Truesdell and Toupin 1960, p. 437):

$$\mathbf{v}^* = \frac{d\mathbf{z}_0^*}{dt} + \boldsymbol{\omega} \wedge [\mathbf{Q} \cdot (\mathbf{z} - \mathbf{z}_0)] + \mathbf{Q} \cdot \mathbf{v} \tag{1.2.2-13}$$

The material in this section is drawn from Truesdell and Noll (1965, p. 42) and Truesdell (1966a, p. 22).

**Exercise 1.2.2-1**

   i) Show that velocity is not frame indifferent.

   ii) Show that at any position in euclidean point space a difference in velocities with respect to the same frame is frame indifferent.

**Exercise 1.2.2-2** *Acceleration*

   i) Determine that (Truesdell 1966a, p. 24)

$$\frac{d_{(m)}\mathbf{v}^*}{dt} = \frac{d^2\mathbf{z}_0^*}{dt^2} + 2\mathbf{A}\cdot\left(\mathbf{v}^* - \frac{d\mathbf{z}_0^*}{dt}\right)$$
$$+ \left(\frac{d\mathbf{A}}{dt} - \mathbf{A}\cdot\mathbf{A}\right)\cdot\left(\mathbf{z}^* - \mathbf{z}_0^*\right) + \mathbf{Q}\cdot\frac{d_{(m)}\mathbf{v}}{dt}$$

   ii) Prove that (Truesdell and Toupin 1960, p. 440)

$$\frac{d_{(m)}\mathbf{v}^*}{dt} = \frac{d^2\mathbf{z}_0^*}{dt^2} + \left(\frac{d_{(m)}\mathbf{A}}{dt} + \mathbf{A}\cdot\mathbf{A}\right)\cdot\mathbf{Q}\cdot(\mathbf{z} - \mathbf{z}_0)$$
$$+ 2\mathbf{A}\cdot\mathbf{Q}\cdot\mathbf{v} + \mathbf{Q}\cdot\frac{d_{(m)}\mathbf{v}}{dt}$$

   iii) Prove that

$$\boldsymbol{\omega}\wedge\left[\boldsymbol{\omega}\wedge\left(\mathbf{z}^* - \mathbf{z}_0^*\right)\right] = \mathbf{A}\cdot\mathbf{A}\cdot\left(\mathbf{z}^* - \mathbf{z}_0^*\right)$$

$$\frac{d\boldsymbol{\omega}}{dt}\wedge\wedge\left(\mathbf{z}^* - \mathbf{z}_0^*\right) = \frac{d\mathbf{A}}{dt}\cdot\left(\mathbf{z}^* - \mathbf{z}_0^*\right)$$

   and

$$\boldsymbol{\omega}\wedge(\mathbf{Q}\cdot\mathbf{v}) = \mathbf{A}\cdot\mathbf{Q}\cdot\mathbf{v}$$

   iv) Conclude that (Truesdell and Toupin 1960, p. 438)

$$\frac{d_{(m)}\mathbf{v}^*}{dt} = \frac{d^2\mathbf{z}_0^*}{dt^2} + \frac{d\boldsymbol{\omega}}{dt}\wedge\cdot[\mathbf{Q}\cdot(\mathbf{z} - \mathbf{z}_0)]$$
$$+ \boldsymbol{\omega}\wedge\{\boldsymbol{\omega}\wedge[\mathbf{Q}\cdot(\mathbf{z} - \mathbf{z}_0)]\}$$
$$+ 2\boldsymbol{\omega}\wedge\cdot(\mathbf{Q}\cdot\mathbf{v}) + \mathbf{Q}\cdot\frac{d_{(m)}\mathbf{v}}{dt}$$

**Exercise 1.2.2-3**   Give an example of a scalar that is not frame indifferent.

*Hint:*   What vector is not frame indifferent?

**Exercise 1.2.2-4** *Motion of a rigid body*   Determine that the velocity distribution in a rigid body may be expressed as

$$\mathbf{v}^* = \frac{d\mathbf{z}_0^*}{dt} + \boldsymbol{\omega} \wedge \left(\mathbf{z}^* - \mathbf{z}_0^*\right)$$

What is the relation of the unstarred frame to the body in this case?

---

## 1.3    Mass

### 1.3.1    Conservation of Mass

This discussion of mechanics is based upon several postulates. The first is

**Conservation of mass**   The mass of a body is independent of time.

Physically, this means that, if we follow a portion of a material body through any number of translations, rotations, and deformations, the mass associated with it will not vary as a function of time. If $\rho$ is the *mass density* of the body, the mass may be represented as

$$\int_{R_{(m)}} \rho \, dV$$

Here $dV$ denotes that a volume integration is to be performed over the region $R_{(m)}$ of space occupied by the body in its current configuration; in general $R_{(m)}$, or the limits on this integration, is a function of time. The postulate of conservation of mass says that

$$\frac{d}{dt} \int_{R_{(m)}} \rho \, dV = 0 \tag{1.3.1-1}$$

Notice that, like the material particle introduced in Section 1.1, *mass* is a primitive concept. Rather than defining mass, we describe its properties. We have just examined its most important property: It is conserved. In addition, I will require that

$$\rho > 0 \tag{1.3.1-2}$$

and that the mass density be a frame-indifferent scalar,

$$\rho^* = \rho \tag{1.3.1-3}$$

Our next objective will be to determine a relationship that expresses the idea of conservation of mass at each point in a material. To do this, we will find it necessary to interchange the operations of differentiation and integration in (1.3.1-1). Yet the limits on this integral describe the boundaries of the body in its current configuration and generally are functions of time. The next section explores this problem in more detail.

### 1.3.2    Transport Theorem

Let us consider the operation

$$\frac{d}{dt} \int_{R_{(m)}} \Psi \, dV$$

Here $\Psi$ is any scalar-, vector-, or tensor-valued function of time and position. Again, we should expect $R_{(m)}$, or the limits on this integration, to be a function of time.

If we look at this volume integration in the reference configuration $\kappa$, the limits on the volume integral are no longer functions of time; the limits are expressed in terms of the material coordinates of the bounding surface of the body. This means that we may interchange differentiation and integration in the above operation. In terms of a rectangular Cartesian coordinate system, let $(z_1, z_2, z_3)$ denote the current coordinates of a material point and let $(z_{\kappa 1}, z_{\kappa 2}, z_{\kappa 3})$ be the corresponding material coordinates. Using the results of Section A.11.3, we may say that

$$
\frac{d}{dt} \int_{R_{(m)}} \Psi \, dV = \frac{d}{dt} \int_{R_{(m)\kappa}} \Psi J \, dV
$$

$$
= \int_{R_{(m)\kappa}} \left( \frac{d_{(m)}\Psi}{dt} + \frac{\Psi}{J} \frac{d_{(m)}J}{dt} \right) J \, dV
$$

$$
= \int_{R_{(m)}} \left( \frac{d_{(m)}\Psi}{dt} + \frac{\Psi}{J} \frac{d_{(m)}J}{dt} \right) dV \tag{1.3.2-1}
$$

where (see Exercise A.11.3-1)

$$
J \equiv \left| \det \left( \frac{\partial z_i}{\partial z_{\kappa j}} \right) \right|
$$

$$
= \left| \det \left( \frac{\partial \chi_{\kappa i}}{\partial z_{\kappa j}} \right) \right|
$$

$$
= |\det \mathbf{F}| \tag{1.3.2-2}
$$

Here

$$
\mathbf{F} \equiv \frac{\partial \chi_{\kappa i}}{\partial z_{\kappa j}} \mathbf{e}_i \mathbf{e}_j
$$

$$
\equiv \frac{\partial z_i}{\partial z_{\kappa j}} \mathbf{e}_i \mathbf{e}_j \tag{1.3.2-3}
$$

is the *deformation gradient*. The quantity $J$ may be thought of as the volume in the current configuration per unit volume in the reference configuration. It will generally be a function of both time and position. Here $R_{(m)\kappa}$ indicates that the integration is to be performed over the region of space occupied by the body in its reference configuration $\kappa$.

What follows is the most satisfying development of (1.3.2-8). If this is your first time through this material, I recommend that you replace (1.3.2-3) through (1.3.2-7) by the alternative development of (1.3.2-8) in Appendix B.

From Exercise A.10.1-5,

$$
\frac{d_{(m)}J}{dt} = J \operatorname{tr} \left( \mathbf{F}^{-1} \cdot \frac{d_{(m)}\mathbf{F}}{dt} \right) \tag{1.3.2-4}
$$

It is easy to show that

$$
\mathbf{F}^{-1} = \frac{\partial z_{\kappa m}}{\partial z_n} \mathbf{e}_m \mathbf{e}_n \tag{1.3.2-5}
$$

Using the definition of velocity from Section 1.1, we have

$$\frac{d_{(m)}\mathbf{F}}{dt} = \frac{\partial v_i}{\partial z_{Kj}}\mathbf{e}_i\mathbf{e}_j = \frac{\partial v_i}{\partial z_r}\frac{\partial z_r}{\partial z_{Kj}}\mathbf{e}_i\mathbf{e}_j \tag{1.3.2-6}$$

Consequently,

$$\begin{aligned}
\text{tr}\left(\mathbf{F}^{-1}\cdot\frac{d_{(m)}\mathbf{F}}{dt}\right) &= \frac{\partial z_{Kj}}{\partial z_i}\frac{\partial v_i}{\partial z_r}\frac{\partial z_r}{\partial z_{Kj}} \\
&= \frac{\partial v_i}{\partial z_i} \\
&= \text{div } \mathbf{v}
\end{aligned} \tag{1.3.2-7}$$

and

$$\frac{1}{J}\frac{d_{(m)}J}{dt} = \text{div } \mathbf{v} \tag{1.3.2-8}$$

Equation (1.3.2-8) allows us to write (1.3.2-1) as

$$\frac{d}{dt}\int_{R_{(m)}}\Psi\,dV = \int_{R_{(m)}}\left(\frac{d_{(m)}\Psi}{dt} + \Psi\,\text{div } \mathbf{v}\right)dV \tag{1.3.2-9}$$

This may also be expressed as

$$\frac{d}{dt}\int_{R_{(m)}}\Psi\,dV = \int_{R_{(m)}}\left[\frac{\partial\Psi}{\partial t} + \text{div}(\Psi\mathbf{v})\right]dV \tag{1.3.2-10}$$

or, by Green's transformation (Section A.11.2), we may say

$$\frac{d}{dt}\int_{R_{(m)}}\Psi\,dV = \int_{R_{(m)}}\frac{\partial\Psi}{\partial t}\,dV + \int_{S_{(m)}}\Psi\mathbf{v}\cdot\mathbf{n}\,dA \tag{1.3.2-11}$$

By $S_{(m)}$, I mean the closed bounding surface of $R_{(m)}$; like $R_{(m)}$, it will in general be a function of time. Equations (1.3.2-9) to (1.3.2-11) are three forms of the *transport theorem* (Truesdell and Toupin 1960, p. 347).

We will have occasion to ask about the derivative with respect to time of a quantity while following a system that is not necessarily a material body. For example, let us take as our system the air in a child's balloon and ask for the derivative with respect to time of the volume associated with the air as the balloon is inflated. Since material (air) is being continuously added to the balloon, we are not following a set of material particles as a function of time. However, there is nothing to prevent us from defining a particular set of fictitious *system particles* to be associated with our system. The only restriction we shall make upon this set of imaginary system particles is that the normal component of velocity of any system particle at the boundary of the system be equal to the normal component of velocity of the boundary of the system. Equations (1.3.2-8) to (1.3.2-11) remain valid if we replace

1) derivatives with respect to time while following material particles, $d_{(m)}/dt$, by derivatives with respect to time while following fictitious system particles, $d_{(s)}/dt$; and
2) the velocity vector for a material particle, $\mathbf{v}$, by the velocity vector for a fictitious system particle, $\mathbf{v}_{(s)}$.

This means that

$$\frac{d}{dt} \int_{R_{(s)}} \Psi \, dV = \int_{R_{(s)}} \frac{\partial \Psi}{\partial t} \, dV + \int_{S_{(s)}} \Psi \mathbf{v}_{(s)} \cdot \mathbf{n} \, dA \qquad (1.3.2\text{-}12)$$

Here $R_{(s)}$ signifies that region of space currently occupied by the system; $S_{(s)}$ is the closed bounding surface of the system. We will refer to (1.3.2-12) as the *generalized transport theorem* (Truesdell and Toupin 1960, p. 347).

For an alternative discussion of the transport theorem, see Appendix B.

**Exercise 1.3.2-1** Show that

$$\frac{d}{dt} \int_{R_{(m)}} dV = \int_{R_{(m)}} \operatorname{div} \mathbf{v} \, dV = \int_{S_{(m)}} \mathbf{v} \cdot \mathbf{n} \, dA$$

Here $S_{(m)}$ is the (time-dependent) closed bounding surface of $R_{(m)}$.

### 1.3.3 Differential Mass Balance

Going back to the postulate of conservation of mass in Section 1.3.1,

$$\frac{d}{dt} \int_{R_{(m)}} \rho \, dV = 0 \qquad (1.3.3\text{-}1)$$

and employing the transport theorem in the form of (1.3.2-9), we have that

$$\int_{R_{(m)}} \left( \frac{d_{(m)}\rho}{dt} + \rho \operatorname{div} \mathbf{v} \right) dV = 0 \qquad (1.3.3\text{-}2)$$

But this statement is true for any body or for any portion of a body, since a portion of a body is a body (see Exercise 1.3.3-4). We conclude that the integrand itself must be identically zero:

$$\frac{d_{(m)}\rho}{dt} + \rho \operatorname{div} \mathbf{v} = 0 \qquad (1.3.3\text{-}3)$$

By Exercise 1.1.0-1, this may also be written as

$$\frac{\partial \rho}{\partial t} + \operatorname{div}(\rho \mathbf{v}) = 0 \qquad (1.3.3\text{-}4)$$

Equations (1.3.3-3) and (1.3.3-4) are two forms of the *equation of continuity* or *differential mass balance*. These equations express the requirement that mass be conserved at every point in the continuous material.

The differential mass balance is presented in Table 2.4.1-1 for rectangular Cartesian, cylindrical, and spherical coordinates.

If the density following a fluid particle does not change as a function of time, (1.3.3-3) reduces to

$$\operatorname{div} \mathbf{v} = 0 \qquad (1.3.3\text{-}5)$$

Such a motion is said to be *isochoric*. If, for the flow under consideration, density is independent of both time and position, we will say that the fluid is *incompressible*. A sufficient, though not necessary, condition for an isochoric motion is that the fluid is incompressible.

**Exercise 1.3.3-1** *Another form of the transport theorem*    Show that, if we assume that mass is conserved, for any $\hat{\Psi}$

$$\frac{d}{dt}\int_{R_{(m)}}\rho\hat{\Psi}\,dV = \int_{R_{(m)}}\rho\frac{d_{(m)}\hat{\Psi}}{dt}\,dV$$

**Exercise 1.3.3-2**    Derive the forms of the differential mass balance shown in Table 2.4.1-1.

**Exercise 1.3.3-3**

i) From (1.3.2-1) and the postulate of conservation of mass, determine that

$$\frac{d_{(m)}}{dt}[\ln(\rho J)] = 0$$

ii) Integrate this equation to conclude that

$$J \equiv |\det \mathbf{F}|$$
$$= \frac{\rho_0}{\rho}$$

where $\rho_0$ denotes the density distribution in the reference configuration.

**Exercise 1.3.3-4** *When volume integral over arbitrary body is zero, the integrand is zero.*    Let us examine the argument that must be supplied in going from (1.3.3-2) to (1.3.3-3).

We can begin by considering the analogous problem in one dimension. It is clear that

$$\int_0^{2\pi}\sin\theta\,d\theta = 0$$

does not imply that $\sin\theta$ is identically zero. But

$$\int_0^x f(y)\,dy = 0 \qquad\qquad (1.3.3\text{-}6)$$

does imply that

$$f(y) = 0 \qquad\qquad (1.3.3\text{-}7)$$

*Proof:*    The Leibnitz rule for the derivative of an integral states that (Kaplan 1952, p. 220)

$$\frac{d}{dx}\int_{a(x)}^{b(x)} g(x,y)\,dy = g(x,b(x))\frac{db}{dx} - g(x,a(x))\frac{da}{dx} + \int_{a(x)}^{b(x)}\frac{\partial g}{\partial x}\,dy$$

If we apply the Leibnitz rule to

$$\frac{d}{dx}\int_0^x f(y)\,dy$$

Equation (1.3.3-7) follows immediately.

Let us consider the analogous problem for

$$\int_{\zeta_1}^{\zeta_2}\int_{\eta_1(z)}^{\eta_2(z)}\int_{\xi_1(y,z)}^{\xi_2(y,z)} g(x,y,z)\,dx\,dy\,dz = 0 \qquad\qquad (1.3.3\text{-}8)$$

where $\xi_1(y, z)$, $\xi_2(y, z)$, $\eta_1(z)$, $\eta_2(z)$, $\zeta_1$, and $\zeta_2$ are completely arbitrary. Prove that this implies

$$g(x, y, z) = 0 \qquad\qquad\qquad\qquad\qquad (1.3.3\text{-}9)$$

**Exercise 1.3.3-5** *Frame indifference of differential mass balance*  Prove that the differential mass balance takes the same form in every frame of reference.

## 1.3.4  Phase Interface

A *phase interface* is that region separating two phases in which the properties or behavior of the material differ from those of the adjoining phases. There is considerable evidence that density is appreciably different in the neighborhood of an interface (Defay et al. 1966, p. 29). As the critical point is approached, density is observed to be a continuous function of position in the direction normal to the interface (Hein 1914; Winkler and Maass 1933; Maass 1938; McIntosh et al. 1939; Palmer 1952). This suggests that the phase interface might be best regarded as a three-dimensional region, the thickness of which may be several molecular diameters or more.

Although it is appealing to regard the interface as a three-dimensional region, there is an inherent difficulty. Except in the neighborhood of the critical point where the interface is sufficiently thick for instrumentation to be inserted, the density and velocity distributions in the interfacial region can be observed only indirectly through their influence upon the adjacent phases.

Gibbs (1928, p. 219) proposed that a phase interface at rest or at equilibrium be regarded as a hypothetical two-dimensional *dividing surface* that lies within or near the interfacial region and separates two homogeneous phases. He suggested that the cumulative effects of the interface upon the adjoining phases be taken into account by the assignment to the dividing surface of any excess mass or energy not accounted for by the adjoining homogeneous phases.

Gibbs's concept may be extended to include dynamic phenomena if we define a homogeneous phase to be one throughout which each description of material behavior applies uniformly. In what follows, we will represent phase interfaces as dividing surfaces. In most problems of engineering significance, we can neglect the effects of excess mass, momentum, energy, and entropy associated with this dividing surface. For this reason, they will be neglected here. For a more detailed discussion of these interfacial effects, see Slattery (1990).

## 1.3.5  Transport Theorem for a Region Containing a Dividing Surface

As described in the preceding section, we will be representing the phase interface as a dividing surface, a surface at which one or more quantities such as density and velocity are discontinuous. In general, a dividing surface is not material; it is common for mass to be transferred across it. As an ice cube melts, as water evaporates, or as solid carbon dioxide sublimes, material is transferred across it and the phase interface moves through the material. We assume that this dividing surface may be in motion through the material with an arbitrary speed of displacement. If $\mathbf{u}$ denotes the velocity of a point on the surface, $\mathbf{u} \cdot \boldsymbol{\xi}^+$ is the *speed of displacement* of the surface measured in the direction $\boldsymbol{\xi}^+$ in Figure 1.3.5-1, and $\mathbf{u} \cdot \boldsymbol{\xi}^-$ is the speed of displacement of the surface measured in the direction $\boldsymbol{\xi}^-$ (Truesdell and Toupin 1960, p. 499).

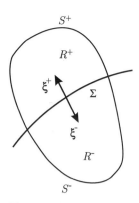

**Figure 1.3.5-1.**    Region containing a dividing surface $\Sigma$.

A typical material region exhibiting a dividing surface $\Sigma$ is illustrated in Figure 1.3.5-1. The quantities $\Psi$ and $\mathbf{v}$ are assumed to be continuously differentiable in the regions $R^+$ and $R^-$. Since in general the dividing surface $\Sigma$ is not material, the regions $R^+$ and $R^-$ are not material. We may write

$$\frac{d}{dt} \int_{R_{(m)}} \Psi \, dV = \frac{d}{dt} \int_{R^+} \Psi \, dV + \frac{d}{dt} \int_{R^-} \Psi \, dV \tag{1.3.5-1}$$

To each term on the right of (1.3.5-1), we may apply the generalized transport theorem of Section 1.3.2 to obtain

$$\frac{d}{dt} \int_{R^+} \Psi \, dV = \int_{R^+} \frac{\partial \Psi}{\partial t} \, dV + \int_{S^+} \Psi \mathbf{v} \cdot \mathbf{n} \, dA - \int_{\Sigma} \Psi^+ \mathbf{u} \cdot \boldsymbol{\xi}^+ \, dA \tag{1.3.5-2}$$

and

$$\frac{d}{dt} \int_{R^-} \Psi \, dV = \int_{R^-} \frac{\partial \Psi}{\partial t} \, dV + \int_{S^-} \Psi \mathbf{v} \cdot \mathbf{n} \, dA - \int_{\Sigma} \Psi^- \mathbf{u} \cdot \boldsymbol{\xi}^- \, dA \tag{1.3.5-3}$$

By $\Psi^+$ and $\Psi^-$, I mean the limits of the function $\Psi$ obtained as any point $\mathbf{z}$ approaches a point $\mathbf{z}_0$ on $\Sigma$ while remaining within $R^+$ and $R^-$, respectively.

Substituting these expressions into (1.3.5-1), we conclude that

$$\frac{d}{dt} \int_{R_{(m)}} \Psi \, dV = \int_{R_{(m)}} \frac{\partial \Psi}{\partial t} \, dV + \int_{S_{(m)}} \Psi \mathbf{v} \cdot \mathbf{n} \, dA - \int_{\Sigma} \left( \Psi^+ \mathbf{u} \cdot \boldsymbol{\xi}^+ + \Psi^- \mathbf{u} \cdot \boldsymbol{\xi}^- \right) dA$$

$$= \int_{R_{(m)}} \frac{\partial \Psi}{\partial t} \, dV + \int_{S_{(m)}} \Psi \mathbf{v} \cdot \mathbf{n} \, dA - \int_{\Sigma} [\![ \Psi \mathbf{u} \cdot \boldsymbol{\xi} ]\!] \, dA \tag{1.3.5-4}$$

where the boldface brackets denote the jump of the quantity enclosed across the interface:

$$[\![ A \boldsymbol{\xi} ]\!] \equiv A^+ \boldsymbol{\xi}^+ + A^- \boldsymbol{\xi}^- \tag{1.3.5-5}$$

Finally, we can use Green's transformation (Section A.11.2) to rewrite (1.3.5-4) as

$$\frac{d}{dt} \int_{R_{(m)}} \Psi \, dV = \int_{R_{(m)}} \left[ \frac{\partial \Psi}{\partial t} + \mathrm{div}(\Psi \mathbf{v}) \right] dV + \int_{\Sigma} [\![ \Psi (\mathbf{v} - \mathbf{u}) \cdot \boldsymbol{\xi} ]\!] \, dA \tag{1.3.5-6}$$

or

$$\frac{d}{dt} \int_{R_{(m)}} \Psi \, dV = \int_{R_{(m)}} \left( \frac{d_{(m)} \Psi}{dt} + \Psi \, \mathrm{div} \, \mathbf{v} \right) dV + \int_{\Sigma} \llbracket \Psi (\mathbf{v} - \mathbf{u}) \cdot \boldsymbol{\xi} \rrbracket \, dA \qquad (1.3.5\text{-}7)$$

We will refer to (1.3.5-7) as the *transport theorem for a region containing a dividing surface*. This discussion is based upon that of Truesdell and Toupin (1960, p. 525).

### 1.3.6 Jump Mass Balance for Phase Interface

In Section 1.3.3, we found that the differential mass balance expresses the requirement that mass is conserved at every point within a continuous material. We wish to examine the implications of mass conservation at a phase interface. As described in Section 1.3.4, we will represent the phase interface as a dividing surface, rather than as a three-dimensional region of some thickness.

Equation (1.3.1-1) again applies to the multiphase body drawn in Figure 1.3.5-1:

$$\frac{d}{dt} \int_{R_{(m)}} \rho \, dV = 0 \qquad (1.3.6\text{-}1)$$

By the transport theorem for a region containing a dividing surface given in Section 1.3.5, Equation (1.3.6-1) may be written as

$$\frac{d}{dt} \int_{R_{(m)}} \rho \, dV = \int_{R_{(m)}} \left( \frac{d_{(m)} \rho}{dt} + \rho \, \mathrm{div} \, \mathbf{v} \right) dV + \int_{\Sigma} \llbracket \rho (\mathbf{v} - \mathbf{u}) \cdot \boldsymbol{\xi} \rrbracket \, dA \qquad (1.3.6\text{-}2)$$

Since the differential mass balance developed in Section 1.3.3 applies everywhere within each phase, (1.3.6-2) reduces to

$$\int_{\Sigma} \llbracket \rho (\mathbf{v} - \mathbf{u}) \cdot \boldsymbol{\xi} \rrbracket \, dA = 0 \qquad (1.3.6\text{-}3)$$

This must be true for any portion of a body containing a phase interface, no matter how large or small the body is. We conclude that the integrand itself must be zero:

$$\llbracket \rho (\mathbf{v} - \mathbf{u}) \cdot \boldsymbol{\xi} \rrbracket = 0 \qquad (1.3.6\text{-}4)$$

This is known as the *jump mass balance* for a phase interface, which is represented by a dividing surface, when all interfacial effects are neglected (Slattery 1990).

**Exercise 1.3.6-1** Discuss how one concludes that, when there is no mass transfer across the phase interface, (1.3.6-4) reduces to

$$\mathbf{u} \cdot \boldsymbol{\xi} = \mathbf{v}^+ \cdot \boldsymbol{\xi} = \mathbf{v}^- \cdot \boldsymbol{\xi}$$

**Exercise 1.3.6-2** *Alternative derivation of jump mass balance* Write the postulate of conservation of mass for a material region that instantaneously contains a phase interface. Employ the transport theorem for a region containing a dividing surface in the form of (1.3.5-4). Deduce the jump mass balance (1.3.6-4) by allowing the material region to shrink around the phase interface (surface of discontinuity) (Truesdell and Toupin 1960, p. 526).

**Exercise 1.3.6-3** *Alternative form of transport theorem for regions containing a dividing surface*   Show that, since mass is conserved, for any $\hat{\Psi}$

$$\frac{d}{dt} \int_{R_{(m)}} \rho \hat{\Psi} \, dV = \int_{R_{(m)}} \rho \frac{d_{(m)} \hat{\Psi}}{dt} \, dV + \int_{\Sigma} \left[ \rho \hat{\Psi} \left( \mathbf{v} - \mathbf{u} \right) \cdot \boldsymbol{\xi} \right] dA$$

**Exercise 1.3.6-4** *Frame indifference of jump mass balance*   Prove that the jump mass balance takes the same form in every frame of reference.

## 1.3.7   Stream Functions

By a two-dimensional motion, we mean here one such that in some coordinate system the velocity field has only two nonzero components.

Let us further restrict our discussion to incompressible fluids, so that the differential mass balance (1.3.3-3) reduces to

$$\text{div } \mathbf{v} = 0 \tag{1.3.7-1}$$

Consider a two-dimensional motion such that in spherical coordinates

$$v_r = v_r(r, \theta), \qquad v_\theta = v_\theta(r, \theta), \qquad v_\varphi = 0 \tag{1.3.7-2}$$

From Table 2.4.1-1, Equation (1.3.7-1) takes the form

$$\frac{1}{r^2} \frac{\partial}{\partial r}(r^2 v_r) + \frac{1}{r \sin\theta} \frac{\partial}{\partial \theta}(v_\theta \sin\theta) = 0 \tag{1.3.7-3}$$

Multiplying by $r^2 \sin\theta$, we may also write this as

$$\frac{\partial}{\partial r}(v_r r^2 \sin\theta) = \frac{\partial}{\partial \theta}(-v_\theta r \sin\theta) \tag{1.3.7-4}$$

Upon comparing (1.3.7-4) with

$$\frac{\partial^2 \psi}{\partial r \, \partial \theta} = \frac{\partial^2 \psi}{\partial \theta \, \partial r} \tag{1.3.7-5}$$

we see that we may define a *stream function* $\psi$ such that

$$v_r = \frac{1}{r^2 \sin\theta} \frac{\partial \psi}{\partial \theta} \tag{1.3.7-6}$$

and

$$v_\theta = -\frac{1}{r \sin\theta} \frac{\partial \psi}{\partial r} \tag{1.3.7-7}$$

The advantage of such a stream function $\psi$ is that it can be used in this way to satisfy identically the differential mass balance for the flow described by (1.3.7-2).

Expressions for velocity components in terms of a stream function are presented for several situations in Tables 2.4.2-1.

**Exercise 1.3.7-1**   Using arguments similar to those employed in this section, for each velocity distribution in Table 2.4.2-1, express the nonzero components of velocity in terms of a stream function.

**Exercise 1.3.7-2** *Stream function is a constant along a streamline.* Prove that, in a two-dimensional flow, the stream function is a constant along a streamline. This allows one to readily plot streamlines as in Figures 1.1.0-2 and 3.4.2-1.

*Hint:* Let $\alpha$ be a parameter measured along a streamline. Begin by recognizing that

$$\frac{d\mathbf{p}}{d\alpha} \wedge \mathbf{v} = 0.$$

**Exercise 1.3.7-3** *Another view of the stream function* Show that the differential mass balance for an incompressible fluid in a two-dimensional flow can be expressed as

$$\frac{\partial}{\partial x^1}(-\sqrt{g}\,v^1) = \frac{\partial}{\partial x^2}(\sqrt{g}\,v^2)$$

This is a necessary and sufficient condition for the existence of a stream function $\psi$ such that

$$\sqrt{g}\,v^1 = -\frac{\partial \psi}{\partial x^2}$$

$$\sqrt{g}\,v^2 = \frac{\partial \psi}{\partial x^1}$$

For further discussion of these ideas as well as an extension of the concept of a stream function to steady compressible flows, see Truesdell and Toupin (1960, p. 477).

# 2

---

# Foundations for Momentum Transfer

I N WHAT FOLLOWS, the principal tools for studying fluid mechanics are developed. We begin by introducing the concept of force. Notice that force is not defined; it is a primitive concept in the same sense as is mass and the material particle in Chapter 1. This forms the basis for introducing our second and third postulates: the momentum balance and the moment of momentum balance. The stress tensor is introduced in order to derive the equations that describe at each point in a material the local balances for momentum and moment of momentum. The differential mass and momentum balances together with the symmetry of the stress tensor form the foundation for fluid mechanics.

We conclude our discussion with an outline of what must be said about real material behavior if we are to analyze any practical problems. It is especially at this point that statistical mechanics, based upon the molecular viewpoint of real materials, can be used to supplement the concepts developed in continuum mechanics. In continuum mechanics, we can indicate a number of rules that constitutive equations for the stress tensor must satisfy (the principle of determinism, the principle of local action, the principle of material frame indifference, . . . ), but from first principles we cannot derive an explicit relationship between stress and deformation. If we work strictly within the bounds of continuum mechanics, we can derive such a relationship only by making some sort of assumption about its form. Our feelings are that the most interesting advances in describing material behavior result from predictions based upon simple molecular models that are generalized through the use of the statements about material behavior that have been postulated in continuum mechanics. For an excellent brief summary of what can be said about material behavior from a molecular point of view, see Bird et al. (1960, Chap. 1).

Although our direct concerns here are for momentum transfer, practically all of the ideas developed will be applied again in examining energy and mass transfer. We firmly believe that the best foundation for energy and mass-transfer studies is a clear understanding of fluid mechanics.

---

## 2.1  Force

Like the *material particle*, *force* is a primitive concept. It is not defined. Instead we describe its attributes in a series of five axioms.

Corresponding to each body $B$, there is a distinct set of bodies $B^e$ such that the mass of the union of these bodies is the mass of the universe. We refer to $B^e$ as the exterior or the surroundings of the body $B$.

a) A system of forces is a vector-valued function $\mathbf{f}(B, C)$ of pairs of bodies. The value of $\mathbf{f}(B, C)$ is called the force exerted on the body $B$ by the body $C$.
b) For a specified body $B$, $\mathbf{f}(C, B^e)$ is an additive function defined over the subbodies $C$ of $B$.
c) Conversely, for a specified body $B$, $\mathbf{f}(B, C)$ is an additive function defined over the subbodies $C$ of $B^e$.
d) In any particular problem, we regard the forces exerted upon a body as being given a priori to all observers; all observers would assume the same set of forces in a given problem. In prescribing these forces, we specify a particular dynamic problem. We consequently assume that all forces are independent of the observer or are *frame indifferent* (Truesdell 1966a, p. 27) (see Section 1.2.1):

$$\mathbf{f}^* = \mathbf{Q} \cdot \mathbf{f} \qquad (2.1.0\text{-}1)$$

There are three types of forces with which we may be concerned:

**External forces**   These arise at least in part from outside the body and act upon the material particles of which the body is composed. One example is the uniform force of gravity. Another example would be the electrostatic force between two charged bodies. Let $P$ indicate a portion of a body $B$ as illustrated in Figure 2.1.0-1. Taking $\mathbf{f}_e$ to be external force per unit mass that the surroundings $B^e$ exert on the body $B$, we write the total external force acting on $P$ in terms of a volume integral over the region occupied by $P$:

$$\int_{R_P} \rho \mathbf{f}_e \, dV$$

In general, the external force per unit mass is a function of position, and $\mathbf{f}_e$ should be regarded as a spatial vector field.

**Mutual forces**   These arise within a body and act upon pairs of material particles. The long-range intermolecular forces acting between a thin film of liquid and the solid upon which it rests are mutual forces. We can imagine a body in which there is a distribution of electrostatic charge; we would speak of the electrostatic force between one portion of the body with a net positive charge and some other element of the body with a net negative charge as being a

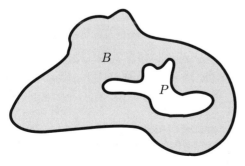

**Figure 2.1.0-1.** The body $B$ of which $P$ is a portion.

mutual force. Let $\mathbf{f}_m$ be the mutual force per unit mass that $B - P$[1] exerts upon $P$; the total mutual force acting upon $P$ may be represented as an integral over the volume of $P$:[2]

$$\int_{R_P} \rho \mathbf{f}_m \, dV$$

We should expect the mutual force per unit mass generally to be a function of position within the material; $\mathbf{f}_m$ should be viewed as a spatial vector field.

**Contact forces**    These forces are not assignable as functions of position but are to be imagined as acting upon the bounding surface of a portion of material in such a way as to be equivalent to the force exerted by one portion of the material upon another beyond that accounted for through mutual forces. In typing you exert a contact force upon the keys of the typewriter. If we deform some putty in our hands, during the deformation any one portion of the putty exerts a contact force upon the remainder at their common boundary. Let $\mathbf{t} = \mathbf{t}(\mathbf{z}, P)$ represent the *stress* vector or force per unit area that $B - P$ exerts upon the boundary of $P$ at the position $\mathbf{z}$. This force per unit area $\mathbf{t}$ is usually referred to as *stress*. The total contact force that $B - P$ exerts upon $P$ may be written as an integral over the bounding surface of $P$:

$$\int_{S_P} \mathbf{t} \, dA$$

The fifth axiom, the stress principle, specifies the nature of the contact load.

e) **Stress principle**    There is a vector-valued function $\mathbf{t}(\mathbf{z}, \mathbf{n})$ defined for all points $\mathbf{z}$ in a body $B$ and for all unit vectors $\mathbf{n}$ such that the stress that $B - P$ exerts upon any portion $P$ of $B$ is given by

$$\mathbf{t}(\mathbf{z}, P) = \mathbf{t}(\mathbf{z}, \mathbf{n}) \tag{2.1.0-2}$$

Here $\mathbf{n}$ is the unit normal that is outwardly directed with respect to the closed bounding surface of $P$. The spatial vector $\mathbf{t} = \mathbf{t}(\mathbf{z}, \mathbf{n})$ is referred to as the stress vector at the position $\mathbf{z}$ acting upon the oriented surface element with normal $\mathbf{n}$; $\mathbf{n}$ points into the material that exerts the stress $\mathbf{t}$ upon the surface element.

The material in this section is drawn from Truesdell (1966b, p. 97), Truesdell and Toupin (1960, pp. 531 and 536) and Truesdell and Noll (1965, p. 39).

---

## 2.2    Additional Postulates

### 2.2.1    Momentum and Moment of Momentum Balance

In Section 1.3.1, we introduced our first postulate, conservation of mass. Our second postulate is (Truesdell 1966b, p. 97; Truesdell and Toupin 1960, pp. 531 and 537; Truesdell and Noll 1965, p. 39)[3]

---

[1] We define $B - P$ to be such that $B = (B - P) \cup P$ and $(B - P) \cap P = 0$.

[2] We recognize here that the sum of the mutual forces exerted by any two parts of $P$ upon each other is zero (Truesdell and Toupin 1960, p. 533).

[3] Truesdell and Toupin (1960, pp. 531 and 534) point out that "the laws of Newton ... are neither unequivocally stated nor sufficiently general to serve as a foundation for continuum mechanics."

**Momentum balance** The time rate of change of the momentum of a body relative to an inertial frame of reference is equal to the sum of the forces acting on the body.

Let the volume and closed bounding surface of a body or any portion of a body be denoted, respectively, as $R_{(m)}$ and $S_{(m)}$. Referring to our discussion of forces in the introduction to Section 2.1, in an inertial frame of reference we may express the momentum balance as

$$\frac{d}{dt} \int_{R_{(m)}} \rho \mathbf{v}\, dV = \int_{S_{(m)}} \mathbf{t}\, dA + \int_{R_{(m)}} \rho \mathbf{f}\, dV \tag{2.2.1-1}$$

Here $\mathbf{f}$ is the field of external and mutual forces per unit mass:

$$\mathbf{f} \equiv \mathbf{f}_e + \mathbf{f}_m \tag{2.2.1-2}$$

In most cases, the effect of mutual forces can be neglected with respect to external forces, one of the primary exceptions being thin films, where long-range, intermolecular forces exerted by the adjoining phases must be taken into account. Hereafter we will assume that mutual forces have been dismissed, and we will refer to $\mathbf{f}$ as the field of external forces per unit mass.

Our understanding is that (2.2.1-1) is written with respect to an *inertial frame of reference*. In reality, we define an inertial frame of reference to be one in which (2.2.1-1) and (2.2.1-3) are valid. Somewhat more casually we describe an inertial frame of reference to be one that is stationary with respect to the *fixed stars*.

Our third postulate is (Truesdell 1966b, p. 97; Truesdell and Toupin 1960, pp. 531 and 537; Truesdell and Noll 1965, p. 39).

**Moment of momentum balance** The time rate of change of the moment of momentum of a body relative to an inertial frame of reference is equal to the sum of the moments of all the forces acting on the body.

In an *inertial frame of reference*, the moment of momentum balance assumes the form

$$\frac{d}{dt} \int_{R_{(m)}} \rho(\mathbf{p} \wedge \mathbf{v})\, dV = \int_{S_{(m)}} \mathbf{p} \wedge \mathbf{t}\, dA + \int_{R_{(m)}} \rho(\mathbf{p} \wedge \mathbf{f})\, dV \tag{2.2.1-3}$$

In writing the moment of momentum balance in this manner we confine our attention to the so-called *nonpolar*[4] case [i.e., we assume that all torques acting on the body are the result of forces acting on the body (Truesdell and Toupin 1960, pp. 538 and 546; Curtiss 1956; Livingston and Curtiss 1959; Dahler and Scriven 1961)]. For example, it is possible to induce a local source of moment of momentum by use of a suitable rotating electric field (Lertes 1921a,b,c; Grossetti 1958, 1959). In such a case it might also be necessary to account for the flux of moment of momentum at the bounding surface of the body. Effects of this type have not been investigated thoroughly, but they are thought to be negligibly small for all but unusual situations. Consequently, they are neglected here.

---

[4] When molecules are referred to as *nonpolar*, it indicates that their dipole moment is zero. This is an entirely different use of the word than that intended here, where *nonpolar* means that all torques acting on the material are the result of forces.

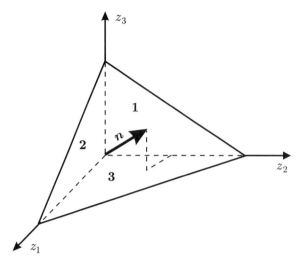

**Figure 2.2.2-1.** A material body in the form of a tetrahedron.

**Exercise 2.2.1-1** *Cauchy's lemma*   Consider two neighboring portions of a continuous body. Apply the momentum balance to each portion and to their union. Deduce that on their common boundary

$$\mathbf{t}(\mathbf{z}, \mathbf{n}) = -\mathbf{t}(\mathbf{z}, -\mathbf{n})$$

or

**Cauchy's lemma**   The stress vectors acting upon opposite sides of the same surface at a given point are equal in magnitude and opposite in direction.

### 2.2.2  Stress Tensor

We ask here how $\mathbf{t}(\mathbf{z}, \mathbf{n})$ varies as the position $\mathbf{z}$ is held fixed and $\mathbf{n}$ changes.

At any point in a body, consider the tetrahedron shown in Figure 2.2.2-1. Three sides are mutually orthogonal and coincide with a set of rectangular Cartesian coordinate planes intersecting at $\mathbf{z}$; the fourth side has an outwardly directed normal $\mathbf{n}$. Let the altitude of the tetrahedron be $h$; let the area of the inclined face be $A$. In terms of the rectangular Cartesian basis fields, we may write $\mathbf{n} = n_i \mathbf{e}_i$.

Let us apply the momentum balance (2.2.1-1) to the material in the tetrahedron at time $t$ to obtain

$$\frac{d}{dt} \int_{R_{(m)}} \rho \mathbf{v} \, dV = \int_{S_{(m)}} \mathbf{t} \, dA + \int_{R_{(m)}} \rho \mathbf{f} \, dV \tag{2.2.2-1}$$

Applying the form of the transport theorem introduced in Exercise 1.3.3-1 to the term on the left and using the theorem of mean value to evaluate the surface integral, we have

$$\int_{R_{(m)}} \left( \rho \frac{d_{(m)} \mathbf{v}}{dt} - \rho \mathbf{f} \right) dV = A \left( \mathbf{t}^* + \frac{A_1}{A} \mathbf{t}_1^* + \frac{A_2}{A} \mathbf{t}_2^* + \frac{A_3}{A} \mathbf{t}_3^* \right) \tag{2.2.2-2}$$

Here the asterisk denotes a mean value of the function, $t^*$ denotes the contact stress that the surroundings exert upon the inclined face, and $\mathbf{t}_i^*$ denotes the contact stress that the surroundings exert upon the face whose inwardly directed unit normal is $\mathbf{e}_i$. Applying the theorem of mean value to the term on the left, we write

$$\left(\rho \frac{d_{(m)}\mathbf{v}}{dt} - \rho\mathbf{f}\right)^* \frac{hA}{3} = A\left(\mathbf{t}^* + n_1\mathbf{t}_1^* + n_2\mathbf{t}_2^* + n_3\mathbf{t}_3^*\right) \tag{2.2.2-3}$$

Since the cosines of the angles between the Cartesian coordinate planes and the inclined plane are $n_1$, $n_2$, and $n_3$, respectively, the areas of the faces of the tetrahedron lying in the coordinate planes are $n_1A$, $n_2A$, and $n_3A$. Consider a sequence of geometrically similar tetrahedrons; in the limit as $h$ approaches zero, we obtain

$$\mathbf{t} = -(\mathbf{t}_1 n_1 + \mathbf{t}_2 n_2 + \mathbf{t}_3 n_3) \tag{2.2.2-4}$$

where all stress vectors are evaluated at the point $\mathbf{z}$. From their definitions, the quantities $\mathbf{t}_1$, $\mathbf{t}_2$, and $\mathbf{t}_3$ do not depend upon $\mathbf{n}$.

With the convention that $T_{km}$ is the $k$th component of the stress vector acting upon the *positive* side of the plane $z_m = $ constant (such that the unit normal is $\mathbf{e}_m$), by *Cauchy's lemma* (Exercise 2.2.1-1) we write

$$-\mathbf{t}_1 = T_{i1}\mathbf{e}_i$$
$$-\mathbf{t}_2 = T_{i2}\mathbf{e}_i \tag{2.2.2-5}$$
$$-\mathbf{t}_3 = T_{i3}\mathbf{e}_i$$

This permits us to express (2.2.2-4) as

$$\mathbf{t} = n_j T_{ij}\mathbf{e}_i \tag{2.2.2-6}$$

The matrix $[T_{ij}]$ defines a second-order *stress tensor* $\mathbf{T}$:

$$\mathbf{T} \equiv T_{ij}\mathbf{e}_i\mathbf{e}_j \tag{2.2.2-7}$$

and (2.2.2-6) becomes

$$\mathbf{t} = \mathbf{T} \cdot \mathbf{n} \tag{2.2.2-8}$$

Remember in using (2.2.2-8) that $\mathbf{n}$ is the unit normal directed into the material that is exerting the force per unit area $\mathbf{t}(\mathbf{z}, \mathbf{n})$ at the position $\mathbf{z}$ on the surface.

**Exercise 2.2.2-1**  Show that the stress tensor is frame indifferent.

**Exercise 2.2.2-2**  In going from (2.2.2-2) to (2.2.2-3), prove that

$$n_i = \frac{A_i}{A}, \quad i = 1, 2, 3$$

## 2.2.3  Differential and Jump Momentum Balances

A modification of the transport theorem that takes into account the postulate that mass is conserved (see Exercise 1.3.6-3) may be used to express the momentum balance (2.2.1-1) as

$$\int_{R_{(m)}} \rho \frac{d_{(m)}\mathbf{v}}{dt} dV + \int_{\Sigma} \left[\rho\mathbf{v}(\mathbf{v} - \mathbf{u}) \cdot \boldsymbol{\xi}\right] dA = \int_{S_{(m)}} \mathbf{t}\, dA + \int_{R_{(m)}} \rho\mathbf{f}\, dV \tag{2.2.3-1}$$

If we express the stress vector **t** in terms of the stress tensor **T** as suggested in Section 2.2.2, the first term on the right of this equation may be rearranged by an application of Green's transformation (Section A.11.2):

$$\int_{S_{(m)}} \mathbf{t}\, dA = \int_{S_{(m)}} \mathbf{T} \cdot \mathbf{n}\, dA$$

$$= \int_{R_{(m)}} \operatorname{div} \mathbf{T}\, dV + \int_{\Sigma} [\![\mathbf{T} \cdot \boldsymbol{\xi}]\!]\, dA \qquad (2.2.3\text{-}2)$$

Substituting (2.2.3-2) into (2.2.3-1), we have

$$\int_{R_{(m)}} \left( \rho \frac{d_{(m)}\mathbf{v}}{dt} - \operatorname{div} \mathbf{T} - \rho \mathbf{f} \right) dV + \int_{\Sigma} [\![\rho \mathbf{v}\,(\mathbf{v} - \mathbf{u}) \cdot \boldsymbol{\xi} - \mathbf{T} \cdot \boldsymbol{\xi}]\!]\, dA = 0 \qquad (2.2.3\text{-}3)$$

Remember that (2.2.3-1) is written for an arbitrary portion $P$ of a body. Whether $P$ is very large or arbitrarily small or whether $P$ consists of one phase or many phases, (2.2.3-3) remains valid. This implies that the *differential momentum balance* (*Cauchy's first "law"*),

$$\rho \frac{d_{(m)}\mathbf{v}}{dt} = \operatorname{div} \mathbf{T} + \rho \mathbf{f} \qquad (2.2.3\text{-}4)$$

must be satisfied at each point within each phase and that the *jump momentum balance*,

$$[\![\rho \mathbf{v}\,(\mathbf{v} - \mathbf{u}) \cdot \boldsymbol{\xi} - \mathbf{T} \cdot \boldsymbol{\xi}]\!] = 0 \qquad (2.2.3\text{-}5)$$

must be obeyed at each point on each phase interface. Here $\mathbf{u} \cdot \boldsymbol{\xi}$ is the speed of displacement of the phase interface; the boldface bracket notation is defined in Section 1.3.5.

**Exercise 2.2.3-1** *Archimedes principles*   Prove the following two theorems of Archimedes (Rouse and Ince 1957, p. 17):

> If a solid lighter than a fluid be forcibly immersed in it, the solid will be driven upwards by a force equal to the difference between its weight and the weight of the fluid displaced.

> A solid heavier than a fluid will, if placed in it, descend to the bottom of the fluid, and the solid will, when weighed in the fluid, be lighter than its true weight by the weight of the fluid displaced.

*Hint:*   Apply the momentum balance and extend the definition of fluid pressure into the region of space occupied by the solid, recognizing that pressure in the fluid is a constant in any horizontal plane.

**Exercise 2.2.3-2** *Another theorem of Archimedes*   Prove (Rouse and Ince 1957, p. 17):

> Any solid lighter than a fluid will, if placed in the fluid, be so far immersed that the weight of the solid will be equal to the weight of the fluid displaced.

*Hint:*   Proceed as in Exercise 2.2.3-1, making use of the jump momentum balance (2.2.3-5).

### Exercise 2.2.3-3 *Another frame of reference*

i) Use the result of Exercise 1.2.2-2 to write the differential momentum balance for an arbitrary frame of reference:

$$\rho \left\{ \frac{d_{(m)}\mathbf{v}^*}{dt} - \frac{d^2\mathbf{z}^*_{(0)}}{dt^2} - \frac{d\boldsymbol{\omega}}{dt} \wedge \left( \mathbf{z}^* - \mathbf{z}^*_{(0)} \right) + \boldsymbol{\omega} \wedge \left[ \boldsymbol{\omega} \wedge \left( \mathbf{z}^* - \mathbf{z}^*_{(0)} \right) \right] \right.$$

$$\left. - 2\boldsymbol{\omega} \wedge \left( \mathbf{v}^* - \frac{d\mathbf{z}^*_{(0)}}{dt} \right) \right\} = \operatorname{div} \mathbf{T}^* + \rho \mathbf{f}^*$$

ii) Determine that the differential momentum balance assumes the form of (2.2.3-4) in every frame of reference that moves at a constant velocity (without rotation) relative to the inertial frame of reference.

## 2.2.4 Symmetry of Stress Tensor

We again restrict ourselves to a two-phase body in which all quantities are smooth functions of position and time as described at the beginning of Section 2.2.3.

After an application of the transport theorem (see Exercise 1.3.6-3) and the differential mass balance, the moment of momentum balance (2.2.1-3) may be written as

$$\int_{R_{(m)}} \rho \frac{d_{(m)}(\mathbf{p} \wedge \mathbf{v})}{dt} \, dV + \int_{\Sigma} \left[ \mathbf{p} \wedge \mathbf{v}(\mathbf{v} - \mathbf{u}) \cdot \boldsymbol{\xi} \right] dA$$

$$= \int_{S_{(m)}} \mathbf{p} \wedge (\mathbf{T} \cdot \mathbf{n}) \, dA + \int_{R_{(m)}} \rho(\mathbf{p} \wedge \mathbf{f}) \, dV \qquad (2.2.4\text{-}1)$$

In writing this equation we have also expressed the stress vector in terms of the stress tensor as described in Section 2.2.2. Let us consider these expressions individually.

From the left-hand term, we have

$$\frac{d_{(m)}(\mathbf{p} \wedge \mathbf{v})}{dt} = \frac{d_{(m)}\mathbf{p}}{dt} \wedge \mathbf{v} + \mathbf{p} \wedge \frac{d_{(m)}\mathbf{v}}{dt} \qquad (2.2.4\text{-}2)$$

But since we have defined (Section 1.1)

$$\mathbf{v} \equiv \frac{d_{(m)}\mathbf{p}}{dt} \qquad (2.2.4\text{-}3)$$

we are left with

$$\frac{d_{(m)}}{dt}(\mathbf{p} \wedge \mathbf{v}) = \mathbf{p} \wedge \frac{d_{(m)}\mathbf{v}}{dt} \qquad (2.2.4\text{-}4)$$

By an application of Green's transformation (Section A.11.2), the first term on the right of (2.2.4-1) becomes

$$\int_{S_{(m)}} \mathbf{p} \wedge (\mathbf{T} \cdot \mathbf{n}) dA = \int_{R_{(m)}} \frac{\partial}{\partial z_m} \left( e_{ijk} z_j T_{km} \right) dV \, \mathbf{e}_i + \int_{\Sigma} \left[ \mathbf{p} \wedge (\mathbf{T} \cdot \boldsymbol{\xi}) \right] dA \qquad (2.2.4\text{-}5)$$

We find that

$$\frac{\partial}{\partial z_m}(e_{ijk} z_j T_{km}) = e_{ijk} \frac{\partial z_j}{\partial z_m} T_{km} + e_{ijk} z_j \frac{\partial T_{km}}{\partial z_m} \qquad (2.2.4\text{-}6)$$

But

$$\frac{\partial z_j}{\partial z_m} = \delta_{jm} \tag{2.2.4-7}$$

Equations (2.2.4-6) and (2.2.4-7) allow us to conclude that

$$\frac{\partial}{\partial z_m}(e_{ijk}z_j T_{km})\mathbf{e}_i = e_{ijk}T_{kj}\mathbf{e}_i + \mathbf{p} \wedge (\mathrm{div}\,\mathbf{T}) \tag{2.2.4-8}$$

Equations (2.2.4-4), (2.2.4-5), and (2.2.4-8) enable us to express (2.2.4-1) in the form

$$\int_{R_{(m)}} \left[ \mathbf{p} \wedge \left( \rho\frac{d_{(m)}\mathbf{v}}{dt} - \mathrm{div}\,\mathbf{T} - \rho\mathbf{f} \right) - e_{ijk}T_{kj}\mathbf{e}_i \right] dV$$
$$+ \int_{\Sigma} \left[ \mathbf{p} \wedge \mathbf{v}(\mathbf{v} - \mathbf{u}) \cdot \boldsymbol{\xi} - \mathbf{T} \cdot \boldsymbol{\xi} \right] dA$$
$$= 0 \tag{2.2.4-9}$$

In view of the differential momentum balance and the jump momentum balance (Section 2.2.3), this becomes

$$\int_{R_{(m)}} e_{ijk}T_{kj}\mathbf{e}_i\, dV = 0 \tag{2.2.4-10}$$

But (2.2.4-1) was written for a portion of a body with the understanding that the portion considered might be arbitrarily large or small. This implies from (2.2.4-10) that

$$e_{ijk}T_{kj}\mathbf{e}_i = 0 \tag{2.2.4-11}$$

Since the rectangular Cartesian basis fields are linearly independent, we have

$$e_{ijk}T_{kj} = 0 \tag{2.2.4-12}$$

We may also write (see Exercise A.2.1-2)

$$e_{imn}e_{ijk}T_{kj} = 0$$
$$T_{nm} - T_{mn} = 0 \tag{2.2.4-13}$$

Equation (2.2.4-13) expresses

**Symmetry of stress tensor**    A necessary and sufficient condition for the moment of momentum balance to be satisfied at every point within a phase is that the stress tensor be symmetric (*Cauchy's second "law"*):

$$\mathbf{T} = \mathbf{T}^T \tag{2.2.4-14}$$

as long as the differential momentum balance is satisfied and the body conforms to the nonpolar case (Section 2.2.1).

Hereafter, we need not worry about satisfying the moment of momentum balance, as long as we take the components of the stress tensor to be symmetric.

Exercise 2.2.4-1 *Orientation of suspended body* An irregular solid body is suspended in a fluid from a swivel that permits it to assume an arbitrary orientation. The density of the solid may be either less than or greater than the density of the fluid. If $\mathbf{z}_{swivel}$ is the position vector locating the point at which the swivel is attached to the body and if

$$\mathbf{z}_c \equiv \int_{R^{(s)}} \left( \rho^{(s)} - \rho^{(f)} \right) \mathbf{z} \, dV \left[ \int_{R^{(s)}} \left( \rho^{(s)} - \rho^{(f)} \right) dV \right]^{-1}$$

prove that the orientation of the body will be such that $\mathbf{z}_c - \mathbf{z}_{swivel}$ is parallel to gravity.

Exercise 2.2.4-2 *Orientation of floating body* An irregular solid body is floating at an interface between two fluids, $A$ and $B$. Let $R_{(A)}$ be the region of the solid surrounded by fluid $A$ and the $A$–$B$ interface; let $R_{(B)}$ be the region of the solid surrounded by fluid $B$ and the $A$–$B$ interface. If

$$\mathbf{z}_{cA} \equiv \int_{R^{(A)}} \left( \rho^{(s)} - \rho^{(f)} \right) \mathbf{z} \, dV \left[ \int_{R^{(A)}} \left( \rho^{(s)} - \rho^{(f)} \right) dV \right]^{-1}$$

and

$$\mathbf{z}_{cB} \equiv \int_{R^{(B)}} \left( \rho^{(s)} - \rho^{(f)} \right) \mathbf{z} \, dV \left[ \int_{R^{(B)}} \left( \rho^{(s)} - \rho^{(f)} \right) dV \right]^{-1}$$

prove that the orientation of the floating solid will be such that $\mathbf{z}_{cA} - \mathbf{z}_{cB}$ is parallel to gravity.

## 2.3  Behavior of Materials

### 2.3.1  Some General Principles

It should occur to you that we have said nothing as yet about the behavior of materials. Mass conservation, the momentum balance, and the moment of momentum balance are stated for all materials. Yet our experience tells us that under similar circumstances air and steel respond to forces in drastically different manners. Somewhere in our theoretical structure we must incorporate this information.

To confirm this intuitive feeling, let us consider the mathematical structure we have developed. For simplicity, assume that the material is incompressible so that mass density $\rho$ is a known constant. Let us also assume that the description of the external force field $\mathbf{f}$ is given; for example, we are commonly concerned with physical situations in an essentially uniform gravitational field. As unknowns in some arbitrary coordinate system we are left with the three components of the velocity vector $\mathbf{v}$ and the six components of the symmetric stress tensor $\mathbf{T}$ (Section 2.2.4). As equations, we have the differential mass balance (Section 1.3.3) and the three components of the differential momentum balance (Section 2.2.3). This means we have four equations in nine unknowns. This reinforces our intuitive feeling that further information is required.

While we have assumed the nature of the external force is known, we have said nothing about the character of the force that one portion of a body exerts on its neighboring portion. We indicated in Section 2.2.1 that we would neglect mutual forces. This leaves only the contact force. We must describe how the contact forces in a body depend upon the motion and deformation of the body. More specifically, we must say how the stress tensor **T** varies with the motion and deformation of the body.

Before plunging ahead to relate the stress tensor to motion, let us see if our everyday experience in observing materials in deformations and motions will help us to lay down some rules governing such a relation. For example, it seems obvious that what happens to a body in the future is going to have no influence on the present stress tensor field. This suggests stating (Truesdell 1966b, p. 6; Truesdell and Noll 1965, p. 56)

**The principle of determinism**    The stress in a body is determined by the history of the motion that the body has undergone.

Our experience is that motion in one portion of a body does not necessarily have any effect on the state of stress in another portion of the body. For example, if we lay down a bead of caulking compound, we may shape one portion of the bead with a putty knife without disturbing the rest. From a somewhat different point of view, the physical idea of a contact force suggests that the circumstances in the immediate neighborhood of the point in question determine it. We may state this as (Truesdell 1966b, p. 6; Truesdell and Noll 1965, p. 56)

**The principle of local action**    The motion of the material outside an arbitrarily small neighborhood of the material point $\zeta$ may be ignored in determining the stress at this material point.

Let us consider an experiment in which a series of weights are successively added to one end of a spring, the other end having been attached to the ceiling of a laboratory. Two experimentalists observe this experiment, one standing on the floor of the laboratory near the spring and the other standing across the room on a turntable that rotates with some angular velocity. The frame of reference for the first observer might be his backbone, his shoulders, and his nose or perhaps the walls of the laboratory. The frame of reference for the second observer consists of the turntable's axis and a series of lines painted upon the turntable; to this observer, the spring and weights appear to be revolving in a circle. Yet we expect both observers to come to the same conclusions regarding the behavior of the spring under stress. Referring to Sections 1.2.1 and 1.2.2 and to Exercise 2.2.2-1, we can summarize our feeling here with (Truesdell and Noll 1965, p. 44)[5]

**The principle of frame indifference**    Descriptions of material behavior must be invariant under changes of frame of reference.

If a description of stress-deformation behavior is satisfied for a process in which the stress tensor and motion are given by

$$\mathbf{T} = \mathbf{T}(\mathbf{z}_\kappa, t) \tag{2.3.1-1}$$

---

[5] In writing this text, I have made an effort to refer the reader to the more lucid reference rather than the historical "first" when a choice has been necessary. A choice was necessary here, because the essential idea of the principle of material frame indifference had been stated by several authors prior to Noll (1958). Oldroyd (1950) in particular attracted considerable interest with his viewpoint. Truesdell and Noll (1965, p. 45) have gone back to the seventeenth century to trace the development of this idea through the literature.

and

$$\mathbf{z} = \chi_\kappa(\mathbf{z}_\kappa, t) \tag{2.3.1-2}$$

then it must also be satisfied for any equivalent process described with respect to another frame of reference. In particular, the description of material behavior must be satisfied for a process in which the stress tensor and motion are given by

$$\mathbf{T}^* = \mathbf{T}^*(\mathbf{z}_\kappa^*, t^*)$$
$$= \mathbf{Q}(t) \cdot \mathbf{T}(\mathbf{z}_\kappa, t) \cdot \mathbf{Q}^T(t) \tag{2.3.1-3}$$

$$\mathbf{z}^* = \chi_\kappa^*(\mathbf{z}_\kappa^*, t^*)$$
$$= \mathbf{z}_0^*(t) + \mathbf{Q}(t) \cdot \left[\chi_\kappa(\mathbf{z}_\kappa, t) - \mathbf{z}_0\right] \tag{2.3.1-4}$$

and

$$t^* = t - a \tag{2.3.1-5}$$

It is possible to make further statements in much the same manner as above[6] (Truesdell and Toupin 1960, p. 700; Truesdell and Noll 1965, p. 101). These principles may be used to help in the construction of particular constitutive equations for the stress tensor. The type of argument involved is illustrated in the following section.

## 2.3.2 Simple Constitutive Equation for Stress

In the preceding section we discussed three principles that every description of stress-deformation behavior should satisfy. Let us propose a simple stress-deformation relationship that is consistent with these principles.

We can satisfy the principle of determinism by requiring the stress to depend only upon a description of the present state of motion in the material. Both the principle of determinism and the principle of local action are satisfied if we assume that the stress at a point is a function of the velocity and velocity gradient at that point:

$$\mathbf{T} = \mathbf{H}(\mathbf{v}, \nabla\mathbf{v}) \tag{2.3.2-1}$$

It is understood that stress may also depend upon local thermodynamic state variables, but since this dependence is not of primary concern as yet, it is not shown explicitly.

Every second-order tensor can be written as the sum of a symmetric tensor and a skew-symmetric tensor. For example, the velocity gradient may be expressed as

$$\nabla\mathbf{v} = \frac{1}{2}[\nabla\mathbf{v} + (\nabla\mathbf{v})^T] + \frac{1}{2}[\nabla\mathbf{v} - (\nabla\mathbf{v})^T]$$
$$= \mathbf{D} + \mathbf{W} \tag{2.3.2-2}$$

---

[6] It is traditional for discussions rooted in the *thermodynamics of irreversible processes* to apply Curie's "law" and the Onsager–Casimir reciprocal relations in developing linear descriptions of material behavior (Bird 1993). Usually these relations are employed in place of the principle of frame indifference and the entropy inequality (the second law of thermodynamics, discussed in Sections 5.2 and 5.3) (Bird 1993). Truesdell (1969, p. 134) follows the historical evolution of Curie's law and the Onsager–Casimir reciprocal relations, and he finds them to be without merit.

where

$$\mathbf{D} \equiv \frac{1}{2}[\nabla\mathbf{v} + (\nabla\mathbf{v})^T]  \qquad (2.3.2\text{-}3)$$

is the *rate of deformation tensor* and

$$\mathbf{W} \equiv \frac{1}{2}[\nabla\mathbf{v} - (\nabla\mathbf{v})^T]  \qquad (2.3.2\text{-}4)$$

is the *vorticity tensor*. This allows us to rewrite (2.3.2-1) as

$$\mathbf{T} = \mathbf{H}(\mathbf{v}, \mathbf{D} + \mathbf{W})  \qquad (2.3.2\text{-}5)$$

The principle of frame indifference discussed in Section 2.3.1 requires that

$$\mathbf{T}^* = \mathbf{Q} \cdot \mathbf{T} \cdot \mathbf{Q}^T = \mathbf{H}(\mathbf{v}^*, \mathbf{D}^* + \mathbf{W}^*)  \qquad (2.3.2\text{-}6)$$

From Section 1.2.2, Exercise 2.3.2-1, and Equation (2.3.2-5), we find that the function $\mathbf{H}$ must be such that

$$\mathbf{Q} \cdot \mathbf{H}(\mathbf{v}, \mathbf{D} + \mathbf{W}) \cdot \mathbf{Q}^T = \mathbf{H}\left( \frac{d\mathbf{z}_0^*}{dt} + \frac{d\mathbf{Q}}{dt} \cdot (\mathbf{z} - \mathbf{z}_0) + \mathbf{Q} \cdot \mathbf{v}, \mathbf{Q} \cdot \mathbf{D} \cdot \mathbf{Q}^T \right.$$
$$\left. + \mathbf{Q} \cdot \mathbf{W} \cdot \mathbf{Q}^T + \frac{d\mathbf{Q}}{dt} \cdot \mathbf{Q}^T \right)  \qquad (2.3.2\text{-}7)$$

Let us choose a particular change of frame such that

$$\frac{d\mathbf{z}_0^*}{dt} = -\frac{d\mathbf{Q}}{dt} \cdot (\mathbf{z} - \mathbf{z}_0) - \mathbf{Q} \cdot \mathbf{v}  \qquad (2.3.2\text{-}8)$$

and

$$\frac{d\mathbf{Q}}{dt} = -\mathbf{Q} \cdot \mathbf{W}  \qquad (2.3.2\text{-}9)$$

With this change of frame, we have from (2.3.2-6) and (2.3.2-7)

$$\mathbf{T}^* = \mathbf{G}(\mathbf{D}^*)$$
$$\equiv \mathbf{H}(\mathbf{0}, \mathbf{Q} \cdot \mathbf{D} \cdot \mathbf{Q}^T + \mathbf{0})  \qquad (2.3.2\text{-}10)$$

Applying the principle of frame indifference, we conclude from (2.3.2-10) that

$$\mathbf{T} = \mathbf{G}(\mathbf{D})  \qquad (2.3.2\text{-}11)$$

Equation (2.3.2-7) requires that the function $\mathbf{G}$ must satisfy

$$\mathbf{Q} \cdot \mathbf{G}(\mathbf{D}) \cdot \mathbf{Q}^T = \mathbf{G}(\mathbf{Q} \cdot \mathbf{D} \cdot \mathbf{Q}^T)  \qquad (2.3.2\text{-}12)$$

Let us try to give a physical interpretation to our use of the principle of frame indifference in the preceding paragraphs. We begin with an observer at the position $\mathbf{z}$ in an arbitrary frame of reference who assumes that the stress tensor depends upon velocity, the rate of deformation tensor, and the vorticity tensor: (2.3.2-5). Equations (2.3.1-4), (2.3.2-8), and (2.3.2-9) describe another observer at the position $\mathbf{z}^*$ in a new frame of reference. The observer at $\mathbf{z}^*$ rotates and translates with the material in such a way that for him the velocity $\mathbf{v}^*$ and the vorticity tensor $\mathbf{W}^*$ are zero. He can see no dependence of the stress tensor $\mathbf{T}^*$

upon $\mathbf{v}^*$ and $\mathbf{W}^*$; he sees a dependence of $\mathbf{T}^*$ on $\mathbf{D}^*$ alone in (2.3.2-10). But the principle of frame indifference requires that all observers come to the same conclusions about the behavior of materials. We consequently conclude that (2.3.2-5) reduces to (2.3.2-11).

The most general form that (2.3.2-11) can take in view of (2.3.2-12) is (Truesdell and Noll 1965, p. 32)

$$\mathbf{T} = \kappa_0 \mathbf{I} + \kappa_1 \mathbf{D} + \kappa_2 \mathbf{D} \cdot \mathbf{D} \tag{2.3.2-13}$$

where

$$\kappa_k = \kappa_k(I_D, II_D, III_D) \tag{2.3.2-14}$$

Here $I_D, II_D$, and $III_D$ are the three principal invariants of the rate of deformation tensor (i.e., the coefficients in the equation for the principal values of $\mathbf{D}$):

$$\det(\mathbf{D} - m\mathbf{I}) = -m^3 + I_D m^2 - II_D m + III_D$$
$$= 0 \tag{2.3.2-15}$$

where

$$I_D \equiv \operatorname{tr} \mathbf{D}$$
$$= \operatorname{div} \mathbf{v} \tag{2.3.2-16}$$

$$II_D \equiv \frac{1}{2}[(I_D)^2 - \bar{II}_D] \tag{2.3.2-17}$$

$$\bar{II}_D \equiv \operatorname{tr}(\mathbf{D} \cdot \mathbf{D}) \tag{2.3.2-18}$$

$$III_D \equiv \det \mathbf{D} \tag{2.3.2-19}$$

Equation (2.3.2-13) was first obtained by Reiner (1945) and Prager (1945) for functions $\mathbf{G}(\mathbf{D})$ in the form of a tensor power series (Truesdell and Noll 1965, p. 33).

Notice that this constitutive equation for stress automatically satisfies the symmetry of the stress tensor, since the rate of deformation tensor is symmetric.

It follows immediately from (2.3.2-13) that the most general *linear* relation between the stress tensor and the rate of deformation tensor that is consistent with the principle of material frame indifference is

$$\mathbf{T} = (\alpha + \lambda \operatorname{div} \mathbf{v})\mathbf{I} + 2\mu \mathbf{D} \tag{2.3.2-20}$$

In Section 5.3.4 (Truesdell and Noll 1965, p. 357), we find that this reduces to the *Newtonian fluid*,

$$\mathbf{T} = (-P + \lambda \operatorname{div} \mathbf{v})\mathbf{I} + 2\mu \mathbf{D} \tag{2.3.2-21}$$

where $P$ is the thermodynamic pressure (Section 5.3.1),

$$\mu > 0 \tag{2.3.2-22}$$

is the *shear viscosity*, and

$$\lambda > -\frac{2}{3}\mu \tag{2.3.2-23}$$

It is often stated that the *bulk viscosity*

$$\kappa \equiv \lambda + \frac{2}{3}\mu$$
$$= 0$$

(2.3.2-24)

and that this has been substantiated for low-density monatomic gases. In fact, this result is implicitly assumed in that theory (Truesdell 1952, Sec. 61A). To our knowledge, experimental measurements indicate that $\lambda$ is positive and that for many fluids it is orders of magnitude greater than $\mu$ (Truesdell 1952, Sec. 61A; Karim and Rosenhead 1952).

Another special case of (2.3.2-20) is the *incompressible Newtonian fluid*:

$$\mathbf{T} = -p\mathbf{I} + 2\mu\mathbf{D}$$

(2.3.2-25)

Equation (2.3.2-25) is sometimes described as a special case of (2.3.2-21). (Note that the thermodynamic pressure $P$ is not defined for an incompressible fluid.) The quantity $p$ is known as the *mean pressure*. From (2.3.2-25), we see that we may take as its definition

$$p \equiv -\frac{1}{3}\operatorname{tr}\mathbf{T}$$

(2.3.2-26)

When discussing stress-deformation behavior, it is common to speak in terms of the *viscous portion of the stress tensor*,

$$\mathbf{S} \equiv \mathbf{T} + P\mathbf{I}$$

(2.3.2-27)

In this way, the strictly thermodynamic quantity $P$ is separated from those effects arising from deformation. For incompressible fluids, we define the *viscous portion of the stress tensor* as

$$\mathbf{S} \equiv \mathbf{T} + p\mathbf{I}$$

(2.3.2-28)

where $p$ is the mean pressure (2.3.2-26).

Although the Newtonian fluid (2.3.2-21) has been found to be useful in describing the behavior of gases and low-molecular-weight liquids, it has not been established that any fluid requires a nonzero value for $\kappa_2$ or a dependence of $\kappa_1$ upon $III_D$ in (2.3.2-13). However, a number of empirical relations based upon limiting forms of (2.3.2-13) have been found to be of some engineering value. A few of these are discussed in the next section.

**Exercise 2.3.2-1**    (Truesdell 1966a, p. 25)

i) Let us define the *deformation gradient* as

$$\mathbf{F} \equiv \operatorname{grad}\chi_\kappa$$
$$= \frac{\partial \chi_{\kappa i}}{\partial z_{\kappa j}}\mathbf{e}_i\mathbf{e}_j$$

where $\chi_\kappa$ is the deformation of the body (see Section 1.1),

$$\mathbf{z} = \chi_\kappa(\mathbf{z}_\kappa, t)$$

and grad denotes the gradient operation in the reference configuration of the material. Note that we have used the same rectangular Cartesian coordinate system (or set of basis vectors) both in the current configuration of the body and in the reference configuration of the body. Use the definition of the material derivative (1.1.0-12), the definition of the velocity vector (1.1.0-13), and the chain rule to show that

$$\frac{d_{(m)}\mathbf{F}}{dt} = (\nabla\mathbf{v}) \cdot \mathbf{F}$$

and that

$$\nabla\mathbf{v} = \frac{d_{(m)}\mathbf{F}}{dt} \cdot \mathbf{F}^{-1}$$

ii) Let $\mathbf{Q}$ be a time-dependent orthogonal transformation associated with a change of frame (Section 1.2.1) and let the motions $\chi$ and $\chi^*$ be referred to the same reference configuration. Starting with (2.3.1-4), show that

$$\mathbf{F}^* = \mathbf{Q} \cdot \mathbf{F}$$

iii) Take the material derivative of this equation and show that

$$(\nabla\mathbf{v})^* \cdot \mathbf{F}^* = \mathbf{Q} \cdot (\nabla\mathbf{v}) \cdot \mathbf{Q}^T \cdot \mathbf{F}^* + \frac{d\mathbf{Q}}{dt} \cdot \mathbf{Q}^T \cdot \mathbf{F}^*$$

Here the asterisk indicates an association with the new frame.

iv) The decomposition of a second-order tensor into skew-symmetric and symmetric portions is unique. Make use of this fact to show that

$$\mathbf{D}^* = \mathbf{Q} \cdot \mathbf{D} \cdot \mathbf{Q}^T$$

and

$$\mathbf{W}^* = \mathbf{Q} \cdot \mathbf{W} \cdot \mathbf{Q}^T + \mathbf{A}$$

We conclude that the rate of deformation tensor $\mathbf{D}$ is frame indifferent, whereas the vorticity tensor $\mathbf{W}$ is not. Note that the angular velocity tensor $\mathbf{A}$ defined by (1.2.2-8) is shown to be skew symmetric in (1.2.2-10).

**Exercise 2.3.2-2**   Starting with tr $\mathbf{D}$, show that

$$\text{tr}\,\mathbf{D} = \text{div}\,\mathbf{v}$$

**Exercise 2.3.2-3** *The existence of a hydrostatic pressure*   Consider a fluid described by (2.3.2-11) and (2.3.2-12). We see from (2.3.2-13) that, when there is no flow,

$$\mathbf{T} = k_0\mathbf{I}$$

Prove this result directly from (2.3.2-11) and (2.3.2-12) by means of Exercise A.5.2-4.

## 2.3.3   Generalized Newtonian Fluid

At the present time it has not been established that any of the many fluids for which the Newtonian fluid is inadequate can be described by (2.3.2-13). But empirical models based upon (2.3.2-13) may have some utility, in that they predict some aspects of real fluid behavior in a restricted class of flows known as *viscometric flows*.

In a viscometric flow, a material particle is subjected to a constant deformation history, so that memory effects are wiped out. Examples of viscometric flows are flow through a tube, Couette flow, and flow in a cone–plate viscometer under conditions such that inertial effects can be neglected. Unsteady-state flows, flow in a periodically constricted tube, flow through a porous rock, and flow through a pump are examples of nonviscometric flows.

The most common class of empirical models for incompressible fluids based upon (2.3.2-13) is the *generalized Newtonian fluid*, which can be written in two forms.

## Primary Form

The primary form of the generalized Newtonian fluid can be expressed as

$$\mathbf{S} \equiv \mathbf{T} + p\mathbf{I}$$
$$= 2\eta(\gamma)\mathbf{D} \tag{2.3.3-1}$$

where

$$\gamma \equiv \sqrt{2\bar{\bar{II}}_D}$$
$$= \sqrt{2\,\mathrm{tr}\,(\mathbf{D} \cdot \mathbf{D})} \tag{2.3.3-2}$$

We refer to $\mathbf{S}$ as the *viscous portion of the stress tensor*; it is common to call $\eta(\gamma)$ the *apparent viscosity* by analogy with (2.3.2-25). From the differential entropy inequality, we can show that (Exercise 5.3.4-1)

$$\eta(\gamma) > 0 \tag{2.3.3-3}$$

It is well to keep in mind that $\eta(\gamma)$ should include a dependence upon all of the parameters required to describe a fluid's behavior. Truesdell (1964; Truesdell and Noll 1965, p. 65) argues that, no matter how many parameters describe a fluid's behavior, only two of them have dimensions, a characteristic viscosity $\mu_0$ and a characteristic (relaxation) time $s_0$:

$$\eta = \eta\,(\gamma, \mu_0, s_0) \tag{2.3.3-4}$$

With this assumption, the Buckingham–Pi theorem (Brand 1957) requires that

$$\frac{\eta}{\mu_0} = \eta^\star\,(s_0\gamma) \tag{2.3.3-5}$$

A number of suggestions have been made for the definitions of $\mu_0$ and $s_0$ (Bird et al. 1977). For example, $\mu_0$ could be identified as the viscosity of the fluid in the limit $\gamma < \gamma_0$ (all fluids are Newtonian in the limit $\gamma \to 0$).

One of the most common two-parameter generalized Newtonian models is the Ostwald–de Waele model or *power-law fluid* (Reiner 1960, p. 243):

$$\eta(\gamma) = \mu_0 m^\star\,(s_0\gamma)^{n-1}$$
$$= m(\gamma)^{n-1} \tag{2.3.3-6}$$

Here

$$m \equiv \mu_0 m^\star s_0^{\,n-1} \tag{2.3.3-7}$$

and $n$ are parameters that must be determined empirically. (Since $s_0$ and $\mu_0 m^\star$ appear only in combination, we don't count them as independent parameters.) When $n = 1$ and $\mu_0 m^\star = \mu$, the power-law fluid reduces to the incompressible Newtonian fluid (2.3.2-25). Since this model is relatively simple, it has been used widely in calculations. Its disadvantage is that it does not reduce to Newtonian behavior either in the limit $\gamma \to 0$ or in the limit $\gamma \to \infty$ as we currently believe all real fluids do. For most polymers and polymer solutions, $n$ is less than unity. For this case, (2.3.3-6) predicts an infinite viscosity in the limit of zero rate of deformation and a zero viscosity as the rate of deformation becomes unbounded.

Hermes and Fredrickson (1967) proposed one possible superposition of Newtonian and power-law behavior:

$$\eta(\gamma) = \frac{\mu_0}{1 + a^\star (s_0\gamma)^{1-n}} \tag{2.3.3-8}$$

where $a^\star$, $\mu_0$, and $n$ are experimentally determined parameters. (Since $a^\star$ and $s_0$ appear only in combination, we do not consider them as independent parameters.) If we assume that $n$ is less than unity, it predicts a lower-limiting viscosity $\mu_0$ as $\gamma \to 0$. [Equation (2.3.3-8) often may prove to be more useful than the Ellis fluid (described below), since the stress tensor is given as an explicit function of the rate of deformation tensor.]

The Sisko fluid (Bird 1965b; Sisko 1958) is another superposition of Newtonian and power-law behavior:

$$\text{for } \gamma < \gamma_0 : \ \eta(\gamma) = \mu_0 \left[ 1 - \left( \frac{\gamma}{\gamma_0} \right)^{\alpha-1} \right] \tag{2.3.3-9}$$

Here $\mu_0$, $\gamma_0 = 1/s_0$, and $\alpha$ are parameters whose values depend upon the particular material being described. It properly predicts a lower-limiting viscosity $\mu_0$ as $\gamma \to 0$, but it cannot be used for $\gamma > \gamma_0$, since $\alpha$ is usually between 1 and 3 (Bird 1965b).

The Bingham plastic (Reiner 1960, p. 114) is of historical interest but of limited current practical value. It describes a material that behaves as a rigid solid until the stress has exceeded some critical value:

$$\text{for } \tau > \tau_0 : \ \eta(\gamma) = \eta_0 + \frac{\tau_0}{\gamma} \tag{2.3.3-10}$$

$$\text{for } \tau < \tau_0 : \ \mathbf{D} = \mathbf{0} \tag{2.3.3-11}$$

This model contains two parameters: $\eta_0$ and $\tau_0$. It was originally proposed to represent the behavior of paint. The idea of a critical stress $\tau_0$ at which the rigid solid yielded and began to flow probably was postulated on the basis of inadequate data in the limit $\tau \to 0$. Though later work has failed to establish that any materials are true Bingham plastics, the concept is firmly established in the older literature.

### Alternative Form

Let us define

$$\tau \equiv \sqrt{\frac{1}{2} \overline{II}_s}$$

$$= \sqrt{\frac{1}{2} \text{tr}(\mathbf{S} \cdot \mathbf{S})} \tag{2.3.3-12}$$

From (2.3.3-1),

$$\tau = \tau(\gamma)$$

$$= \eta(\gamma)\gamma \tag{2.3.3-13}$$

We assume that $\eta(\gamma)$ is a differentiable function, which means

$$\left.\frac{d\tau}{d\gamma}\right|_{\gamma=0} = \lim_{\gamma\to0} \frac{\tau(\gamma)-\tau(0)}{\gamma}$$

$$= \lim_{\gamma\to0} \frac{\tau(\gamma)}{\gamma} = \eta(0) \tag{2.3.3-14}$$

Equations (2.3.3-3) and (2.3.3-14) require that

$$\left.\frac{d\tau}{d\gamma}\right|_{\gamma=0} > 0 \tag{2.3.3-15}$$

If the derivative $d\tau/d\gamma$ is continuous, it must be positive in some neighborhood of $\gamma = 0$. In this neighborhood, $\tau(\gamma)$ will be a strictly increasing function of $\gamma$ and, for this reason, it will have an inverse:

$$\gamma = \lambda(\tau) \tag{2.3.3-16}$$

Equation (2.3.3-16) follows from (2.3.3-1) for $\gamma$ sufficiently close to zero. It is possible that $\tau(\gamma)$ ceases to increase when $\gamma$ exceeds some value, but such behavior has not been observed experimentally.

From (2.3.3-13) and (2.3.3-16),

$$\frac{\gamma}{\tau} = \frac{1}{\eta(\gamma)} = \frac{\lambda(\tau)}{\tau} = \frac{1}{\eta(\lambda(\tau))} \tag{2.3.3-17}$$

This suggests that we may write (2.3.3-1) in the alternative form

$$2\mathbf{D} = \varphi(\tau)\mathbf{S} \tag{2.3.3-18}$$

where

$$\varphi(\tau) \equiv \frac{1}{\eta(\lambda(\tau))} \tag{2.3.3-19}$$

is referred to as the *fluidity*.

By analogy with (2.3.3-4) and (2.3.3-5), if we assume

$$\varphi = \varphi(\tau, \mu_0, s_0) \tag{2.3.3-20}$$

the Buckingham–Pi theorem (Brand 1957) requires that

$$\mu_0\varphi = \varphi^\star\left(\frac{s_0\tau}{\mu_0}\right) \tag{2.3.3-21}$$

The power-law fluid (2.3.3-6) can be written in this alternative form:

$$\varphi(\tau) = \frac{1}{\mu_0} m^{\star-1/n} \left(\frac{\tau s_0}{\mu_0}\right)^{(1-n)/n}$$

$$= m^{-1/n} \tau^{(1-n)/n} \tag{2.3.3-22}$$

It still does not reduce to Newtonian behavior in the limit $\tau \to 0$ or in the limit $\tau \to \infty$ as real fluids do.

The Ellis fluid (Reiner 1960, p. 246; Bird 1965b) is still another superposition of Newtonian and power-law behavior:

$$\varphi(\tau) = \frac{1}{\mu_0}\left[1 + \left(\frac{\tau}{\tau_{1/2}}\right)^{\alpha-1}\right] \qquad (2.3.3\text{-}23)$$

where $\mu_0$, $\tau_{1/2}$, and $\alpha$ are parameters to be fixed by comparison with experimental data. It includes the power-law fluid as a special case corresponding to the limit

$$\frac{1}{\mu_0} \to 0$$

$$\begin{aligned}
\tau_{1/2}{}^{1-\alpha} &\to m^{\star-1/n}\left(\frac{s_0}{\mu_0}\right)^{(1-n)/n} \\
&= \mu_0 m^{-1/n}
\end{aligned} \qquad (2.3.3\text{-}24)$$

$$\alpha \to \frac{1}{n}$$

For polymers and their solutions, $\alpha$ is usually between 1 and 3 (Bird 1965b), which means that it properly predicts a lower-limiting viscosity $\mu_0$ as $\tau \to 0$. Equation (2.3.3-23) may be one of the more useful three-parameter models of the class of generalized Newtonian fluids.

## Summary

Many more models of the form of (2.3.3-1) and (2.3.3-18) have been proposed in addition to those mentioned here (Reiner 1960; Bird 1965b; Bird, Armstrong, and Hassager 1987, p. 169). They often have been published in a one-dimensional form. The reader interested in applying to another situation a model that has been presented in this fashion should first express the model in a form consistent with either (2.3.3-1) or (2.3.3-18).

**Exercise 2.3.3-1**   Show that $p$ in (2.3.3-1) must be the *mean pressure* defined by (2.3.2-26).

**Exercise 2.3.3-2**   Starting from (2.3.3-1), derive (2.3.3-13).

**Exercise 2.3.3-3** *Reiner–Philippoff fluid*   One of the classic descriptions of stress-deformation behavior from the pre-1945 literature is the Reiner–Philippoff fluid (Philippoff 1935):

$$\frac{dv_1}{dz_2} = \left[\mu_\infty + \frac{\mu_0 - \mu_\infty}{1 + (S_{12}/\tau_0)^2}\right]^{-1} S_{12}$$

Following the usual practice of that period, we have stated the model in a form appropriate for a one-dimensional flow in rectangular, Cartesian coordinates:

$$v_1 = v_1(z_2)$$
$$v_2 = 0$$
$$v_3 = 0$$

Here $\mu_0$, $\mu_\infty$, and $\tau_0$ are three material parameters, constants for a given material and a given set of thermodynamic state variables.

i) How would you generalize this model so that it could be applied to a totally different, multidimensional flow?

ii) Show that this model correctly predicts Newtonian behavior for both low and high stresses.

**Exercise 2.3.3-4** *Eyring fluid*    The Eyring fluid is another classic description of stress-deformation behavior (Bird et al. 1960). For a one-dimensional flow in rectangular, Cartesian coordinates, it takes the form

$$S_{12} = A \operatorname{arcsinh} \left( \frac{1}{B} \frac{dv_1}{dz_2} \right)$$

Here $A$ and $B$ are material parameters, constants for a given material and a given set of thermodynamic state variables.

i) How would you generalize this model so that it could be applied to a totally different, multidimensional flow?
ii) Are there any unpleasant features to this model?

**Exercise 2.3.3-5** *Knife in a jar of peanut butter*    You all will have had an experience similar to
- finding that a kitchen knife could be supported vertically in a jar of peanut butter without the knife touching the bottom of the jar or
- finding that a screwdriver could be supported vertically in a can of grease without the screwdriver touching the bottom of the can.

To better understand such observations, consider a knife that is allowed to slip slowly into a Bingham plastic, until it comes to rest without being in contact with any of the bounding surfaces of the system. Relate the depth $H$ to which the knife is submerged and the total weight of the knife to the properties of the Bingham plastic. You may assume that you know both the mass density $\rho^{(m)}$ of the metal and the mass density $\rho^{(f)}$ of the fluid, you may neglect *end effects* at the intersection of the knife with the fluid–fluid interface, and you may approximate the knife as parallel planes ignoring its edges.

*Hint:*    Make the same assumption about the pressure that you did in solving Exercise 2.2.3-1.

## 2.3.4    Noll Simple Fluid

Many commercial processes involve viscoelastic fluids, ranging from polymers and polymer solutions to food products. (Viscoelastic is used here in the sense that, following a material particle, the stress depends upon the history of the deformation to which the immediate neighborhood of the material particle has been subjected. These fluids exhibit a finite relaxation time and normal stresses in viscometric flows [Coleman et al. 1966, p. 47].) The behavior of these fluids is generally much more complex than we have suggested in Sections 2.3.2 and 2.3.3. Sometimes the simple models discussed in Section 2.3.3 are adequate for representing the principal aspects of material behavior to be observed in a particular experiment. More often they are not. Noll and coworkers (Noll 1958; Coleman, Markovitz, and Noll 1966; Coleman and Noll 1961; Truesdell and Noll 1965) have suggested a description of stress-deformation behavior that apparently can be used to explain all aspects of the behavior of viscoelastic liquids that have been observed experimentally.

Before trying to say exactly what is meant by a Noll simple fluid, let us pause for a little background. Let $\xi$ be the position at time $t - s$ ($0 \leq s < \infty$) of the material particle that at time $t$ occupies the position $z$:

$$\xi = \chi_t(\mathbf{z}, t - s) \tag{2.3.4-1}$$

We call $\chi_t$ the *relative deformation function* because material particles are named or identified by their positions in the current configuration. Equation (2.3.4-1) describes the motion that took place in the material at all times $t - s$ prior to the time $t$. The gradient with respect to $\mathbf{z}$ of the relative deformation function is called the *relative deformation gradient* (see also Exercise 2.3.2-1):

$$\mathbf{F}_t(t - s) \equiv \nabla \chi_t(\mathbf{z}, t - s) \tag{2.3.4-2}$$

The *right relative Cauchy–Green strain tensor* is defined as

$$\mathbf{C}_t(t - s) \equiv \mathbf{F}_t^T(t - s) \cdot \mathbf{F}_t(t - s) \tag{2.3.4-3}$$

Noll (Noll 1958, Coleman et al. 1966, Coleman and Noll 1961, Truesdell and Noll 1965) defines an *incompressible simple fluid* as one for which the extra stress tensor $\mathbf{S}$ at the position $\mathbf{z}$ and time $t$ is specified by the history of the relative right Cauchy–Green strain tensor for the material that is within an arbitrarily small neighborhood of $\mathbf{z}$ at time $t$:

$$\mathbf{S} = \frac{\mu_0}{s_0} \overset{\infty}{\underset{\sigma=0}{\mathcal{H}}}{}^{\star} (\mathbf{C}_t(t - s_0\sigma)) \tag{2.3.4-4}$$

Here we follow Truesdell's discussion of the dimensional indifference of the definition of a simple material (Truesdell 1964; Truesdell and Noll 1965, p. 65). The quantity $\mathcal{H}_{\sigma=0}^{\infty}{}^{\star}$ is a dimensionally invariant tensor-valued functional (an operator that maps tensor-valued functions into a tensor). The constants $\mu_0$ and $s_0$ are the characteristic viscosity and the characteristic (relaxation) time of the fluid. Like any characteristic quantities introduced in defining dimensionless variables, the definitions for $\mu_0$ and $s_0$ are arbitrary. The advantages and disadvantages of particular definitions for $\mu_0$ and $s_0$ have been discussed elsewhere (Slattery 1968a).

Equation (2.3.4-4) clearly satisfies the principles of determination and local action (Section 2.3.1). That it also satisfies the principle of frame indifference is less obvious (Truesdell 1966a, pp. 39, 58, and 63).

Since the form of the functional $\mathcal{H}_{\sigma=0}^{\infty}{}^{\star}$ is left unspecified, it is clear that the Noll simple fluid incorporates a great deal of flexibility. It is for exactly this reason that many workers believe the Noll simple fluid to be capable of explaining all(?) manifestations of fluid behavior that have been observed experimentally to date. It should be viewed as representing an entire class of constitutive equations or an entire class of fluid behaviors.

But the generality of the Noll simple fluid is also its weakness. Only two classes of flows have been shown to be dynamically possible for *every* simple fluid (Coleman et al. 1966; Coleman and Noll 1959, 1961, 1962; Truesdell and Noll 1965; Coleman 1962; Noll 1962; Slattery 1964): the viscometric flows and the extensional flows. Most flows of engineering interest cannot be analyzed without first specifying a particular form for the functional $\mathcal{H}_{\sigma=0}^{\infty}{}^{\star}$.

This does not mean that the Noll simple fluid is of no significance to those of us interested in practical problems. It actually is a very simple model for fluid behavior in the sense that it incorporates, at most, two dimensional parameters (Truesdell 1964; Truesdell and Noll 1965, p. 65): $\mu_0$ and $s_0$. This dimensional simplicity, together with its capacity for representing a wide range of fluid behaviors, makes the Noll simple fluid ideal for use in preparing

**Table 2.4.1-1.** The differential mass balance in three coordinate systems

---

*Rectangular Cartesian coordinates* $(z_1, z_2, z_3)$:

$$\frac{\partial \rho}{\partial t} + \frac{\partial}{\partial z_1}(\rho v_1) + \frac{\partial}{\partial z_2}(\rho v_2) + \frac{\partial}{\partial z_3}(\rho v_3) = 0$$

*Cylindrical coordinates* $(r, \theta, z)$:

$$\frac{\partial \rho}{\partial t} + \frac{1}{r}\frac{\partial}{\partial r}(\rho r v_r) + \frac{1}{r}\frac{\partial}{\partial \theta}(\rho v_\theta) + \frac{\partial}{\partial z}(\rho v_z) = 0$$

*Spherical coordinates* $(r, \theta, \varphi)$:

$$\frac{\partial \rho}{\partial t} + \frac{1}{r^2}\frac{\partial}{\partial r}(\rho r^2 v_r) + \frac{1}{r \sin \theta}\frac{\partial}{\partial \theta}(\rho v_\theta \sin \theta) + \frac{1}{r \sin \theta}\frac{\partial}{\partial \varphi}(\rho v_\varphi) = 0$$

---

dimensionless correlations of experimental data. The limitation upon correlations formed in this way is that, since material behavior in the form of the functional $\mathcal{H}^\infty_{\sigma=0}{}^*$ has not been fully specified, correlations of experimental data can be made for only one fluid at a time. Although this is a serious limitation, it is not necessary to have all of the data that would be required in order to describe the behavior of the material under study. For more on scale-ups and data correlations for viscoelastic fluids, see Slattery (1965, 1968a).

There have been alternative descriptions for the complex behavior observed in real fluids. Of these, Oldroyd's (1965) *generalized elasticoviscous fluid* has attracted perhaps the most interest.

If you wish to learn more about the behavior of real materials and their description, you are fortunate to have several excellent texts available (Fredrickson 1964, Lodge 1964, Truesdell and Noll 1965, Coleman et al. 1966, Truesdell 1966a, Leigh 1968, Bird et al. 1977).

---

## 2.4　Summary

### 2.4.1　Differential Mass and Momentum Balances

We would like to summarize here some of the most common relationships in rectangular Cartesian, cylindrical, and spherical coordinates.

Table 2.4.1-1 presents the differential mass balance (1.3.3-4)

$$\frac{\partial \rho}{\partial t} + \text{div}(\rho \mathbf{v}) = 0 \tag{2.4.1-1}$$

for these coordinate systems.

It is generally more convenient to work with the differential momentum balance (2.2.3-4) in terms of the viscous portion of the stress tensor $\mathbf{S} \equiv \mathbf{T} + P\mathbf{I}$:

$$\rho \left[ \frac{\partial \mathbf{v}}{\partial t} + (\nabla \mathbf{v}) \cdot \mathbf{v} \right] = -\nabla P + \text{div}\, \mathbf{S} + \rho \mathbf{f} \tag{2.4.1-2}$$

It is the components of this equation that are presented in Tables 2.4.1-2, 2.4.1-4, and 2.4.1-6.

**Table 2.4.1-2.** Differential momentum balance in rectangular Cartesian coordinates

$z_1$ *component*:

$$\rho \left( \frac{\partial v_1}{\partial t} + v_1 \frac{\partial v_1}{\partial z_1} + v_2 \frac{\partial v_1}{\partial z_2} + v_3 \frac{\partial v_1}{\partial z_3} \right) = -\frac{\partial P}{\partial z_1} + \frac{\partial S_{11}}{\partial z_1} + \frac{\partial S_{12}}{\partial z_2} + \frac{\partial S_{13}}{\partial z_3} + \rho f_1$$

$z_2$ *component*:

$$\rho \left( \frac{\partial v_2}{\partial t} + v_1 \frac{\partial v_2}{\partial z_1} + v_2 \frac{\partial v_2}{\partial z_2} + v_3 \frac{\partial v_2}{\partial z_3} \right) = -\frac{\partial P}{\partial z_2} + \frac{\partial S_{21}}{\partial z_1} + \frac{\partial S_{22}}{\partial z_2} + \frac{\partial S_{23}}{\partial z_3} + \rho f_2$$

$z_3$ *component*:

$$\rho \left( \frac{\partial v_3}{\partial t} + v_1 \frac{\partial v_3}{\partial z_1} + v_2 \frac{\partial v_3}{\partial z_2} + v_3 \frac{\partial v_3}{\partial z_3} \right) = -\frac{\partial P}{\partial z_3} + \frac{\partial S_{31}}{\partial z_1} + \frac{\partial S_{32}}{\partial z_2} + \frac{\partial S_{33}}{\partial z_3} + \rho f_3$$

**Table 2.4.1-3.** Differential momentum balance in rectangular Cartesian coordinates for a Newtonian fluid with constant $\rho$ and $\mu$, the Navier–Stokes equation

$z_1$ *component*:

$$\rho \left( \frac{\partial v_1}{\partial t} + v_1 \frac{\partial v_1}{\partial z_1} + v_2 \frac{\partial v_1}{\partial z_2} + v_3 \frac{\partial v_1}{\partial z_3} \right) = -\frac{\partial p}{\partial z_1} + \mu \left( \frac{\partial^2 v_1}{\partial z_1^2} + \frac{\partial^2 v_1}{\partial z_2^2} + \frac{\partial^2 v_1}{\partial z_3^2} \right) + \rho f_1$$

$z_2$ *component*:

$$\rho \left( \frac{\partial v_2}{\partial t} + v_1 \frac{\partial v_2}{\partial z_1} + v_2 \frac{\partial v_2}{\partial z_2} + v_3 \frac{\partial v_2}{\partial z_3} \right) = -\frac{\partial p}{\partial z_2} + \mu \left( \frac{\partial^2 v_2}{\partial z_1^2} + \frac{\partial^2 v_2}{\partial z_2^2} + \frac{\partial^2 v_2}{\partial z_3^2} \right) + \rho f_2$$

$z_3$ *component*:

$$\rho \left( \frac{\partial v_3}{\partial t} + v_1 \frac{\partial v_3}{\partial z_1} + v_2 \frac{\partial v_3}{\partial z_2} + v_3 \frac{\partial v_3}{\partial z_3} \right) = -\frac{\partial p}{\partial z_3} + \mu \left( \frac{\partial^2 v_3}{\partial z_1^2} + \frac{\partial^2 v_3}{\partial z_2^2} + \frac{\partial^2 v_3}{\partial z_3^2} \right) + \rho f_3$$

Commonly, the only external force to be considered is a uniform gravitational field, which we may represent as

$$\mathbf{f} = -\nabla \phi \tag{2.4.1-3}$$

For an incompressible fluid, (2.4.1-3) allows us to express (2.4.1-2) as

$$\rho \left[ \frac{\partial \mathbf{v}}{\partial t} + (\nabla \mathbf{v}) \cdot \mathbf{v} \right] = -\nabla \mathcal{P} + \operatorname{div} \mathbf{S} \tag{2.4.1-4}$$

where

$$\mathcal{P} \equiv p + \rho \phi \tag{2.4.1-5}$$

is referred to as the *modified pressure*. The components of (2.4.1-4) are easily found from Tables 2.4.1-2, 2.4.1-4, and 2.4.1-6 by deleting the components of $\mathbf{f}$ and replacing $P$ with $\mathcal{P}$.

**Table 2.4.1-4.** Differential momentum balance in cylindrical coordinates

*r component*:

$$\rho \left( \frac{\partial v_r}{\partial t} + v_r \frac{\partial v_r}{\partial r} + \frac{v_\theta}{r} \frac{\partial v_r}{\partial \theta} - \frac{v_\theta{}^2}{r} + v_z \frac{\partial v_r}{\partial z} \right)$$

$$= -\frac{\partial P}{\partial r} + \frac{1}{r} \frac{\partial}{\partial r}(r S_{rr}) + \frac{1}{r} \frac{\partial S_{r\theta}}{\partial \theta} - \frac{S_{\theta\theta}}{r} + \frac{\partial S_{rz}}{\partial z} + \rho f_r$$

*θ component*:

$$\rho \left( \frac{\partial v_\theta}{\partial t} + v_r \frac{\partial v_\theta}{\partial r} + \frac{v_\theta}{r} \frac{\partial v_\theta}{\partial \theta} + \frac{v_r v_\theta}{r} + v_z \frac{\partial v_\theta}{\partial z} \right)$$

$$= -\frac{1}{r} \frac{\partial P}{\partial \theta} + \frac{1}{r^2} \frac{\partial}{\partial r}(r^2 S_{\theta r}) + \frac{1}{r} \frac{\partial S_{\theta\theta}}{\partial \theta} + \frac{\partial S_{\theta z}}{\partial z} + \rho f_\theta$$

*z component*:

$$\rho \left( \frac{\partial v_z}{\partial t} + v_r \frac{\partial v_z}{\partial r} + \frac{v_\theta}{r} \frac{\partial v_z}{\partial \theta} + v_z \frac{\partial v_z}{\partial z} \right)$$

$$= -\frac{\partial P}{\partial z} + \frac{1}{r} \frac{\partial}{\partial r}(r S_{zr}) + \frac{1}{r} \frac{\partial S_{z\theta}}{\partial \theta} + \frac{\partial S_{zz}}{\partial z} + \rho f_z$$

**Table 2.4.1-5.** Differential momentum balance in cylindrical coordinates for a Newtonian fluid with constant $\rho$ and $\mu$, the Navier–Stokes equation

*r component*:

$$\rho \left( \frac{\partial v_r}{\partial t} + v_r \frac{\partial v_r}{\partial r} + \frac{v_\theta}{r} \frac{\partial v_r}{\partial \theta} - \frac{v_\theta{}^2}{r} + v_z \frac{\partial v_r}{\partial z} \right)$$

$$= -\frac{\partial p}{\partial r} + \mu \left[ \frac{\partial}{\partial r} \left( \frac{1}{r} \frac{\partial}{\partial r}(r v_r) \right) + \frac{1}{r^2} \frac{\partial^2 v_r}{\partial \theta^2} - \frac{2}{r^2} \frac{\partial v_\theta}{\partial \theta} + \frac{\partial^2 v_r}{\partial z^2} \right] + \rho f_r$$

*θ component*:

$$\rho \left( \frac{\partial v_\theta}{\partial t} + v_r \frac{\partial v_\theta}{\partial r} + \frac{v_\theta}{r} \frac{\partial v_\theta}{\partial \theta} + \frac{v_r v_\theta}{r} + v_z \frac{\partial v_\theta}{\partial z} \right)$$

$$= -\frac{1}{r} \frac{\partial p}{\partial \theta} + \mu \left[ \frac{\partial}{\partial r} \left( \frac{1}{r} \frac{\partial}{\partial r}(r v_\theta) \right) + \frac{1}{r^2} \frac{\partial^2 v_\theta}{\partial \theta^2} + \frac{2}{r^2} \frac{\partial v_r}{\partial \theta} + \frac{\partial^2 v_\theta}{\partial z^2} \right] + \rho f_\theta$$

*z component*:

$$\rho \left( \frac{\partial v_z}{\partial t} + v_r \frac{\partial v_z}{\partial r} + \frac{v_\theta}{r} \frac{\partial v_z}{\partial \theta} + v_z \frac{\partial v_z}{\partial z} \right)$$

$$= -\frac{\partial p}{\partial z} + \mu \left[ \frac{1}{r} \frac{\partial}{\partial r} \left( r \frac{\partial v_z}{\partial r} \right) + \frac{1}{r^2} \frac{\partial^2 v_z}{\partial \theta^2} + \frac{\partial^2 v_z}{\partial z^2} \right] + \rho f_z$$

**Table 2.4.1-6.** Differential momentum balance in spherical coordinates

*r component*:

$$\rho\left(\frac{\partial v_r}{\partial t} + v_r\frac{\partial v_r}{\partial r} + \frac{v_\theta}{r}\frac{\partial v_r}{\partial \theta} + \frac{v_\varphi}{r\sin\theta}\frac{\partial v_r}{\partial \varphi} - \frac{v_\theta^2 + v_\varphi^2}{r}\right)$$

$$= -\frac{\partial P}{\partial r} + \frac{1}{r^2}\frac{\partial}{\partial r}(r^2 S_{rr}) + \frac{1}{r\sin\theta}\frac{\partial}{\partial \theta}(S_{r\theta}\sin\theta) + \frac{1}{r\sin\theta}\frac{\partial S_{r\varphi}}{\partial \varphi} - \frac{S_{\theta\theta} + S_{\varphi\varphi}}{r} + \rho f_r$$

*θ component*:

$$\rho\left(\frac{\partial v_\theta}{\partial t} + v_r\frac{\partial v_\theta}{\partial r} + \frac{v_\theta}{r}\frac{\partial v_\theta}{\partial \theta} + \frac{v_\varphi}{r\sin\theta}\frac{\partial v_\theta}{\partial \varphi} + \frac{v_r v_\theta}{r} - \frac{v_\varphi^2\cot\theta}{r}\right)$$

$$= -\frac{1}{r}\frac{\partial P}{\partial \theta} + \frac{1}{r^3}\frac{\partial}{\partial r}(r^3 S_{\theta r}) + \frac{1}{r\sin\theta}\frac{\partial}{\partial \theta}(S_{\theta\theta}\sin\theta) + \frac{1}{r\sin\theta}\frac{\partial S_{\theta\varphi}}{\partial \varphi} - \frac{\cot\theta}{r}S_{\varphi\varphi} + \rho f_\theta$$

*φ component*:

$$\rho\left(\frac{\partial v_\varphi}{\partial t} + v_r\frac{\partial v_\varphi}{\partial r} + \frac{v_\theta}{r}\frac{\partial v_\varphi}{\partial \theta} + \frac{v_\varphi}{r\sin\theta}\frac{\partial v_\varphi}{\partial \varphi} + \frac{v_\varphi v_r}{r} + \frac{v_\theta v_\varphi}{r}\cot\theta\right)$$

$$= -\frac{1}{r\sin\theta}\frac{\partial P}{\partial \varphi} + \frac{1}{r^3}\frac{\partial}{\partial r}(r^3 S_{\varphi r}) + \frac{1}{r\sin^2\theta}\frac{\partial}{\partial \theta}(S_{\varphi\theta}\sin^2\theta) + \frac{1}{r\sin\theta}\frac{\partial S_{\varphi\varphi}}{\partial \varphi} + \rho f_\varphi$$

In Sections 2.3.1 to 2.3.4, we pointed out the need for information beyond the differential mass and momentum balances and the symmetry of the stress tensor, and we discussed several possible descriptions of stress-deformation behavior. One of these was the Newtonian fluid, (2.3.2-21). The divergence of the stress tensor for a Newtonian fluid may be written as

$$\text{div }\mathbf{T} = \text{div}(-P\mathbf{I} + \lambda[\text{div }\mathbf{v}]\mathbf{I} + 2\mu\mathbf{D}) \tag{2.4.1-6}$$

If we assume that $\lambda$ and $\mu$ are constants with respect to position, (2.4.1-6) becomes

$$\text{div }\mathbf{T} = -\nabla P + \lambda\nabla(\text{div }\mathbf{v}) + 2\mu\,\text{div }\mathbf{D} \tag{2.4.1-7}$$

Since

$$\text{div }\mathbf{D} = \frac{1}{2}\text{div}\,(\nabla\mathbf{v}) + \frac{1}{2}\nabla(\text{div }\mathbf{v}) \tag{2.4.1-8}$$

we have

$$\text{div }\mathbf{T} = -\nabla P + (\lambda + \mu)\nabla(\text{div }\mathbf{v}) + \mu\,\text{div}\,(\nabla\mathbf{v}) \tag{2.4.1-9}$$

With (2.4.1-9), the differential momentum balance (2.2.3-4) becomes for a Newtonian fluid

$$\rho\frac{d_{(m)}\mathbf{v}}{dt} = -\nabla P + (\lambda + \mu)\nabla(\text{div }\mathbf{v}) + \mu\,\text{div}\,(\nabla\mathbf{v}) + \rho\mathbf{f} \tag{2.4.1-10}$$

Stress-deformation behavior of an incompressible Newtonian fluid is described by (2.3.2-25). The differential momentum balance for this case becomes

$$\rho\frac{d_{(m)}\mathbf{v}}{dt} = -\nabla p + \mu\,\text{div}\,(\nabla\mathbf{v}) + \rho\mathbf{f} \tag{2.4.1-11}$$

**Table 2.4.1-7.** Differential momentum balance in spherical coordinates for a Newtonian fluid with constant $\rho$ and $\mu$, the Navier–Stokes equation

*r component:*

$$\rho\left(\frac{\partial v_r}{\partial t} + v_r\frac{\partial v_r}{\partial r} + \frac{v_\theta}{r}\frac{\partial v_r}{\partial \theta} + \frac{v_\varphi}{r\sin\theta}\frac{\partial v_r}{\partial \varphi} - \frac{v_\theta{}^2 + v_\varphi{}^2}{r}\right)$$

$$= -\frac{\partial p}{\partial r} + \mu\left(\mathcal{H}v_r - \frac{2}{r^2}v_r - \frac{2}{r^2}\frac{\partial v_\theta}{\partial \theta} - \frac{2}{r^2}v_\theta\cot\theta - \frac{2}{r^2\sin\theta}\frac{\partial v_\varphi}{\partial \varphi}\right) + \rho f_r$$

*θ component:*

$$\rho\left(\frac{\partial v_\theta}{\partial t} + v_r\frac{\partial v_\theta}{\partial r} + \frac{v_\theta}{r}\frac{\partial v_\theta}{\partial \theta} + \frac{v_\varphi}{r\sin\theta}\frac{\partial v_\theta}{\partial \varphi} + \frac{v_r v_\theta}{r} - \frac{v_\varphi{}^2\cot\theta}{r}\right)$$

$$= -\frac{1}{r}\frac{\partial p}{\partial \theta} + \mu\left(\mathcal{H}v_\theta + \frac{2}{r^2}\frac{\partial v_r}{\partial \theta} - \frac{v_\theta}{r^2\sin^2\theta} - \frac{2\cos\theta}{r^2\sin^2\theta}\frac{\partial v_\varphi}{\partial \varphi}\right) + \rho f_\theta$$

*φ component:*

$$\rho\left(\frac{\partial v_\varphi}{\partial t} + v_r\frac{\partial v_\varphi}{\partial r} + \frac{v_\theta}{r}\frac{\partial v_\varphi}{\partial \theta} + \frac{v_\varphi}{r\sin\theta}\frac{\partial v_\varphi}{\partial \varphi} + \frac{v_r v_\varphi}{r} + \frac{v_\theta v_\varphi}{r}\cot\theta\right)$$

$$= -\frac{1}{r\sin\theta}\frac{\partial p}{\partial \varphi} + \mu\left(\mathcal{H}v_\varphi - \frac{v_\varphi}{r^2\sin^2\theta} + \frac{2}{r^2\sin\theta}\frac{\partial v_r}{\partial \varphi} + \frac{2\cos\theta}{r^2\sin^2\theta}\frac{\partial v_\theta}{\partial \varphi}\right) + \rho f_\varphi$$

where

$$\mathcal{H} \equiv \frac{1}{r^2}\frac{\partial}{\partial r}\left(r^2\frac{\partial}{\partial r}\right) + \frac{1}{r^2\sin\theta}\frac{\partial}{\partial \theta}\left(\sin\theta\frac{\partial}{\partial \theta}\right) + \frac{1}{r^2\sin^2\theta}\left(\frac{\partial^2}{\partial \varphi^2}\right)$$

which is the *Navier–Stokes equation*. The components of the Navier–Stokes equation are presented in Tables 2.4.1-3, 2.4.1-5, and 2.4.1-7.

We deal only with physical components of spatial vector fields and tensor fields when discussing curvilinear coordinate systems in most of this text. For this reason, in Tables 2.4.1-8 through 2.4.1-10 we adopt a somewhat simpler notation for physical components in cylindrical and spherical coordinates than that suggested in Appendix A. We denote the physical components of spatial vector fields in cylindrical coordinates as $v_r$, $v_\theta$, and $v_z$ rather than $v_{(1)}$, $v_{(2)}$, and $v_{(3)}$; the physical components of second-order tensor field are indicated as $D_{rr}$, $D_{r\theta}$, $D_{\theta z}$, etc. The notation used in spherical coordinates is very similar.

Because of this change in notation, we do *not* employ the summation convention hereafter with the physical components of spatial vector fields and second-order tensor fields. The quantity $D_{rr}$ is a single physical component of the second-order tensor field $\mathbf{D}$. When used in context, there should be no occasion to misinterpret it as the sum of three rectangular Cartesian components. When we have occasion to discuss physical components with respect to other curvilinear coordinate systems, we revert to the notation introduced in Appendix A.

**Exercise 2.4.1-1**    In a convenient rectangular Cartesian coordinate system, derive an expression for $\phi$ and show that $\phi$ is arbitrary to a constant.

**Exercise 2.4.1-2**    Starting with div $\mathbf{D}$, derive (2.4.1-8).

**Table 2.4.1-8.** Components of rate of deformation tensor in rectangular Cartesian coordinates

$$D_{11} = \frac{\partial v_1}{\partial z_1}$$

$$D_{22} = \frac{\partial v_2}{\partial z_2}$$

$$D_{33} = \frac{\partial v_3}{\partial z_3}$$

$$D_{12} = D_{21} = \frac{1}{2}\left(\frac{\partial v_1}{\partial z_2} + \frac{\partial v_2}{\partial z_1}\right)$$

$$D_{13} = D_{31} = \frac{1}{2}\left(\frac{\partial v_1}{\partial z_3} + \frac{\partial v_3}{\partial z_1}\right)$$

$$D_{23} = D_{32} = \frac{1}{2}\left(\frac{\partial v_2}{\partial z_3} + \frac{\partial v_3}{\partial z_2}\right)$$

**Table 2.4.1-9.** Components of the rate of deformation tensor in cylindrical coordinates

$$D_{rr} = \frac{\partial v_r}{\partial r}$$

$$D_{\theta\theta} = \frac{1}{r}\frac{\partial v_\theta}{\partial \theta} + \frac{v_r}{r}$$

$$D_{zz} = \frac{\partial v_z}{\partial z}$$

$$D_{r\theta} = D_{\theta r} = \frac{1}{2}\left[r\frac{\partial}{\partial r}\left(\frac{v_\theta}{r}\right) + \frac{1}{r}\frac{\partial v_r}{\partial \theta}\right]$$

$$D_{rz} = D_{zr} = \frac{1}{2}\left(\frac{\partial v_r}{\partial z} + \frac{\partial v_z}{\partial r}\right)$$

$$D_{\theta z} = D_{z\theta} = \frac{1}{2}\left(\frac{\partial v_\theta}{\partial z} + \frac{1}{r}\frac{\partial v_z}{\partial \theta}\right)$$

## 2.4.2 Stream Function and the Navier–Stokes Equation

In Section 1.3.7, we expressed the velocity components for a two-dimensional motion of an incompressible fluid in terms of a stream function $\psi$. In this way, the differential mass balance is automatically satisfied. Here we examine the result of the introduction of a stream function upon the equation of motion for an incompressible Newtonian fluid, where the external force may be expressed in terms of a potential as described in Section 2.4.1.

When the external force is representable as the gradient of a scalar potential, we may introduce the modified pressure of Section 2.4.1 into the Navier–Stokes equation (2.4.1-11)

**Table 2.4.1-10.** Components of the rate of deformation tensor in spherical coordinates

$$D_{rr} = \frac{\partial v_r}{\partial r}$$

$$D_{\theta\theta} = \frac{1}{r}\frac{\partial v_\theta}{\partial \theta} + \frac{v_r}{r}$$

$$D_{\varphi\varphi} = \frac{1}{r \sin\theta}\frac{\partial v_\varphi}{\partial \varphi} + \frac{v_r}{r} + \frac{v_\theta \cot\theta}{r}$$

$$D_{r\theta} = D_{\theta r} = \frac{1}{2}\left[ r\frac{\partial}{\partial r}\left(\frac{v_\theta}{r}\right) + \frac{1}{r}\frac{\partial v_r}{\partial \theta}\right]$$

$$D_{r\varphi} = D_{\varphi r} = \frac{1}{2}\left[\frac{1}{r \sin\theta}\frac{\partial v_r}{\partial \varphi} + r\frac{\partial}{\partial r}\left(\frac{v_\varphi}{r}\right)\right]$$

$$D_{\theta\varphi} = D_{\varphi\theta} = \frac{1}{2}\left[\frac{\sin\theta}{r}\frac{\partial}{\partial \theta}\left(\frac{v_\varphi}{\sin\theta}\right) + \frac{1}{r \sin\theta}\frac{\partial v_\theta}{\partial \varphi}\right]$$

to obtain

$$\rho\frac{\partial \mathbf{v}}{\partial t} + \rho(\nabla \mathbf{v})\cdot \mathbf{v} = -\nabla\mathcal{P} + \mu\,\mathrm{div}\,(\nabla \mathbf{v}) \tag{2.4.2-1}$$

If we take the curl of this equation, modified pressure $\mathcal{P}$ is eliminated to yield

$$\frac{\partial(\mathrm{curl}\,\mathbf{v})}{\partial t} + \mathrm{curl}([\nabla \mathbf{v}]\cdot \mathbf{v}) = \nu\,\mathrm{div}(\nabla[\mathrm{curl}\,\mathbf{v}]) \tag{2.4.2-2}$$

where

$$\nu \equiv \frac{\mu}{\rho} \tag{2.4.2-3}$$

is the *kinematic viscosity*. In any coordinate system for which the velocity vector has only two nonzero components, Equation (2.4.2-2) has only one nonzero component. The nonzero component expressed in terms of the stream function is presented for several situations in Table 2.4.2-1.

The differential equations of Table 2.4.2-1 can also be derived by recognizing that in any two-dimensional flow (2.4.2-1) will have only two nonzero components. Modified pressure may be eliminated between these two equations by recognizing that

$$\frac{\partial^2 \mathcal{P}}{\partial x^i\,\partial x^j} = \frac{\partial^2 \mathcal{P}}{\partial x^j\,\partial x^i} \tag{2.4.2-4}$$

The velocity components in the resulting differential equation may be expressed in terms of a stream function.

**Table 2.4.2-1.** The stream function

| Coordinate system[a] | Assumed form of velocity distribution | Velocity components | Nonzero component of (2.4.2-2)[b] | Operator |
|---|---|---|---|---|
| Rectangular Cartesian | $v_3 = 0$ <br> $v_1 = v_1(z_1, z_2)$ <br><br> $v_2 = v_2(z_1, z_2)$ | $v_1 = \dfrac{\partial \psi}{\partial z_2}$ <br><br><br> $v_2 = -\dfrac{\partial \psi}{\partial z_1}$ | $\dfrac{\partial}{\partial t}(E^2\psi) - \dfrac{\partial(\psi, E^2\psi)}{\partial(z_1, z_2)}$ <br> $= \nu E^4 \psi$ | $E^2 = \dfrac{\partial^2}{\partial z_1{}^2} + \dfrac{\partial^2}{\partial z_2{}^2}$ <br> $E^4\psi = E^2(E^2\psi)$ <br> $= \left(\dfrac{\partial^4}{\partial z_1{}^4} + 2\dfrac{\partial^4}{\partial z_1{}^2\partial z_2{}^2}\right.$ <br> $\left. + \dfrac{\partial^4}{\partial z_2{}^4}\right)\psi$ |
| Cylindrical | $v_z = 0$ <br> $v_r = v_r(r, \theta)$ <br><br> $v_\theta = v_\theta(r, \theta)$ | $v_r = \dfrac{1}{r}\dfrac{\partial \psi}{\partial \theta}$ <br><br><br> $v_\theta = -\dfrac{\partial \psi}{\partial r}$ | $\dfrac{\partial}{\partial t}(E^2\psi) - \dfrac{1}{r}\dfrac{\partial(\psi, E^2\psi)}{\partial(r, \theta)}$ <br> $= \nu E^4 \psi$ | $E^2 = \dfrac{\partial^2}{\partial r^2} + \dfrac{1}{r}\dfrac{\partial}{\partial r}$ <br><br> $+ \dfrac{1}{r^2}\dfrac{\partial^2}{\partial \theta^2}$ |
| Cylindrical | $v_\theta = 0$ <br> $v_r = v_r(r, z)$ <br><br> $v_z = v_z(r, z)$ | $v_r = \dfrac{1}{r}\dfrac{\partial \psi}{\partial z}$ <br><br><br> $v_z = -\dfrac{1}{r}\dfrac{\partial \psi}{\partial r}$ | $\dfrac{\partial}{\partial t}(E^2\psi) - \dfrac{1}{r}\dfrac{\partial(\psi, E^2\psi)}{\partial(r, z)}$ <br> $-\dfrac{2}{r^2}\dfrac{\partial \psi}{\partial z}E^2\psi = \nu E^4 \psi$ | $E^2 = \dfrac{\partial^2}{\partial r^2} - \dfrac{1}{r}\dfrac{\partial}{\partial r} + \dfrac{\partial^2}{\partial z^2}$ |
| Spherical | $v_\varphi = 0$ <br> $v_r = v_r(r, \theta)$ <br><br><br><br> $v_\theta = v_\theta(r, \theta)$ | $v_r = \dfrac{1}{r^2 \sin\theta}\dfrac{\partial \psi}{\partial \theta}$ <br><br><br><br> $v_\theta = -\dfrac{1}{r\sin\theta}\dfrac{\partial \psi}{\partial r}$ | $\dfrac{\partial}{\partial t}(E^2\psi) - \dfrac{1}{r^2 \sin\theta}$ <br> $\times \dfrac{\partial(\psi, E^2\psi)}{\partial(r, \theta)} + \dfrac{2E^2\psi}{r^2 \sin\theta}$ <br> $\times \left(\dfrac{\partial \psi}{\partial r}\cos\theta - \dfrac{1}{r}\dfrac{\partial \psi}{\partial \theta}\sin\theta\right)$ <br> $= \nu E^4 \psi$ | $E^2 = \dfrac{\partial^2}{\partial r^2}$ <br><br> $+ \dfrac{\sin\theta}{r^2}\dfrac{\partial}{\partial \theta}\left(\dfrac{1}{\sin\theta}\dfrac{\partial}{\partial \theta}\right)$ |

[a]This table is taken from Bird et al. (1960, Table 4.2-1) and from Goldstein (1938, p. 114). Goldstein also presents relations for axisymmetric flows with a nonzero component of velocity around the axis.
[b]The Jacobian notation signifies

$$\frac{\partial(f, g)}{\partial(x, y)} = \begin{vmatrix} \dfrac{\partial f}{\partial x} & \dfrac{\partial f}{\partial y} \\ \dfrac{\partial g}{\partial x} & \dfrac{\partial g}{\partial y} \end{vmatrix}$$

### 2.4.3    Interfacial Tension and the Jump Mass and Momentum Balances

To this point, when we have used a dividing surface to represent a phase interface, we have not recognized that there might be mass, momentum, or stresses associated with the interface. While we will continue to neglect mass and momentum associated with the interface, the imbalance of long-range intermolecular forces at a deformed interface can be taken into account in the jump momentum balance through the introduction of stresses that act tangent to the dividing surface. In the majority of problems, the magnitude of these stresses can be described in terms of the *interfacial tension* $\gamma$ and the jump momentum balance becomes (Slattery 1990, p. 237)

$$\nabla_{(\sigma)}\gamma + 2H\gamma\boldsymbol{\xi} + \left[-\rho\mathbf{v}\left(\mathbf{v} - \mathbf{u}\right) \cdot \boldsymbol{\xi} + \mathbf{T} \cdot \boldsymbol{\xi}\right] = 0 \qquad (2.4.3\text{-}1)$$

Here $H = (\kappa_1 + \kappa_2)/2$ is the *mean curvature* (Slattery 1990, p. 1116) of the surface; $\kappa_1$ and $\kappa_2$ are the principal curvatures of the surface (Slattery 1990, p. 1119). The surface gradient operator $\nabla_{(\sigma)}$ defines a gradient with respect to position $(y^1, y^2)$ on the surface (Slattery 1990, p. 1075). The surface coordinates $y^1$ and $y^2$ in general define a curvilinear coordinate system on the surface (Slattery 1990, p. 1065). It is only in the case of a planar surface that a rectangular Cartesian coordinate system can be introduced and the surface gradient operation takes a familiar form. The surface gradient $\nabla_{(\sigma)}\gamma$ of interfacial tension appears because $\gamma$ will often be a function of position on the surface through its dependence upon temperature and concentration.

For most problems, you will have to express the jump mass balance (Section 1.3.6)

$$\left[\rho\left(\mathbf{v} - \mathbf{u}\right) \cdot \boldsymbol{\xi}\right] = 0 \qquad (2.4.3\text{-}2)$$

and the jump momentum balance (2.4.3-1) in forms appropriate for the configuration of the surface. Many problems involve interfaces that we are willing to describe as planes, cylinders, spheres, two-dimensional surfaces, or axially symmetric surfaces. For these configurations, you will find your work already finished in Tables 2.4.3-1 through 2.4.3-8.

For a further introduction to interfacial behavior, see Slattery (1990).

**Exercise 2.4.3-1** *Floating sphere*    As you can simply demonstrate for yourself, it is easy to make a small needle float at a water–air interface. Just rub the needle with your fingers before carefully placing it on the surface of the water. The oils from your skin ensure that the solid surface is not wet by the water, and the needle is supported by the force of interfacial tension.

Determine the maximum diameter $d_{\max}$ of a solid sphere that will float at an interface, assuming that the contact angle measured through the gas is zero:

$$d_{\max} = \left\{ \frac{6\gamma}{g\left[\rho^{(s)} - \left(\rho^{(l)} + \rho^{(g)}\right)/2\right]} \right\}^{1/2}$$

**Table 2.4.3-1.** Stationary plane dividing surface viewed in a rectangular Cartesian coordinate system

---

*Dividing surface*
$$z_3 = \text{a constant}$$

*Surface coordinates*
$$y^1 \equiv z_1$$
$$y^2 \equiv z_2$$

*Jump mass balance*
$$\left[\rho v_3 \xi_3\right] = 0$$

*Jump momentum balance*
  *$z_1$ component*
$$\frac{\partial \gamma}{\partial z_1} + \left[T_{13}\xi_3\right] = 0$$

  *$z_2$ component*
$$\frac{\partial \gamma}{\partial z_2} + \left[T_{23}\xi_3\right] = 0$$

  *$z_3$ component*
$$\left[-\rho v_3{}^2\xi_3 + T_{33}\xi_3\right] = 0$$

---

**Table 2.4.3-2.** Stationary plane dividing surface viewed in a cylindrical coordinate system

---

*Dividing surface*
$$z = \text{a constant}$$

*Surface coordinates*
$$y^1 \equiv r$$
$$y^2 \equiv \theta$$

*Jump mass balance*
$$\left[\rho v_z \xi_z\right] = 0$$

*Jump momentum balance*

  *r component*
$$\frac{\partial \gamma}{\partial r} + \left[T_{rz}\xi_z\right] = 0$$

  *θ component*
$$\frac{1}{r}\frac{\partial \gamma}{\partial \theta} + \left[T_{\theta z}\xi_z\right] = 0$$

  *z component*
$$\left[-\rho v_z{}^2\xi_z + T_{zz}\xi_z\right] = 0$$

---

**Table 2.4.3-3.** Alternative form for stationary plane dividing surface viewed in a cylindrical coordinate system

---

*Dividing surface*

$$\theta = a \, constant$$

*Surface coordinates*

$$y^1 \equiv r$$
$$y^2 \equiv z$$

*Jump mass balance*

$$\left[\rho v_\theta \xi_\theta\right] = 0$$

*Jump momentum balance*
  *r component*

$$\frac{\partial \gamma}{\partial r} + \left[T_{r\theta}\xi_\theta\right] = 0$$

  *θ component*

$$\left[-\rho v_\theta^2 \xi_\theta + T_{\theta\theta}\xi_\theta\right] = 0$$

  *z component*

$$\frac{\partial \gamma}{\partial z} + \left[T_{z\theta}\xi_\theta\right] = 0$$

---

**Table 2.4.3-4.** Cylindrical dividing surface viewed in a cylindrical coordinate system

---

*Dividing surface*

$$r = R(t)$$

*Surface coordinates*

$$y^1 \equiv \theta$$
$$y^2 \equiv z$$

*Unit normal*

$$\xi_r = 1$$
$$\xi_\theta = \xi_z = 0$$

*Mean curvature*

$$H = -\frac{1}{2R}$$

*Speed of displacement of surface*

$$\mathbf{u} \cdot \boldsymbol{\xi} = \frac{dR}{dt}$$

*Jump mass balance*

$$\left[\rho v_r \xi_r\right] = 0$$

*Jump momentum balance*

 *r component*

$$-\frac{\gamma}{R} + \left[-\rho v_r{}^2\xi_r + T_{rr}\xi_r\right] = 0$$

 *θ component*

$$\frac{1}{R}\frac{\partial \gamma}{\partial \theta} + \left[T_{\theta r}\xi_r\right] = 0$$

 *z component*

$$\frac{\partial \gamma}{\partial z} + \left[T_{zr}\xi_r\right] = 0$$

**Table 2.4.3-5.** Spherical dividing surface viewed in a spherical coordinate system

*Dividing surface*

$$r = R(t)$$

*Surface coordinates*

$$y^1 \equiv \theta$$

$$y^2 \equiv \phi$$

*Unit normal*

$$\xi_r = 1$$

$$\xi_\theta = \xi_\phi = 0$$

*Mean curvature*

$$H = -\frac{1}{R}$$

*Speed of displacement of surface*

$$\mathbf{u} \cdot \boldsymbol{\xi} = \frac{dR}{dt}$$

*Jump mass balance*

$$\left[\rho v_r \xi_r\right] = 0$$

*Jump momentum balance*
 *r component*

$$-\frac{2\gamma}{R} + \left[-\rho v_r{}^2\xi_r + T_{rr}\xi_r\right] = 0$$

 *θ component*

$$\frac{1}{R}\frac{\partial \gamma}{\partial \theta} + \left[T_{\theta r}\xi_r\right] = 0$$

 *φ component*

$$\frac{1}{R\sin\theta}\frac{\partial \gamma}{\partial \phi} + \left[T_{\phi r}\xi_r\right] = 0$$

**Table 2.4.3-6.** Two-dimensional surface viewed in a rectangular Cartesian coordinate system

---

*Dividing surface*

$$z_3 = h(z_1, t)$$

*Surface coordinates*

$$y^1 \equiv z_1$$

$$y^2 \equiv z_2$$

*Unit normal*

$$\xi_1 = -\frac{\partial h}{\partial z_1}\left[1 + \left(\frac{\partial h}{\partial z_1}\right)^2\right]^{-1/2}$$

$$\xi_2 = 0$$

$$\xi_3 = \left[1 + \left(\frac{\partial h}{\partial z_1}\right)^2\right]^{-1/2}$$

*Mean curvature*

$$H = \frac{1}{2}\frac{\partial^2 h}{\partial z_1^2}\left[1 + \left(\frac{\partial h}{\partial z_1}\right)^2\right]^{-3/2}$$

*Speed of displacement of surface*

$$\mathbf{u} \cdot \boldsymbol{\xi} = \frac{\partial h}{\partial t}\left[1 + \left(\frac{\partial h}{\partial z_1}\right)^2\right]^{-1/2}$$

*Jump mass balance*

$$\left[\rho(\mathbf{v} \cdot \boldsymbol{\xi} - \mathbf{u} \cdot \boldsymbol{\xi})\right] = 0$$

*Jump momentum balance*

　$z_1$ *component*

$$\left[1 + \left(\frac{\partial h}{\partial z_1}\right)^2\right]^{-1}\frac{\partial \gamma}{\partial z_1} + 2H\gamma\xi_1 + \left[-\rho(\mathbf{v} \cdot \boldsymbol{\xi} - \mathbf{u} \cdot \boldsymbol{\xi})^2\xi_1 + T_{11}\xi_1 + T_{13}\xi_3\right] = 0$$

　$z_2$ *component*

$$\left[T_{21}\xi_1 + T_{23}\xi_3\right] = 0$$

　$z_3$ *component*

$$\left[1 + \left(\frac{\partial h}{\partial z_1}\right)^2\right]^{-1}\frac{\partial h}{\partial z_1}\frac{\partial \gamma}{\partial z_1} + 2H\gamma\xi_3 + \left[-\rho(\mathbf{v} \cdot \boldsymbol{\xi} - \mathbf{u} \cdot \boldsymbol{\xi})^2\xi_3 + T_{31}\xi_1 + T_{33}\xi_3\right] = 0$$

---

**Table 2.4.3-7.** Axially symmetric surface viewed in a cylindrical coordinate system

---

*Dividing surface*

$$z = h(r, t)$$

*Surface coordinates*

$$y^1 \equiv r$$

$$y^2 \equiv \theta$$

*Unit normal*

$$\xi_r = -\frac{\partial h}{\partial r}\left[1 + \left(\frac{\partial h}{\partial r}\right)^2\right]^{-1/2}$$

$$\xi_\theta = 0$$

$$\xi_z = \left[1 + \left(\frac{\partial h}{\partial r}\right)^2\right]^{-1/2}$$

*Mean curvature*

$$H = \frac{1}{2r}\left[r\frac{\partial^2 h}{\partial r^2} + \frac{\partial h}{\partial r} + \left(\frac{\partial h}{\partial r}\right)^3\right]\left[1 + \left(\frac{\partial h}{\partial r}\right)^2\right]^{-3/2}$$

$$= \frac{1}{2r}\frac{\partial}{\partial r}\left\{r\frac{\partial h}{\partial r}\left[1 + \left(\frac{\partial h}{\partial r}\right)^2\right]^{-1/2}\right\}$$

*Speed of displacement of surface*

$$\mathbf{u} \cdot \boldsymbol{\xi} = \frac{\partial h}{\partial t}\left[1 + \left(\frac{\partial h}{\partial r}\right)^2\right]^{-1/2}$$

*Jump mass balance*

$$\left[\rho(\mathbf{v} \cdot \boldsymbol{\xi} - \mathbf{u} \cdot \boldsymbol{\xi})\right] = 0$$

*Jump momentum balance*

 *r component*

$$\left[1 + \left(\frac{\partial h}{\partial r}\right)^2\right]^{-1}\frac{\partial \gamma}{\partial r} + 2H\gamma\xi_r$$

$$+ \left[-\rho(\mathbf{v} \cdot \boldsymbol{\xi} - \mathbf{u} \cdot \boldsymbol{\xi})^2\xi_r + T_{rr}\xi_r + T_{rz}\xi_z\right] = 0$$

 *θ component*

$$\left[T_{\theta r}\xi_r + T_{\theta z}\xi_z\right] = 0$$

 *z component*

$$\left[1 + \left(\frac{\partial h}{\partial r}\right)^2\right]^{-1}\frac{\partial h}{\partial r}\frac{\partial \gamma}{\partial r} + 2H\gamma\xi_z$$

$$+ \left[-\rho(\mathbf{v} \cdot \boldsymbol{\xi} - \mathbf{u} \cdot \boldsymbol{\xi})^2\xi_z + T_{zr}\xi_r + T_{zz}\xi_z\right] = 0$$

---

**Table 2.4.3-8.** Alternative form for axially symmetric surface viewed in a cylindrical coordinate system

---

*Dividing surface*

$$r = c(z, t)$$

*Surface coordinates*

$$y^1 \equiv z$$

$$y^2 \equiv \theta$$

*Unit normal*

$$\xi_r = \left[ 1 + \left( \frac{\partial c}{\partial z} \right)^2 \right]^{-1/2}$$

$$\xi_\theta = 0$$

$$\xi_z = -\frac{\partial c}{\partial z} \left[ 1 + \left( \frac{\partial c}{\partial z} \right)^2 \right]^{-1/2}$$

*Mean curvature*

$$H = \frac{1}{2c} \left[ c \frac{\partial^2 c}{\partial z^2} - \left( \frac{\partial c}{\partial z} \right)^2 - 1 \right] \left[ 1 + \left( \frac{\partial c}{\partial z} \right)^2 \right]^{-3/2}$$

*Speed of displacement of surface*

$$\mathbf{u} \cdot \boldsymbol{\xi} = \frac{\partial c}{\partial t} \left[ 1 + \left( \frac{\partial c}{\partial z} \right)^2 \right]^{-1/2}$$

*Jump mass balance*

$$\left[ \rho (\mathbf{v} \cdot \boldsymbol{\xi} - \mathbf{u} \cdot \boldsymbol{\xi}) \right] = 0$$

*Jump momentum balance*

 *r component*

$$\left[ 1 + \left( \frac{\partial c}{\partial z} \right)^2 \right]^{-1} \frac{\partial c}{\partial z} \frac{\partial \gamma}{\partial z} + 2H\gamma \xi_r$$

$$+ \left[ -\rho (\mathbf{v} \cdot \boldsymbol{\xi} - \mathbf{u} \cdot \boldsymbol{\xi})^2 \xi_r + T_{rr} \xi_r + T_{rz} \xi_z \right] = 0$$

 *θ component*

$$\left[ T_{\theta r} \xi_r + T_{\theta z} \xi_z \right] = 0$$

 *z component*

$$\left[ 1 + \left( \frac{\partial c}{\partial z} \right)^2 \right]^{-1} \frac{\partial \gamma}{\partial z} + 2H\gamma \xi_z$$

$$+ \left[ -\rho (\mathbf{v} \cdot \boldsymbol{\xi} - \mathbf{u} \cdot \boldsymbol{\xi})^2 \xi_z + T_{zr} \xi_r + T_{zz} \xi_z \right] = 0$$

---

# 3

## Differential Balances
## in Momentum Transfer

A FTER CONSIDERABLE PREPARATION (Appendix A, Chapters 1 and 2), we
are ready to analyze the detailed motions of materials in particular geometries. Try
not to be discouraged if some of our initial examples appear to be too simple. The simple
problems are there in order to allow you to gain both facility and confidence.

The objection is sometimes raised that studying fluid mechanics in the detail described
here is useless because the problems that can be solved are trivial. A statement such as
this is normally made as an exaggeration to make a point. The point is that the interesting
problems require numerical solutions or that the interesting problems are susceptible only
to approximate solutions.

There is an element of truth here. Our feeling is that the best way to be introduced to this
material is through the use of problems that can be solved with a minimum of programming.
In this way, concepts and techniques can be polished rapidly. These problems are required
preparation for those who wish to study more sophisticated problems, to develop limiting
cases as checks on the validity of numerical work, or to develop approximate solutions using
the integral techniques of Chapter 4.

What about the statement that the most interesting problems are susceptible only to
approximate solutions? There are at least four classes of approximations that can be easily
distinguished:

1. Sometimes the physical problem in which we are primarily interested is too difficult for
   us to handle. One answer is to replace it by a problem that has most of the important
   features of our original problem, but which is sufficiently simple for us to analyze. A
   good example here is provided by flow through a tube. From a practical point of view,
   we are always concerned with flow through finite tubes. But sometimes the entrance and
   exit regions are of lesser importance to us and the real problem can be replaced by an
   idealized one in which entrance and exit effects are negligible: flow through a tube of
   infinite length.
2. Even after such an idealization of our original physical problem, it may still be too difficult.
   We may wish to consider a limiting case in which one or more terms in a differential
   equation are neglected. We have tried to place special emphasis in this text upon the
   way in which one should argue to arrive at such an approximation. Note in particular the
   discussions of creeping flow, nonviscous flow, and boundary-layer theory.

3. Many times our requirements do not demand detailed solutions of the differential balances. Perhaps we are only interested in some type of integral average. The approach to integral averages is explored in some detail in Chapter 4.

4. Finally, there is the question of mathematical approximations. This is really the problem with which you are primarily concerned in carrying out numerical solutions. A mathematical approximation is applied repeatedly in order to arrive at a solution for a differential equation. The approximate solution presented by the computer can be made to approach the exact solution as closely as desired, perhaps by using smaller and smaller step sizes. We do not directly address numerical analysis in this text.

In summary, this chapter begins by examining those problems for which solutions to the original differential equation can be developed simply. We then look at three limiting cases that are themselves worthy of textbooks: creeping flow, potential flow, and boundary-layer theory.

Our purpose here is to show you how physical problems can be described in mathematical terms, not to teach you how to solve differential equations either analytically or numerically. For this reason, we suggest that some of the more difficult exercises be solved using one of the symbolic manipulator programs such as Mathematica (1993). In some cases, these programs may help you develop analytic solutions; in others, numerical solutions.

## 3.1    Philosophy

In Chapters 1 and 2, we developed the differential and jump balances describing mass conservation, the momentum balance, and the moment of momentum balance. Our intention here is to discuss how one uses this structure to solve problems.

The first step is to decide just what problem it is that you wish to solve. One must choose a description of the stress-deformation behavior of the fluid, for example, an incompressible Newtonian fluid. One must indicate the geometry through which the material is to move and the applied forces that cause the fluid to move. We might wish to study the flow of our incompressible fluid through a horizontal tube in a uniform gravitational field, when the stresses indicated by two pressure gauges mounted on the tube at different axial positions are given. (Hereafter, we assume the external force is due to a uniform gravitational field, unless we specifically state otherwise.)

The specification of stress at two points on the tube wall is one example of boundary conditions. By a *boundary condition*, all we mean is that one of the variables in the problem is specified or restricted in some fashion at some point or on some portion of the boundary. There are several common types of boundary conditions for which one should look in considering an unfamiliar problem.

1. We shall always assume that tangential components of velocity are continuous at a phase interface. This appears to be an excellent assumption, even though we might be dealing with a liquid–solid interface such that the liquid does not *wet* the surface. For example, there is no evidence of slip at the wall when mercury flows through glass tubes (Goldstein 1938, p. 676).

   Continuity of the tangential components of velocity at a phase interface is suggested by visualizing that, in a sense, local equilibrium is established at a phase boundary. Continuity of the tangential components of velocity across a phase interface is a necessary condition for equilibrium (Slattery 1990, p. 842).

2. The jump mass balance introduced in Section 1.3.6 must be satisfied at every phase interface. This may be used to relate the normal components of velocity in each phase at an interface.

3. The jump momentum balance (2.2.3-5) must be satisfied at every phase interface. It is common in solving for the velocity distribution within a fluid to ignore the stress distribution in bounding solid walls. When this is done, we employ the jump momentum balance at fluid–fluid phase interfaces exclusively.

4. The velocities and stresses are finite at all points in the fluid.

Having specified what it is that we wish to describe, we are faced with a strictly mathematical problem of solving several partial differential equations simultaneously to find a solution that is consistent with the boundary conditions. This is not easy, even in the relatively simple case of an incompressible Newtonian fluid, since no general solution of the Navier–Stokes equation (Section 2.4.1) consistent with the differential mass balance of Section 1.3.3 is known. We will begin each problem by assuming the form of solution that we are seeking. If we are successful in finding a solution of this form that satisfies all of the boundary conditions, we know that our initial assumption was correct, in the sense that we have found our solution to the problem. If not, we must start over with a revised assumption for the form of the solution.

One point is worth keeping in mind in reading this chapter. We shall take the approach here of most workers and merely ask for a solution. Experiments may suggest that the solution we find is unique, but this is not always the case. We will not investigate the uniqueness of the solution.

## 3.2   Complete Solutions

### 3.2.1   Flow in a Tube

As our first example of a problem for which an exact solution can be found, consider the steady-state flow of an incompressible Newtonian fluid through the horizontal tube of radius $R$ shown in Figure 3.2.1-1. It is assumed that we are working with a section of the tube that is so removed from the entrance and exit that disturbances originating at the entrance and exit can be neglected. In terms of a cylindrical coordinate system whose $z$ axis coincides with the axis of the tube, we have two pressure gauges mounted on the tube that measure the $r$ component of the force per unit area that the fluid exerts on the tube wall, $-T_{rr}$:

$$\text{at } z = 0, \; r = R, \; \theta = 0: \quad t_r = -T_{rr}$$
$$= p - S_{rr}$$
$$= P_0 \tag{3.2.1-1}$$

$$\text{at } z = L, \; r = R, \; \theta = 0: \quad t_r = -T_{rr}$$
$$= p - S_{rr}$$
$$= P_L \tag{3.2.1-2}$$

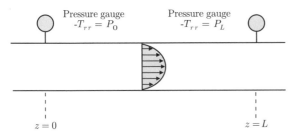

**Figure 3.2.1-1.** Flow through a tube.

As further boundary conditions, continuity of the tangential components of velocity and the jump mass balance require that, at the tube wall,

$$\text{at } r = R : \quad \mathbf{v} = \mathbf{0} \qquad (3.2.1\text{-}3)$$

As mentioned in Section 3.1, we must start with an assumption about the form of solution for the velocity distribution. Usually one's intuition can be assisted by a simple sketch. For example, in Figure 3.2.1-1 the flow is in the axial direction and, if the velocity is to be zero at the tube wall, it must reach a maximum at the centerline of the tube. This suggests that we might assume

$$v_r = v_\theta = 0 \qquad (3.2.1\text{-}4)$$

$$v_z = v_z(r, z) \qquad (3.2.1\text{-}5)$$

and see whether we can find a solution to the Navier–Stokes equation and the differential mass balance that satisfies the boundary conditions in (3.2.1-1) to (3.2.1-3).

Remembering that the fluid is incompressible, we see from Table 2.4.1-1 that the differential mass balance requires in cylindrical coordinates

$$\frac{\partial v_z}{\partial z} = 0 \qquad (3.2.1\text{-}6)$$

or

$$v_z = v_z(r) \qquad (3.2.1\text{-}7)$$

In view of (3.2.1-4), (3.2.1-7), and Table (2.4.1-9), the only nonzero physical components of the viscous portion of the stress tensor for an incompressible Newtonian fluid (Section 2.3.2) are

$$S_{rz} = S_{zr} = \mu \frac{dv_z}{dr} \qquad (3.2.1\text{-}8)$$

From Table 2.4.1-4, the three components of the differential momentum balance for a uniform gravitational field in Section 2.4.1 become

$$0 = -\frac{\partial \mathcal{P}}{\partial r} \qquad (3.2.1\text{-}9)$$

$$0 = -\frac{1}{r}\frac{\partial \mathcal{P}}{\partial \theta} \qquad (3.2.1\text{-}10)$$

and

$$\frac{\partial \mathcal{P}}{\partial z} = \frac{1}{r}\frac{d}{dr}(r S_{rz}) \qquad (3.2.1\text{-}11)$$

Here $\mathcal{P}$ is the modified pressure:

$$\mathcal{P} = p + \rho\phi \tag{3.2.1-12}$$

and $\phi$ is the gravitational potential. Since $\phi$ is arbitrary to a constant (Exercise 2.4.1-1), let us define

$$\text{at } r = R, \ \theta = 0 : \ \phi = 0 \tag{3.2.1-13}$$

Equations (3.2.1-9) and (3.2.1-10) imply that $\mathcal{P}$ is a function only of $z$. But the term on the right of (3.2.1-11) is a function only of $r$, whereas the term on the left is a function only of $z$. This can be true only if

$$\frac{d\mathcal{P}}{dz} = A = \text{constant} \tag{3.2.1-14}$$

Integrating this equation using (3.2.1-1) and (3.2.1-2) [remember that $S_{rr}$ is zero as the result of (3.2.1-4) and (3.2.1-7)] or using

$$\text{at } z = 0 : \ \mathcal{P} = P_0 \tag{3.2.1-15}$$

$$\text{at } z = L : \ \mathcal{P} = P_L \tag{3.2.1-16}$$

we have

$$-A = \frac{P_0 - P_L}{L} \tag{3.2.1-17}$$

Having eliminated $d\mathcal{P}/dz$ between (3.2.1-11) and (3.2.1-17), we may integrate:

$$-\frac{(P_0 - P_L)}{L} \int_0^r r \, dr = \int_0^{r S_{rz}} d(r S_{rz})$$

$$-\frac{(P_0 - P_L)}{L} \frac{r}{2} = S_{rz} \tag{3.2.1-18}$$

Our only assumption here is that $S_{rz}$ is finite at $r = 0$. Substituting for $S_{rz}$ from (3.2.1-8), we may again integrate using (3.2.1-3) as the boundary condition:

$$-\frac{(P_0 - P_L)}{2\mu L} \int_R^r r \, dr = \int_0^{v_z} dv_z$$

$$v_z = \frac{(P_0 - P_L)R^2}{4\mu L} \left[ 1 - \left(\frac{r}{R}\right)^2 \right] \tag{3.2.1-19}$$

This equation confirms the velocity distribution that we sketched in Figure 3.2.1-1 on the basis of our intuition. This means that our initial assumptions in (3.2.1-5) were justified and that there is a solution to the problem of the form assumed.

The maximum velocity occurs on the centerline of the tube, where

$$v_{z(\text{max})} = \frac{(P_0 - P_L)R^2}{4\mu L} \tag{3.2.1-20}$$

The volume rate of flow $Q$ is easily computed to be

$$Q = \int_0^{2\pi} \int_0^R v_z r \, dr \, d\theta$$

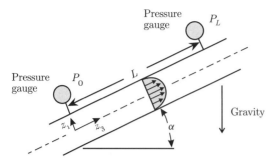

**Figure 3.2.1-2.** Flow through an inclined tube.

$$= 2\pi R^2 \frac{(P_0 - P_L)R^2}{4\mu L} \int_0^1 (1 - x^2)x\, dx$$

$$= \frac{\pi(P_0 - P_L)R^4}{8\mu L} \tag{3.2.1-21}$$

Equation (3.2.1-21) is often referred to as Poiseuille's law. Its general form was published in 1841 by Jean Louis Poiseuille, a physician interested in experimental physiology. He presented it as an empirical relationship correlating data for the flow of water through glass capillary tubes (Rouse and Ince 1957, p. 160).

**Exercise 3.2.1-1**   Show how the discussion in this section is modified when the tube is inclined at an angle $\alpha$ with respect to the horizon, as illustrated in Figure 3.2.1-2.

*Answer:*

$$-\frac{d\mathcal{P}}{dz} = -A$$

$$= \frac{P_0 - P_L - \rho g L \sin \alpha}{L}$$

**Exercise 3.2.1-2** *Flow in a tube of a power-law fluid*   Derive the analog of Poiseuille's law for a fluid described by the power-law model.

*Answer:*

$$v_z = \frac{n}{1+n} \left( \frac{-AR^{1+n}}{2m} \right)^{1/n} \left[ 1 - \left( \frac{r}{R} \right)^{(1+n)/n} \right]$$

$$Q = \frac{\pi n}{(1+3n)} \left( \frac{-AR^{1+3n}}{2m} \right)^{1/n}$$

**Exercise 3.2.1-3** *Flow of an Ellis fluid through a tube*   Derive the analog of Poiseuille's law for a fluid described by the Ellis fluid.

*Answer:*

$$v_z = \frac{-AR^2}{4\mu_0}\left[1 - \left(\frac{r}{R}\right)^2\right] + \left(\frac{-A}{2\tau_{1/2}}\right)^\alpha \frac{\tau_{1/2}R^{\alpha+1}}{\mu_0(\alpha+1)}\left[1 - \left(\frac{r}{R}\right)^{\alpha+1}\right]$$

$$Q = \frac{-A\pi R^4}{8\mu_0} + \frac{\pi \tau_{1/2} R^{\alpha+3}}{\mu_0(\alpha+3)}\left(\frac{-A}{2\tau_{1/2}}\right)^\alpha$$

Exercise 3.2.1-4 *Flow of a Bingham plastic through a tube*   Derive the analog of Poiseuille's law for a fluid described by the Bingham plastic model.

*Answer:*

for $\dfrac{-AR}{2} < \tau_0$ : no flow

for $\dfrac{-Ar}{2} > \tau_0$ : $v_z = \dfrac{-AR^2}{4\eta_0}\left[1 - \left(\dfrac{r}{R}\right)^2\right] - \dfrac{\tau_0 R}{\eta_0}\left(1 - \dfrac{r}{R}\right)$

for $\dfrac{-AR}{2} > \tau_0$ : $Q = \dfrac{-A\pi R^4}{8\eta_0}\left[1 - \left(\dfrac{2\tau_0}{-AR}\right)^4\right] - \dfrac{\pi R^3 \tau_0}{3\eta_0}\left[1 - \left(\dfrac{2\tau_0}{-AR}\right)^3\right]$

Exercise 3.2.1-5 *Newtonian flow in a wire-coating die* (Bird et al. 1960, p. 65; Paton et al. 1959) A somewhat simplified picture of a wire-coating die is shown in Figure 3.2.1-3a. The wire is assumed to be coaxial with the cylindrical die and moving axially with a speed $V$. The reservoir at the left is assumed to be filled with a liquid coating, taken here to be a Newtonian fluid. We wish to determine the steady-state velocity distribution, the volume rate of flow of the fluid in the annular region, and the force per unit length required to pull the wire through the die.

This is a difficult problem to analyze as we have presented it. So let us neglect the *end effects* and determine the quantities requested above for the idealized flow shown in Figure 3.2.1-3b.

Exercise 3.2.1-6 *More about Newtonian flow in a wire-coating die*   Repeat Exercise 3.2.1-5 with the assumption that there is a specified pressure difference $P_0 - P_L$ across a die of length $L$.

Exercise 3.2.1-7 *Power-law model flow in a wire-coating die*   Repeat Exercise 3.2.1-5 for a fluid described by the power-law model.

Exercise 3.2.1-8 *Flow of an Ellis fluid in a wire-coating die*   Repeat Exercise 3.2.1-5 for a fluid described by as an Ellis fluid, but assume that the axial component $F_z$ of the force per unit length required to pull the wire through the die is given and the corresponding speed $V$ of the wire is to be determined.

How does your result simplify for a Newtonian fluid (Exercise 3.2.1-5) and for a fluid described by the power-law model (Exercise 3.2.1-7)?

*Answer:*

$$V = \frac{F_z}{2\pi L \mu_0}\ln\left(\frac{1}{\kappa}\right) + \frac{1}{\mu_0(1-\alpha)}\left(\frac{F_z}{2\pi L}\right)^\alpha (\kappa R \tau_{1/2})^{1-\alpha}\left[\left(\frac{1}{\kappa}\right)^{1-\alpha} - 1\right]$$

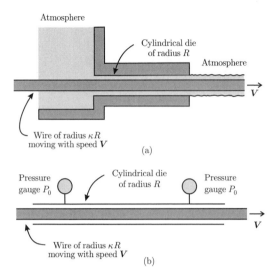

**Figure 3.2.1-3.** (a) Flow in a wire-coating die; (b) flow in an idealized wire-coating die.

**Exercise 3.2.1-9** *Flow of a Bingham plastic in a wire-coating die*    Repeat Exercise 3.2.1-5 for a fluid described as a Bingham plastic. Let us assume that the parameters for the Bingham plastic model are known from another experiment. What is the axial component of the minimum force to be applied to the wire, in order to move it through the die?

**Exercise 3.2.1-10** *Newtonian flow through an annulus*    An incompressible Newtonian fluid flows through the inclined annulus shown in Figure 3.2.1-4. Determine the velocity distribution in the annular region and the volume rate of flow through the annulus.

*Answer:*

$$v_z = \frac{-AR^2}{4\mu}\left[1 - \left(\frac{r}{R}\right)^2 + \frac{1 - \kappa^2}{\ln(1/\kappa)}\ln\frac{r}{R}\right]$$

$$Q = \frac{-A\pi R^4}{8\mu}\left[1 - \kappa^4 - \frac{(1-\kappa^2)^2}{\ln(1/\kappa)}\right]$$

$$-A = \frac{P_0 - P_L - \rho g L \sin\alpha}{L}$$

**Exercise 3.2.1-11** *Newtonian flow through a channel*    An incompressible Newtonian fluid flows through the channel of width $2b$ shown in Figure 3.2.1-5 (two parallel planes of infinite extent separated by a distance $2b$). Determine the velocity distribution and the volume rate of flow per unit width of channel.

*Answer:*

$$v_3 = \frac{-Ab^2}{2\mu}\left[1 - \left(\frac{z_1}{b}\right)^2\right]$$

$$Q = -\frac{2}{3}\frac{Ab^3}{\mu}$$

$$-A \equiv \frac{P_0 - P_L}{L} - \rho g \sin\alpha$$

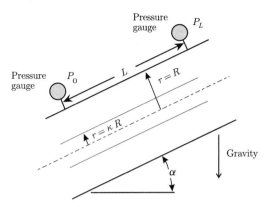

**Figure 3.2.1-4.** Flow through an inclined annulus.

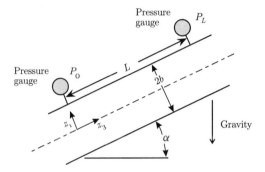

**Figure 3.2.1-5.** Flow through an inclined channel.

**Exercise 3.2.1-12** *Power-law fluid through a channel*   Repeat Exercise 3.2.1-11 for a power-law fluid.

**3.2.2**   Flow of a Generalized Newtonian Fluid in a Tube

Here we wish to repeat the problem of Section 3.2.1 for an incompressible generalized Newtonian fluid (Section 2.3.3). We assume that entrance and exit effects may be neglected, but we take the tube to be inclined with respect to the horizon as shown in Figure 3.2.1-2.

We begin by looking for a velocity distribution of the form

$$v_r = v_\theta = 0$$
$$v_z = v_z(r) \tag{3.2.2-1}$$

which automatically satisfies the differential mass balance.

From (3.2.2-1) and Table 2.4.1-9, we see that there is only one nonzero component of the rate of deformation tensor **D** in cylindrical coordinates:

$$D_{rz} = \frac{1}{2}\frac{dv_z}{dr} \tag{3.2.2-2}$$

From Section 2.3.3, we observe that

$$\mathbf{S} = 2\eta(\gamma)\mathbf{D} \tag{3.2.2-3}$$

which means that there can be only one nonzero component of the viscous portion of the stress tensor, $S_{rz}$. From (3.2.1-17) and (3.2.1-18),

$$\frac{d\mathcal{P}}{dz} = A$$
$$= \text{a constant} \tag{3.2.2-4}$$

and

$$S_{rz} = \frac{Ar}{2} \tag{3.2.2-5}$$

We see from Figure 3.2.1-2 that, with respect to the rectangular Cartesian coordinate system indicated,

$$-\frac{\partial \phi}{\partial z_1} = f_1$$
$$= -g \cos \alpha \tag{3.2.2-6}$$

and

$$-\frac{\partial \phi}{\partial z_3} = f_3$$
$$= -g \sin \alpha \tag{3.2.2-7}$$

The potential energy $\phi$ may be determined from this information by a line integration:

$$\phi = \int_0^\phi d\phi = \int_R^{z_1} \left.\frac{\partial \phi}{\partial z_1}\right|_{z_3=0} dz_1 + \int_0^{z_3} \frac{\partial \phi}{\partial z_3} dz_3$$
$$= -(R - z_1)g \cos \alpha + z_3 g \sin \alpha \tag{3.2.2-8}$$

where we have assumed that

$$\text{at } r = R, \ \theta = 0, \ z = 0 : \ \phi = 0 \tag{3.2.2-9}$$

In terms of cylindrical coordinates, (3.2.2-8) becomes

$$\phi = -(R - r \cos \theta)g \cos \alpha + zg \sin \alpha \tag{3.2.2-10}$$

We are now in a position to integrate (3.2.2-4) in order to determine $A$. Since $S_{rr} = 0$, the boundary conditions on stress, given in Section 3.2.1, reduce to

$$\text{at } z = 0, \ r = R, \ \theta = 0 : \ p = P_0 \tag{3.2.2-11}$$

and

$$\text{at } z = L, \ r = R, \ \theta = 0 : \ p = P_L \tag{3.2.2-12}$$

Equation (3.2.2-10) allows us to write these as boundary conditions on modified pressure:

$$\text{at } z = 0, \ r = R, \ \theta = 0 : \ \mathcal{P} = P_0 \tag{3.2.2-13}$$

$$\text{at } z = L, \ r = R, \ \theta = 0 : \ \mathcal{P} = P_L + \rho g L \sin \alpha \tag{3.2.2-14}$$

Equations (3.2.2-13) and (3.2.2-14) may be used in integrating (3.2.2-4) to yield

$$-A = \frac{P_0 - P_L - \rho g L \sin \alpha}{L} \qquad (3.2.2\text{-}15)$$

Compare this with the similar result in Section 3.2.1.

From Section 2.3.3, we have

$$2\mathbf{D} = \varphi(\tau)\mathbf{S} \qquad (3.2.2\text{-}16)$$

For this flow,

$$\tau \equiv \sqrt{\frac{1}{2}\text{tr}\,\mathbf{S}^2} = |S_{rz}| \qquad (3.2.2\text{-}17)$$

With the help of (3.2.2-2), (3.2.2-5), and (3.2.2-17), we may write the only nonzero component of (3.2.2-16) as

$$\frac{dv_z}{dr} = \varphi\left(\frac{|A|r}{2}\right)\frac{Ar}{2} \qquad (3.2.2\text{-}18)$$

Using the boundary condition for velocity at the wall of the tube described in Section 3.2.1, we may integrate this to obtain for the velocity distribution

$$v_z = \int_r^R \varphi\left(\frac{|A|r}{2}\right)\left[-\frac{Ar}{2}\right]dr \qquad (3.2.2\text{-}19)$$

If we require flow to be in the positive $z$ direction as indicated in Figure 3.2.1-2, it is clear that $-A$ in (3.2.2-15) must be positive, so that (3.2.2-19) becomes

$$v_z = \int_r^R \varphi\left(\frac{-Ar}{2}\right)\left[\frac{-Ar}{2}\right]dr \qquad (3.2.2\text{-}20)$$

Equations (3.2.2-4), (3.2.2-15), and (3.2.2-20) satisfy the differential mass and momentum balances, a description of stress-deformation behavior, and the required boundary conditions on velocity and stress. Thus we have found a solution to this problem that is consistent with our initial assumptions concerning the velocity distribution.

It is easy to measure experimentally the pressure gradient and the volume rate of flow:

$$Q = 2\pi \int_0^R v_z r\, dr \qquad (3.2.2\text{-}21)$$

It would be useful to be able to determine directly from these measurements the function $\varphi(\tau)$ of (3.2.2-16) or the apparent viscosity function $\eta(\gamma)$ in the equivalent description of stress deformation behavior, (3.2.2-3). As we showed in Section 2.3.3, we may derive from (3.2.2-3) and (3.2.2-16)

$$\tau = \eta(\gamma)\gamma \qquad (3.2.2\text{-}22)$$

and

$$\gamma = \varphi(\tau)\tau \qquad (3.2.2\text{-}23)$$

Equations (3.2.2-5) and (3.2.2-15) indicate that we may interpret our pressure measurements in terms of $\tau$ evaluated at the wall of the tube:

$$
\begin{aligned}
\tau_R &\equiv \tau \big|_{r=R} \\
&= -S_{rz} \big|_{r=R} \\
&= -\frac{AR}{2}
\end{aligned}
\tag{3.2.2-24}
$$

Our problem is solved if we can measure $\gamma$ at the tube wall. An integration by parts allows us to express (3.2.2-21) as

$$
Q = -\pi \int_0^R \frac{dv_z}{dr} r^2 \, dr
\tag{3.2.2-25}
$$

Let us make the change of variable

$$
\tau = \frac{r}{R} \tau_R
\tag{3.2.2-26}
$$

in (3.2.2-25), which we can now write as

$$
Q = -\pi \left( \frac{R}{\tau_R} \right)^3 \int_0^{\tau_R} \frac{dv_z}{dr} \tau^2 \, d\tau
\tag{3.2.2-27}
$$

This in turn may be differentiated with respect to $\tau_R$ to find, using the Leibnitz rule (see Exercise 1.3.3-4),

$$
\frac{d(Q\tau_R{}^3)}{d\tau_R} = -\pi R^3 \tau_R{}^2 \frac{dv_z}{dr} \bigg|_{r=R}
\tag{3.2.2-28}
$$

or

$$
\gamma_R \equiv \gamma \big|_{r=R} = -\frac{dv_z}{dr} \bigg|_{r=R} = \frac{4Q}{\pi R^3} \left( \frac{3}{4} + \frac{1}{4} \frac{d \ln Q}{d \ln \tau_R} \right)
\tag{3.2.2-29}
$$

From Section 3.2.1, we find that for an incompressible Newtonian fluid

$$
\gamma_R = \frac{4Q}{\pi R^3}
\tag{3.2.2-30}
$$

This means that we can interpret the term in parentheses on the right side of (3.2.2-29) as a correction for generalized Newtonian behavior.

Equations (3.2.2-15), (3.2.2-24), and (3.2.2-29) allow us to prepare a plot of $\tau_R$ as a function of $\gamma_R$ from experimental measurements of pressure drop as a function of volume rate of flow. This plot of $\tau_R$ versus $\gamma_R$ may be compared with either (3.2.2-22) or (3.2.2-23) to determine the function $\eta(\gamma)$ or the function $\varphi(\tau)$.

- If, as indicated in Figure 3.2.2-1a, a log–log plot of the experimental data is a straight line whose slope is unity, the fluid can be described as an incompressible Newtonian fluid.
- Figure 3.2.2-1b is meant to depict a plot of experimental data that can be represented by a straight line whose slope is other than unity. Such behavior is well described by the power-law model.

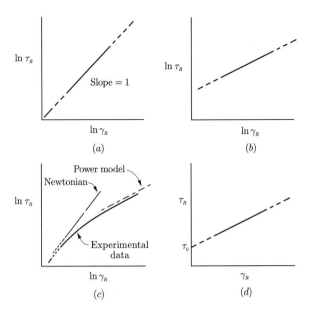

**Figure 3.2.2-1.** (a) Incompressible Newtonian fluid; (b) power-law model; (c) Ellis fluid or Sisko fluid; (d) Bingham plastic fluid.

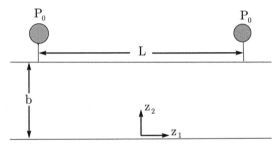

**Figure 3.2.2-2.** Couette flow between parallel plates. The upper plate moves in the $z_1$ direction with a constant speed $V$; the lower plate is stationary.

- In Figure 3.2.2-1c, the experimental data appear to be asymptotic to power-law model behavior for large values of $\gamma_R$. One should attempt to fit such data with the Ellis fluid or Sisko fluid. Ashare et al. (1965) have suggested a convenient manner for fitting the three-parameter Ellis fluid to experimental data.
- Occasionally, a limited range of experimental data will appear to be well described by a linear relationship between $\tau_R$ and $\gamma_R$ as depicted in Figure 3.2.2-1d. The Bingham plastic model will describe these experimental data, though one should be cautious in extrapolating the results to either larger or smaller values of $\gamma_R$.

The analysis given here was first reported by Herzog and Weissenberg (1928) and by Rabinowitsch (1929). It can be extended to a more general description of material behavior, the Noll simple fluid introduced in Section 2.3.4. For a discussion of the Noll simple fluid

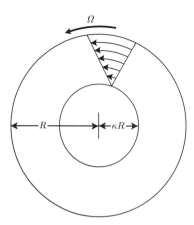

**Figure 3.2.3-1.** Tangential annular
flow.

and an important class of problems that can be analyzed for this model of material behavior,
see an excellent book by Coleman et al. (1966).

**Exercise 3.2.2-1** *Couette flow*   Prove that the velocity profile in Couette flow shown in Figure
   3.2.2-2 assumes the same form for any member of the class of generalized Newtonian fluids
   (Section 2.3.3).

### 3.2.3   Tangential Annular Flow

An incompressible Ellis fluid is trapped between the two concentric cylinders in Figure
3.2.3-1. The outer cylinder rotates with a constant angular velocity, while the inner cylinder
is held stationary. As in the preceding sections, we neglect end effects – we treat the flow as
though the cylinders were infinitely long, even though they are a finite length $L$.

   This geometry is the basis for one common type of viscometer, an instrument used to
study the stress-deformation behavior of fluids or to measure the shear viscosity of Newtonian
fluids. Two measurements are recorded: the angular velocity of the outer cylinder and the
axial component of the torque (moment of the force) that the fluid exerts on the inner cylinder.
Here we assume that torque is given and that the corresponding angular velocity of the outer
cylinder is to be computed.

   Let us begin with the boundary conditions to be satisfied. The inner cylinder is stationary,
so that in cylindrical coordinates

$$\text{at } r = \kappa R : \quad \mathbf{v} = \mathbf{0} \tag{3.2.3-1}$$

The torque $\mathcal{T}$ that the fluid exerts on the inner cylinder may be expressed as

$$\text{at } r = \kappa R : \quad \mathcal{T} = \int_0^{2\pi} \int_0^L \mathbf{p} \wedge (\mathbf{T} \cdot \mathbf{n}) \kappa R \, dz \, d\theta \tag{3.2.3-2}$$

where $\mathbf{p}$ is the position vector with respect to some point on the axis of the cylinders.
As pointed out in Section A.11.1, integrals of vectors are defined only in a rectangular
Cartesian coordinate system. Yet here we have a problem where the major portion of our
computations are most naturally done in cylindrical coordinates. In terms of rectangular

Cartesian coordinates, the axial component or $z_3$ component of the torque that the fluid exerts upon the inner cylinder may be expressed as (see Exercise A.4.1-4)

$$\text{at } r = \kappa R : \quad \mathcal{T}_3 = \int_0^{2\pi} \int_0^L \bar{a}_3 \kappa R \, dz \, d\theta \tag{3.2.3-3}$$

Here we define

$$\mathbf{a} \equiv \mathbf{p} \wedge (\mathbf{T} \cdot \mathbf{n}) \tag{3.2.3-4}$$

and use overbars to denote rectangular Cartesian coordinates. The relation between $\bar{a}_3$ and the cylindrical components of $\mathbf{a}$ is simple (Exercise A.4.1-8):

$$\bar{a}_3 = a_z \tag{3.2.3-5}$$

From Exercise A.9.2-1,

$$a_z \equiv a_{\langle 3 \rangle} = e_{3jk} p_{\langle j \rangle} T_{\langle km \rangle} n_{\langle m \rangle} \tag{3.2.3-6}$$

Since the only nonzero physical component of $\mathbf{n}$ in cylindrical coordinates is

$$n_r = n_{\langle 1 \rangle} = 1 \tag{3.2.3-7}$$

and since the physical cylindrical components of the position vector are (Exercise A.4.1-10)

$$\begin{aligned} p_{\langle 1 \rangle} &= r \\ p_{\langle 2 \rangle} &= 0 \\ p_{\langle 3 \rangle} &= z \end{aligned} \tag{3.2.3-8}$$

Equation (3.2.3-6) reduces to

$$\begin{aligned} a_z &= e_{312} p_{\langle 1 \rangle} T_{\langle 21 \rangle} n_{\langle 1 \rangle} \\ &= r T_{\theta r} \\ &= r S_{\theta r} \end{aligned} \tag{3.2.3-9}$$

Equations (3.2.3-5) and (3.2.3-9) allow us to express (3.2.3-3) as

$$\text{at } r = \kappa R : \quad \mathcal{T}_3 = \int_0^{2\pi} \int_0^L S_{\theta r} (\kappa R)^2 \, dz \, d\theta \tag{3.2.3-10}$$

Next we must make an assumption about the form of the solution we are seeking. In Figure 3.2.3-1, we have sketched what we might imagine the velocity distribution to be like. This suggests assuming in cylindrical coordinates that

$$\begin{aligned} v_r &= v_z = 0 \\ v_\theta &= v_\theta(r, \theta) \end{aligned} \tag{3.2.3-11}$$

Remember that these are assumptions based upon our physical grasp of the problem. We have no guarantee that a solution of this form exists or that, if such a solution exists, it is a unique solution. (In reality, it is not a unique solution, as we shall see in Section 3.6.)

From Table 2.4.1-1 the differential mass balance for an incompressible fluid moving with a velocity distribution given by (3.2.3-11) requires

$$\frac{\partial v_\theta}{\partial \theta} = 0 \tag{3.2.3-12}$$

or

$$v_\theta = v_\theta(r) \tag{3.2.3-13}$$

From Table 2.4.1-9 as well as Equations (3.2.3-11) and (3.2.3-13), we see that the only nonzero component of the rate of deformation tensor is

$$D_{r\theta} = \frac{r}{2} \frac{d}{dr} \left( \frac{v_\theta}{r} \right) \tag{3.2.3-14}$$

For the Ellis fluid (Section 2.3.3), we have that

$$2\mathbf{D} = \frac{1}{\mu_0} \left[ 1 + \left( \frac{\tau}{\tau_{1/2}} \right)^{\alpha-1} \right] \mathbf{S} \tag{3.2.3-15}$$

which means that the only nonzero component of the viscous portion of the stress tensor is

$$S_{r\theta} = S_{r\theta}(r) \tag{3.2.3-16}$$

Accordingly, from Table 2.4.1-4 the three components of the differential momentum balance reduce to

$$\frac{\rho(v_\theta)^2}{r} = \frac{\partial \mathcal{P}}{\partial r} \tag{3.2.3-17}$$

$$\frac{\partial \mathcal{P}}{\partial \theta} = \frac{1}{r} \frac{d}{dr}(r^2 S_{r\theta}) \tag{3.2.3-18}$$

$$0 = \frac{\partial \mathcal{P}}{\partial z} \tag{3.2.3-19}$$

Here we have assumed that the external force per unit mass may be expressed in terms of a potential, and we have introduced the modified pressure $\mathcal{P}$ defined in Section 2.4.1.

Equations (3.2.3-17) through (3.2.3-19) require that

$$\mathcal{P} = \mathcal{P}(r, \theta)$$
$$= \theta f(r) + g(r) \tag{3.2.3-20}$$

Though we did not mention it as a boundary condition, it is clear that all quantities must be periodic functions of $\theta$. In particular,

$$\mathcal{P}(r, \theta) = \mathcal{P}(r, \theta + 2\pi) \tag{3.2.3-21}$$

This means that

$$\frac{\partial \mathcal{P}}{\partial \theta} = f(r)$$
$$= 0 \tag{3.2.3-22}$$

and (3.2.3-18) reduces to

$$\frac{d}{dr}(r^2 S_{r\theta}) = 0 \tag{3.2.3-23}$$

In view of (3.2.3-16), Equation (3.2.3-10) may be simplified to

$$\text{at } r = \kappa R : \quad \mathcal{T}_3 = 2\pi L (\kappa R)^2 S_{r\theta} \tag{3.2.3-24}$$

This is a boundary condition on $S_{r\theta}$, and it may be used to integrate (3.2.3-23):

$$\int_{\kappa^2 R^2 S_0}^{r^2 S_{r\theta}} d(r^2 S_{r\theta}) = 0 \tag{3.2.3-25}$$

$$S_{r\theta} = \frac{\kappa^2 R^2}{r^2} S_0$$

where

$$S_0 \equiv \frac{T_3}{2\pi L \kappa^2 R^2} \tag{3.2.3-26}$$

Since the only nonzero component of the viscous portion of the stress tensor is $S_{r\theta}$, we have

$$\tau \equiv \left[ \frac{1}{2} \text{tr} \, (\mathbf{S} \cdot \mathbf{S}) \right]^{1/2}$$

$$= |S_{r\theta}|$$

$$= S_{r\theta} \tag{3.2.3-27}$$

Here we have made use of (3.2.3-15) to say that $S_{r\theta}$ must have the same sign as $D_{r\theta}$: positive. From (3.2.3-14), (3.2.3-15), (3.2.3-25), and (3.2.3-27), we arrive at

$$r \frac{d}{dr} \left( \frac{v_\theta}{r} \right) = \frac{1}{\mu_0} \frac{\kappa^2 R^2}{r^2} S_0 + \frac{\tau_{1/2}}{\mu_0} \left( \frac{\kappa^2 R^2 S_0}{r^2 \tau_{1/2}} \right)^\alpha \tag{3.2.3-28}$$

Using boundary condition (3.2.3-1), we may integrate this last expression to obtain

$$\frac{v_\theta}{r} = \frac{S_0}{2\mu_0} \left[ 1 - \left( \frac{\kappa R}{r} \right)^2 \right] + \frac{\tau_{1/2}}{2\alpha\mu_0} \left( \frac{S_0}{\tau_{1/2}} \right)^\alpha \left[ 1 - \left( \frac{\kappa R}{r} \right)^{2\alpha} \right] \tag{3.2.3-29}$$

In answer to our original question, the angular velocity of the outer cylinder is, from (3.2.3-26) and (3.2.3-29),

$$\Omega = \frac{T_3(1 - \kappa^2)}{4\pi L \kappa^2 R^2 \mu_0} + \frac{\tau_{1/2}}{2\alpha\mu_0} \left( \frac{T_3}{2\pi L \kappa^2 R^2 \tau_{1/2}} \right)^\alpha (1 - \kappa^{2\alpha}) \tag{3.2.3-30}$$

Finally, (3.2.3-17) can be used to determine $P$ as a function of $r$.

In a manner somewhat similar to that used in Section 3.2.2, it is possible to analyze the tangential annular flow of a generalized Newtonian fluid (Section 2.3.3) without assuming a specific functional form for either $\eta(\gamma)$ or $\varphi(\tau)$ (Krieger and Elrod 1953, Pawlowski 1953). Without much additional difficulty, tangential annular flow of a Noll simple fluid (Section 2.3.4) can be treated; Coleman et al. (1966) give an excellent discussion of this problem and related topics.

For a survey of the problems that have been solved using the Ellis fluid, see Matsuhisa and Bird (1965).

**Exercise 3.2.3-1** *Tangential annular flow of a power-law fluid* Repeat this analysis for an incompressible fluid described by the power-law model.

**Exercise 3.2.3-2** *Tangential annular flow of a Newtonian fluid*   Repeat this analysis for an incompressible Newtonian fluid, but assume that the angular velocity of the outer cylinder is given and the axial component of the moment of the force that the fluid exerts on the inner cylinder is to be calculated.

**Exercise 3.2.3-3** *Helical flow of a Newtonian fluid*   Let us consider the helical flow of an incompressible Newtonian fluid through an inclined annulus. Figure 3.2.1-4 again applies, with the understanding that the inner cylinder rotates with an angular velocity $\Omega$ while the outer cylinder is stationary. Therefore, in cylindrical coordinates, there are *two* nonzero components of velocity, $v_\theta$ and $v_z$, and the particle paths are helices.

Compute the components of the velocity field in the annulus, the volume rate of flow through the annulus, and the torque that the fluid exerts on the inner cylinder (and that the motor driving the inner cylinder at the angular velocity $\Omega$ must overcome).

*Answer:*

$$\frac{v_\theta}{\kappa R \Omega} = \frac{r/R - R/r}{\kappa - 1/\kappa}$$

$$v_z = -\frac{AR^2}{4\mu}\left[1 - \left(\frac{r}{R}\right)^2 + \frac{1 - \kappa^2}{\ln(1/\kappa)}\ln\frac{r}{R}\right]$$

$$-A = \frac{P_0 - P_L - \rho g L \sin\alpha}{L}$$

$$Q = -\frac{A\pi R^4}{8\mu}\left[1 - \kappa^4 - \frac{(1 - \kappa^2)^2}{\ln(1/\kappa)}\right]$$

$$\text{at } r = \kappa R : \ T_3 = \frac{4\pi\mu L R^2 \Omega}{1 - (1/\kappa)^2}$$

*Hint:*   From the $r$ component of the differential momentum balance,

$$\mathcal{P} = h(r) + g(z, \theta)$$

From the $\theta$ component,

$$\frac{\partial g}{\partial \theta} = B = \text{a constant}$$

From the $z$ component,

$$\frac{\partial g}{\partial z} = A = \text{a constant}$$

We conclude that

$$\mathcal{P} = h(r) + Az + B\theta + C$$

where $C$ is a constant.

**Exercise 3.2.3-4** *Still more about Newtonian flow in a wire-coating die*   Repeat Exercise 3.2.1-5, assuming that the wire rotates with a constant angular velocity $\Omega$. Is the force on the wire altered by rotating the wire?

**Exercise 3.2.3-5** *More about flow of a power-law fluid in a wire-coating die*   Go as far as you can in repeating Exercise 3.2.1-5 for a fluid described by the power-law model, again assuming that the wire rotates with a constant angular velocity $\Omega$. Is the force on the wire altered by rotating the wire?

**Exercise 3.2.3-6**   For the problem discussed in the text, show that the other two components of the torque that the fluid exerts on the inner cylinder are zero.

**Exercise 3.2.3-7** *Tangential annular flow of an incompressible power-law fluid*   For the apparatus described in this section, what is the relation between $T_3$ and $\Omega$ for an incompressible fluid described by the power-law fluid?

**Exercise 3.2.3-8** *Cylinder rotates in an infinite fluid*   A long vertical circular cylinder of radius $R$ rotates with an angular velocity $\Omega$ in an infinite incompressible Newtonian fluid. Determine the velocity distribution and the torque exerted upon the cylinder by the fluid.

**Exercise 3.2.3-9** *Cylinder rotates in an infinite Ellis fluid*   Repeat Exercise 3.2.3-8 for an Ellis fluid.

**Exercise 3.2.3-10** *Tangential annular flow*   This experiment can be used to determine the apparent viscosity function (Krieger and Maron 1952, 1954; Krieger and Elrod 1953). One of the simplest suggestions can be understood as follows:

i) Assume that the fluid behavior is described by the class of generalized Newtonian fluids in the form (see Section 2.3.3)

$$2\mathbf{D} = \varphi(\tau)\mathbf{S}$$

Prompted by the discussion in this section, determine that the velocity distribution has the form

$$v_r = v_z = 0$$

$$\frac{v_\theta}{r\Omega} = \int_{\kappa R}^{r} \frac{1}{r^3} \varphi\left(\frac{\kappa^2 R^2 S_0}{r^2}\right) dr \left[\int_{\kappa R}^{R} \frac{1}{r^3} \varphi\left(\frac{\kappa^2 R^2 S_0}{r^2}\right) dr\right]^{-1}$$

This proves that there is a solution to the differential mass and momentum balances that has the form of (3.2.3-11) and that satisfies all of the required boundary conditions.

ii) Regarding $v_\theta/r$ as a function of $\tau$, determine that

$$\Omega = \frac{1}{2} \int_{\kappa^2 S_0}^{S_0} \varphi(\tau) d\tau$$

iii) Use this result to say

$$\varphi\left(\kappa^2 S_0\right) = \frac{1}{\kappa^2 S_0} \left(\frac{\partial\Omega}{\partial \ln\frac{1}{\kappa}}\right)_{S_0}$$

iv) Explain how experimental data could be used to construct a plot of the function $\varphi(\tau)$. Krieger and Maron (1952, 1954) have shown that sufficient accuracy can usually be obtained with just two inner cylinders (two values of $\kappa$).

### 3.2.4  Wall That Is Suddenly Set in Motion

A semi-infinite stationary body of an incompressible Newtonian fluid is bounded on one side by a wall coinciding with the $z_1 z_3$ plane. Gravity acts in the negative $z_2$ direction. At time $t = 0$, the wall is suddenly set in motion in the positive $z_1$ direction, with a constant magnitude of velocity $V$. We wish to determine the velocity distribution in the fluid as a function of time.

We visualize that this body of fluid is actually a deep pool with a horizontal phase interface in contact with the atmosphere. We approximate this experimental situation by saying that, as $z_2$ approaches infinity, pressure is no longer a function of $z_1$ and $z_3$.

The boundary conditions on velocity and pressure that must be satisfied are

$$\text{for all } t \text{ as } z_2 \to \infty: \quad \frac{\partial p}{\partial z_1} = \frac{\partial p}{\partial z_3} = 0 \tag{3.2.4-1}$$

$$\text{for all } z_2 \text{ at } t = 0: \quad \mathbf{v} = \mathbf{0} \tag{3.2.4-2}$$

$$\text{for } t > 0 \text{ at } z_2 = 0: \quad v_1 = V, \ v_2 = v_3 = 0 \tag{3.2.4-3}$$

Some authors feel that as an additional boundary condition one should take the velocity to be zero as $z_2$ approaches infinity for all values of time (Schlichting 1979, p. 90; Bird et al. 1960, p. 125). We take the view here that velocity is not directly constrained to be zero at infinity for time greater than zero; it is only necessary that velocity be finite as $z_2$ approaches infinity.

The initial and boundary conditions suggest that we seek a solution of the form

$$v_1 = v_1(z_2, t)$$
$$v_2 = v_3 = 0 \tag{3.2.4-4}$$

which automatically satisfies the differential mass balance. With this assumption, the rectangular Cartesian components of the Navier–Stokes equation (Section 2.4.1) written in terms of modified pressure become (see Table 2.4.1-3)

$$\frac{\partial \mathcal{P}}{\partial z_1} = -\rho \frac{\partial v_1}{\partial t} + \mu \frac{\partial^2 v_1}{\partial z_2^2} \tag{3.2.4-5}$$

and

$$\frac{\partial \mathcal{P}}{\partial z_2} = \frac{\partial \mathcal{P}}{\partial z_3} = 0 \tag{3.2.4-6}$$

In view of (3.2.4-6), the left side of (3.2.4-5) is a function of $t$ and $z_1$. The right side of this equation can at most be a function of $t$ and $z_2$. This means that each side of this equation is a function of time alone:

$$\frac{\partial \mathcal{P}}{\partial z_1} = A(t) \tag{3.2.4-7}$$

But (3.2.4-7) is consistent with boundary condition (3.2.4-1) only if

$$A(t) = 0 \tag{3.2.4-8}$$

Our problem has reduced to finding a solution for

$$\frac{\partial v_1}{\partial t} = \frac{\mu}{\rho} \frac{\partial^2 v_1}{\partial z_2^2} \tag{3.2.4-9}$$

that satisfies

for all $z_2$ at $t = 0$: $v_1 = 0$ (3.2.4-10)

and

for $t > 0$ at $z_2 = 0$: $v_1 = V$ (3.2.4-11)

Sometimes a partial differential equation such as this may be solved by transforming it into an ordinary differential equation. For example, let us seek a solution of the form

$$\frac{v_1}{V} = g(\eta) \tag{3.2.4-12}$$

where

$$\eta \equiv a z_2 t^c \tag{3.2.4-13}$$

(or any convenient power of $a z_2 t^c$). If we introduce the change of variable (3.2.4-13) in (3.2.4-9), we find

$$\frac{\partial v_1}{\partial t} = \frac{V c \eta}{t} \frac{dg}{d\eta}$$

$$\frac{\partial v_1}{\partial z_2} = \frac{V \eta}{z_2} \frac{dg}{d\eta}$$

$$\frac{\partial^2 v_1}{\partial z_2^2} = V \left( \frac{\eta}{z_2} \right)^2 \frac{d^2 g}{d\eta^2}$$

and

$$c \frac{dg}{d\eta} = \frac{\mu}{\rho} \frac{\eta t}{z_2^2} \frac{d^2 g}{d\eta^2}$$

By inspection, an ordinary differential equation for $g(\eta)$ results, if $c \equiv -1/2$. The choice of

$$a \equiv \left( \frac{4\mu}{\rho} \right)^{-1/2}$$

allows the resulting ordinary differential equation to be further simplified. Specifically, if we take

$$\eta = \frac{z_2}{\sqrt{4\mu t / \rho}} \tag{3.2.4-14}$$

Equation (3.2.4-9) becomes

$$g'' + 2\eta g' = 0 \tag{3.2.4-15}$$

This prime notation is used here to denote differentiation with respect to $\eta$. With this change of variable, the initial condition (3.2.4-10) becomes

as $\eta \to \infty$: $g \to 0$ (3.2.4-16)

and boundary condition (3.2.4-11) transforms to

at $\eta = 0$ :  $g = 1$ (3.2.4-17)

Equation (3.2.4-15) is a first-order separable differential equation in $g'$, which may be integrated to give

$$g' = C_1 e^{-\eta^2}$$ (3.2.4-18)

Here $C_1$ is a constant of integration. This in turn may be integrated using (3.2.4-17):

$$g - 1 = C_1 \int_0^\eta e^{-\eta^2} d\eta$$ (3.2.4-19)

Finally, we may determine $C_1$ using (3.2.4-16):

$$-\frac{1}{C_1} = \int_0^\infty e^{-\eta^2} d\eta = \frac{\sqrt{\pi}}{2}$$ (3.2.4-20)

As a final result we have

$$\frac{v_1}{V} = g = 1 - \frac{2}{\sqrt{\pi}} \int_0^\eta e^{-\eta^2} d\eta$$ (3.2.4-21)

or

$$\frac{v_1}{V} = 1 - \mathrm{erf}\left(\frac{z_2}{\sqrt{4\mu t/\rho}}\right)$$ (3.2.4-22)

Here we have introduced the error function

$$\mathrm{erf}(x) \equiv \frac{2}{\sqrt{\pi}} \int_0^x e^{-t^2} dt$$ (3.2.4-23)

This solution is shown in Figure 3.2.4-1.

This discussion was inspired by Schlichting's (1979, p. 90) development.

**Exercise 3.2.4-1**  Starting with (3.2.4-12) and (3.2.4-13), show how one argues to deduce that the choice $c = -1/2$ reduces (3.2.4-9) to the simplest possible ordinary differential equation.

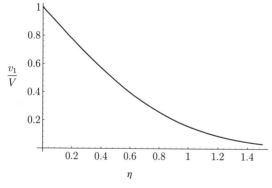

**Figure 3.2.4-1.** Velocity distribution $v_1/V$ as a function of $\eta$ for flow of a Newtonian fluid in the neighborhood of a wall suddenly set in motion.

**Exercise 3.2.4-2** Take the Laplace transformation of (3.2.4-9) to obtain a linear second-order differential equation. Solve with the required boundary conditions to find (3.2.4-22).

**Exercise 3.2.4-3** *An oscillating wall*    Consider a situation similar to that described in the text except that a periodic oscillation is superimposed upon the motion of the wall:

$$\text{at } z_2 = 0 : \quad v_1 = V + A\cos(\omega t - \epsilon)$$

Let us determine the velocity distribution in the fluid as $t \to \infty$.

After formulating the problem and discussing the pressure distribution, seek a solution of the form

$$v_1^\star \equiv \frac{v_1 - V}{A} = u(z_2)\exp[i(\omega t - \epsilon)]$$

Only the real portion of this solution is of interest to us, but we must allow for the possibility that $u(z_2)$ is a complex function. Determine that

$$v_1^\star = \exp(-Kz_2)\cos(\omega t - Kz_2 - \epsilon)$$

where

$$K \equiv \sqrt{\frac{\omega\rho}{2\mu}}$$

*Hint:*   Observe that $i = (1+i)^2/2$.

**Exercise 3.2.4-4** *More on an oscillating wall*    Consider a modification of the system described in Exercise 3.2.4-3 in which a liquid–gas interface is located at $z_2 = L$. You may assume that the gas phase is at a uniform atmospheric pressure $p_o$.

i) Under what circumstances can viscous effects in the gas phase be neglected with respect to those in the liquid phase?

ii) Determine the velocity distribution in the liquid phase as $t \to \infty$.

**Exercise 3.2.4-5** *Oscillatory flow between two flat plates* (Bird 1965b)    An incompressible Newtonian fluid is contained between the two parallel plates shown in Figure 3.2.4-2. The lower plate oscillates periodically:

$$\text{at } z_2 = 0 : \quad v_1 = v_0 \cos \omega t$$

while the upper plate is held stationary. Determine the velocity distribution in the fluid.

*Answer:*

$$v_1^\star \equiv \frac{v_1}{v_0}$$

$$= \left\{ \begin{bmatrix} \sinh a(1-\xi)\cos a(1-\xi)\sinh a \cos a \\ +\sin a(1-\xi)\cosh a(1-\xi)\sin a \cosh a \end{bmatrix} \cos \omega t \right.$$

$$\left. + \begin{bmatrix} \sinh a(1-\xi)\cos a(1-\xi)\sin a \cosh a \\ -\sin a(1-\xi)\cosh a(1-\xi)\sinh a \cos a \end{bmatrix} \sin \omega t \right\}$$

$$\times \left\{ \sinh^2 a \cos^2 a + \sin^2 a \cosh^2 a \right\}^{-1}$$

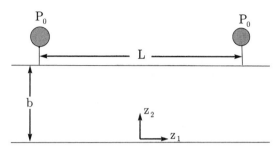

**Figure 3.2.4-2.** Oscillatory flow between two flat plates.

where

$$\xi \equiv \frac{z_2}{b}$$

$$a \equiv \sqrt{\frac{\omega \rho b^2}{2\mu}}$$

*Hint:*   See Exercise 3.2.4-3.

**Exercise 3.2.4-6** *Unsteady-state flow in a tube*    The inclined tube shown in Figure 3.2.1-2 is filled with a stationary incompressible Newtonian fluid. At time $t = 0$, the pressure gradient indicated in the figure is imposed and the fluid begins to flow. We wish to determine the velocity distribution in the fluid as a function of time.

i) Argue that the equation of motion in the form

$$\rho \frac{\partial v_z}{\partial t} = -A + \mu \frac{1}{r} \frac{\partial}{\partial r} \left( r \frac{\partial v_z}{\partial r} \right)$$

is to be solved consistent with the boundary conditions

at $t = 0: \ v_z = 0$

and

at $r = R : \ v_z = 0$

Here

$$-A \equiv \frac{P_0 - P_L}{L} - \rho g \sin \alpha$$

ii) Let us introduce as dimensionless variables

$$v_z^\star \equiv \frac{4\mu}{R^2(-A)} v_z$$

$$r^\star \equiv \frac{r}{R}$$

$$t^\star \equiv \frac{\mu t}{\rho R^2}$$

We shall attempt to find a solution of the form

$$v_z^\star \equiv v_{z\infty}^\star - \Phi(r^\star, t^\star)$$

where $v_{z\infty}^\star$ is the steady-state velocity distribution. Determine that $\Phi(r^\star, t^\star)$ is a solution to

$$\frac{\partial \Phi}{\partial t^\star} = \frac{1}{r^\star} \frac{\partial}{\partial r^\star}\left(r^\star \frac{\partial \Phi}{\partial r^\star}\right)$$

that is consistent with the boundary conditions

at $t^\star = 0 : \quad \Phi = 1 - r^{\star 2}$

and

at $r^\star = 1 : \quad \Phi = 0$

iii) Solve the problem posed in (ii) to find

$$v_z^\star = 1 - r^{\star 2} - \sum_{n=1}^{\infty} \frac{8}{(\alpha_n)^3 J_1(\alpha_n)} J_0(\alpha_n r^\star) \exp\left(-\alpha_n^2 t^\star\right)$$

Here the $\alpha_n (n = 1, 2, \ldots)$ are roots of

$$J_0(\alpha_n) = 0$$

(Irving and Mullineux 1959, p. 130). This solution is shown in Figure 3.2.4-3.

iv) From a practical point of view, we may be more interested in the volume flow rate $Q$ as a function of time:

$$\frac{8\mu Q}{\pi R^4(-A)} = 1 - \sum_{n=1}^{\infty} \frac{32}{(\alpha_n)^4} \exp\left(-\alpha_n^2 t^\star\right)$$

This result is plotted in Figure 4.2.1-1.

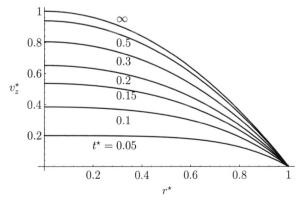

**Figure 3.2.4-3.** Velocity distribution $v_z^\star$ as a function of $r^\star$ at $t^\star = 0.05, 0.1, 0.15, 0.2, 0.3, 0.5,$ and $\infty$ (from bottom to top) for unsteady-state flow from rest in a circular tube (Szymanski 1932).

For another view of this problem, see Section 4.2.1.

*Hint:* From Irving and Mullineux (1959, p. 161) we know

$$\int_0^1 r^\star(1 - r^{\star 2})J_0(\alpha_m r^\star)\, dr^\star = \frac{2}{(\alpha_m)^2}J_2(\alpha_m)$$

**Exercise 3.2.4-7** *More about unsteady-state flow in a tube*    Steady-state flow of an incompressible Newtonian fluid through a tube has been established. Repeat Exercise 3.2.4-6 assuming that at $t = 0$ the pressure drop between $z = 0$ and $z = L$ is increased by a factor of $x$.

**Exercise 3.2.4-8** *Unsteady-state flow through an annulus*    The analysis outlined in Exercise 3.2.4-6 may be used as a guide in treating flow from rest of an incompressible Newtonian fluid in a coaxial annulus. Determine the velocity distribution in unsteady-state axial flow (Müller 1936).

**Exercise 3.2.4-9** *Unsteady-state tangential flow in an annulus*    Determine the velocity distribution in unsteady-state tangential flow in an annulus for an incompressible Newtonian fluid, assuming that for $t > 0$ the outer cylinder rotates with an angular velocity $\Omega$ (Bird and Curtiss 1959).

**Exercise 3.2.4-10** *Suddenly accelerated wire in an unbounded Newtonian fluid*    A straight, rigid, horizontal wire is at rest in an unbounded, incompressible, Newtonian fluid. At time $t = 0$, the wire is suddenly set in motion in the positive $z$ direction (the direction of its axis) with a constant magnitude of velocity $V$. Determine the velocity distribution in the fluid and the force per unit length required to pull the wire through the fluid as functions of time.

*Hint:* Use the Laplace transformation to arrive at a form of Bessel's equation (Carslaw and Jaeger 1959, p. 335).

### 3.2.5    Rotating Meniscus

The right circular cylinder shown in Figure 3.2.5-1 is partially filled with an incompressible Newtonian fluid and rotates with a constant angular velocity $\Omega$. We wish to determine the shape of the gas–liquid phase interface.

In this discussion, we shall neglect viscous effects in the gas phase and assume that the pressure in the gas phase has a uniform value $p_0$. The analysis will be done in two steps. In the first step, we will neglect the effects of interfacial tension and contact angle (see Section 2.4.3); in the second, we will take these effects into account.

#### Neglecting the Effects of Interfacial Tension and Contact Angle

Most readers will immediately recognize that the fluid in the bucket must rotate as a solid body and that the velocity distribution in the fluid takes the following form in cylindrical coordinates (see also Exercise 3.2.5-1):

$$\frac{v_\theta}{r} = \Omega$$

$$v_r = v_z = 0 \tag{3.2.5-1}$$

For this velocity distribution, the differential mass balance is satisfied identically and the rate of deformation tensor is zero (since the fluid rotates as a solid body).

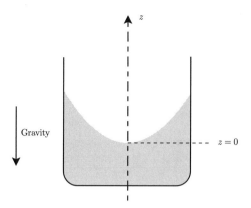

**Figure 3.2.5-1.** The rotating bucket.

Therefore, the viscous portion of the stress tensor **S** is zero as well and the three components of the differential momentum balance reduce to

$$\frac{\partial \mathcal{P}}{\partial r} = \rho r \Omega^2 \tag{3.2.5-2}$$

and

$$\frac{\partial \mathcal{P}}{\partial \theta} = \frac{\partial \mathcal{P}}{\partial z} = 0 \tag{3.2.5-3}$$

Because there is no mass transfer across the phase interface, the normal component of velocity of the fluid at the phase interface must be the same as the speed of displacement of the phase interface (see Section 1.3.5):

$$\mathbf{v} \cdot \boldsymbol{\xi} = \mathbf{u} \cdot \boldsymbol{\xi} \tag{3.2.5-4}$$

This assures that the jump mass balance for the phase interface (Section 1.3.6) is identically satisfied and the jump momentum balance (2.2.3-5) reduces to

$$[\![ \mathbf{T} \cdot \boldsymbol{\xi} ]\!] = 0 \tag{3.2.5-5}$$

It is easy to show that, in view of (3.2.5-1), the $\theta$ component of (3.2.5-5) is satisfied identically and the $r$ and $z$ components require only that, at the phase interface,

$$p = p_0 \tag{3.2.5-6}$$

Let us define our coordinate system such that the phase interface passes through the origin ($r = 0, z = 0$) and let us define the external force potential $\phi$ such that

$$\phi = gz \tag{3.2.5-7}$$

In view of (3.2.5-3), this allows us to say that

$$\text{at } r = 0 : \quad \mathcal{P} = p_0 \tag{3.2.5-8}$$

We may integrate (3.2.5-2) using (3.2.5-8) as a boundary condition to determine that the pressure at any point in the fluid is given by

$$p - p_0 = \rho \frac{\Omega^2}{2} r^2 - \rho g z \tag{3.2.5-9}$$

The phase interface is that surface on which (3.2.5-6) is satisfied. From (3.2.5-9) the phase interface is the parabolic surface

$$z = h(r)$$

$$= \left(\frac{\Omega^2}{2g}\right) r^2 \tag{3.2.5-10}$$

or

$$h^\star = \left(\frac{\Omega^2 R}{2g}\right) r^{\star 2} \tag{3.2.5-11}$$

where

$$r^\star \equiv \frac{r}{R}$$

$$h^\star \equiv \frac{h}{R} \tag{3.2.5-12}$$

As an example, let us consider a rotating tube of water such that

$$\rho = 998 \, \text{kg/m}^3$$

$$g = 9.80 \, \text{m/s}^2$$

$$R = 10^{-2} \, \text{m} \tag{3.2.5-13}$$

$$\Omega = 2\pi \, \text{radians/s}$$

Mathematica (1993) was used to plot (3.2.5-11) as the lower curve in Figure 3.2.5-2.

### Including the Effects of Interfacial Tension and Contact Angle

In order to include the effects of interfacial tension, we must use the more complete form of the jump momentum balance discussed in Section 2.4.3. With the recognition that at steady state the interfacial tension $\gamma$ will be independent of position, we find from Table 2.4.3-7

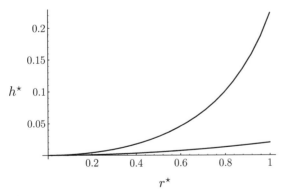

**Figure 3.2.5-2.** Interface configuration $h^\star (r^\star)$ for water in a rotating tube. The lower curve is the prediction of (3.2.5-11). The upper curve is the complete solution of (3.2.5-18) for $\gamma = 70 \times 10^{-3}$ N/m and $\theta = \pi/4$.

that, in contrast with (3.2.5-6), the $r$ and $z$ components of the jump momentum balance require

$$p - p_0 = -2H\gamma$$

$$= -\frac{\gamma}{r}\left[r\frac{d^2h}{dr^2} + \frac{dh}{dr} + \left(\frac{dh}{dr}\right)^3\right]\left[1 + \left(\frac{dh}{dr}\right)^2\right]^{-3/2} \tag{3.2.5-14}$$

Alternatively, in terms of the dimensionless variables (3.2.5-12), Equation (3.2.5-14) becomes

$$p - p_0 = -\frac{\gamma}{Rr^\star}\left[r^\star\frac{d^2h^\star}{dr^{\star 2}} + \frac{dh^\star}{dr^\star} + \left(\frac{dh^\star}{dr^\star}\right)^3\right]\left[1 + \left(\frac{dh^\star}{dr^\star}\right)^2\right]^{-3/2} \tag{3.2.5-15}$$

If $p_{\text{ref}}$ is defined to be the pressure in liquid at $r = 0$ adjacent to the phase interface, we can write in place of (3.2.5-8)

$$\text{at } r = 0: \quad \mathcal{P} = p_{\text{ref}} \tag{3.2.5-16}$$

This can be used in integrating (3.2.5-2) to find

$$p - p_{\text{ref}} = \rho\frac{\Omega^2}{2}r^2 - \rho g z \tag{3.2.5-17}$$

Subtracting (3.2.5-17) from (3.2.5-15) to eliminate $p$, we find

$$\frac{1}{r^\star}\left[r^\star\frac{d^2h^\star}{dr^{\star 2}} + \frac{dh^\star}{dr^\star} + \left(\frac{dh^\star}{dr^\star}\right)^3\right]\left[1 + \left(\frac{dh^\star}{dr^\star}\right)^2\right]^{-3/2}$$

$$+ A + Br^{\star 2} - Ch^\star = 0 \tag{3.2.5-18}$$

where

$$A \equiv \frac{(p_{\text{ref}} - p_0)R}{\gamma}$$

$$B \equiv \frac{\rho R^3\Omega^2}{2\gamma} \tag{3.2.5-19}$$

$$C \equiv \frac{\rho g R^2}{\gamma}$$

This second-order differential equation must be integrated consistent with the boundary conditions

$$\text{at } r^\star = 0: \quad h^\star = 0 \tag{3.2.5-20}$$

$$\text{at } r^\star = 0: \quad \frac{dh^\star}{dr^\star} = 0 \tag{3.2.5-21}$$

$$\text{at } r^\star = 1: \quad \frac{dh^\star}{dr^\star} = (\tan\theta)^{-1} \tag{3.2.5-22}$$

Here $\theta$ is the *contact angle* measured through the liquid phase. Three boundary conditions are required, because $p_{\text{ref}}$, and therefore $A$, is unknown.

As an example, let us consider a rotating tube of water for the conditions described by (3.2.5-13) and $\gamma = 70 \times 10^{-3}$ N/m. Mathematica (1993) was used to solve this problem. The results for $\theta = \pi/4$ are shown as the upper curve in Figure 3.2.5-2.

These results are consistent with those first developed by Wasserman and Slattery (1964).

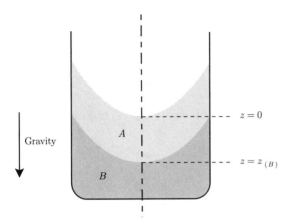

**Figure 3.2.5-3.** A rotating bucket filled with two immiscible, incompressible Newtonian fluids.

**Exercise 3.2.5-1**    Repeat the analysis of this section beginning with the assumption that the velocity distribution is of the form

$$\frac{v_\theta}{r} = \omega(r), \quad v_r = v_z = 0$$

**Exercise 3.2.5-2** *Two rotating menisci*    Repeat the analysis of this section assuming that the bucket is filled with two immiscible, incompressible Newtonian fluids as shown in Figure 3.2.5-3. You may neglect the effects of contact angles and interfacial tensions. Determine that the shape of the liquid–liquid phase interface is given by

$$z - z_{(B)} = \frac{\Omega^2 r^2}{2g}$$

**Exercise 3.2.5-3** *A rotating annulus*    Repeat the problem discussed in the text, substituting a rotating annulus for the rotating bucket. You may neglect the effects of contact angles and interfacial tension.

**Exercise 3.2.5-4** *More about rotating meniscus*    Repeat the discussion in the text, assuming that the bucket rotates around an axis outside the bucket. You may neglect the effects of contact angles and interfacial tension.

**Exercise 3.2.5-5** *Flow in a film*    In the manufacture of photographic film and x-ray film, several stratified emulsion layers flow down an inclined plane to be deposited on a moving continuous strip that forms the backing for the film.

As a first step in understanding this problem, let us consider the flow of an incompressible Newtonian fluid down an inclined plane as shown in Figure 3.2.5-4. Neglect viscous effects in the gas phase and assume that the pressure is a constant $p_0$. Determine the velocity distribution and volume rate of flow in this film, assuming that the thickness of the film is a constant $\delta$.

The volume rate of flow is determined by the speed of the film backing and by the required thickness of the coating. In practice, one would probably wish to match the average velocity

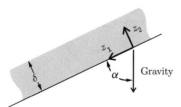

**Figure 3.2.5-4.** Flow down an inclined plane.

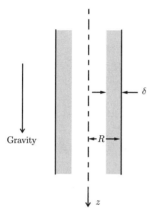

**Figure 3.2.5-5.** Flow in a film on the interior of a vertical pipe.

of the fluid to the speed of the film backing, to minimize the disturbance created as the fluid pours onto the backing. This results in two equations for the two unknowns: $\delta$ and $\cos \alpha$.

*Answer:*

$$v_1 = \frac{\rho g \delta^2 \cos \alpha}{2\mu} \left[ 2\frac{z_2}{\delta} - \left(\frac{z_2}{\delta}\right)^2 \right]$$

**Exercise 3.2.5-6** *Flow in a cylindrical film* An incompressible Newtonian fluid flows down the inside of a vertical pipe as shown in Figure 3.2.5-5. Neglect viscous effects in the gas phase and take the pressure in the gas phase to have a uniform value $p_0$. Find the velocity distribution in this film, assuming that the thickness of the film is a constant $\delta$.

*Answer:*

$$v_z = \frac{\rho g (R - \delta)^2}{2\mu} \left[ \ln\left(\frac{r}{R}\right) - \frac{1}{2}\left(\frac{r}{R-\delta}\right)^2 + \frac{1}{2}\left(\frac{R}{R-\delta}\right)^2 \right]$$

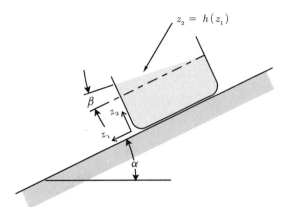

**Figure 3.2.5-6.** Acceleration of tank is constant as it slides down the plane.

**Exercise 3.2.5-7**   A tank containing a liquid slides down an inclined plane as shown in Figure 3.2.5-6. The coefficient of friction between the tank and the plane is $\kappa$. Assuming that the acceleration of the tank is independent of time, determine that the angle $\beta = \tan^{-1}\kappa$.

*Hint:*   There are three points to note. (1) The fluid will move as a solid body. (2) Its movement is best described in a moving frame of reference in which the fluid is stationary (see Exercise 2.2.3-3). (3) The acceleration can be determined by an application of the momentum balance to the tank as a body.

**Exercise 3.2.5-8** *Flow in a vertical film* (Whitaker 1968, p. 183)    A film of water flows down a vertical wall as shown in Figure 3.2.5-7. How large must the air gap $h_{(2)}$ be to allow us to neglect the effect of the air on the water stream?

i) Assume in both the air and water that

$$v_1 = v_1(z_2)$$
$$v_2 = v_3 = 0$$

Assume that the difference in pressure in the gas phase between $z_1 = 0$ and $z_1 = L$ is that which would exist in a static situation or

$$\frac{\partial \mathcal{P}_{(air)}}{\partial z_1} = 0$$

Prove that

$$\frac{\partial \mathcal{P}_{(water)}}{\partial z_1} = -g\left(\rho_{(water)} - \rho_{(air)}\right)$$

ii) Determine the velocity distribution in the water film to be

$$v_{1(water)} = \left[\frac{g(\rho_{(water)} - \rho_{(air)})h_{(1)}^2}{2\mu_{(water)}}\right]\left[1 - \left(\frac{z_2}{h_1}\right)^2 - \frac{1 + z_2/h_{(1)}}{1 + \left(\mu_{(water)}h_{(2)}/\mu_{(air)}h_{(1)}\right)}\right]$$

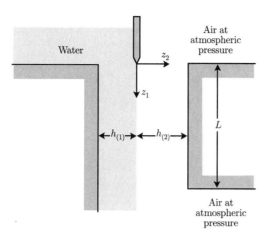

**Figure 3.2.5-7.** The effect of air on a vertical water film.

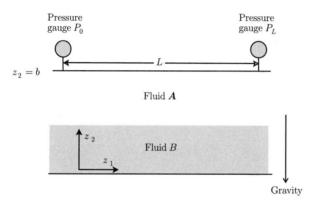

**Figure 3.2.5-8.** Stratified two-phase flow between horizontal parallel plates.

iii) At 20°C and atmospheric pressure,

$$\mu_{(air)} = 1.8 \times 10^{-5} \, \text{Pa s}$$
$$\mu_{(water)} = 1.3 \times 10^{-3} \, \text{Pa s}$$

Conclude that the air gap has a negligible effect provided that

$$\frac{h_{(2)}}{h_{(1)}} \gg 2 \times 10^{-2}$$

**Exercise 3.2.5-9** Repeat Exercise 3.2.5-5 for a power-law fluid.

**Exercise 3.2.5-10** *Stratified flow* Two incompressible Newtonian fluids $A$ and $B$ are pumped at equal volumetric flow rates between two horizontal parallel plates as indicated in Figure 3.2.5-8. Determine the location $z_2 = f$ of the phase interface for fully developed laminar flow as a function of the ratio of the viscosities as shown in Figure 3.2.5-9.

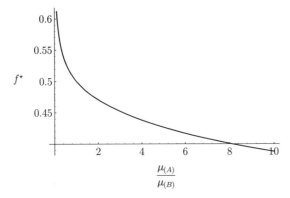

**Figure 3.2.5-9.** $f^\star \equiv f/b$ as a function of $\mu_{(A)}/\mu_{(B)}$ for stratified two-phase flow between horizontal parallel plates.

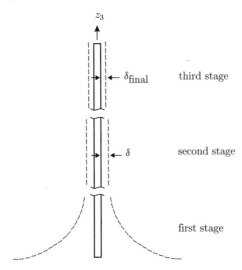

**Figure 3.2.5-10.** Three stages in the continuous coating of a flat sheet.

Exercise 3.2.5-11 *Two-phase, annular flow in a vertical pipe*    Two phases are being pumped through a vertical pipe of radius $R$ under conditions such that the fluid–fluid interface can be represented as a cylinder of radius $\kappa R$. Determine the volume rate of flow of each phase as a function of pressure gauge readings at two different axial positions on the wall of the pipe as well as the appropriate physical properties.

Exercise 3.2.5-12 *Continuous coating*    A flat sheet is to be continuously coated with an incompressible Newtonian fluid in a process shown schematically in Figure 3.2.5-10. This process may be idealized as consisting of three stages.

In the first stage of the process, the liquid film is formed on the rising sheet. This portion of the process determines the thickness $\delta$ of the liquid film observed in stage 2. It will be a function of the speed $V$ of the rising sheet and of the physical properties of the fluid, including the interfacial tension. Although this is the most important stage of the process, it

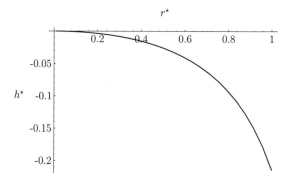

**Figure 3.2.5-11.** Configuration ($h^\star$ as a function of $r^\star$) of a sessile drop of water on a horizontal plane for the contact angle $\theta = \pi/4$. These results were obtained using Mathematica (1993).

is also the most difficult, and it has received adequate attention only recently in the literature. We will not be concerned with this stage of the process here.

In the second stage of the process, a liquid film of uniform thickness $\delta$ is observed.

In the third stage of the process, a solid film of uniform thickness $\delta_{final}$ is observed.

i)  Determine the velocity distribution in the film in stage 2.
ii) For stage 2, determine $Q$, the volume rate of flow of fluid in the film.
iii) Determine the relation between $\delta$ (the film thickness in stage 2) and $\delta_{final}$ (the film thickness in stage 3).

**Exercise 3.2.5-13** *Continuous coating of a wire*   Repeat Exercise 3.2.5-12 for the continuous coating of a wire.

**Exercise 3.2.5-14** *Sessile drop*   A drop of fluid resting on the upper side of a horizontal plane is usually referred to as a *sessile drop*. Determine the configuration of a sessile drop, assuming that the surface tension $\gamma$ and contact angle $\theta$ are given. A numerical solution of the resulting differential equation will be required for each value of $\gamma$ and $\theta$ chosen. (Figure 3.2.5-11)

**Exercise 3.2.5-15** *Hanging drop*   Repeat Exercise 3.2.5-14 for a *hanging drop* on the lower side of a horizontal plane.

**Exercise 3.2.5-16** *Capillary rise*   A vertical tube is inserted in a liquid–gas phase interface. How high will the interface rise in the capillary? Determine the configuration of the interface.

**Exercise 3.2.5-17** *More about flow in a cylindrical film*   Repeat Exercise 3.2.5-6 for an incompressible power-law fluid.

## 3.3   Creeping Flows

In the preceding sections we see examples of problems that can be solved exactly. Most problems are not of this nature; after we have made what seem to be reasonable initial assumptions, the partial differential equations with which we are faced may be very difficult to solve. Before beginning an extensive numerical solution, one is advised to consider what might be learned from limiting cases.

Let us look at the Navier–Stokes equation (Section 2.4.1), which is the expression of the differential momentum balance appropriate to an incompressible Newtonian fluid with a constant viscosity. Define the following dimensionless variables:

$$z_i^\star \equiv \frac{z_i}{L_0}$$

$$\mathbf{v}^\star \equiv \frac{\mathbf{v}}{v_0}$$

$$\mathcal{P}^\star \equiv \frac{\mathcal{P}}{\mathcal{P}_0} \tag{3.3.0-1}$$

$$t^\star \equiv \frac{t}{t_0}$$

where $v_0$ is a characteristic magnitude of velocity, $L_0$ a characteristic length, $\mathcal{P}_0$ a characteristic modified pressure, and $t_0$ a characteristic time. By a characteristic quantity, we mean one that occurs as a parameter in one of the governing equations or boundary conditions of that problem. We assume that the external force per unit mass is representable by a potential as discussed in Section 2.4.1. Under this condition and in terms of these dimensionless variables, the Navier–Stokes equation may easily be rearranged to the form

$$\frac{1}{N_{St}} \frac{\partial \mathbf{v}^\star}{\partial t^\star} + (\nabla \mathbf{v}^\star) \cdot \mathbf{v}^\star = -\frac{1}{N_{Ru}} \nabla \mathcal{P}^\star + \frac{1}{N_{Re}} \mathrm{div}(\nabla \mathbf{v}^\star) \tag{3.3.0-2}$$

Here $N_{St}$, $N_{Ru}$, and $N_{Re}$ are the Strouhal number, the Ruark number, and the Reynolds number:

$$N_{St} \equiv \frac{t_0 v_0}{L_0}$$

$$N_{Ru} \equiv \frac{\rho v_0{}^2}{\mathcal{P}_0} \tag{3.3.0-3}$$

$$N_{Re} \equiv \frac{L_0 v_0 \rho}{\mu}$$

One limiting case corresponds to $N_{Re} \ll 1$. In a given geometry with a given fluid, this is accomplished by letting $v_0$ approach zero; hence, it is termed *creeping flow*. If $N_{St}$ and $N_{Ru}$ take arbitrary values in the limit $N_{Re} \ll 1$, we intuitively expect the convective inertial terms $(\nabla \mathbf{v}^\star \cdot \mathbf{v}^\star)$ to become negligibly small with respect to the viscous terms $N_{Re}{}^{-1} \mathrm{div}(\nabla \mathbf{v}^\star)$. This suggests that in the limit $N_{Re} \ll 1$, Equation (3.3.0-2) reduces to

$$\frac{1}{N_{St}} \frac{\partial \mathbf{v}^\star}{\partial t^\star} = -\frac{1}{N_{Ru}} \nabla \mathcal{P}^\star + \frac{1}{N_{Re}} \mathrm{div}(\nabla \mathbf{v}^\star) \tag{3.3.0-4}$$

or

$$\rho \frac{\partial \mathbf{v}}{\partial t} = -\nabla \mathcal{P} + \mu \mathrm{div}(\nabla \mathbf{v}) \tag{3.3.0-5}$$

The Strouhal number normally will take a value other than unity in a periodic flow. For a nonperiodic flow, we often may be willing to define

$$t_0 \equiv \frac{L_0}{v_0} \tag{3.3.0-6}$$

or

$$N_{St} = 1 \tag{3.3.0-7}$$

With this understanding, in the limit $N_{Re} \ll 1$, Equation (3.3.0-2) further reduces to

$$0 = -\frac{1}{N_{Ru}}\nabla \mathcal{P}^\star + \frac{1}{N_{Re}}\text{div}(\nabla \mathbf{v}^\star) \tag{3.3.0-8}$$

or

$$0 = -\nabla \mathcal{P} + \mu \, \text{div}(\nabla \mathbf{v}) \tag{3.3.0-9}$$

It is important to realize that we have used an intuitive argument in the above discussion. There is no theorem that says that, in a partial differential equation, a term with a small coefficient will have a small effect on the solution. Stokes's paradox (Birkhoff 1955, p. 33) calls our attention to a situation in which the inertial terms are not negligibly small compared with the viscous terms in (3.3.0-2) for $N_{Re} \ll 1$:

**Stokes's Paradox**   No steady, two-dimensional, creeping flow of an incompressible Newtonian fluid past an infinite circular cylinder is possible.

A little later we see another example of a term in the Navier–Stokes equation with a very small coefficient that yields a finite effect (see the introduction to Section 3.4). The point to keep in mind is that intuitive arguments such as this are successful the vast majority of times. The situations where they fail are well documented and pose no surprises.

**Exercise 3.3.0-1** *Equation for modified pressure*   Prove that, for creeping flow of an incompressible Newtonian fluid, modified pressure $\mathcal{P}$ satisfies Laplace's equation

$$\text{div}(\nabla \mathcal{P}) = 0$$

## 3.3.1   Flow in a Cone–Plate Viscometer

An incompressible power-law fluid is continuously deformed between the rotating plate and stationary cone shown in Figure 3.3.1-1. We neglect *edge effects*. This is similar to saying that the cone and plate extend to infinity, though we shall see later that there is at least one important difference.

This geometry is the basis for another type of viscometer. Two measurement are made: the angular velocity of the plate and the axial component of the torque that the fluid exerts

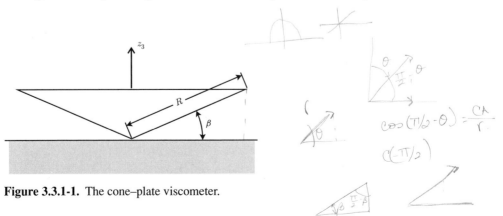

**Figure 3.3.1-1.** The cone–plate viscometer.

on the cone. We assume here that the angular velocity $\Omega$ of the plate is given and that the axial component of the torque (moment of force) that the fluid exerts on the cone is to be computed.

The boundary conditions to be satisfied in terms of spherical coordinates are

$$\text{at } \theta = \frac{\pi}{2} - \beta : \mathbf{v} = 0 \tag{3.3.1-1}$$

$$\text{at } \theta = \frac{\pi}{2} : \quad v_r = v_\theta = 0, \quad v_\varphi = \Omega r \sin\theta \tag{3.3.1-2}$$

These boundary conditions and symmetry suggest that we assume our velocity distribution is of the form

$$\begin{aligned} v_r &= v_\theta = 0 \\ v_\varphi &= v_\varphi(r, \theta) \end{aligned} \tag{3.3.1-3}$$

With a little more reflection, we might guess that the angular velocity of the fluid in the gap is a function only of $\theta$:

$$\frac{v_\varphi}{r \sin\theta} = \omega = \omega(\theta) \tag{3.3.1-4}$$

The differential mass balance in spherical coordinates (Table 2.4.1-1) is satisfied identically by (3.3.1-3) and (3.3.1-4).

From Table 2.4.1-10, the only nonzero components of the rate of deformation tensor are

$$D_{\theta\varphi} = D_{\varphi\theta} = \frac{1}{2}\sin\theta \frac{d\omega}{d\theta} \tag{3.3.1-5}$$

For a power-law fluid (Section 2.3.3),

$$\mathbf{S} = 2m\gamma^{n-1}\mathbf{D} \tag{3.3.1-6}$$

where

$$\gamma \equiv \sqrt{2\,\mathrm{tr}\,\mathbf{D}\cdot\mathbf{D}} = \sin\theta \left|\frac{d\omega}{d\theta}\right| \tag{3.3.1-7}$$

This means that the only nonzero components of the viscous portion of the stress tensor are

$$S_{\theta\varphi} = S_{\varphi\theta} = m\left(\sin\theta \frac{d\omega}{d\theta}\right)^n \tag{3.3.1-8}$$

In writing this last, we have observed that $d\omega/d\theta$ is positive in this situation.

From Table 2.4.1-6, the three components of the differential momentum balance can be written in terms of the modified pressure $\mathcal{P}$ as

$$\rho\frac{v_\varphi^2}{r} = \frac{\partial \mathcal{P}}{\partial r} \tag{3.3.1-9}$$

$$\rho v_\varphi^2 \cot\theta = \frac{\partial \mathcal{P}}{\partial \theta} \tag{3.3.1-10}$$

$$\frac{\partial \mathcal{P}}{\partial \varphi} = \frac{1}{\sin\theta}\frac{d}{d\theta}(S_{\theta\varphi}\sin^2\theta) \tag{3.3.1-11}$$

From (3.3.1-11),

$$\begin{aligned} \mathcal{P} &= \mathcal{P}(r, \theta, \varphi) \\ &= \varphi f(\theta) + g(r, \theta) \end{aligned} \tag{3.3.1-12}$$

But we require $\mathcal{P}$ to be a periodic function of $\varphi$ with period $2\pi$:

$$\mathcal{P}(r, \theta, \varphi) = \mathcal{P}(r, \theta, \varphi + 2\pi) \tag{3.3.1-13}$$

which means that

$$\frac{\partial \mathcal{P}}{\partial \varphi} = f(\theta)$$
$$= 0 \tag{3.3.1-14}$$

This leaves us with

$$S_{\theta\varphi} = \frac{A}{\sin^2 \theta} \tag{3.3.1-15}$$

where $A$ is a constant of integration. This can be used together with (3.3.1-8) to solve for the velocity distribution. All we have left to do is to check (3.3.1-9) and (3.3.1-10) for consistency. We require that

$$\frac{\partial^2 \mathcal{P}}{\partial r \, \partial \theta} = \frac{\partial^2 \mathcal{P}}{\partial \theta \, \partial r} \tag{3.3.1-16}$$

But this is not satisfied by (3.3.1-9) and (3.3.1-10) (see Exercise 3.3.1-7).

What does this mean? Our assumptions in (3.3.1-3) and (3.3.1-4) have led us to a contradiction. There is not a steady-state solution of this form to the differential mass and momentum balances for an incompressible power-law fluid (or for that matter any model of the type discussed in Section 2.3.3). This could mean that the $r$ dependence assumed in (3.3.1-4) is incorrect; this possibility is investigated in Exercise 3.3.1-5 and found to be without merit. This probably means that a solution will have *three* nonzero components of the velocity vector. Suddenly the problem has become very difficult.

At this point we can revise our assumptions in (3.3.1-3) and (3.3.1-4) and attempt a solution of the resulting problem, which would require a numerical solution. Or we can stop and ask about limiting cases. Let us take the latter alternative and ask about the creeping flow limit discussed in the introduction to Section 3.3.

We may write the differential momentum balance in terms of the modified pressure for an incompressible power-law fluid (2.3.3-6) as

$$\rho \frac{\partial \mathbf{v}}{\partial t} + \rho(\nabla \mathbf{v}) \cdot \mathbf{v} = -\nabla \mathcal{P} + \operatorname{div}\left[2\mu_0 m^\star (s_0 \gamma)^{n-1} \mathbf{D}\right] \tag{3.3.1-17}$$

In terms of the dimensionless variables introduced in (3.3.0-1), this becomes

$$\frac{1}{N_{St}} \frac{\partial \mathbf{v}^\star}{\partial t^\star} + (\nabla \mathbf{v}^\star) \cdot \mathbf{v}^\star = -\frac{1}{N_{Ru}} \nabla \mathcal{P}^\star + \frac{1}{N_{Re\,PL}} \operatorname{div}\left(2\gamma^{\star n-1} \mathbf{D}^\star\right) \tag{3.3.1-18}$$

where

$$\gamma^\star \equiv \frac{L_0}{v_0} \gamma$$

$$\mathbf{D}^\star \equiv \frac{L_0}{v_0} \mathbf{D} \tag{3.3.1-19}$$

$$N_{Re\,PL} \equiv \frac{\rho v_0 L_0}{\mu_0} \left(\frac{s_0 v_0}{L_0}\right)^{1-n} \left(\frac{1}{m^\star}\right)$$

By analogy with our discussion in the introduction to Section 3.3, the creeping flow limit corresponds to the limit $N_{Re\,PL} \ll 1$. In this limit it appears that inertial effects may be negligibly small with respect to viscous effects in the differential momentum balance, and (3.3.1-18) reduces to

$$\frac{1}{N_{St}}\frac{\partial \mathbf{v}^\star}{\partial t^\star} = -\frac{1}{N_{Ru}}\nabla \mathcal{P}^\star + \frac{1}{N_{Re\,PL}}\mathrm{div}(2\gamma^{\star n-1}\mathbf{D}^\star) \tag{3.3.1-20}$$

or

$$\rho\frac{\partial \mathbf{v}}{\partial t} = -\nabla \mathcal{P} + \mathrm{div}(2m\gamma^{n-1}\mathbf{D}) \tag{3.3.1-21}$$

Here [see (2.3.3-7)]

$$m \equiv \mu_0 m^\star s_0^{\,n-1} \tag{3.3.1-22}$$

If we look at (3.3.1-9) and (3.3.1-10) in the limit of steady-state, creeping flow, they reduce to

$$0 = \frac{\partial \mathcal{P}}{\partial r} \tag{3.3.1-23}$$

$$0 = \frac{\partial \mathcal{P}}{\partial \theta} \tag{3.3.1-24}$$

and condition (3.3.1-16) is satisfied.

We have not yet identified the characteristic length and velocity used in (3.3.1-19). This is a somewhat unusual problem in that, when we neglect edge effects, we find that no lengths or velocities appear in the boundary conditions, Equations (3.3.1-1) and (3.3.1-2). If we truly wished to analyze flow between an infinite cone and an infinite plate (the mathematical problem to be solved when edge effects are neglected), we would have little choice but to define $v_0 \equiv L_0\Omega$ and $L_0$ such that $N_{Re\,PL} = 1$, which would be inconsistent with the creeping-flow argument.

Our point of view here is that we wish to discuss the finite geometry pictured in Figure 3.3.1-1 under conditions such that edge effects are negligibly small. This suggests that we take

$$v_0 \equiv R\Omega\cos\beta$$
$$L_0 \equiv R\sin\beta \tag{3.3.1-25}$$
$$N_{Re\,PL} \equiv \frac{\rho R^2\Omega\sin\beta\cos\beta}{\mu_0}\left(\frac{s_0\Omega\cos\beta}{\sin\beta}\right)^{1-n}\left(\frac{1}{m^\star}\right)$$

If $\beta$ is sufficiently small, small values of $N_{Re\,PL}$ may be achieved even though $\Omega$ may be large and $0 < n \le 1$ (the more common situation).[1]

We may now determine the velocity distribution in the limit of creeping flow. The component $S_{\theta\varphi}$ may be eliminated between (3.3.1-8) and (3.3.1-15) to obtain

$$m\left(\sin\theta\frac{d\omega}{d\theta}\right)^n = \frac{A}{\sin^2\theta} \tag{3.3.1-26}$$

---

[1] For more general viscoelastic behavior such as that represented by the Noll simple fluid (Section 2.3.4), it is necessary to require $\beta \to 0$ in order that the compatibility condition (3.3.1-16) be satisfied (Coleman et al. 1966, p. 51).

This may be rearranged as

$$\frac{d\omega}{d\theta} = \frac{1}{\sin\theta}\left(\frac{A}{m\sin^2\theta}\right)^{1/n} \tag{3.3.1-27}$$

and integrated to satisfy boundary condition (3.3.1-2):

$$\omega - \Omega = \left(\frac{A}{m}\right)^{1/n}\int_{\pi/2}^{\theta}\frac{d\theta}{(\sin\theta)^{(n+2)/n}} \tag{3.3.1-28}$$

Applying boundary condition (3.3.1-1), we have

$$\Omega = \left(\frac{A}{m}\right)^{1/n}\int_{\pi/2-\beta}^{\pi/2}\frac{d\theta}{(\sin\theta)^{(n+2)/n}} \tag{3.3.1-29}$$

Together these equations imply that

$$\frac{\Omega-\omega}{\Omega} = \int_{\theta}^{\pi/2}\frac{d\theta}{(\sin\theta)^{(n+2)/n}}\left[\int_{\pi/2-\beta}^{\pi/2}\frac{d\theta}{(\sin\theta)^{(n+2)/n}}\right]^{-1} \tag{3.3.1-30}$$

These integrals cannot be evaluated analytically, but they pose no numerical difficulty.

Our original object was to compute the axial component of $\mathcal{T}$, the torque that the fluid exerts on the stationary cone:

at $\theta = \dfrac{\pi}{2} - \beta$ :

$$\mathcal{T} = \int_0^{2\pi}\int_0^R \mathbf{p}\wedge(\mathbf{T}\cdot\mathbf{n})r\sin\left(\frac{\pi}{2}-\beta\right)dr\,d\varphi \tag{3.3.1-31}$$

In terms of a rectangular Cartesian coordinate system defined by the relations given in Exercise A.4.1-5, the axial component or $z_3$ component of $\mathcal{T}$ may be expressed as

at $\theta = \dfrac{\pi}{2} - \beta$ : $\quad \mathcal{T}_3 = \displaystyle\int_0^{2\pi}\int_0^R a_3 r\,\sin\left(\frac{\pi}{2}-\beta\right)dr\,d\varphi$ (3.3.1-32)

Here we define

$$\mathbf{a} \equiv \mathbf{p}\wedge(\mathbf{T}\cdot\mathbf{n}) \tag{3.3.1-33}$$

The relation between the rectangular Cartesian component $a_3$ and the spherical components of $\mathbf{a}$ is (Exercise A.4.1-8)

$$a_3 = a_r\cos\theta - a_\theta\sin\theta \tag{3.3.1-34}$$

From Exercise A.9.2-1,

$$a_r \equiv a_{\langle 1\rangle} = e_{1jk}p_{\langle j\rangle}T_{\langle km\rangle}n_{\langle m\rangle} \tag{3.3.1-35}$$

Since the only nonzero physical component of $\mathbf{n}$ in spherical coordinates is

$$n_\theta \equiv n_{\langle 2\rangle} = 1 \tag{3.3.1-36}$$

and since the physical spherical components of the position vector are (Exercise A.4.1-10)

$$\begin{aligned}p_{\langle 1\rangle} &= r\\ p_{\langle 2\rangle} &= p_{\langle 3\rangle} = 0\end{aligned} \tag{3.3.1-37}$$

Equation (3.3.1-35) reduces to

$$a_r = e_{11k} r T_{(k2)} = 0 \tag{3.3.1-38}$$

In the same manner, we get

$$
\begin{aligned}
a_\theta &\equiv a_{(2)} \\
&= e_{2jk} p_{(j)} T_{(km)} n_{(m)} \\
&= e_{213} r T_{(32)} \\
&= -r T_{\varphi\theta} \\
&= -r S_{\varphi\theta} \tag{3.3.1-39}
\end{aligned}
$$

With (3.3.1-34), (3.3.1-38), and (3.3.1-39), Equation (3.3.1-32) can be expressed as

$$\text{at } \theta = \frac{\pi}{2} - \beta: \quad \mathcal{T}_3 = \int_0^{2\pi} \int_0^R S_{\varphi\theta} r^2 \sin^2\left(\frac{\pi}{2} - \beta\right) dr\, d\varphi \tag{3.3.1-40}$$

From (3.3.1-15),

$$
\begin{aligned}
\mathcal{T}_3 &= \int_0^{2\pi} \int_0^R A r^2 \, dr\, d\varphi \\
&= \frac{2\pi A R^3}{3} \tag{3.3.1-41}
\end{aligned}
$$

where $A$ is given by (3.3.1-29).

In a manner somewhat similar to that used in Section 3.2.2, creeping flow between a rotating cone and plate may be analyzed for the generalized Newtonian fluid discussed in Section 2.3.3 without assuming a specific functional form for either $\eta(\gamma)$ or $\varphi(\tau)$ (Slattery 1961). It is hardly more difficult to treat the flow of a Noll simple fluid (Section 2.3.4) in this geometry; see Coleman et al. (1966). for an excellent discussion.

**Exercise 3.3.1-1**    Starting with (3.3.1-9) and (3.3.1-10) and with the realization that the boundary conditions require $\omega$ to be a function of $\theta$, conclude that (3.3.1-16) cannot be satisfied by a velocity distribution described by (3.3.1-3) and (3.3.1-4).

**Exercise 3.3.1-2**    For the discussion in the text, determine that the other two components of the torque that the fluid exerts on the cone are zero.

**Exercise 3.3.1-3** *Ellis fluid in a cone–plate viscometer*    Repeat the analysis of this section for a fluid described by the Ellis fluid, but assume that the axial component of the moment of the force that the fluid exerts on the cone is given and that the angular velocity of the plate is to be calculated.

**Exercise 3.3.1-4** *Incompressible Newtonian fluid in a cone–plate viscometer*    Repeat Exercise 3.3.1-3 for an incompressible Newtonian fluid.

**Exercise 3.3.1-5** *More about Newtonian flow in a cone–plate viscometer*    (Slattery 1959, p. 175)

i) If (3.3.1-4) is assumed rather than (3.3.1-3), does this remove the contradiction found in the text between the $r$ and $\theta$ components of the differential momentum balance?

ii) Let us consider the creeping flow of an incompressible Newtonian fluid in the cone–plate viscometer described in the text and shown in Figure 3.3.1-1 and let us assume that the velocity distribution is described by (3.3.1-3).

iii) Show that the differential momentum balance yields

$$\frac{\partial}{\partial r}\left(r^2\frac{\partial v_\varphi}{\partial r}\right) + \frac{1}{\sin\theta}\frac{\partial}{\partial\theta}\left(\sin\theta\frac{\partial v_\varphi}{\partial\theta}\right) - \frac{v_\varphi}{\sin^2\theta} = 0$$

which must be solved consistent with

$$\text{at } \theta = \frac{\pi}{2} - \beta : \ v_\varphi = 0$$

$$\text{at } \theta = \frac{\pi}{2} : \ v_\varphi = r\Omega$$

$$\text{at } r = 0 : \ v_\varphi = 0$$

iv) Look for separable solutions that have the form

$$v_\varphi = R(r)\Theta(\theta)$$

Find that the differential equations to be solved for $R(r)$ and $\Theta(\theta)$ are

$$\frac{d}{dr}\left(r^2\frac{dR}{dr}\right) = mR$$

and

$$\frac{d}{dx}\left[(1-x^2)\frac{d\Theta}{dx}\right] + \left(m - \frac{1}{1-x^2}\right)\Theta = 0$$

where

$$x \equiv \cos\theta$$

and $m$ is a constant. If we take

$$m \equiv n(n+1)$$

where $n$ is a positive integer, $\Theta$ is a solution of Legendre's associated equation of the first order and $n$th degree. A solution for the partial differential equation of (iii) is

$$v_\varphi = \sum_{n=1}^{\infty}\left\{r^n\left[A_nP_n^1(\cos\theta) + B_nQ_n^1(\cos\theta)\right] + r^{-n-1}\left[C_nP_n^1(\cos\theta) + D_nQ_n^1(\cos\theta)\right]\right\}$$

Here $P_n^1(\cos\theta)$ is Legendre's function of the first kind, first order, and $n$th degree; $Q_n^1(\cos\theta)$ is Legendre's function of the second kind, first order, and $n$th degree; $A_n, B_n, C_n, D_n$ are constants to be determined in such a way that this expression for $v_\varphi$ satisfies the required boundary conditions.

v) Apply the boundary conditions to find

$$\frac{v_\varphi}{\Omega r \sin\theta} = 1 - \left[\ln\frac{1+\cos\theta}{1-\cos\theta} + \frac{\cos\theta}{\sin^2\theta}\right]$$

$$\times\left[\ln\frac{1+\cos(\pi/2-\beta)}{1-\cos(\pi/2-\beta)} + 2\frac{\cos(\pi/2-\beta)}{\sin^2(\pi/2-\beta)}\right]^{-1}$$

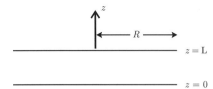

**Figure 3.3.1-2.** Tangential flow between parallel disks.

**Exercise 3.3.1-6** *Still more about Newtonian flow in a cone–plate viscometer*   As indicated in Figure 3.3.1-1, the cone–plate viscometer has finite dimensions. In reality, there is a liquid–gas interface that intersects both the cone and the plate at $r = R$.

i) Determine the pressure distribution in the fluid. You may neglect the effects of gravity, of interfacial tension, and of contact angles on both the cone and the plate.
ii) Determine the axial component of the force that the fluid exerts on the cone.
iii) Determine the axial component of the force that the fluid exerts on the plate.

**Exercise 3.3.1-7** *Tangential flow between parallel disks*   The gap between the two parallel plates shown in Figure 3.3.1-2 is filled with an incompressible Newtonian fluid. The upper plate rotates with a constant angular velocity $\Omega$; the lower plate is stationary. Determine the velocity distribution in the fluid and the axial component of the torque that the fluid exerts upon the upper plate.

**Exercise 3.3.1-8** *Sphere rotating in an unbounded fluid*   A sphere of radius $R$ rotates with a constant angular velocity in an unbounded incompressible Newtonian fluid. Determine the velocity distribution in the fluid as well as the axial component of the torque required to maintain the motion of the sphere.

*Answer:*

$$\frac{v_\varphi}{r\Omega \sin\theta} = \left(\frac{R}{r}\right)^3$$

$$T_3 = -8\pi\mu\Omega R^3 \quad \text{exerted by fluid on the sphere}$$

**Exercise 3.3.1-9** *Simplified analysis for the screw extruder*   In Figure 3.3.1-3, we have sketched a single-flighted, single-screw extruder. The screw rotates with a constant angular velocity whose magnitude is $\Omega$; the barrel of the extruder is stationary. Pellets of solid polymer are fed to the screw on the left and forced to move to the right by the rotating screw. As they move toward the right, they are heated by viscous forces as they deform, and they finally melt. We will consider only that portion of the extruder to the right in the figure, where the polymer flows as a liquid. To keep things simple, we will assume that the polymer can be described as an incompressible Newtonian fluid. For the moment, we will also assume that the heat transfer to the cooling jacket of the extruder is perfect and that the temperature of the system is uniform.

If the ratio of the height of the thread to the diameter of the screw is very small, the curvature of the channel can be neglected. (A little man standing in one of the channels formed by the screw flights will not be aware of the curvature of the channel.) For that

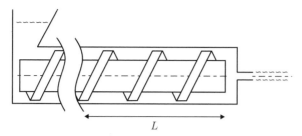

**Figure 3.3.1-3.** Screw extruder.

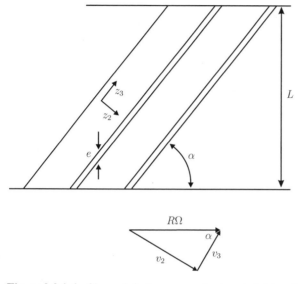

**Figure 3.3.1-4.** Channels in the screw extruder, straightened and viewed from above.

reason, we will replace the geometry shown in Figure 3.3.1-3 by the straight channels of Figure 3.3.1-4.

The characteristic dimension of this geometry is the depth of the channel $a$ as depicted in Figure 3.3.1-5. The depth $a$ is normally so small that under realistic operating conditions the Reynolds number is much less than unity, and inertial effects can be neglected.

Because we are neglecting inertial effects, we are able to switch to a rotating frame of reference without changing the form of the differential mass and momentum balances (see Exercise 2.2.3-3). The problem will be mathematically more convenient if we view the screw extruder in a rotating frame of reference in which the screw is stationary and the barrel rotates with a constant angular velocity $\Omega$.

Under these conditions, the flow within a single channel of the screw can be described by the boundary conditions

$$\text{at } z_1 = 0: \quad \mathbf{v} = \mathbf{0}$$
$$\text{at } z_1 = a: \quad v_1 = 0$$
$$v_2 = R\Omega \sin \alpha$$
$$v_3 = R\Omega \cos \alpha$$

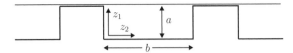

**Figure 3.3.1-5.** Channels in the screw extruder, straightened and viewed in cross section.

at $z_2 = 0$ : $\mathbf{v} = \mathbf{0}$

at $z_2 = b$ : $\mathbf{v} = \mathbf{0}$

The screw forces the molten polymer through an orifice or die, the dimensions and configuration of which are determined by the desired product. Because the polymer must flow through the die, a pressure gradient develops:

at $z_3 = 0$, $z_1 = a$, $z_2 = 0$ : $-T_{11} = P_0$

at $z_3 = \dfrac{L}{\sin \alpha}$, $z_1 = a$, $z_2 = 0$ : $-T_{11} = P_L$

This pressure gradient is specified by the dimensions of the die and the volume rate of flow $Q$ through the die:

$$P_L - P_0 = \frac{\mu}{K} Q$$

Here $\mu$ is the viscosity of the fluid, $K$ the die coefficient, and $n$ the number of channels formed by the screw flights. In effect, we are describing flow through the die as flow through a short channel.

We can further considerably simplify the analysis if we also assume that $a/b \ll 1$. This permits us to ignore the boundary conditions at the edges of each channel (at $z_2 = 0$ and at $z_2 = b$) and to seek a solution that has the form

$v_1 = 0$

$v_2 = v_2(z_1)$

$v_3 = v_3(z_1)$

i) Calculate the volume rate of flow $Q/n$ through a single channel.
ii) To maximize the area available for flow with a given value of the radius $R$ of the barrel of the extruder, what must $n$ be?
iii) What values must $\alpha$ and $a^\star \equiv a/R$ take if $Q^\star \equiv Q/\left(nR^3\Omega\right)$ is to be maximized? You may assume $e^\star \equiv e/R \ll 1$.
iv) Determine $v_2$ as a function of $z_1$. In carrying out this computation, you will want to note that there can be no net volume rate of flow in the $z_2$ direction:

$$\int_0^a v_2 \, dz_1 = 0$$

v) Assuming that we are to use the preceding analysis to design a screw extruder, determine that there are four parameters that have not yet been specified? What are they?
vi) To determine these four parameters, we have two equations. What are they?

In summary, we are missing two relationships required to fully specify a screw extruder within the context of the analysis outlined here.

- The maximum temperature developed in the flow must be a specified fraction of the *scorch temperature* of the polymer. At the scorch temperature, the polymer begins to char and black flecks appear in the product.
- A specified linear combination of capital and operating costs per unit time must be minimized.

Exercise 3.3.1-10 *A rotating cup of tea*    A cylindrical cup partially filled with tea has been spinning for some time at a constant angular velocity $\Omega$. At $t = 0$, you pick up the cup, forcing it to stop rotating. Determine the velocity distribution in the tea as it decays with time.

You may assume a creeping flow, and you may neglect the effects of surface tension and the bottom of the cylindrical cup.

Exercise 3.3.1-11 *More on a rotating cup of tea*    The tea in a cup is at rest. At time $t = 0$, the cup begins to rotate at a constant angular velocity $\Omega$. Starting with the solution of Exercise 3.3.1-10 and switching to a rotating frame of reference, write down the solution.

## 3.3.2  Flow Past a Sphere

Consider a sphere that falls at a constant speed $v_\infty$ along the axis of a cylinder filled with an incompressible Newtonian fluid. We wish to calculate the force that the fluid exerts on the sphere.

To simplify the problem, let us neglect effects attributable to the bounding surfaces of the container and assume that the sphere is moving in the direction of gravity through a relatively unbounded expanse of fluid.

In a laboratory frame of reference, this is an unsteady-state problem, since the position of the sphere is changing as a function of time. It becomes a steady-state problem if we view the flow in a coordinate system that is fixed with respect to the sphere. Let us adopt a spherical coordinate system whose origin coincides with the center of the sphere. Referring to Figure 3.3.2-1, we see that the boundary conditions for our problem become

$$\text{at } r = R : \quad \mathbf{v} = \mathbf{0} \tag{3.3.2-1}$$

and

$$\text{as } r \to \infty : \quad \mathbf{v} \to v_\infty \mathbf{e}_3 \tag{3.3.2-2}$$

It seems reasonable to assume that the velocity distribution in spherical coordinates is of the form

$$v_r = v_r(r, \theta)$$

$$v_\theta = v_\theta(r, \theta) \tag{3.3.2-3}$$

$$v_\varphi = 0$$

In Section 1.3.7, we show how the differential mass balance can be satisfied identically in such a two-dimensional motion by the introduction of a stream function $\psi$. In Table 2.4.2-1,

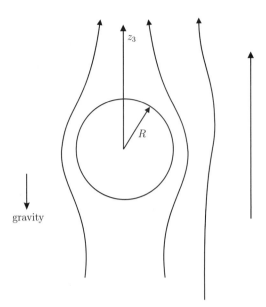

**Figure 3.3.2-1.** Flow past a sphere.

we find that we may represent the velocity components in terms of this stream function by

$$v_r = \frac{1}{r^2 \sin \theta} \frac{\partial \psi}{\partial \theta}$$

$$v_\theta = -\frac{1}{r \sin \theta} \frac{\partial \psi}{\partial r}$$

(3.3.2-4)

For a steady-state motion in which the creeping-flow assumption can be made, the Navier–Stokes equation simplifies considerably when written in terms of the stream function (see Table 2.4.2-1):

$$\left[ \frac{\partial^2}{\partial r^2} + \frac{\sin \theta}{r^2} \frac{\partial}{\partial \theta} \left( \frac{1}{\sin \theta} \frac{\partial}{\partial \theta} \right) \right]^2 \psi = 0$$

(3.3.2-5)

From (3.3.2-1), the corresponding boundary conditions are

$$\text{at } r = R : \quad \frac{1}{r^2 \sin \theta} \frac{\partial \psi}{\partial \theta} = 0$$

(3.3.2-6)

and

$$\text{at } r = R : \quad \frac{1}{r \sin \theta} \frac{\partial \psi}{\partial r} = 0$$

(3.3.2-7)

From (3.3.2-2) and Exercise A.4.1-9, we find that

$$\text{as } r \to \infty : \quad v_r = \frac{1}{r^2 \sin \theta} \frac{\partial \psi}{\partial \theta}$$

$$\to v_\infty \cos \theta$$

(3.3.2-8)

The stream function $\psi$ is arbitrary to a constant and we are free to require

$$\text{as } r \to \infty \text{ for } \theta = 0 : \quad \psi \to 0 \tag{3.3.2-9}$$

This allows us to say

$$
\begin{aligned}
\text{as } r \to \infty : \quad \psi &= \int_0^\psi d\psi \\
&= \int_0^\theta \frac{\partial \psi}{\partial \theta} d\theta \\
&\to r^2 v_\infty \int_0^\theta \sin \theta \cos \theta \, d\theta \\
&\to \frac{1}{2} r^2 v_\infty \sin^2 \theta
\end{aligned}
\tag{3.3.2-10}
$$

The boundary condition (3.3.2-10) suggests that the stream function might be of the form

$$\psi = f(r) \sin^2 \theta \tag{3.3.2-11}$$

With this transformation of variable, (3.3.2-5) becomes

$$\frac{d^4 f}{dr^4} - \frac{4}{r^2} \frac{d^2 f}{dr^2} + \frac{8}{r^3} \frac{df}{dr} - \frac{8f}{r^4} = 0 \tag{3.3.2-12}$$

Boundary conditions (3.3.2-6), (3.3.2-7), and (3.3.2-10) become

$$\text{at } r = R : \quad f = 0 \tag{3.3.2-13}$$

$$\text{at } r = R : \quad \frac{df}{dr} = 0 \tag{3.3.2-14}$$

and

$$\text{as } r \to \infty : \quad f \to \frac{1}{2} r^2 v_\infty \tag{3.3.2-15}$$

The fourth-order differential equation (3.3.2-12) is linear and homogeneous. One form of solution is

$$f(r) = ar^n \tag{3.3.2-16}$$

This implies

$$f(r) = \frac{A}{r} + Br + Cr^2 + Dr^4 \tag{3.3.2-17}$$

Boundary conditions (3.3.2-13) to (3.3.2-15) require

$$
\begin{aligned}
A &= \frac{1}{4} v_\infty R^3 \\
B &= -\frac{3}{4} v_\infty R \\
C &= \frac{1}{2} v_\infty \\
D &= 0
\end{aligned}
\tag{3.3.2-18}
$$

In summary, (3.3.2-4), (3.3.2-11), (3.3.2-17), and (3.3.2-18) tell us that the components of velocity within the fluid are

$$\frac{v_r}{v_\infty} = \left[1 - \frac{3}{2}\frac{R}{r} + \frac{1}{2}\left(\frac{R}{r}\right)^3\right]\cos\theta \tag{3.3.2-19}$$

and

$$\frac{v_\theta}{v_\infty} = -\left[1 - \frac{3}{4}\frac{R}{r} - \frac{1}{4}\left(\frac{R}{r}\right)^3\right]\sin\theta \tag{3.3.2-20}$$

Let us next determine the pressure distribution within the fluid. It is most convenient to work in terms of the modified pressure

$$\mathcal{P} \equiv p + \rho\phi \tag{3.3.2-21}$$

where

$$\phi = gz_3 = gr\cos\theta \tag{3.3.2-22}$$

In order to completely specify pressure within the fluid (pressure in an incompressible fluid is determined by the motion up to a constant), we require

$$\text{as } r \to \infty \text{ for } \theta = \frac{\pi}{2}: \quad \mathcal{P} = p_0 \tag{3.3.2-23}$$

From the $r$ and $\theta$ components of the Navier–Stokes equation (see Table 2.4.1-7), as well as (3.3.2-19) and (3.3.2-20), we have

$$\frac{\partial\mathcal{P}}{\partial r} = 3\mu v_\infty R\frac{\cos\theta}{r^3} \tag{3.3.2-24}$$

and

$$\frac{\partial\mathcal{P}}{\partial\theta} = \frac{3}{2}\mu v_\infty R\frac{\sin\theta}{r^2} \tag{3.3.2-25}$$

This allows us to determine the pressure distribution by the line integration

$$\mathcal{P} - p_0 = \int_{p_0}^{\mathcal{P}} d\mathcal{P}$$

$$= \int_\infty^r \left.\frac{\partial\mathcal{P}}{\partial r}\right|_{\theta=\pi/2} dr + \int_{\pi/2}^\theta \frac{\partial\mathcal{P}}{\partial\theta} d\theta \tag{3.3.2-26}$$

as

$$p = p_0 - \rho gr\cos\theta - \frac{3}{2}\mu v_\infty R\frac{\cos\theta}{r^2} \tag{3.3.2-27}$$

The $z_3$ component of the force $\mathcal{F}$ that the fluid exerts upon the sphere can be expressed as (an alternative expression for $\mathcal{F}_3$ is found in Exercise 4.4.8-13)

$$\mathcal{F}_3 = R^2 \int_0^{2\pi} \int_0^\pi (T_{rr}\cos\theta - T_{\theta r}\sin\theta)|_{r=R}\sin\theta\, d\theta d\varphi \tag{3.3.2-28}$$

In view of (3.3.2-19), we have

$$\text{at } r = R : \quad S_{rr} = 0 \tag{3.3.2-29}$$

and (3.3.2-28) reduces to

$$\mathcal{F}_3 = R^2 \int_0^{2\pi} \int_0^{\pi} (-p \cos\theta - T_{\theta r} \sin\theta)|_{r=R} \sin\theta \, d\theta \, d\varphi \tag{3.3.2-30}$$

After carrying out the required integrations, we find

$$\mathcal{F}_3 = \frac{4}{3}\pi R^3 \rho g + 6\pi \mu R v_\infty \tag{3.3.2-31}$$

The first term on the right describes the buoyant force that the fluid exerts upon the sphere; the second term is the result of the motion in the fluid. Equation (3.3.2-31) is known as *Stokes's law*.

It is very common to express the force that a fluid exerts on a submerged object beyond the force attributable to the ambient pressure and to the hydrostatic pressure in terms of a drag coefficient $c_D$ (see Section 4.4.4). For flow past a sphere,

$$c_D \equiv \mathcal{F}_3 - \frac{4}{3}\pi R^3 \rho g \left(\frac{1}{2}\rho v_\infty^2 \pi R^2\right)^{-1} \tag{3.3.2-32}$$

Stokes's law may be expressed in terms of this drag coefficient as

$$c_D = \frac{6\pi \mu R v_\infty}{\frac{1}{2}\rho v_\infty^2 \pi R^2} = \frac{24}{N_{Re}} \tag{3.3.2-33}$$

where

$$N_{Re} \equiv \frac{2\rho v_\infty R}{\mu} \tag{3.3.2-34}$$

In arriving at (3.3.2-33), we restrict ourselves to the limit as $N_{Re} \ll 1$ in order to justify neglecting the inertial terms with respect to the viscous terms in the equation of motion. A comparison with experimental data (Schlichting 1979, p. 17) indicates that Stokes's law is an excellent representation of reality for $N_{Re} < 0.5$.

For a sphere falling at a constant velocity under the action of gravity, the force that the fluid exerts upon the sphere should be equal in magnitude and opposite in direction to the force of gravity upon the sphere:

$$\mathcal{F}_3 = \frac{4}{3}\pi R^3 \rho^{(s)} g \tag{3.3.2-35}$$

Here $\rho^{(s)}$ is the density of the sphere. This gives us an explicit expression for the drag coefficient:

$$c_D = \frac{8}{3}\frac{Rg}{v_\infty^2}\frac{\rho^{(s)} - \rho}{\rho} \tag{3.3.2-36}$$

When Stokes's law is applicable, Equations (3.3.2-33) and (3.3.2-36) represent a useful relationship between the properties of the sphere and the properties of the fluid. If the density of the fluid is larger than the density of the sphere, (3.3.2-33) and (3.3.2-36) are still applicable; it merely means that $v_\infty$ must be negative and the sphere rises rather than falls.

For more on the solution of steady-state creeping flows, see Sampson (1891), Proudman and Pearson (1957), Taylor and Acrivos (1964), Acrivos and Taylor (1964)), Lamb (1945, Sections 335 and 336), and Happel and Brenner (1965, Eq. 3-2.3).

**Exercise 3.3.2-1** *The pressure distribution in flow past a sphere*   Derive (3.3.2-24), (3.3.2-25), and (3.3.2-27).

**Exercise 3.3.2-2** *Force that the fluid exerts upon the sphere*

i) Starting with

$$\mathcal{F}_3 = R^2 \int_0^{2\pi} \int_0^{\pi} t_3 \sin\theta \, d\theta d\varphi$$

where $t_3$ is the $z_3$ component of the force per unit area that the fluid exerts on the sphere, derive (3.3.2-28).

ii) Starting with (3.3.2-30), arrive at (3.3.2-31).

### 3.3.3   Thin Draining Films

The squeezing flow between two parallel disks provides a simplified model for film drainage during the coalescence of a bubble or drop at an interface (Edwards et al. 1991, Sec. 11.3). This problem is normally analyzed using a scaling argument (the lubrication approximation) to neglect terms in the mass and momentum balances (Landau and Lifshitz 1987, p. 66; Bird et al. 1987, p. 20).

Bird et al. (1977, Example 1.2-6) presented an interesting and different approach to this problem for an incompressible Newtonian fluid that avoided the use of a scaling argument. With reference to Figure 3.3.3-1, they assumed creeping flow and postulated that

$$v_r = v_r(r, z, t)$$
$$v_z = v_z(z, t)$$
$$(3.3.3\text{-}1)$$

(Bird et al. (1977, Example 1.2-6) use a frame of reference that is fixed with respect to the center of the film rather than the one used here, which is fixed with respect to the lower interface). Unfortunately, they also assumed $p = p(r, t)$, which was inconsistent with the $z$ component of the differential momentum balance. They apparently recognized this later, because in the second edition of their book (Bird et al. 1987, p. 20) a traditional scaling argument was used.

Here we follow Vaughn and Slattery (1995) in extending the core of the original Bird et al. (1977, Example 1.2-6) argument both to a draining film bounded by immobile interfaces (large surface tension gradients or large surface viscosities as the result of soluble surfactants in the system) and to a draining film bounded by mobile interfaces (uniform surface tension and vanishingly small surface viscosities in the absence of soluble surfactants). As they did, we will consider the creeping flow of an incompressible Newtonian fluid, neglecting the effects of gravity.

Note that these films are deformable, and their shape is specified by the normal component of the jump momentum balance (2.4.3). Experimental evidence suggests that nearly plane-parallel films are not uncommon.

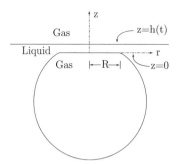

**Figure 3.3.3-1.** Idealized film formed as a small bubble rises through a continuous liquid to an interface between the liquid and another gas. The film is observed in a frame of reference in which the interface between the bubble and the liquid is stationary.

## Immobile Interfaces

In view of (3.3.3-1), the differential mass balance reduces to

$$\frac{1}{r}\frac{\partial}{\partial r}(r v_r) + \frac{\partial v_z}{\partial z} = 0 \tag{3.3.3-2}$$

and the $r$ and $z$ components of the differential momentum balance become

$$-\frac{\partial p}{\partial r} + \mu \left[ \frac{\partial}{\partial r}\left( \frac{1}{r}\frac{\partial}{\partial r}(r v_r) \right) + \frac{\partial^2 v_r}{\partial z^2} \right] = 0 \tag{3.3.3-3}$$

and

$$-\frac{\partial p}{\partial z} + \mu \frac{\partial^2 v_z}{\partial z^2} = 0 \tag{3.3.3-4}$$

These equations are to be solved consistent with the boundary conditions

at $z = 0$ : $v_r = 0$, $v_z = 0$ $\tag{3.3.3-5}$

at $z = h(t)$ : $v_r = 0$, $v_z = \dfrac{dh}{dt}$ $\tag{3.3.3-6}$

at $r = 0$ : $v_r = 0$ $\tag{3.3.3-7}$

where $h(t)$ is the position of the upper interface.

Bird et al. (1977, Example 1.2-6) observed that (3.3.3-1) and (3.3.3-2) implied that the derivative with respect to $r$ in (3.3.3-3) was zero and that the $r$ component of the differential momentum balance became

$$-\frac{\partial p}{\partial r} + \mu \frac{\partial^2 v_r}{\partial z^2} = 0 \tag{3.3.3-8}$$

This together with (3.3.3-1) and (3.3.3-4) require that

$$p = f_1(z, t) + f_2(r, t) \tag{3.3.3-9}$$

In view of (3.3.3-9), we can integrate (3.3.3-8) consistent with boundary conditions (3.3.3-5) and (3.3.3-6) to obtain

$$v_r = \frac{1}{2\mu} \frac{\partial p}{\partial r} \left( z^2 - hz \right) \tag{3.3.3-10}$$

Referring to (3.3.3-1), the differential mass balance (3.3.3-2) can be integrated consistent with (3.3.3-7) to find

$$v_r = -\frac{\partial v_z}{\partial z} \frac{r}{2} \tag{3.3.3-11}$$

Eliminating $v_r$ between (3.3.3-11) and (3.3.3-10), we can integrate the result consistent with boundary conditions (3.3.3-5) and (3.3.3-6) to conclude

$$v_z = \frac{1}{h^3} \frac{dh}{dt} \left( 3hz^2 - 2z^3 \right) \tag{3.3.3-12}$$

and

$$\frac{\partial p}{\partial r} = \frac{6\mu r}{h^3} \frac{dh}{dt} \tag{3.3.3-13}$$

This last equation permits us to express (3.3.3-10) more conveniently as

$$v_r = -\frac{3r}{h^3} \frac{dh}{dt} \left( hz - z^2 \right) \tag{3.3.3-14}$$

Note that the jump momentum balance has not been employed in this analysis. The tangential components of the jump momentum balance could be used to discuss the surface tension gradients required to create immobile interfaces. The role of the normal components of the jump momentum balance would be to determine the configurations of the interfaces, which we have arbitrarily represented as planes.

It is common to impose the condition

$$\text{at } r = R: \quad p = p_h \tag{3.3.3-15}$$

where $p_h$ is the hydrostatic pressure. This clearly contradicts (3.3.3-4), (3.3.3-9), and (3.3.3-13). If we ignore this *edge effect*, we find that the force that the fluid exerts on the bubble is

$$\mathcal{F}_z = \int_0^{2\pi} \int_0^R (p - p_0 - S_{zz})|_{z=0} \, r \, dr \, d\theta$$

$$= -(p_0 - p_h) \pi R^2 - \frac{3\pi R^4 \mu}{2h^3} \frac{dh}{dt} \tag{3.3.3-16}$$

where $p_0$ is the pressure in the bubble. Note that the normal stress $S_{zz}$ on an immobile interface is always zero (Bird et al. 1987, p. 12).

**Immobile Interfaces: Lubrication Approximation**

Let us begin by introducing dimensionless variables in (3.3.3-2) through (3.3.3-4):

$$\frac{1}{r^\star}\frac{\partial}{\partial r^\star}(r^\star v_r^\star) + \frac{\partial v_z^\star}{\partial z^\star} = 0 \tag{3.3.3-17}$$

$$-\frac{1}{N_{Ru}}\frac{\partial p^\star}{\partial r^\star} + \frac{1}{N_{Re}}\left[\frac{\partial}{\partial r^\star}\left(\frac{1}{r^\star}\frac{\partial}{\partial r^\star}(r^\star v_r^\star)\right) + \frac{\partial^2 v_r^\star}{\partial z^{\star 2}}\right] = 0 \tag{3.3.3-18}$$

$$-\frac{1}{N_{Ru}}\frac{\partial p^\star}{\partial z^\star} + \frac{1}{N_{Re}}\frac{\partial^2 v_z^\star}{\partial z^{\star 2}} = 0 \tag{3.3.3-19}$$

where

$$r^\star \equiv \frac{r}{R}$$
$$z^\star \equiv \frac{z}{R}$$
$$\mathbf{v}^\star \equiv \frac{\mathbf{v}}{v_0} \tag{3.3.3-20}$$
$$p^\star \equiv \frac{p}{\mathcal{P}_0}$$

and

$$N_{St} \equiv \frac{t_0 v_0}{R}$$
$$N_{Ru} \equiv \frac{\rho v_0^{\,2}}{\mathcal{P}_0} \tag{3.3.3-21}$$
$$N_{Re} \equiv \frac{R v_0 \rho}{\mu}$$

Referring to Figure 3.3.3-1, it is clear that not all of the dimensionless derivatives in (3.3.3-17) through (3.3.3-19) are of the same order of magnitude, since the thickness of the film is so much smaller than its radius.

We will argue that dimensionless forms of these derivatives can be made to have the same order of magnitude with the introduction of more appropriate dimensionless variables:

$$z^{\star\star} \equiv \frac{z}{\delta_0}$$
$$v_z^{\star\star} \equiv \frac{R v_z}{\delta_0 v_0} \tag{3.3.3-22}$$

In terms of these variables, (3.3.3-17) through (3.3.3-19) become

$$\frac{1}{r^\star}\frac{\partial}{\partial r^\star}\left(r^\star v_r^\star\right) + \frac{\partial v_z^{\star\star}}{\partial z^{\star\star}} = 0 \tag{3.3.3-23}$$

$$-\frac{1}{N_{Ru}}\frac{\partial p^\star}{\partial r^\star} + \frac{1}{N_{Re}}\left[\frac{\partial}{\partial r^\star}\left(\frac{1}{r^\star}\frac{\partial}{\partial r^\star}\left(r^\star v_r^\star\right)\right) + \left(\frac{R}{\delta_0}\right)^2\frac{\partial^2 v_r^\star}{\partial z^{\star\star 2}}\right] = 0 \tag{3.3.3-24}$$

$$-\frac{1}{N_{Ru}}\frac{\partial p^\star}{\partial z^{\star\star}} + \frac{1}{N_{Re}}\frac{\partial^2 v_z^{\star\star}}{\partial z^{\star\star 2}} = 0 \tag{3.3.3-25}$$

In the limit

$$\left(\frac{R}{\delta_0}\right)^2 \gg 1 \tag{3.3.3-26}$$

Equation (3.3.3-24) reduces to

$$-\frac{1}{N_{Ru}}\frac{\partial p^\star}{\partial r^\star} + \frac{1}{N_{Re}}\left(\frac{R}{\delta_0}\right)^2\frac{\partial^2 v_r^\star}{\partial z^{\star2}} = 0 \tag{3.3.3-27}$$

and we conclude that

$$\frac{\partial p^\star}{\partial z^{\star\star}} \ll \frac{\partial p^\star}{\partial r^\star} \tag{3.3.3-28}$$

In view of (3.3.3-28), the common conclusion is that (3.3.3-9) reduces to

$$p = f_2(r, t). \tag{3.3.3-29}$$

Although one still obtains (3.3.3-12) through (3.3.3-14), there is a clear contradiction to (3.3.3-25).

## Mobile Interfaces

When the surface is mobile (or partially mobile, as in the drainage of a thin film that is stabilized by a soluble surfactant), the situation is different. From the tangential component of the jump momentum balance, we have

$$\text{at } z = 0: \quad \frac{\partial v_r}{\partial z} = 0 \tag{3.3.3-30}$$

$$\text{at } z = h(t): \quad \frac{\partial v_r}{\partial z} = 0 \tag{3.3.3-31}$$

The normal component of the jump momentum balance is not satisfied because we have arbitrarily represented the interfaces as planes. In addition, we must recognize the kinematic conditions

$$\text{at } z = 0: \quad v_z = 0 \tag{3.3.3-32}$$

$$\text{at } z = h(t): \quad v_z = \frac{dh}{dt} \tag{3.3.3-33}$$

In the same way that we arrived at (3.3.3-12) through (3.3.3-14) for the case of the immobile interfaces, we conclude here that

$$v_r = -\frac{1}{2}\frac{r}{h}\frac{dh}{dt} \tag{3.3.3-34}$$

$$v_z = \frac{z}{h}\frac{dh}{dt} \tag{3.3.3-35}$$

$$p = p_h \tag{3.3.3-36}$$

where we have employed (3.3.3-15).

Noting that

$$
\begin{aligned}
S_{zz} &= 2\mu \frac{\partial v_z}{\partial z} \\
&= \frac{2\mu}{h} \frac{dh}{dt}
\end{aligned}
\tag{3.3.3-37}
$$

we have the $z$ component of the force that the fluid exerts on the upper interface:

$$
\begin{aligned}
\mathcal{F}_z &= \int_0^{2\pi} \int_0^{R(t)} (p_h - p_0 - S_{zz}|_{z=h}) r\, dr\, d\theta \\
&= -(p_0 - p_h)\pi R^2 - S_{zz}|_{z=h}\, \pi R^2 \\
&= -(p_0 - p_h)\pi R^2 - 2\pi R^2 \mu \frac{1}{h} \frac{dh}{dt}
\end{aligned}
\tag{3.3.3-38}
$$

or

$$
\frac{\mathcal{F}_z}{\pi R^2 p_0} = -\left(1 - \frac{p_h}{p_0}\right) - \frac{2\mu}{p_0 h} \frac{dh}{dt}
\tag{3.3.3-39}
$$

## Mobile Interfaces: Lubrication Approximation

For mobile interfaces, the lubrication approximation, applied as explained above, again leads to (3.3.3-34) through (3.3.3-36), but without the contradiction found above with immobile interfaces.

## Conclusions

The viscous normal stress $S_{zz}$ affects the force that the fluid exerts on the bubble only in the case of mobile (or partially mobile) interfaces. In the case of small bubbles, it is easy to argue that the effect of $S_{zz}$ can be neglected with respect to the first term on the right in (3.3.3-15), as it has been common to assume (Lin and Slattery 1982a). The effect of $S_{zz}$ is likely to be more important in the case of larger bubbles, where the first term on the right of (3.3.3-15) approaches zero.

The lubrication approximation gives the same velocity and pressure distribution as the complete theory for both immobile and mobile interfaces, although in the case of immobile interfaces the $z$ component of the differential momentum balance is not satisfied.

## 3.3.4 Melt Spinning

Melt spinning, illustrated in Figure 3.3.4-1, is the process in which a cylindrical liquid thread of molten polymer is continuously extruded vertically through an orifice or die. After initially swelling (a phenomenon called "die swell"), the thread is drawn down to a smaller diameter $R_f$ by an axial force created in winding the finished product, and it undergoes a phase change as the result of the energy lost to the surrounding gas. The extension of the thread tends to align the molecules along the axis, creating a stronger structure after solidification. During this extension, the thread undergoes a phase transition from a fluid to a solid.

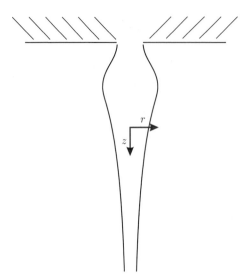

**Figure 3.3.4-1.** Extrusion of a monofilament thread.

Two classes of experimental studies have been reported for this process, depending upon the value of the Froude number

$$N_{Fr} \equiv \frac{V^2}{gR_0} \tag{3.3.4-1}$$

Here $V$ is the axial component of velocity and $R_0$ is the radius of the thread immediately following the die swell (see below); $g$ is the acceleration of gravity. Several experiments have been reported with low-speed data for which $N_{Fr} < 1$ (Spearot and Metzner 1972, Zeichner 1973). The only high-speed data for which $N_{Fr} \gg 1$ are those of George (1982) and Shimizu et al. (1985).

Unfortunately for present purposes, the data of Shimizu et al. (1985, figures 19 and 20) are primarily concerned with the region downstream of the onset of crystallization. In this region, there are discontinuities in the thread diameter that are not well understood. Since we will be concerned only with the region upstream of the phase transition, we will not attempt to compare our results with these data.

Ziabicki (1976) summarized our knowledge of melt spinning through the early 1970s. Kase and Matsuo (1965) analyzed the process for a nonisothermal, Newtonian fluid. Matovich and Pearson (1969) presented analyses both for Newtonian and generalized Newtonian (no dependence upon deformation history) fluids. These authors worked in terms of an area-averaged temperature and a film coefficient for energy transfer based upon the difference between this area-averaged temperature and the bulk temperature of the surrounding gas stream. The measurements of average temperature reported by George (1982) are in good agreement with their predictions. Unfortunately, the polymers used commercially in melt spinning are not accurately represented as either Newtonian or generalized Newtonian fluids.

Slattery (1966) analyzed melt spinning for a Noll simple fluid (Noll 1958, Coleman et al. 1966, Coleman and Noll 1961, Truesdell and Noll 1965) undergoing an extensional flow (Coleman and Noll 1962). Unfortunately he did not discuss the change of properties as the thread cooled, nor did he provide any comparison with experimental data.

Denn et al. (1975) used the generalized upper-convected Maxwell model with a single relaxation time to show that viscoelastic behavior could be responsible for the experimentally observed dependence of the axial component of velocity upon axial position. They also recognized that this particular model for viscoelastic behavior limits the draw ratios that can be observed to values less than those seen in commercial processes.

Phan-Thien (1978) used the Phan-Thien and Tanner (PTT) model with multiple relaxation times to describe low-speed experimental data. Gagon and Denn (1981) used this same model to describe the high-speed data of George (1982) for polyethylene terephthalate (PET).

In what follows, we will summarize the analysis of Slattery (1966) for the extensional flow of the liquid thread following the die swell and prior to solidification. We will show that this analysis is valid only for high-speed, commercial processes in which $N_{Fr} \gg 1$, and we will address the effects of the temperature changes in the thread. We will demonstrate that, with little change, the results are equally valid for Newtonian fluids and for the Noll simple fluid, a very general class of viscoelastic fluids. We will conclude with a comparison with the data of George (1982), the only data currently available that satisfy our constraint on $N_{Fr}$.

### Analysis

Our objective in the analysis that follows is to determine the radius of the liquid thread as a function of axial position and the axial component of the force that must be exerted in elongating the thread to achieve the desired final radius $R_f$. In carrying out this analysis, we will make the following assumptions:

1. The polymer can be represented as an incompressible fluid. We will examine two cases: a Newtonian fluid and a Noll simple fluid (Section 2.3.4).
2. The effects of gravity can be neglected, since

$$N_{Fr} \gg 1 \tag{3.3.4-2}$$

The extruder is mounted vertically and the diameter of the liquid thread is very small.

3. We will ignore exit effects immediately downstream of the die (orifice or *spinneret*) such as *die swell*, which is observed only in viscoelastic fluids (see Figure 3.3.4-1). The position $z = 0$ is chosen downstream of the die swell, perhaps where the radius of the thread is equal to the radius of the die, $R_0$. Physically, this will be very close to the die. We also will assume that, at this cross section, the axial component of velocity $V$ and temperature are independent of position.
4. For commercial high-speed spinning, the *capillary number*

$$N_{ca} \equiv \frac{\mu V}{\gamma} \gg 1 \tag{3.3.4-3}$$

which means that the effect of interfacial tension can be neglected with respect to viscous effects in the jump momentum balance. Here $\mu$ is the viscosity of the liquid and $\gamma$ is the interfacial tension.

5. We will assume that the liquid thread is axisymmetric and that it does not rotate.

Referring to Figure 3.3.4-1, we will say that in cylindrical coordinates

$$\text{at } z = 0: \quad R = R_0$$
$$v_z = V \tag{3.3.4-4}$$

$$\text{at } z = L: \quad R = R_f \tag{3.3.4-5}$$

The position $z = L$ denotes where the phase change begins; $R = R(z)$ describes the radius of the thread as a function of axial position $z$.

We will seek a solution in the form of an *extensional flow*, a subclass of potential flows such that, in a rectangular Cartesian coordinate system,

$$\Phi = -\frac{1}{2}\left(a_1 z_1{}^2 + a_2 z_2{}^2 + a_3 z_3{}^2\right) - b_1 z_1 - b_2 z_2 - b_3 z_3 \tag{3.3.4-6}$$

The corresponding velocity distribution is $\mathbf{v} = -\nabla\Phi$ or

$$
\begin{aligned}
v_1 &= a_1 z_1 + b_1 \\
v_2 &= a_2 z_2 + b_2 \\
v_3 &= a_3 z_3 + b_3
\end{aligned}
\tag{3.3.4-7}
$$

Here the $a_i$ and $b_i$ are constants to be determined. This means that, in cylindrical coordinates,

$$
\begin{aligned}
v_r &= a_1 r \, \cos^2\theta + b_1 \cos\theta + a_2 r \, \sin^2\theta + b_2 \sin\theta \\
v_\theta &= -a_1 r \, \sin\theta \cos\theta - b_1 \sin\theta + a_2 r \, \sin\theta \cos\theta + b_2 \cos\theta \\
v_z &= a_3 z + b_3
\end{aligned}
\tag{3.3.4-8}
$$

To give (3.3.4-7) and (3.3.4-8) some perspective, we usually begin the solution of a fluid mechanics problem with a statement about the functional form of the velocity distribution that we are seeking, perhaps in this case

$$
\begin{aligned}
v_r &= v_r(r, z) \\
v_\theta &= v_\theta(r, z) \\
v_z &= v_z(r, z)
\end{aligned}
\tag{3.3.4-9}
$$

Equation (3.3.4-7) is simply a more specific assumption that results in (3.3.4-18), a special case of (3.3.4-9).

In rectangular Cartesian coordinates, the differential mass balance requires that

$$a_1 + a_2 + a_3 = 0 \tag{3.3.4-10}$$

From (3.3.4-4), we conclude that

$$b_3 = V \tag{3.3.4-11}$$

By assumption 5,

$$\text{at } r = 0 : \quad v_r = 0 \tag{3.3.4-12}$$

and

$$\text{everywhere}: \quad v_\theta = 0 \tag{3.3.4-13}$$

which imply

$$b_1 = b_2 = 0 \tag{3.3.4-14}$$

and

$$a_1 = a_2 \tag{3.3.4-15}$$

In view of (3.3.4-10) and (3.3.4-15), we conclude that

$$a_1 = a_2$$
$$= -\frac{a}{2} \tag{3.3.4-16}$$

in which we have introduced

$$a \equiv a_3 \tag{3.3.4-17}$$

In summary, in view of (3.3.4-7), (3.3.4-11), (3.3.4-14), (3.3.4-16), and (3.3.4-17), Equation (3.3.4-8) reduces to

$$v_r = -\frac{a}{2}r$$
$$v_\theta = 0 \tag{3.3.4-18}$$
$$v_z = az + V$$

The path lines in the liquid–gas interface determine the configuration of the thread. In rectangular Cartesian coordinates, we have for $i = 1, 2, 3$

$$\frac{dz_i}{dt} = a_i z_i + b_i \quad \text{(no sum on } i\text{)} \tag{3.3.4-19}$$

This equation can be integrated with the boundary condition

$$\text{at } t = 0 : \ z_i = z_{(0)i} \tag{3.3.4-20}$$

to find

$$\frac{z_1}{z_{(0)1}} = \exp\left(-\frac{at}{2}\right)$$
$$\frac{z_2}{z_{(0)2}} = \exp\left(-\frac{at}{2}\right) \tag{3.3.4-21}$$
$$\frac{az_3 + V}{V} = \exp(at)$$

or, eliminating $t$,

$$\frac{R}{R_0} = \left(\frac{az}{V} + 1\right)^{-1/2} \tag{3.3.4-22}$$

in which $R = R(z)$ is the radius of the thread. From (3.3.4-22) and Table 2.4.3-8, the three components of the unit normal to the polymer–gas phase interface (directed into the gas phase) are

$$\xi_r = \left[1 + \frac{1}{4}\left(\frac{aR_0}{V}\right)^2 \left(\frac{r}{R_0}\right)^6\right]^{-1/2}$$
$$\xi_\theta = 0 \tag{3.3.4-23}$$
$$\xi_z = \left[\frac{1}{2}\left(\frac{r}{R_0}\right)^3 \frac{aR_0}{V}\right]\left[1 + \frac{1}{4}\left(\frac{aR_0}{V}\right)^2 \left(\frac{r}{R_0}\right)^6\right]^{-1/2}$$

which will be required shortly.

## Isothermal, Incompressible Newtonian Fluid

Given (3.3.4-18), we find that for an isothermal, incompressible Newtonian fluid the three components of the differential momentum balance reduce to

$$\rho \frac{a^2}{4} r = -\frac{\partial p}{\partial r}$$

$$0 = -\frac{\partial p}{\partial \theta}$$

$$\rho a(az + V) = -\frac{\partial p}{\partial z}$$

(3.3.4-24)

In terms of the dimensionless variables

$$r^\star \equiv \frac{r}{R_0}, \quad z^\star \equiv \frac{z}{R_0}$$

$$v_r^\star \equiv \frac{v_r}{V}, \quad v_z^\star \equiv \frac{v_z}{V}, \quad p^\star \equiv \frac{p}{p_0}$$

(3.3.4-25)

Equation (3.3.4-24) reduces to

$$\frac{a^{\star 2}}{4} r^\star = -\frac{1}{N_{Ru}} \frac{\partial p^\star}{\partial r^\star}$$

$$0 = -\frac{\partial p^\star}{\partial \theta}$$

$$a^\star \left(a^\star z^\star + 1\right) = -\frac{1}{N_{Ru}} \frac{\partial p^\star}{\partial z^\star}$$

(3.3.4-26)

where

$$a^\star \equiv \frac{a R_0}{V}$$

(3.3.4-27)

$$N_{Ru} \equiv \frac{\rho V^2}{p_0}$$

(3.3.4-28)

and $p_0$ is atmospheric pressure.

From (3.3.4-5) and (3.3.4-22), if

$$a^\star = \frac{R_0}{L} \left[ \left(\frac{R_0}{R_f}\right)^2 - 1 \right]$$

$$\ll 1$$

(3.3.4-29)

Equation (3.3.4-24) requires that

$$p = \text{a constant}$$

(3.3.4-30)

George (1982) notes that, for commercial high-speed spinning of PET, typically $R_0 = 3.5 \times 10^{-5}$ m, $R_f = 1.1 \times 10^{-5}$ m, $L = 1.2$ m, and

$$a^\star = 8.2 \times 10^{-5}$$

(3.3.4-31)

From Table 2.4.3-8, the $r$ component of the jump momentum balance requires that

$$(T_{rr} + p_0)\,\xi_r = 0 \tag{3.3.4-32}$$

or

$$(-p - \mu a + p_0)\,\xi_r = 0 \tag{3.3.4-33}$$

This means that, for an isothermal process, (3.3.4-30) becomes

$$p = p_0 - \mu a \tag{3.3.4-34}$$

The $\theta$ component is satisfied identically. The $z$ component takes the form

$$(-p + 2\mu a + p_0)\,\xi_z = 0 \tag{3.3.4-35}$$

which is valid in the limit (3.3.4-29), since (3.3.4-23) indicates that $\xi_z$ is small.

Our conclusion is that the extensional flow velocity distribution (3.3.4-7) is a solution for the commercially significant limiting case described by (3.3.4-2), (3.3.4-3), and (3.3.4-29).

Finally, the axial component of the force that the liquid thread exerts on the solidified thread (beyond the opposing force of atmospheric pressure) is

$$\begin{aligned}
\mathcal{F}_z &= \int_0^{2\pi} \int_0^{R_f} (t_3 - p_0)\, r\, dr\, d\theta \\
&= \int_0^{2\pi} \int_0^{R_f} (-T_{zz} - p_0)\, r\, dr\, d\theta \\
&= \int_0^{2\pi} \int_0^{R_f} (p - 2\mu a - p_0)\, r\, dr\, d\theta \\
&= -\int_0^{2\pi} \int_0^{R_f} 3\mu a\, r\, dr\, d\theta \\
&= -3\pi \mu a R_f^{\,2}
\end{aligned} \tag{3.3.4-36}$$

Remember that this force is equal in magnitude and opposite in direction to the force that the solidified thread exerts on the liquid thread.

## Isothermal, Incompressible Noll Simple Fluid

For an incompressible Noll simple fluid, the extra stress tensor $S$ can be understood to be either a functional of the history of the right relative Cauchy–Green tensor (Noll 1958; Coleman and Noll 1961) or simply the relative deformation gradient (Coleman et al. 1966, p. 17). Given the particle paths (3.3.4-21), we conclude that both the relative deformation gradient and the right relative Cauchy–Green strain tensor are independent of position (Coleman and Noll 1962, Slattery 1966). This means that the components of $S$ for a Noll simple fluid are also independent of position (Coleman and Noll 1962, Slattery 1966) except for the dependence upon temperature of the physical parameters used to describe a particular member of this class of fluids. Once again the three components of the differential momentum balance reduce to (3.3.4-26). With the same arguments used above for the Newtonian fluid, we conclude that (3.3.4-30) again holds and that the three components of the jump momentum balance are satisfied.

Again, our conclusion is that the extensional flow velocity distribution (3.3.4-7) is a solution for the commercially important limiting case described by (3.3.4-2), (3.3.4-3), and (3.3.4-29).

Finally, instead of (3.3.4-36), we have

$$
\begin{aligned}
\mathcal{F}_z &= \int_0^{2\pi} \int_0^{R_f} (p - S_{zz} - p_0) r \, dr \, d\theta \\
&= \int_0^{2\pi} \int_0^{R_f} (S_{rr} - S_{zz}) r \, dr \, d\theta \\
&= \pi \, (S_{rr} - S_{zz}) \, R_f{}^2
\end{aligned}
\tag{3.3.4-37}
$$

We know that, in view of (3.3.4-31) (Slattery 1966, Eqs. 23 and 24),

$$
(S_{rr} - S_{zz}) \sim a
\tag{3.3.4-38}
$$

but the proportionality factor depends upon the particular member of the class of Noll simple fluids with which one is working. It would have to be measured in an extensional flow.

### Nonisothermal, Incompressible Newtonian Fluid

Since the viscosity $\mu$ is a function of temperature, (3.3.4-24) must be replaced by

$$
\begin{aligned}
\rho \frac{a^2}{4} r &= -\frac{\partial p}{\partial r} + \frac{\partial}{\partial r}(\mu a) \\
0 &= -\frac{\partial p}{\partial \theta} \\
\rho a (az + V) &= -\frac{\partial p}{\partial z} + \frac{\partial}{\partial z}(2\mu a)
\end{aligned}
\tag{3.3.4-39}
$$

Arguing as we did in the isothermal case, we can write this in terms of dimensionless variables as

$$
\begin{aligned}
\frac{a^{\star 2}}{4} r^\star &= -\frac{1}{N_{Ru}} \frac{\partial p^\star}{\partial r^\star} + \frac{1}{N_{RE}} \frac{\partial}{\partial r^\star} \left( \mu^\star a^\star \right) \\
0 &= -\frac{\partial p^\star}{\partial \theta} \\
a^\star (a^\star z^\star + 1) &= -\frac{1}{N_{Ru}} \frac{\partial p^\star}{\partial z^\star} + \frac{1}{N_{RE}} \frac{\partial}{\partial r^\star} \left( 2\mu^\star a^\star \right)
\end{aligned}
\tag{3.3.4-40}
$$

Here,

$$
N_{Re} \equiv \frac{\rho V R_0}{\mu_0}
\tag{3.3.4-41}
$$

is the Reynolds number and

$$
\mu^\star \equiv \frac{\mu}{\mu_0}
\tag{3.3.4-42}
$$

For the data of George (1982), $\rho = 980 \text{ kg/m}^3$, $V = 5 \text{ m/s}$, $R_0 = 3.5 \times 10^{-5} \text{ m}$, and $\mu_0 = 360$ Pa s, which means $N_{Re} = 4.9 \times 10^{-4}$. In the limit (3.3.4-29), we again can neglect the inertial terms. Assuming that temperature is independent of position at $z = 0$, we can do a line integration to conclude that

$$
p = 2\mu a + \text{a constant}
\tag{3.3.4-43}
$$

This raises a question as to whether the jump momentum balance (3.3.4-33) is still satisfied, since $\mu$ is a function of $T$. If the temperature of the polymer–gas interface is a constant, $\mu$ is independent of position in the interface, and the three components of the jump momentum balance are satisfied as before. There are two cases to be discussed.

In the first case, we assume that there is forced convection in the adjoining gas phase beyond that induced by the movement of the thread and that this convection is sufficiently complex to defy a detailed analysis. Problems of this character are attacked in two different ways. The preferred manner would be to express the energy flux from the surface of the solid in terms of an empirical film coefficient of heat transfer. Unfortunately the only correlation for energy transfer from a polymer thread currently available was developed in terms of the area-averaged temperature of the thread (Kase and Matsuo 1965); it cannot be used to determine the surface temperature of the thread. A generally less accurate but commonly employed boundary condition is to assume that the surface temperature is the same as the temperature of the gas at some distance from the interface, generally a constant.

In the second case, all forced convection in the adjoining gas phase is the result of the movement of the thread. In Section 6.7.7, we analyze the temperature distribution in the surrounding gas phase, and we conclude that, in at least one solution, the temperature of the polymer–gas interface is independent of position. Although we have solved one limiting case, we have not determined the complete temperature distribution in the thread.

Assuming that one of these arguments is applicable, our conclusion is that the extensional flow velocity distribution (3.3.4-7) is a solution for the commercially important limiting case described by (3.3.4-2), (3.3.4-3), and (3.3.4-29).

Finally, in using (3.3.4-36), we must replace $\mu$ in the last line by its area-averaged value.

### Nonisothermal, Incompressible Noll Simple Fluid

For the nonisothermal case, we must recognize that the physical parameters describing a particular member of the class of Noll simple fluids are functions of position as the result of their dependence upon temperature. Since all nondiagonal components of $\mathbf{S}$ are zero and $S_{rr} = S_{\theta\theta}$ (Slattery 1966), the three components of the differential momentum balance reduce to

$$\rho \frac{a^2}{4} r = -\frac{\partial p}{\partial r} + \frac{1}{r} \frac{\partial}{\partial r} (r S_{rr}) - \frac{S_{\theta\theta}}{r}$$

$$0 = -\frac{\partial p}{\partial \theta} \tag{3.3.4-44}$$

$$\rho a (az + V) = -\frac{\partial p}{\partial z} + \frac{\partial S_{zz}}{\partial z}$$

Arguing as we did in the isothermal case, we can neglect the inertial terms. Assuming that temperature is independent of position at $z = 0$, we can do a line integration to conclude that

$$p = S_{zz} + \text{a constant} \tag{3.3.4-45}$$

Following the discussion of the nonisothermal, incompressible Newtonian fluid above, we will again assume that temperature is independent of position in the polymer–gas interface and that the jump momentum balance is satisfied. Our conclusion is that the extensional flow velocity distribution (3.3.4-7) is a solution for the commercially important limiting case described by (3.3.4-2), (3.3.4-3), and (3.3.4-29).

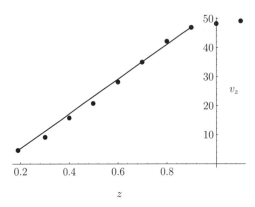

**Figure 3.3.4-2.** Comparison of the predictions of (3.3.4-18) with measurements reported by George (1982) for PET: $v_z$ (m/s) as a function of $z$ (m). One must allow for the uncertainty in locating the cross-section $z = 0$ and eliminate the region in which the phase transition occurs.

## Comparison with Experimental Data

To the extent that the temperature dependence of density can be neglected, we have found that the extensional flow velocity distribution (3.3.4-18) satisfies the differential mass balance, the differential momentum balance, as well as the jump mass and momentum balances at the polymer–gas interface for either Newtonian fluids or Noll simple fluids. The results are valid for both isothermal and nonisothermal processes, so long as the argument that the temperature of the polymer–air interface is either approximately or exactly a constant. For these reasons, (3.3.4-18) should describe a commercial process within the limits (3.3.4-2), (3.3.4-3), and (3.3.4-29).

Figure 3.3.4-2 compares (3.3.4-18) with measurements reported by George (1982, Fig. 5) for $v_z$ as a function of $z$ in the spinning of PET. In view of the uncertainty in locating the cross-section $z = 0$, we are forced to choose $z = 0$ as the location of his first data point. We also must eliminate the region in which the phase transition occurs, which means that we will eliminate his last two points. Within the context of these decisions, $V = 4.7$ m/s, $L = 0.71$ m, and

$$\left(\frac{R_0}{R_f}\right)^2 = \frac{v_z|_{z=L}}{V}$$
$$= 10.0$$

which permits us to compute $a = 59.3$ s$^{-1}$ from (3.3.4-22).

**Exercise 3.3.4-1** *Extrusion of a two-dimensional film*    Repeat the discussion in the text for the extrusion of a thin polymer film.

**Exercise 3.3.4-2** *A simple drawing experiment*    Let us assume that we have a very viscous, nearly solid-like material that we are willing to describe as an incompressible Newtonian fluid. We wish to measure its viscosity. Because the material is so viscous, it is not practical to measure its viscosity in a cone–plate viscometer or in flow through a tube.

It is proposed that the viscosity may be measured by casting the material in the form of a cylindrical rod, fixing the position of one end of the rod, and slowly drawing out the other end at a constant speed $V$. The axial component $F_1$ of the force required to draw the sample out at this constant speed will vary as a function of time $t$, but it should be related to the viscosity $\mu$ of the sample.

Develop an argument similar to that given in text to determine the relationship between $F_1$, $V$, $\mu$, and $t$. Neglect any end effects attributable to the devices used to grip the rod.

**Exercise 3.3.4-3** *Another simple drawing experiment*   As a variation on the experiment described in Exercise 3.3.4-2, assume that both ends of the rod are slowly moved in opposite directions at a constant speed $V$.

---

## 3.4    Nonviscous Flows

Let us confine our attention to an incompressible Newtonian fluid.

In the introduction to Section 3.3, we examine the limit of creeping flow or $N_{Re} \ll 1$. Let us now examine the limit $N_{Re} \gg 1$. Referring again to the dimensionless form of the Navier–Stokes equation (3.3.0-2), we have

$$\frac{1}{N_{St}} \frac{\partial \mathbf{v}^\star}{\partial t^\star} + (\nabla \mathbf{v}^\star) \cdot \mathbf{v}^\star = -\frac{1}{N_{Ru}} \nabla \mathcal{P}^\star + \frac{1}{N_{Re}} \operatorname{div}(\nabla \mathbf{v}^\star) \tag{3.4.0-1}$$

For $N_{Re} \gg 1$, it appears that the Navier–Stokes equation reduces to

$$\frac{1}{N_{St}} \frac{\partial \mathbf{v}^\star}{\partial t^\star} + (\nabla \mathbf{v}^\star) \cdot \mathbf{v}^\star = -\frac{1}{N_{Ru}} \nabla \mathcal{P}^\star \tag{3.4.0-2}$$

or

$$\rho \frac{\partial \mathbf{v}}{\partial t} + \rho(\nabla \mathbf{v}) \cdot \mathbf{v} = -\nabla \mathcal{P} \tag{3.4.0-3}$$

Our next task is to study the solutions of (3.4.0-3) that are consistent with the differential mass balance. A velocity distribution of the form

$$\mathbf{v} = -\nabla \Phi \tag{3.4.0-4}$$

suggests itself, since in this case the Navier–Stokes equation

$$\rho \frac{\partial \mathbf{v}}{\partial t} + \rho(\nabla \mathbf{v}) \cdot \mathbf{v} = -\nabla \mathcal{P} + \mu \operatorname{div}(\nabla \mathbf{v}) \tag{3.4.0-5}$$

reduces to (3.4.0-3) for all values of $N_{Re}$:

$$\begin{aligned}
\operatorname{div}(\nabla \mathbf{v}) &= \frac{\partial^2 v_i}{\partial z_j \partial z_j} \mathbf{e}_i \\
&= -\frac{\partial^3 \Phi}{\partial z_j \partial z_j \partial z_i} \mathbf{e}_i \\
&= \frac{\partial^2 v_j}{\partial z_i \partial z_j} \mathbf{e}_i \\
&= \nabla(\operatorname{div} \mathbf{v}) \\
&= 0
\end{aligned} \tag{3.4.0-6}$$

**Table 3.4.0-1.** Laplace's equation, div $(\nabla \Phi) = 0$, in three coordinate systems

---

*Rectangular Cartesian coordinates:*

$$\frac{\partial^2 \Phi}{\partial z_1^2} + \frac{\partial^2 \Phi}{\partial z_2^2} + \frac{\partial^2 \Phi}{\partial z_3^2} = 0$$

*Cylindrical coordinates:*

$$\frac{1}{r}\frac{\partial}{\partial r}\left(r\frac{\partial \Phi}{\partial r}\right) + \frac{1}{r^2}\frac{\partial^2 \Phi}{\partial \theta^2} + \frac{\partial^2 \Phi}{\partial z^2} = 0$$

*Spherical coordinates:*

$$\frac{\partial}{\partial r}\left(r^2\frac{\partial \Phi}{\partial r}\right) + \frac{1}{\sin\theta}\frac{\partial}{\partial \theta}\left(\sin\theta\frac{\partial \Phi}{\partial \theta}\right) + \frac{1}{\sin^2\theta}\frac{\partial^2 \Phi}{\partial \varphi^2} = 0$$

---

In view of the differential mass balance for an incompressible fluid, the equation that must be solved for the potential $\Phi$ is Laplace's equation:

$$\text{div}\,(\nabla \Phi) = 0 \tag{3.4.0-7}$$

Laplace's equation is presented in Table 3.4.0-1 for rectangular Cartesian, cylindrical, and spherical coordinates. This equation has received considerable attention in the literature (Kellogg 1929; Churchill 1960, Chaps. 9 and 10 and Appendix 2). Equation (3.4.0-3) is used only to determine the modified pressure distribution.

When the velocity distribution is given by an equation of the form of (3.4.0-4), we say that the velocity **v** is representable by a potential $\Phi$. We refer to such a flow as a *potential flow*. For a potential flow, it is easy to see that the vorticity vector (Exercise A.9.1-1)

$$\mathbf{w} \equiv \text{curl}\,\mathbf{v}$$
$$= 0 \tag{3.4.0-8}$$

A flow in which the vorticity vector vanishes everywhere is said to be an *irrotational flow*. The physical picture is that there is no local angular motion.

In summary, incompressible potential flows (or incompressible irrotational flows) form one class of flows in which the viscous terms in the Navier–Stokes equation are negligibly small with respect to the convective inertial terms. They are not the only flows with this property. It is only because of their mathematical simplicity that they have received the most attention.

Note that the viscous terms in the differential momentum balance for an incompressible fluid described by (2.3.2-13) or the generalized Newtonian fluid (2.3.3-1) are not automatically zero in a potential flow (Slattery 1962).

Upon reflection, it should be clear that a serious problem has developed in arriving at (3.4.0-3): All of the second derivatives have dropped out of the differential momentum balance. This implies that we will not be able to satisfy all of the required boundary conditions. This is a case in which a casual argument about the order of magnitude of terms in a differential equation fails. An apparently small term has a very large effect. The error occurred when we assumed that div $(\nabla \mathbf{v}^\star) \sim O(1)$.

For potential flows in particular, a unique solution is available if we eliminate the requirement that the tangential components of velocity be continuous at phase interfaces (Kellogg 1929, pp. 211, 216, 311). In these problems, we satisfy only the constraints placed upon the normal component of velocity by the jump mass balance.

We argue there that, for sufficiently small values of the Reynolds number $N_{Re}$, the convective inertial terms, $(\nabla \mathbf{v}^*) \cdot \mathbf{v}^*$, might be neglected with respect to the viscous terms. This intuitive argument seems to work in three-dimensional flows such as flow in a cone–plate viscometer discussed in Section 3.3.1, but we pointed out that it fails in plane motions. A velocity field is defined to be *plane* if there is some rectangular Cartesian coordinate system such that $v_1 = v_1(z_1, z_2)$, $v_2 = v_2(z_1, z_2)$, and $v_3 = 0$.

It should not be too surprising that an argument that suggests one term in a differential equation may be neglected with respect to another occasionally fails. There is no mathematical basis for such a step. It is reasonable to neglect one term with respect to another in the *solution* of a differential equation, but it is an entirely different matter in the differential equation used to obtain that solution. Rather than being surprised when an overly simple argument such as this fails, we should be grateful that it is so often helpful.

You may start to wonder whether flows in which viscous effects are neglected with respect to inertial effects have any importance. They do. Their true value will be better appreciated in the context of boundary-layer theory in Section 3.5.

### 3.4.1 Bernoulli Equation

In this section, let us assume that we are concerned with a compressible fluid and that we have made an intuitive argument to neglect the viscous terms in the differential momentum balance. For example, we might start with a compressible Newtonian fluid (2.3.2-21) and argue that, in the limit where both $L_0 v_0 \rho / \mu \gg 1$ and $L_0 v_0 \rho / \lambda \gg 1$, the viscous terms should become negligibly small compared with the convective inertial terms. As a result, the differential momentum balance reduces to

$$\rho \frac{\partial \mathbf{v}}{\partial t} + \rho (\nabla \mathbf{v}) \cdot \mathbf{v} = -\nabla P + \rho \mathbf{f} \tag{3.4.1-1}$$

Let us further assume that the external force per unit mass is representable by a potential:

$$\mathbf{f} = -\nabla \phi \tag{3.4.1-2}$$

It is easily shown that (see Exercise 3.4.1-1)

$$\mathbf{v} \wedge \mathbf{w} = \nabla \left( \frac{1}{2} v^2 \right) - (\nabla \mathbf{v}) \cdot \mathbf{v} \tag{3.4.1-3}$$

where the vorticity vector is defined by (3.4.0-8). Equations (3.4.1-2) and (3.4.1-3) allow us to rewrite (3.4.1-1) as (Milne-Thomson 1955, p. 75)

$$\frac{\partial \mathbf{v}}{\partial t} - \mathbf{v} \wedge \mathbf{w} = -\frac{1}{\rho} \nabla P - \nabla \left( \frac{1}{2} v^2 \right) - \nabla \phi$$
$$= -\nabla \chi \tag{3.4.1-4}$$

where

$$\chi \equiv \int_{P_0}^{P} \frac{dP}{\rho} + \frac{1}{2}v^2 + \phi \qquad (3.4.1\text{-}5)$$

Notice that in arriving at (3.4.1-4) we have assumed that density is a function only of pressure and not of temperature (see Exercise 3.4.1-3).

Let $s$ be a parameter along an arbitrary curve in space. At any point along this arbitrary curve, $d\mathbf{p}/ds$ is a unit tangent vector to the curve. Consider the component of (3.4.1-4) in the direction $d\mathbf{p}/ds$:

$$\frac{\partial \mathbf{v}}{\partial t} \cdot \frac{d\mathbf{p}}{ds} - (\mathbf{v} \wedge \mathbf{w}) \cdot \frac{d\mathbf{p}}{ds} = -\frac{d\mathbf{p}}{ds} \cdot \nabla \chi$$

$$= -\frac{d\chi}{ds} \qquad (3.4.1\text{-}6)$$

Since

$$\frac{d}{ds} \int_{s_0}^{s} \left( \frac{\partial \mathbf{v}}{\partial t} \cdot \frac{d\mathbf{p}}{ds} \right) ds = \frac{\partial \mathbf{v}}{\partial t} \cdot \frac{d\mathbf{p}}{ds} \qquad (3.4.1\text{-}7)$$

we may express (3.4.1-6) as (Sabersky, Acosta, and Hauptmann 1989, p. 86)

$$-(\mathbf{v} \wedge \mathbf{w}) \cdot \frac{d\mathbf{p}}{ds} = -\frac{dX}{ds} \qquad (3.4.1\text{-}8)$$

where

$$X \equiv \int_{P_0}^{P} \frac{dP}{\rho} + \frac{1}{2}v^2 + \phi + \int_{s_0}^{s} \left( \frac{\partial \mathbf{v}}{\partial t} \cdot \frac{d\mathbf{p}}{ds} \right) ds \qquad (3.4.1\text{-}9)$$

If, at any point in time, $d\mathbf{p}/ds$ is tangent to a streamline,

$$\frac{dX}{ds} = (\mathbf{v} \wedge \mathbf{w}) \cdot \frac{d\mathbf{p}}{ds}$$

$$= 0. \qquad (3.4.1\text{-}10)$$

We conclude that, along a streamline, $X = $ a constant.

If, at any point in time, $d\mathbf{p}/ds$ is tangent to a vortex line, (3.4.1-10) is again valid, and we conclude that, along a vortex line, $X = $ a constant. (The *vortex lines* for time $t$ form that family of curves to which the vorticity vector field is everywhere tangent. The parametric equations for the vortex lines are solutions of the differential equations

$$\frac{d\mathbf{p}}{d\alpha} = \mathbf{w} \qquad (3.4.1\text{-}11)$$

where $\alpha$ is an arbitrary parameter measured along the curves and time $t$ is a constant. (See the discussion of streamlines in Exercise 1.1.0-7.)

If we are dealing with an irrotational flow ($\mathbf{w} = \mathbf{0}$) and $d\mathbf{p}/ds$ is the tangent vector to *any* curve in the flowfield, Equation (3.4.1-10) still holds. We deduce that, in an irrotational flow, $X = $ constant along any curve in the flowfield.

A more interesting result for an irrotational flow may be obtained by starting directly with (3.4.1-4). Because this must be a potential flow, we may write

$$
\nabla \left( \int_{P_0}^{P} \frac{dP}{\rho} + \frac{1}{2} v^2 + \phi - \frac{\partial \Phi}{\partial t} \right) = 0
\tag{3.4.1-12}
$$

which means that

$$
\Psi \equiv \int_{P_0}^{P} \frac{dP}{\rho} + \frac{1}{2} v^2 + \phi - \frac{\partial \Phi}{\partial t}
$$
$$
= F(t)
\tag{3.4.1-13}
$$

The form of $F(t)$ has no physical significance, for we can define

$$
\tilde{\Phi} \equiv \Phi + \int_{0}^{t} F(t) \, dt
\tag{3.4.1-14}
$$

such that

$$
\int_{P_0}^{P} \frac{dP}{\rho} + \frac{1}{2} v^2 + \phi - \frac{\partial \tilde{\Phi}}{\partial t} = 0
\tag{3.4.1-15}
$$

Yet $\tilde{\Phi}$ retains the important physical significance

$$
\nabla \tilde{\Phi} = \nabla \Phi = -\mathbf{v}
\tag{3.4.1-16}
$$

For this reason, it is customary to write for an irrotational flow

$$
\Psi = \text{constant}
\tag{3.4.1-17}
$$

In summary,

$$
X \equiv \int_{P_0}^{P} \frac{dP}{\rho} + \frac{1}{2} v^2 + \phi + \int_{s_0}^{s} \left( \frac{\partial \mathbf{v}}{\partial t} \cdot \frac{d\mathbf{p}}{ds} \right) ds
$$
$$
= \text{a constant}
\tag{3.4.1-18}
$$

is *Bernoulli's equation*, which is valid along streamlines and vortex lines in any flow such that viscous effects can be neglected in the differential momentum balance. In any irrotational flow such that viscous effects can be neglected in the differential momentum balance, (3.4.1-18) is valid along every curve in the flowfield; more importantly, *Bernoulli's equation for a potential flow* holds throughout the flowfield:

$$
\Psi \equiv \int_{P_0}^{P} \frac{dP}{\rho} + \frac{1}{2} v^2 + \phi - \frac{\partial \Phi}{\partial t}
$$
$$
= \text{a constant}
\tag{3.4.1-19}
$$

**Exercise 3.4.1-1**  Derive (3.4.1-3).

**Exercise 3.4.1-2**  Starting with the result that $X = $ a constant along every curve in an irrotational flowfield such that viscous effects can be neglected in the differential momentum balance, derive the result that $\Psi$ is independent of position.

**Exercise 3.4.1-3** *More about Bernoulli's equation*   In deriving (3.4.1-4), we assumed that the density $\rho$ is a function only of pressure. In this way we restrict ourselves to an isothermal (constant composition) flow. Where was this assumption introduced?

**Exercise 3.4.1-4**   A vertical tube of length $L$ is filled with an incompressible fluid and a plate is held over the lower end. At time $t = 0$, the plate is removed and the fluid is allowed to run out of the tube.

Use Bernoulli's equation to derive an expression for the time required to empty the tube. (See also Exercises 3.4.2-4 and 4.4.8-4.)

*Hint:*   Don't hesitate to make a reasonable approximation in evaluating the line integral.

**Exercise 3.4.1-5**   Let us assume that the lower end of the vertical tube in Exercise 3.4.1-4 is finished with a rounded orifice. Use Bernoulli's equation to derive a differential equation that describes (approximately) the height of water in the tube as a function of time. Give the required boundary conditions.

*Answer:*

$$f \frac{d^2 f}{dt^2} + \frac{1}{2}\left[1 - \left(\frac{A_{(t)}}{A_{(o)}}\right)^2\right]\left(\frac{df}{dt}\right)^2 + gf = 0$$

$$\text{at } t = 0: \ f = L \text{ and } \frac{df}{dt} = 0$$

where $A_{(t)}$ and $A_{(o)}$ are the cross-sectional areas of the tube and orifice, respectively, and $f$ is the height of fluid at time $t$.

### 3.4.2   Potential Flow Past a Sphere

Let us consider steady-state potential flow past a stationary sphere such that a fluid at a very large distance from the sphere moves with a uniform velocity in the positive $z_3$ direction:

$$\text{as } r \rightarrow \infty: \ \mathbf{v} \rightarrow V\mathbf{e}_3 \tag{3.4.2-1}$$

The origins of the spherical and rectangular Cartesian coordinate systems referred to here coincide with the center of the sphere; the relationship between these coordinate systems is that adopted in Exercise A.4.1-5. We wish to determine the velocity potential $\Phi$ for this flow and the corresponding velocity distribution

$$\mathbf{v} = -\nabla\Phi \tag{3.4.2-2}$$

Boundary condition (3.4.2-1) may be interpreted as a restriction on the velocity potential,

$$\text{as } r \rightarrow \infty: \ \Phi \rightarrow -Vz_3$$
$$= -Vr\cos\theta \tag{3.4.2-3}$$

Our discussion in the introduction to Section 3.4 indicates that only the constraint placed on the normal component of velocity by the jump mass balance can be satisfied by a potential flow. If $a$ denotes the radius of a sphere, this means that

$$\text{at } r = a: \ v_r = -\frac{\partial\Phi}{\partial r} = 0 \tag{3.4.2-4}$$

where we have used the physical components of the gradient of a scalar found in Exercise A.4.3-6.

Since a sphere is a body of revolution and since boundary condition (3.4.2-3) is axially symmetric, let us look for a velocity distribution of the form

$$v_r = v_r(r, \theta)$$
$$v_\theta = v_\theta(r, \theta) \tag{3.4.2-5}$$
$$v_\varphi = 0$$

Hence we are seeking a velocity potential that is independent of the spherical coordinate $\varphi$:

$$\Phi = \Phi(r, \theta) \tag{3.4.2-6}$$

From Table 3.4.0-1, Laplace's equation becomes in spherical coordinates

$$\frac{\partial}{\partial r}\left(r^2 \frac{\partial \Phi}{\partial r}\right) + \frac{1}{\sin\theta}\frac{\partial}{\partial\theta}\left(\sin\theta\frac{\partial \Phi}{\partial\theta}\right) = 0 \tag{3.4.2-7}$$

We wish to find a solution to this equation that satisfies boundary conditions (3.4.2-3) and (3.4.2-4).

If we use the method of separation of the variables, it is easy to show that (3.4.2-7) has a solution of the form (Irving and Mullineux 1959, p. 30)

$$\Phi = \sum_{n=0}^{\infty}\left\{r^n[A_n P_n(\cos\theta) + B_n Q_n(\cos\theta)] + r^{-n-1}[C_n P_n(\cos\theta) + D_n Q_n(\cos\theta)]\right\} \tag{3.4.2-8}$$

where $P_n(x)$ is the Legendre function of the first kind, $Q_n(x)$ is the Legendre function of the second kind, and $A_n$, $B_n$, $C_n$, and $D_n$ are constants. Since

$$\text{as } \cos\theta \to \pm 1: \quad Q_n(\cos\theta) \to \infty \tag{3.4.2-9}$$

we take

$$B_n = D_n = 0 \text{ for } n \geq 0 \tag{3.4.2-10}$$

Recognizing that

$$P_1(\cos\theta) = \cos\theta \tag{3.4.2-11}$$

we see that (3.4.2-8) must reduce to

$$\Phi = -Vr\, P_1(\cos\theta) + \sum_{n=0}^{\infty} r^{-n-1}C_n P_n(\cos\theta) \tag{3.4.2-12}$$

in order that it satisfy (3.4.2-3). Applying boundary condition (3.4.2-4) to (3.4.2-12), we find that

$$V P_1(\cos\theta) + \sum_{n=0}^{\infty}(n+1)a^{-n-2}C_n P_n(\cos\theta) = 0 \tag{3.4.2-13}$$

The Legendre functions of the first kind are orthogonal on the interval $-1 \leq x \leq 1$ (Jahnke and Emde 1945, p. 116):

$$\int_{-1}^{+1} P_m(x)P_n(x)\, dx = 0, \quad m \neq n \tag{3.4.2-14}$$

$$\int_{-1}^{+1} [P_n(x)]^2 dx = \frac{2}{2n+1} \tag{3.4.2-15}$$

These orthogonality relations allow us to conclude from (3.4.2-13) that

$$C_1 = -\frac{Va^3}{2}$$
$$C_n = 0 \text{ for } n \neq 1$$
(3.4.2-16)

In summary, we find that the velocity potential is given by

$$\Phi = -V\left(r + \frac{a^3}{2r^2}\right)\cos\theta$$
(3.4.2-17)

Using Exercise A.4.3-6, we learn that the corresponding components of velocity are

$$v_r = -\frac{\partial\Phi}{\partial r}$$

$$= V\left(1 - \frac{a^3}{r^3}\right)\cos\theta$$
(3.4.2-18)

$$v_\theta = -\frac{1}{r}\frac{\partial\Phi}{\partial\theta}$$

$$= -V\left(1 + \frac{a^3}{2r^3}\right)\sin\theta$$
(3.4.2-19)

$$v_\varphi = 0$$
(3.4.2-20)

This confirms our initial assumption that there is at least one solution to this problem of the form assumed in (3.4.2-6). It is shown elsewhere that this solution is unique (Milne-Thomson 1955, p. 95).

To help in visualizing what we have found here, let us calculate the stream function $\psi$ for this solution. From Table 2.4.2-1,

$$v_r = \frac{1}{r^2\sin\theta}\frac{\partial\psi}{\partial\theta}$$
(3.4.2-21)

and

$$v_\theta = -\frac{1}{r\sin\theta}\frac{\partial\psi}{\partial r}$$
(3.4.2-22)

Equations (3.4.2-18) and (3.4.2-19) allow us to carry out a line integration to determine the stream function:

$$\psi = \int_0^\psi d\psi$$

$$= \int_a^r \frac{\partial\psi}{\partial r}\bigg|_{\theta=0} dr + \int_0^\theta \frac{\partial\psi}{\partial\theta} d\theta$$

$$= \frac{Va^2}{2}\left(\frac{r^2}{a^2} - \frac{a}{r}\right)\sin^2\theta$$
(3.4.2-23)

Since the stream function is arbitrary to a constant, we have chosen in this integration

$$\text{at } r = a, \ \theta = 0: \ \psi = 0$$
(3.4.2-24)

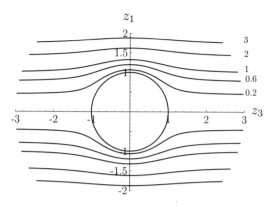

**Figure 3.4.2-1.** Streamlines for potential flow past a sphere corresponding to $2\psi/(Va^2) = 0.2, 0.6, 1, 2, 3$.

Noting that $\psi$ is a constant on a streamline (Exercise 1.3.7-2), we have plotted various streamlines corresponding to different values of $2\psi/(Va^2)$ in Figure 3.4.2-1.

**Exercise 3.4.2-1** *More on potential flow past a sphere*

i) Using the results of this section and Section 3.4.1, determine the pressure distribution in the fluid.

ii) Determine that the total force exerted upon the sphere is the buoyant force.

**Exercise 3.4.2-2** *Potential flow past a cylinder*  Consider steady-state, plane potential flow past a stationary cylinder such that

$$\text{as } r \to \infty : \ v \to V\mathbf{e}_1$$

and such that the viscous terms may be neglected in the equation of motion. By plane flow, we mean that, with respect to a rectangular Cartesian coordinate system,

$$v_1 = v_1(z_1, z_2), \quad v_2 = v_2(z_1 z_2), \quad v_3 = 0$$

i) Determine the velocity potential $\Phi$ for this flow and the corresponding velocity components with respect to a cylindrical coordinate system.

ii) Verify that the stream function $\psi$ for this flow in the absence of circulation is proportional to the constant $C$ given in Exercise 1.1.0-8. (See also Exercise 1.3.7-2.)

*Answer:*

$$\Phi = -V\left(r + \frac{a^2}{r}\right)\cos\theta + \Lambda\theta$$

where $\Lambda$ is an arbitrary constant. For a limiting case of this result corresponding to no circulation, see Figure 1.1.0-2.

**Exercise 3.4.2-3** *More on potential flow past a cylinder*

i) Using the results of Exercise 3.4.2-2 and Section 3.4.1, determine the pressure distribution in the fluid.

ii) Calculate both the $z_1$ and $z_2$ components of the force exerted upon the cylinder by the fluid beyond the buoyant force.

**Exercise 3.4.2-4**    A vertical tube of length $L$ is filled with an incompressible fluid and a plate is held over the lower end. At time $t = 0$, the plate is removed and the fluid is allowed to run out of the tube. Derive an expression for the time required to empty the tube.

You may assume that this is a potential flow, neglect viscous effects in the differential momentum balance, and assume that the top surface of the fluid remains a horizontal plane as the tube empties.

See also Exercises 3.4.1-5 and 4.4.8-4.

**Exercise 3.4.2-5** *Flow in convergent or divergent channels*    Consider two plane walls oriented in such a way that they form surfaces of constant $\theta$ in cylindrical coordinates. An incompressible fluid moves in either the positive or negative $r$ direction between these two planes. This is known as flow in either a divergent or a convergent channel. Determine the velocity distribution for the corresponding steady-state potential flow, assuming that the viscous terms may be neglected in the differential momentum balance.

### 3.4.3    Plane Potential Flow in a Corner

Consider the plane flow whose streamlines are pictured in Figure 3.4.3-1. Fluid approaches the plane $z_2 = 0$ from the $z_2$ direction, is turned to one side by the corner, and departs in the $z_1$ direction. By *plane flow*, we mean that

$$v_1 = v_1(z_1, z_2)$$
$$v_2 = v_2(z_1, z_2) \qquad\qquad\qquad (3.4.3\text{-}1)$$
$$v_3 = 0$$

Let us determine the velocity potential and the velocity distribution for the corresponding potential flow.

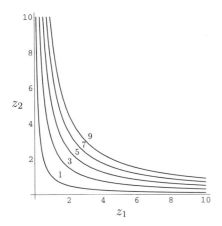

**Figure 3.4.3-1.** Streamlines for plane potential flow in the neighborhood of a corner.

Rather than trying to describe the fluid movements in this entire semi-infinite region, let us restrict our attention to the immediate neighborhood of the corner where the velocity potential $\Phi$ can be represented by a Taylor expansion about the origin:

$$\Phi = \Phi(z_1, z_2)$$

$$= \Phi(0, 0) + z_1 \frac{\partial \Phi}{\partial z_1}(0, 0) + z_2 \frac{\partial \Phi}{\partial z_2}(0, 0) + \frac{1}{2}(z_1)^2 \frac{\partial^2 \Phi}{\partial z_1^2}(0, 0)$$

$$+ z_1 z_2 \frac{\partial^2 \Phi}{\partial z_1 \partial z_2}(0, 0) + \frac{1}{2}(z_2)^2 \frac{\partial^2 \Phi}{\partial z_2^2}(0, 0) + \cdots \qquad (3.4.3\text{-}2)$$

We require as boundary conditions that

$$\text{at } z_1 = 0: \quad v_1 = -\frac{\partial \Phi}{\partial z_1} = 0 \qquad (3.4.3\text{-}3)$$

and

$$\text{at } z_2 = 0: \quad v_2 = -\frac{\partial \Phi}{\partial z_2} = 0 \qquad (3.4.3\text{-}4)$$

These boundary conditions mean that in (3.4.3-2)

$$\frac{\partial \Phi}{\partial z_1}(0, 0) = \frac{\partial \Phi}{\partial z_2}(0, 0)$$

$$= \frac{\partial^2 \Phi}{\partial z_1 \partial z_2}(0, 0)$$

$$= 0 \qquad (3.4.3\text{-}5)$$

Laplace's equation must hold everywhere, in particular at the origin:

$$\frac{\partial^2 \Phi}{\partial z_2^2}(0, 0) = -\frac{\partial^2 \Phi}{\partial z_1^2}(0, 0)$$

$$= k$$

$$= \text{a constant} \qquad (3.4.3\text{-}6)$$

Since the velocity potential is arbitrary to a constant, we choose to define

$$\Phi(0, 0) = 0 \qquad (3.4.3\text{-}7)$$

Equations (3.4.3-5) to (3.4.3-7) allow us to express (3.4.3-2) as

$$\Phi = \frac{k}{2}\left(z_2^2 - z_1^2\right) + \cdots \qquad (3.4.3\text{-}8)$$

The corresponding velocity components are

$$v_1 = -\frac{\partial \Phi}{\partial z_1}$$

$$= k z_1 + \cdots \qquad (3.4.3\text{-}9)$$

and

$$v_2 = -\frac{\partial \Phi}{\partial z_2}$$

$$= -k z_2 + \cdots \qquad (3.4.3\text{-}10)$$

It is helpful in visualizing these results to compute the stream function and to plot the streamlines. From Table 2.4.2-1,

$$v_1 = \frac{\partial \psi}{\partial z_2}$$

$$v_2 = -\frac{\partial \psi}{\partial z_1} \tag{3.4.3-11}$$

Equations (3.4.3-9) to (3.4.3-11) allow us to carry out a line integration to determine the stream function:

$$\psi = \int_0^\psi d\psi$$

$$= \int_0^{z_1} \frac{\partial \psi}{\partial z_1}\bigg|_{z_2=0} dz_1 + \int_0^{z_2} \frac{\partial \psi}{\partial z_2} dz_2$$

$$= k z_1 z_2 + \cdots \tag{3.4.3-12}$$

Since the stream function is arbitrary to a constant, we have chosen in this integration

$$\text{at } z_1 = z_2 = 0: \ \psi = 0 \tag{3.4.3-13}$$

Various streamlines corresponding to different values of $\psi/k$ are shown in Figure 3.4.3-1.

It is important to realize that the first terms of the velocity potential given by (3.4.3-8) cannot possibly describe a real flow at large distances from the origin. Equations (3.4.3-9) and (3.4.3-10) indicate that the magnitude of velocity is proportional to the distance from the origin:

$$|\mathbf{v}| = k\,|\mathbf{z}| \tag{3.4.3-14}$$

It is unbounded from above in the field of flow. This reinforces the viewpoint suggested when we expanded the velocity potential in a Taylor series earlier in this discussion: The results found here should be thought of as describing a potential flow in the neighborhood of the corner.

Our method of seeking a solution here (expansion of $\Phi$ in a Taylor series about the origin) implies that the first two terms given in (3.4.3-8) cannot possibly represent a unique potential flow for this geometry. But to make the point more plainly, consider (Churchill 1960, p. 211)

$$\Phi = -Be^{z_1^2-z_2^2}\cos(2z_1 z_2) \tag{3.4.3-15}$$

This is another solution to Laplace's equation that satisfies boundary conditions (3.4.3-3) and (3.4.3-4).

As an added benefit, the solution found here for potential flow in the neighborhood of a corner also represents potential flow in the neighborhood of a stagnation point on the plane $z_2 = 0$. Fluid approaching from the positive $z_2$ direction is turned aside by the plane $z_2 = 0$; it subsequently moves in the positive and negative $z_1$ directions. This flow has a plane of symmetry at $z_1 = 0$ along which the $z_1$ component of velocity is zero. Since the velocity is zero at the origin, it is known as a *stagnation point*. In the neighborhood of this stagnation point, the stream function is again given by (3.4.3-12).

In Section 3.5.4 we give another analysis of plane stagnation flow, taking viscous effects into account.

Exercise 3.4.3-1 *Axially symmetric potential flow in the neighborhood of a stagnation point*   Repeat the analysis in the text for axially symmetric potential flow in the neighborhood of a stagnation point.

## 3.5    Boundary-Layer Theory

Boundary-layer theory will be developed in three steps. In this section, we will confine our attention to momentum transfer. In Section 6.7, we will consider both momentum and energy transfer. Finally, in Section 9.5, we will add mass transfer.

### 3.5.1    Plane Flow Past a Flat Plate

In the introduction to Section 3.4, we see that a simple argument also cannot be used to neglect the viscous terms with respect to the convective inertial terms in the limit $N_{Re} \gg 1$. The problem is that not all of the required boundary conditions can be satisfied in this case. For potential flows in particular, a unique solution can be obtained, if we satisfy the constraint placed upon the normal component of velocity at a phase boundary by the jump mass balance, neglecting the requirement that the tangential components of velocity be continuous.

With this thought in mind, let us reexamine the problem of flows for $N_{Re} \gg 1$. It helps to be acquainted with what one might observe experimentally. To be specific, we restrict our attention to plane flow past a *thin* flat plate. The concepts we develop by looking at this particular flow may be readily extended to other situations.

In the time-lapse photograph shown in Figure 3.5.1-1, we see particle paths in flow past a flat plate, made visible by air bubbles in water. Outside a thin region next to the plate, the flow is essentially undisturbed. The thickness of the disturbed region increases with distance from the leading edge of the plate. In Figure 3.5.1-2, we illustrate on an exaggerated scale the velocity distribution in the immediate neighborhood of the plate. If we observed this flow as a function of the $N_{Re}$, we would find that the thickness of this region of retarded velocity decreased as the $N_{Re}$ increased.

Outside a very thin region next to a body, the flow is essentially that predicted by a nonviscous flow. This is the case for relatively thin streamlined bodies, such as airfoils. More generally, this thin boundary-layer region, where viscous effects are very important, can appear to *separate* from a body (Schlichting 1979, p. 28), and vortices can form at some distance from the body. See, for example, the photograph showing particle paths in flow past a circular cylinder in Figure 3.5.1-3. Here the flow at some distance from the body is obviously rotational, which means that the flow outside the boundary layer on the leading surface of the cylinder would not be precisely that predicted by a potential flow (Schlichting 1979, p. 21). It is perhaps safer to say that viscous effects may be ignored outside the boundary layer. A potential flow represents one particular type of nonviscous flow for a given set of boundary conditions on velocity, but it does not form a *unique solution*.

From what we have observed and suggested above, it seems clear that viscous effects are just as important as inertial effects in the Navier–Stokes equation for $N_{Re} \gg 1$, at least in the boundary layer. The argument we presented in the introduction to Section 3.4 is not applicable to this region, since it says that viscous effects are negligibly small compared with

**Figure 3.5.1-1.** Flow from left to right past a flat plate (with beveled edges) at zero incidence, made visible by air bubbles in water (Werlé 1974), as presented by VanDyke (1982, Plate 29). At a Reynolds number of 10,000 based on the length of the plate, the uniform stream is disturbed only in the immediate neighborhood of the plate (the dark region). The thickness of the disturbed region increases with distance measured from the leading edge.

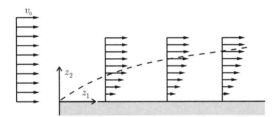

**Figure 3.5.1-2.** The character of the velocity distribution in flow past a flat plate.

convective inertial effects; apparently, this treatment must be reserved for the flow outside the boundary layer. If a similar approximation is to be used for the boundary layer, we must revise the definitions of our dimensionless variables so as to prevent the disappearance of all the viscous terms. Since the thickness of the boundary layer decreases as $N_{Re}$ increases, it seems reasonable that we should magnify the thickness of the boundary layer by introducing into the discussion of flow past a flat plate

$$z_2^{\star\star} \equiv N_{Re}{}^a z_2^{\star}$$

$$= N_{Re}{}^a \frac{z_2}{L_0} \tag{3.5.1-1}$$

while retaining the definitions for the other two dimensionless variables introduced in the introduction to Section 3.3.

**Figure 3.5.1-3.** Flow from left to right past a circular cylinder at $N_{Re} = 26$, as photographed by Sadatoshi Taneda (VanDyke 1982, Plate 42). The cylinder, moving through a tank of water containing aluminum powder, is illuminated by a sheet of light below the free surface. The boundary layer has separated from the surface, and two vortices have formed in the cylinder's wake.

In the context of the definition of the velocity vector as the material derivative of the position vector (Section 1.1), the dimensionless form of the differential momentum balance given in the introduction to Section 3.3,

$$\frac{1}{N_{St}}\frac{\partial \mathbf{v}^\star}{\partial t^\star} + (\nabla \mathbf{v}^\star)\cdot\mathbf{v}^\star = -\frac{1}{N_{Ru}}\nabla \mathcal{P}^\star + \frac{1}{N_{Re}}\text{div}(\nabla\mathbf{v}^\star) \tag{3.5.1-2}$$

suggests as a reasonable definition for the dimensionless $z_2$ component of velocity

$$v_2^{\star\star} = N_{Re}{}^{a} v_2^\star \tag{3.5.1-3}$$

Alternatively, if we require that each term in the differential mass balance have equal weight,

$$\frac{\partial v_1^\star}{\partial z_1^\star} + N_{Re}{}^{a}\frac{\partial v_2^\star}{\partial z_2^{\star\star}} + \frac{\partial v_3^\star}{\partial z_3^\star} = 0 \tag{3.5.1-4}$$

we are again led to (3.5.1-3).

Let us assume that our plate is relatively wide so that we may reasonably take the flow to be planar:

$$v_1^\star = v_1^\star(z_1^\star, z_2^{\star\star}, t^\star)$$
$$v_2^{\star\star} = v_2^{\star\star}(z_1^\star, z_2^{\star\star}, t^\star) \tag{3.5.1-5}$$
$$v_3^\star = 0$$

Under these conditions, the components of (3.5.1-2) are

$$\frac{1}{N_{St}}\frac{\partial v_1^\star}{\partial t^\star} + v_1^\star\frac{\partial v_1^\star}{\partial z_1^\star} + v_2^{\star\star}\frac{\partial v_1^\star}{\partial z_2^{\star\star}}$$
$$= -\frac{1}{N_{Ru}}\frac{\partial \mathcal{P}^\star}{\partial z_1^\star} + \frac{1}{N_{Re}}\frac{\partial^2 v_1^\star}{\partial z_1^{\star 2}} + N_{Re}{}^{2a-1}\frac{\partial^2 v_1^\star}{\partial z_2^{\star\star 2}} \tag{3.5.1-6}$$

$$\frac{1}{N_{St}}\frac{\partial v_2^{\star\star}}{\partial t^\star} + v_1^\star \frac{\partial v_2^{\star\star}}{\partial z_1^\star} + v_2^{\star\star}\frac{\partial v_2^{\star\star}}{\partial z_2^{\star\star}}$$

$$= -\frac{(N_{Re})^{2a}}{N_{Ru}}\frac{\partial \mathcal{P}^\star}{\partial z_2^{\star\star}} + \frac{1}{N_{Re}}\frac{\partial^2 v_2^{\star\star}}{\partial z_1^{\star 2}} + N_{Re}{}^{2a-1}\frac{\partial^2 v_2^{\star\star}}{\partial z_2^{\star\star 2}} \tag{3.5.1-7}$$

and

$$\frac{\partial \mathcal{P}^\star}{\partial z_3^\star} = 0 \tag{3.5.1-8}$$

If any viscous terms are to be the same order of magnitude as the convective inertial terms for $N_{Re} \gg 1$, we must identify

$$a \equiv \frac{1}{2} \tag{3.5.1-9}$$

In the limit $N_{Re} \gg 1$, Equation (3.5.1-6) reduces to

$$\frac{1}{N_{St}}\frac{\partial v_1^\star}{\partial t^\star} + v_1^\star \frac{\partial v_1^\star}{\partial z_1^\star} + v_2^{\star\star}\frac{\partial v_1^\star}{\partial z_2^{\star\star}} = -\frac{1}{N_{Ru}}\frac{\partial \mathcal{P}^\star}{\partial z_1^\star} + \frac{\partial^2 v_1^\star}{\partial z_2^{\star\star 2}} \tag{3.5.1-10}$$

and (3.5.1-7) reduces to

$$\frac{\partial \mathcal{P}^\star}{\partial z_2^{\star\star}} = 0 \tag{3.5.1-11}$$

Equations (3.5.1-8) and (3.5.1-11) mean that

$$\mathcal{P}^\star = \mathcal{P}^\star\left(z_1^\star, t^\star\right) \tag{3.5.1-12}$$

In summary, by the above arguments for flow past a thin flat plate, the differential mass and momentum balances have been reduced, respectively, to

$$\frac{\partial v_1^\star}{\partial z_1^\star} + \frac{\partial v_2^{\star\star}}{\partial z_2^{\star\star}} = 0 \tag{3.5.1-13}$$

and

$$\frac{1}{N_{St}}\frac{\partial v_1^\star}{\partial t^\star} + v_1^\star \frac{\partial v_1^\star}{\partial z_1^\star} + v_2^{\star\star}\frac{\partial v_1^\star}{\partial z_2^{\star\star}} = -\frac{1}{N_{Ru}}\frac{\partial \mathcal{P}^\star}{\partial z_1^\star} + \frac{\partial^2 v_1^\star}{\partial z_2^{\star\star 2}} \tag{3.5.1-14}$$

These are two equations in three unknowns: $v_1^\star$, $v_2^{\star\star}$, and $\mathcal{P}^\star$. Fortunately we have some information that we have not yet employed.

Outside the boundary-layer region, viscous effects can be neglected in the equation of motion. Let

$$\mathbf{v}^{(e)\star} = \mathbf{v}^{(e)\star}\left(z_1^\star, z_2^\star, t^\star\right)$$

$$= \mathbf{v}^{(e)\star}\left(z_1^\star, \frac{z_2^{\star\star}}{\sqrt{N_{Re}}}, t^\star\right) \tag{3.5.1-15}$$

and

$$\mathcal{P}^{(e)\star} = \mathcal{P}^{(e)\star}(z_1^\star, z_2^\star, t^\star)$$
$$= \mathcal{P}^{(e)\star}\left(z_1^\star, \frac{z_2^{\star\star}}{\sqrt{N_{Re}}}, t^\star\right) \tag{3.5.1-16}$$

denote the dimensionless velocity and modified pressure in the nonviscous external flow. Within a region where both the boundary-layer solution and the external nonviscous flow are valid,

$$\tilde{\mathbf{v}}^\star \equiv \lim N_{Re} \gg 1 \text{ for } z_1^\star \text{ and } z_2^{\star\star} \text{ fixed}: \; \mathbf{v}^{(e)\star}\left(z_1^\star, \frac{z_2^{\star\star}}{\sqrt{N_{Re}}}, t^\star\right)$$
$$= \mathbf{v}^{(e)\star}\left(z_1^\star, 0, t^\star\right) \tag{3.5.1-17}$$

and

$$\tilde{\mathcal{P}}^\star \equiv \lim N_{Re} \gg 1 \text{ for } z_1^\star \text{ and } z_2^{\star\star} \text{ fixed}: \; \mathcal{P}^{(e)\star}\left(z_1^\star, \frac{z_2^{\star\star}}{\sqrt{N_{Re}}}, t^\star\right)$$
$$= \mathcal{P}^{(e)\star}(z_1^\star, 0, t^\star) \tag{3.5.1-18}$$

As $z_2^{\star\star} \to \infty$, we must require that the $z_1$ component of velocity from the boundary-layer solution approach asymptotically the corresponding velocity component for the nonviscous external flow evaluated at $z_2^\star = 0$:

$$\text{as } z_2^{\star\star} \to \infty: \; v_1^\star \to \tilde{v}_1^\star$$
$$\mathcal{P}^\star \to \tilde{\mathcal{P}}^\star \tag{3.5.1-19}$$

(Because the fluid in the boundary layer moves more slowly than that in the main stream, fluid is continuously displaced from the boundary-layer region, and $v_2^{\star\star}$ does not approach zero as $z_2^{\star\star} \to \infty$.) In view of (3.5.1-12), we conclude that in (3.5.1-14) we may identify

$$-\frac{1}{N_{Ru}}\frac{\partial \mathcal{P}^\star}{\partial z_1^\star} = -\frac{1}{N_{Ru}}\frac{\partial \tilde{\mathcal{P}}^\star}{\partial z_1^\star}$$
$$\equiv \frac{1}{N_{St}}\frac{\partial \tilde{v}_1^\star}{\partial t^\star} + \tilde{v}_1^\star \frac{\partial \tilde{v}_1^\star}{dz_1^\star} \tag{3.5.1-20}$$

and write (3.5.1-14) as

$$\frac{1}{N_{St}}\frac{\partial v_1^\star}{\partial t^\star} + v_1^\star \frac{\partial v_1^\star}{\partial z_1^\star} + v_2^{\star\star}\frac{\partial v_1^\star}{\partial z_2^{\star\star}} = \frac{1}{N_{St}}\frac{\partial \tilde{v}_1^\star}{\partial t^\star} + \tilde{v}_1^\star \frac{\partial \tilde{v}_1^\star}{\partial z_1^\star} + \frac{\partial^2 v_1^\star}{\partial z_2^{\star\star 2}} \tag{3.5.1-21}$$

Since we assume that $\tilde{\mathbf{v}}^\star$ is known a priori, Equations (3.5.1-13) and (3.5.1-21) are to be solved simultaneously for the two unknowns $v_1^\star$ and $v_2^{\star\star}$. These equations are commonly referred to as the *boundary-layer equations* for plane flow past a flat plate at zero incidence.

In the next section, we carry through to a solution the ideas developed here for flow past a flat plate.

For an alternative view, consider a perturbation solution for the differential mass and momentum balances, in which the perturbation parameter is taken to be $N_{Re}^{-1}$. This case is a *singular perturbation* problem, in the sense that all of the second-order terms drop out of the differential momentum balance, with the result that the no-slip boundary condition can no longer be satisfied. This nonviscous flow describes the zeroth-order term in an asymptotic expansion in terms of $N_{Re}^{-1}$ for the *outer problem* (outside the immediate neighborhood of a phase boundary), where viscous effects are negligible. To describe the *inner problem* (the boundary layer), variables are rescaled in such a way that not all of the viscous terms are lost at the zeroth order. The asymptotic expansions for the inner and outer problems are matched in a region where both are valid. For more on the theory of matched asymptotic expansions, see Cole (1968).

**Exercise 3.5.1-1** *Power-law fluid in flow past a flat plate*    An incompressible fluid that is described by the power-law fluid (see Section 2.3.3) is plane flow past a flat plate. Construct an argument that suggests that, in the limit where modified Reynolds number $N_{Re,Pl} \gg 1$, the appropriate boundary-layer equations are (3.5.1-13) and

$$\frac{1}{N_{St}}\frac{\partial v_1^\star}{\partial t^\star} + v_1^\star \frac{\partial v_1^\star}{\partial z_1^\star} + v_2^{\star\star}\frac{\partial v_1^\star}{\partial z_2^{\star\star}} = \frac{1}{N_{St}}\frac{\partial \tilde{v}_1^\star}{\partial t^\star} + \tilde{v}_1^\star\frac{\partial \tilde{v}_1^\star}{\partial z_1^\star} + \frac{\partial}{\partial z_2^{\star\star}}\left[ \left| \frac{\partial v_1^\star}{\partial z_2^{\star\star}} \right|^{n-1} \frac{\partial v_1^\star}{\partial z_2^{\star\star}} \right]$$

Here

$$z_2^{\star\star} = (N_{Re,Pl})^{1/(n+1)} z_2^\star$$

$$v_2^{\star\star} = (N_{Re,Pl})^{1/(n+1)} v_2^\star$$

and

$$N_{Re,Pl} \equiv \frac{\rho v_0^{\,2-n} L_0^{\,n}}{m}$$

As before, $v_0$ and $L_0$ represent a magnitude of velocity and a length, respectively, that are characteristic of the flow.

For further reading on boundary-layer flows of viscoelastic fluids, see Schowalter (1960), Acrivos et al. (1960), Yau and Tien (1963), Gutfinger and Shinnar (1964), White and Metzner (1965), Acrivos et al. (1965), and Hermes and Fredrickson (1967).

### 3.5.2    More on Plane Flow Past a Flat Plate

We wish to discuss a particular case to which the theory of the preceding section is applicable: steady-state plane flow past a flat plate at zero incidence of an incompressible Newtonian fluid.

Let us assume that the velocity field outside the boundary layer might be representable in terms of a potential as Figures 3.5.1-1 and 3.5.1-2 suggest. We neglect any displacement of the potential flow by the boundary layer and by the thin plate. Consequently, the velocity field outside the boundary layer is uniform in magnitude and direction, the positive $z_1$ direction in Figure 3.5.1-2.

From Section 3.5.1, the equations to be solved for the velocity distribution in the boundary layer become

$$\frac{\partial v_1^\star}{\partial z_1^\star} + \frac{\partial v_2^{\star\star}}{\partial z_2^{\star\star}} = 0 \tag{3.5.2-1}$$

and

$$v_1^\star \frac{\partial v_1^\star}{\partial z_1^\star} + v_2^{\star\star}\frac{\partial v_1^\star}{\partial z_2^{\star\star}} = \frac{\partial^2 v_1^\star}{\partial z_2^{\star\star 2}} \tag{3.5.2-2}$$

The corresponding boundary conditions are

$$\text{at } z_2^{\star\star} = 0: \quad v_1^\star = v_2^{\star\star} = 0 \tag{3.5.2-3}$$

and

$$\text{as } z_2^{\star\star} \to \infty: \quad v_1^\star \to 1 \tag{3.5.2-4}$$

Here

$$
\begin{aligned}
v_1^\star &\equiv \frac{v_1}{\tilde{v}_1} \\
v_2^{\star\star} &\equiv \sqrt{N_{Re}}\,\frac{v_2}{\tilde{v}_1} \\
z_1^\star &\equiv \frac{z_1}{L} \\
z_2^{\star\star} &\equiv \sqrt{N_{Re}}\,\frac{z_2}{L}
\end{aligned}
\tag{3.5.2-5}
$$

where $L$ is the length of the plate and

$$N_{Re} \equiv \frac{\tilde{v}_1 L \rho}{\mu} \tag{3.5.2-6}$$

We will notice later that the quantity $L$ drops out of the solution, which is consistent with our intuitive view that the solution we obtain should be valid for the semi-infinite flat plate.

Equation (3.5.2-1) may be satisfied identically by expressing the velocity components in terms of a stream function as suggested in Section 1.3.7. Referring to Table 2.4.2-1, we introduce a dimensionless stream function $\psi$ by requiring

$$v_1^\star = \frac{\partial \psi}{\partial z_2^{\star\star}} \tag{3.5.2-7}$$

and

$$v_2^{\star\star} = -\frac{\partial \psi}{\partial z_1^\star} \tag{3.5.2-8}$$

Introduction of the stream function into (3.5.2-2) yields

$$\frac{\partial \psi}{\partial z_2^{\star\star}}\frac{\partial^2 \psi}{\partial z_1^\star \partial z_2^{\star\star}} - \frac{\partial \psi}{\partial z_1^\star}\frac{\partial^2 \psi}{\partial z_2^{\star\star2}} = \frac{\partial^3 \psi}{\partial z_2^{\star\star3}} \tag{3.5.2-9}$$

We must solve one partial differential equation for $\psi$, (3.5.2-9). It sometimes helps in a situation like this to make a change of independent and dependent variables with the aim of transforming the partial differential equation into an ordinary differential equation. For example, if we postulate here that

$$\psi = \sqrt{z_1^\star}\,f(\eta) \tag{3.5.2-10}$$

where

$$\eta \equiv \frac{z_2^{\star\star}}{\sqrt{z_1^\star}} \tag{3.5.2-11}$$

then

$$v_1^\star \equiv \frac{v_1}{\tilde{v}_1}$$

$$= f'$$

$$v_2^{\star\star} \equiv \sqrt{N_{Re}}\,\frac{v_2}{\tilde{v}_1} \tag{3.5.2-12}$$

$$= \frac{1}{2\sqrt{z_1^\star}}(\eta f' - f)$$

and Equation (3.5.2-9) becomes

$$f f'' + 2 f''' = 0 \tag{3.5.2-13}$$

The primes denote differentiation with respect to $\eta$. Boundary conditions (3.5.2-3) and (3.5.2-4) may be expressed as

$$\text{at } \eta = 0: \ f = f' = 0 \tag{3.5.2-14}$$

and

$$\text{as } \eta \to \infty: \ f' \to 1 \tag{3.5.2-15}$$

Note that $L$ has dropped out of the definition for $\eta$ and out of the solution for this problem. As we suggested earlier, this is to be expected, since this geometry has no inherent characteristic length.

We have arrived at a completely defined mathematical problem, since (3.5.2-13) is a third-order ordinary differential equation to be solved consistent with the two boundary conditions (3.5.2-14) and (3.5.2-15). Notice that we never said anything about $v_2^{\star\star}$ as $z_2^{\star\star} \to \infty$; it is different from zero as the edge of the boundary layer is approached (Schlichting 1979, Fig. 7.8) and is determined by the other conditions placed upon the flow.

This problem was originally solved by Blasius (Blasius 1908; Schlichting 1979, p. 135), though one of the most accurate solutions was given by Howarth (1938). Today, Equation (3.5.2-13) is readily solved numerically consistent with boundary conditions (3.5.2-14) and (3.5.2-15). The results for $v_1^\star$ and $\sqrt{z_1^\star}v_2^{\star\star}$ are shown in Figures 3.5.2-1 and 3.5.2-2. Note that, as suggested in Section 3.5.1, $v_2^{\star\star}$ does not approach zero as $\eta \to \infty$.

There have been at least two particularly notable experimental studies confirming these results. Figure 3.5.2-1 shows the velocity distribution measured by Nikuradse (1942). Figure 3.5.2-3 compares the dimensionless stress on the wall as a function of Reynolds number:

$$S_{12}^\star \equiv \frac{S_{12}}{\rho \tilde{v}_1^2}$$

$$= N_{Re}^{-1/2}\,\frac{dv_1^\star}{dz_2^{\star\star}}\bigg|_{z_2^{\star\star}=0}$$

$$= N_{Re}z_1^{\star-1/2}\,f''\big|_{\eta=0}$$

$$= 0.332 N_{Re}z_1^{\star-1/2} \tag{3.5.2-16}$$

**Exercise 3.5.2-1**   Given (3.5.2-10) and (3.5.2-11), derive (3.5.2-13) and boundary conditions (3.5.2-14) and (3.5.2-15).

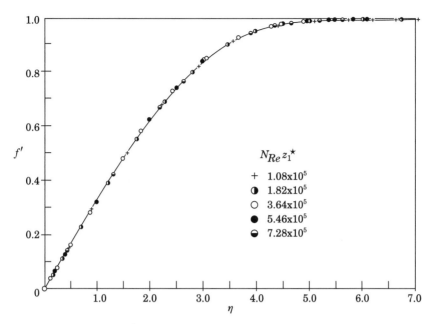

**Figure 3.5.2-1.** $v_1^\star = f'$ as a function of $\eta$ for boundary-layer flow past a flat plate, with experimental data by Nikuradse (1942). Taken from Schlichting (1979, Fig. 7.9).

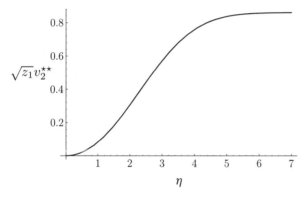

**Figure 3.5.2-2.** $\sqrt{z_1^\star}\, v_2^{\star\star}$ as a function of $\eta$ for boundary-layer flow past a flat plate.

**Exercise 3.5.2-2** Starting with

$$\psi = \left(z_1^\star\right)^a f(\eta)$$

and

$$\eta \equiv \left(z_1^\star\right)^b \left(z_2^{\star\star}\right)^c$$

show how one might argue to arrive at the transformations indicated by (3.5.2-10) and (3.5.2-11).

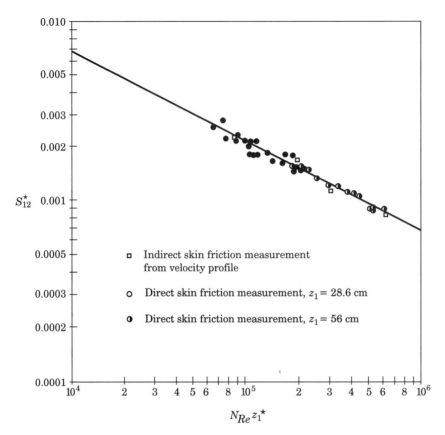

**Figure 3.5.2-3.** $S_{12}^\star$ as a function of $N_{Re}z_1^\star$ for boundary-layer flow past a flat plate, with experimental data by Liepmann and Dhawan (1951) and by Dhawan (1953). Taken from Schlichting (1979, Fig. 7.10).

**Exercise 3.5.2-3** *Boundary layer external to an extruding polymer film* (Sakiadis 1961a, b)    In Exercise 3.3.4-1, we examined extrusion of a polymer film. In preparation for later discussions of energy and mass transfer from an extruding film, let us construct the velocity and pressure distributions for the boundary layer created in the adjacent gas. You may neglect any motion in the gas very far away from the film, you may assume that the polymer enters the gas through a slit in a solid wall at $z_1 = 0$, and, for the purposes of this discussion, you may assume that the polymer–air phase interfaces are parallel planes that do not stretch. Compare your results with those for flow past a flat plate as described in the text.

**Exercise 3.5.2-4** *More on power-law fluid in flow past a flat plate*    Complete the discussion of boundary-layer flow of a power-law fluid that was begun in Exercise 3.5.1-1.

*Hint:*    In place of (3.5.2-10) and (3.5.2-11), define

$$\psi = z_1^{\star 1/(n+1)} f(\eta)$$

$$\eta \equiv \frac{z_2^{\star\star}}{z_1^{\star 1/(n+1)}}$$

**Exercise 3.5.2-5** *Thickness of film flowing down an inclined plate* A liquid film exposed to air flows down an inclined plane. At the entrance to this flow, $v_1 = V = $ a constant and the volume rate of flow per unit width of plate is $Q$. Assume that the film thickness changes as the result of the development of a boundary layer in the liquid.

i) Determine the film thickness $h$ as a function of $z_1$ measured from the entrance.
ii) Estimate the distance measured from the entrance at which the film profile is fully developed (see Exercise 3.2.5-5).

### 3.5.3 Plane Flow Past a Curved Wall

In Section 3.5.1, we discussed plane flow past a flat plate. In the limit, for $N_{Re} \gg 1$, we found that the Navier–Stokes equation may be simplified for the boundary layer. This simplification involved neglecting some (though not all) of the viscous terms with respect to the convective inertial terms.

In this section we wish to consider the boundary layer formed by an incompressible Newtonian fluid in plane flow past a curved wall. Intuitively, it seems clear that the fluid in a sufficiently thin boundary layer behaves in the same manner whether the wall is curved or flat. Our object here is to show in what sense this intuitive feeling is correct.

A portion of this wall is shown in Figure 3.5.3-1. With respect to the rectangular Cartesian coordinate system indicated, the equation of this surface is

$$z_2 = f(z_1) \tag{3.5.3-1}$$

To better compare flow past a curved wall with flow past a flat plate, let us view this problem in terms of an orthogonal curvilinear coordinate system such that

$x \equiv x^1$ is defined to be arc length measured along the wall in a plane of constant $z$,
$y \equiv x^2$ is defined to be arc length measured along straight lines that are normal to the wall,
$z \equiv x^3 \equiv z_3$ is the coordinate normal to the plane of flow.

By *plane flow* we mean here that

$$
\begin{aligned}
v_x &\equiv v_{(1)} \\
&= v_x(x, y, t) \\
v_y &\equiv v_{(2)} \\
&= v_y(x, y, t) \\
v_z &\equiv v_{(3)} \\
&= 0
\end{aligned}
\tag{3.5.3-2}
$$

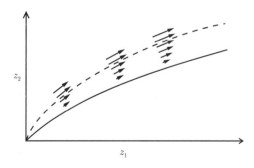

**Figure 3.5.3-1.** Plane flow past a curved wall.

With these restrictions, the differential mass balance and the three components of the Navier–Stokes equation may be written as (Goldstein 1938, p. 119; see also Exercise 3.5.3-1)

$$\frac{\partial v_x}{\partial x} + (1 + \kappa y)\frac{\partial v_y}{\partial y} + \kappa v_y = 0 \tag{3.5.3-3}$$

$$(1 + \kappa y)\frac{\partial v_x}{\partial t} + v_x\frac{\partial v_x}{\partial x} + (1 + \kappa y)v_y\frac{\partial v_x}{\partial y} + \kappa v_x v_y$$

$$= -\frac{1}{\rho}\frac{\partial \mathcal{P}}{\partial x} + \frac{\mu}{\rho}\left[\frac{1}{1+\kappa y}\frac{\partial^2 v_x}{\partial x^2} + \frac{v_y}{(1+\kappa y)^2}\frac{\partial \kappa}{\partial x} + \frac{2\kappa}{1+\kappa y}\frac{\partial v_y}{\partial x}\right.$$

$$\left. - \frac{y}{(1+\kappa y)^2}\frac{d\kappa}{dx}\frac{\partial v_x}{\partial x} - \frac{\kappa^2}{1+\kappa y}v_x + \kappa\frac{\partial v_x}{\partial y} + (1+\kappa y)\frac{\partial^2 v_x}{\partial y^2}\right] \tag{3.5.3-4}$$

$$\frac{\partial v_y}{\partial t} + \frac{v_x}{1+\kappa y}\frac{\partial v_y}{\partial x} + v_y\frac{\partial v_y}{\partial y} - \frac{\kappa}{1+\kappa y}v_x{}^2$$

$$= -\frac{1}{\rho}\frac{\partial \mathcal{P}}{\partial y} + \frac{\mu}{\rho}\left[\frac{1}{(1+\kappa y)^2}\frac{\partial^2 v_y}{\partial x^2} - \frac{y}{(1+\kappa y)^3}\frac{d\kappa}{dx}\frac{\partial v_y}{\partial x} + \frac{\kappa}{1+\kappa y}\frac{\partial v_y}{\partial y}\right.$$

$$\left. - \frac{2\kappa}{(1+\kappa y)^2}\frac{\partial v_y}{\partial x} - \frac{\kappa^2}{(1+\kappa y)^2}v_y - \frac{v_x}{(1+\kappa y)^3}\frac{d\kappa}{dx} + \frac{\partial^2 v_y}{\partial y^2}\right] \tag{3.5.3-5}$$

and

$$\frac{\partial \mathcal{P}}{\partial z} = 0 \tag{3.5.3-6}$$

where we define

$$\kappa \equiv \frac{-f''}{[1 + (f')^2]^{3/2}}$$
$$= \kappa(x) \tag{3.5.3-7}$$

The primes here indicate differentiation with respect to $z_1$.

In addition to the dimensionless velocity, dimensionless modified pressure, and dimensionless time introduced at the beginning of Section 3.3, let us define

$$x^\star \equiv \frac{x}{L}$$
$$y^\star \equiv \frac{y}{L} \tag{3.5.3-8}$$
$$\kappa^\star \equiv \kappa L$$

If we extend the arguments in Section 3.5.1 to this geometry, we are motivated to express our results for the boundary layer in terms of

$$y^{\star\star} \equiv N_{Re}{}^a y^\star$$
$$v_y^{\star\star} \equiv N_{Re}{}^a v_y^\star \tag{3.5.3-9}$$

If any viscous terms in the Navier–Stokes equation are to be of the same order of magnitude as the convective inertial terms for $N_{Re} \gg 1$, we are again forced to specify that

$$a = \frac{1}{2}$$

With this understanding, for $N_{Re} \gg 1$, Equations (3.5.3-3), (3.5.3-4), and (3.5.3-5) reduce to

$$\frac{\partial v_x^\star}{\partial x^\star} + (1 + \kappa^{\star\star} y^{\star\star}) \frac{\partial v_y^{\star\star}}{\partial y^{\star\star}} + \kappa^{\star\star} v_y^{\star\star} = 0 \tag{3.5.3-10}$$

$$\frac{1}{N_{St}}(1 + \kappa^{\star\star} y^{\star\star}) \frac{\partial v_x^\star}{\partial t^\star} + v_x^\star \frac{\partial v_x^\star}{\partial x^\star} + (1 + \kappa^{\star\star} y^{\star\star}) v_y^{\star\star} \frac{\partial v_x^\star}{\partial y^{\star\star}} + \kappa^{\star\star} v_x^\star v_y^{\star\star}$$

$$= -\frac{1}{N_{St}} \frac{\partial \mathcal{P}}{\partial x^\star} - \frac{\kappa^{\star\star 2}}{1 + \kappa^{\star\star} y^{\star\star}} v_x^\star + \kappa^{\star\star} \frac{\partial v_x^\star}{\partial y^{\star\star}} + (1 + \kappa^{\star\star} y^{\star\star}) \frac{\partial^2 v_x^\star}{\partial y^{\star\star 2}} \tag{3.5.3-11}$$

and

$$\frac{\kappa^{\star\star}}{1 + \kappa^{\star\star} y^{\star\star}} v_x^{\star\star 2} = \frac{1}{N_{Ru}} \frac{\partial \mathcal{P}}{\partial y^{\star\star}} \tag{3.5.3-12}$$

where we define

$$\kappa^{\star\star} \equiv \kappa^\star N_{Re}^{-1/2} \tag{3.5.3-13}$$

For a fixed wall configuration, $\kappa^{\star\star} \to 0$ in the limit $N_{Re} \gg 1$. Equations (3.5.3-10), (3.5.3-11), and (3.5.3-12) simplify under these conditions to

$$\frac{\partial v_x^\star}{\partial x^\star} + \frac{\partial v_y^{\star\star}}{\partial y^{\star\star}} = 0 \tag{3.5.3-14}$$

$$\frac{1}{N_{St}} \frac{\partial v_x^\star}{\partial t^\star} + v_x^\star \frac{\partial v_x^\star}{\partial x^\star} + v_y^{\star\star} \frac{\partial v_x^\star}{\partial y^{\star\star}} = -\frac{1}{N_{Ru}} \frac{\partial \mathcal{P}^\star}{\partial x^\star} + \frac{\partial^2 v_x^\star}{\partial y^{\star\star 2}} \tag{3.5.3-15}$$

and

$$\frac{\partial \mathcal{P}^\star}{\partial y^{\star\star}} = 0 \tag{3.5.3-16}$$

Equations (3.5.3-6) and (3.5.3-16) mean that

$$\mathcal{P}^\star = \mathcal{P}^\star(x^\star) \tag{3.5.3-17}$$

It can be shown that $-\kappa = 2H$, where $H$ is the mean curvature of the surface; it is also the only nonzero principal curvature (Slattery 1990, pp. 1116 and 1119) of the surface.

As we suggested in Section 3.5.1,

$$\text{as } y^{\star\star} \to \infty : \quad v_x^\star \to \tilde{v}_x^\star \tag{3.5.3-18}$$

where $\tilde{v}_x^\star$ is the dimensionless $x$ component of velocity at the curved wall from the nonviscous flow solution. In view of (3.5.3-17), we conclude that we may identify

$$-\frac{1}{N_{Ru}} \frac{\partial \mathcal{P}^\star}{\partial x^\star} = -\frac{1}{N_{Ru}} \frac{d\tilde{\mathcal{P}}^\star}{dx^\star}$$

$$= \frac{1}{N_{St}} \frac{\partial \tilde{v}_x^\star}{\partial t^\star} + \tilde{v}_x^\star \frac{d\tilde{v}_x^\star}{dx^\star} \tag{3.5.3-19}$$

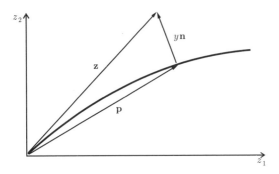

**Figure 3.5.3-2.** The relation among the position vector **p**, the unit normal **n**, and any point **z** in the boundary layer.

and write (3.5.3-15) as

$$\frac{1}{N_{St}}\frac{\partial v_x^\star}{\partial t^\star} + v_x^\star\frac{\partial v_x^\star}{\partial x^\star} + v_y^{\star\star}\frac{\partial v_x^\star}{\partial y^{\star\star}} = \frac{1}{N_{St}}\frac{\partial \tilde{v}_x^\star}{\partial t^\star} + \tilde{v}_x^\star\frac{d\tilde{v}_x^\star}{dx^\star} + \frac{\partial^2 v_x^\star}{\partial y^{\star\star 2}} \tag{3.5.3-20}$$

Since we assume that $\tilde{v}_x^\star$ is known a priori, (3.5.3-14) and (3.5.3-20) are to be solved simultaneously for the two unknowns $v_x^\star$ and $v_y^{\star\star}$. These equations are commonly referred to as the *boundary-layer equations* for plane flow past a curved wall. As our intuition suggested, they have the same form as the boundary-layer equations developed in Section 3.5.1 for plane flow past a flat plate.

**Exercise 3.5.3-1** *Derivation of (3.5.3-3) to (3.5.3-6)*

i) In Figure 3.5.3-2, **p** denotes the position vector of any point on the curved wall:

$$\mathbf{p} = z_1\mathbf{e}_1 + f(z_1)\mathbf{e}_2 + z_3\mathbf{e}_3$$

Determine that

$$\frac{dz_1}{dx} = [1 + (f')^2]^{-1/2}$$

where the prime denotes differentiation with respect to $z_1$.

ii) In Figure 3.5.3-2, **z** denotes the position vector of any point in the boundary layer:

$$\mathbf{z} = \mathbf{p} + y\mathbf{n}$$

By **n** we mean the unit normal to the wall directed into the fluid.

iii) For the curvilinear coordinate system described in the text of this section, find that

$$\mathbf{g}_1 \equiv \frac{\partial \mathbf{z}}{\partial x} = \frac{1 + \kappa y}{[1 + (f')^2]^{1/2}}(\mathbf{e}_1 + f'\mathbf{e}_2)$$

and

$$\mathbf{g}_2 \equiv \frac{\partial \mathbf{z}}{\partial y} = \mathbf{n} = \frac{-f'\mathbf{e}_1 + \mathbf{e}_2}{[1 + (f')^2]^{1/2}}$$

where $\kappa$ is defined by (3.5.3-7).

iv) Determine that

$$g_{11} = (1 + \kappa y)^2$$

and

$$g_{22} = g_{33} = 1$$

v) Show that the only nonzero Christoffel symbols of the second kind are

$$\begin{Bmatrix} 1 \\ 1 \ 1 \end{Bmatrix} = \frac{y}{1 + \kappa y} \frac{d\kappa}{dx}$$

$$\begin{Bmatrix} 1 \\ 1 \ 2 \end{Bmatrix} = \begin{Bmatrix} 1 \\ 2 \ 1 \end{Bmatrix} = \frac{\kappa}{1 + \kappa y}$$

$$\begin{Bmatrix} 2 \\ 1 \ 1 \end{Bmatrix} = -\kappa(1 + \kappa y)$$

vi) Derive (3.5.3-3) through (3.5.3-6).

**Exercise 3.5.3-2** *Derivation of (3.5.3-10) to (3.5.3-12)*

i) Introduce in (3.5.3-3) to (3.5.3-6) the dimensionless velocity, dimensionless modified pressure, and dimensionless time defined in the introduction to Section 3.3 as well as those dimensionless variables defined in (3.5.3-8).

ii) Express these equations in terms of the variables defined in (3.5.3-9) and construct the reasoning that leads to the definition $a = 1/2$ and to Equations (3.5.3-10) to (3.5.3-12).

**Exercise 3.5.3-3** *Generalized Newtonian fluids* Let us restrict our attention to incompressible generalized Newtonian fluids whose behavior can be represented by one of the simple empirical models discussed in Section 2.3.3:

$$\mathbf{S} = 2\eta(\gamma)\mathbf{D}$$

Let us further require $\eta(\gamma)$ to be a homogeneous function of degree $p$. By this we mean that, for a fixed value of $p$,

$$t^p \eta(\gamma) = \eta(t\gamma)$$

no matter what value $t$ assumes.

Argue that, for plane boundary-layer flow past a curved wall, the differential momentum balance requires

$$\frac{1}{N_{St}} \frac{\partial v_x^\star}{\partial t^\star} + v_x^\star \frac{\partial v_x^\star}{\partial x^\star} + v_y^{\star\star} \frac{\partial v_x^\star}{\partial y^{\star\star}}$$

$$= \frac{1}{N_{St}} \frac{\partial \tilde{v}_x^\star}{\partial t^\star} + \tilde{v}_x^\star \frac{\partial \tilde{v}_x^\star}{\partial x^\star} + \frac{\partial}{\partial y^{\star\star}} \left[ \eta^\star \left( \left| \frac{\partial v_x^\star}{\partial y^{\star\star}} \right| \right) \frac{\partial v_x^\star}{\partial y^{\star\star}} \right] \quad (3.5.3\text{-}21)$$

and

$$\frac{\partial \mathcal{P}^\star}{\partial y^{\star\star}} = 0$$

Here we have defined

$$\eta^\star \left( \left| \frac{\partial v_x^\star}{\partial y^{\star\star}} \right| \right) \equiv \frac{1}{\mu_0} (N_{Re})^{(-p)/(p+2)} \eta \left( \left| \frac{\partial v_x}{\partial y} \right| \right)$$

$$y^{\star\star} \equiv y^\star N_{Re}^{1/(p+2)}$$

$$v_y^{\star\star} \equiv v_y^\star N_{Re}^{1/(p+2)}$$

$$N_{Re} \equiv \frac{L_0 v_0 \rho}{\mu_0}$$

By $\mu_0$, we mean a viscosity characteristic of the fluid.

For a fluid described by the power-law model,

$$\eta \left( \left| \frac{\partial v_x}{\partial y} \right| \right) = m \left( \frac{v_0}{L_0} \right)^{n-1} (N_{Re\,PL})^{(n-1)/(n+1)} \left| \frac{\partial v_x^\star}{\partial y^{\star\star}} \right|^{n-1}$$

we see that

$$p = n - 1$$

and that

$$\eta^\star \left( \left| \frac{\partial v_x^\star}{\partial y^{\star\star}} \right| \right) \equiv \left| \frac{\partial v_x^\star}{\partial y^{\star\star}} \right|^{n-1}$$

If we define

$$\mu_0 \equiv m \left( \frac{v_0}{L_0} \right)^{n-1}$$

then

$$N_{Re\,PL} \equiv \frac{\rho v_0^{2-n} L_0^n}{m}$$

With these definitions, (3.5.3-21) is consistent with the result found in Exercise 3.5.1-1 for flow past a flat plate.

### 3.5.4 Solutions Found by Combination of Variables

In Sections 3.5.1 and 3.5.3, we found that the same differential equations govern plane boundary-layer flow past either a flat wall or a curved wall: (3.5.3-14) and (3.5.3-20). Although (3.5.3-20) is a simplification of the Navier–Stokes equations, it is still a nonlinear partial differential equation. Our object here is to determine the external flows for which the solution of these equations reduces to the solution of a single ordinary differential equation. We shall focus our attention upon steady-state flows, following the discussion given by Schlichting (1979, p. 152).

As discussed in Section 1.3.7, Equation (3.5.3-14) may be satisfied identically by expressing the velocity components in terms of a stream function. Referring to Table 2.4.2-1, we introduce a dimensionless stream function $\psi$ by requiring

$$v_x^\star = \frac{\partial \psi}{\partial y^{\star\star}} \qquad (3.5.4\text{-}1)$$

and

$$v_y^{\star\star} = -\frac{\partial \psi}{\partial x^\star} \tag{3.5.4-2}$$

These expressions allow us to write (3.5.3-20) for a steady-state flow as

$$\frac{\partial \psi}{\partial y^{\star\star}} \frac{\partial^2 \psi}{\partial x^\star \partial y^{\star\star}} - \frac{\partial \psi}{\partial x^\star} \frac{\partial^2 \psi}{\partial y^{\star\star 2}} = \tilde{v}_x^\star \frac{d\tilde{v}_x^\star}{dx^\star} + \frac{\partial^3 \psi}{\partial y^{\star\star 3}} \tag{3.5.4-3}$$

Since it is assumed that $\tilde{v}_x^\star$ is known a priori, (3.5.4-3) is a nonlinear partial differential equation which must be solved for $\psi$. In what follows we find that there are several classes of problems for which this equation may be expressed as an ordinary differential equation.

Our approach is to introduce new independent and dependent variables defined as

$$\eta \equiv \frac{y^{\star\star}}{g(x^\star)} \tag{3.5.4-4}$$

and

$$f(\eta) \equiv \frac{\psi}{h(x^\star)} \tag{3.5.4-5}$$

The functions $g$ and $h$ are to be determined by the requirement that (3.5.4-3) be expressed as an ordinary differential equation for $f$ as a function of $\eta$. In terms of these variables, (3.5.4-3) becomes

$$f''' + g\frac{dh}{dx^\star}ff'' - \left(g\frac{dh}{dx^\star} - h\frac{dg}{dx^\star}\right)f'^2 + \frac{g^3}{h}\tilde{v}_x^\star\frac{d\tilde{v}_x^\star}{dx^\star} = 0 \tag{3.5.4-6}$$

where the prime denotes differentiation with respect to $\eta$. This equation suggests that we define $h$ by requiring

$$g\frac{dh}{dx^\star} - h\frac{dg}{dx^\star} = \frac{g^3}{h}\tilde{v}_x^\star\frac{d\tilde{v}_x^\star}{dx^\star} \tag{3.5.4-7}$$

This may easily be rearranged to read

$$\frac{d}{dx^\star}\left(\frac{h^2}{g^2}\right) = \frac{d\tilde{v}_x^{\star 2}}{dx^\star} \tag{3.5.4-8}$$

from which we have the definition

$$h \equiv g\tilde{v}_x^\star \tag{3.5.4-9}$$

This allows us to express (3.5.4-6) as

$$f''' + \alpha ff'' + \beta[1 - f'^2] = 0 \tag{3.5.4-10}$$

where

$$\alpha \equiv g\frac{d}{dx^\star}\left(g\tilde{v}_x^\star\right) \tag{3.5.4-11}$$

and

$$\beta \equiv g^2\frac{d\tilde{v}_x^\star}{dx^\star} \tag{3.5.4-12}$$

In order that $f$ be a function of $x^\star$ only through its dependence upon $\eta$, we must require that $\alpha$ and $\beta$ be independent of $x^\star$. These two requirements determine the functions $g$ and $\tilde{v}_x^\star$. When we determine the function $\tilde{v}_x^\star$, we specify those geometries for which (3.5.4-3) may be reduced to an ordinary differential equation by the changes of variables defined in (3.5.4-4) and (3.5.4-5).

## Case 1: $2\alpha - \beta \neq 0$

From (3.5.4-11) and (3.5.4-12) we have that

$$2\alpha - \beta = \frac{d}{dx^\star}\left(g^2\tilde{v}_x^\star\right) \tag{3.5.4-13}$$

and

$$\alpha - \beta = g\tilde{v}_x^\star \frac{dg}{dx^\star} \tag{3.5.4-14}$$

This latter equation may be rearranged as

$$\frac{\alpha - \beta}{\tilde{v}_x^\star}\frac{d\tilde{v}_x^\star}{dx^\star} = g\frac{d\tilde{v}_x^\star}{dx^\star}\frac{dg}{dx^\star}$$

$$= \frac{\beta}{g}\frac{dg}{dx^\star} \tag{3.5.4-15}$$

and integrated to give

$$\tilde{v}_x^{\star\,\alpha-\beta} = Kg^\beta \tag{3.5.4-16}$$

where $K$ is a constant of integration. Equation (3.5.4-13) may also be integrated to yield

$$(2\alpha - \beta)x^\star = g^2\tilde{v}_x^\star + C \tag{3.5.4-17}$$

Here $C$ indicates a constant of integration. We may eliminate $g$ between (3.5.4-16) and (3.5.4-17) to obtain

$$(2\alpha - \beta)x^\star = \frac{\tilde{v}_x^{\star\,(2\alpha-\beta)/\beta}}{K^{2/\beta}} + C \tag{3.5.4-18}$$

If we assume that $2\alpha - \beta \neq 0$, this result may be solved for $\tilde{v}_x^\star$:

$$\tilde{v}_x^\star = K^{2/(2\alpha-\beta)}[(2\alpha - \beta)x^\star - C]^{\beta/(2\alpha-\beta)} \tag{3.5.4-19}$$

The corresponding form for $g$ may be obtained by eliminating $\tilde{v}_x^\star$ between (3.5.4-16) and (3.5.4-19):

$$g = K^{-1/(2\alpha-\beta)}[(2\alpha - \beta)x^\star - C]^{(\alpha-\beta)/(2\alpha-\beta)} \tag{3.5.4-20}$$

## Case 1a: $\alpha \neq 0$, $2\alpha - \beta \neq 0$

It is clear from (3.5.4-19) that the result is independent of any common factor of $\alpha$ and $\beta$, since any common factor may be included in $g$. We may assume that $\alpha \neq 0$ and put $\alpha = 1$ without loss of generality. Equations (3.5.4-19) and (3.5.4-20) then become

$$\tilde{v}_x^\star = K^{2/(2-\beta)}[(2 - \beta)x^\star - C]^{\beta/(2-\beta)} \tag{3.5.4-21}$$

and

$$g = K^{-1/(2-\beta)}[(2-\beta)x^\star - C]^{(1-\beta)/(2-\beta)} \tag{3.5.4-22}$$

If we define

$$m \equiv \frac{\beta}{2-\beta} \tag{3.5.4-23}$$

these results take the somewhat simpler forms

$$\tilde{v}_x^\star = K^{1+m}\left[\left(\frac{2}{1+m}\right)x^\star - C\right]^m \tag{3.5.4-24}$$

and

$$g = K^{-(1+m)/2}\left[\left(\frac{2}{1+m}\right)x^\star - C\right]^{(1-m)/2} \tag{3.5.4-25}$$

When $C = 0$, Equation (3.5.4-24) describes the potential flow velocity distribution at the wall near the leading edge of a wedge whose included angle is (Schlichting 1979, p. 156)

$$\pi\beta = \frac{2\pi m}{1+m} \tag{3.5.4-26}$$

For $\beta = m = 1$, we have plane stagnation flow (see Section 3.4.3). For $\beta = m = 0$, we have flow past a flat plate at zero angle of incidence. For a summary of the solutions available describing flow past a wedge, see Schlichting (1979, p. 164) as well as Section 6.7.4.

**Case 1b: $\alpha = 0$, $2\alpha - \beta \neq 0$**

In the event that $\alpha = 0$, (3.5.4-19) and (3.5.4-20) become

$$\tilde{v}_x^\star = K^{-2/\beta}(-\beta x^\star - C)^{-1} \tag{3.5.4-27}$$

and

$$g = K^{1/\beta}(-\beta x^\star - C) \tag{3.5.4-28}$$

It is convenient to take $\beta = \pm 1$, depending upon the sign of $d\tilde{v}_x^\star/dx^\star$ in (3.5.4-12).

When $C = 0$, Equation (3.5.4-27) may be interpreted as flow in either a flat-walled convergent or divergent channel (see Exercise 3.4.2-5). For a convergent channel, $\tilde{v}_x^\star$ is negative and $d\tilde{v}_x^\star/dx^\star$ is positive; (3.5.4-12) and (3.5.4-27) require that we define $\beta = +1$. For a divergent channel, $\tilde{v}_x^\star$ is positive and $d\tilde{v}_x^\star/dx^\star$ is negative; $\beta = -1$. Flow in a convergent channel is discussed in Section 3.5.5.

The full Navier–Stokes equations have been solved for plane flow in convergent and divergent channels with flat walls. Goldstein (1938, p. 105) and Schlichting (1979, p. 107) give interesting discussions of the available solutions.

**Case 2: $2\alpha - \beta = 0$**

In the event that $2\alpha - \beta = 0$, we have from (3.5.4-13) that

$$\tilde{v}_x^\star = \frac{K}{g^2} \tag{3.5.4-29}$$

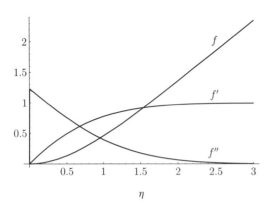

**Figure 3.5.4-1.** For plane stagnation flow, $f$, $f'$, and $f''$ as functions of $\eta$.

where $K$ is a constant of integration. Equation (3.5.4-14), consequently, reduces to

$$-\alpha = \frac{K}{g} \frac{dg}{dx^\star} \tag{3.5.4-30}$$

which may be integrated to give

$$g = \exp\left(-\frac{\alpha}{K}x^\star + C\right) \tag{3.5.4-31}$$

Here $C$ denotes another constant of integration. This last expression allows us to write (3.5.4-29) as

$$\tilde{v}_x^\star = \frac{K}{\exp(-[2\alpha/K]x^\star + 2C)} \tag{3.5.4-32}$$

**Exercise 3.5.4-1** *Plane stagnation flow*   Determine (numerically) the velocity distribution for plane stagnation flow (see also Section 3.4.3) of an incompressible Newtonian fluid in the neighborhood of the origin. This problem was first solved by Hiemenz (1911) and later by Howarth (1934). The results are shown in Figure 3.5.4-1.

### 3.5.5  Flow in a Convergent Channel

Consider the steady-state flow of an incompressible Newtonian fluid in a convergent channel. The convergent channel should be thought of as two plane walls oriented in such a way that they form surfaces of constant $\theta$ in cylindrical coordinates; the fluid moves in the negative $r$ direction. This discussion is only approximately applicable to a real channel of finite length, since we have made no attempt to account for end effects.

In Exercise 3.4.2-5, we analyzed the corresponding potential flow and found

$$\begin{aligned} v_r^\star &= -\frac{1}{r^\star} \\ v_\theta^\star &= v_z^\star = 0 \end{aligned} \tag{3.5.5-1}$$

In this section, we wish to solve the corresponding boundary-layer flow pictured in Figure 3.5.5-1. One reason for discussing this particular flow is that it is a rare case where an analytic solution to the boundary-layer equations can be determined.

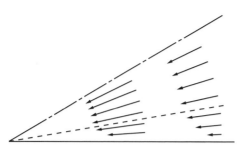

**Figure 3.5.5-1.** Flow in the neighborhood of
the wall in a convergent channel.

In the notation of Section 3.5.3, the potential flow described by (3.5.5-1) specifies the
condition that the boundary-layer equation must satisfy at infinity:

$$\text{as } y^{\star\star} \to \infty: \quad \tilde{v}_x^\star \to -\frac{1}{x^\star} \tag{3.5.5-2}$$

All of the boundary layers we have considered up to this point have developed downstream
of a leading edge, entrance, or stagnation point. This is not true here. The coordinate $x^\star$ is
measured from the imaginary sink for a convergent channel rather than from the leading
edge of a finite channel. But the interpretation of the boundary layer is the same. In a flow
for which the Reynolds number is large, it is that region (normally within the immediate
neighborhood of a phase interface) in which viscous effects cannot be neglected with respect
to inertial effects.

We say in Section 3.5.4 (Case 1b) that boundary-layer flow in a convergent channel is one
of the situations for which the boundary-layer equations may be simplified by the method
of combination of variables. Having expressed the dimensionless velocity components in
terms of a stream function $\psi$,

$$v_x^\star = \frac{\partial \psi}{\partial y^{\star\star}} \tag{3.5.5-3}$$

$$v_y^{\star\star} = -\frac{\partial \psi}{\partial x^\star} \tag{3.5.5-4}$$

we find that

$$\begin{aligned}
\psi &= h(x^\star)f(\eta) \\
&= g\tilde{v}_x^\star f(\eta) \\
&= K^{-1}f(\eta)
\end{aligned} \tag{3.5.5-5}$$

where $f = f(\eta)$ is a solution to

$$f''' - f'^2 + 1 = 0 \tag{3.5.5-6}$$

The primes in this last expression indicate differentiation with respect to $\eta$. On comparison
of (3.5.5-2) with the expression for $\tilde{v}_x^\star$ appropriate to this case in Section 3.5.4, we find that
the constant $K$ in (3.5.5-5) must be either $+1$ or $-1$. The variable $\eta$ is defined as

$$\eta \equiv \frac{y^{\star\star}}{g} = -\frac{y^{\star\star}}{Kx^\star} = \frac{y^{\star\star}}{x^\star} \tag{3.5.5-7}$$

where we have chosen to define

$$K \equiv -1 \tag{3.5.5-8}$$

In addition to (3.5.5-2), the boundary conditions to be satisfied by this boundary-layer flow are

$$\text{at } y^{**} = 0: \quad v_x^* = v_y^{**} = 0 \tag{3.5.5-9}$$

From (3.5.5-3), (3.5.5-4), (3.5.5-5), (3.5.5-7), and (3.5.5-8) we find that

$$v_x^* = -\frac{f'}{x^*} \tag{3.5.5-10}$$

and

$$v_y^{**} = -\frac{f'\eta}{x^*} \tag{3.5.5-11}$$

This means that boundary conditions (3.5.5-2) and (3.5.5-9) may be expressed as follows:

$$\text{as } \eta \to \infty: \quad f' \to 1 \tag{3.5.5-12}$$

and

$$\text{at } \eta = 0: \quad f' = 0 \tag{3.5.5-13}$$

If we multiply (3.5.5-6) by $2f''$, we may integrate it once:

$$2f'' f''' + 2f''(1 - f'^2) = 0$$
$$\frac{d}{d\eta}\left[ f''^2 - \frac{2}{3}(1 - f')^2(f' + 2) \right] = 0 \tag{3.5.5-14}$$
$$f''^2 - \frac{2}{3}(1 - f')^2(f' + 2) = a$$

Here, $a$ is a constant of integration. If our primary interest is in the velocity distribution, we may view this last as a separable first-order ordinary differential equation in $f'$ with two boundary conditions, (3.5.5-12) and (3.5.5-13). The problem as it stands is well posed mathematically, and there would be no particular difficulty in obtaining a numerical solution for $f'$.

However, with a little insight we may obtain an analytic expression for $f'$ (Schlichting 1979, p. 166). Our discussion of boundary conditions in Sections 3.5.1 and 3.5.3 suggests that

$$\text{as } y^{**} \to \infty: \quad \frac{\partial v_x^*}{\partial y^{**}} = -\frac{f''}{x^{*2}} \to 0 \tag{3.5.5-15}$$

or

$$\text{as } \eta \to \infty: \quad f'' \to 0 \tag{3.5.5-16}$$

Equations (3.5.5-12) and (3.5.5-16) suggest that we take

$$a = 0 \tag{3.5.5-17}$$

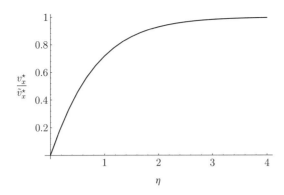

**Figure 3.5.5-2.** Velocity distribution for boundary-layer flow in a convergent channel.

which allows us to write (3.5.5-14) as

$$\frac{f''}{\sqrt{2}(1 - f')(f'/3 + 2/3)^{1/2}} = 1$$

$$\sqrt{2}\frac{d}{d\eta}\left[\text{arctanh}\left(f'/3 + 2/3\right)^{1/2}\right] = 1 \tag{3.5.5-18}$$

This last expression may be integrated once using boundary condition (3.5.5-13) to give

$$\int_0^\eta d\eta = \sqrt{2}\int_{f'=0}^{f'} d\left(\text{arctanh}\sqrt{f'/3 + 2/3}\right)$$

$$\eta = \sqrt{2}\left(\text{arctanh}\sqrt{f'/3 + 2/3} - \text{arctanh}\sqrt{2/3}\right) \tag{3.5.5-19}$$

This then satisfies boundary condition (3.5.5-12) as was originally required.

To repeat, (3.5.5-16) should not be viewed as a boundary condition for this problem. Rather, this is a condition that we suggested as being helpful in seeking an analytic solution to (3.5.5-14).

Equation (3.5.5-19) can be solved for $f'$:

$$\frac{v_x^\star}{\tilde{v}_x^\star} = f' = 3\,\tanh^2\left(\frac{\eta}{\sqrt{2}} + 1.146\right) - 2 \tag{3.5.5-20}$$

In arriving at this expression, we have noted that $\text{arctanh}\,(2/3)^{1/2} = 1.146$. Equation 3.5.5-20) is plotted in Figure 3.5.5-2.

### 3.5.6 Flow Past a Body of Revolution

In Sections 3.5.1 and 3.5.3, we discussed plane flow past a flat plate and plane flow past a curved wall. We found that the boundary layer could be described by the same set of equations in both cases.

In what follows, we consider the boundary layer formed by an incompressible Newtonian fluid flowing past a body of revolution. In some sense we expect the fluid in a sufficiently thin boundary layer to behave in the same manner on a body of revolution as it would on a flat wall. We wish to determine here in what sense this intuitive feeling is correct.

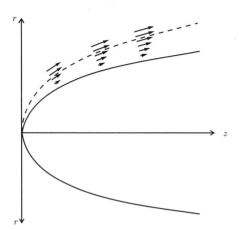

**Figure 3.5.6-1.** Flow past a body of revolution.

A plane section through the axis of symmetry of the body is shown in Figure 3.5.6-1. With respect to the cylindrical coordinate system indicated, the equation of the axially symmetric surface is

$$r = f(z) \tag{3.5.6-1}$$

To better compare flow past a body of revolution with flow past a flat plate, let us view this problem in terms of an orthogonal curvilinear coordinate system such that

$x \equiv x^3$ is defined to be arc length measured along the wall (in the direction of flow) in a plane of constant $\theta$,
$y \equiv x^1$ is defined to be arc length measured along straight lines that are normal to the wall,
$\theta \equiv x^2$ is the azimuthal cylindrical coordinate (measured around the axis of the body).

The shape of the wall suggests that

$$
\begin{aligned}
v_x &\equiv v_{(3)} \\
&= v_x(x, y, t) \\
v_y &\equiv v_{(1)} \\
&= v_y(x, y, t) \\
v_\theta &\equiv v_{(2)} \\
&= 0
\end{aligned}
\tag{3.5.6-2}
$$

With these restrictions, the differential mass balance and the three components of the Navier–Stokes equation may be written as (see Exercise 3.5.6-1)

$$\frac{1}{1 + \kappa y} \frac{\partial v_x}{\partial x} + \frac{1}{1 + \kappa y} \frac{\partial}{\partial y}[(1 + \kappa y)v_y] + \frac{f'}{g}v_x + \frac{1}{g}v_y = 0 \tag{3.5.6-3}$$

$$
(1 + \kappa y)\frac{\partial v_x}{\partial t} + v_x\frac{\partial v_x}{\partial x} + (1 + \kappa y)v_y\frac{\partial v_x}{\partial y} + \kappa v_x v_y = -\frac{1}{\rho}\frac{\partial \mathcal{P}}{\partial x}
$$
$$
+ \frac{\mu}{\rho}\left[\frac{1}{1 + \kappa y}\frac{\partial^2 v_x}{\partial x^2} + \frac{v_y}{(1 + \kappa y)^2}\frac{d\kappa}{dx} + \frac{2\kappa}{1 + \kappa y}\frac{\partial v_y}{\partial x}\right.
$$

$$
-\frac{y}{(1+\kappa y)^2}\frac{d\kappa}{dx}\frac{\partial v_x}{\partial x} - \frac{\kappa^2}{1+\kappa y}v_x + \kappa\frac{\partial v_x}{\partial y} + (1+\kappa y)\frac{\partial^2 v_x}{\partial y^2}
$$

$$
-\frac{f'(1+\kappa y)}{g^2}v_y + \frac{\kappa f'}{g}(1+f'^2)^{1/2}v_y
$$

$$
\left. +\frac{f'}{g}(1+f'^2)^{1/2}\frac{\partial v_x}{\partial x} + \frac{(1+\kappa y)}{g}\frac{\partial v_x}{\partial y}\right]
\tag{3.5.6-4}
$$

$$
\frac{\partial v_y}{\partial t} + \frac{v_x}{1+\kappa y}\frac{\partial v_y}{\partial x} + v_y\frac{\partial v_y}{\partial y} - \frac{\kappa}{1+\kappa y}v_x{}^2 = -\frac{1}{\rho}\frac{\partial \mathcal{P}}{\partial y}
$$

$$
+\frac{\mu}{\rho}\left[\frac{1}{(1+\kappa y)^2}\frac{\partial^2 v_y}{\partial x^2} - \frac{y}{(1+\kappa y)^3}\frac{d\kappa}{dx}\frac{\partial v_y}{\partial x} + \frac{\kappa}{1+\kappa y}\frac{\partial v_y}{\partial y}\right.
$$

$$
-\frac{2\kappa}{(1+\kappa y)^2}\frac{\partial v_x}{\partial x} - \frac{\kappa^2}{(1+\kappa y)^2}v_y - \frac{v_x}{(1+\kappa y)^3}\frac{d\kappa}{dx} + \frac{\partial^2 v_y}{\partial y^2}
$$

$$
\left. -\frac{v_y}{g^2} + \frac{1}{g}\frac{\partial v_y}{\partial y} + \frac{f'}{(1+\kappa y)g}(1+f'^2)^{1/2}\left(\frac{\partial v_y}{\partial x} - \kappa v_x\right)\right]
\tag{3.5.6-5}
$$

and

$$
\frac{\partial \mathcal{P}}{\partial \theta} = 0
\tag{3.5.6-6}
$$

where we define

$$
\kappa \equiv \frac{-f''}{\left(1+f'^2\right)^{1/2}} = \kappa(x)
\tag{3.5.6-7}
$$

and

$$
g \equiv f\left(1+f'^2\right)^{1/2} + y
\tag{3.5.6-8}
$$

We may refer to $-\kappa$ as one of the principal curvatures (Slattery 1990, p. 1119) of the surface. The primes here indicate differentiation with respect to the cylindrical coordinate $z$ measured along the axis of revolution.

In addition to the dimensionless velocity, dimensionless modified pressure, and dimensionless time introduced in the introduction to Section 3.3, let us define

$$
x^\star \equiv \frac{x}{L}, \quad y^\star \equiv \frac{y}{L}, \quad f^\star \equiv \frac{f}{L}, \quad \kappa^\star \equiv \kappa L
\tag{3.5.6-9}
$$

If we extend the arguments of Section 3.5.1 to this geometry, we are motivated to express our result in terms of

$$
y^{\star\star} \equiv N_{Re}{}^a y^\star \quad \text{and} \quad v_y^{\star\star} \equiv N_{Re}{}^a v_y^\star
\tag{3.5.6-10}
$$

To ensure that some viscous terms in the equation of motion have the same order of magnitude as the inertial terms for $N_{Re} \gg 1$, we are again forced to specify that $a = 1/2$. With this understanding, for $N_{Re} \gg 1$, Equations (3.5.3-3) to (3.5.3-5) reduce to

$$\frac{1}{1 + \kappa^{**} y^{**}} \frac{\partial v_x \star}{\partial x^{\star}} + \frac{1}{1 + \kappa^{**} y^{**}} \frac{\partial}{\partial y^{**}} \left[ (1 + \kappa^{**} y^{**}) v_y^{**} \right]$$

$$+ \frac{f^{\star\prime}}{f^{\star} \left( 1 + f^{\star\prime 2} \right)^{1/2}} v_x^{\star} = 0 \tag{3.5.6-11}$$

$$\frac{1}{N_{St}} (1 + \kappa^{**} y^{**}) \frac{\partial v_x^{\star}}{\partial t^{\star}} + v_x^{\star} \frac{\partial v_x^{\star}}{\partial x^{\star}} + (1 + \kappa^{**} y^{**}) v_y^{**} \frac{\partial v_x^{\star}}{\partial y^{**}} + \kappa^{**} v_x^{\star} v_y^{**}$$

$$= -\frac{1}{N_{Ru}} \frac{\partial \mathcal{P}^{\star}}{\partial x^{\star}} - \frac{\kappa^{**2}}{1 + \kappa^{**} y^{**}} v_x^{\star} + \kappa^{**} \frac{\partial v_x^{\star}}{\partial y^{**}} + (1 + \kappa^{**} y^{**}) \frac{\partial^2 v_x^{\star}}{\partial y^{**2}} \tag{3.5.6-12}$$

and

$$\frac{\kappa^{**}}{1 + \kappa^{**} y^{**}} v_x^{**2} = \frac{1}{N_{Ru}} \frac{\partial \mathcal{P}^{\star}}{\partial y^{**}} \tag{3.5.6-13}$$

where we define

$$\kappa^{**} \equiv \kappa^{\star} N_{Re}^{-1/2} \tag{3.5.6-14}$$

The primes in (3.5.6-11) now indicate differentiation with respect to $z^{\star}$.

For a fixed-wall configuration, $\kappa^{**} \to 0$ in the limit $N_{Re} \gg 1$. Equations (3.5.6-11) to (3.5.6-13) simplify under these conditions to

$$\frac{\partial v_x^{\star}}{\partial x^{\star}} + \frac{\partial v_y^{**}}{\partial y^{**}} + \frac{f^{\star\prime}}{f^{\star} \left( 1 + f^{\star\prime 2} \right)^{1/2}} v_x^{\star} = 0 \tag{3.5.6-15}$$

$$\frac{1}{N_{St}} \frac{\partial v_x^{\star}}{\partial t^{\star}} + v_x^{\star} \frac{\partial v_x^{\star}}{\partial x^{\star}} + v_y^{**} \frac{\partial v_x^{\star}}{\partial y^{**}} = -\frac{1}{N_{Ru}} \frac{\partial \mathcal{P}^{\star}}{\partial x^{\star}} + \frac{\partial^2 v_x^{\star}}{\partial y^{**2}} \tag{3.5.6-16}$$

and

$$\frac{\partial \mathcal{P}^{\star}}{\partial y^{**}} = 0 \tag{3.5.6-17}$$

Since (see Exercise 3.5.6-1)

$$\frac{dz^{\star}}{dx^{\star}} = \left( 1 + f^{\star\prime 2} \right)^{-1/2} \tag{3.5.6-18}$$

(3.5.6-15) may also be written as

$$\frac{\partial \left( f^{\star} v_x^{\star} \right)}{\partial x^{\star}} + \frac{\partial \left( f^{\star} v_y^{**} \right)}{\partial y^{**}} = 0 \tag{3.5.6-19}$$

Equations (3.5.6-6) and (3.5.6-17) mean that

$$\mathcal{P}^{\star} = \mathcal{P}^{\star}(x^{\star}) \tag{3.5.6-20}$$

As we suggested in Section 3.5.1,

$$\text{as } y^{**} \to \infty : \quad v_x^{\star} \to \tilde{v}_x^{\star} \tag{3.5.6-21}$$

where $\tilde{v}_x^\star$ is the dimensionless $x$ component of velocity at the curved wall from the nonviscous flow solution. In view of (3.5.6-20), we conclude that we may identify

$$-\frac{1}{N_{Ru}}\frac{\partial \mathcal{P}^\star}{\partial x^\star} = -\frac{1}{N_{Ru}}\frac{d\tilde{\mathcal{P}}^\star}{\partial x^\star}$$

$$= \tilde{v}_x^\star \frac{d\tilde{v}_x^\star}{dx^\star} \tag{3.5.6-22}$$

and write (3.5.6-16) as

$$\frac{1}{N_{St}}\frac{\partial v_x^\star}{\partial t^\star} + v_x^\star \frac{\partial v_x^\star}{\partial x^\star} + v_y^{\star\star}\frac{\partial v_x^\star}{\partial y^{\star\star}} = \tilde{v}_x^\star \frac{d\tilde{v}_x^\star}{dx^\star} + \frac{\partial^2 v_x^\star}{\partial y^{\star\star2}} \tag{3.5.6-23}$$

Since we assume that $\tilde{v}_x^\star$ is known a priori, (3.5.6-19) and (3.5.6-23) are to be solved simultaneously for the two unknowns $v_x^\star$ and $v_y^{\star\star}$. These equations may be referred to as the *boundary-layer equations* for flow past a body of revolution.

Mangler (1979, p. 245; see also Exercises 3.5.6-3 and 3.5.6-4) has suggested an interesting change of variables. Define

$$\bar{x}^\star \equiv \int_0^{x^\star} f^{\star 2}dx^\star \tag{3.5.6-24}$$

$$\bar{y}^{\star\star} \equiv f^\star y^{\star\star} \tag{3.5.6-25}$$

$$\bar{t}^\star \equiv f^{\star 2}t^\star \tag{3.5.6-26}$$

and

$$\bar{v}_y^{\star\star} \equiv \frac{1}{f^\star}v_y^{\star\star} + \frac{f^{\star\prime}}{f^{\star 2}\left(1 + f^{\star\prime 2}\right)^{1/2}}y^{\star\star}v_x^\star \tag{3.5.6-27}$$

With this change of variables, (3.5.6-19) and (3.5.6-23) become

$$\frac{\partial v_x^\star}{\partial x^\star} + \frac{\partial \bar{v}_y^{\star\star}}{\partial \bar{y}^{\star\star}} = 0 \tag{3.5.6-28}$$

and

$$\frac{1}{N_{St}}\frac{\partial v_x^\star}{\partial \bar{t}^\star} + v_x^\star \frac{\partial v_x^\star}{\partial \bar{x}^\star} + \bar{v}_y^{\star\star}\frac{\partial v_x^\star}{\partial \bar{y}^{\star\star}} = \tilde{v}_x^\star \frac{d\tilde{v}_x^\star}{dx^\star} + \frac{\partial^2 v_x^\star}{\partial \bar{y}^{\star\star2}} \tag{3.5.6-29}$$

These equations have the same form as the boundary-layer equations found in Sections 3.5.1 and 3.5.3. They may be referred to as *Mangler's boundary-layer equations* for flow past a body of revolution. It is in this sense that our original intuitive feelings are confirmed: The mathematical problems that describe flow past a body of revolution and plane flow past a curved wall have the same form.

**Exercise 3.5.6-1** *Derivation of Equations (3.5.6-3) to (3.5.6-6)*

i) In Figure 3.5.6-2, **p** denotes the position vector of any point on the surface of revolution. With respect to a cylindrical coordinate basis such that $z$ is measured along the axis of the body of revolution,

$$\mathbf{p} = f(z)\mathbf{g}_r + z\mathbf{g}_z$$

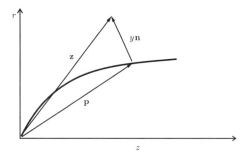

**Figure 3.5.6-2.** Relation between the position vectors for a point on the surface and the corresponding point in the fluid.

With respect to a rectangular Cartesian coordinate basis such that $z_3 \equiv z$,

$$\mathbf{p} = [f(z_3)\cos\theta]\mathbf{e}_1 + [f(z_3)\sin\theta]\mathbf{e}_3 + z_3\mathbf{e}_3$$

Determine that

$$\frac{dz_3}{dx} = \left(1 + f'^2\right)^{-1/2}$$

where the prime denotes differentiation with respect to $z_3(\equiv z)$.

ii) In Figure 3.5.6-2, $\mathbf{z}$ denotes the position vector of any point in the boundary layer:

$$\mathbf{z} = \mathbf{p} + y\mathbf{n}.$$

By $\mathbf{n}$ we mean the unit normal to the wall directed into the fluid. For the curvilinear coordinate system described in the text of this section, find that

$$\mathbf{g}_1 \equiv \frac{\partial \mathbf{z}}{\partial y} = \mathbf{n}$$
$$= [1 + (f')^2]^{-\frac{1}{2}}(\cos\theta\mathbf{e}_1 + \sin\theta\mathbf{e}_2 - f'\mathbf{e}_3)$$
$$\mathbf{g}_2 \equiv \frac{\partial \mathbf{z}}{\partial \theta}$$
$$= \{f + y[1 + (f')^2]^{-\frac{1}{2}}\}(-\sin\theta\mathbf{e}_1 + \cos\theta\mathbf{e}_2)$$

and

$$\mathbf{g}_3 \equiv \frac{\partial \mathbf{z}}{\partial x}$$
$$= (1 + \kappa y)[1 + (f')^2]^{-\frac{1}{2}}(f'\cos\theta\mathbf{e}_1 + f'\sin\theta\mathbf{e}_2 + \mathbf{e}_3)$$

where $\kappa$ is defined by (3.5.6-7).

iii) Determine that

$$g_{11} = 1$$
$$g_{22} = \left[f + y\left(1 + f'^2\right)^{-1/2}\right]^2$$

and

$$g_{33} = (1 + \kappa y)^2$$

iv) Show that the only nonzero Christoffel symbols of the second kind are

$$\left\{ \begin{matrix} 1 \\ 2 \ 2 \end{matrix} \right\} = -\frac{g}{1 + f'^2}$$

$$\left\{ \begin{matrix} 1 \\ 3 \ 3 \end{matrix} \right\} = -\kappa(1 + \kappa y)$$

$$\left\{ \begin{matrix} 2 \\ 1 \ 2 \end{matrix} \right\} = \frac{1}{g},$$

$$\left\{ \begin{matrix} 2 \\ 2 \ 3 \end{matrix} \right\} = \frac{f'(1 + \kappa y)}{g}$$

$$\left\{ \begin{matrix} 3 \\ 1 \ 3 \end{matrix} \right\} = \frac{\kappa}{1 + \kappa y}$$

$$\left\{ \begin{matrix} 3 \\ 2 \ 2 \end{matrix} \right\} = -f'g(1 + \kappa y)^{-1}\left(1 + f'^2\right)^{-1/2}$$

$$\left\{ \begin{matrix} 3 \\ 3 \ 3 \end{matrix} \right\} = \frac{y}{1 + \kappa y}\frac{d\kappa}{dx}$$

where $g$ is defined by (3.5.6-8).

v) Derive (3.5.6-3) to (3.5.6-6).

**Exercise 3.5.6-2** *Derivation of (3.5.6-11) to (3.5.6-13)*

i) Introduce in (3.5.6-3) to (3.5.6-6) the dimensionless velocity, dimensionless modified pressure, and dimensionless time defined in the introduction to Section 3.3, as well as those dimensionless variables defined in (3.5.6-9).

ii) Express these equations in terms of the variables defined in (3.5.6-10) and construct the reasoning that leads to the definition $a = 1/2$ and to Equations (3.5.6-11) to (3.5.6-13).

**Exercise 3.5.6-3** *Mangler's transformation* (Schlichting 1979, p. 245)

i) If $A$ is any scalar and if $\overline{x}^\star$ and $\overline{y}^{\star\star}$ are defined by (3.5.6-24) and (3.5.6-25), prove that

$$\frac{\partial A}{\partial x^\star} = f^{\star 2}\frac{\partial A}{\partial \overline{x}^\star} + \frac{f^{\star\prime}}{f^\star\left(1 + f^{\star\prime 2}\right)^{1/2}}\overline{y}^{\star\star}\frac{\partial A}{\partial \overline{y}^{\star\star}}$$

and

$$\frac{\partial A}{\partial y^{\star\star}} = f^\star\frac{\partial A}{\partial \overline{y}^{\star\star}}$$

ii) Given (3.5.6-19) and (3.5.6-23), make the change of variables indicated by (3.5.6-24) to (3.5.6-27) to arrive at (3.5.6-28) and (3.5.6-29).

**Exercise 3.5.6-4** (Schlichting 1979, p. 247) Consider a body of revolution such that

$$f^\star = f^\star(z^\star)$$

$$= x^\star$$

and

$$\tilde{v}_x^\star = ax^\star$$

where $a$ is a dimensionless constant. Show that for this case Mangler's boundary-layer equations, (3.5.6-28) and (3.5.6-29), have the same form as the boundary-layer equations for plane flow past a wedge such that

$$\tilde{v}_x^\star = b\overline{x}^{\star 1/3}$$

Here $b$ is another dimensionless constant.

**Exercise 3.5.6-5** *Generalized Newtonian fluids*    Let us restrict our attention to incompressible fluids whose behavior can be represented by one of the simple empirical models discussed in Section 2.3.3:

$$\mathbf{S} = 2\eta(\gamma)\mathbf{D}$$

We require $\eta(\gamma)$ to be a homogeneous function of degree $p$.

Construct an argument to conclude that, for boundary-layer flow past a body of revolution, the differential momentum balance assumes the same form as that found in Exercise 3.5.3-3.

**Exercise 3.5.6-6** *Boundary layer external to an extruding monofilament thread* (Sakiadis 1961a,b)    In Section 3.3.4, we examined the spinning or extrusion of a monofilament thread. In preparation for later discussions of energy and mass transfer from an extruding thread, let us construct the velocity and pressure distributions for the boundary layer created in the adjacent gas. You may neglect any motion in the gas very far away from the film, and you may assume that the thread enters the gas through a hole in a solid wall at $z_1 = 0$. Compare your results with those for flow past a flat plate as described in Section 3.5.2 and with the boundary layer external to an extruding film in Exercise 3.5.2-3.

**Exercise 3.5.6-7** *Axially symmetric boundary layer in the neighborhood of a stagnation point*    Describe the velocity and pressure distributions created within a boundary layer by an axially symmetric flow in the neighborhood of a stagnation point (see Section 3.4.3).

---

## 3.6    Stability

Up to this point, we have been assuming that the flows being discussed were stable, but unstable flows are commonly observed.

We know that there is a difference in the stability of tangential annular flow (Section 3.2.3) as a function of the Reynolds number, depending upon the boundary conditions. If the inner cylinder is rotating and the outer cylinder is stationary, one observes two transitions: a transition to the laminar Taylor vortices shown in Figures 3.6.0-1 and 3.6.0-2 and, at a higher Reynolds number, a transition to turbulence. If the inner cylinder is stationary and the outer cylinder is rotating, one observes only a transition to turbulence.

We know that, although laminar flow through a tube has been observed at Reynolds numbers greater than $10^4$ (Goldstein 1938, p. 69; Schlichting 1979, p. 450), for Reynolds

**Figure 3.6.0-1.** Axisymmetric laminar Taylor vortices in tangential annular flow between rotating cylinders (Burkhalter and Koschmieder 1974), as presented by VanDyke (1982, Plate 127). Machine oil containing aluminum powder fills the gap between a fixed outer glass cylinder and a rotating inner metal one, whose relative radius is 0.727. The top and bottom plates are fixed. The rotation speed is 1.16 times the critical speed at which the transition from the one-dimensional flow of Section 3.2.3 is observed. The flow is radially inward on the heavier dark horizontal rings and outward on the finer ones. The motion was started impulsively, giving narrower vortices than would result from a smooth start.

numbers greater than 2,000 this flow is unstable. It becomes turbulent (Section 4.1) if it is subjected to small perturbations in its boundary conditions: a tap on the tube with a hammer, walking across the laboratory floor, an electric motor running in the building. . . .

Laminar flow through a tube has been subjected to a *linear stability analysis*, in which the effects of vanishingly small disturbances are observed (Schlichting 1979, p. 457). Success has been achieved, in the sense that flow through a pipe can be shown to be unstable above a critical value of the Reynolds number. Unfortunately, the predicted critical value of the Reynolds number is much smaller than that observed experimentally. A linear stability analysis allows one to determine when an instability first appears, but it does not permit one to discuss the development of the instability. The initiation of an infinitesimal instability does not imply that the instability will grow to fill the entire cross section of the pipe at some finite distance downstream.

The transition to turbulence in a pipe illustrates the use of linear stability analyses to compute conservative limits, below which a flow is stable.

Sometimes a linear stability analysis can be used to characterize some aspect of the more stable state, as in the problem described in the next section.

**Figure 3.6.0-2.** Laminar Taylor vortices in tangential annular flow between rotating cylinders (Koschmieder 1979), as presented by VanDyke (1982, Plate 128). The ratio of the radius of the inner cylinder to that of the outer cylinder is 0.896. Again only the inner cylinder rotates. The rotation speed is 8.5 times the critical speed at which the transition from the one-dimensional flow of Section 3.2.3 is observed. The flow is doubly periodic, with six waves around the circumference, drifting with the rotation.

**3.6.1**   Stability of a Liquid Thread

The stability of a long cylindrical thread of liquid was investigated by Rayleigh (1878) who neglected viscous effects, by Bohr (1909) and Weber (1931) who neglected the effects of the outer phase, and by Tomotika (1935) who included the viscous effects of both phases. All of these studies took a uniform surface tension into account. Hajiloo et al. (1987) considered a more detailed description of interfacial stress-deformation behavior appropriate for systems containing surfactants. They argued that the effects of mass transfer could be ignored and that the interfacial tension could be assumed to be uniform. Here we will follow the analysis of Hajiloo et al. (1987) for the case considered by Tomotika (1935): viscous effects in both phases and a uniform interfacial tension.

Referring to Figure 3.6.1-1, we will make several assumptions in this analysis:

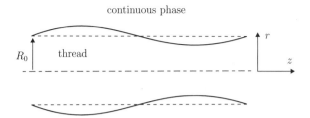

continuous phase

**Figure 3.6.1-1.** The liquid thread viewed in cylindrical co-ordinates.

1) Both fluids are incompressible and Newtonian.
2) Mass transfer to and from the adjoining phases can be ignored, and the interfacial tension is a constant.
3) The inner and outer phases are unbounded in the $z$ direction, and the outer phase is unbounded in the $r$ direction.
4) The Bond number

$$N_{Bo} \equiv \frac{\left(\rho^{(1)} - \rho^{(2)}\right) g R_0^2}{\gamma}$$

$$\ll 1 \tag{3.6.1-1}$$

where $\ldots^{(1)}$ refers to the inner phase, $\ldots^{(2)}$ refers to the outer phase, $g$ is the magnitude of the acceleration of gravity, $R_0$ the initial radius of the thread, and $\gamma$ the interfacial tension. This neglects the effects of gravity. In our analysis, we will not assume that the densities of the two phases are equal, but we will assume that, in a frame of reference fixed with respect to the thread, there is no relative motion of the two phases in the absence of disturbances.
5) Prompted by assumption 3.6.1, we will further assume that the liquid thread is axially symmetric and that it does not rotate.

The differential mass balance requires for each phase $j = 1, 2$

$$\text{div } \mathbf{v}^{(j)\star} = 0 \tag{3.6.1-2}$$

The differential momentum balance demands

$$\frac{\partial \mathbf{v}^{(1)\star}}{\partial t^\star} + \nabla \mathbf{v}^{(1)\star} \cdot \mathbf{v}^{(1)\star} = -\nabla \mathcal{P}^{(1)\star} + \text{div}\left(\nabla \mathbf{v}^{(1)\star}\right) \tag{3.6.1-3}$$

$$\frac{\partial \mathbf{v}^{(2)\star}}{\partial t^\star} + \nabla \mathbf{v}^{(2)\star} \cdot \mathbf{v}^{(2)\star} = -\nabla \mathcal{P}^{(2)\star} + \frac{N_\mu}{N_\rho} \text{div}\left(\nabla \mathbf{v}^{(2)\star}\right) \tag{3.6.1-4}$$

where

$$\mathcal{P}^{(j)} \equiv p^{(j)} + \rho^{(j)}\phi \tag{3.6.1-5}$$

with the understanding that the acceleration of gravity

$$\mathbf{b} = -\nabla\phi \tag{3.6.1-6}$$

and $\phi$ is the potential energy per unit mass. Here we have introduced as dimensionless variables

$$\mathbf{v}^{(j)\star} \equiv \frac{\rho^{(1)} R_0 \mathbf{v}^{(j)}}{\mu^{(1)}}$$

$$\mathcal{P}^{(j)\star} \equiv \frac{\mathcal{P}^{(j)}}{\rho^{(j)}} \left( \frac{\rho^{(1)} R_0}{\mu^{(1)}} \right)^2$$

$$t^\star \equiv \frac{\mu^{(1)} t}{\rho^{(1)} R_0^2} \tag{3.6.1-7}$$

$$r^\star \equiv \frac{r}{R_0}$$

$$z^\star \equiv z R_0$$

and the dimensionless parameters

$$N_\mu \equiv \frac{\mu^{(2)}}{\mu^{(1)}}$$

$$\tag{3.6.1-8}$$

$$N_\rho \equiv \frac{\rho^{(2)}}{\rho^{(1)}}$$

The jump mass balance is satisfied identically by assumption 3.6.1. In view of assumption 3.6.1, the jump momentum balance requires (Section 2.2.3)

$$2H^\star \boldsymbol{\xi} + N_{ca} \left( \mathcal{P}^{(1)\star} - N_\rho \mathcal{P}^{(2)\star} \right) \boldsymbol{\xi} + 2N_\mu N_{ca} \mathbf{D}^{(2)\star} \cdot \boldsymbol{\xi} - 2N_{ca} \mathbf{D}^{(1)\star} \cdot \boldsymbol{\xi} = 0 \tag{3.6.1-9}$$

in which

$$H^\star \equiv R_0 H \tag{3.6.1-10}$$

the *capillary number* is defined by

$$N_{ca} \equiv \frac{\mu^{(1)2}}{\gamma \rho^{(1)} R_0} \tag{3.6.1-11}$$

$H$ is the mean curvature of the interface (Table 2.4.3-8), and $\boldsymbol{\xi}$ is the unit normal to the dividing surface pointing into the outer phase.

Finally, with reference to Table 2.4.3-8, the configuration of the interface will be described by

$$r^\star = c^\star(z^\star, t^\star) \tag{3.6.1-12}$$

Let $\epsilon$ characterize the magnitude of the very small (undefined) disturbances to the system. We can visualize that, if we had a complete solution to this problem with well-defined boundary conditions (including the disturbances), we could expand $c^\star$, $\mathcal{P}^{(j)\star}$, and $\mathbf{v}^{(j)\star}$ in

Taylor series with respect to $\epsilon$:

$$c^\star = c_{(0)}^\star \left(r^\star, z^\star, t^\star\right) + \epsilon c_{(1)}^\star \left(r^\star, z^\star, t^\star\right) + \cdots$$

$$\mathcal{P}^{(j)\star} = \mathcal{P}_{(0)}^{(j)\star} \left(r^\star, z^\star, t^\star\right) + \epsilon \mathcal{P}_{(1)}^{(j)\star} \left(r^\star, z^\star, t^\star\right) + \cdots$$

$$v_r^{(j)\star} = v_{(0)r}^{(j)\star} \left(r^\star, z^\star, t^\star\right) + \epsilon v_{(1)r}^{(j)\star} \left(r^\star, z^\star, t^\star\right) + \cdots \qquad (3.6.1\text{-}13)$$

$$v_z^{(j)\star} = v_{(0)z}^{(j)\star} \left(r^\star, z^\star, t^\star\right) + \epsilon v_{(1)z}^{(j)\star} \left(r^\star, z^\star, t^\star\right) + \cdots$$

$$v_\theta^{(j)\star} = 0$$

Note that in writing these series we have taken advantage of assumption 3.6.1.

The quantities $c_{(0)}^\star$, $\mathcal{P}_{(0)}^{(j)\star}$, and $\mathbf{v}_{(0)}^{(j)\star}$ are referred to as the zeroth perturbations of these quantities. Since they represent the flow in the absence of a disturbance, we can immediately write

$$c_{(0)}^\star = 1$$

$$\mathcal{P}_{(0)}^{(j)\star} = \text{a constant within phase } j \qquad (3.6.1\text{-}14)$$

$$\mathbf{v}_{(0)}^{(j)\star} = \mathbf{0}$$

The zeroth perturbations of the modified pressures in each phase are related by the normal component of (3.6.1-11):

$$N_{ca} \left(\mathcal{P}^{(1)\star} - N_\rho \mathcal{P}^{(2)\star}\right) = 1 \qquad (3.6.1\text{-}15)$$

Substituting (3.6.1-13) into (3.6.1-2) through (3.6.1-4) and (3.6.1-9) and retaining only the coefficients of $\epsilon$, we arrive at the boundary-value problem to be solved for the first perturbations: $c_{(1)}^\star$, $\mathcal{P}_{(1)}^{(j)\star}$, and $\mathbf{v}_{(1)}^{(j)\star}$. In a linear stability analysis, we seek no more than these first-order terms in the Taylor series (3.6.1-13). Our objective is to determine under what conditions a system becomes unstable. We will make no attempt to determine how these instabilities develop as a function of time.

The first perturbation of (3.6.1-2) for each phase,

$$\frac{1}{r^\star} \frac{\partial}{\partial r^\star} \left(r^\star v_{(1)r}^{(j)\star}\right) + \frac{\partial v_{(1)z}^{(j)\star}}{\partial z^\star} = 0 \qquad (3.6.1\text{-}16)$$

can be satisfied automatically with the introduction of stream functions (Table 2.4.2-1):

$$v_{(1)r}^{(j)\star} = \frac{1}{r^\star} \frac{\partial \psi^{(j)}}{\partial z^\star}$$

$$\qquad (3.6.1\text{-}17)$$

$$v_{(1)z}^{(j)\star} = -\frac{1}{r^\star} \frac{\partial \psi^{(j)}}{\partial r^\star}$$

In terms of these stream functions, (3.6.1-3) and (3.6.1-4) may be expressed as (Table 2.4.2-1)

$$\left(\frac{\partial}{\partial t^\star} - E^2\right) E^2 \psi^{(1)} = 0 \qquad (3.6.1\text{-}18)$$

and

$$\left(\frac{\partial}{\partial t^\star} - \frac{N_\mu}{N_\rho} E^2\right) E^2 \psi^{(2)} = 0 \qquad (3.6.1-19)$$

where

$$E^2 \equiv \frac{\partial^2}{\partial r^{\star 2}} - \frac{1}{r^\star}\frac{\partial}{\partial r^\star} + \frac{\partial^2}{\partial z^{\star 2}} \qquad (3.6.1-20)$$

In view of assumption 3.6.1, the jump mass balance and continuity of the tangential components of velocity require at $r^\star = 1$

$$\frac{\partial \psi^{(1)}}{\partial z^\star} = \frac{\partial \psi^{(2)}}{\partial z^\star}$$

$$= \frac{\partial c_{(1)}^\star}{\partial t^\star} \qquad (3.6.1-21)$$

and

$$\frac{\partial \psi^{(1)}}{\partial r^\star} = \frac{\partial \psi^{(2)}}{\partial r^\star} \qquad (3.6.1-22)$$

The $r$ and $z$ components of (3.6.1-9) become at $r^\star = 1$ (see Table 2.4.3-8)

$$\frac{\partial c_{(1)}^\star}{\partial z^{\star 2}} + c_{(1)}^\star + N_{ca}\left(\mathcal{P}_{(1)}^{(1)\star} - N_\rho \mathcal{P}_{(1)}^{(2)\star}\right) + 2N_\mu N_{ca}\left(\frac{\partial^2 \psi^{(2)}}{\partial r^\star \partial z^\star} - \frac{\partial \psi^{(2)}}{\partial z^\star}\right)$$

$$- 2N_{ca}\left(\frac{\partial^2 \psi^{(1)}}{\partial r^\star \partial z^\star} - \frac{\partial \psi^{(1)}}{\partial z^\star}\right) = 0 \qquad (3.6.1-23)$$

and

$$N_\mu\left(\frac{\partial \psi^{(2)}}{\partial r^\star} - \frac{\partial^2 \psi^{(2)}}{\partial r^{\star 2}} + \frac{\partial^2 \psi^{(2)}}{\partial z^{\star 2}}\right) - \frac{\partial \psi^{(1)}}{\partial r^\star} + \frac{\partial^2 \psi^{(1)}}{\partial r^{\star 2}} - \frac{\partial^2 \psi^{(1)}}{\partial z^{\star 2}} = 0 \qquad (3.6.1-24)$$

Since the stream functions are determined only within arbitrary constants, we will say

as $r \to \infty: \ \psi^{(2)} \to 0$ \qquad (3.6.1-25)

at $r = 0: \ \psi^{(1)}$ is finite \qquad (3.6.1-26)

We will seek a solution for the first perturbation of the form

$$c_{(1)}^\star = \hat{c}_{(1)}^\star (r^\star) \exp(im^\star z^\star) \exp(n^\star t^\star)$$

$$\mathcal{P}_{(1)}^{(j)\star} = \hat{\mathcal{P}}_{(1)}^{(j)\star}(r^\star) \exp(im^\star z^\star) \exp(n^\star t^\star) \qquad (3.6.1-27)$$

$$\psi^{(j)\star} = \hat{\psi}^{(j)\star}(r^\star) \exp(im^\star z^\star) \exp(n^\star t^\star)$$

where $m^\star$ and $n^\star$ are real. We conclude from (3.6.1-18) and (3.6.1-19) that

$$\left(n - E_{(1)}^{\ 2}\right) E_{(1)}^{\ 2} \hat{\psi}^{(1)\star} = 0 \qquad (3.6.1-28)$$

$$\left(n - \frac{N_\mu}{N_\rho} E_{(1)}^{\ 2}\right) E_{(1)}^{\ 2} \hat{\psi}^{(2)\star} = 0 \qquad (3.6.1-29)$$

in which

$$E_{(1)}^2 \equiv \frac{d^2}{dr^{\star 2}} - \frac{1}{r^\star}\frac{d}{dr^\star} - m^{\star 2} \tag{3.6.1-30}$$

Equations (3.6.1-21) through (3.6.1-24) require at $r^\star = 1$

$$\hat{\psi}^{(1)\star} = \hat{\psi}^{(2)\star}$$

$$= \frac{n^\star}{im^\star}\hat{c}_{(1)}^\star \tag{3.6.1-31}$$

$$\frac{d\hat{\psi}^{(1)}}{dr^\star} = \frac{d\hat{\psi}^{(2)}}{dr^\star} \tag{3.6.1-32}$$

$$\left(1 - m^{\star 2}\right)\hat{c}^\star + N_{ca}\left(\hat{\mathcal{P}}^{(1)} - N_\rho\hat{\mathcal{P}}^{(2)}\right)$$

$$+ 2im^\star N_{ca}(N_\mu - 1)\left(\frac{d\hat{\psi}^{(1)}}{dr^\star} - \hat{\psi}^{(1)}\right) = 0 \tag{3.6.1-33}$$

$$(N_\mu - 1)\left(m^{\star 2}\hat{\psi}^{(1)} - \frac{d\hat{\psi}^{(1)}}{dr^\star}\right) + \left(N_\mu\frac{d^2\hat{\psi}^{(2)}}{dr^{\star 2}} - \frac{d^2\hat{\psi}^{(1)}}{dr^{\star 2}}\right) = 0 \tag{3.6.1-34}$$

Similarly, (3.6.1-25) and (3.6.1-26) demand

$$\text{as } r \to \infty : \ \hat{\psi}^{(2)} \to 0 \tag{3.6.1-35}$$

$$\text{at } r = 0 : \ \hat{\psi}^{(1)} \text{ is finite} \tag{3.6.1-36}$$

Finally, from the $z$ components of (3.6.1-3) and (3.6.1-4), we can observe that

$$im^\star\hat{\mathcal{P}}^{(1)} = \left(\frac{d^2}{dr^{\star 2}} + \frac{1}{r^\star}\frac{d}{dr^\star} - m^{\star 2} - n^\star\right)\left(-\frac{1}{r^\star}\frac{d\hat{\psi}^{(1)}}{dr^\star}\right) \tag{3.6.1-37}$$

and

$$im^\star N_\rho\hat{\mathcal{P}}^{(2)} = \left(N_\mu\frac{d^2}{dr^{\star 2}} + N_\mu\frac{1}{r^\star}\frac{d}{dr^\star} - N_\mu m^{\star 2} - N_\rho n^\star\right)\left(-\frac{1}{r^\star}\frac{d\hat{\psi}^{(2)}}{dr^\star}\right) \tag{3.6.1-38}$$

Solutions to (3.6.1-28) and (3.6.1-29) consistent with (3.6.1-35) and (3.6.1-36) are

$$\hat{\psi}^{(1)} = Ar^\star I_1(m^\star r^\star) + Br^\star I_1\left(m_1^\star r^\star\right) \tag{3.6.1-39}$$

$$\hat{\psi}^{(2)} = Cr^\star K_1(m^\star r^\star) + Dr^\star K_1\left(m_2^\star r^\star\right) \tag{3.6.1-40}$$

in which $I_1$ and $K_1$ are the modified Bessel functions of order one and

$$m_1 \equiv (m^{\star 2} + n^\star)^{1/2}$$

$$m_2 \equiv \left(m^{\star 2} + n^\star\frac{N_\rho}{N_\mu}\right)^{1/2} \tag{3.6.1-41}$$

Boundary conditions (3.6.1-31) through (3.6.1-34) reduce to

$$I_1\left(m^\star\right)A + I_1\left(m_1^\star\right)B - K_1\left(m^\star\right)C - K_1\left(m_2^\star\right)D = 0 \tag{3.6.1-42}$$

$$m^\star I_0\left(m^\star\right)A + m_1 I_0\left(m_1^\star\right)B + m^\star K_0\left(m^\star\right)C + m_2 I_0\left(m_2^\star\right)D = 0 \tag{3.6.1-43}$$

$$\left\{ m^\star\left(-m^{\star 2}\right)I_1(m^\star) - n^{\star 2}N_{ca}I_0(M^\star) + 2m^\star n^\star[m^\star I_0(m^\star) - I_1(m^\star)]N_{ca}(N_\mu - 1)\right\} A$$

$$+ \left\{ m^\star\left(-m^{\star 2}\right)I_1\left(m_1^\star\right) + 2m^\star n^\star\left[m_1^\star I_0\left(m_1^\star\right) - I_1\left(m_1^\star\right)\right]N_{ca}(N_\mu - 1)\right\} B$$

$$- n^{\star 2}N_{ca}N_\rho K_0(m^\star)C = 0 \tag{3.6.1-44}$$

$$2m^{\star 2}I_1\left(m^\star\right) A + \left(m_1^{\star 2} + m^{\star 2}\right)I_1\left(m_1^\star\right) B - 2m^{\star 2}N_\mu K_1(m^\star)C$$

$$- \left(m_2^{\star 2} + m^{\star 2}\right)N_\mu K_1\left(m_2^\star\right) D = 0 \tag{3.6.1-45}$$

In arriving at (3.6.1-45), we have employed (3.6.1-42) and (3.6.1-43). [I recommend that you use Mathematica (1993) or some similar program to arrive at (3.6.1-44) and (3.6.1-45).]

To ensure that the solution will be nontrivial, the determinant of the $4 \times 4$ matrix whose entries are the coefficients of $A$, $B$, $C$, and $D$ in (3.6.1-42) through (3.6.1-45) must be equal to zero. The solution to this equation is the dimensionless growth rate $n^\star$ for a given dimensionless wave number $m^\star$. The wavelength corresponding to $m^\star$ is

$$\lambda = \frac{2\pi}{m} = \frac{2\pi R_0}{m^\star} \tag{3.6.1-46}$$

We note immediately from (3.6.1-44) that, for neutral stability ($n^\star = 0$), the determinant of the coefficients of $A$, $B$, $C$, and $D$ in (3.6.1-42) through (3.6.1-45) reduces to zero when $m^\star = 1$. The thread is stable ($n^\star < 0$) for $m > 1$. The mode of maximum instability is the dimensionless wave number $m_{max}^\star$ that belongs to the largest possible growth rate $n_{max}^\star$:

$$\lambda^\star \equiv \frac{\lambda}{R_0} = \frac{2\pi}{m_{max}^\star} \tag{3.6.1-47}$$

It is by this mode that the system will tend to break up, when the system is subjected to infinitesimal perturbations. It is clear from (3.6.1-42) through (3.6.1-45) that $m_{max}^\star$ and $n_{max}^\star$ are functions of $N_{ca}$, $N_\mu$, and $N_\rho$.

Rumscheidt and Mason (1962) studied the stability of a liquid thread experimentally. In Table 3.6.1-1, we compare our results obtained using Mathematica (1993) with their observations of their system 2, for which they provide somewhat more detailed information. It is interesting that, for these cases, changing $R_0$ and $N_{ca}$ affected $\lambda^\star$ in only the fourth significant figure (not shown). At least for these parameters, the results are sensitive to the values of the densities. Some of the deviation from experimental observation may be attributable to our estimated densities, since the actual densities were not supplied by Rumscheidt and Mason (1962).

## Liquid Jet

The results described above can be applied to the stability of a laminar jet of liquid in a gas with two further assumptions approximations:

6) Viscous effects in the gas can be neglected with respect to those in the liquid.
7) In a frame of reference in which the average velocity of the liquid is zero relatively far from the orifice, end effects can be neglected.

**Table 3.6.1-1.** Comparison of (3.6.1-47) with the experimental observations of Rumscheidt and Mason (1962) for their system 2. For this system, the continuous phase is silicone oil 5000 (Dow Corning) for which $\mu^{(C)} = 5.26$ Pa s, and the liquid thread is Pale 4 (oxidized castor oil, Baker Castor Oil Co., New York) for which $\mu^{(T)} = 6.0$ Pa s. For this system, they report $\gamma = 4.8 \times 10^{-3}$ N/m. In the absence of information, we assume that $\rho^{(C)} = \rho^{(T)} = 950$ kg/m$^3$

| $R_0$ ($\mu$m) | $\lambda^\star$ | $\lambda^\star_{measured}$ |
|---|---|---|
| 70 | 11.2 | 14.3 |
| 75 | 11.2 | 14.7 |
| 125 | 11.2 | 13.6 |
| 125 | 11.2 | 14.4 |
| 210 | 11.2 | 13.3 |

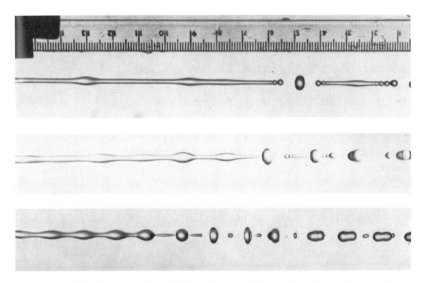

**Figure 3.6.1-2.** Capillary instability of a liquid jet. Water, forced from a 4 mm tube into air, is perturbed at various frequencies by a loudspeaker (Rutland and Jameson 1971), as presented by VanDyke (1982, Plate 122). Beginning with the top figure, $\lambda^\star = 84$, 25, and 9.2.

For the jets of water in air shown in Figure 3.6.1-2, we can say

$$N_{ca} \ll 1$$
$$m^\star_{max} \approx 1 \qquad (3.6.1\text{-}48)$$
$$\lambda^\star \approx 2\pi$$

The deviation between theory and experimental observation may be attributable to assumptions 6 and 7.

# 4

## Integral Averaging in Momentum Transfer

I MENTIONED IN MY INTRODUCTION to Chapter 3 that not every interesting problem should be attacked by directly solving the differential mass and momentum balances. Some problems are really too difficult to be solved in this manner. In other cases, the amount of effort required for such a solution is not justified, when the end purpose for which the solution is being developed is taken into account.

In the majority of momentum transfer problems, the quantity of ultimate interest is an integral. Perhaps it is an average velocity, a volume flow rate, or a force on a surface. This suggested that I set aside an entire chapter in order to exploit approaches to problems in which the independent variables are integrals or integral averages.

I begin by approaching turbulence in terms of time-averaged variables. Then I look at some problems that are normally explained in terms of area-averaged variables. The random geometry encountered in flow through porous media suggests the use of a local volume-averaged variable. The chapter concludes with the relatively well-known integral balances for arbitrary systems.

Again I encourage those of you who feel you are primarily interested in energy and mass transfer to pay close attention to this chapter. The ideas developed here are taken over, almost without change, and applied to energy transfer in Chapter 7 and to mass transfer in Chapter 10.

### 4.1    Time Averaging

The most common example of time averaging is in the context of turbulence. *Turbulence* is defined to be a motion that varies randomly with time over at least a portion of the flow field such that statistically distinct average values can be discerned. (More precisely, we should require the motion to vary randomly with time in all possible frames of reference. We certainly would not wish to say that a rigid solid was in a turbulent motion merely because it was subject to random rotations and translations.) Any fluid motion that is not turbulent is termed *laminar*. In thinking about turbulence you should carefully distinguish between a complex-appearing laminar flow and a true turbulent flow. A laminar flow may to the eye have a very complex dependence upon position in space and time, but it is only the turbulent flow that exhibits the random variations with time.

From a practical point of view, turbulent flows are probably more important than laminar flows. When a fluid is pumped through a pipe in a commercial process, it is likely in turbulent flow. You might ask why, if turbulent flows are so important, we have waited until now to mention them. The difficulty is that I can tell you nothing about exact solutions for turbulent flows. It is the averaging techniques discussed in Chapters 4, 7, and 9 that have proved more helpful in our attempts to analyze practical problems.

In reading the literature, it is helpful to have a few commonly accepted definitions in mind. Turbulence that is generated and continuously affected by fixed walls is referred to as *wall turbulence*. It is wall turbulence that we observe in flow through a tube. In the absence of walls, we speak of *free turbulence*. Aircraft encounter free turbulence sometimes in apparently clear skies.

If the turbulence has quantitatively the same structure everywhere in the flowfield, it is said to be *homogeneous*. If its statistical features do not depend upon direction, the turbulence is called *isotropic*. [The word isotropic is overworked. We have isotropic functions (Truesdell and Noll 1965, p. 22) and isotropic materials (Truesdell 1966a, p. 60), not to mention isotropic porous media (see Example 1 of Section 4.3.5). I would prefer to talk about *oriented* and *nonoriented* turbulence.] Where the mean velocity shows a gradient, we speak of the turbulence as being nonisotropic or anisotropic. Wall turbulence will always be anisotropic.

The approach to turbulence that is outlined in the next few sections is very old. Everyone recognizes now that it can never by itself lead to a detailed understanding of the phenomena. We recommend this approach, at least for an introduction, because it has been the most fruitful in terms of engineering results. For more detailed studies of turbulence from a statistical point of view see Lin and Reid (1963), Corrsin (1963), Hinze (1959), Batchelor (1959), and Townsend (1956). See also the introduction to Section 3.6.

### 4.1.1  Time Averages

Consider a constant volume rate of flow through a tube. Although the Reynolds number $N_{Re} > 5,000$ and the flow must be turbulent, a pressure gauge mounted on the wall of the tube shows a reading that is independent of time. If we examine the velocity distribution using a Pitot tube, we see that only the axial component of velocity differs from zero and that it also appears to be independent of time. We appear to have a contradiction, because the flow is turbulent and yet the velocity and pressure distributions appear to be independent of time.

There is no contradiction. Both the Bourdon-tube pressure gauge and the Pitot tube damp out the high-frequency variations with time. The readings they give us are time averages of pressure and velocity.

Since our ordinary instruments measure time averages, at least for engineering purposes we might work exclusively in terms of time-averaged variables. Let $B$ be any scalar, vector, or tensor. We will define its time average as

$$\overline{B}(t) \equiv \frac{1}{\Delta t} \int_t^{t+\Delta t} B(t') \, dt' \tag{4.1.1-1}$$

By $\Delta t$, we mean a finite time interval that is large with respect to the period or timescale of the random fluctuations of this variable but small compared with the period or timescale of any slow variations in the field of flow that is not associated with turbulence. There is a degree of arbitrariness in the choice of the fluctuations attributed to turbulence. In practice,

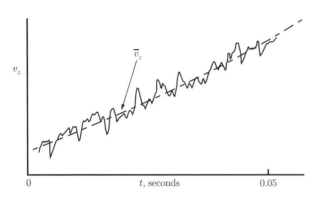

**Figure 4.1.1-1.** Random variation of velocity component from a mean value as might be measured with a hot-wire anemometer.

a choice can usually be made without too much difficulty. Figure 4.1.1-1 suggests how a velocity component might vary randomly from its mean value, even though the mean value itself is a function of time.

If we are to talk about turbulent flows in terms of their time-averaged distributions for velocity, pressure, density, ..., let us begin by taking the time average of the differential mass balance:

$$\frac{1}{\Delta t} \int_{t}^{t+\Delta t} \left[ \frac{\partial \rho}{\partial t'} + \text{div}\,(\rho \mathbf{v}) \right] dt' = 0 \tag{4.1.1-2}$$

The first term on the left may be integrated directly and rearranged using the Leibnitz rule for differentiation of an integral (Exercise 1.3.3-4):

$$\frac{1}{\Delta t} \int_{t}^{t+\Delta t} \frac{\partial \rho}{\partial t'} dt' = \frac{\partial \overline{\rho}}{\partial t} \tag{4.1.1-3}$$

In the second term on the left of (4.1.1-2), the divergence operation commutes with time averaging. As a result, (4.1.1-2) becomes

$$\frac{\partial \overline{\rho}}{\partial t} + \text{div}\,(\overline{\rho \mathbf{v}}) = 0 \tag{4.1.1-4}$$

This will be referred to as the *time average of the differential mass balance.* For an incompressible fluid, this equation reduces to

$$\text{div}\,\overline{\mathbf{v}} = 0 \tag{4.1.1-5}$$

In looking at (4.1.1-4), it is important to realize that in general the time average of a product is not the same as the product of the time averages:

$$\overline{\rho \mathbf{v}} \neq \overline{\rho}\,\overline{\mathbf{v}} \tag{4.1.1-6}$$

This point will arise repeatedly throughout our discussion of averaging operations.

Let us take the time average of the differential momentum balance:

$$\frac{1}{\Delta t} \int_{t}^{t+\Delta t} \left[ \frac{\partial \rho \mathbf{v}}{\partial t'} + \text{div}(\rho \mathbf{v}\mathbf{v}) + \nabla P - \text{div}\,\mathbf{S} - \rho \mathbf{f} \right] dt' = 0 \tag{4.1.1-7}$$

In a manner directly analogous to that used in going from (4.1.1-2) to (4.1.1-4), we can express this as

$$\frac{\partial \overline{\rho \mathbf{v}}}{\partial t} + \operatorname{div}\left(\overline{\rho}\,\overline{\mathbf{v}}\,\overline{\mathbf{v}}\right) = -\nabla \overline{P} + \operatorname{div}\left(\overline{\mathbf{S}} + \mathbf{T}^{(t)}\right) + \overline{\rho}\mathbf{f} \tag{4.1.1-8}$$

where

$$\mathbf{T}^{(t)} \equiv \overline{\rho}\,\overline{\mathbf{v}}\,\overline{\mathbf{v}} - \overline{\rho \mathbf{v}\mathbf{v}} \tag{4.1.1-9}$$

is the *Reynolds stress tensor*. It is introduced to allow us to express the convective inertial terms in (4.1.1-8) as the divergence of a product of averages rather than as a divergence of an average of products. For an incompressible fluid, (4.1.1-8) and (4.1.1-9) simplify to

$$\rho\left(\frac{\partial \overline{\mathbf{v}}}{\partial \mathbf{t}} + \nabla \overline{\mathbf{v}} \cdot \overline{\mathbf{v}}\right) = -\nabla \overline{p} + \operatorname{div}\left(\overline{\mathbf{S}} + \mathbf{T}^{(t)}\right) + \rho\mathbf{f} \tag{4.1.1-10}$$

and

$$\mathbf{T}^{(t)} = \rho\left(\overline{\mathbf{v}}\,\overline{\mathbf{v}} - \overline{\mathbf{v}\mathbf{v}}\right). \tag{4.1.1-11}$$

In arriving at (4.1.1-8) and (4.1.1-10), we have taken $\mathbf{f}$ to be independent of time, as in the case of gravity.

For the sake of simplicity, in the sections that follow we restrict our attention to incompressible fluids. We will be interested in solving (4.1.1-5) and (4.1.1-10) consistent with appropriate boundary conditions and descriptions of material behavior.

It is clear that merely saying we have an incompressible Newtonian fluid will not be sufficient to specify $\mathbf{T}^{(t)}$. We have lost some detail in time averaging the differential momentum balance. We will replace that lost information with an empirical data correlation for $\mathbf{T}^{(t)}$.

**Exercise 4.1.1-1** *An incompressible Newtonian fluid*   Determine that, for an incompressible Newtonian fluid, the time average of the differential momentum balance becomes

$$\rho\left(\frac{\partial \overline{\mathbf{v}}}{\partial \mathbf{t}} + \nabla \overline{\mathbf{v}} \cdot \overline{\mathbf{v}}\right) = -\nabla \overline{p} + \operatorname{div}\left(\mu \nabla \overline{\mathbf{v}} + \mathbf{T}^{(t)}\right) + \rho\mathbf{f}$$

## 4.1.2   The Time Average of a Time-Averaged Variable

In the next section and repeatedly in the literature, we are asked to identify the time average of a time-averaged variable,

$$\overline{\overline{B}} \equiv \frac{1}{\Delta t}\int_{t}^{t+\Delta t} \overline{B}\, dt' \tag{4.1.2-1}$$

with simply the time average of that variable. It should be understood that $B$ can be any scalar, vector, or tensor function of time (and usually position as well). This seems intuitively reasonable as long as we are averaging over sufficiently small increments $\Delta t$ in time. Our purpose here is to show in what sense this intuitive feeling is confirmed.

By way of orientation, let us consider a somewhat simpler problem. Given some function $f(x)$, let us ask about the value of

$$B \equiv \frac{1}{R_2 R_1}\int_{0}^{R_2}\int_{0}^{R_1} f(x + X)\, dX\, dx \tag{4.1.2-2}$$

where the constants $R_1$ and $R_2$ are known. If we expand $f(x + X)$ in a Taylor series, (4.1.2-2) may be expressed as

$$
\begin{aligned}
B &= \frac{1}{R_2 R_1} \int_0^{R_2} \int_0^{R_1} \left[ f(x) + X \frac{\partial f(x)}{\partial x} + \frac{1}{2} X^2 \frac{\partial^2 f(x)}{\partial x^2} + \cdots \right] dX \, dx \\
&= \frac{1}{R_2} \int_0^{R_2} f(x) \, dx + \frac{1}{2} \frac{R_1}{R_2} \int_0^{R_2} \frac{\partial f(x)}{\partial x} \, dx + \cdots \\
&= \frac{1}{R_2} \int_0^{R_2} f(x) \, dx + \frac{1}{2} \frac{R_1}{R_2} [f(R_2) - f(0)] + \cdots
\end{aligned}
\tag{4.1.2-3}
$$

This suggests that, as $R_1 / R_2 \to 0$,

$$
\frac{1}{R_2 R_1} \int_0^{R_2} \int_0^{R_1} f(x + X) \, dX \, dx \to \frac{1}{R_2} \int_0^{R_2} f(x) \, dx
\tag{4.1.2-4}
$$

This indicates how we might look at the time average of a time-averaged variable:

$$
\begin{aligned}
\overline{\overline{B}} &= \frac{1}{(\Delta t)^2} \int_t^{t+\Delta t} \int_T^{T+\Delta t} B(T') \, dT' \, dT \\
&= \frac{1}{(\Delta t)^2} \int_t^{t+\Delta t} \int_0^{\Delta t} B(T + \tau) \, d\tau \, dT \\
&= \frac{1}{(\Delta t)^2} \int_t^{t+\Delta t} \int_0^{\Delta t} \left[ B(T) + \tau \frac{\partial B}{\partial T} + \frac{1}{2} \tau^2 \frac{\partial^2 B}{\partial T^2} + \cdots \right] d\tau \, dT \\
&= \frac{1}{\Delta t} \int_t^{t+\Delta t} \left[ B(T) + \frac{1}{2} \Delta t \frac{\partial B}{\partial T} + \frac{(\Delta t)^2}{6} \frac{\partial^2 B}{\partial T^2} + \cdots \right] dT
\end{aligned}
\tag{4.1.2-5}
$$

It is convenient to introduce as a dimensionless variable

$$
T^\star \equiv \frac{T}{t_0}
\tag{4.1.2-6}
$$

where $t_0$ is characteristic of the timescale of any slow variations in the field of flow that we do not wish to regard as belonging to the turbulence. In terms of this dimensionless time, (4.1.2-5) becomes

$$
\overline{\overline{B}} = \frac{t_0}{\Delta t} \int_{t^\star}^{t^\star + \Delta t / t_0} \left[ B(T^\star) + \frac{1}{2} \frac{\Delta t}{t_0} \frac{\partial B}{\partial T^\star} + \frac{1}{6} \left( \frac{\Delta t}{t_0} \right)^2 \frac{\partial^2 B}{\partial T^{\star 2}} + \cdots \right] dT^\star
\tag{4.1.2-7}
$$

This motivates our saying that

$$
\text{as } \frac{\Delta t}{t_0} \to 0 : = \overline{\overline{B}} \to \overline{B}
\tag{4.1.2-8}
$$

### 4.1.3  Empirical Correlations for $\mathbf{T}^{(t)}$

In this section, we illustrate how empirical data correlations for the Reynolds stress tensor $\mathbf{T}^{(t)}$ can be formulated. We base this discussion on three points.

1) If we limit ourselves to changes of frame such that

$$\overline{\mathbf{Q}} \doteq \mathbf{Q} \tag{4.1.3-1}$$

we may use the result of Section 4.1.2 to conclude that $\mathbf{T}^{(t)}$ is frame indifferent:

$$\begin{aligned}
\mathbf{T}^{(t)*} &\equiv \rho \left( \overline{\mathbf{v}^* \, \mathbf{v}^*} - \overline{\mathbf{v}^*}\,\overline{\mathbf{v}^*} \right) \\
&= \rho \left( \overline{(\mathbf{v}^* - \overline{\mathbf{v}^*})(\mathbf{v}^* - \overline{\mathbf{v}^*})} \right) \\
&= \rho \mathbf{Q} \cdot \overline{(\overline{\mathbf{v}} - \mathbf{v})(\mathbf{v} - \overline{\mathbf{v}})} \cdot \mathbf{Q}^T \\
&= \mathbf{Q} \cdot \mathbf{T}^{(t)} \cdot \mathbf{Q}^T
\end{aligned} \tag{4.1.3-2}$$

Here $\mathbf{Q}$ is a (possibly) time-dependent orthogonal second-order tensor. We make use of the fact that a velocity difference is frame indifferent (see Exercise 1.2.2-1).

2) We assume that the principle of frame indifference introduced in Section 2.3.1 applies to any empirical correlations developed for $\mathbf{T}^{(t)}$, provided the change of frame considered satisfies (4.1.3-1).

3) The Buckingham–Pi theorem (Brand 1957) serves to further limit the form of any expression for $\mathbf{T}^{(t)}$.

## Example 1: Prandtl's Mixing-Length Theory

Let us attempt to develop an empirical correlation for $\mathbf{T}^{(t)}$ appropriate to wall turbulence. Before beginning, it is advisable to examine available experimental evidence, so as to establish the qualitative character of the phenomena to be described.

Most of our experimental evidence concerns flow through a tube. Although the direction of the velocity fluctuations is nearly random at the center of a tube, in the immediate neighborhood of the tube wall the magnitude of the velocity fluctuation in the axial direction is greater than that in the radial direction. All velocity fluctuations approach zero in the limit of the wall itself. This changing character of the velocity fluctuations suggests that three regimes of wall turbulence might be recognized.

- Referring to Figure 4.1.3-1, we can visualize that there is a thin layer next to the wall, where the viscous stresses are far more important than the Reynolds stresses. This is called the *laminar sublayer*.
- Outside this laminar sublayer is an intermediate region, where the viscous stresses are of the same order of magnitude as the Reynolds stresses. This is called the *buffer zone*.
- Beyond this buffer zone, the turbulent flow is said to be *fully developed*, and the Reynolds stresses dominate the viscous stresses.

We will fix our attention here on the fully developed flow regime.

If we think of the Reynolds stress tensor as being in some way similar to the stress tensor $\mathbf{T}$, we might attack the problem of empirical correlations for $\mathbf{T}^{(t)}$ in the same way that we approached the problem of constitutive equations for $\mathbf{T}$ in Section 2.3.2. For example, we might assume

$$\mathbf{T}^{(t)} = \mathbf{T}^{(t)}(\rho, l, \nabla \overline{\mathbf{v}}) \tag{4.1.3-3}$$

We specifically do not include viscosity as an independent variable, because we are considering the fully developed turbulent flow regime. Arguing as we did in Section 2.3.2, we

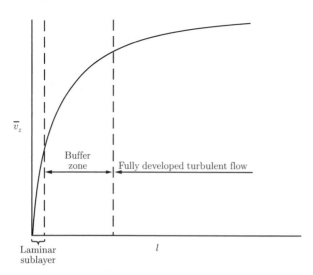

**Figure 4.1.3-1.** Time-averaged velocity as a function of distance measured from the wall.

conclude that the most general form that (4.1.3-3) can take, if it is to satisfy our restricted form of the principle of frame indifference, is

$$\mathbf{T}^{(t)} = \eta_0\mathbf{I} + \eta_1\overline{\mathbf{D}} + \eta_2\overline{\mathbf{D}} \cdot \overline{\mathbf{D}} \tag{4.1.3-4}$$

where

$$\eta_k = \eta_k\left(\rho, l, \operatorname{div}\overline{\mathbf{v}}, \operatorname{tr}\left(\overline{\mathbf{D}} \cdot \overline{\mathbf{D}}\right), \det\overline{\mathbf{D}}\right) \tag{4.1.3-5}$$

Our experience in formulating simple empirical models for $\mathbf{T}$ (see Section 2.3.3) suggests that we restrict our attention to the special case

$$\mathbf{T}^{(t)} = \eta_0\mathbf{I} + \eta_1\overline{\mathbf{D}} \tag{4.1.3-6}$$

Since we are considering only incompressible fluids,

$$-p^{(t)} \equiv \frac{1}{3}\operatorname{tr}\mathbf{T}^{(t)}$$

$$= \eta_0 \tag{4.1.3-7}$$

$$\eta_1 = \eta_1\left(\rho, l, \operatorname{tr}\left(\overline{\mathbf{D}} \cdot \overline{\mathbf{D}}\right), \det\overline{\mathbf{D}}\right) \tag{4.1.3-8}$$

and (4.1.3-6) becomes

$$\mathbf{T}^{(t)} + p^{(t)}\mathbf{I} = \eta_1\overline{\mathbf{D}} \tag{4.1.3-9}$$

The coefficient $\eta_1$ is referred to as the *eddy viscosity*.

An early proposal by Boussinesq (1877) was that

$$\eta_1 = \text{a constant} \tag{4.1.3-10}$$

This has not proved to be realistic. Experimentally we find that $\eta_1$ is a function of position in a flowfield.

Let us investigate the special case of (4.1.3-9) for which

$$\eta_1 = \eta_1 \left( \rho, l, \operatorname{tr} \left( \overline{\mathbf{D}} \cdot \overline{\mathbf{D}} \right) \right) \tag{4.1.3-11}$$

By the Buckingham–Pi theorem (Brand 1957), we find

$$\frac{\eta_1}{\rho l^2 \sqrt{2 \operatorname{tr} \left( \overline{\mathbf{D}} \cdot \overline{\mathbf{D}} \right)}} = 2\eta_1^\star \tag{4.1.3-12}$$

where $\eta_1^\star$ is a dimensionless constant. In view of (4.1.3-7) and (4.1.3-12), Equation (4.1.3-4) becomes

$$\mathbf{T}^{(t)} + p^{(t)}\mathbf{I} = 2\eta_1^\star \rho l^2 \sqrt{2 \operatorname{tr} \left( \overline{\mathbf{D}} \cdot \overline{\mathbf{D}} \right)} \tag{4.1.3-13}$$

This should be viewed as the tensorial form of *Prandtl's mixing-length theory* (Hinze 1959, p. 277; Prandtl 1925; Schlichting 1979, p. 579).

It is perhaps worth emphasizing that we should not expect the Prandtl mixing length theory to be appropriate to the laminar sublayer or buffer zone. We assumed at the beginning that we were constructing a representation for the Reynolds stress tensor in the fully developed turbulent flow regime. We see in the next section that it provides an excellent representation of the time-averaged velocity distribution within the fully developed regime for turbulent flow through tubes.

There are also some similarities between this result and Taylor's vorticity transport theory (Hinze 1959, p. 281; Taylor 1932; Goldstein 1938, p. 209). Von Karman's similarity hypothesis (Schlichting 1979, p. 5585) has a considerably different character. His development is in the context of a special two-dimensional flow, but he expected the Reynolds stress tensor to depend upon second derivatives of $\overline{\mathbf{v}}$. The tensorial representation of his result given by Bird et al. (1960, p. 161) suggests that his approach may not have been entirely reasonable. If we were to approach his expression for the Reynolds stress tensor in a manner similar to that suggested above, instead of (4.1.3-3) we would start with

$$\mathbf{T}^{(t)} = \mathbf{T}^{(t)}(\rho, l, \overline{\mathbf{D}}, \Omega) \tag{4.1.3-14}$$

where

$$\Omega \equiv \nabla \mathbf{w} + (\nabla \mathbf{w})^T \tag{4.1.3-15}$$

and $\mathbf{w}$ is the vorticity vector (3.4.0-8). Although it is true that the Bird et al. (1960) tensorial expression for Von Karman's result does obey the principle of frame indifference, $\Omega$ seems to be an unlikely independent variable in (4.1.3-14), since it is not a frame-indifferent tensor [see, for example, Truesdell and Noll (1965, p. 24)].

### Example 2: Deissler's Expression for the Region near the Wall

If we focus our attention on that portion of the turbulent flow of an incompressible Newtonian fluid in the immediate vicinity of a bounding wall, the laminar sublayer, and the buffer zone, it might appear reasonable to propose

$$\mathbf{T}^{(t)} = \mathbf{T}^{(t)} \left( \rho, \mu, l, \overline{\mathbf{v}} - \mathbf{v}_{(s)}, \nabla \overline{\mathbf{v}} \right) \tag{4.1.3-16}$$

By $\mathbf{v}_{(s)}$, we mean the velocity of the bounding wall. The most general expression of this form that is consistent both with the principle of frame indifference and the Buckingham–Pi theorem is exceedingly difficult (Spencer and Rivlin 1959, Sec. 7; Smith 1965). Let us instead look at a special case of (4.1.3-16) that satisfies the principle of frame indifference:

$$\mathbf{T}^{(t)} + p^{(t)}\mathbf{I} = \kappa\left(\rho, \mu, l, \left|\overline{\mathbf{v}} - \mathbf{v}_{(s)}\right|\right)\overline{\mathbf{D}} \tag{4.1.3-17}$$

After an application of the Buckingham–Pi theorem, we conclude that

$$\mathbf{T}^{(t)} + p^{(t)}\mathbf{I} = 2\eta^\star \rho l \left|\overline{\mathbf{v}} - \mathbf{v}_{(s)}\right|\overline{\mathbf{D}} \tag{4.1.3-18}$$

Here

$$\eta^\star = \eta^\star(N) \tag{4.1.3-19}$$

and

$$N \equiv \frac{\rho l \left|\overline{\mathbf{v}} - \mathbf{v}_{(s)}\right|}{\mu} \tag{4.1.3-20}$$

Deissler (1955) has proposed on empirical grounds that

$$\mathbf{T}^{(t)} + p^{(t)}\mathbf{I} = 2n^2 \rho l \left|\overline{\mathbf{v}} - \mathbf{v}_{(s)}\right|[1 - \exp\left(-n^2 N\right)]\overline{\mathbf{D}} \tag{4.1.3-21}$$

We see in the next section that this represents well the time-averaged velocity distribution within the laminar sublayer and buffer zone for turbulent flow through tubes.

### 4.1.4  Wall Turbulence in the Flow Through a Tube

Our purpose is to discuss wall turbulence in the flow of an incompressible Newtonian fluid through a long inclined tube of radius $R$ as shown in Figure 3.2.1-2. We will break the discussion into two parts:

1) fully developed turbulent flow immediately outside the laminar sublayer and buffer zone and
2) the laminar sublayer and buffer zone.

#### Fully Developed Turbulent Flow Immediately Outside the Laminar Sublayer and Buffer Zone

For the moment, we will focus our attention on that portion of the flow in which the turbulence can be considered to be fully developed. We will use the Prandtl mixing-length theory for $\mathbf{T}^{(t)}$ [Equation (4.1.3-13)].

Let us begin by assuming that the time-averaged velocity distribution has the form

$$\overline{v}_z = \overline{v}_z(r)$$
$$\overline{v}_r = \overline{v}_\theta = 0 \tag{4.1.4-1}$$

This means that the viscous and Reynolds stress tensors have only one nonzero shear component:

$$\overline{S}_{rz} + T_{rz}^{(t)} = \left[\mu + \eta_1^\star \rho(R - r)^2 \left|\frac{d\overline{v}_z}{dr}\right|\right]\frac{d\overline{v}_z}{dr} \tag{4.1.4-2}$$

The time-averaged differential mass balance for an incompressible fluid is satisfied identically by (4.1.4-1). The three components of the time-averaged differential momentum balance reduce to

$$\frac{\partial \mathcal{P}}{\partial r} = \frac{\partial \mathcal{P}}{\partial \theta} = 0 \tag{4.1.4-3}$$

and

$$\frac{\partial \mathcal{P}}{\partial z} = \frac{1}{r} \frac{d}{dr} \left[ r \left( \overline{S}_{rz} + T_{rz}^{(t)} \right) \right] \tag{4.1.4-4}$$

Arguing in much the same manner as we did in Section 3.2.1, we conclude that

$$-\frac{\partial \mathcal{P}}{\partial z} = \frac{1}{L} \left( P_0 - P_L - \rho g L \, \sin \alpha \right) \tag{4.1.4-5}$$

and

$$\overline{S}_{rz} + T_{rz}^{(t)} = -S_0 \frac{r}{R} \tag{4.1.4-6}$$

where

$$S_0 \equiv \frac{\left( P_0 - P_L - \rho g L \, \sin \alpha \right) R}{2L} \tag{4.1.4-7}$$

When we recognize that

$$\left| \frac{d\overline{v}_z}{dr} \right| = -\frac{d\overline{v}_z}{dr} \tag{4.1.4-8}$$

Equations (4.1.4-2) and (4.1.4-6) may be combined to say

$$\left[ \mu - \eta_1^\star \rho (R - r)^2 \frac{d\overline{v}_z}{dr} \right] \frac{d\overline{v}_z}{dr} = -\frac{S_0 r}{R} \tag{4.1.4-9}$$

This can be expressed in terms of a dimensionless distance measured from the tube wall

$$s^\star \equiv \frac{R - r}{R} = 1 - \frac{r}{R} \tag{4.1.4-10}$$

and a dimensionless velocity

$$v^\star \equiv \frac{\overline{v}_z}{v_0} \tag{4.1.4-11}$$

as

$$\left( 1 + \frac{\rho v_0 R}{\mu} \eta_1^\star s^{\star 2} \frac{dv^\star}{ds^\star} \right) \frac{dv^\star}{ds^\star} = \frac{R S_0}{\mu v_0} (1 - s^\star) \tag{4.1.4-12}$$

If we identify the characteristic speed as

$$v_0 \equiv \sqrt{\frac{S_0}{\rho}} \tag{4.1.4-13}$$

(4.1.4-12) simplifies to

$$\left( 1 + N_{(t)} \eta_1^\star s^{\star 2} \frac{dv^\star}{ds^\star} \right) \frac{dv^\star}{ds^\star} = N_{(t)} (1 - s^\star) \tag{4.1.4-14}$$

where we define

$$N_{(t)} \equiv \frac{\rho v_0 R}{\mu}$$

$$= \frac{R\sqrt{\rho S_0}}{\mu} \tag{4.1.4-15}$$

In the immediate neighborhood of the wall, $s^\star \ll 1$ and (4.1.4-14) reduces to

$$\left(1 + \eta_1^\star s^{\star\star2} \frac{dv^\star}{ds^{\star\star}}\right) \frac{dv^\star}{ds^{\star\star}} = 1 \tag{4.1.4-16}$$

The expanded variable

$$s^{\star\star} \equiv N_{(t)} s^\star$$

$$= \frac{s\sqrt{\rho S_0}}{\mu} \tag{4.1.4-17}$$

has been introduced here for simplicity.

We began by saying that we would confine our attention to that portion of the flow that can be considered to be fully developed. We suggested in Section 4.1.3 that, in the fully developed portion of the flow, the effect of the viscous stresses was negligible compared with that of the Reynolds stresses. Consequently, we will assume that

$$\eta_1^\star s^{\star\star2} \frac{dv^\star}{ds^{\star\star}} \gg 1 \tag{4.1.4-18}$$

and that, in the fully developed portion of the flow, (4.1.4-16) reduces to

$$\eta_1^\star s^{\star\star2} \left(\frac{dv^\star}{ds^{\star\star}}\right)^2 = 1 \tag{4.1.4-19}$$

In a moment, we will return to check inequality (4.1.4-18).

Equation (4.1.4-19) can be integrated to find

$$\text{for } s^{\star\star} \geq s_1^{\star\star} : \quad v^\star - v_1^\star = \frac{1}{\sqrt{\eta_1^\star}} \ln \frac{s^{\star\star}}{s_1^{\star\star}} \tag{4.1.4-20}$$

where it is convenient to interpret $s_1^{\star\star}$ as the outer edge of the buffer zone and $v_1^\star$ as the dimensionless velocity at this position.

As a result of the comparison with experimental data shown in Figure (4.1.4-1), Deissler (1955) recommends that we take $\sqrt{\eta_1^\star} = 0.36$, $s_1^{\star\star} = 26$, and $v_1^\star = 12.85$. With these values, (4.1.4-20) becomes

$$\text{for } s^{\star\star} \geq 26 : \quad v^\star = \frac{1}{0.36} \ln s^{\star\star} + 3.8 \tag{4.1.4-21}$$

This means that

$$\text{for } s^{\star\star} \geq 26 : \quad \eta_1^\star s^{\star\star2} \frac{dv^\star}{ds^{\star\star}} \geq 9.4 \tag{4.1.4-22}$$

and inequality (4.1.4-18) appears to be justified. Equation (4.1.4-21) is shown in Figure 4.1.4-1.

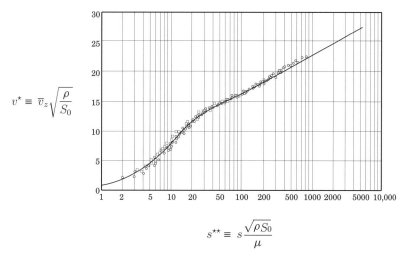

$$s^{\star\star} \equiv s\,\frac{\sqrt{\rho S_0}}{\mu}$$

**Figure 4.1.4-1.** Velocity distribution for wall turbulence in isothermal flow in tubes as reported by Deissler (1955, p. 3). For $s^{\star\star} \leq 5$, the data are described by (4.1.4-28); for $s^{\star\star} \geq 26$, by (4.1.4-21). For $5 < s^{\star\star} < 26$ and $n = 0.124$, a numerical integration of (4.1.4-25) consistent with (4.1.4-26) is in excellent agreement with the data. The experimental data shown are those of Deissler (1950) and Laufer (1953), indicated by circles and squares, respectively.

Notice in Figure 4.1.4-1 that $dv^{\star}/ds^{\star\star} \neq 0$ at the center of the tube, in contrast with our intuition. But our intuition has not failed us. Rather, as explained previously, we have restricted ourselves to the region near the wall (although outside the laminar sublayer and buffer zone) in deriving (4.1.4-16) and (4.1.4-21). We should not expect it to be valid for the region near the axis of the tube.

### In the Laminar Sublayer and Buffer Zone

We will use the empirical proposal of Deissler (1955) to describe the velocity distribution in the laminar sublayer and in the buffer zone.

If we again start with the assumption that the time-averaged velocity distribution has the form indicated in (4.1.4-1), the viscous and Reynolds stress tensors have only one nonzero component:

$$\overline{S}_{rz} + T_{rz}^{(t)} = \left( \mu + n^2 \rho \overline{v}_z (R - r) \times \left\{ 1 - \exp\left( \frac{-n^2 \rho \overline{v}_z [R - r]}{\mu} \right) \right\} \right) \frac{d\overline{v}_z}{dr} \qquad (4.1.4\text{-}23)$$

In view of (4.1.4-6), this means that

$$\{1 + n^2 v^{\star} s^{\star\star} [1 - \exp\left(-n^2 v^{\star} s^{\star\star}\right)]\} \frac{dv^{\star}}{ds^{\star\star}} = 1 - \frac{s^{\star\star}}{N_{(t)}} \qquad (4.1.4\text{-}24)$$

where, for convenience, we have introduced the dimensionless variables defined by (4.1.4-11) and (4.1.4-13). Since we are interested here in the laminar sublayer and buffer zone, we can restrict ourselves to the immediate neighborhood of the wall:

$$s^{\star} = \frac{s^{\star\star}}{N_{(t)}} \ll 1 : \left\{1 + n^2 v^{\star} s^{\star\star} \left[1 - \exp\left(-n^2 v^{\star} s^{\star\star}\right)\right]\right\} \frac{dv^{\star}}{ds^{\star\star}} = 1 \qquad (4.1.4\text{-}25)$$

This can be integrated numerically consistent with the constraint that

$$\text{at } s^{**} = 0: \quad v^* = 0 \tag{4.1.4-26}$$

The result for $n = 0.124$, recommended by Deissler (1955), is shown in Figure 4.1.4-1. For $s^{**} \leq 26$, it represents very well the velocity distribution in both the laminar sublayer and in the buffer zone.

If we are primarily interested in the laminar sublayer, we should examine the limit of (4.1.4-25) as the wall is approached:

$$\text{as } s^{**} \to 0: \quad \frac{dv^*}{ds^{**}} \to 1 \tag{4.1.4-27}$$

Integrating, we find that the velocity distribution in the laminar sublayer should have the form

$$v^* = s^{**} \tag{4.1.4-28}$$

Referring to Figure 4.1.4-1, we see that this relationship provides a very good representation for $s^{**} \leq 5$. The laminar sublayer for flow through very long tubes is defined to be that region in which (4.1.4-28) describes the time-averaged velocity distribution.

**Exercise 4.1.4-1** *Turbulent flow between two flat plates*   Repeat the analysis of this section for turbulent flow through the inclined channel shown in Figure 3.2.1-5. Determine that (4.1.4-20) and (4.1.4-25) again apply with the understanding that, here,

$$S_0 \equiv \frac{(P_0 - P_L - \rho g L \sin \alpha)b}{L}$$

and

$$v^* \equiv \frac{\overline{v}_1}{v_0}$$

---

## 4.2    Area Averaging

For most engineering purposes, complete velocity distributions are not required. We are usually concerned with estimating some macroscopic aspect of a problem such as a volume flow rate or a force on a wall.

When the dependence of the velocity distribution upon the directions normal to the macroscopic flow do not appear to be of prime interest, it may be wise to average the equation of motion over the cross section normal to the flow. This can lead to a considerable simplification. For this reason, it is a particularly desirable approach when the original problem posed requires considerable time and money for solution. You must make a judgment in the context of the application with which you are concerned. If you must have an answer accurate within 1 percent, a complete solution is required. If you are willing to accept as much as a 20 or 25 percent error (and no rigorous error bounds), an integral averaging technique, such as area averaging, may be useful.

With any of the integral averaging techniques, information is lost that must be replaced by an empiricism or an approximation. In time averaging, we found that it was necessary to supply an empirical data correlation for the Reynolds stress tensor $\mathbf{T}^{(t)}$. In area averaging, there are two ways in which this empiricism can be introduced.

In the first class of problems, as illustrated in Section 4.2.1, we concern ourselves primarily with an area-averaged variable, perhaps a volume flow rate. Normally, an approximation is made concerning the force per unit area or stress at a bounding wall.

The second and more highly developed class of problems is often referred to as *approximate boundary-layer theory* (Schlichting 1979, p. 206; Slattery 1981, p. 186). In approximate boundary-layer theory, the form of the velocity distribution is assumed in terms of a function $\delta(x)$ of the arc length $x$. This function is often referred to as the (approximate) boundary-layer thickness. The area-averaged equation of motion yields an ordinary differential equation for $\delta$. Approximate boundary-layer theory was originally developed to avoid difficult numerical integrations. With better computers and better numerical methods, it finds relatively little use today.

### 4.2.1  Flow from Rest in a Circular Tube

To test the accuracy of this approach, let us consider flow from rest in an inclined tube of an incompressible Newtonian fluid, since we can compare with the exact solution for this problem developed in Exercise 3.2.4-6. We will assume that we are primarily concerned with determining the volume rate of flow $Q$ through the tube as a function of time. In other words, we would like to find the area-averaged axial component of velocity as a function of time.

Having assumed in Exercise 3.2.4-6 that there was only one nonzero component of velocity,

$$v_z = v_z(r, t)$$
$$v_r = v_\theta = 0 \tag{4.2.1-1}$$

we found that the differential momentum balance implied for an incompressible Newtonian fluid that

$$\frac{\partial v_z}{\partial t} = -A + \frac{1}{r}\frac{\partial}{\partial r}(r S_{rz}) \tag{4.2.1-2}$$

where

$$-A \equiv \frac{P_0 - P_L}{L} - \rho g \sin \alpha \tag{4.2.1-3}$$

Equation (4.2.1-2) and the appropriate constitutive equation for $S_{rz}$ were solved simultaneously, consistent with the initial and boundary conditions

$$\text{at } t = 0: \ v_z = 0 \tag{4.2.1-4}$$

and

$$\text{at } r = R: \ v_z = 0 \tag{4.2.1-5}$$

But if we are primarily interested in the area-averaged velocity

$$\bar{v}_z \equiv \frac{1}{\pi R^2} \int_0^{2\pi} \int_0^R v_z r \, dr \, d\theta \tag{4.2.1-6}$$

we might consider averaging (4.2.1-2) over the cross section of the tube normal to flow:

$$\rho \frac{d\bar{v}_z}{dt} = -A + \frac{1}{\pi R^2} \int_0^{2\pi} \int_0^R \frac{\partial}{\partial r}(r\,S_{rz})\,dr\,d\theta$$

$$= -A + \frac{2}{R}\,S_{rz}|_{r=R} \tag{4.2.1-7}$$

There is a problem here. We do not know *a priori* $S_{rz}|_{r=R}$. We have lost some information in averaging, just as we did in Section 4.1.1. We need either an empirical data correlation or an approximation, in order to evaluate $S_{rz}|_{r=R}$. Perhaps the simplest thing to do is to say that the relationship between $S_{rz}|_{r=R}$ and $\bar{v}_z$ can be approximated as that found for steady-state flow in Section 3.2.1:

$$S_{rz}|_{r=R} = -\frac{4\mu}{R}\bar{v}_z \tag{4.2.1-8}$$

In applying this approximation, we find that (4.2.1-7) yields an ordinary differential equation for the area-averaged velocity:

$$\rho \frac{d\bar{v}_z}{dt} = -A - \frac{8\mu}{R^2}\bar{v}_z \tag{4.2.1-9}$$

This can easily be integrated consistent with initial condition (4.2.1-4) in the form of

$$\text{at } t = 0: \ \bar{v}_z = 0 \tag{4.2.1-10}$$

to find

$$\frac{8\mu Q}{\pi R^4(-A)} = 1 - \exp\left(-8t^\star\right) \tag{4.2.1-11}$$

where

$$t^\star \equiv \frac{\mu t}{\rho R^2} \tag{4.2.1-12}$$

In Figure 4.2.1-1, we compare this last expression with the exact solution given in Exercise 3.2.4-6. The error introduced in using (4.2.1-11) is acceptable for most purposes for $t^\star > 0.5$.

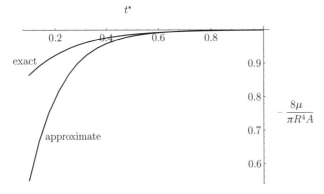

**Figure 4.2.1-1.** Comparison of area-averaged analysis with exact solution for flow from rest in a circular tube. The upper curve is from Exercise 3.2.4-6; the lower curve is from (4.2.1-11).

## **4.3** Local Volume Averaging

The movement of gases and liquids through porous media is common to many industrial processes. Distillation and absorption columns are often filled with beads or packing in a variety of shapes. A chemical reactor may be filled with porous pellets impregnated with a catalyst. Filters are employed in most chemical processes. The movements of water and oil through porous rock are important in water conservation and oil exploration.

Darcy (1856) made the first serious study of this problem. As a correlation of experimental data for water moving axially with a volume flow rate $Q$ through a cylindrical packed bed of cross-sectional area $A$ and length $L$ under the influence of a pressure difference $\Delta p$, he proposed (Scheidegger 1963, p. 634)

$$\frac{\Delta(p + \rho\phi)}{L} = b\frac{Q}{A} \tag{4.3.0-1}$$

It was later observed that $b$ is proportional to the viscosity for an incompressible Newtonian fluid,

$$\frac{\Delta(p + \rho\phi)}{L} = \frac{\mu}{k}\frac{Q}{A} \tag{4.3.0-2}$$

This is usually referred to as *Darcy's law*, and the coefficient $k$ is referred to as the *permeability*.

Equation (4.3.0-2) prompts the question: How should one analyze flow in other geometries or under the influence of other boundary conditions? The standard answer has been to say that a differential equation, inspired by (4.3.0-2) and commonly referred to as Darcy's law, describes the flow at each point in the porous medium (Scheidegger 1963, p. 634):

$$-\nabla p + \rho\mathbf{f} = \frac{\mu}{k}\mathbf{w} \tag{4.3.0-3}$$

A major difficulty with this equation has been that, since originally it was not derived, the average pressure $p$ and the average velocity $\mathbf{w}$ were not defined in terms of the local pressure and velocity distribution in the pores. It is in attempting to avoid this difficulty that local volume averaging will be introduced.

The concept of local volume averaging was intuitively obvious to many. The formal derivation of the local volume-averaged differential mass, momentum, and energy balances was made practical by the development of the theorem for the local volume average of a gradient, which was described independently in 1967 by Anderson and Jackson (1967), Marle (1967), Slattery (1967), and Whitaker (1967b). Drew (1971) and Bear (1972, p. 90) proposed local volume averaging, but they made no mention of the theorem for the local volume average of a gradient. Drew (1971) suggested multiple volume averages, in order to ensure that the averaged variables were sufficiently differentiable. Although this is generally not necessary, Quintard and Whitaker (1994) have shown that a double average is necessary for the treatment of spatially periodic systems. Instead of the theorem for the local volume average of a gradient, Bachmat (1972) derived a theorem for the local volume average of a material derivative that leads to somewhat less explicit forms of the local volume-averaged differential balances.

In the sections that follow, I restrict my attention to a porous structure that is rigid, stationary, and filled with a single fluid. For this reason, the discussion will not apply for example to filter cakes, which are normally compressible. Because we will be confining our attention to flows in porous media, we will also assume that the fluid is in laminar flow. But keep in mind that these restrictions are not necessary; they are made for convenience in introducing the subject.

For more on the development and application of local volume averaging, I suggest that you examine the literature (Whitaker 1966; Slattery 1967, 1968b, 1969, 1970, 1974; Whitaker 1969, 1973; Patel, Hedge, and Slattery 1972; Gray 1975; Gray and O'Neil 1976; Sha and Slattery 1980; Lin and Slattery 1982b; Jiang et al. 1987). I also encourage you to examine the extensions to local area averaging (Wallis 1969; Delhaye 1977a, 1981b) and to local space-time averaging (Drew 1971; Delhaye 1977a, 1981b; Sha and Slattery 1980).

### 4.3.1  The Concept

From one point of view, no special equations are required to describe a dispersed flow. The implications of conservation of mass, the momentum balance, and the moment of momentum balance derived in Chapters 1 and 2 are sufficient.

Let us think for the moment about the flow of water through a permeable rock. The first problem is to describe geometrically the configuration of the pore walls bounding the flow. The configuration of the pore space in a permeable rock, in a bed of sand, or in an irregular bed of spheres will normally be, at least in part, a random function of position in space, in which case we will not have an equation available with which to describe it. Even if we consider a regular arrangement of spheres, the description of the pore geometry may be too troublesome for most purposes. Somehow we would like to speak about the movement of the fluid without having to describe in detail the pores through which it is moving.

For most purposes, we are not particularly concerned with the detailed velocity distribution of the water within a single pore. We may not care even about its average velocity within a single pore. We are more concerned with the variation of the average velocity of the water as it moves through the rock over distances that are large compared with the average diameter of a pore. Our primary concern is with averages defined at every position in the porous media. The approach taken is to speak in terms of local volume averages at each point in the structure.

Our initial objective is to associate with every point in a porous medium a local volume average of the differential mass balance for the fluid:

$$\frac{\partial \rho}{\partial t} + \mathrm{div}(\rho \mathbf{v}) = 0 \tag{4.3.1-1}$$

When I say every point in the porous medium, I include the solid phase as well as the fluid phase and the solid–fluid phase interface.

Referring to Figure 4.3.1-1, let us begin by thinking of a particular point $\mathbf{z}$ in the porous medium. It makes no difference whether this point is located in the solid phase, the fluid phase, or on the solid–fluid phase interface; the argument remains unchanged. Let us associate with this point a closed surface $S$. I have chosen a sphere in Figure 4.3.1-1.

We will associate this averaging surface $S$ with every point in the porous medium by a simple translation of $S$ without rotation. The diameter of $S$ should be sufficiently large that averages over the pore space enclosed by $S$ vary smoothly with position. Whenever possible, the diameter of $S$ should be so small as to be negligible with respect to a characteristic dimension of the macroscopic porous body. Yet it should not be so small that $S$ encloses

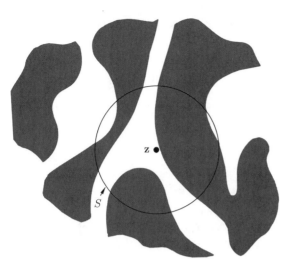

**Figure 4.3.1-1.** The averaging surface $S$ to be associated
with every point $\mathbf{z}$ in the porous medium.

only solid or only fluid at many points in the porous structure. The minimum size of $S$ will
be discussed shortly.

Let $R$ be the region enclosed by $S$; $V$ is its volume. Let $R^{(f)}$ denote the pores that contain
fluid in the interior of $S$; the volume of $R^{(f)}$ as well as its shape in general will change from
point to point in the porous medium. The closed boundary surface $S^{(f)}$ of $R^{(f)}$ is the sum of
$S_e$ and $S_w$ : $S_e$ coincides with $S$ and $S_w$ coincides with the pore walls. We may think of $S_e$ as
the entrance and exit surfaces of $R^{(f)}$ through which fluid passes in and out.

Let us write a mass balance for the fluid contained within this closed surface $S$. The most
convenient way of doing this is to integrate the differential mass balance (4.3.1-1) over $R^{(f)}$,
the region of space occupied by the fluid within $S$:

$$\int_{R^{(f)}} \left[ \frac{\partial \rho}{\partial t} + \text{div}(\rho \mathbf{v}) \right] dV = 0 \tag{4.3.1-2}$$

We can immediately interchange the operations of volume integration and differentiation
with respect to time in the first term on the left to find

$$\frac{\partial \overline{\rho}^{(f)}}{\partial t} + \frac{1}{V} \int_{R^{(f)}} \text{div}(\rho \mathbf{v}) \, dV = 0 \tag{4.3.1-3}$$

where

$$\overline{\rho}^{(f)} \equiv \frac{1}{V} \int_{R^{(f)}} \rho \, dV \tag{4.3.1-4}$$

Assume that $B$ is some scalar, vector, or tensor associated with phase $i$. We will have
occasion to speak of

$$\overline{B}^{(i)} \equiv \frac{1}{V} \int_{R^{(i)}} B \, dV \tag{4.3.1-5}$$

as the *superficial volume average* for phase $i$ of $B$ (the mean value of $B^{(i)}$ in $R$),

$$\langle B \rangle^{(i)} \equiv \frac{1}{V^{(i)}} \int_{R^{(i)}} B \, dV \tag{4.3.1-6}$$

as the *intrinsic volume average* for phase $i$ of $B$ (the mean value of $B$ in $R^{(i)}$), and

$$\langle B \rangle \equiv \frac{1}{V} \int_R B \, dV$$

$$= \sum_{i=1}^{M} \overline{B}^{(i)} \tag{4.3.1-7}$$

as the *total volume average* of $B$ over all $M$ phases present (the mean value of $B$ in $R$).

Let $L_0$ be a characteristic dimension of $S$. The minimum acceptable size of $S$ or the minimum acceptable value of $L_0$ is such that $\overline{B}^{(f)}$ is nearly independent of position over distances of the same order of magnitude. This implies that

$$\langle \overline{B}^{(f)} \rangle^{(f)} = \overline{B}^{(f)} \tag{4.3.1-8}$$

$$\langle \langle B \rangle^{(f)} \rangle^{(f)} = \langle B \rangle^{(f)} \tag{4.3.1-9}$$

It would be nice if we could interchange the volume integration with the divergence operation in the second term on the left of (4.3.1-3). But the limits on this volume integration depend upon the pore geometry enclosed by $S$, and they must be functions of position $\mathbf{z}$. The next section explores this problem in more detail.

## 4.3.2  Theorem for the Local Volume Average of a Gradient

Let $B$ be any scalar, spatial vector, or second-order tensor associated with the fluid. Given

$$\overline{\nabla B}^{(f)} \equiv \frac{1}{V} \int_{R^{(f)}} \nabla B \, dV \tag{4.3.2-1}$$

let us ask in what sense we might interchange the volume average with the gradient operation to obtain

$$\nabla \overline{B}^{(f)} \equiv \nabla \left( \frac{1}{V} \int_{R^{(f)}} B \, dV \right) \tag{4.3.2-2}$$

Let us associate the averaging surface $S$, which was introduced in Section 4.3.1, with every point in the porous medium. We will do this by a simple translation of $S$ without rotation. As an example, if $S$ is a unit sphere, the center of which coincides with the point initially considered, we center upon each point in the porous medium a unit sphere. If $S$ is small compared with the average pore diameter, it may enclose only solid or only fluid at many points; if it is large, many pores may intersect $S$, the intersections serving as entrances and exits to the fluid enclosed by $S$.

Consider any arbitrary curve running through the porous medium as shown in Figure 4.3.2-1. Let $s$ be a parameter such as arc length measured along this curve. We can identify with each point along this curve a region denoted by $R^{(f)}$, composed of the pores containing fluid enclosed by surface $S$. We may think of $R^{(f)}$ as a function of the parameter $s$ along this

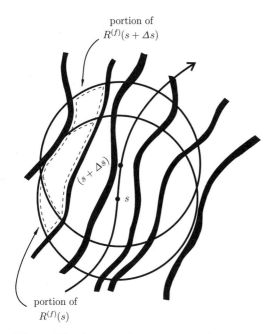

portion of
$R^{(f)}(s + \Delta s)$

$(s + \Delta s)$

$s$

portion of
$R^{(f)}(s)$

**Figure 4.3.2-1.** An arbitrary curve running through
the porous medium, where $s$ is a parameter measured along this curve.

curve. If we simply replace the parameter time by $s$ in the generalized transport theorem of
Section 1.3.2, we have

$$\frac{d}{ds} \int_{R^{(f)}} B \, dV = \int_{R^{(f)}} \frac{\partial B}{\partial s} dV + \int_{S^{(f)}} B \frac{d\mathbf{p}}{ds} \cdot \mathbf{n} \, dA \qquad (4.3.2\text{-}3)$$

Here $\mathbf{p}$ is the position vector field.

Let us further restrict ourselves to quantities $B$ that are explicit functions of position (and
time) only:

$$\frac{\partial B}{\partial s} = 0 \qquad (4.3.2\text{-}4)$$

(By $\partial B/\partial s$, we mean a derivative with respect to $s$ holding position and time fixed.) As a
function of $s$, the artificial system particles clearly move with the normal component of the
velocity of $S_w$:

$$\text{on } S_w : \quad \frac{d\mathbf{p}}{ds} \cdot \mathbf{n} = 0 \qquad (4.3.2\text{-}5)$$

Let $\mathbf{r}_0(s)$ be the position vector locating the point $s$ on the arbitrary curve, and let $\mathbf{r}(s)$ be
the position vector locating points on $S^{(f)}$ relative to this point $s$ (which is at the center of
the sphere for the case illustrated in Figures 4.3.2-1 and 4.3.2-2). Provided $S$ is translated
without rotation along this arbitrary curve in identifying it with every point in the porous
medium, we can say

$$\text{on } S_e : \quad \frac{d\mathbf{p}}{ds} = \frac{d\mathbf{r}_0}{ds} + \frac{d\mathbf{r}}{ds}$$

$$= \frac{d\mathbf{r}_0}{ds} \qquad (4.3.2\text{-}6)$$

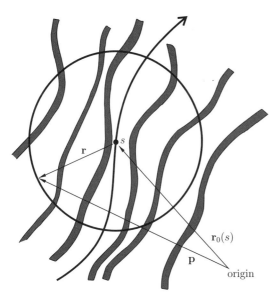

**Figure 4.3.2-2.** The vector $\mathbf{r}_0(s)$ denotes the position of the point $s$ along the curve; $\mathbf{r}$ denotes position on $S^{(f)}$ with respect to the point $s$.

Equations (4.3.2-4) through (4.3.2-6) allow us to rewrite (4.3.2-3) as

$$\frac{d}{ds}\int_{R^{(f)}} B\,dV = \frac{d\mathbf{r}_0}{ds}\cdot\nabla\int_{R^{(f)}} B\,dV$$

$$= \int_{S_e} B\frac{d\mathbf{r}_0}{ds}\cdot\mathbf{n}\,dA \qquad (4.3.2\text{-}7)$$

or, since $d\mathbf{r}_0/ds$ is independent of position on $S_e$,

$$\frac{d\mathbf{r}_0}{ds}\cdot\nabla\int_{R^{(f)}} B\,dV = \left(\int_{S_e} B\mathbf{n}\,dA\right)\cdot\frac{d\mathbf{r}_0}{ds} \qquad (4.3.2\text{-}8)$$

Because we have been concerned with any arbitrary curve running through the porous medium, this implies

$$\nabla\int_{R^{(f)}} B\,dV = \int_{S_e} B\mathbf{n}\,dA \qquad (4.3.2\text{-}9)$$

An application of Green's transformation allows us to express (4.3.2-1) as

$$\frac{1}{V}\int_{R^{(f)}}\nabla B\,dV = \frac{1}{V}\int_{S_e+S_w} B\mathbf{n}\,dA \qquad (4.3.2\text{-}10)$$

In view of (4.3.2-9), we have the useful result

$$\overline{\nabla B}^{\,(f)} \equiv \frac{1}{V}\int_{R^{(f)}}\nabla B\,dV = \nabla\left(\frac{1}{V}\int_{R^{(f)}} B\,dV\right) + \frac{1}{V}\int_{S_w} B\mathbf{n}\,dA$$

$$= \nabla\overline{B}^{\,(f)} + \frac{1}{V}\int_{S_w} B\mathbf{n}\,dA \qquad (4.3.2\text{-}11)$$

We may refer to this as the *theorem for the volume average of a gradient*.

A special case of (4.3.2-11) is

$$\overline{\operatorname{div}^{(f)} \mathbf{B}} \equiv \frac{1}{V} \int_{R^{(f)}} \operatorname{div} \mathbf{B} \, dV$$

$$= \operatorname{div} \overline{\mathbf{B}}^{(f)} + \frac{1}{V} \int_{S_w} \mathbf{B} \cdot \mathbf{n} \, dA \tag{4.3.2-12}$$

Here $\mathbf{B}$ should be interpreted as a spatial vector field or second-order tensor field. Equation (4.3.2-12) may be referred to as the *theorem for the volume average of the divergence*.

Let us apply these results in obtaining the local volume average of the differential mass balance.

### 4.3.3 The Local Volume Average of the Differential Mass Balance

In Section 4.3.1, we found that the local volume average of the differential mass balance could be written as

$$\frac{\partial \overline{\rho}^{(f)}}{\partial t} + \frac{1}{V} \int_{R^{(f)}} \operatorname{div}(\rho \mathbf{v}) \, dV = 0 \tag{4.3.3-1}$$

An application of the theorem of Section 4.3.2 allows us to express this as

$$\frac{\partial \overline{\rho}^{(f)}}{\partial t} + \operatorname{div}\left(\overline{\rho \mathbf{v}}^{(f)}\right) = 0 \tag{4.3.3-2}$$

In arriving at this result, we have observed that the velocity of the fluid is zero on the pore walls $S_w$. It is unfortunate that the superficial volume average $\overline{\rho v}^{(f)}$ occurs rather than the product of the superficial volume averages $\overline{\rho}^{(f)} \, \overline{v}^{(f)}$.

For the special case of an incompressible fluid, we have the simpler result

$$\operatorname{div} \overline{\mathbf{v}}^{(f)} = 0 \tag{4.3.3-3}$$

It is for this reason that incompressible fluids are easier to work with in considering flows through porous media.

**Exercise 4.3.3-1** Equation (4.3.3-3) can be obtained immediately by taking a local volume average of the differential mass balance for an incompressible fluid. How does one obtain the same result from (4.3.3-2)?

*Hint:* Note that

$$\overline{\rho}^{(f)} = \rho \frac{V^{(f)}}{V}$$

### 4.3.4 The Local Volume Average of the Differential Momentum Balance

We can start as we did in Section 4.3.1, where we began considering the local volume average of the differential mass balance. Let us think of a particular point $\mathbf{z}$ in the porous medium, and let us integrate the differential momentum balance over $R^{(f)}$, the region of space occupied by the fluid within $S$ associated with $\mathbf{z}$:

$$\frac{1}{V} \int_{R^{(f)}} \left[ \frac{\partial (\rho \mathbf{v})}{\partial t} + \operatorname{div}(\rho \mathbf{v} \mathbf{v}) - \operatorname{div} \mathbf{T} - \rho \mathbf{f} \right] dV = 0 \tag{4.3.4-1}$$

The operations of volume integration and differentiation with respect to time may be interchanged in the first term on the left:

$$\frac{1}{V} \int_{R^{(f)}} \frac{\partial(\rho \mathbf{v})}{\partial t} dV = \frac{\partial \overline{\rho \mathbf{v}}^{(f)}}{\partial t} \tag{4.3.4-2}$$

The theorem of Section 4.3.2 can be used to express the second and third terms on the left of (4.3.4-1) as

$$\frac{1}{V} \int_{R^{(f)}} \mathrm{div}(\rho \mathbf{v}\mathbf{v}) \, dV = \mathrm{div}\left(\overline{\rho \mathbf{v}\mathbf{v}}^{(f)}\right) \tag{4.3.4-3}$$

and

$$\frac{1}{V} \int_{R^{(f)}} \mathrm{div}\, \mathbf{T}\, dV = \mathrm{div}\, \overline{\mathbf{T}}^{(f)} + \frac{1}{V} \int_{S_w} \mathbf{T} \cdot \mathbf{n}\, dA \tag{4.3.4-4}$$

In arriving at (4.3.4-3), we have observed that the velocity vector must be zero at the fluid–solid phase interface $S_w$. In view of (4.3.4-2) through (4.3.4-4), Equation (4.3.4-1) becomes

$$\frac{\partial \overline{\rho \mathbf{v}}^{(f)}}{\partial t} + \mathrm{div}\left(\overline{\rho \mathbf{v}\mathbf{v}}^{(f)}\right) = \mathrm{div}\, \overline{\mathbf{T}}^{(f)} + \overline{\rho \mathbf{f}}^{(f)} + \frac{1}{V} \int_{S_w} \mathbf{T} \cdot \mathbf{n}\, dA \tag{4.3.4-5}$$

Throughout the remainder of our discussion of flow through porous media, we will restrict ourselves to incompressible fluids, and we will assume that all inertial effects may be neglected in the local volume average of the differential momentum balance. We shall also find it convenient to assume that the external force per unit mass $\mathbf{f}$ may be represented by a scalar potential $\varphi$:

$$\mathbf{f} = -\nabla \varphi \tag{4.3.4-6}$$

With these restrictions, (4.3.4-5) simplifies to

$$\mathrm{div}\left(\overline{\mathbf{T}}^{(f)} - \rho \overline{\varphi}^{(f)} \mathbf{I}\right) + \frac{1}{V} \int_{S_w} (\mathbf{T} - \rho \varphi \mathbf{I}) \cdot \mathbf{n}\, dA = 0 \tag{4.3.4-7}$$

or

$$\nabla \overline{(\mathcal{P} - p_0)}^{(f)} - \mathrm{div}\, \overline{\mathbf{S}}^{(f)} + \mathbf{g} = 0 \tag{4.3.4-8}$$

Here $\mathcal{P}$ is the modified pressure

$$\mathcal{P} \equiv p + \rho \varphi \tag{4.3.4-9}$$

and $\mathbf{S}$ is the viscous portion of the stress tensor. A constant reference or ambient pressure $p_0$ is introduced here so that we may identify

$$\mathbf{g} \equiv -\frac{1}{V} \int_{S_w} [\mathbf{T} + (p_0 - \rho \varphi)\mathbf{I}] \cdot \mathbf{n}\, dA \tag{4.3.4-10}$$

as the force per unit volume that the fluid exerts upon the pore walls contained within $S$ beyond the hydrostatic force and beyond any force attributable to the ambient pressure. This force per unit volume $\mathbf{g}$ is assignable to the motion of the fluid.

For an incompressible Newtonian fluid,

$$\overline{\mathbf{S}}^{(f)} = \mu \left[ \overline{\nabla \mathbf{v}}^{(f)} + \overline{(\nabla \mathbf{v})^T}^{(f)} \right] \tag{4.3.4-11}$$

The theorem of Section 4.3.2 and the fact that the velocity of the fluid is zero at the pore walls allow us to say

$$
\overline{\nabla \mathbf{v}}^{(f)} \equiv \frac{1}{V} \int_{R^{(f)}} \nabla \mathbf{v} \, dV
$$

$$
= \frac{1}{V} \int_{S_e + S_w} \mathbf{v} \mathbf{n} \, dA
$$

$$
= \nabla \overline{\mathbf{v}}^{(f)} \tag{4.3.4-12}
$$

Exactly the same argument may be used to show

$$
\overline{(\nabla \mathbf{v})^T}^{(f)} = \left( \nabla \overline{\mathbf{v}}^{(f)} \right)^T \tag{4.3.4-13}
$$

Consequently,

$$
\overline{\mathbf{S}}^{(f)} = \mu \left[ \nabla \overline{\mathbf{v}}^{(f)} + \left( \nabla \overline{\mathbf{v}}^{(f)} \right)^T \right] \tag{4.3.4-14}
$$

and

$$
\operatorname{div} \overline{\mathbf{S}}^{(f)} = \mu \operatorname{div} \left( \nabla \overline{\mathbf{v}}^{(f)} \right) \tag{4.3.4-15}
$$

In summary, when we neglect all inertial effects and assume that the external force per unit mass can be represented by the gradient of a scalar potential, the local volume average of the differential momentum balance for an incompressible Newtonian fluid can be written as

$$
\nabla \overline{(\mathcal{P} - p_0)}^{(f)} - \mu \operatorname{div} (\nabla \overline{\mathbf{v}}^{(f)}) + \mathbf{g} = 0 \tag{4.3.4-16}
$$

In the next section we discuss the preparation of empirical correlations for $\mathbf{g}$.

## 4.3.5 Empirical Correlations for $\mathbf{g}$

In this section, we use three examples to indicate how experimental data can be used to prepare correlations for $\mathbf{g}$, introduced in Section 4.3.4. We base this discussion upon four points:

1) The force per unit volume $\mathbf{g}$ is frame indifferent:

$$
\mathbf{g}^* = -\frac{1}{V} \int_{S_w} [\mathbf{T}^* + (p_0 - \rho \varphi) \mathbf{I}^*] \cdot \mathbf{n}^* \, dA
$$

$$
= -\frac{1}{V} \int_{S_w} \mathbf{Q} \cdot [\mathbf{T} + (p_0 - \rho \varphi) \mathbf{I}] \cdot \mathbf{n} \, dA
$$

$$
= \mathbf{Q} \cdot \mathbf{g} \tag{4.3.5-1}
$$

Here $\mathbf{Q}$ is a (possibly) time-dependent, orthogonal, second-order tensor.

2) We assume that the principle of frame indifference introduced in Section 2.3.1 applies to any empirical correlation developed for $\mathbf{g}$.

3) The Buckingham–Pi theorem serves to further restrict the form of any expression for $\mathbf{g}$.

4) The averaging surface $S$ is sufficiently large that $\mathbf{g}$ may be assumed *not* to be an *explicit* function of position in the porous structure, although it very well may be an *implicit* function of position as the result of its dependence upon other quantities.

## Example 1: Flow of Newtonian Fluid in a Nonoriented Medium

By an *oriented* porous structure (the term *anisotropic* is in common use), I mean one that has a direction or a set of directions intrinsically associated with the pore geometry. For example, in a naturally occurring stratified rock there is often a gradient in "particle diameter" in the direction of gravity or in what was originally the direction of gravity, when the sediment was deposited. A *nonoriented* porous structure (often referred to as *isotropic* has no such direction intrinsically associated with it.

For the moment, let us assume that **g** is a function of the difference between the intrinsic volume-averaged velocity of the fluid $\langle \mathbf{v} \rangle^{(f)}$ and the intrinsic volume-average velocity of the solid:

$$\langle \mathbf{u} \rangle^{(s)} \equiv \frac{1}{V_{(s)}} \int_{R_{(s)}} \mathbf{u} \, dV \tag{4.3.5-2}$$

Here **u** is the velocity distribution within the solid (which may be undergoing a rigid-body rotation and translation); $V_{(s)}$ is the volume occupied by the solid within the averaging surface $S$. Equivalently, we can assume **g** is a function of the difference between the superficial volume-averaged velocity of the fluid $\overline{\mathbf{v}}^{(f)}$ and $\Psi \langle \mathbf{u} \rangle^{(s)}$:

$$\mathbf{g} = \hat{\mathbf{g}}(\overline{\mathbf{v}}^{(f)} - \Psi \langle \mathbf{u} \rangle^{(s)}) \tag{4.3.5-3}$$

where

$$\Psi \equiv \frac{R^{(f)}}{V} \tag{4.3.5-4}$$

denotes the local *porosity* of the structure (assuming all of the pores are filled with fluid).

By the principle of frame indifference, the functional relationship between these variables should be the same in every frame of reference. This means that

$$
\begin{aligned}
\mathbf{g}^* &= \mathbf{Q} \cdot \mathbf{g} \\
&= \mathbf{Q} \cdot \hat{\mathbf{g}} \left( \overline{\mathbf{v}}^{(f)} - \Psi \langle \mathbf{u} \rangle^{(s)} \right) \\
&= \hat{\mathbf{g}} \left[ \mathbf{Q} \cdot \left( \overline{\mathbf{v}}^{(f)} - \Psi \langle \mathbf{u} \rangle^{(s)} \right) \right]
\end{aligned} \tag{4.3.5-5}
$$

or **g** is an isotropic function (Truesdell and Noll 1965, p. 22):

$$\hat{\mathbf{g}} \left( \overline{\mathbf{v}}^{(f)} - \Psi \langle \mathbf{u} \rangle^{(s)} \right) = \mathbf{Q}^T \cdot \hat{\mathbf{g}} \left[ \mathbf{Q} \cdot \left( \overline{\mathbf{v}}^{(f)} - \Psi \langle \mathbf{u} \rangle^{(s)} \right) \right] \tag{4.3.5-6}$$

By a representation theorem for a vector-valued isotropic function of one vector (Truesdell and Noll 1965, p. 35), we may write

$$
\begin{aligned}
\mathbf{g} &= \hat{\mathbf{g}} \left( \overline{\mathbf{v}}^{(f)} - \Psi \langle \mathbf{u} \rangle^{(s)} \right) \\
&= R \left[ \overline{\mathbf{v}}^{(f)} - \Psi \langle \mathbf{u} \rangle^{(s)} \right]
\end{aligned} \tag{4.3.5-7}
$$

It is to be understood here that the resistance coefficient $R$ is a function of the magnitude of the local volume-averaged velocity of the fluid relative to the local volume-averaged velocity of the solid $\left| \overline{\mathbf{v}}^{(f)} - \Psi \langle \mathbf{u} \rangle^{(s)} \right|$ and a function of the viscosity of the fluid $\mu$, the porosity $\Psi$, as well as a characteristic length $l_0$ of the porous medium:

$$R = R \left( \left| \overline{\mathbf{v}}^{(f)} - \Psi \langle \mathbf{u} \rangle^{(s)} \right|, \mu, \Psi, l_0 \right) \tag{4.3.5-8}$$

We have not considered the fluid density here, since it does not appear in the local volume-averaged differential momentum balance (4.3.4-8) or the local volume-averaged differential mass balance for an incompressible fluid (4.3.3-3). By the Buckingham–Pi theorem (Brand 1957), Equation (4.3.5-8) can be written in terms of a dimensionless permeability $k_0^\star$, which is a function of $\Psi$ only:

$$R = \frac{\Psi \mu}{l_0^2 k_0^\star} \tag{4.3.5-9}$$

In summary, Equations (4.3.5-7) and (4.3.5-9) may be used to describe the force per unit volume that an incompressible Newtonian fluid exerts on a nonoriented porous structure (beyond the hydrostatic force and the force attributable to the ambient pressure).

### Example 2: Flow of Viscoelastic Fluid in a Nonoriented Medium

Let us repeat Example 1 for an incompressible viscoelastic fluid. In order that our results have a wide range of applicability, let us assume that the behavior of this viscoelastic fluid can be described by the Noll simple fluid discussed in Section 2.3.4.

The initial argument given in Example 1 and concluding with (4.3.5-7) is again applicable here. The only modification necessary is to say that the resistance coefficient $R$ is a function of a characteristic viscosity $\mu_0$ and characteristic time $s_0$ of the fluid as well as $\left| \overline{\mathbf{v}}^{(f)} - \Psi \langle \mathbf{u} \rangle^{(s)} \right|$, the porosity $\Psi$, and the characteristic length $l_0$:

$$R = R \left( \left| \overline{\mathbf{v}}^{(f)} - \Psi \langle \mathbf{u} \rangle^{(s)} \right|, \mu_0, s_0, \Psi, l_0 \right) \tag{4.3.5-10}$$

The Buckingham–Pi theorem (Brand 1957) allows us to conclude that

$$R = \frac{\Psi \mu_0}{l_0^2 k^\star} \tag{4.3.5-11}$$

where $k^\star$ is a function of the local Weissenberg number $N_{Wi}$ and $\Psi$:

$$k^\star = k^\star (N_{Wi}, \Psi) \tag{4.3.5-12}$$

$$N_{Wi} \equiv \frac{s_0 \left| \overline{\mathbf{v}}^{(f)} - \Psi \langle \mathbf{u} \rangle^{(s)} \right|}{l_0} \tag{4.3.5-13}$$

Anticipating the result of Section 4.3.9, we can postulate as an alternative to (4.3.5-10)

$$R = R \left( \left| \nabla \left( \overline{\mathcal{P} - p_0}^{(f)} \right) \right|, \mu_0, s_0, \Psi, l_0 \right) \tag{4.3.5-14}$$

By the Buckingham–Pi theorem (Brand 1957), we conclude that (4.3.5-11) still applies, but now

$$k^\star = k^\star \left( \frac{\left| \nabla \overline{(\mathcal{P} - p_0)}^{(f)} \right| l_0 s_0}{\mu_0}, \Psi \right) \tag{4.3.5-15}$$

In summary, Equations (4.3.5-7) and (4.3.5-11) describe the force per unit volume that an incompressible viscoelastic fluid exerts upon a nonoriented porous structure, provided the behavior of the fluid is representable by the Noll simple fluid. The dimensionless permeability $k^\star$ must be considered to be a function of the Weissenberg number $N_{Wi}$. As mentioned in Section 2.3.4 [see also Slattery (1965, 1968a)], an empirical correlation of this type can be prepared for only one viscoelastic fluid at a time, since the functional $\mathcal{H}_{\sigma=0}^{\infty}{}^\star$ describing the simple fluid has not been fully specified. However, if we are interested in only a single fluid, it is not necessary to have particular values for the characteristic viscosity $\mu_0$ and the characteristic time $s_0$. As far as this empirical correlation is concerned, we can avoid an extensive (and perhaps somewhat indeterminate) study of material behavior in a set of viscometers.

### Example 3: Flow of Newtonian Fluid in an Oriented Medium

One should not expect (4.3.5-7) to be applicable to the flow of an incompressible Newtonian fluid through a porous structure in which particle diameter $l$ is a function of position. For such a structure, (4.3.5-2) must be altered to include a dependence upon additional vector and possibly tensor quantities. For example, one might postulate a dependence of $\mathbf{g}$ upon the local gradient of particle diameter as well as $\overline{\mathbf{v}}^{(f)} - \Psi\langle\mathbf{u}\rangle^{(s)}$:

$$\mathbf{g} = \hat{\mathbf{g}}\left(\overline{\mathbf{v}}^{(f)} - \Psi\langle\mathbf{u}\rangle^{(s)}, \nabla l\right) \tag{4.3.5-16}$$

For the moment, we leave the additional dependence of $\mathbf{g}$ upon $\mu$ and $l$ understood.

The principle of frame indifference again requires $\hat{\mathbf{g}}$ to be an isotropic function:

$$\begin{aligned}
&\hat{\mathbf{g}}\left(\overline{\mathbf{v}}^{(f)} - \Psi\langle\mathbf{u}\rangle^{(s)}, \nabla l\right) \\
&= \mathbf{Q}^T \cdot \hat{\mathbf{g}}\left[\mathbf{Q} \cdot \left(\overline{\mathbf{v}}^{(f)} - \Psi\langle\mathbf{u}\rangle^{(s)}\right), \mathbf{Q} \cdot \nabla l\right]
\end{aligned} \tag{4.3.5-17}$$

By representation theorems of Spencer and Rivlin (1959, Sec. 7) and of Smith (1965), the most general polynomial isotropic vector function of two vectors has the form[1]

$$\mathbf{g} = \varphi_{(1)}\left(\overline{\mathbf{v}}^{(f)} - \Psi\langle\mathbf{u}\rangle^{(s)}\right) + \varphi_{(2)}\nabla l \tag{4.3.5-18}$$

Here $\varphi_{(1)}$ and $\varphi_{(2)}$ are scalar-valued polynomials:

$$\varphi_{(i)} = \varphi_{(i)}\left[\left|\overline{\mathbf{v}}^{(f)} - \Psi\langle\mathbf{u}\rangle^{(s)}\right|, |\nabla l|, \left(\overline{\mathbf{v}}^{(f)} - \Psi\langle\mathbf{u}\rangle^{(s)}\right) \cdot \nabla l, \mu, \Psi, l\right] \tag{4.3.5-19}$$

An application of the Buckingham–Pi theorem (Brand 1957) allows us to conclude that

$$\varphi_{(1)} = \frac{\Psi\mu}{l^2 k^\star_{(1)}} \tag{4.3.5-20}$$

and

$$\varphi_{(2)} = \frac{\mu\left|\overline{\mathbf{v}}^{(f)} - \Psi\langle\mathbf{u}\rangle^{(s)}\right|}{l^2 k^\star_{(2)}} \tag{4.3.5-21}$$

where

$$k^\star_{(i)} = k^\star_{(i)}\left(|\nabla l|, \frac{\overline{\mathbf{v}}^{(f)} - \Psi\langle\mathbf{u}\rangle^{(s)}}{\left|\overline{\mathbf{v}}^{(f)} - \Psi\langle\mathbf{u}\rangle^{(s)}\right|} \cdot \nabla l, \Psi\right) \tag{4.3.5-22}$$

As we would expect, $\varphi_{(2)} = 0$ for $\left|\overline{\mathbf{v}}^{(f)} - \Psi\langle\mathbf{u}\rangle^{(s)}\right| = 0$, in order that $\mathbf{g} = \mathbf{0}$ in this limit.

---

[1] In applying the theorem of Spencer and Rivlin, we identify a vector $\mathbf{b}$ that has covariant components $b_i$ with the skew-symmetric tensor that has contravariant components $\epsilon^{ijk}b_i$. Their theorem requires an additional term in (4.3.5-18) proportional to the vector product $\left[\left(\overline{\mathbf{v}}^{(f)} - \Psi\langle\mathbf{u}\rangle^{(s)}\right) \wedge \nabla l\right]$. This term is not consistent with the requirement that $\mathbf{g}$ be isotropic (Truesdell and Noll 1965, p. 24) and consequently is dropped.

In summary, (4.3.5-18) and (4.3.5-20) through (4.3.5-22) may be used to represent the force per unit volume that an incompressible Newtonian fluid exerts upon an oriented structure such that the orientation of the structure is fully described by the local gradient of particle diameter. The resulting expression for **g** is somewhat more complicated than that which we found for a nonoriented structure in Example 1.

### 4.3.6 Summary of Results for an Incompressible Newtonian Fluid

In Section 4.3.3, we found that the local volume average of the differential mass balance requires

$$\operatorname{div} \bar{\mathbf{v}}^{(f)} = 0 \tag{4.3.6-1}$$

Under conditions such that inertial effects can be neglected and the external force per unit mass can be represented as the gradient of a scalar potential, Sections 4.3.4 and 4.3.5 indicate that the local volume average of the differential momentum balance for an incompressible Newtonian fluid flowing through a nonoriented porous medium has the form

$$\nabla \left[ \Psi \left( \langle P \rangle^{(f)} - p_0 \right) \right] - \mu \operatorname{div} \left( \nabla \bar{\mathbf{v}}^{(f)} \right) + \frac{\Psi \mu}{k} \bar{\mathbf{v}}^{(f)} = 0 \tag{4.3.6-2}$$

Here $p_0$ is a reference or ambient pressure,

$$k \equiv l_0^2 k_0^{\star}$$

is the *permeability*, $l_0$ is a characteristic length of the porous medium, and $k_0^{\star}$ is a dimensionless function of porosity. In writing (4.3.6-2), we have recognized that, in the frame of reference for which (4.3.4-16) is appropriate, the velocity of the porous medium is assumed to be zero. With the exception of the second term on the left, (4.3.6-2) is similar to Darcy's law for flow (4.3.0-3). Interestingly, Brinkman (1949a,b) proposed a similar relationship without derivation.

In Section 4.3.5, we considered the flow of an incompressible Newtonian fluid through a porous medium such that the orientation of the gradient of the local pore diameter $l$ is significant. Under conditions such that inertial effects can be neglected and the external force per unit mass can be represented in terms of a scalar potential, Sections 4.3.4 and 4.3.5 suggest that the local volume average of the differential momentum balance is

$$\nabla \left[ \Psi \left( \langle P \rangle^{(f)} - p_0 \right) \right] - \mu \operatorname{div} \left( \nabla \bar{\mathbf{v}}^{(f)} \right) + \frac{\Psi \mu}{l^2 k_{(1)}^{\star}} \bar{\mathbf{v}}^{(f)} + \frac{\mu |\bar{\mathbf{v}}^{(f)}|}{l^2 k_{(2)}^{\star}} \nabla l = \mathbf{0} \tag{4.3.6-3}$$

where

$$k_{(i)}^{\star} = k_{(i)}^{\star} \left( |\nabla l|, \frac{\bar{\mathbf{v}}^{(f)}}{|\bar{\mathbf{v}}^{(f)}|} \cdot \nabla l, \Psi \right) \tag{4.3.6-4}$$

In stating these results in the form of (4.3.6-3) and (4.3.6-4), we have recognized that, in the frame of reference being considered, the porous structure is stationary.

With the exception of the second term on the left, (4.3.6-2) has the same form as the extended Darcy law for a nonoriented porous structure. Equation (4.3.6-3) might also be written as (Whitaker 1969)

$$\nabla \left[ \Psi \left( \langle P \rangle^{(f)} - p_0 \right) \right] - \mu \operatorname{div} \left( \nabla \bar{\mathbf{v}}^{(f)} \right) + \frac{\Psi \mu}{l^2} \mathbf{R}^{\star} \cdot \bar{\mathbf{v}}^{(f)} = \mathbf{0} \tag{4.3.6-5}$$

in which

$$\mathbf{R}^{\star} \equiv \frac{1}{k_{(1)}^{\star}} \mathbf{I} + \frac{|\overline{\mathbf{v}}^{(f)}|}{\Psi k_{(2)}^{\star} \left(\overline{\mathbf{v}}^{(f)} \cdot \nabla l\right)} \nabla l \nabla l \tag{4.3.6-6}$$

In Section 4.3.9, we further explore these similarities.

Before illustrating how we can use these equations to analyze flows through porous media, let us stop and ask how local volume-averaged variables can be used to calculate some of the macroscopic area and volume averages of practical interest.

### 4.3.7  Averages of Volume-Averaged Variables

One commonly used type of packed bed is prepared by filling a cylindrical tube with small particles (sand, glass beads, catalyst pellets, ... ). In the next section, we analyze the flow through such a bed to determine the local volume-averaged velocity. From a practical point of view, we are more interested in the average of velocity over the cross section of the tube. Intuitively, we feel that the average of velocity over the cross section is equal to the average over the cross section of the local volume-averaged velocity, so long as the pores in the structure are sufficiently small compared with the diameter of the tube. The purpose of this section is to confirm these intuitive feelings.

More generally, we would like to discuss under what circumstances

$$\frac{1}{\mathcal{A}} \int_{\mathcal{S}} \overline{f}^{(f)} \, dA = \frac{1}{\mathcal{A}} \int_{\mathcal{S}} f \, dA \tag{4.3.7-1}$$

where $f$ is some quantity (scalar, vector, or second-order tensor field) associated with the fluid, $\mathcal{S}$ is some macroscopic surface, and $\mathcal{A}$ is the area of $\mathcal{S}$.

Let us rearrange (4.3.7-1) as

$$\frac{1}{\mathcal{A}} \int_{\mathcal{S}} \overline{f}^{(f)} \, dA = \frac{1}{\mathcal{A}} \int_{\mathcal{S}} \left[ f + \left( \overline{f}^{(f)} - f \right) \right] dA \tag{4.3.7-2}$$

and fix our attention upon the last two terms on the right, writing

$$\frac{1}{\mathcal{A}} \int_{\mathcal{S}} \left( \overline{f}^{(f)} - f \right) dA = \frac{1}{\mathcal{A}} \sum_{n} \int_{\mathcal{S}_n} \left( \overline{f}^{(f)} - f \right) dA \tag{4.3.7-3}$$

The understanding here is that

$$\mathcal{S} = \sum_{n} \mathcal{S}_n \tag{4.3.7-4}$$

The only limitation placed upon the subsurface $\mathcal{S}_n$ is that the characteristic dimension of each be of the order $L_0$, the characteristic dimension of the averaging surface $S$ (see Section 4.3.1). With this limitation, it seems reasonable to approximate the mean value of $f$ on $\mathcal{S}_n$ by

$$\overline{f}^{(f)} \doteq \frac{1}{\mathcal{A}_n} \int_{\mathcal{S}_n} f \, dA \tag{4.3.7-5}$$

and to say, since $\overline{f}^{(f)}$ is nearly independent of position over distances of order $L_0$

(Section 4.3.1),

$$\int_{\mathcal{S}_n} \left( \overline{f}^{(f)} - f \right) dA \doteq 0 \tag{4.3.7-6}$$

By $\mathcal{A}_n$, I mean the area of $\mathcal{S}_n$.

Equation (4.3.7-1) follows from (4.3.7-2), (4.3.7-3), and (4.3.7-6). The limitations on (4.3.7-1) are these:

1) The characteristic dimension $L_0$ of the averaging surface $S$ is chosen such that $\overline{f}^{(f)}$ is nearly independent of position over distances of the same order.
2) The characteristic dimension of $S$ must be greater than or equal to $L_0$.

**Exercise 4.3.7-1**  Let $\mathcal{R}$ be some macroscopic region in space whose volume is $\mathcal{V}$. Construct an argument similar to that given in the text to conclude that

$$\frac{1}{\mathcal{V}} \int_{\mathcal{R}} \overline{f}^{(f)} \, dV = \frac{1}{\mathcal{V}} \int_{\mathcal{R}} f \, dV$$

Assumption (1) of the text again applies as well as

2′) The characteristic dimension of $\mathcal{R}$ must be greater than or equal to $L_0$.

## 4.3.8  Flow Through a Packed Tube

An incompressible Newtonian fluid flows through a nonoriented permeable structure of uniform porosity

$$\Psi \equiv \frac{V^{(f)}}{V} \tag{4.3.8-1}$$

bounded by a cylindrical tube of radius $r_0$. We wish to determine the local volume-averaged velocity distribution for the fluid as well as the corresponding volume rate of flow through the tube.

Referring to Figure 4.3.8-1, we say that

$$\text{at } r = r_0, \ \theta = 0, \ z = 0 : \ \langle p \rangle^{(f)} = P_L \tag{4.3.8-2}$$

and

$$\text{at } r = r_0, \ \theta = 0, \ z = L : \ \langle p \rangle^{(f)} = P_0 \tag{4.3.8-3}$$

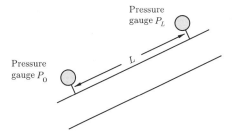

**Figure 4.3.8-1.** Flow through a packed tube.

Here we express our belief that experimentalists measure more nearly (see Exercise 4.3.8-1)

$$\langle P \rangle^{(f)} \equiv \frac{1}{V^{(f)}} \int_{V_{(f)}} p \, dV \tag{4.3.8-4}$$

Because of the finite size of the averaging surface $S$, the local volume-averaged velocity $\overline{\mathbf{v}}^{(f)}$ will not be zero at the tube wall. But as we advance into the pipe wall, $\overline{\mathbf{v}}^{(f)}$ decreases since less of the averaging surface $S$ intercepts the porous structure that contains the fluid. At some distance $\epsilon$ inside the impermeable wall, $\overline{\mathbf{v}}^{(f)}$ goes to zero (this distance $\epsilon$ will depend upon the averaging surface $S$ chosen):

$$\text{at } r = r_0 + \epsilon : \quad \overline{\mathbf{v}}^{(f)} = 0 \tag{4.3.8-5}$$

A solution of the form

$$\overline{v}_r^{(f)} = \overline{v}_\theta^{(f)}$$
$$= 0$$
$$\overline{v}_z^{(f)} = \overline{v}_z^{(f)}(r) \tag{4.3.8-6}$$

satisfies the local volume-averaged differential mass balance for an incompressible fluid, (4.3.3-3). The three components of (4.3.6-2) reduce to

$$\frac{\partial \left( \langle \mathcal{P} \rangle^{(f)} - p_0 \right)}{\partial r} = \frac{\partial \left( \langle \mathcal{P} \rangle^{(f)} - p_0 \right)}{\partial \theta}$$
$$= 0 \tag{4.3.8-7}$$

and

$$\Psi \frac{\partial \left( \langle \mathcal{P} \rangle^{(f)} - p_0 \right)}{\partial z} = \frac{\mu}{r} \frac{d}{dr} \left( r \frac{d\overline{v}_z^{(f)}}{dr} \right) - \frac{\Psi \mu}{k} \overline{v}_z^{(f)} \tag{4.3.8-8}$$

These equations imply that

$$\Psi \frac{\partial \left( \langle \mathcal{P} \rangle^{(f)} - p_0 \right)}{\partial z} = \frac{\mu}{r} \frac{d}{dr} \left( r \frac{d\overline{v}_z^{(f)}}{dr} \right) - \frac{\Psi \mu}{k} \overline{v}_z^{(f)}$$
$$= C$$
$$= \text{a constant} \tag{4.3.8-9}$$

We can integrate (4.3.8-9) with boundary conditions (4.3.8-2) and (4.3.8-3) to find

$$C = -\frac{1}{L} \left\{ \Psi \left( P_0 - P_L \right) + \rho \Psi \left[ \langle \varphi \rangle^{(f)} (r_0, 0, 0) - \langle \varphi \rangle^{(f)} (r_0, 0, L) \right] \right\} \tag{4.3.8-10}$$

where $\varphi = \varphi(r, \theta, z)$. If in (4.3.8-9) we introduce

$$u^\star \equiv -\frac{\overline{v}_z^{(f)} \Psi \mu}{Ck} - 1 \tag{4.3.8-11}$$

and

$$r^\star = \frac{r}{r_0} \tag{4.3.8-12}$$

we obtain a form of Bessel's equations,

$$\frac{d^2 u^\star}{dr^{\star 2}} + \frac{1}{r^\star}\frac{du^\star}{dr^\star} - N^2 u^\star = 0 \tag{4.3.8-13}$$

Here

$$N \equiv \left[\frac{\Psi r_0{}^2}{k}\right]^{1/2} \tag{4.3.8-14}$$

The boundary conditions for (4.3.8-13) are that $u^\star$ remains finite at $r^\star = 0$ and

$$\text{at } r^\star = 1 + \frac{\epsilon}{r_0}: \quad u^\star = -1 \tag{4.3.8-15}$$

The required solution is

$$u^\star = -\frac{I_0(Nr^\star)}{I_0(N[1 + \epsilon/r_0])} \tag{4.3.8-16}$$

or

$$\bar{v}_z^{(f)} = -\frac{Ck}{\Psi\mu}\left[1 - \frac{I_0(Nr^\star)}{I_0(N[1 + \epsilon/r_0])}\right] \tag{4.3.8-17}$$

By $I_0$, we mean the zero-order modified Bessel function of the first kind (Irving and Mullineux 1959, p. 143).

For sufficiently large values of $N$, Equation (4.3.8-17) tells us that $\bar{v}_z^{(f)}$ is essentially constant over the cross section, except in the immediate vicinity of the wall, where it approaches zero. Equation (4.3.8-19) indicates that as $N \to \infty$ the volume rate of flow through the tube is essentially that found by multiplying the centerline velocity by the cross-sectional area.

To get a better feeling for the magnitude of the wall effect, let us consider an example. Say that we have a tube, 2 cm in diameter, packed with spherical particles of diameter 2 mm such that the void fraction or porosity $\Psi = 0.3$. We may use the Blake–Kozeny equation (Bird et al. 1960, p. 199) to estimate $k = 1.5 \times 10^{-5}$ cm$^2$. We conclude that

$$N = 141 \tag{4.3.8-18}$$

In this case, there is a small wall effect in (4.3.8-19).

Finally, the volume rate of flow through the packed tube can be calculated using (4.3.8-17):

$$Q = 2\pi r_0{}^2 \int_0^{1+\epsilon/r_0} \bar{v}_z^{(f)} r^\star \, dr^\star$$

$$= -\frac{\pi r_0{}^2 (1 + \epsilon/r_0)^2 \, kC}{\Psi\mu}\left[1 - \frac{2}{N}\frac{I_1(N[1 + \epsilon/r_0])}{I_0(N[1 + \epsilon/r_0])}\right] \tag{4.3.8-19}$$

For sufficient large values of $N$, this reduces to

$$Q = -\frac{\pi r_0{}^2 (1 + \epsilon/r_0)^2 \, kC}{\Psi\mu} \tag{4.3.8-20}$$

Usually, we will be willing to say that $\epsilon/r_0 \ll 1$, in which case this further reduces to

$$Q = -\frac{\pi r_0{}^2 kC}{\Psi\mu} \tag{4.3.8-21}$$

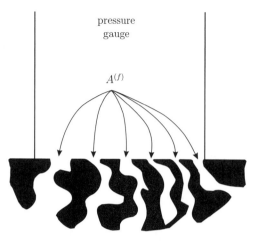

**Figure 4.3.8-2.** Detail of a somewhat idealized pressure-gauge probe.

Exercise 4.3.8-1 *The pressure-gauge measurement*    Figure 4.3.8-2 shows the detail of a somewhat idealized pressure-gauge probe. Let $A^{(f)}$ indicate the area at which the fluid acts upon the probe; $A$ denotes the total cross-sectional area of the probe.

i) If $P_0$ is the pressure-gauge reading, argue that

$$P_0 = \frac{A}{A^{(f)}} \overline{p}^{(f)}$$

ii) Let us define the void-volume distribution function

$$\alpha(\mathbf{z}) = \begin{cases} 1 & \text{if } \mathbf{z} \text{ lies in the fluid} \\ 0 & \text{if } \mathbf{z} \text{ lies in the solid} \end{cases}$$

Estimate that (Whitaker 1969)

$$\frac{A^{(f)}}{A} = \frac{V^{(f)}}{V}$$
$$= \Psi$$

This allows us to conclude that

$$P_0 = \Psi^{-1} \overline{p}^{(f)}.$$

Exercise 4.3.8-2 *Channel flow* (Slattery 1969)    Let us use Figure 4.3.8-1 to describe flow of an incompressible Newtonian fluid through a nonoriented permeable structure of uniform porosity bounded by two infinite parallel planes. The distance between the planes is $2b$. Determine the local volume-averaged velocity distribution corresponding to the fluid as well as the volume rate of flow of the fluid through the channel.

Exercise 4.3.8-3 *Radial flow* (Slattery 1969)    A problem closely related to radial flow to or from a well bore may be described in cylindrical coordinates by a local volume-averaged velocity distribution of the form

$$\overline{v}_r^{(f)} = \overline{v}_r^{(f)}(r)$$
$$\overline{v}_\theta^{(f)} = \overline{v}_z^{(f)}$$
$$= 0$$

with an associated local volume-averaged pressure distribution that satisfies

$$\text{at } r = r_1, \; \theta = 0, \; z = 0: \; \langle p \rangle^{(f)} = P_1$$

and

$$\text{at } r = r_2, \; \theta = 0, \; z = 0: \; \langle p \rangle^{(f)} = P_2$$

Assume that the cylindrical coordinate system is oriented with the $z$ axis in the direction opposite to the action of gravity. Determine the radial component of the local volume-averaged velocity distribution as well as a volume rate of flow through the cylindrical surface $r = r_1$.

## 4.3.9 Neglecting the Divergence of the Local Volume-Averaged Extra Stress

Our discussion of flow through a packed tubed in Section 4.3.8 suggests that under many circumstances we may be able to neglect the effect of the divergence of the local volume-averaged extra stress in the equation of motion for a porous medium, (4.3.4-8). In this section, we justify neglecting this term for the flow of incompressible Newtonian and incompressible viscoelastic fluids through nonoriented porous media. The extension to oriented permeable structures follows along the same lines (see Exercise 4.3.9-1).

Equation (4.3.6-2) describes the flow of an incompressible Newtonian fluid through a nonoriented porous structure. In terms of dimensionless variables, it says

$$\frac{P_0 k}{L_0 \Psi \mu v_0} \nabla \left[ \Psi \left( \langle P \rangle^{(f)} - p_0 \right) \right]^\star - \frac{k}{L_0{}^2 \Psi} \text{div} \left( \nabla \overline{\mathbf{v}}^{(f)\star} \right) + \overline{\mathbf{v}}^{(f)\star} = \mathbf{0} \tag{4.3.9-1}$$

Here $P_0$ is a characteristic pressure, $v_0$ is a characteristic magnitude of velocity, and $L_0$ is a length characteristic of the macroscopic geometry. The quantity $k/\left(L_0{}^2 \Psi\right)$ would be a very small number for common porous media problems. This suggests that the second term on the left of (4.3.9-1) be neglected with respect to the third term to obtain Darcy's law [compare with (4.3.0-3)]

$$\nabla \left[ \Psi \left( \langle P \rangle^{(f)} - p_0 \right) \right] + \frac{\Psi \mu}{k} \overline{\mathbf{v}}^{(f)} = \mathbf{0} \tag{4.3.9-2}$$

But the reader is cautioned that this is an intuitive argument that need not be true in every situation. There is no theorem that says that, because a term of a differential equation is multiplied by a small parameter, the term can be neglected.

This same argument can be repeated for a viscoelastic fluid as represented by the Noll simple fluid model discussed in Section 2.3.4. For flow through a nonoriented porous structure, (4.3.4-8), (4.3.5-7), and (4.3.5-11) tell us that

$$\nabla \left[ \Psi \left( \langle P \rangle^{(f)} - p_0 \right) \right] - \text{div} \, \mathbf{S} + \frac{\Psi \mu_0}{l_0{}^2 k^\star} \overline{v}^{(f)} = 0 \tag{4.3.9-3}$$

where $k^\star$ is a function of the Weissenberg number

$$N_{Wi} \equiv \frac{s_0 \left|\overline{\mathbf{v}}^{(f)}\right|}{l_0} \tag{4.3.9-4}$$

Let us introduce the same dimensionless variables as we did above, with the additional definition

$$\overline{\mathbf{S}}^{(f)\star} \equiv \frac{s_0 \overline{\mathbf{S}}^{(f)}}{\mu_0} \tag{4.3.9-5}$$

This allows us to write (4.3.9-3) in a dimensionless form as

$$\frac{P_0 l_0^2 k^\star}{\Psi L_0 \mu_0 v_0} \nabla \left[ \Psi \left( \langle P \rangle^{(f)} - p_0 \right) \right] - \frac{l_0^2 k^\star}{\Psi s_0 L_0 v_0} \operatorname{div} \overline{\mathbf{S}}^{(f)\star} + \overline{\mathbf{v}}^{(f)\star} = 0 \tag{4.3.9-6}$$

To get a feeling for the magnitude of the parameter multiplying the second term on the left of (4.3.9-6), let us consider an example using some typical values for the various characteristic quantities appropriate to flow in an oil-bearing rock structure:

$$\Psi \approx 0.3$$

$$l_0^2 k^\star \approx 250 \text{ millidarcys or } 2.5 \times 10^{-9} \text{cm}^2$$

$$s_0 = 10^{-2} \text{ s}$$

$$v_0 = 1 \text{ ft/day or } 3.5 \times 10^{-4} \text{ cm/s}$$

$$\frac{l_0^2 k^\star}{\Psi s_0 L_0 v_0} \approx \frac{2 \times 10^{-3}}{L_0} \tag{4.3.9-7}$$

Here we have judged that a characteristic time of $10^{-2}$ s appears to be reasonable for some viscoelastic fluids (Shertzer and Metzner 1965, Gin and Metzner 1965). If we remember that $L_0$ is a length characteristic of the macroscopic geometry, this suggests that the second term on the left of (4.3.9-6) may be neglected with respect to the third term to obtain

$$\nabla \left[ \Psi \left( \langle P \rangle^{(f)} - p_0 \right) \right] + \frac{\Psi \mu_0}{l_0^2 k^\star} \overline{\mathbf{v}}^{(f)} = 0 \tag{4.3.9-8}$$

In dropping out the divergence of the local volume-averaged extra stress to obtain (4.3.9-2) and (4.3.9-8), there is a corresponding reduction in order of the differential equation. This is most obvious in comparing (4.3.9-1) with (4.3.9-2) for an incompressible Newtonian fluid. With this reduction in order, we lose our ability to satisfy boundary conditions on the tangential components of the local volume-averaged velocity vector. This is similar to the problem we ran into in discussing potential flow in the introduction to Section 3.4. So long as we are not very interested in the velocity distribution in the immediate neighborhood of an impermeable boundary to a porous medium, the approximation suggested here should be entirely satisfactory.

As an example of the type of problem where one might get into trouble by neglecting the divergence of the local volume-averaged extra stress, consider flow through a porous-walled tube. The description of the local volume-averaged velocity distribution in the immediate neighborhood of the boundary of the porous medium would appear to be important in determining the proper boundary conditions for the fluid flowing through the tube. Since the tangential component of velocity would not necessarily go to zero at the tube wall, one might expect to see in the experimental data an apparent slip at the wall.

In summary, as long as we can neglect inertial effects and represent the external force vector in terms of a scalar potential, we will almost always be justified in writing the local volume average of the differential momentum balance as

$$\nabla \left[ \Psi \left( \langle P \rangle^{(f)} - p_0 \right) \right] + \mathbf{g} = \mathbf{0} \tag{4.3.9-9}$$

where the force per unit volume $\mathbf{g}$ that the fluid exerts upon the porous structure must be determined from an empirical correlation of the form suggested in Section 4.3.5.

What about the relation of these results to the force balances that have been in use in the literature for some time? For the flow of an incompressible Newtonian fluid through a nonoriented porous structure, (4.3.9-9) takes the form of (4.3.9-2). Upon comparison of (4.3.9-2) with (4.3.0-3), it becomes clear that we have derived here an extended form of Darcy's law with a clear interpretation for the variables being used. For the flow of an incompressible Newtonian fluid through an oriented porous structure summarized in Section 4.3.6, Equation (4.3.9-9) becomes

$$\nabla \left[ \Psi \left( \langle P \rangle^{(f)} - p_0 \right) \right] + \frac{\Psi \mu}{l^2} \mathbf{R}^\star \cdot \bar{\mathbf{v}}^{(f)} = 0 \tag{4.3.9-10}$$

where

$$\mathbf{R}^\star \equiv \frac{1}{k_{(1)}^\star} \mathbf{I} + \frac{\left| \bar{\mathbf{v}}^{(f)} \right|}{\Psi k_{(2)}^\star \left( \bar{\mathbf{v}}^{(f)} \cdot \nabla l \right)} \nabla l \nabla l \tag{4.3.9-11}$$

and

$$k_{(i)}^\star = k_{(i)}^\star \left( |\nabla l|, \frac{\bar{\mathbf{v}}^{(f)}}{|\bar{\mathbf{v}}^{(f)}|} \cdot \nabla l, \Psi \right) \tag{4.3.9-12}$$

**Exercise 4.3.9-1** *Flow of an incompressible Newtonian fluid through an oriented porous medium* Justify the use of (4.3.9-10) to describe the flow of an incompressible Newtonian fluid through an oriented porous structure.

---

## 4.4 Integral Balances

In Chapters 1 and 2, we discussed mass, momentum, and moment-of-momentum balances for material bodies. Here we wish to develop the corresponding balances for more general systems. The system may be the fluid in a surge tank or the fuel in a jet aircraft that is moving along an arbitrary curve in space. In these examples, the system is not a material body, since fluid may be entering or leaving the surge tank and fuel is being consumed and gases are being exhausted by the jet engine.

We frequently use integral balances in situations where we wish to make a statement about the system as a whole without worrying about a detailed description of the motions of the fluids within its interior. For example, in relating the thrust developed by a rocket engine to the average velocity of its exhaust gases, it may not be important to do a detailed study of the atomization of the liquid fuel. This atomization process may be important to the subsequent combustion, and in this way it may directly affect the rocket's performance. But once we have been given the average velocity of the rocket's exhaust, we will find that we have sufficient information to estimate the thrust developed by the engine.

Integral balances are one of the most commonly used techniques in engineering, since they result in algebraic equations or relatively simple differential equations. Their importance cannot be overemphasized. Yet their simplicity is misleading, since one is often forced to make a series of approximations based upon intuitive judgments or related experimental knowledge. It becomes more difficult to say whether a particular analysis will describe an experimental observation within a prescribed error. One often hears remarks to the effect that "a good engineer develops a facility for making intuitive judgments." This certainly is true of the engineer who is successful in applying integral balances.

Many of the ideas associated with integral balances that we present here are due to Bird (1957).

### 4.4.1   The Integral Mass Balance

The differential mass balance developed in Section 1.3.3,

$$\frac{\partial \rho}{\partial t} + \operatorname{div}(\rho \mathbf{v}) = 0 \tag{4.4.1-1}$$

describes mass conservation at any point in a body. To develop a mass balance for an arbitrary system, we must integrate this equation over the system

$$\int_{R_{(s)}} \left( \frac{\partial \rho}{\partial t} + \operatorname{div}(\rho \mathbf{v}) \right) dV = 0 \tag{4.4.1-2}$$

Here $R_{(s)}$ is the region of space occupied by the system. After an application of Green's transformation to the second term, we obtain

$$\int_{R_{(s)}} \frac{\partial \rho}{\partial t} dV = - \int_{S_{(s)}} \rho \mathbf{v} \cdot \mathbf{n} \, dA \tag{4.4.1-3}$$

where $S_{(s)}$ is the closed bounding surface of $R_{(s)}$.

The generalized transport theorem developed in Section 1.3.2 allows us to say

$$\frac{d}{dt} \int_{R_{(s)}} \rho \, dV = \int_{R_{(s)}} \frac{\partial \rho}{\partial t} dV + \int_{S_{(s)}} \rho \mathbf{v}_{(s)} \cdot \mathbf{n} \, dA \tag{4.4.1-4}$$

in which $\mathbf{v}_{(s)}$ is the velocity of $S_{(s)}$, which in general will be a function of position on $S_{(s)}$. Using this, we may rewrite (4.4.1-3) as

$$\frac{d}{dt} \int_{R_{(s)}} \rho \, dV = \int_{S_{(s)}} \rho \left( \mathbf{v} - \mathbf{v}_{(s)} \right) \cdot (-\mathbf{n}) \, dA \tag{4.4.1-5}$$

In words, this equation tells us that the time rate of change of mass in the system is equal to the net rate at which mass enters the system. Notice how naturally the velocity of the fluid relative to the boundary of the system, $(\mathbf{v} - \mathbf{v}_{(s)})$, enters. Since $(\mathbf{v} - \mathbf{v}_{(s)})$ is different from zero only on the entrance and exit portions of $S_{(s)}$, $S_{(\text{ent ex})}$, we may write (4.4.1-5) as

$$\frac{d}{dt} \int_{R_{(s)}} \rho \, dV = \int_{S_{(\text{ent ex})}} \rho \left( \mathbf{v} - \mathbf{v}_{(s)} \right) \cdot (-\mathbf{n}) \, dA \tag{4.4.1-6}$$

This is the *integral mass balance* for a single-phase system.

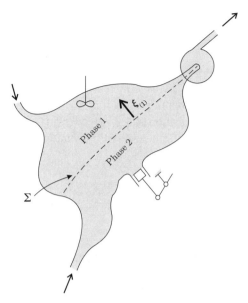

**Figure 4.4.1-1.** A system consisting of two phases.

It is important to realize that, as (4.4.1-6) has been derived, it applies only to a single-phase system. More often than not in practice we are concerned with multiphase systems. For example, if we should wish to talk about the mass of an accelerating rocket as a function of time, the most natural choice of systems would be the rocket and all its contents. This system would consist of millions of distinct phases: gases, liquids, and solids. Let us derive for such a system a relation analogous to (4.4.1-6).

For simplicity, let us assume that our arbitrary system consists of only two phases. Each phase may itself be regarded as a system to which (4.4.1-6) applies. Referring to Figure 4.4.1-1, if we denote the region occupied by phase $i$ in the system as $R_{(i)}$, the region occupied by the entire system may be indicated as

$$R_{(s)} = R_{(1)} + R_{(2)}$$

The entrance and exit surfaces of $R_{(i)}$ are $S_{(\text{ent ex } i)}$. Some of these entrance and exit surfaces may coincide with a portion of the entrance and exit surface for the system as a whole, $S_{(\text{ent ex})}$; others will coincide with a phase interface (dividing surface) $\Sigma$, since we wish to allow for the possibility of interphase mass transfer. Adding the statements of (4.4.1-6) made for each phase, we have

$$\frac{d}{dt} \int_{R_{(s)}} \rho \, dV = \frac{d}{dt} \int_{R_{(1)}} \rho \, dV + \frac{d}{dt} \int_{R_{(2)}} \rho \, dV$$

$$= \int_{S_{(\text{ent ex } 1)}} \rho \left( \mathbf{v} - \mathbf{v}_{(s)} \right) \cdot (-\mathbf{n}) \, dA + \int_{S_{(\text{ent ex } 2)}} \rho \left( \mathbf{v} - \mathbf{v}_{(s)} \right) \cdot (-\mathbf{n}) \, dA$$

or

$$\frac{d}{dt} \int_{R_{(s)}} \rho \, dV = \int_{S_{(\text{ent ex})}} \rho \left( \mathbf{v} - \mathbf{v}_{(s)} \right) \cdot (-\mathbf{n}) \, dA + \int_{\Sigma} \left[ \rho (\mathbf{v} - \mathbf{u}) \cdot \boldsymbol{\xi} \right] dA \qquad (4.4.1\text{-}7)$$

Here we have employed the notation first introduced in Section 1.3.5. If $\Psi_{(i)}$ denotes the value of $\Psi$ obtained in the limit as $\Sigma$ is approached from the interior of phase $i$,

$$\text{at } \Sigma : \quad \llbracket \Psi \xi \rrbracket = \Psi_{(1)}\xi_{(1)} + \Psi_{(2)}\xi_{(2)} \tag{4.4.1-8}$$

Here $\xi$ is the unit normal to $\Sigma$ directed into phase $i$. This argument for two-phase systems is readily extended to systems containing any number of phases.

Given the jump mass balance of Section 1.3.6, Equation (4.4.1-7) reduces to (4.4.1-6). The integral mass balance (4.4.1-6) applies equally well to single-phase and multiphase systems.

## 4.4.2 The Integral Mass Balance for Turbulent Flows

Some of the most important and interesting applications of the integral balances are to systems, portions of which are in turbulent flow.

For example, we may wish to determine the pressure drop between pumping stations in a pipeline that transports natural gas from the Southwest to Chicago. It is convenient to choose the natural gas in the pipeline between the pumping stations as our system. Our first step is to determine the restrictions placed upon this system by the integral mass balance. To do this, we must make some statement about the time rate of change of the mass of the system. Is this a steady-state problem or an unsteady-state problem? Since the gas is in turbulent flow, this must by definition be an unsteady-state situation. Yet, when the pipeline superintendent speaks about the volume rate of flow at a particular cross section in the pipeline, he ignores any random fluctuations caused by turbulence. He has little choice, since whatever instruments he uses to infer the volume flow rate report only time-averaged readings. The integral mass balance derived in Section 4.4.1 is not directly applicable, because it relates the instantaneous mass of the system to the instantaneous mass flow rates through the entrances and exits.

Let us limit ourselves to single-phase or multiphase systems that do not involve fluid–fluid phase interfaces. Starting with the time-averaged differential mass balance found in Section 4.1.1,

$$\frac{\partial \overline{\rho}}{\partial t} + \text{div } (\overline{\rho \mathbf{v}}) = 0 \tag{4.4.2-1}$$

the derivation of Section 4.4.1 may be repeated to find

$$\frac{d}{dt} \int_{R_{(s)}} \overline{\rho} \, dV = \int_{S_{(\text{ent ex})}} \left( \overline{\rho \mathbf{v}} - \overline{\rho} \mathbf{v}_{(s)} \right) \cdot (-\mathbf{n}) \, dA + \int_{\Sigma} \left[ \left( \overline{\rho \mathbf{v}} - \overline{\rho} \mathbf{u} \right) \cdot \xi \right] dA \tag{4.4.2-2}$$

The time-averaged jump mass balance of Exercise 4.4.2-1 simplifies this to

$$\frac{d}{dt} \int_{R_{(s)}} \overline{\rho} \, dV = \int_{S_{(\text{ent ex})}} \left( \overline{\rho \mathbf{v}} - \overline{\rho} \mathbf{v}_{(s)} \right) \cdot (-\mathbf{n}) \, dA \tag{4.4.2-3}$$

This says that the time rate of change of the time-averaged mass of the system equals the net time-averaged rate at which mass is brought into the system through the entrances and exits. Equation (4.4.2-3) is the *integral mass balance* appropriate to a turbulent single-phase or multiphase system that does not involve fluid–fluid phase interfaces.

I do not recommend this approach when discussing systems that contain or are bounded by fluid–fluid phase interfaces, since the positions of these phase interfaces will in general be random functions of time. Even after a definition for a time-averaged phase interface is agreed upon, we find that it is not the time average of the jump mass balance (Section 1.3.6) that applies. There are important interfacial effects that are directly attributable to turbulence.

For single-phase or multiphase systems that include one or more fluid–fluid phase interfaces, I recommend time averaging the integral mass balance obtained in Section 4.4.1:

$$\frac{d}{dt}\int_{R_{(s)}} \rho \, dV = \int_{S_{(\text{ent ex})}} \rho \left(\mathbf{v} - \mathbf{v}_{(s)}\right) \cdot (-\mathbf{n}) \, dA \tag{4.4.2-4}$$

to find

$$\overline{\frac{d}{dt}\int_{R_{(s)}} \rho \, dV} = \overline{\int_{S_{(\text{ent ex})}} \rho \left(\mathbf{v} - \mathbf{v}_{(s)}\right) \cdot (-\mathbf{n}) \, dA} \tag{4.4.2-5}$$

Here we have observed that

$$\overline{\frac{d}{dt}\int_{R_{(s)}} \rho \, dV} \equiv \frac{1}{\Delta t}\int_t^{t+\Delta t}\left(\frac{d}{dt'}\int_{R_{(s)}} \rho \, dV\right) dt'$$

$$= \frac{1}{\Delta t}\left(\int_{R_{(s)}} \rho \, dV\Big|_{t+\Delta t} - \int_{R_{(s)}} \rho \, dV\Big|_t\right)$$

$$= \frac{d}{dt}\left(\frac{1}{\Delta t}\int_t^{t+\Delta t}\int_{R_{(s)}} \rho \, dV \, dt'\right)$$

$$= \frac{d}{dt}\overline{\int_{R_{(s)}} \rho \, dV} \tag{4.4.2-6}$$

Equation (4.4.2-5) tells us that the time rate of change of the time-averaged mass of the system is equal to the sum of the time-averaged mass flow rates through the entrances of the system minus the time-averaged mass flow rates through the exits of the system.

As we discussed in the introduction to Section 4.1, the interval $\Delta t$ for time averaging is chosen to be large with respect to a time characteristic of a random fluctuation in turbulent flow but small with respect to a time characteristic of the macroscopic process being studied. This suggests that, if we restrict ourselves to single-phase or multiphase systems that do not involve fluid–fluid phase interfaces, for any quantity $\Psi$

$$\overline{\int_{R_{(s)}} \Psi \, dV} = \int_{R_{(s)}} \overline{\Psi} \, dV, \tag{4.4.2-7}$$

$$\overline{\int_{S_{(\text{ent ex})}} \Psi \, dA} = \int_{S_{(\text{ent ex})}} \overline{\Psi} \, dA \tag{4.4.2-8}$$

and

$$\overline{\mathbf{v}_{(s)} \cdot \mathbf{n}} = \mathbf{v}_{(s)} \cdot \mathbf{n} \tag{4.4.2-9}$$

Under these circumstances, (4.4.2-5) reduces to (4.4.2-3).

**Exercise 4.4.2-1** *Time-averaged jump mass balance* Determine that the time-averaged jump mass balance applicable to solid–fluid phase interfaces that bound turbulent flows is identical to the balance found in Section 1.3.6:

$$\left[ (\overline{\rho}\mathbf{v} - \overline{\rho}\mathbf{u}) \cdot \boldsymbol{\xi} \right] = \left[ \rho(\mathbf{v} - \mathbf{u}) \cdot \boldsymbol{\xi} \right]$$
$$= 0$$

### 4.4.3  The Integral Momentum Balance

Our objective is to derive a momentum balance for an arbitrary system. Let us take the same approach that we used in arriving at the integral mass balance.

Let us begin by integrating the differential momentum balance of Section 2.2.3,

$$\frac{\partial(\rho\mathbf{v})}{\partial t} + \operatorname{div}(\rho\mathbf{v}\mathbf{v}) = \operatorname{div}\mathbf{T} + \rho\mathbf{f} \tag{4.4.3-1}$$

over the region occupied by an arbitrary system:

$$\int_{R_{(s)}} \left[ \frac{\partial(\rho\mathbf{v})}{\partial t} + \operatorname{div}(\rho\mathbf{v}\mathbf{v}) - \operatorname{div}\mathbf{T} - \rho\mathbf{f} \right] dV = \mathbf{0} \tag{4.4.3-2}$$

The first integral on the left may be expressed more conveniently using the generalized transport theorem,

$$\frac{d}{dt} \int_{R_{(s)}} \rho\mathbf{v}\, dV = \int_{R_{(s)}} \frac{\partial(\rho\mathbf{v})}{\partial t} dV + \int_{S_{(s)}} \rho\mathbf{v}\big(\mathbf{v}_{(s)} \cdot \mathbf{n}\big)\, dA \tag{4.4.3-3}$$

The fourth integral on the left of (4.4.3-2) can be left unchanged. The second and third integrals have clearer physical meaning after an application of Green's transformation:

$$\int_{R_{(s)}} \operatorname{div}(\rho\mathbf{v}\mathbf{v})\, dV = \int_{S_{(s)}} \rho\mathbf{v}(\mathbf{v} \cdot \mathbf{n})\, dA \tag{4.4.3-4}$$

$$\int_{R_{(s)}} \operatorname{div}\mathbf{T}\, dV = \int_{S_{(s)}} \mathbf{T} \cdot \mathbf{n}\, dA \tag{4.4.3-5}$$

Equations (4.4.3-3) through (4.4.3-5) allow us to rewrite (4.4.3-2) as

$$\frac{d}{dt} \int_{R_{(s)}} \rho\mathbf{v}\, dV = \int_{S_{(s)}} \rho\mathbf{v}\big(\mathbf{v} - \mathbf{v}_{(s)}\big) \cdot (-\mathbf{n})\, dA$$
$$- \int_{S_{(s)}} \mathbf{T} \cdot (-\mathbf{n})\, dA + \int_{R_{(s)}} \rho\mathbf{f}\, dV \tag{4.4.3-6}$$

The first term on the right is the net rate at which momentum is brought into the system with whatever material is crossing the boundaries. The second term is the force that the material in the system exerts on the bounding surfaces of the system. This is the total force, whereas we usually speak in terms of the force in excess of the force of atmospheric pressure. To correct for the effect of atmospheric pressure $p_0$, it is easiest to return to the differential momentum balance and note that

$$\operatorname{div}\mathbf{T} = \operatorname{div}(\mathbf{T} + p_0\mathbf{I}) \tag{4.4.3-7}$$

This means that (4.4.3-6) may be replaced by

$$\frac{d}{dt} \int_{R_{(s)}} \rho\mathbf{v}\, dV = \int_{S_{(s)}} \rho\mathbf{v}\big(\mathbf{v} - \mathbf{v}_{(s)}\big) \cdot (-\mathbf{n})\, dA$$
$$- \int_{S_{(s)}} (\mathbf{T} + p_0\mathbf{I}) \cdot (-\mathbf{n})\, dA + \int_{R_{(s)}} \rho\mathbf{f}\, dV \tag{4.4.3-8}$$

If we recognize that $(\mathbf{v} - \mathbf{v}_{(s)}) \cdot \mathbf{n}$ is different from zero only on the exit and entrance surfaces $S_{(ent\ ex)}$, we may rewrite (4.4.3-8) as

$$\frac{d}{dt} \int_{R_{(s)}} \rho \mathbf{v} \, dV = \int_{S_{(ent\ ex)}} \rho \mathbf{v} \left(\mathbf{v} - \mathbf{v}_{(s)}\right) \cdot (-\mathbf{n}) \, dA$$

$$- \int_{S_{(ent\ ex)}} (\mathbf{T} + p_0 \mathbf{I}) \cdot (-\mathbf{n}) \, dA - \mathcal{F} + \int_{R_{(s)}} \rho \mathbf{f} \, dV \qquad (4.4.3\text{-}9)$$

where we define

$$\mathcal{F} \equiv \int_{S_{(s)} - S_{(ent\ ex)}} (\mathbf{T} + p_0 \mathbf{I}) \cdot (-\mathbf{n}) \, dA \qquad (4.4.3\text{-}10)$$

Physically, $\mathcal{F}$ denotes the force that the system exerts upon the impermeable portion of its bounding surface beyond the force attributable to the ambient pressure $p_0$. Equation (4.4.3-9) is the *integral momentum balance* applicable to a single-phase system.

Commonly, one chooses a system for an application of the integral momentum balance in such a way that viscous forces on the entrance and exit surfaces $S_{(ent\ ex)}$ may be neglected. This might be done by selecting entrance and exit surfaces as cross sections normal to flow in long straight pipes and by thinking of the flow in these straight pipes as being approximately represented by Poiseuille flow (Section 3.2.1). In most problems, one is forced to do this for lack of sufficient information to do otherwise. Under these conditions, (4.4.3-9) becomes

$$\frac{d}{dt} \int_{R_{(s)}} \rho \mathbf{v} \, dV = \int_{S_{(ent\ ex)}} \rho \mathbf{v} \left(\mathbf{v} - \mathbf{v}_{(s)}\right) \cdot (-\mathbf{n}) \, dA$$

$$- \int_{S_{(ent\ ex)}} (P - p_0)\mathbf{n} \, dA - \mathcal{F} + \int_{R_{(s)}} \rho \mathbf{f} \, dV \qquad (4.4.3\text{-}11)$$

This form of the integral momentum balance is more commonly employed than (4.4.3-9).

When dealing with incompressible fluids under circumstances such that $\mathbf{f}$ is representable by a potential,

$$\mathbf{f} = -\nabla \varphi \qquad (4.4.3\text{-}12)$$

it may be more convenient to express (4.4.3-11) as

$$\frac{d}{dt} \int_{R_{(s)}} \rho \mathbf{v} \, dV = \int_{S_{(ent\ ex)}} \rho \mathbf{v} \left(\mathbf{v} - \mathbf{v}_{(s)}\right) \cdot (-\mathbf{n}) \, dA$$

$$- \int_{S_{(ent\ ex)}} (\mathcal{P} - p_0)\mathbf{n} \, dA - \mathcal{G} \qquad (4.4.3\text{-}13)$$

where we define

$$\mathcal{G} \equiv \int_{S_{(s)} - S_{(ent\ ex)}} [\mathbf{T} + (p_0 - \rho\varphi)\mathbf{I}] \cdot (-\mathbf{n}) \, dA \qquad (4.4.3\text{-}14)$$

By $\mathcal{G}$, we mean the force that the system exerts upon the impermeable portion of its bounding surface beyond the force attributable to the ambient pressure $p_0$ and to the hydrostatic pressure.

As (4.4.3-11) and (4.4.3-13) have been derived, they are applicable only to a single-phase system. As we pointed out in Section 4.4.1, we are more commonly concerned with multiphase systems. Using the approach and notation of Section 4.4.1 and making no additional assumptions, we find that the integral momentum balance appropriate to a multiphase system is

$$\frac{d}{dt} \int_{R_{(s)}} \rho \mathbf{v} \, dV = \int_{S_{(\text{ent ex})}} \rho \mathbf{v} \left( \mathbf{v} - \mathbf{v}_{(s)} \right) \cdot (-\mathbf{n}) \, dA$$

$$- \int_{S_{(\text{ent ex})}} (\mathbf{T} + p_0 \mathbf{I}) \cdot (-\mathbf{n}) \, dA - \mathcal{F} + \int_{R_{(s)}} \rho \mathbf{f} \, dV$$

$$+ \int_{\Sigma} \left[ \rho \mathbf{v}(\mathbf{v} - \mathbf{u}) \cdot \boldsymbol{\xi} - \mathbf{T} \cdot \boldsymbol{\xi} \right] dA \tag{4.4.3-15}$$

Recognizing the jump momentum balance (Section 2.2.3), this simplifies to

$$\frac{d}{dt} \int_{R_{(s)}} \rho \mathbf{v} \, dV = \int_{S_{(\text{ent ex})}} \rho \mathbf{v} \left( \mathbf{v} - \mathbf{v}_{(s)} \right) \cdot (-\mathbf{n}) \, dA$$

$$- \int_{S_{(\text{ent ex})}} (\mathbf{T} + p_0 \mathbf{I}) \cdot (-\mathbf{n}) \, dA - \mathcal{F} + \int_{R_{(s)}} \rho \mathbf{f} \, dV \tag{4.4.3-16}$$

For a multiphase system composed of incompressible materials, this also may be written as

$$\frac{d}{dt} \int_{R_{(s)}} \rho \mathbf{v} \, dV = \int_{S_{(\text{ent ex})}} \rho \mathbf{v} \left( \mathbf{v} - \mathbf{v}_{(s)} \right) \cdot (-\mathbf{n}) \, dA$$

$$- \int_{S_{(\text{ent ex})}} [\mathbf{T} + (p_0 - \rho \varphi)\mathbf{I}] \cdot (-\mathbf{n}) \, dA - \mathcal{G}$$

$$+ \int_{\Sigma} \left[ \rho \boldsymbol{\xi} \right] \varphi \, dA \tag{4.4.3-17}$$

There are three common types of problems in which the integral momentum balance is applied: The force $\mathcal{F}$ (or $\mathcal{G}$) may be neglected, it may be the unknown to be determined, or it may be known from previous experimental data. In this last case, one employs an empirical correlation of data for $\mathcal{F}$ (or $\mathcal{G}$). In Section 4.4.4, we discuss the form that these empirical correlations should take.

**Exercise 4.4.3-1**   Derive (4.4.3-15) and (4.4.3-17).

**Exercise 4.4.3-2** *The integral momentum balance for turbulent flows*   I recommend following the discussion in Section 4.4.2 in developing forms of the integral momentum balance for turbulent flows.

i) Show that, for single-phase or multiphase systems that do not involve fluid–fluid phase interfaces, the derivation of Section 4.4.3 may be repeated to find

$$\frac{d}{dt} \int_{R_{(s)}} \overline{\rho \mathbf{v}} \, dV = \int_{S_{(\text{ent ex})}} \left( \overline{\rho} \, \overline{\mathbf{v}} \, \overline{\mathbf{v}} - \overline{\rho \mathbf{v}} \mathbf{v}_{(s)} \right) \cdot (-\mathbf{n}) \, dA$$

$$- \int_{S_{(\text{ent ex})}} (\overline{P} - p_0) \mathbf{n} \, dA - \mathcal{F} + \int_{R_{(s)}} \overline{\rho} \mathbf{f} \, dV$$

Here we have recognized that $S_{(s)} - S_{(\text{ent ex})}$ and $\Sigma$ must be composed of fluid–solid and solid–solid phase interfaces on which the turbulent fluctuations are identically zero, and we have taken advantage of Exercise 4.4.3-3. This is the form of the *integral momentum balance* recommended for turbulent single-phase or multiphase systems that do not involve fluid–fluid phase interfaces. Note that in arriving at this result we have neglected any viscous or turbulent forces acting at the entrances and exits to the system.

ii) Show that, for single-phase or multiphase systems that do not involve fluid–fluid phase interfaces and that are composed only of incompressible material,

$$\frac{d}{dt} \int_{R_{(s)}} \rho \bar{\mathbf{v}} \, dV = \int_{S_{(ent\ ex)}} \rho \left( \bar{\mathbf{v}}\bar{\mathbf{v}} - \bar{\mathbf{v}}\mathbf{v}_{(s)} \right) \cdot (-\mathbf{n}) \, dA$$

$$- \int_{S_{(ent\ ex)}} (\bar{\mathcal{P}} - p_0)\mathbf{n} \, dA - \mathcal{G} + \int_{\Sigma} [\rho \boldsymbol{\xi}] \varphi \, dA$$

iii) For single-phase or multiphase systems that include one or more fluid–fluid phase interfaces, I recommend simple time averages of the results derived in the text.

**Exercise 4.4.3-3** *Time-averaged jump momentum balance*   Find that the time-averaged jump momentum balance applicable to solid–fluid phase interfaces that bound turbulent flows is identical to (2.2.3-5):

$$\left[ (\bar{\rho}\,\bar{\mathbf{v}}\bar{\mathbf{v}} - \overline{\rho\mathbf{v}\mathbf{u}}) \cdot \boldsymbol{\xi} - \mathbf{T} \cdot \boldsymbol{\xi} \right] = \left[ \rho\mathbf{v}(\mathbf{v} - \mathbf{u}) \cdot \boldsymbol{\xi} - \mathbf{T} \cdot \boldsymbol{\xi} \right]$$

$$= 0$$

### 4.4.4   Empirical Correlations for $\mathcal{F}$ and $\mathcal{G}$

In this section, we use two examples to illustrate how empirical data correlations for $\mathcal{F}$ and $\mathcal{G}$ (or, when we are dealing with turbulent flows, $\bar{\mathcal{F}}$ and $\bar{\mathcal{G}}$), introduced in Section 4.4.3, can be formulated. We base this discussion on three points:

1) The forces $\mathcal{F}$ and $\mathcal{G}$ are frame indifferent. For example,

$$\mathcal{G}^* = \int_{S_{(s)} - S_{(ent\ ex)}} \left[ \mathbf{T}^* + (p_0 - \rho\varphi)\mathbf{I} \right] \cdot (-\mathbf{n}^*) \, dA$$

$$= \int_{S_{(s)} - S_{(ent\ ex)}} \mathbf{Q} \cdot [\mathbf{T} + (p_0 - \rho\varphi)\mathbf{I}] \cdot (-\mathbf{n}) \, dA$$

$$= \mathbf{Q} \cdot \mathcal{G} \tag{4.4.4-1}$$

where $\mathbf{Q}$ is a (possibly) time-dependent, orthogonal, second-order tensor.

2) We assume that the principle of frame indifference introduced in Section 2.3.1 applies to any empirical correlation developed for either $\mathcal{F}$ or $\mathcal{G}$.

3) The Buckingham–Pi theorem (Brand 1957) serves to further restrict the form of any expression for $\mathcal{F}$ and $\mathcal{G}$.

### Example 1: Flow Past a Sphere

As our first example, let us consider a sphere of radius $a$ that is in relative motion with respect to a large body of an incompressible Newtonian fluid. In a frame of reference that is fixed with respect to the laboratory, we observe that the sphere translates without rotation at a constant velocity $\mathbf{v}_0$ and that at a very large distance from the sphere the fluid moves with a uniform velocity $\mathbf{v}_\infty$. It seems reasonable to say that $\mathcal{G}$ should be a function of the fluid's density $\rho$ and viscosity $\mu$, the sphere's radius $a$, and the velocity difference $\mathbf{v}_\infty - \mathbf{v}_0$:

$$\mathcal{G} = \mathbf{h}(\rho, \mu, a, \mathbf{v}_\infty - \mathbf{v}_0) \tag{4.4.4-2}$$

Our reason for choosing $\mathbf{v}_\infty - \mathbf{v}_0$ as an independent variable in this expression, rather than $\mathbf{v}_0$ and $\mathbf{v}_\infty$ separately, is that velocity is not frame indifferent, whereas a velocity difference is (see Exercise 1.2.2-1).

For the moment, let the dependence of $\mathcal{G}$ upon $\rho, \mu$, and $a$ be understood, and let us concentrate our attention upon $(\mathbf{v}_\infty - \mathbf{v}_0)$:

$$\mathcal{G} = \hat{\mathbf{h}}(\mathbf{v}_\infty - \mathbf{v}_0) \tag{4.4.4-3}$$

By the principle of frame indifference, the functional relationship between these variables should be the same in every frame of reference. This means that

$$\begin{aligned}
\mathcal{G}^* &= \mathbf{Q} \cdot \mathcal{G} \\
&= \mathbf{Q} \cdot \hat{\mathbf{h}}(\mathbf{v}_\infty - \mathbf{v}_0) \\
&= \hat{\mathbf{h}}[\mathbf{Q} \cdot (\mathbf{v}_\infty - \mathbf{v}_0)]
\end{aligned} \tag{4.4.4-4}$$

or $\hat{\mathbf{h}}$ is an isotropic function (Truesdell and Noll 1965, p. 22):

$$\hat{\mathbf{h}}(\mathbf{v}_\infty - \mathbf{v}_0) = \mathbf{Q}^T \cdot \hat{\mathbf{h}}[\mathbf{Q} \cdot (\mathbf{v}_\infty - \mathbf{v}_0)] \tag{4.4.4-5}$$

By a representation theorem for a vector-valued isotropic function of one vector (Truesdell and Noll 1965, p. 35), we may write

$$\begin{aligned}
\mathcal{G} &= \hat{\mathbf{h}}(\mathbf{v}_\infty - \mathbf{v}_0) \\
&= \mathcal{G}\frac{\mathbf{v}_\infty - \mathbf{v}_0}{|\mathbf{v}_\infty - \mathbf{v}_0|}
\end{aligned} \tag{4.4.4-6}$$

where $\mathcal{G} \equiv |\mathcal{G}|$ is a function of the magnitude of the undisturbed fluid relative to the sphere, $|\mathbf{v}_\infty - \mathbf{v}_0|$, as well as a function of $\rho, \mu$, and $a$:

$$\mathcal{G} = \mathcal{G}(\rho, \mu, a, |\mathbf{v}_\infty - \mathbf{v}_0|) \tag{4.4.4-7}$$

It is customary to express $\mathcal{G}$ in terms of a drag coefficient or friction factor $c$. The drag coefficient or friction factor is introduced in many contexts, but it is almost always defined as the ratio of $\mathcal{G}$ to an area $A$ that is characteristic of $S_{(s)} - S_{(\text{ent ex})}$ and to a kinetic energy per unit volume that is characteristic of the flow,

$$\text{in general}: \quad c \equiv \frac{\mathcal{G}}{\frac{1}{2}\rho u^2 A} \tag{4.4.4-8}$$

For flow past a sphere, we define $A = \pi a^2$ and $u = |\mathbf{v}_\infty - \mathbf{v}_0|$ so that

$$\text{for flow past a sphere}: \quad c \equiv \frac{\mathcal{G}}{\left(\frac{1}{2}\rho \, |\mathbf{v}_\infty - \mathbf{v}_0|^2\right)\pi a^2} \tag{4.4.4-9}$$

The drag coefficient for a sphere is, consequently, a dimensionless function,

$$c = c(\rho, \mu, a, |\mathbf{v}_\infty - \mathbf{v}_0|) \tag{4.4.4-10}$$

By the Buckingham–Pi theorem (Brand 1957), we find that $c$ is a function of the Reynolds number,

$$c = \hat{c}(N_{Re}) \tag{4.4.4-11}$$

where

$$N_{Re} \equiv \frac{a\,|\mathbf{v}_\infty - \mathbf{v}_0|\,\rho}{\mu} \tag{4.4.4-12}$$

Therefore, we find for this situation

$$\mathcal{G} = c\left(\frac{1}{2}\rho\,|\mathbf{v}_\infty - \mathbf{v}_0|^2\right)\pi a^2 \frac{\mathbf{v}_\infty - \mathbf{v}_0}{|\mathbf{v}_\infty - \mathbf{v}_0|} \tag{4.4.4-13}$$

where the drag coefficient $c$ is to be determined from a correlation of experimental data in the form of (4.4.4-11) (Schlichting 1979, p. 17).

If we had wished to develop an empirical correlation for $\mathcal{F}$ instead, we would have had to include the additional dependence of $\mathcal{F}$ upon the external force vector $\mathbf{f}$ (see Exercise 4.4.4-1). Because of this additional complication with empirical correlations for $\mathcal{F}$, it is not common to find them being used in the literature.

### Example 2: Plane Flow Past a Cylindrical Body

An infinitely long cylindrical body (the surface of which is traced by a straight line moving parallel to a fixed straight line and intersecting a fixed closed curve) is in relative motion with respect to a large body of an incompressible Newtonian fluid. In a frame of reference that is fixed with respect to the earth, the cylindrical body translates without rotation at a constant velocity $\mathbf{v}_0$ and the fluid at a very large distance from the body moves with a uniform velocity $\mathbf{v}_\infty$. The vectors $\mathbf{v}_0$ and $\mathbf{v}_\infty$ are normal to the axis of the cylinder so that we may expect that the fluid moves in a plane flow. One unit vector $\boldsymbol{\alpha}$ is sufficient to describe the orientation of the cylinder. Following the discussion of the previous example, we postulate that $\mathcal{G}$ is a function of $\rho$, $\mu$, a length $L$ that is characteristic of the cylinder's cross section, $\mathbf{v}_\infty - \mathbf{v}_0$, and $\boldsymbol{\alpha}$:

$$\mathcal{G} = \mathbf{h}(\rho, \mu, L, \mathbf{v}_\infty - \mathbf{v}_0, \boldsymbol{\alpha}) \tag{4.4.4-14}$$

Let us concentrate our attention upon the independent variables $\mathbf{v}_\infty - \mathbf{v}_0$ and $\boldsymbol{\alpha}$:

$$\mathcal{G} = \hat{\mathbf{h}}(\mathbf{v}_\infty - \mathbf{v}_0, \boldsymbol{\alpha}) \tag{4.4.4-15}$$

By the principle of frame indifference, we conclude that $\hat{\mathbf{h}}$ is a vector-valued isotropic function of two vectors:

$$\hat{\mathbf{h}}(\mathbf{v}_\infty - \mathbf{v}_0, \boldsymbol{\alpha}) = \mathbf{Q}^T \cdot \hat{\mathbf{h}}(\mathbf{Q} \cdot [\mathbf{v}_\infty - \mathbf{v}_0], \mathbf{Q} \cdot \boldsymbol{\alpha}) \tag{4.4.4-16}$$

Again, $\mathbf{Q}$ is an orthogonal second-order tensor. The representation theorems of Spencer and Rivlin (1959, Sec. 7) and of Smith (1965) tell us that the most general polynomial isotropic vector function of two vectors has the form

$$\begin{aligned}
\mathcal{G} &= \hat{\mathbf{h}}(\mathbf{v}_\infty - \mathbf{v}_0, \boldsymbol{\alpha}) \\
&= \varphi_{(1)}\frac{\mathbf{v}_\infty - \mathbf{v}_0}{|\mathbf{v}_\infty - \mathbf{v}_0|} + \varphi_{(2)}\boldsymbol{\alpha}
\end{aligned} \tag{4.4.4-17}$$

where $\varphi_{(1)}$ and $\varphi_{(2)}$ are scalar-valued polynomials of $\rho$, $\mu$, and $L$ as well as $|\mathbf{v}_\infty - \mathbf{v}_0|$ and $(\mathbf{v}_\infty - \mathbf{v}_0) \cdot \boldsymbol{\alpha}$. (In applying the theorem of Spencer and Rivlin, we identify a vector $\mathbf{b}$ that has covariant components $b_i$ with the skew-symmetric tensor that has contravariant components $\epsilon^{ijk} b_i$. Their theorem requires an additional term in (4.4.4-17) proportional to the vector product $(\mathbf{v}_\infty - \mathbf{v}_0) \wedge \boldsymbol{\alpha}$. This term is not consistent with the requirement that $\hat{\mathbf{h}}$ be isotropic (Truesdell and Noll 1965, p. 24), and it is dropped.) We expect $\varphi_{(2)} = 0$ for $(\mathbf{v}_\infty - \mathbf{v}_0) = 0$ in order that $\mathcal{G} = 0$ in this limit.

Equation (4.4.4-17) is not commonly seen in the literature. Rather, $\mathcal{G}$ is expressed as a linear combination of the direction of the relative motion

$$\frac{\mathbf{v}_\infty - \mathbf{v}_0}{|\mathbf{v}_\infty - \mathbf{v}_0|}$$

and the direction orthogonal to the relative motion $\boldsymbol{\lambda}$:

$$\mathcal{G} = \mathcal{D}\left(\frac{\mathbf{v}_\infty - \mathbf{v}_0}{|\mathbf{v}_\infty - \mathbf{v}_0|}\right) + \mathcal{L}\boldsymbol{\lambda} \tag{4.4.4-18}$$

We refer to $\mathcal{D}$ as the *drag* component of $\mathcal{G}$ and $\mathcal{L}$ as the *lift* component.

Current practice is to express $\mathcal{D}$ and $\mathcal{L}$ in terms of drag and lift coefficients defined in the manner of (4.4.4-8):

$$c_{\mathcal{D}} \equiv \frac{\mathcal{D}}{\frac{1}{2}\rho\,|\mathbf{v}_\infty - \mathbf{v}_0|^2\,A_{\mathcal{D}}} \tag{4.4.4-19}$$

and

$$c_{\mathcal{L}} \equiv \frac{\mathcal{L}}{\frac{1}{2}\rho\,|\mathbf{v}_\infty - \mathbf{v}_0|^2\,.A_{\mathcal{L}}} \tag{4.4.4-20}$$

The characteristic areas for drag and lift are not necessarily the same. The area of the body projected on the plane normal to the direction of flow is usually chosen for $A_{\mathcal{D}}$. In the case of airfoil sections, $A_{\mathcal{L}}$ is taken to be the product of the chord length and wing length.

The Buckingham–Pi theorem (Brand 1957) tells us finally that the dimensionless drag and lift coefficients are functions of two Reynolds numbers:

$$c_{\mathcal{D}} = c_{\mathcal{L}}\left(N_{Re(1)}, N_{Re(2)}\right)$$
$$c_{\mathcal{L}} = c_{\mathcal{L}}\left(N_{Re(1)}, N_{Re(2)}\right) \tag{4.4.4-21}$$

where

$$N_{Re(1)} \equiv \frac{L\,|\mathbf{v}_\infty - \mathbf{v}_0|\,\rho}{\mu}$$
$$N_{Re(2)} \equiv \frac{L\,[(\mathbf{v}_\infty - \mathbf{v}_0) \cdot \boldsymbol{\alpha}]\,\rho}{\mu} \tag{4.4.4-22}$$

In summary,

$$\mathcal{G} = c_{\mathcal{D}} A_{\mathcal{D}}\left(\frac{1}{2}\rho\,|\mathbf{v}_\infty - \mathbf{v}_0|^2\right)\frac{\mathbf{v}_\infty - \mathbf{v}_0}{|\mathbf{v}_\infty - \mathbf{v}_0|} + c_{\mathcal{L}} A_{\mathcal{L}}\left(\frac{1}{2}\rho\,|\mathbf{v}_\infty - \mathbf{v}_0|^2\right)\boldsymbol{\lambda} \tag{4.4.4-23}$$

where $c_{\mathcal{D}}$ and $c_{\mathcal{L}}$ are to be determined from empirical correlations of experimental data. These empirical correlations should be expected to have the form of (4.4.4-21) or their equivalent (Schlichting 1979, p. 22).

**Exercise 4.4.4-1**  Repeat Example 1 to obtain the form of an empirical data correlation for $\mathcal{F}$.

**Exercise 4.4.4-2**  Repeat Example 1 for a power-law fluid (see Section 2.3.3).

**Exercise 4.4.4-3**  Repeat Example 1 for any body of revolution.

**Exercise 4.4.4-4**  Repeat Example 1 for a spinning sphere. Assume that the orientation of the axis of rotation may be a function of time.

### 4.4.5  The Mechanical Energy Balance

In Section 3.4.1, we developed the Bernoulli equation by means of which we were able to relate differences in pressure between two points in a fluid to differences in kinetic and potential energy. But the use of the Bernoulli equation is restricted to situations such that the viscous terms may be neglected in the differential momentum balance. Let us try to develop an equation that tells us something about the kinetic energy of an arbitrary system without making any assumption about viscous effects.

Although we will find that the practical results from this section are for incompressible fluids, I suggest that we refrain from imposing this restriction until we are nearly finished. In this way, our intermediate results will be directly applicable to a further discussion of the integral mechanical energy balance in Section 7.4.3.

Let us begin by taking the scalar product of the differential momentum balance with the velocity vector,

$$\mathbf{v} \cdot \left( \rho \frac{d_{(m)} \mathbf{v}}{dt} - \operatorname{div} \mathbf{T} - \rho \mathbf{f} \right) = 0 \tag{4.4.5-1}$$

and rearrange the first term that appears on the left:

$$\rho \mathbf{v} \cdot \frac{d_{(m)} \mathbf{v}}{dt} = \rho \frac{d_{(m)}}{dt} \left( \frac{1}{2} v^2 \right)$$

$$= \frac{d_{(m)}}{dt} \left( \frac{1}{2} \rho v^2 \right) - \frac{1}{2} v^2 \frac{d_{(m)} \rho}{dt}$$

$$= \frac{d_{(m)}}{dt} \left( \frac{1}{2} \rho v^2 \right) + \frac{1}{2} \rho v^2 \operatorname{div} \mathbf{v}$$

$$= \frac{\partial}{\partial t} \left( \frac{1}{2} \rho v^2 \right) + \operatorname{div} \left( \frac{1}{2} \rho v^2 \mathbf{v} \right) \tag{4.4.5-2}$$

This result, together with Green's transformation, may be used to express the integral of (4.4.5-1) over our system as

$$\int_{R_{(s)}} \frac{\partial}{\partial t} \left( \frac{1}{2} \rho v^2 \right) dV = - \int_{S_{(s)}} \frac{1}{2} \rho v^2 (\mathbf{v} \cdot \mathbf{n}) \, dA + \int_{R_{(s)}} (\mathbf{v} \cdot \operatorname{div} \mathbf{T} + \rho \mathbf{v} \cdot \mathbf{f}) \, dV \tag{4.4.5-3}$$

The generalized transport theorem of Section 1.3.2 can be used to write the term on the left of (4.4.5-3) as

$$\int_{R_{(s)}} \frac{\partial}{\partial t} \left( \frac{1}{2} \rho v^2 \right) dV = \frac{d}{dt} \int_{R_{(s)}} \frac{1}{2} \rho v^2 \, dV - \int_{S_{(s)}} \frac{1}{2} \rho v^2 \left( \mathbf{v}_{(s)} \cdot \mathbf{n} \right) dA \tag{4.4.5-4}$$

But in any given problem the value of the first integral on the right will not be immediately obvious. The difficulty is essentially the same as those we encountered in Sections 4.4.1 and 4.4.3.

The second integral on the right of (4.4.5-3) has no direct physical significance as it stands, which suggests that we rearrange it by an application of Green's transformation. In terms of a rectangular Cartesian coordinate system, we have

$$
\begin{aligned}
\int_{R_{(s)}} \mathbf{v} \cdot \operatorname{div} \mathbf{T} \, dV &= \int_{R_{(s)}} v_i \frac{\partial T_{ij}}{\partial z_j} \, dV \\
&= \int_{R_{(s)}} \left[ \frac{\partial}{\partial z_j} \left( v_i T_{ij} \right) - T_{ij} \frac{\partial v_i}{\partial z_j} \right] dV \\
&= \int_{S_{(s)}} v_i T_{ij} n_j \, dA - \int_{R_{(s)}} \operatorname{tr} \left( \mathbf{T}^T \cdot \nabla \mathbf{v} \right) dV \\
&= \int_{S_{(s)}} \mathbf{v} \cdot (\mathbf{T} \cdot \mathbf{n}) \, dA - \int_{R_{(s)}} \operatorname{tr} (\mathbf{T} \cdot \nabla \mathbf{v}) \, dV
\end{aligned}
\tag{4.4.5-5}
$$

Note that we use the symmetry of the stress tensor in arriving at this last expression. We find it convenient to observe further that

$$
\begin{aligned}
\int_{R_{(s)}} \operatorname{tr} \left( \mathbf{T} \cdot \nabla \mathbf{v} \right) dV &= \int_{R_{(s)}} \operatorname{tr} \left( [-P\mathbf{I} + \mathbf{S}] \cdot \nabla \mathbf{v} \right) dV \\
&= - \int_{R_{(s)}} P \operatorname{div} \mathbf{v} \, dV + \int_{R_{(s)}} \operatorname{tr} \left( \mathbf{S} \cdot \nabla \mathbf{v} \right) dV
\end{aligned}
\tag{4.4.5-6}
$$

This allows us to express (4.4.5-5) as

$$
\begin{aligned}
\int_{R_{(s)}} \mathbf{v} \cdot \operatorname{div} \mathbf{T} \, dV &= \int_{S_{(s)}} \mathbf{v} \cdot (\mathbf{T} \cdot \mathbf{n}) \, dA + \int_{R_{(s)}} P \operatorname{div} \mathbf{v} \, dV \\
&\quad - \int_{R_{(s)}} \operatorname{tr} \left( \mathbf{S} \cdot \nabla \mathbf{v} \right) dV
\end{aligned}
\tag{4.4.5-7}
$$

The third term on the right of (4.4.5-3) may be left unchanged for many purposes. A somewhat more familiar form is obtained if the external force can be expressed in terms of a potential energy per unit mass $\varphi$:

$$
\mathbf{f} = -\nabla \varphi
\tag{4.4.5-8}
$$

This means that

$$
\begin{aligned}
\int_{R_{(s)}} \rho \mathbf{v} \cdot \mathbf{f} \, dV &= - \int_{R_{(s)}} \rho \mathbf{v} \cdot \nabla \varphi \, dV \\
&= - \int_{R_{(s)}} \operatorname{div} (\rho \varphi \mathbf{v}) \, dV + \int_{R_{(s)}} \varphi \operatorname{div} (\rho \mathbf{v}) \, dV \\
&= - \int_{S_{(s)}} \rho \varphi (\mathbf{v} \cdot \mathbf{n}) \, dA + \int_{R_{(s)}} \varphi \operatorname{div} (\rho \mathbf{v}) \, dV
\end{aligned}
\tag{4.4.5-9}
$$

From the differential mass balance,

$$\int_{R_{(s)}} \varphi \operatorname{div}(\rho \mathbf{v}) \, dV = -\int_{R_{(s)}} \varphi \frac{\partial \rho}{\partial t} \, dV \tag{4.4.5-10}$$

Let us restrict ourselves to an external force potential that is independent of time. For most situations, including the case where gravity is the only external force, this is completely satisfactory. We are now in position for an application of the generalized transport theorem:

$$\int_{R_{(s)}} \varphi \operatorname{div}(\rho \mathbf{v}) \, dV = -\int_{R_{(s)}} \frac{\partial(\rho \varphi)}{\partial t} \, dV$$

$$= -\frac{d}{dt} \int_{R_{(s)}} \rho \varphi \, dV + \int_{S_{(s)}} \rho \varphi \left( \mathbf{v}_{(s)} \cdot \mathbf{n} \right) \, dA \tag{4.4.5-11}$$

Equations (4.4.5-9) and (4.4.5-11) give

$$\int_{R_{(s)}} \rho \mathbf{v} \cdot \mathbf{f} \, dV = -\frac{d}{dt} \int_{R_{(s)}} \rho \varphi \, dV - \int_{S_{(s)}} \rho \varphi \left( \mathbf{v} - \mathbf{v}_{(s)} \right) \cdot \mathbf{n} \, dA \tag{4.4.5-12}$$

We may use (4.4.5-4), (4.4.5-7), and (4.4.5-12) to express (4.4.5-3) as

$$\frac{d}{dt} \int_{R_{(s)}} \rho \left( \frac{1}{2} v^2 + \varphi \right) dV = \int_{S_{(s)}} \rho \left( \frac{1}{2} v^2 + \varphi \right) (\mathbf{v} - \mathbf{v}_{(s)}) \cdot (-\mathbf{n}) \, dA$$

$$+ \int_{R_{(s)}} P \operatorname{div} \mathbf{v} \, dV - \int_{S_{(s)}} \mathbf{v} \cdot [\mathbf{T} \cdot (-\mathbf{n})] \, dA - \mathcal{E} \tag{4.4.5-13}$$

in which

$$\mathcal{E} \equiv \int_{R_{(s)}} \operatorname{tr}(\mathbf{S} \cdot \nabla \mathbf{v}) \, dV \tag{4.4.5-14}$$

The third term on the right of (4.4.5-13) is to be interpreted as the rate at which work is done by the system on the surroundings. Intuitively, we realize that there should be no work due to the action of the ambient pressure $p_0$. As an obvious extension of our discussion in Section 4.4.3, we may write (4.4.5-13) as

$$\frac{d}{dt} \int_{R_{(s)}} \rho \left( \frac{1}{2} v^2 + \varphi \right) dV = \int_{S_{(\text{ent ex})}} \rho \left( \frac{1}{2} v^2 + \varphi \right) (\mathbf{v} - \mathbf{v}_{(s)}) \cdot (-\mathbf{n}) \, dA$$

$$+ \int_{R_{(s)}} (P - p_0) \operatorname{div} \mathbf{v} \, dV - \int_{S_{(s)}} \mathbf{v} \cdot [(\mathbf{T} + p_0 \mathbf{I}) \cdot (-\mathbf{n})] \, dA - \mathcal{E} \tag{4.4.5-15}$$

or

$$\frac{d}{dt} \int_{R_{(s)}} \rho \left( \frac{1}{2} v^2 + \varphi \right) dV$$

$$= \int_{S_{(\text{ent ex})}} \rho \left( \frac{1}{2} v^2 + \varphi + \frac{P - p_0}{\rho} \right) (\mathbf{v} - \mathbf{v}_{(s)}) \cdot (-\mathbf{n}) \, dA$$

$$+ \int_{R_{(s)}} (P - p_0) \operatorname{div} \mathbf{v} \, dV - \mathcal{W} - \mathcal{E}$$

$$+ \int_{S_{(\text{ent ex})}} \left[ -(P - p_0) \mathbf{v}_{(s)} \cdot \mathbf{n} + \mathbf{v} \cdot (\mathbf{S} \cdot \mathbf{n}) \right] dA \tag{4.4.5-16}$$

where we define

$$W \equiv \int_{S_{(s)}-S_{(\text{ent ex})}} \mathbf{v} \cdot [(\mathbf{T} + p_0\mathbf{I}) \cdot (-\mathbf{n})] \, dA \tag{4.4.5-17}$$

We should think of $W$ as the rate at which work is done by the system on the surroundings at the moving impermeable surfaces of the system (beyond any work done on these surfaces by the ambient pressure $p_0$). Equation (4.4.5-16) can be thought of as one *general form of the mechanical energy balance*.

The difficulty with (4.4.5-16) is that the value of the second integral on the right is not immediately obvious for most situations. More will be said on this point in Section 7.4.3.

For the moment, we are primarily interested in the special case of an incompressible fluid. By the differential mass balance,

$$\int_{R_{(s)}} (P - p_0) \operatorname{div} \mathbf{v} \, dV = 0 \tag{4.4.5-18}$$

and (4.4.5-16) reduces to

$$\frac{d}{dt} \int_{R_{(s)}} \rho \left( \frac{1}{2} v^2 + \varphi \right) dV$$

$$= \int_{S_{(\text{ent ex})}} \rho \left( \frac{1}{2} v^2 + \varphi + \frac{p - p_0}{\rho} \right) (\mathbf{v} - \mathbf{v}_{(s)}) \cdot (-\mathbf{n}) \, dA - W - \mathcal{E}$$

$$+ \int_{S_{(\text{ent ex})}} \left[ -(p - p_0)\mathbf{v}_{(s)} \cdot \mathbf{n} + \mathbf{v} \cdot (\mathbf{S} \cdot \mathbf{n}) \right] dA \tag{4.4.5-19}$$

Notice that in writing this last expression, we recognize that the pressure being used can no longer be the thermodynamic pressure $P$ (see Section 2.3.2). We shall refer to this equation as a general form of the *mechanical energy balance* for incompressible materials.

In words, the term on the left of (4.4.5-19) denotes the time rate of change of the kinetic and potential energy (often referred to as the *mechanical energy*) associated with the system. On the right of this equation,

$$\int_{S_{(\text{ent ex})}} \rho \left( \frac{1}{2} v^2 + \varphi \right) (\mathbf{v} - \mathbf{v}_{(s)}) \cdot (-\mathbf{n}) \, dA$$

is the net rate at which kinetic and potential energy is brought into the system with any material that moves across the boundary. By

$$\int_{S_{(\text{ent ex})}} \rho \left( \frac{p - p_0}{\rho} \right) (\mathbf{v} - \mathbf{v}_{(s)}) \cdot (-\mathbf{n}) \, dA + \int_{S_{(\text{ent ex})}} -(p - p_0)\mathbf{v}_{(s)} \cdot \mathbf{n} \, dA$$

$$= \int_{S_{(\text{ent ex})}} (p - p_0)(-\mathbf{v} \cdot \mathbf{n}) \, dA$$

we mean the net rate at which pressure forces (beyond the reference or ambient pressure $p_0$) do work on the system at the entrances and exits. We have already defined $W$ to be the rate at which work is done by the system on the surroundings at the moving impermeable surfaces of the system (beyond any work done on the surfaces by the ambient pressure $p_0$); see (4.4.5-17). The rate at which mechanical energy is dissipated by the action of viscous forces is denoted by $\mathcal{E}$, defined by (4.4.5-14). We will see in our discussion of the energy balance in Section 5.1.3 that we may also interpret $\mathcal{E}$ as the rate of production of internal energy by the action of viscous forces. The last term on the right,

$$\int_{S_{(ent\ ex)}} \mathbf{v} \cdot (\mathbf{S} \cdot \mathbf{n}) \, dA$$

represents the rate at which work is done on the system at the entrances and exits by the viscous forces.

One usually attempts to choose a system for an application of the mechanical energy balance in such a way that work done by viscous forces at entrances and exits may be neglected. In this way, we may write (4.4.5-19) as

$$\frac{d}{dt} \int_{R_{(s)}} \rho \left( \frac{1}{2} v^2 + \varphi \right) dV$$

$$= \int_{S_{(ent\ ex)}} \rho \left[ \frac{1}{2} v^2 + \varphi + \frac{(p - p_0)}{\rho} \right] (\mathbf{v} - \mathbf{v}_{(s)}) \cdot (-\mathbf{n}) \, dA - \mathcal{W} - \mathcal{E}$$

$$- \int_{S_{(ent\ ex)}} (p - p_0) (\mathbf{v}_{(s)} \cdot \mathbf{n}) \, dA \qquad (4.4.5\text{-}20)$$

This would be rigorous if the entrances and exits could be chosen as normal to the motion in very long straight pipes in which the flow is laminar (see Exercise 4.4.5-3).

As (4.4.5-20) has been derived, it is applicable only to a single-phase system. We are more commonly concerned with multiphase systems. Using the approach and notation of Section 4.4.1 and neglecting the work done by viscous forces at entrances and exits, we find that the *mechanical energy balance* for a multiphase system of incompressible materials is

$$\frac{d}{dt} \int_{R_{(s)}} \rho \left( \frac{1}{2} v^2 + \varphi \right) dV$$

$$= \int_{S_{(ent\ ex)}} \rho \left[ \frac{1}{2} v^2 + \varphi + \frac{(p - p_0)}{\rho} \right] (\mathbf{v} - \mathbf{v}_{(s)}) \cdot (-\mathbf{n}) \, dA - \mathcal{W} - \mathcal{E}$$

$$- \int_{S_{(ent\ ex)}} (p - p_0) (\mathbf{v}_{(s)} \cdot \mathbf{n}) \, dA$$

$$+ \int_{\Sigma} \left[ \rho \left( \frac{1}{2} v^2 + \varphi \right) (\mathbf{v} - \mathbf{u}) \cdot \boldsymbol{\xi} - \mathbf{v} \cdot [(\mathbf{T} + p_0 \mathbf{I}) \cdot \boldsymbol{\xi}] \right] dA \qquad (4.4.5\text{-}21)$$

When there is no mass transfer across internal phase interfaces and when the jump mass and momentum balances of Sections 1.3.6 and 2.2.3 apply, this equation reduces to (4.4.5-20). For this reason, it is (4.4.5-20) with which we will be primarily concerned.

There are three common categories of problems in which the integral mechanical energy balances apply: $\mathcal{E}$ may be neglected, it may be the unknown to be determined, or it may be known from previous experimental data. In this last case, one employs an empirical correlation of data for $\mathcal{E}$. In Section 4.4.6, we discuss the form that these empirical data correlations should take.

For more about the mechanical energy balance, see Section 7.4.3. An extensive compilation of alternative forms that it can take is given in Tables 7.4.3-1 through 7.4.3-3.

**Exercise 4.4.5-1**    Derive (4.4.5-18).

**Exercise 4.4.5-2** *Relation to Bernoulli's equation*    Show that for a steady-state flow through a system with one entrance and one exit, (4.4.5-16) reduces to a form that is similar to Bernoulli's equation.

**Exercise 4.4.5-3**    Prove that, if the entrances and exits are located in very long straight pipes in which the flow is laminar,

$$\int_{S_{(\text{ent ex})}} \mathbf{v} \cdot (\mathbf{S} \cdot \mathbf{n}) \, dA = 0$$

**Exercise 4.4.5-4** *The mechanical energy balance for incompressible materials in turbulent flow*    I recommend following the discussion in Section 4.4.2 in developing forms of the mechanical energy balance for turbulent flows.

i) Let us restrict our attention here to incompressible materials. Show that, for single-phase or multiphase systems that do not involve fluid–fluid phase interfaces,

$$\frac{d}{dt} \int_{R_{(s)}} \rho \left( \frac{1}{2} \overline{v}^2 + \varphi \right) dV$$

$$= \int_{S_{(\text{ent ex})}} \rho \left[ \frac{1}{2} \overline{v}^2 + \varphi + \frac{(\overline{p} - p_0)}{\rho} \right] (\overline{\mathbf{v}} - \mathbf{v}_{(s)}) \cdot (-\mathbf{n}) \, dA$$

$$- \int_{S_{(\text{ent ex})}} (\overline{p} - p_0) \left( \mathbf{v}_{(s)} \cdot \mathbf{n} \right) \, dA - \mathcal{W} - \mathcal{E}^{(t)}$$

$$+ \int_{\Sigma} \left[ \rho \left( \frac{1}{2} \overline{v}^2 + \varphi \right) (\overline{\mathbf{v}} - \mathbf{u}) \cdot \boldsymbol{\xi} - \overline{\mathbf{v}} \cdot \left[ (\overline{\mathbf{T}} + p_0 \mathbf{I}) \cdot \boldsymbol{\xi} \right] \right] dA \qquad (4.4.5\text{-}22)$$

where

$$\mathcal{E}^{(t)} \equiv \int_{R_{(s)}} \text{tr} \left[ (\overline{\mathbf{S}} + \mathbf{T}^{(t)}) \cdot \nabla \overline{\mathbf{v}} \right] dV$$

In arriving at this result, we have neglected the rate at which work is done by viscous and turbulent forces on the entrances and exits of the system, and we have noted that for this case $S_{(s)} - S_{(\text{ent ex})}$ and $\Sigma$ must be composed of fluid–solid and solid–solid phase interfaces on which the turbulent fluctuations are identically zero. When there is no mass transfer across internal phase interfaces, the jump mass and momentum balances of Exercises 4.4.2-1 and 4.4.3-3 permit the last integral in (4.4.5-22) to be neglected.

ii) For single-phase or multiphase systems that include one or more fluid–fluid phase interfaces, I recommend simple time averages of the results derived in the text.

## 4.4.6    Empirical Correlations for

In formulating empirical data correlations for $\mathcal{E}$ ($\overline{\mathcal{E}}$ when concerned with turbulent flows), one proceeds in much the same manner as we described in Section 4.4.3, where we discussed correlations for $\mathcal{F}$ and $\mathcal{G}$. Three principal thoughts should be kept in mind.

1) The total rate of dissipation of mechanical energy is frame indifferent:

$$
\begin{aligned}
\mathcal{E}^{\star} &\equiv \int_{R_{(s)}} \operatorname{tr}\left(\mathbf{S}^{\star} \cdot \nabla \mathbf{v}^{\star}\right) dV \\
&= \int_{R_{(s)}} \operatorname{tr}\left(\mathbf{S} \cdot \nabla \mathbf{v}\right) dV \\
&= \mathcal{E}
\end{aligned}
\tag{4.4.6-1}
$$

2) We assume that the principle of frame indifference introduced in Section 2.3.1 applies to any empirical correlation developed for $\mathcal{E}$.

3) The form of any expression for $\mathcal{E}$ must satisfy the Buckingham–Pi theorem (Brand 1957).

The most common class of problems where we employ correlations for $\mathcal{E}$ includes flow through a conduit and flow through a wide range of pipe or tubing fittings (valves, elbows, tees, nozzles, etc.). As an illustration of the general approach, let us consider the flow of an incompressible Newtonian fluid through a valve (of a specified design) mounted in a run of tubing on a jet aircraft.

Let us say that we have already described the contribution to $\mathcal{E}$ as a result of viscous dissipation in the tubing. We wish to consider the additional contribution to $\mathcal{E}$ resulting from the presence of the valve. Let us define

$$
u \equiv \frac{4G}{\rho \pi D^2}
\tag{4.4.6-2}
$$

where $D$ denotes the inside diameter of the tubing, $\rho$ the density of the fluid, and $G$ the mass flow rate of fluid through the tube. (Note that $G$, and therefore $u$ as well, is a frame-indifferent scalar, since it is based upon a velocity measured with respect to the tubing, which is a difference of velocities. See Exercise 1.2.2-1.) If $\mu$ denotes the viscosity of the fluid, it seems reasonable to assume that

$$
\mathcal{E} = \mathcal{E}(\rho, \mu, D, u)
\tag{4.4.6-3}
$$

which automatically satisfies the principle of frame indifference.

The Buckingham–Pi theorem (Brand 1957) requires that this last expression be of the form

$$
e = e(N_{Re})
\tag{4.4.6-4}
$$

where $e$ is known as the *energy loss coefficient*:

$$
e \equiv \frac{\mathcal{E}}{\frac{1}{2} u^2 G}
\tag{4.4.6-5}
$$

and $N_{Re}$ is a Reynolds number:

$$
N_{Re} \equiv \frac{D u \rho}{\mu}
\tag{4.4.6-6}
$$

The details of this discussion apply equally well to any pipeline fitting, in either laminar or turbulent flow (for a turbulent flow, we merely replace $\mathcal{E}$ by $\overline{\mathcal{E}}$). The dependence of the friction loss coefficient $e$ for a given fitting upon a Reynolds number has been confirmed experimentally, at least for laminar flows (Kittredge and Rowley 1957). For turbulent flows, $e$ appears to be approximately a constant (Lapple 1949; Bird et al. 1960, p. 217).

### 4.4.7  Integral Moment-of-Momentum Balance

Corresponding to the differential mass and momentum balances, we have written the integral mass and momentum balances. This suggests that we develop an integral moment-of-momentum balance. From a different point of view, it could be useful to have an equation that would allow us to estimate the torque exerted on a solid surface as the result of a flow.

Considering our intention of developing an integral moment-of-momentum balance, we are prompted to consider the integral over an arbitrary system of the vector product of the position vector field $\mathbf{p}$ and the differential momentum balance:

$$\int_{R_{(s)}} \mathbf{p} \wedge \left( \rho \frac{d_{(m)}\mathbf{v}}{dt} - \operatorname{div} \mathbf{T} - \rho \mathbf{f} \right) dV = \mathbf{0} \tag{4.4.7-1}$$

Let us consider the terms in this equation individually.

The integrand of the first term in (4.4.7-1) may be expressed as

$$\mathbf{p} \wedge \left( \rho \frac{d_{(m)}\mathbf{v}}{dt} \right) = \mathbf{p} \wedge \left[ \frac{\partial(\rho \mathbf{v})}{\partial t} + \operatorname{div}(\rho \mathbf{v}\mathbf{v}) \right] \tag{4.4.7-2}$$

in terms of rectangular Cartesian components. The first term on the right of this last equation becomes

$$
\begin{aligned}
\mathbf{p} \wedge \frac{\partial(\rho \mathbf{v})}{\partial t} &= e_{ijk} z_j \frac{\partial(\rho v_k)}{\partial t} \mathbf{e}_i \\
&= \frac{\partial}{\partial t}(\mathbf{p} \wedge \rho \mathbf{v})
\end{aligned}
\tag{4.4.7-3}
$$

The second term on the right of (4.4.7-2) may be expressed as

$$
\begin{aligned}
\mathbf{p} \wedge \operatorname{div}(\rho \mathbf{v}\mathbf{v}) &= e_{ijk} z_j \frac{\partial}{\partial z_m}(\rho v_k v_m)\mathbf{e}_i \\
&= \frac{\partial}{\partial z_m}(e_{ijk} z_j \rho v_k v_m)\mathbf{e}_i - e_{ijk}\delta_{jm}\rho v_k v_m \mathbf{e}_i \\
&= \frac{\partial}{\partial z_m}(e_{ijk} z_j \rho v_k v_m)\mathbf{e}_i \\
&= \operatorname{div}([\mathbf{p} \wedge \rho \mathbf{v}]\mathbf{v})
\end{aligned}
\tag{4.4.7-4}
$$

Equations (4.4.7-3) and (4.4.7-4) allow us to express the first term on the left of (4.4.7-1) as

$$\int_{R_{(s)}} \mathbf{p} \wedge \left( \rho \frac{d_{(m)}\mathbf{v}}{dt} \right) dV = \int_{R_{(s)}} \left\{ \frac{\partial}{\partial t}(\mathbf{p} \wedge \rho \mathbf{v}) + \operatorname{div}[(\mathbf{p} \wedge \rho \mathbf{v})\mathbf{v}] \right\} dV \tag{4.4.7-5}$$

or, after an application of Green's transformation,

$$
\begin{aligned}
&\int_{R_{(s)}} \mathbf{p} \wedge \left( \rho \frac{d_{(m)}\mathbf{v}}{dt} \right) dV \\
&= \int_{R_{(s)}} \frac{\partial}{\partial t}(\mathbf{p} \wedge \rho \mathbf{v})\, dV + \int_{S_{(s)}} (\mathbf{p} \wedge \rho \mathbf{v})(\mathbf{v} \cdot \mathbf{n})\, dA \\
&= \frac{d}{dt} \int_{R_{(s)}} \mathbf{p} \wedge \rho \mathbf{v}\, dV + \int_{S_{(s)}} (\mathbf{p} \wedge \rho \mathbf{v})\left(\mathbf{v} - \mathbf{v}_{(s)}\right) \cdot \mathbf{n}\, dA
\end{aligned}
\tag{4.4.7-6}
$$

In this last line, we have used the generalized transport theorem of Section 1.3.2.

Rearrangement of the second term on the left of (4.4.7-1) is again best seen in terms of rectangular Cartesian components:

$$-\int_{R_{(s)}} \mathbf{p} \wedge (\operatorname{div} \mathbf{T}) \, dV = -\int_{R_{(s)}} e_{ijk} z_j \frac{\partial T_{km}}{\partial z_m} \mathbf{e}_i \, dV$$

$$= -\int_{R_{(s)}} \frac{\partial}{\partial z_m} (e_{ijk} z_j T_{km}) \mathbf{e}_i \, dV + \int_{R_{(s)}} e_{ijk} \delta_{jm} T_{km} \mathbf{e}_i \, dV$$

$$= -\int_{R_{(s)}} \frac{\partial}{\partial z_m} (e_{ijk} z_j T_{km}) \mathbf{e}_i \, dV$$

$$= -\int_{R_{(s)}} \operatorname{div} (\mathbf{p} \wedge \mathbf{T}) \, dV \qquad (4.4.7\text{-}7)$$

We have taken advantage of the symmetry of the stress tensor in declaring the second term on the right of (4.4.7-7) to be zero. After an application of Green's transformation, we are left with

$$-\int_{R_{(s)}} \mathbf{p} \wedge (\operatorname{div} \mathbf{T}) \, dV = -\int_{S_{(s)}} e_{ijk} z_j T_{km} n_m \mathbf{e}_i \, dA$$

$$= -\int_{S_{(s)}} \mathbf{p} \wedge (\mathbf{T} \cdot \mathbf{n}) \, dA \qquad (4.4.7\text{-}8)$$

Equations (4.4.7-6) and (4.4.7-8) allow us to express (4.4.7-1) as

$$\frac{d}{dt} \int_{R_{(s)}} \mathbf{p} \wedge \rho \mathbf{v} \, dV = \int_{S_{(s)}} (\mathbf{p} \wedge \rho \mathbf{v})(\mathbf{v} - \mathbf{v}_{(s)}) \cdot (-\mathbf{n}) \, dA$$

$$- \int_{S_{(s)}} \mathbf{p} \wedge [\mathbf{T} \cdot (-\mathbf{n})] \, dA + \int_{R_{(s)}} \mathbf{p} \wedge \rho \mathbf{f} \, dV \qquad (4.4.7\text{-}9)$$

The first term on the right of this equation is the net rate at which moment-of-momentum is brought into the system with whatever material is crossing the boundaries. The second term is the torque or moment of the force that the material in the system exerts on the bounding surface of the system. This is the total torque, whereas we usually speak in terms of the torque in excess of that attributable to an ambient or atmosphere pressure $p_0$. Repeating the argument we used in Section 4.4.3 to account for $p_0$ and neglecting viscous effects at the entrances and exits of the system, we can finally write (Slattery and Gaggioli 1962)

$$\frac{d}{dt} \int_{R_{(s)}} \mathbf{p} \wedge \rho \mathbf{v} \, dV = \int_{S_{(\text{ent ex})}} (\mathbf{p} \wedge \rho \mathbf{v})(\mathbf{v} - \mathbf{v}_{(s)}) \cdot (-\mathbf{n}) \, dA$$

$$- \int_{S_{(\text{ent ex})}} \mathbf{p} \wedge [(\mathbf{T} + p_0 \mathbf{I}) \cdot (-\mathbf{n})] \, dA - \boldsymbol{\mathcal{T}} + \int_{R_{(s)}} \mathbf{p} \wedge \rho \mathbf{f} \, dV \qquad (4.4.7\text{-}10)$$

where we define

$$\boldsymbol{\mathcal{T}} \equiv \int_{S_{(s)} - S_{(\text{ent ex})}} \mathbf{p} \wedge [(\mathbf{T} + p_0 \mathbf{I}) \cdot (-\mathbf{n})] \, dA \qquad (4.4.7\text{-}11)$$

Physically, $\boldsymbol{\mathcal{T}}$ denotes the torque that the system exerts upon the impermeable portion of its bounding surface beyond the torque attributable to the ambient pressure $p_0$. Equation

(4.4.7-10) is the *general integral moment-of-momentum balance* applicable to a single-phase system.

We generally try to choose a system for application of the integral moment-of-momentum balance in such a way that the torque attributable to viscous forces at the entrance and exit surfaces may be neglected. This might be done by selecting entrance and exit surfaces as cross sections normal to the flow in long straight pipes and by thinking of the flow in these straight pipes as being approximately represented by Poiseuille flow (Section 3.2.1). In most problems, one is forced to do this for lack of sufficient information to do otherwise. Under these conditions, (4.4.7-10) reduces to

$$
\frac{d}{dt} \int_{R_{(s)}} \mathbf{p} \wedge \rho \mathbf{v} \, dV = \int_{S_{(\text{ent ex})}} (\mathbf{p} \wedge \rho \mathbf{v})(\mathbf{v} - \mathbf{v}_{(s)}) \cdot (-\mathbf{n}) \, dA
$$

$$
- \int_{S_{(\text{ent ex})}} (P - p_0)(\mathbf{p} \wedge \mathbf{n}) \, dA - \mathcal{T} + \int_{R_{(s)}} \mathbf{p} \wedge \rho \mathbf{f} \, dV \qquad (4.4.7\text{-}12)
$$

This form of the integral moment-of-momentum balance is more commonly employed than (4.4.7-10).

As (4.4.7-10) and (4.4.7-12) have been derived, they are applicable only to a single-phase system. As we pointed out in Section 4.4.1, we are more often concerned with multiphase systems. Using the approach and notation of Section 4.4.1 and making no additional assumptions, we find that the *general integral moment-of-momentum balance* appropriate to a multiphase system is

$$
\frac{d}{dt} \int_{R_{(s)}} \mathbf{p} \wedge \rho \mathbf{v} \, dV = \int_{S_{(\text{ent ex})}} (\mathbf{p} \wedge \rho \mathbf{v})\left(\mathbf{v} - \mathbf{v}_{(s)}\right) \cdot (-\mathbf{n}) \, dA
$$

$$
- \int_{S_{(\text{ent ex})}} \mathbf{p} \wedge [(\mathbf{T} + p_0 \mathbf{I}) \cdot (-\mathbf{n})] \, dA - \mathbf{T} + \int_{R_{(s)}} \mathbf{p} \wedge \rho \mathbf{f} \, dV
$$

$$
+ \int_{\Sigma} \mathbf{p} \wedge \left[ \rho \mathbf{v}(\mathbf{v} - \mathbf{u}) \cdot \boldsymbol{\xi} - \mathbf{T} \cdot \boldsymbol{\xi} \right] dA. \qquad (4.4.7\text{-}13)
$$

If we assume that the jump momentum balance (2.2.3-5) is applicable and if we neglect viscous effects at the entrance and exit surfaces, we find that (4.4.7-13) reduces to the equivalent result for a single-phase system, (4.4.7-12).

We can visualize the three types of problems in which the integral moment-of-momentum balance might be applied: The torque $\mathcal{T}$ might be neglected, it might be unknown and to be determined, or it might be known from previous experimental data. Perhaps because the integral moment-of-momentum balance has not seen as much use in the literature as the integral momentum balance or, more likely, because the need has not generally arisen, very little in the way of empirical data correlations for $\mathcal{T}$ are available. For this reason, we will not devote a special section to empirical correlation for $\mathcal{T}$. For anyone interested, the approach should be similar to that taken for $\mathcal{F}$ and $\mathcal{G}$ in Section 4.4.4.

**Exercise 4.4.7-1** *Integral moment-of-momentum balance for turbulent flow*    I recommend following the discussion in Section 4.4.2 in developing forms of the integral moment-of-momentum balance for turbulent flows.

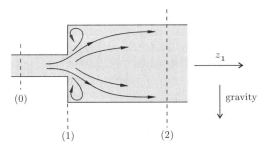

**Figure 4.4.8-1.** Flow through a sudden expansion.

i) Show that, for single-phase or multiphase systems that do not involve fluid–fluid phase interfaces,

$$\frac{d}{dt}\int_{R_{(s)}} \mathbf{p}\wedge\overline{\rho\mathbf{v}}\,dV = \int_{S_{(\text{ent ex})}} \mathbf{p}\wedge\left(\overline{\rho\,\mathbf{v}\,\mathbf{v}} - \overline{\rho\mathbf{v}}\mathbf{v}_{(s)}\right)\cdot(-\mathbf{n})\,dA$$

$$-\int_{S_{(\text{ent ex})}}(\overline{P} - p_0)(\mathbf{p}\wedge\mathbf{n})\,dA - \mathcal{T} + \int_{R_{(s)}}\mathbf{p}\wedge\overline{\rho\mathbf{f}}\,dV$$

$$+\int_{\Sigma}\mathbf{p}\wedge\left[\left(\overline{\rho\,\mathbf{v}\,\mathbf{v}} - \overline{\rho\mathbf{v}}\mathbf{u}\right)\cdot\boldsymbol{\xi} - \overline{\mathbf{T}}\cdot\boldsymbol{\xi}\right]dA \qquad (4.4.7\text{-}14)$$

In arriving at this result, we have observed that, for this case, $S_{(s)} - S_{(\text{ent ex})}$ and $\Sigma$ must be composed of fluid–solid and solid–solid phase interfaces on which the turbulent fluctuations are identically zero. The time-averaged jump momentum balance of Exercise 4.4.3-3 permits us to neglect the last term of (4.4.7-14).

ii) For a single-phase or multiphase system that includes one or more fluid–fluid phase interfaces, I recommend simple time averaging of the integral moment-of-momentum balance of the text.

### 4.4.8 Examples

Our intention here is not only to illustrate how integral balances can be used but also to point out that sometimes the same problem can be analyzed in more than one way. Since in each case somewhat different approximations may be made, it should not be surprising when differing answers are obtained.

To illustrate these points, let us estimate for an incompressible fluid the change in pressure across the sudden expansion in a pipeline pictured in Figure 4.4.8-1. More specifically, we seek the change in pressure between cross sections 0 and 2. It is assumed that the velocity distributions at these cross sections are essentially unaffected by the presence of the sudden expansion, although the pressure difference is primarily attributable to the change in cross-sectional area rather than pipe friction. We contrast results obtained using the integral momentum and mechanical energy balances. We leave it to the reader in Exercise 4.4.8-10 to analyze this same problem using the Bernoulli equation.

#### Integral Momentum Balance

Let us choose as our system all the fluid in the pipe between cross sections 0 and 2.

Let us understand that $\langle \bar{v}_1 \rangle_{(0)}$ denotes the area-averaged $z_1$ component of the time-averaged velocity at cross section 0:

$$\langle \bar{v}_1 \rangle_{(0)} \equiv \frac{1}{A_{(0)}} \int_{S_{(0)}} \bar{v}_1 \, dA \tag{4.4.8-1}$$

The time-averaged integral mass balance tells us that

$$\langle \bar{v}_1 \rangle_{(2)} = \langle \bar{v}_1 \rangle_{(0)} \left( \frac{D_{(0)}}{D_{(2)}} \right)^2 \tag{4.4.8-2}$$

where $D_{(i)}$ denotes the diameter of the pipe at cross section $i$.

For this system we may use the integral momentum balance of Exercise 4.4.3-2, the $z_1$ component of which tells us that

$$\rho \left\langle \bar{v}_1{}^2 \right\rangle_{(0)} \frac{\pi D_{(0)}{}^2}{4} - \rho \left\langle \bar{v}_1{}^2 \right\rangle_{(2)} \frac{\pi D_{(2)}{}^2}{4} + \left( \langle \bar{p} \rangle_{(0)} - p_0 \right) \frac{\pi D_{(0)}{}^2}{4} - \left( \langle \bar{p} \rangle_{(2)} - p_0 \right) \frac{\pi D_{(2)}{}^2}{4}$$
$$+ \left( \langle p \rangle_{(1)} - p_0 \right) \left( \frac{\pi D_{(2)}{}^2}{4} - \frac{\pi D_{(0)}{}^2}{4} \right) = 0 \tag{4.4.8-3}$$

In writing this last equation we have neglected the effect of the turbulent Reynolds stresses at the entrances and exits and we have neglected the viscous contributions to $\mathcal{F}$ on the bounding metal surfaces. Let us assume that $\bar{v}_1$ is nearly independent of position at cross sections 0 and 2 (see Exercise 4.4.8-11), so that we may write

$$\langle \bar{v}_1{}^2 \rangle_{(0)} \doteq \langle \bar{v}_1 \rangle_{(0)}{}^2$$
$$\langle \bar{v}_1{}^2 \rangle_{(2)} \doteq \langle \bar{v}_1 \rangle_{(2)}{}^2 \tag{4.4.8-4}$$

We estimate that

$$\langle p \rangle_{(1)} \doteq \langle \bar{p} \rangle_{(0)} \tag{4.4.8-5}$$

With (4.4.8-2), (4.4.8-4), and (4.4.8-5), equation (4.4.8-3) takes the desired form:

$$\frac{\langle \bar{p} \rangle_{(2)} - \langle \bar{p} \rangle_{(0)}}{\frac{1}{2} \rho \langle \bar{v}_1 \rangle_{(2)}{}^2} = 2 \left[ \left( \frac{D_{(2)}}{D_{(0)}} \right)^2 - 1 \right] \tag{4.4.8-6}$$

## Mechanical Energy Balance

The mechanical energy balance of Exercise 4.4.5-4 may be applied to this system to find

$$\frac{1}{2} \rho \langle \bar{v}_1{}^3 \rangle_{(0)} \frac{\pi D_{(0)}{}^2}{4} + \left( \langle \bar{p} \, \bar{v}_1 \rangle_{(0)} - p_0 \langle \bar{v}_1 \rangle_{(0)} \right) \frac{\pi D_{(0)}{}^2}{4} - \frac{1}{2} \rho \langle \bar{v}_1{}^3 \rangle_{(2)} \frac{\pi D_{(2)}{}^2}{4}$$
$$- \left( \langle \bar{p} \, \bar{v}_1 \rangle_{(2)} - p_0 \langle \bar{v}_1 \rangle_{(2)} \right) \frac{\pi D_{(2)}{}^2}{4} = 0 \tag{4.4.8-7}$$

In arriving at this result, we have neglected all effects of turbulence at the entrances and exits as well as the time-averaged rate of dissipation of mechanical energy $\mathcal{E}^{(t)}$. Again we assume that $\bar{v}_1$ is sufficiently uniform with respect to position at the entrance and exit that we may approximate (see Exercise 4.4.8-11)

$$\langle \overline{v}_1{}^3 \rangle_{(0)} \doteq \langle \overline{v}_1 \rangle_{(0)}{}^3$$

$$\langle \overline{v}_1{}^3 \rangle_{(2)} \doteq \langle \overline{v}_1 \rangle_{(2)}{}^3$$

$$\langle \overline{p}\,\overline{v}_1 \rangle_{(0)} \doteq \langle \overline{p} \rangle_{(0)} \langle \overline{v}_1 \rangle_{(0)}$$

$$\langle \overline{p}\,\overline{v}_1 \rangle_{(2)} \doteq \langle \overline{p} \rangle_{(2)} \langle \overline{v}_1 \rangle_{(2)}$$

(4.4.8-8)

Equations (4.4.8-2) and (4.4.8-8) allow us to express (4.4.8-7) as

$$\frac{\langle \overline{p} \rangle_{(2)} - \langle \overline{p} \rangle_{(0)}}{\frac{1}{2}\rho \langle \overline{v}_1 \rangle_{(2)}{}^2} = \left[ \left( \frac{D_{(2)}}{D_{(0)}} \right)^4 - 1 \right]$$

(4.4.8-9)

### Discussion

Whitaker (1968, p. 242) makes an interesting comparison of (4.4.8-6) and (4.4.8-9) with experimental data. He finds that the result from the integral momentum balance, (4.4.8-6), is in reasonable agreement with the data, whereas the result from the mechanical energy balance, (4.4.8-9), consistently gives values that are too large. If we go back to the mechanical energy balance and include the rate of dissipation of mechanical energy by the action of viscous and turbulent forces, we find

$$\frac{\langle \overline{p} \rangle_{(2)} - \langle \overline{p} \rangle_{(0}}{\frac{1}{2}\rho \langle \overline{v}_1 \rangle_{(2)}{}^2} = \left[ \left( \frac{D_{(2)}}{D_{(0)}} \right)^4 - 1 \right] - \frac{8\mathcal{E}^{(t)}}{\pi D_{(2)}{}^2 \rho \langle \overline{v}_1 \rangle_{(2)}{}^3}$$

(4.4.8-10)

Although it is clear that $\overline{\mathcal{E}} \geq 0$ for an isothermal flow (see Section 7.4.5), a similar statement has not been proved for $\mathcal{E}^{(t)}$. However, the experimental data quoted by Whitaker suggest that, at least for this situation, $\mathcal{E}^{(t)} \geq 0$.

Before the results of these comparisons with experimental data were described to you, it may not have been obvious that the error involved in neglecting $\mathcal{E}^{(t)}$ in the mechanical energy balance was any more serious than the error incurred in neglecting the viscous contribution to $\mathcal{F}$ in the integral momentum balance. I can only tell you that there is no substitute for experience, otherwise known as a well-formed "engineering judgment." This is a good illustration of why one must be wary in applying the integral balances. It is usually necessary to make approximations in order to obtain answers that are in a useful form, but these approximations are made in such a way that it is rarely possible to place error bounds on the results. It is only after a successful comparison with experimental data that most of us can have confidence in using these analyses.

Because (4.4.8-6) does do a reasonable job of representing available experimental data, it and (4.4.8-10) are often used to estimate the friction loss coefficient (introduced in Section 4.4.6) for this flow as

$$\overline{e} \equiv \frac{8\mathcal{E}^{(t)}}{\pi D_{(2)}{}^2 \rho \langle \overline{v}_1 \rangle_{(2)}{}^3} = \left[ \left( \frac{D_{(2)}}{D_{(0)}} \right)^2 - 1 \right]^2$$

(4.4.8-11)

**Exercise 4.4.8-1** *The Borda mouthpiece*  A tank has an orifice as shown in Figure 4.4.8-2. With such an orifice, there is relatively little movement of the fluid in the neighborhood of the tank walls. Neglecting any effects of friction and assuming the fluid is incompressible, find

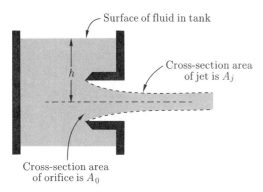

**Figure 4.4.8-2.** The Borda mouthpiece.

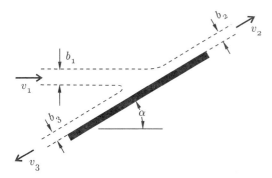

**Figure 4.4.8-3.** Deflection of a two-dimensional jet
by a flat plate.

the contraction coefficient for this nozzle $A_j/A_0$. This orifice is referred to as a *Borda mouthpiece*.

*Hint:*   The Bernoulli equation is useful here.

**Exercise 4.4.8-2** *Deflection of a two-dimensional jet*

i) A two-dimensional jet of incompressible fluid strikes a plane surface. Referring to Figure 4.4.8-3, let $v_i$ ($i = 1, 2, 3$) denote the average magnitude of the velocity of stream $i$ ($i = 1, 2, 3$) and $b_i$ the corresponding width of the stream. Neglecting viscous effects and the effect of gravity, find $v_2$, $v_3$, $b_2$, and $b_3$ in terms of the density of the fluid $\rho$, $v_1$, $b_1$, and $\alpha$.

ii) Find the direction and magnitude of the force per unit width required to hold the plate stationary.

**Exercise 4.4.8-3** *An accelerating rocket*   At time $t = 0$, a rocket, the initial mass of which is $M_0$, is ignited. It accelerates along a straight line in the opposite direction to the action of gravity. The rocket consumes propellant at a constant rate $G$ (mass per unit time). The speed of the exhaust gas with respect to the accelerating rocket is $v_e$. Determine the speed of the rocket as a function of time.

**Figure 4.4.8-4.** The force on a dam.

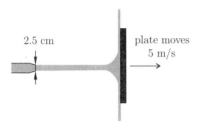

**Figure 4.4.8-5.** Force on a moving plate.

**Exercise 4.4.8-4** *More on emptying a vertical tube* Do Exercise 3.4.2-4 using the integral momentum balance.

**Exercise 4.4.8-5** *The force on a dam* Referring to Figure 4.4.8-4, estimate the $x$ component of the force that the fluid exerts upon the dam per unit width of the dam. You may neglect viscous effects.

**Exercise 4.4.8-6** *A moving frame of reference*

i) Determine that the form of the integral mass balance is independent of the frame of reference.

ii) Prove that the form of the integral momentum balance is unchanged for any frame of reference that moves at a constant velocity with respect to the fixed stars.

iii) Prove that Bernoulli's equation is unchanged for any frame of reference that moves at a constant velocity with respect to the fixed stars.

**Exercise 4.4.8-7** *Force on a moving plate* (Truesdell and Noll 1965, p. 276) Figure 4.4.8-5 shows a cylindrical jet of water issuing from a 2.5-cm (inside diameter) nozzle at a speed of 10 m/s. It strikes a flat plate that is moving away from the jet at a speed of 5 m/s. What force does the jet exert on the plate?

*Hint:* Choose a system and a frame of reference that are fixed with respect to the plate. See Exercise 4.4.8-6.

**Exercise 4.4.8-8** *More about the force on a moving plate* Repeat Exercise 4.4.8-2 assuming now that the plate moves to the right at a constant speed $v_p$.

In carrying out this calculation, let $v_1$, $v_2$, and $v_3$ denote speeds measured with respect to a stationary frame of reference and let $\tilde{v}_1$, $\tilde{v}_2$, and $\tilde{v}_3$ signify speeds measured with respect to the moving frame of reference.

**Table 4.4.8-1.** Dependence of $n$ upon $N_{Re}$

| $n$ | 6 | 7 | 10 |
|---|---|---|---|
| $N_{Re}$ | $4 \times 10^3$ | $1 \times 10^5$ | $3 \times 10^6$ |

*Source:*  Schlichting (1979, p. 600).

**Exercise 4.4.8-9** *A fire hose*    Does the nozzle on a fire hose place the hose in tension or compression? Derive an expression for the force that the fluid exerts upon the nozzle that confirms your intuitive opinion.

**Exercise 4.4.8-10** *Pipe flow* (Bird et al. 1960, p. 216)

i) Consider flow down a vertical pipe under steady-state conditions (either laminar flow or turbulent flow such that the time-averaged variables are independent of time at fixed positions). If $\bar{v}$ is the area-averaged axial component of velocity in the pipe, show that

$$\mathcal{E}^{(t)} = \bar{v}\mathcal{F}$$

where $\mathcal{F}$ is the axial component of the force that the fluid exerts on the walls of the pipe in the direction of flow.

ii) Conclude that

$$e = \frac{4cL}{D}$$

where $c$ is the drag coefficient defined in the manner indicated in Section 4.4.4 and $D$ is the diameter of the pipe.

**Exercise 4.4.8-11** *More on pipe flow* (Whitaker 1968, p. 242)    In Section 3.2.1, we found that the steady-state velocity distribution of an incompressible Newtonian fluid flowing through a very long tube of radius $R$ is given by

$$\frac{v_z}{v_{z,\max}} = 1 - \left(\frac{r}{R}\right)^2$$

The time-averaged velocity distribution for turbulent flow through a pipe is found to be approximately represented by (Schlichting 1979, p. 599)

$$\frac{\bar{v}_z}{\bar{v}_{z,\max}} = \left(1 - \frac{r}{R}\right)^{1/n}$$

where $n$ is a function of the Reynolds number as indicated in Table 4.4.8-1.

Determine that for turbulent flow through a pipe

$$\frac{\langle \bar{v}_z \rangle}{\bar{v}_{z,\max}} = \frac{2n^2}{(1+n)(1+2n)}$$

$$\frac{\langle \bar{v}_z^2 \rangle}{\langle \bar{v}_z \rangle^2} = \frac{(1+n)(1+2n)^2}{4n^2(2+n)}$$

**Table 4.4.8-2.** Deviations from flat velocity profile

| | Turbulent | | | | | Laminar |
|---|---|---|---|---|---|---|
| $n$ | 6 | 7 | 8 | 9 | 10 | Parabolic profile |
| $\dfrac{\langle \overline{v}_z \rangle}{\overline{v}_{z,\max}}$ | 0.76 | 0.82 | 0.84 | 0.85 | 0.86 | 0.50 |
| $\dfrac{\langle \overline{v}_z{}^2 \rangle}{\langle \overline{v}_z \rangle^2}$ | 1.03 | 1.01 | 1.01 | 1.02 | 1.03 | 1.33 |
| $\dfrac{\langle \overline{v}_z{}^3 \rangle}{\langle \overline{v}_z \rangle^3}$ | 1.08 | 1.05 | 1.03 | 1.05 | 1.06 | 2 |

*Source:* Whitaker (1968, p. 243).

and

$$\frac{\langle \overline{v}_z{}^3 \rangle}{\langle \overline{v}_z \rangle^3} = \frac{(1+n)^3(1+2n)^3}{4n^4(3+n)(3+2n)}$$

These results are evaluated for selected values of $n$ in Table 4.4.8-2.

**Exercise 4.4.8-12** *The Egyptian water clock* (Whitaker 1968, p. 277)   The Egyptians used water clocks similar to that illustrated in Figure 4.4.8-6 (Rouse and Ince 1957, p. 7). The radius $r_0$ of the circular bowl is a function of $z$, the distance from the bottom of the bowl. Determine the functional dependence of $r_0$ on $z$ required in order that the depth of liquid be a linear function of time.

  i) Use the integral mass and mechanical energy balances. Do not neglect the time rate of change of potential energy in the mechanical energy balance.
  ii) Use the integral mass balance and the steady-state Bernoulli equation. This is known as a quasi-state analysis.

  *Answer for both cases:*

  $$r_0 = (2g)^{1/4} \left[ \frac{A_{ex}}{-\pi(dz/dt)} \right]^{1/2} z^{1/4}$$

**Exercise 4.4.8-13** *An alternative expression for the force on a sphere*   Consider an infinite expanse of fluid that moves past a sphere fixed in space. Very far away from the sphere, the fluid moves with a constant velocity $\mathbf{v}_\infty$. (Mathematically, this is the same as considering a sphere that moves with a constant velocity $-\mathbf{v}_\infty$ through an infinite expanse of fluid that is stationary very far away from the sphere.)

  i) Use the integral momentum and mechanical energy balances to find

  $$\mathbf{v}_\infty \cdot \mathcal{G} = \int_0^{2\pi} \int_0^{\pi} \int_0^R \text{tr}(\mathbf{S} \cdot \nabla \mathbf{v}) r^2 \sin\theta \, dr \, d\theta \, d\varphi$$

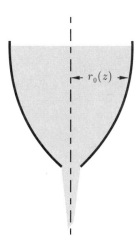

**Figure 4.4.8-6.** Egyptian water clock.

ii) For flow of an incompressible Newtonian fluid past a sphere as discussed in Section 3.3.2, determine that

$$v_\infty \mathcal{F}_3 - \frac{4}{3}\pi R^3 \rho g v_\infty$$

$$= 2\pi\mu \int_R^\infty \int_0^\pi \left\{ 2\left(\frac{\partial v_r}{\partial r}\right)^2 + 2\left(\frac{1}{r}\frac{\partial v_\theta}{\partial \theta} + \frac{v_r}{r}\right)^2 + 2\left(\frac{v_r}{r} + \frac{v_\theta \cot\theta}{r}\right)^2 \right.$$

$$\left. + \left[ r\frac{\partial}{\partial r}\left(\frac{v_\theta}{r}\right) + \frac{1}{r}\frac{\partial v_r}{\partial \theta}\right]^2 \right\} r^2 \sin\theta \, dr \, d\theta$$

where $\mathcal{F}_3$ is the $z_3$ component of the force that the fluid exerts on the sphere.

iii) Use the expression obtained in (ii) as well as the velocity distribution from Section 3.3.2 to arrive at Stokes's law. Compared with the expression used in Section 3.3.2, the result from (ii) has the advantage of not requiring the pressure distribution in the fluid.

**Exercise 4.4.8-14** *The hydraulic ram* (Whitaker 1968, p. 275)   A simple hydraulic brake consists of a cylindrical ram that displaces fluid from a slightly larger cylinder as shown in Figure 4.4.8-7. The speed of the ram is $v_0$. We wish to determine the magnitude $F$ of the force required to maintain this motion.

i) Use the integral mass balance to estimate the area-averaged $z_1$ component of velocity in the annular space between the ram and cylinder as

$$\langle v_1 \rangle_{ex} = \frac{v_0}{(D_1/D_0)^2 - 1}$$

ii) Use the integral momentum balance to suggest

$$F = \frac{\rho v_0^2 \pi D_1^2}{4[(D_1/D_0)^2 - 1]^2}$$

Neglect viscous effects in this analysis.

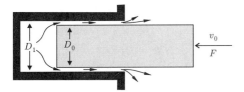

**Figure 4.4.8-7.** The hydraulic ram.

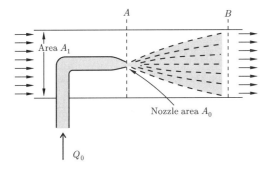

**Figure 4.4.8-8.** A simple ejector pump.

iii) Use the integral mechanical energy balance to say

$$F = \frac{\rho v_0{}^2 \pi D_1{}^2}{8[(D_1/D_0)^2 - 1]^2}$$

Viscous effects should be neglected in this analysis as well.

iv) The integral mechanical energy balance probably gives a better answer here. Why?

**Exercise 4.4.8-15** *Ejector pump*   Figure 4.4.8-8 shows a very simple device that can be used to pump fluids, the ejector pump. Let us assume that both fluids are the same.

i) Derive an expression for the pressure rise between cross section $A$ and cross section $B$.

ii) Derive an equation or set of equations to be solved for $Q_1$ as a function of $Q_0$, $A_0$, and $A_1$.

**Exercise 4.4.8-16** *Drag on an arbitrary body*   Consider an infinite expanse of fluid that moves past an arbitrary body that is fixed in space. Very far away from the body, the fluid moves with a constant velocity $\mathbf{v}_\infty$.

i) Use the integral momentum and mechanical energy balances to find

$$\mathbf{v}_\infty \cdot \mathcal{G} = \mathcal{E}$$

$$\equiv \int_{R_{(\text{fluid})}} \text{tr}(\mathbf{S} \cdot \nabla \mathbf{v}) \, dV$$

ii) Determine that

$$\mathbf{v}_\infty \cdot \mathcal{G} = \mathbf{v}_\infty \cdot \mathcal{F} - \mathcal{M}(\mathbf{v}_\infty \cdot \mathbf{f})$$

where $\mathcal{M}$ is the mass of fluid displaced by the body.

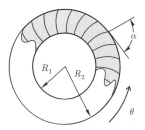

**Figure 4.4.8-9.** Cutaway pump impeller.

iii) For plane flow past a cylindrical body, find that (see Section 4.4.4, Example 2)

$$c_D = \frac{2\mathcal{E}}{A_D \rho \, |\mathbf{v}_\infty|^3}$$

where $c_D$ is the dimensionless drag coefficient.

**Exercise 4.4.8-17** *A stationary pump impeller* (Sabersky et al. 1989, p. 142)  Consider a two-dimensional, frictionless, steady, incompressible flow in Figure 4.4.8-9 from $r = R_1$ to $r = R_2$ such that in cylindrical coordinates

at $r = R_1$ :  $v_r = v_{r(1)}$

$v_\theta = v_{\theta(1)}$

The guide vanes of the impeller align the flow such that

at $r = R_2$ :  $v_r = v_{r(2)}$

$= -v_{\theta(2)} \tan \alpha$

Assuming that the flow is axially symmetric and neglecting any disturbance to the flow due to the thickness of the guide vanes, determine the torque exerted on the guide vanes per unit width.

**Exercise 4.4.8-18** *A rotating pump impeller*  The impeller in Exercise 4.4.8-17 rotates at a constant angular velocity $\Omega$ in the direction of increasing $\theta$. This implies that

$r^* = r$

$\theta^* = \theta + \Omega t$

$z^* = z$

where $(r^*, \theta^*, z^*)$ is a set of cylindrical coordinates in a frame fixed with respect to the stars and $(r, \theta, z)$ is a set of cylindrical coordinates in a frame fixed with respect to the guide vanes on the rotating impeller. These two frames share a common origin on the axis of rotation of the impeller. Consider the velocity components introduced in Exercise 4.4.8-17 to be measured in the cylindrical coordinate system $(r, \theta, z)$.

i) Determine the torque required to turn the impeller for a volume rate of flow $Q$.

ii) What is

$$\mathcal{W} \equiv \int_{S_{(imp)}} \mathbf{v} \cdot (\mathbf{T} + p_0 \mathbf{I}) \cdot \mathbf{n} \, dA$$

the rate at which work is done by the impeller on the fluid? (In writing this expression, our understanding is that $\mathbf{n}$ is the unit normal to the impeller surface directed from the fluid into the metal.)

This problem was suggested by Sabersky et al. (1989, p. 143).

# 5

---

# Foundations for Energy Transfer

THIS CHAPTER is concerned with the foundations for energy transfer. And yet perhaps
that is not sufficiently descriptive of what we are actually going to do here, because the
foundations for energy transfer are really the foundations for the subject we normally think
of as thermodynamics.

By thermodynamics, we do not mean precisely the subject presented to us by Gibbs
(1928), since he was concerned with materials at equilibrium. We are concerned here with
nonequilibrium situations in which momentum and energy are being transferred. (Don't
make the mistake of confusing the terms *steady state* and *equilibrium*.)

There are at least two points to note.

1) Rather than a balance equation for entropy, we have an inequality: the entropy inequality
   or second law of thermodynamics.
2) There are two forms for the entropy inequality. The sole purpose of the differential entropy
   inequality is to place restrictions upon descriptions of material behavior: constitutive
   equations for specific internal energy (the fundamental equation of state relating specific
   internal energy and specific entropy), for the stress tensor, for the energy flux vector,
   . . . . The integral entropy inequality, which will be developed in Section 7.4.4, is more
   familiar to most readers, since it is generally the only form mentioned in undergraduate
   texts on thermodynamics.

---

## 5.1  Energy

### 5.1.1  Energy Balance

We think that you will be better able to visualize our next step if you consider for a moment
a particulate model of a real material. The molecules are in relative motion with respect to
the material that they comprise. They have kinetic energy associated with them beyond the
kinetic energy of the material as a whole. They also possess potential energy as the result
of their positions in the various intermolecular force fields. It is these forms of kinetic and
potential energy that we describe as the internal energy of the material.

But this does not define internal energy in our continuum model for a real material. Like material particle, mass, and force, internal energy is a primitive concept; it is not defined in the context of continuum mechanics. Instead, we describe its properties. We are about to state as a postulate its most important property: the energy balance. In addition, we will require that internal energy per unit mass be positive and frame indifferent:

$$\hat{U} > 0$$
$$\hat{U}^* = \hat{U}$$
(5.1.1-1)

We might be tempted to postulate that the time rate of change of the internal energy of a body is equal to the rate at which work is done on the body by the system of forces acting upon it plus the rate of energy transmission to it. This appears to be simple to put in quantitative terms for single-phase bodies. However, it is an awkward statement for multiphase bodies, when mass transfer is permitted (see Exercise 5.1.3-6).

Instead we take as a postulate applicable to all materials the

**Energy balance**   In an inertial frame of reference, the time rate of change of the internal and kinetic energy of a body is equal to the rate at which work is done on the body by the system of contact, external, and mutual forces acting upon it plus the rate of energy transmission to the body. (We have assumed that all work on the body is the result of forces acting on the body.[1])

The energy balance is also known as the *first law of thermodynamics*.

### 5.1.2   Radiant and Contact Energy Transmission

The rate of energy transmission is described in a manner entirely analogous to that in which force is described (see the introduction to Section 2.1) in the sense that it is a primitive concept. It is not defined. Instead we describe its attributes in a series of five axioms.

Corresponding to each body $B$, there is a distinct set of bodies $B^e$ such that the mass of the union of these bodies is the mass of the universe. We refer to $B^e$ as the exterior or the surroundings of the body $B$.

1) A system of energy transmission rates is a scalar-valued function $\mathcal{Q}(B, C)$ of pairs of bodies. The value of $\mathcal{Q}(B, C)$ is called the rate of energy transmission from body $C$ to body $B$.
2) For a specified body $B$, $\mathcal{Q}(C, B^e)$ is an additive function defined over the subbodies $C$ of $B$.

---

[1] We assume here that all work on the body is the result of forces acting on the body, including, for example, the electrostatic forces exerted by one portion of a body upon another. It is possible to induce in a polar material a local source of moment of momentum with a rotating electric field (Lertes 1921a,b,c; Grossetti 1958, 1959). In such a case, it might also be necessary to account for the work done by the flux of moment of momentum at the bounding surface of the body. Effects of this type have not been investigated thoroughly, but they are thought to be negligibly small for all but unusual situations. They are neglected here.

3) Conversely, for a specified body $B$, $Q(B, C)$ is an additive function defined over the subbodies $C$ of $B^e$.

4) The rate of energy transmission to a body should have nothing to do with the motion of the observer or experimentalist relative to the body. It should be frame indifferent:

$$Q^* = Q \qquad (5.1.2\text{-}1)$$

There are three types of energy transmission with which we are concerned:

**Rate of external radiant energy transmission**   The rate of external radiant energy transmission refers to energy transmission from outside the body to the material particles of which the body is composed. One example is the radiation from the Sun to the gas that composes Earth's atmosphere. Another example is induction heating in which energy is transferred to the polar molecules of a body by means of an alternating magnetic field. It is presumed to be related to the masses of the bodies, and it is described as though it acts directly on each material particle:

$$Q_e = \int_{R_{(m)}} \rho Q_e \, dV \qquad (5.1.2\text{-}2)$$

Here $Q_e$ is the rate of radiant energy transmission per unit mass.

**Rate of mutual radiant energy transmission**   The rate of mutual radiant energy transmission refers to energy transmission between pairs of material particles that are part of the same body. Radiation within a hot colored gas is one example. Let $Q_m$ be the mutual energy transmission rate per unit mass from $B - P$ [define $B - P$ to be such that $B = (B - P) \cup P$ and $(B - P) \cap P = 0$] to $P$; the total mutual energy transmission to $P$ may be represented as an integral over the region occupied by $P$:[2]

$$Q_m = \int_{R_{(m)}} \rho Q_m \, dV \qquad (5.1.2\text{-}3)$$

**Rate of contact energy transmission**   This is energy transmission that is not assignable as a function of position within the body, but which is to be imagined as energy transmission through the bounding surface of a portion of material in such a way as to be equivalent to the energy transmission from the surroundings beyond that accounted for through external and mutual radiant energy transmission. As an example, when we press our hands to a hot metal surface, there is contact energy transfer with the result that our hands may be burned. Let $h = h(\mathbf{z}, P)$ represent the rate of energy transmission per unit area from $B - P$ to the boundary of $P$ at the position $\mathbf{z}$. This rate of energy transmission per unit area $h$ may be referred to as the *contact energy flux*. The total rate of contact energy transfer from $B - P$ to $P$ may be written as an integral over the bounding surface of $P$:

$$Q_c = \int_{S_{(m)}} h \, dA \qquad (5.1.2\text{-}4)$$

---

[2]   We recognize here that the sum of the mutual radiant energy transmission between any two parts of $P$ must be zero; the proof is much the same as that for mutual forces (Truesdell and Toupin 1960, p. 533).

The fifth axiom, the energy flux principle, specifies the nature of the contact energy transmission.

5) **Energy flux principle** There is a frame-indifferent, scalar-valued function $h(\mathbf{z}, \mathbf{n})$ defined for all points $\mathbf{z}$ in a body $B$ and for all unit vectors $\mathbf{n}$ such that the rate of contact energy transmission per unit area from $B - P$ to any portion $P$ of $B$ is given by

$$h(\mathbf{z}, P) = h(\mathbf{z}, \mathbf{n}) \tag{5.1.2-5}$$

Here $\mathbf{n}$ is the unit normal vector that is outwardly directed with respect to the closed bounding surface of $P$. The scalar $h = h(\mathbf{z}, \mathbf{n})$ is referred to as the contact energy flux at the position $\mathbf{z}$ across the oriented surface element with normal $\mathbf{n}$; $\mathbf{n}$ points into the material from which the contact energy flux to the surface element is $h$.

**5.1.3** Differential and Jump Energy Balances

In an inertial frame of reference, the energy balance of Section 5.1.1 says

$$\frac{d}{dt} \int_{R_{(m)}} \rho \left( \hat{U} + \frac{1}{2}v^2 \right) dV = \int_{S_{(m)}} \mathbf{v} \cdot (\mathbf{T} \cdot \mathbf{n}) \, dA + \int_{R_{(m)}} \rho \mathbf{v} \cdot \mathbf{f} \, dV$$
$$+ \int_{S_{(m)}} h \, dA + \int_{R_{(m)}} \rho Q \, dV \tag{5.1.3-1}$$

where $\hat{U}$ denotes *internal energy per unit mass* and

$$Q \equiv Q_e + Q_m \tag{5.1.3-2}$$

is the scalar field that represents the sum of the external and mutual energy transmission rates per unit mass.

Let us assume that we are considering a multiphase body that includes a set of internal phase interfaces $\Sigma$. Under these conditions, (5.1.3-1) implies a differential equation that expresses a balance of energy at every point within a material. The steps are very similar to those used to obtain the differential momentum balance from momentum balance in Section 2.2.3. The alternative form of the transport theorem for regions containing a dividing surface (Exercise 1.3.6-3) permits us to express the left side of (5.1.3-1) as

$$\frac{d}{dt} \int_{R_{(m)}} \rho \left( \hat{U} + \frac{1}{2}v^2 \right) dV = \int_{R_{(m)}} \rho \frac{d_{(m)}}{dt} \left( \hat{U} + \frac{1}{2}v^2 \right) dV$$
$$+ \int_{\Sigma} \left[ \rho \left( \hat{U} + \frac{1}{2}v^2 \right) (\mathbf{v} - \mathbf{u}) \cdot \boldsymbol{\xi} \right] dA \tag{5.1.3-3}$$

The first term on the right of (5.1.3-1) may be expressed in terms of a volume integral by means of Green's transformation (see Section A.11.2):

$$\int_{S_{(m)}} \mathbf{v} \cdot (\mathbf{T} \cdot \mathbf{n}) \, dA = \int_{R_{(m)}} \text{div}(\mathbf{T} \cdot \mathbf{v}) \, dV + \int_{\Sigma} [\mathbf{v} \cdot (\mathbf{T} \cdot \boldsymbol{\xi})] \, dA \tag{5.1.3-4}$$

in which we have taken advantage of the symmetry of the stress tensor. The third term on the right of (5.1.3-1) may be rewritten in terms of a volume integral, only if we can take advantage of Green's transformation. By Exercise 5.1.3-2, we may express the contact energy flux in terms of the *energy flux vector* **q**:

$$h = h(\mathbf{z}, \mathbf{n})$$
$$= -\mathbf{q} \cdot \mathbf{n} \tag{5.1.3-5}$$

and write

$$\int_{S_{(m)}} h\, dA = -\int_{S_{(m)}} \mathbf{q} \cdot \mathbf{n}\, dA$$
$$= -\int_{R_{(m)}} \operatorname{div} \mathbf{q}\, dV - \int_{\Sigma} [\![\mathbf{q} \cdot \boldsymbol{\xi}]\!]\, dA \tag{5.1.3-6}$$

As the result of (5.1.3-3), (5.1.3-4), and (5.1.3-6), Equation (5.1.3-1) becomes

$$\int_{R_{(m)}} \left[ \rho \frac{d_{(m)}}{dt} \left( \hat{U} + \frac{1}{2} v^2 \right) - \operatorname{div}(\mathbf{T} \cdot \mathbf{v}) - \rho(\mathbf{v} \cdot \mathbf{f}) + \operatorname{div} \mathbf{q} - \rho Q \right] dV$$
$$+ \int_{\Sigma} \left[ \rho \left( \hat{U} + \frac{1}{2} v^2 \right) (\mathbf{v} - \mathbf{u}) \cdot \boldsymbol{\xi} + \mathbf{q} \cdot \boldsymbol{\xi} - \mathbf{v} \cdot (\mathbf{T} \cdot \boldsymbol{\xi}) \right] dA$$
$$= 0 \tag{5.1.3-7}$$

Since the size of our body is arbitrary, we conclude that at each point within each phase we must require that the *differential energy balance*,

$$\rho \frac{d_{(m)}}{dt} \left( \hat{U} + \frac{1}{2} v^2 \right) = -\operatorname{div} \mathbf{q} + \operatorname{div}(\mathbf{T} \cdot \mathbf{v}) + \rho(\mathbf{v} \cdot \mathbf{f}) + \rho Q \tag{5.1.3-8}$$

must be satisfied and that at each point on each phase interface we must require that the *jump energy balance*,

$$\left[\!\!\left[ \rho \left( \hat{U} + \frac{1}{2} v^2 \right) (\mathbf{v} - \mathbf{u}) \cdot \boldsymbol{\xi} + \mathbf{q} \cdot \boldsymbol{\xi} - \mathbf{v} \cdot (\mathbf{T} \cdot \boldsymbol{\xi}) \right]\!\!\right] = 0 \tag{5.1.3-9}$$

must be obeyed.

Equation 5.1.3-8 may be simplified by taking advantage of the differential momentum balance. From the scalar product of the velocity vector with the differential momentum balance, we have

$$\mathbf{v} \cdot \left( \rho \frac{d_{(m)}\mathbf{v}}{dt} - \operatorname{div} \mathbf{T} - \rho \mathbf{f} \right) = 0 \tag{5.1.3-10}$$

or

$$\rho \frac{d_{(m)}}{dt} \left( \frac{1}{2} v^2 \right) = \operatorname{div}(\mathbf{T} \cdot \mathbf{v}) - \operatorname{tr}(\mathbf{T} \cdot \nabla \mathbf{v}) + \rho(\mathbf{v} \cdot \mathbf{f}) \tag{5.1.3-11}$$

Subtracting this last equation from (5.1.3-8), we are left with another form of the differential energy balance:

$$\rho \frac{d_{(m)}\hat{U}}{dt} = -\text{div }\mathbf{q} + \text{tr}(\mathbf{T} \cdot \nabla \mathbf{v}) + \rho Q \tag{5.1.3-12}$$

Several different forms of the differential energy balance in common use are given in Table 5.4.0-1.

**Exercise 5.1.3-1**  Consider two neighboring portions of a continuous body. Apply the energy balance to each portion and to their union. Deduce that on their common boundary

$$h(\mathbf{z}, \mathbf{n}) = -h(\mathbf{z}, -\mathbf{n})$$

This says that the contact energy fluxes upon opposite sides of the same surface at a given point are equal in magnitude and opposite in sign.

**Exercise 5.1.3-2** *The energy flux vector*  By a development that parallels that given in Section 2.2.2, show that the contact energy flux may be expressed as

$$h(\mathbf{z}, \mathbf{n}) = -\mathbf{q} \cdot \mathbf{n}$$

where $\mathbf{q}$ is known as the *energy flux vector*.

**Exercise 5.1.3-3** *More about the energy flux vector*  Show that the energy flux vector is frame indifferent.

**Exercise 5.1.3-4** *Alternative form of the differential energy balance*  If the external force per unit mass may be expressed in terms of a potential energy per unit mass $\phi$,

$$\mathbf{f} = -\nabla \phi$$

show that (5.1.3-8) may be written as

$$\rho \frac{d_{(m)}}{dt} \left( \hat{U} + \frac{1}{2}v^2 + \phi \right) = -\text{div }\mathbf{q} + \text{div}(\mathbf{T} \cdot \mathbf{v}) + \rho Q$$

**Exercise 5.1.3-5** *Rigid-body motion*  Show that for a rigid body tumbling in space (see Exercise 2.3.2-1)

$$\text{tr}(\mathbf{T} \cdot \nabla \mathbf{v}) = 0$$

**Exercise 5.1.3-6** *Energy balance for single phase*

i) As a lemma of the energy balance, prove

**Energy balance for a single-phase body**  In an inertial frame of reference, the time rate of change of the internal energy of a body is equal to the rate at which work is done on the body by the system of inertial, contact, external, and mutual forces acting upon it plus the rate of energy transmission to the body.

ii) If instead of a lemma, we adopted this as an axiom for all bodies, prove that the differential energy balance (5.1.3-8) would remain unchanged and that the jump energy balance (5.1.3-9) would take the form

$$\left[ \rho \hat{U}(\mathbf{v} - \mathbf{u}) \cdot \boldsymbol{\xi} + \mathbf{q} \cdot \boldsymbol{\xi} - \mathbf{v} \cdot (\mathbf{T} \cdot \boldsymbol{\xi}) \right] = 0$$

To our knowledge, there is not sufficient experimental evidence to distinguish between these two forms.

## 5.2   Entropy

### 5.2.1   Entropy Inequality

Let us review the physical picture for internal energy in the context of a particulate model for a real material. As the result of their relative motion, the molecules possess kinetic energy with respect to the material as a whole. They also have potential energy as the result of their relative positions in the various force fields acting among the molecules. We think of this kinetic and potential energy as the internal energy of the material.

Still working in the context of a particulate model, we can see that the internal energy of a material is not sufficient to specify its state. Consider two samples of the same material, both having the same internal energy. One has been compressed and cooled; its molecules are in close proximity to one another, and they move slowly. The other has been expanded and heated; the molecules are not very close to one another, and they move rapidly. As we have pictured them, these two samples can be distinguished by their division of internal energy between the kinetic energy and potential energy of the molecules. Alternatively, we can imagine that the molecules in the compressed and cooled material appear in more orderly arrays than do those in the expanded and heated material. There is a difference in the degree of disorder between the samples.

*Entropy* is the term that we use to describe the disorder in a material. Like internal energy, it is a primitive concept; it is not defined in the context of continuum mechanics. Instead, we describe its properties. We are about to state as a postulate its most important property: the entropy inequality. In addition, we will require that entropy per unit mass $\hat{S}$ be frame indifferent:

$$\hat{S}^* = \hat{S} \tag{5.2.1-1}$$

Some familiar observations suggest what we should say about entropy. Let us begin by thinking about some situations in which the surroundings do relatively little work on a body.

Even on a cold day, the air in a closed room becomes noticeably warmer because of the sunshine through the window. During the winter, the room is heated by the energy transmission from a radiator. In the summer, it is cooled by the energy transmission to the coils in an air conditioner. Since the volume of air in the room is a constant, we can say that any energy transferred to the room increases the kinetic energy of the air molecules and the entropy of the air.

On the basis of these observations, we might be inclined to propose as a fundamental postulate that *the time rate of change of the entropy of a body is locally proportional to the rate of energy transmission to the body*. Unfortunately, this is not entirely consistent with other experiments.

Let us consider some experiments in which the energy transfer between the body and the surroundings seems less important. Open a paper clip and repeatedly twist the ends with respect to one another until the metal breaks. The metal is warm to the touch. It is easy to confirm that the grease-packed front-wheel bearings on an automobile (with power transmitted to the rear axle) become hot during a highway trip. Since the paper clip and the grease-packed bearings have roughly constant volumes, we can estimate that the kinetic energy of the molecules has increased, that the molecules are somewhat less ordered, and that the entropy of the system has increased as the result of the systems of forces acting upon it. More important, in every situation that we can recall where there is negligible energy transfer with the surroundings, the entropy of a body always increases as the result of work done. It does not matter whether the work is done by the body on the surroundings or by the surroundings on the body.

It appears that we have two choices open to us in trying to summarize our observations.

We might say that *the time rate of change of the entropy of a body is equal to the rate of entropy transmission to the body plus a multiple of the absolute value of the rate at which work is done on the body by the surroundings*. We cannot say whether this would lead to a self-consistent theory, but it is clear that it would be awkward to work in terms of the absolute value of the rate at which work is done.

As the literature has developed, it appears preferable instead to state as a postulate the

**Entropy inequality**   The minimum time rate of change of the entropy of a body is equal to the rate of entropy transmission to the body.

We realize that this is not a directly useful statement as it stands. To make it useful, we must be able to describe the rate of entropy transmission to the body in terms of the rates of energy transmission to the body.

For a lively and rewarding discussion of the entropy inequality, or the *second law of thermodynamics*, we encourage you to read Truesdell (1969). As will become plain shortly, we have also been influenced here by Gurtin and Vargas (1971).

### 5.2.2   Radiant and Contact Entropy Transmission

The rate of entropy transmission can be described by a set of six axioms that are very similar to those used to describe the rate of energy transmission (see Section 5.1.2).

Corresponding to each body $B$, there is a distinct set of bodies $B^e$ such that the union of these bodies forms the universe. We refer to $B^e$ as the exterior or the surroundings of the body $B$.

1) A system of entropy transmission rates is a scalar-valued function $\mathcal{E}(B, C)$ of pairs of bodies. The value of $\mathcal{E}(B, C)$ is called the *rate of entropy transmission* from body $C$ to body $B$.

2) For a specified body $B$, $\mathcal{E}(C, B^e)$ is an additive function defined over the subbodies $C$ of $B$.

3) Conversely, for a specified body $B$, $\mathcal{E}(B, C)$ is an additive function defined over the subbodies $C$ of $B^e$.

4) The rate of entropy transmission to a body should have nothing to do with the motion of the observer or experimentalist relative to the body. It should be frame indifferent:

$$\mathcal{E}^* = \mathcal{E} \tag{5.2.2-1}$$

There are three types of entropy transmission with which we are concerned:

**Rate of external radiant entropy transmission**    The rate of external radiant entropy transmission refers to entropy transmission from outside the body to the material particles of which the body is composed. It is presumed to be related to the masses of the bodies, and it is described as though it acts directly on each material particle:

$$\mathcal{E}_e = \int_{R_{(m)}} \rho E_e \, dV \tag{5.2.2-2}$$

Here $E_e$ is the rate of external radiant entropy transmission per unit mass to the material.

**Rate of mutual radiant entropy transmission**    The rate of mutual radiant entropy transmission refers to entropy transmission between pairs of material particles that are part of the same body. Let $E_m$ be the mutual entropy transmission rate per unit mass from $B - P$ to $P$; the total mutual entropy transmission to $P$ may be represented as an integral over the region occupied by $P$:[3]

$$\mathcal{E}_m = \int_{R_{(m)}} \rho E_m \, dV \tag{5.2.2-3}$$

**Rate of contact entropy transmission**    This is the entropy transmission that is not assignable as a function of position within the body, but which is to be imagined as entropy transmission through the bounding surface of a portion of material in such a way as to be equivalent to the entropy transmission from the surroundings beyond that accounted for through external and mutual radiant entropy transmission. Let $\eta = \eta(\mathbf{z}, P)$ represent the rate of entropy transmission per unit area from $B - P$ to the boundary of $P$ at the position $\mathbf{z}$. This rate of entropy transmission per unit area $\eta$ may be referred to as the *contact entropy flux*. The total rate of contact entropy transfer from $B - P$ to $P$ may be written as an integral over the bounding surface of $P$:

$$\mathcal{E}_c = \int_{S_{(m)}} \eta \, dA \tag{5.2.2-4}$$

The fifth axiom, the entropy flux principle, specifies the nature of the contact entropy transmission.

5) **Entropy flux principle**    There is a frame-indifferent, scalar-valued function $\eta(\mathbf{z}, \mathbf{n})$ defined for all points $\mathbf{z}$ in a body $B$ and for all unit vectors $\mathbf{n}$ such that the rate of contact entropy transmission per unit area from $B - P$ to any portion $P$ of $B$ is given by

$$\eta(\mathbf{z}, P) = \eta(\mathbf{z}, \mathbf{n}) \tag{5.2.2-5}$$

---

[3]  We recognize here that the sum of the mutual radiant entropy transmission between any two parts of $P$ must be zero; the proof is much the same as that for mutual forces (Truesdell and Toupin 1960, p. 533).

Here $\mathbf{n}$ is the unit normal that is outwardly directed with respect to the closed bounding surface of $P$. The scalar $\eta = \eta(\mathbf{z}, \mathbf{n})$ is referred to as the contact entropy flux at the position $\mathbf{z}$ across the oriented surface element with normal $\mathbf{n}$; $\mathbf{n}$ points into the material from which the contact entropy flux to the surface element is $\eta$.

The experimental observations noted in Section 5.2.1 suggest as a sixth axiom:

6) The rates of radiant energy and entropy transmission have the same sign, and they are proportional:

$$
\begin{aligned}
E &\equiv E_e + E_m \\
&= \frac{Q_e + Q_m}{T}
\end{aligned}
$$

(5.2.2-6)

The proportionality factor $T$ is a positive, frame-indifferent, scalar field known as *temperature*.

## 5.2.3 The Differential and Jump Entropy Inequalities

In an inertial frame of reference, the entropy inequality of Section 5.2.1 says

$$
\text{minimum} \; \frac{d}{dt} \int_{R_{(m)}} \rho \hat{S} \, dV = \int_{S_{(m)}} \eta \, dA + \int_{R_{(m)}} \rho \frac{Q}{T} \, dV
$$

(5.2.3-1)

or

$$
\frac{d}{dt} \int_{R_{(m)}} \rho \hat{S} \, dV \geq \int_{S_{(m)}} \eta \, dA + \int_{R_{(m)}} \rho \frac{Q}{T} \, dV
$$

(5.2.3-2)

where $\hat{S}$ denotes entropy per unit mass and

$$
Q \equiv Q_e + Q_m
$$

(5.2.3-3)

is the scalar field that represents the sum of the external and mutual radiant energy transmission rates per unit mass.

Equation (5.2.3-2) implies a differential inequality that describes the production of entropy at every point within a material. The steps are very similar to those used to obtain the differential energy balance in Section 5.1.3. The alternative form of the transport theorem for regions containing a dividing surface (Exercise 1.3.6-3) permits us to express the left side of (5.2.3-2) as

$$
\frac{d}{dt} \int_{R_{(m)}} \rho \hat{S} \, dV = \int_{R_{(m)}} \rho \frac{d_{(m)}}{dt} \hat{S} \, dV + \int_{\Sigma} \left[ \rho \hat{S}(\mathbf{v} - \mathbf{u}) \cdot \boldsymbol{\xi} \right] dA
$$

(5.2.3-4)

The first term on the right of (5.2.3-2) may be rewritten in terms of a volume integral, only if we can take advantage of Green's transformation. By Exercise 5.2.3-1, we may express the contact entropy flux in terms of the *thermal energy flux vector* $\mathbf{e}$ and the *temperature* $T$,

$$
\begin{aligned}
\eta &= \eta(\mathbf{z}, \mathbf{n}) \\
&= -\frac{1}{T} \mathbf{e} \cdot \mathbf{n}
\end{aligned}
$$

(5.2.3-5)

and write

$$\int_{S_{(m)}} \eta \, dA = -\int_{S_{(m)}} \frac{1}{T} \mathbf{e} \cdot \mathbf{n} \, dA$$

$$= -\int_{R_{(m)}} \text{div}\left(\frac{\mathbf{e}}{T}\right) dV - \int_{\Sigma} \left[\frac{1}{T} \mathbf{e} \cdot \boldsymbol{\xi}\right] dA \tag{5.2.3-6}$$

As the result of (5.2.3-4) and (5.2.3-6), Equation (5.2.3-2) becomes

$$\int_{R_{(m)}} \left[\rho \frac{d_{(m)}\hat{S}}{dt} + \text{div}\left(\frac{\mathbf{e}}{T}\right) - \rho\frac{Q}{T}\right] dV + \int_{\Sigma} \left[\rho\hat{S}(\mathbf{v} - \mathbf{u}) \cdot \boldsymbol{\xi} + \frac{1}{T}\mathbf{e} \cdot \boldsymbol{\xi}\right] dA \geq 0$$

$$\tag{5.2.3-7}$$

Since the size of our body is arbitrary, we conclude that the *differential entropy inequality*,

$$\rho \frac{d_{(m)}\hat{S}}{dt} \geq -\text{div}\left(\frac{\mathbf{e}}{T}\right) + \rho\frac{Q}{T} \tag{5.2.3-8}$$

must be satisfied and that at each point on each phase interface the *jump entropy inequality*,

$$\left[\rho\hat{S}(\mathbf{v} - \mathbf{u}) \cdot \boldsymbol{\xi} + \frac{1}{T}\mathbf{e} \cdot \boldsymbol{\xi}\right] \geq 0 \tag{5.2.3-9}$$

must be obeyed.

**Exercise 5.2.3-1** *The thermal energy flux vector*    By a development that parallels that given in Section 2.2.2, show that the contact entropy flux may be expressed as

$$\eta(\mathbf{z}, \mathbf{n}) = -\frac{1}{T}\mathbf{e} \cdot \mathbf{n}$$

where $\mathbf{e}$ is known as the *thermal energy flux vector* and $T$ is temperature introduced in Section 5.2.2.

**Exercise 5.2.3-2** *More about the thermal energy flux vector*    Show that the thermal energy flux vector is frame indifferent.

**Exercise 5.2.3-3** *More about the jump entropy inequality*    Show that, if we neglect interphase mass transfer and if we assume that temperature is continuous across a phase interface (see Section 6.1), the jump entropy inequality (5.2.3-9) reduces to

$$\left[\mathbf{e} \cdot \boldsymbol{\xi}\right] \geq 0$$

In (5.3.1-25), we find that $\mathbf{e} = \mathbf{q}$. In view of the jump energy balance (5.1.3-9), conclude that

$$\mathbf{v} \cdot \left[\mathbf{T} \cdot \boldsymbol{\xi}\right] \geq 0$$

Because we have placed little emphasis in this text upon interfacial effects, we will have no further use for the jump entropy inequality. To see its role in placing constraints upon interfacial behavior, refer to Slattery (1990).

## 5.3    Behavior of Materials

Up to this point in this chapter, we have been concerned with the form and implications of postulates stated for all materials. But all materials do not behave in the same manner.

The primary idea to be exploited in this section is that any material should be capable of undergoing all processes that are consistent with our fundamental postulates. In particular, we use the differential entropy inequality to restrict the form of descriptions for material behavior. (Contrast this philosophy with one in which these inequalities are used to define the class of processes consistent with a given set of statements about material behavior.)

### 5.3.1    Implications of the Differential Entropy Inequality

Let us begin by investigating the restrictions that the differential entropy inequality (Section 5.2.3),

$$\rho T \frac{d_{(m)}\hat{S}}{dt} \geq -T \operatorname{div}\left(\frac{\mathbf{e}}{T}\right) + \rho Q \tag{5.3.1-1}$$

places upon the form of descriptions for bulk material behavior. The approach is suggested by Gurtin and Vargas (1971).

If we subtract (5.3.1-1) from the differential energy balance (Section 5.1.3),

$$\rho \frac{d_{(m)}\hat{U}}{dt} = -\operatorname{div}\mathbf{q} + \operatorname{tr}(\mathbf{T} \cdot \nabla \mathbf{v}) + \rho Q \tag{5.3.1-2}$$

we have

$$\rho \frac{d_{(m)}\hat{U}}{dt} - \rho T \frac{d_{(m)}\hat{S}}{dt} - \operatorname{tr}(\mathbf{T} \cdot \nabla \mathbf{v}) + \operatorname{div}(\mathbf{q} - \mathbf{e}) + \frac{1}{T}\mathbf{e} \cdot \nabla T \leq 0 \tag{5.3.1-3}$$

We will find it more convenient to work in terms of the *Helmholtz free energy per unit mass*

$$\hat{A} \equiv \hat{U} - T\hat{S} \tag{5.3.1-4}$$

in terms of which (5.3.1-3) becomes

$$\rho \frac{d_{(m)}\hat{A}}{dt} + \rho \hat{S} \frac{d_{(m)}T}{dt} - \operatorname{tr}(\mathbf{T} \cdot \nabla \mathbf{v}) + \operatorname{div}(\mathbf{q} - \mathbf{e}) + \frac{1}{T}\mathbf{e} \cdot \nabla T \leq 0 \tag{5.3.1-5}$$

To make further progress, we must restrict ourselves to a class of material behavior or a class of constitutive equations. Let us assume that

$$\hat{A} = \hat{A}(\mathbf{\Lambda}, \mathbf{D})$$
$$\hat{S} = \hat{S}(\mathbf{\Lambda}, \mathbf{D})$$
$$\mathbf{q} = \mathbf{q}(\mathbf{\Lambda}) \tag{5.3.1-6}$$
$$\mathbf{e} = \mathbf{e}(\mathbf{\Lambda})$$
$$\mathbf{T} = \mathbf{T}(\mathbf{\Lambda}, \mathbf{D})$$

where the set of variables

$$\Lambda \equiv \left( \hat{V}, \, T, \, \nabla \hat{V}, \, \nabla T \right) \tag{5.3.1-7}$$

is a set of independent variables common to all of these constitutive equations and $\mathbf{D}$ is the rate of deformation tensor.

Using the chain rule, we can say from (5.3.1-6) that[4]

$$\frac{d_{(m)}\hat{A}}{dt} = \frac{\partial \hat{A}}{\partial \hat{V}} \frac{d_{(m)}\hat{V}}{dt} + \frac{\partial \hat{A}}{\partial T} \frac{d_{(m)}T}{dt} + \frac{\partial \hat{A}}{\partial \nabla \hat{V}} \cdot \frac{d_{(m)}\nabla \hat{V}}{dt}$$

$$+ \frac{\partial \hat{A}}{\partial \nabla T} \cdot \frac{d_{(m)}\nabla T}{dt} + \mathrm{tr}\left( \frac{\partial \hat{A}}{\partial \mathbf{D}} \cdot \frac{d_{(m)}\mathbf{D}^{\mathbf{T}}}{dt} \right) \tag{5.3.1-8}$$

This together with the differential mass balance (Section 1.3.3)

$$\rho \frac{d_{(m)}\hat{V}}{dt} = \mathrm{div}\, \mathbf{v} \tag{5.3.1-9}$$

allow us to rearrange (5.3.1-5) in the form

$$\rho \left[ \left( \frac{\partial \hat{A}}{\partial T} + \hat{S} \right) \frac{d_{(m)}T}{dt} + \frac{\partial \hat{A}}{\partial \nabla \hat{V}} \cdot \frac{d_{(m)}\nabla \hat{V}}{dt} + \frac{\partial \hat{A}}{\partial \nabla T} \cdot \frac{d_{(m)}\nabla T}{dt} \right.$$

$$\left. + \mathrm{tr}\left( \frac{\partial \hat{A}}{\partial \mathbf{D}} \cdot \frac{d_{(m)}\mathbf{D}^{\mathbf{T}}}{dt} \right) \right] - \mathrm{tr}\left[ \left( \mathbf{T} - \frac{\partial \hat{A}}{\partial \hat{V}}\mathbf{I} \right) \cdot \nabla \mathbf{v} \right] + \mathrm{div}(\mathbf{q} - \mathbf{e}) + \frac{1}{T}\mathbf{e} \cdot \nabla T$$

$$\leq 0 \tag{5.3.1-10}$$

---

[4] Let $f(\mathbf{v})$ be a scalar function of a vector $\mathbf{v}$. The derivative of $f$ with respect to $\mathbf{v}$ is a vector denoted by $\partial f / \partial \mathbf{v}$ and defined by its scalar product with any arbitrary vector $\mathbf{a}$:

$$\frac{\partial f}{\partial \mathbf{v}} \cdot \mathbf{a} \equiv \mathrm{limit}\, s \to 0 : \frac{1}{s}[f(\mathbf{v} + s\mathbf{a}) - f(\mathbf{v})]$$

In a rectangular Cartesian coordinate system, this last expression takes the form (see Section A.3.1)

$$\frac{\partial f}{\partial \mathbf{v}} \cdot \mathbf{a} = \frac{\partial f}{\partial v_i} a_i$$

For the particular case $\mathbf{a} = \mathbf{e}_j$

$$\frac{\partial f}{\partial \mathbf{v}} \cdot \mathbf{e}_j = \frac{\partial f}{\partial v_j}$$

and we conclude that

$$\frac{\partial f}{\partial \mathbf{v}} = \frac{\partial f}{\partial v_i}\mathbf{e}_i$$

In a similar manner, if $f(\mathbf{D})$ is a scalar function of a tensor $\mathbf{D}$, the derivative of $f$ with respect to $\mathbf{D}$ is a tensor denoted by $\partial f / \partial \mathbf{D}$ and defined by

$$\mathrm{tr}\left( \frac{\partial f}{\partial \mathbf{D}} \cdot \mathbf{A}^T \right) \equiv \mathrm{limit}\, s \to 0 : \frac{1}{s}[f(\mathbf{D} + s\mathbf{A}) - f(\mathbf{D})]$$

where $\mathbf{A}$ is an arbitrary tensor. We conclude that

$$\frac{\partial f}{\partial \mathbf{D}} = \frac{\partial f}{\partial D_{ij}}\mathbf{e}_i\mathbf{e}_j$$

It is a simple matter to construct $T$, $\hat{V}$, $\nabla T$, and $\nabla \hat{V}$ fields such that at any given point within a phase at any specified time

$$\frac{d_{(m)}T}{dt}, \ \frac{d_{(m)}\nabla T}{dt}, \ \frac{d_{(m)}\nabla \hat{V}}{dt}, \ \frac{d_{(m)}\mathbf{D}^{\mathbf{T}}}{dt} \tag{5.3.1-11}$$

take arbitrary values. We conclude that

$$\hat{A} = \hat{A}\left(\hat{V}, T\right) \tag{5.3.1-12}$$

$$\hat{S} = \hat{S}\left(\hat{V}, T\right)$$

$$= -\left(\frac{\partial \hat{A}}{\partial T}\right)_{\hat{V}} \tag{5.3.1-13}$$

and

$$-\text{tr}\left\{\left[\mathbf{T} - \left(\frac{\partial \hat{A}}{\partial \hat{V}}\right)_T \mathbf{I}\right] \cdot \nabla \mathbf{v}\right\} + \text{div}(\mathbf{q} - \mathbf{e}) + \frac{1}{T}\mathbf{e} \cdot \nabla T \leq 0 \tag{5.3.1-14}$$

For simplicity, let us introduce

$$\mathbf{k} \equiv \mathbf{q} - \mathbf{e} \tag{5.3.1-15}$$

and write inequality (5.3.1-14) as

$$\text{div } \mathbf{k} + f(\mathbf{\Lambda}, \mathbf{D}) \leq 0 \tag{5.3.1-16}$$

The vector $\mathbf{k}$ is frame indifferent, because $\mathbf{q}$ and $\mathbf{e}$ are frame indifferent by Exercises 5.1.3-3 and 5.2.3-2:

$$\mathbf{k}^* = \mathbf{Q} \cdot \mathbf{k} \tag{5.3.1-17}$$

From (5.3.1-6), we see that $\mathbf{k}$ is a function only of $\mathbf{\Lambda}$:

$$\mathbf{k} = \mathbf{K}(\mathbf{\Lambda}) \tag{5.3.1-18}$$

By the principle of frame indifference (Section 2.3.1), this same form of relationship must hold in every frame of reference:

$$\mathbf{k}^* = \mathbf{K}(\mathbf{\Lambda}^*)$$

$$= \mathbf{K}(\mathbf{Q} \cdot \mathbf{\Lambda}) \tag{5.3.1-19}$$

where the set of variables

$$\mathbf{Q} \cdot \mathbf{\Lambda} \equiv \left(\hat{V}, T, \mathbf{Q} \cdot \nabla \hat{V}, \mathbf{Q} \cdot \nabla T\right) \tag{5.3.1-20}$$

Equations (5.3.1-17) through (5.3.1-19) imply that $\mathbf{K}(\mathbf{\Lambda})$ is an isotropic function:

$$\mathbf{K}(\mathbf{Q} \cdot \mathbf{\Lambda}) = \mathbf{Q} \cdot \mathbf{K}(\mathbf{\Lambda}) \tag{5.3.1-21}$$

Applying the chain rule to (5.3.1-18), we have

$$\text{div } \mathbf{k} = \text{tr}\left(\frac{\partial \mathbf{K}}{\partial \nabla \hat{V}} \cdot \nabla \nabla \hat{V}\right) + \text{tr}\left(\frac{\partial \mathbf{K}}{\partial \nabla T} \cdot \nabla \nabla T\right) + \frac{\partial \mathbf{K}}{\partial T} \cdot \nabla T + \frac{\partial \mathbf{K}}{\partial \hat{V}} \cdot \nabla \hat{V} \tag{5.3.1-22}$$

We can construct compatible $\hat{V}$ and $T$ fields such that at any given point within a phase at any specified time

$$\nabla\hat{V}, \ \nabla T, \ \nabla\nabla\hat{V}, \ \nabla\nabla T$$

take arbitrary values (Gurtin and Vargas 1971). In view of (5.3.1-16) and (5.3.1-22),

$$
\begin{aligned}
\mathrm{tr}\left(\frac{\partial \mathbf{K}}{\partial\nabla\hat{V}} \cdot \nabla\nabla\hat{V}\right) &= \mathrm{tr}\left(\frac{\partial \mathbf{K}}{\partial\nabla T} \cdot \nabla\nabla T\right) \\
&= \frac{\partial \mathbf{K}}{\partial T} \cdot \nabla T \\
&= \frac{\partial \mathbf{K}}{\partial\hat{V}} \cdot \nabla\hat{V} \\
&= 0
\end{aligned}
\tag{5.3.1-23}
$$

which implies that the symmetric parts of

$$\frac{\partial \mathbf{K}}{\partial\nabla\hat{V}}, \ \frac{\partial \mathbf{K}}{\partial\nabla T}$$

are both zero as well as

$$\frac{\partial \mathbf{K}}{\partial\hat{V}} = \frac{\partial \mathbf{K}}{\partial T} = 0$$

Using the principle of frame indifference (Section 2.3.1), Gurtin (1971, Lemma 6.2; Gurtin and Vargas 1971, Lemma 10.2) has proved that, when the symmetric portions of the derivatives of an isotropic vector-valued function $\mathbf{K}(\mathbf{\Lambda})$ with respect to each of the independent vectors are all zero, the function itself is zero. In this case we conclude that

$$\mathbf{k} = 0 \tag{5.3.1-24}$$

or

$$\mathbf{e} = \mathbf{q} \tag{5.3.1-25}$$

This in turn implies that (5.3.1-14) reduces to

$$-\mathrm{tr}\left\{\left[\mathbf{T} - \left(\frac{\partial\hat{A}}{\partial\hat{V}}\right)_T \mathbf{I}\right] \cdot \nabla\mathbf{v}\right\} + \frac{1}{T}\mathbf{q} \cdot \nabla T \leq 0 \tag{5.3.1-26}$$

In what follows, we will find it convenient to define the *extra stress tensor* (or viscous portion of the stress tensor) as

$$\mathbf{S} \equiv \mathbf{T} + P\mathbf{I} \tag{5.3.1-27}$$

and to write (5.3.1-26) as

$$-\mathrm{tr}\left(\left\{\mathbf{S} - \left[P + \left(\frac{\partial\hat{A}}{\partial\hat{V}}\right)_T\right]\mathbf{I}\right\} \cdot \nabla\mathbf{v}\right) + \frac{1}{T}\mathbf{q} \cdot \nabla T \leq 0 \tag{5.3.1-28}$$

Because of the inability of a fluid to support a shear stress at equilibrium (Truesdell 1977, p. 202), we argue that

$$\text{limit } \mathbf{D} \to 0 : \ \mathbf{S}(\Lambda, \mathbf{D}) \to 0 \tag{5.3.1-29}$$

Let us consider two classes of isothermal flows:[5]

1) Let us assume that $P + \left(\frac{\partial \hat{A}}{\partial \hat{V}}\right)_T > 0$. For the class of flows $\text{tr}(\nabla \mathbf{v}) = \text{div } \mathbf{v} > 0$, the extra stress tensor $\mathbf{S}$ can be arbitrarily small, violating (5.3.1-28).

2) Let us assume that $P + \left(\frac{\partial \hat{A}}{\partial \hat{V}}\right)_T < 0$. For the class of flows $\text{tr}(\nabla \mathbf{v}) < 0$, we again can make $\mathbf{S}$ arbitrarily small, violating (5.3.1-28).

Since we are assuming here that appropriate descriptions of material behavior should be consistent with the entropy inequality for all motions, we conclude that the *thermodynamic pressure*

$$P \equiv -\left(\frac{\partial \hat{A}}{\partial \hat{V}}\right)_T \tag{5.3.1-30}$$

This permits us to express (5.3.1-28) as

$$-\text{tr}\left[(\mathbf{T} + P\mathbf{I}) \cdot \nabla \mathbf{v}\right] + \frac{1}{T}\mathbf{q} \cdot \nabla T \leq 0 \tag{5.3.1-31}$$

or

$$-\text{tr}\left[\mathbf{S} \cdot \nabla \mathbf{v}\right] + \frac{1}{T}\mathbf{q} \cdot \nabla T \leq 0 \tag{5.3.1-32}$$

Before we examine the implications of this inequality, we will look at the consequences of (5.3.1-12) in the next section.

**Exercise 5.3.1-1** *Elastic solids* A *solid* has some preferred configuration from which all changes of shape can be detected by experiment. Consequently, all nonorthogonal transformations (*orthogonal transformations* are defined in Section A.5.2) from a preferred configuration for a solid can be detected by experiment (Noll 1958; Truesdell and Noll 1965, p. 81; Truesdell 1966a, p. 61).

An *elastic solid* is one in which the dependence upon specific volume $\hat{V}$ and the rate of deformation $\mathbf{D}$ in (5.3.1-6) is replaced by a dependence upon the deformation gradient $\mathbf{F}$, measured with respect to this preferred configuration and defined by (1.3.2-3) (Truesdell 1966a, p. 98; Truesdell 1977, p. 165).[6]

Let us extend the discussion in the text through (5.3.1-14) to elastic solids. Starting with the result from Exercise 2.3.2-1 that

$$\frac{d_{(m)}\mathbf{F}}{dt} = (\nabla \mathbf{v}) \cdot \mathbf{F}$$

or

$$\frac{d_{(m)}\mathbf{F}^T}{dt} = \mathbf{F}^T \cdot (\nabla \mathbf{v})^T$$

---

[5] This argument was suggested by P. K. Dhori.
[6] Note that, in view of the result of Exercise 1.3.3-3, $\hat{V}$ or $\rho$ is not independent of $\mathbf{F}$.

conclude that (5.3.1-12) and (5.3.1-13) should be replaced by

$$\hat{A} = \hat{A}(T, \mathbf{F})$$
$$\hat{S} = \hat{S}(T, \mathbf{F})$$

and that (5.3.1-14) becomes

$$-\mathrm{tr}\left\{\left[\mathbf{T} - \rho\left(\frac{\partial \hat{A}}{\partial \mathbf{F}}\right)_T \cdot \mathbf{F}^T\right] \cdot (\nabla \mathbf{v})^T\right\} + \mathrm{div}(\mathbf{q} - \mathbf{e}) + \frac{1}{T}\mathbf{e} \cdot \nabla T \leq 0$$

Following a discussion similar to that given in the context, we conclude that

$$\mathbf{q} = \mathbf{e}$$

and

$$\mathrm{tr}\left[\left(\mathbf{T} - \rho\left(\frac{\partial \hat{A}}{\partial \mathbf{F}}\right)_T \cdot \mathbf{F}^T\right) \cdot (\nabla \mathbf{v})^T\right] - \frac{1}{T}\mathbf{q} \cdot \nabla T \geq 0$$

This exercise was written with the help of P. K. Dhori.

**Exercise 5.3.1-2** *Hyperelastic solid*  The inequality developed in the preceding exercise imposes a constraint on the constitutive equations for $\mathbf{T}$ and $\mathbf{q}$. Since a constitutive equation for $\mathbf{T}$ should satisfy this inequality for all processes, for an isothermal process

$$\mathrm{tr}\left[\left(\mathbf{T} - \rho\left(\frac{\partial \hat{A}}{\partial \mathbf{F}}\right)_T \cdot \mathbf{F}^T\right) \cdot (\nabla \mathbf{v})^T\right] \geq 0$$

Since $\mathbf{T}$ and $\hat{A}$ are independent of $\nabla \mathbf{v}$ (they are functions of $\mathbf{F}$), we conclude that

$$\mathbf{T} = \rho\left(\frac{\partial \hat{A}}{\partial \mathbf{F}}\right)_T \cdot \mathbf{F}^T$$

Such a material is known as a *hyperelastic solid* (Truesdell 1966a, p. 182).

**Exercise 5.3.1-3** *Simple fluid*  Several physical ideas have been associated with the term *fluid* (Truesdell 1966a, p. 62). For example, Batchelor (1967, p. 1) feels "A portion of fluid ... does not have a preferred shape . . . ." This might be interpreted to mean that a fluid does not have preferred reference configurations, as long as the density is unchanged (Truesdell and Noll 1965, p. 79).

When a fluid is allowed to relax following a deformation, it does not return to its predeformation configuration but to a new stress-free configuration. For this reason, while dealing with a viscoelastic fluid, the current configuration is taken as its reference configuration and all deformations are measured relative to the current configuration.

We want to know what constraint the entropy inequality imposes on the material behavior of a fluid. We will restrict ourselves to a class of *simple fluids*, the behavior of which are such that[7]

$$\hat{A} = \hat{A}^{\infty}_{s=0} \left( \mathbf{F}^t_t(s), T, \nabla T, \hat{V}, \nabla \hat{V} \right) \tag{5.3.1-33}$$

$$\hat{S} = \hat{S}^{\infty}_{s=0} \left( \mathbf{F}^t_t(s), T, \nabla T, \hat{V}, \nabla \hat{V} \right) \tag{5.3.1-34}$$

$$\mathbf{T} = \mathbf{T}^{\infty}_{s=0} \left( \mathbf{F}^t_t(s), T, \nabla T, \hat{V}, \nabla \hat{V} \right)$$

$$\mathbf{q} = \left( T, \nabla T, \hat{V}, \nabla \hat{V} \right)$$

$$\mathbf{e} = \mathbf{e} \left( T, \nabla T, \hat{V}, \nabla \hat{V} \right)$$

where

$$\mathbf{F}^t_t(s) \equiv \mathbf{F}_t(t - s)$$

$$\equiv \nabla \bar{\chi}_t(z_1, z_2, z_3, \bar{t})$$

$$= \frac{\partial \bar{\chi}_t(z_1, z_2, z_3, \bar{t})}{\partial z_j} \mathbf{e}_j$$

$$= \frac{\partial \bar{z}_i}{\partial z_j} \bar{\mathbf{e}}_i \mathbf{e}_j \tag{5.3.1-35}$$

is the *relative deformation gradient*. The discussion of motion parallels that given in Section 1.1; the definition of the deformation gradient parallels that given in Exercise 2.3.2-1. The relative deformation gradient tells how the position in some past configuration (at some past time $\bar{t} \equiv t - s$) changes as the result of a small change in the current configuration (at current time $t$).

We will also find it convenient to introduce a motion and a relative deformation gradient in which the configuration at some past time $\bar{t}$ is taken to be the reference configuration:

$$\mathbf{F}_{\bar{t}}(t) \equiv \mathbf{F}_{t-s}(t)$$

$$\equiv \bar{\nabla} \chi_{\bar{t}}(\bar{z}_1, \bar{z}_2, \bar{z}_3, t)$$

$$= \frac{\partial \chi_{\bar{t}}(\bar{z}_1, \bar{z}_2, \bar{z}_3, t)}{\partial \bar{z}_j} \bar{\mathbf{e}}_j$$

$$= \frac{\partial z_i}{\partial \bar{z}_j} \mathbf{e}_i \bar{\mathbf{e}}_j \tag{5.3.1-36}$$

i) Use (5.3.1-35) and (5.3.1-36) to show that

$$\mathbf{F}_t(\bar{t}) \cdot \mathbf{F}_{\bar{t}}(t) = \bar{\mathbf{I}} \tag{5.3.1-37}$$

ii) Starting with (5.3.1-36), determine that

$$\left( \frac{d_{(m)} \mathbf{F}_{\bar{t}}(t)}{dt} \right)_{\bar{t}} = \nabla \mathbf{v} \cdot \mathbf{F}_{\bar{t}}(t) \tag{5.3.1-38}$$

---

[7] The behavior of a simple fluid can be further limited by recognizing that a fluid is isotropic or non-oriented, in the sense that it has no natural direction (Truesdell 1977, p. 203).

iii) Differentiate (5.3.1-37) with respect to $t$ holding $\bar{t}$ constant, and use (5.3.1-38) to conclude that

$$\left(\frac{d_{(m)}\mathbf{F}_t(\bar{t})}{dt}\right)_{\bar{t}} = -\mathbf{F}_t(\bar{t}) \cdot \nabla \mathbf{v} \tag{5.3.1-39}$$

iv) Apply the chain rule to (5.3.1-33) to write

$$\frac{d_{(m)}\hat{A}}{dt} = \delta\hat{A}_{s=0}^{\infty}\left(\mathbf{F}_t^t(s), T, \nabla T, \hat{V}, \nabla\hat{V}\left|\frac{d_{(m)}\mathbf{F}_t^t(s)}{dt}\right.\right) + \frac{\partial\hat{A}}{\partial\hat{V}}\frac{d_{(m)}\hat{V}}{dt}$$

$$+ \mathrm{tr}\left(\frac{\partial\hat{A}}{\partial\nabla\hat{V}} \cdot \frac{d_{(m)}\nabla\hat{V}}{dt}\right) + \frac{\partial\hat{A}}{\partial T}\frac{d_{(m)}T}{dt} + \frac{\partial\hat{A}}{\partial\nabla T} \cdot \frac{d_{(m)}\nabla T}{dt} \tag{5.3.1-40}$$

where

$$\delta\hat{A}_{s=0}^{\infty}\left(\cdots\left|\frac{d_{(m)}\mathbf{F}_t^t(s)}{dt}\right.\right)$$

is the first Fréchet derivative of $\hat{A}$ with respect to $\mathbf{F}_t^t(s)$, which is linear in its last argument (Coleman 1964, pp. 12–13)

$$\frac{d_{(m)}\mathbf{F}_t^t(s)}{dt}$$

Recognizing with the help of (5.3.1-35) and (5.3.1-39) that

$$\left(\frac{d_{(m)}\mathbf{F}_t^t(s)}{dt}\right)_s = \left(\frac{d_{(m)}\mathbf{F}_t(\bar{t})}{d\bar{t}}\right)_t \left(\frac{\partial\bar{t}}{\partial t}\right)_s + \left(\frac{d_{(m)}\mathbf{F}_t(\bar{t})}{dt}\right)_{\bar{t}}$$

$$= \nabla\mathbf{v}(t - s) - \mathbf{F}_t^t(s) \cdot \nabla\mathbf{v}$$

we can express the first Fréchet derivative as

$$\delta\hat{A}_{s=0}^{\infty}\left(\mathbf{F}_t^t(s), T, \nabla T, \hat{V}, \nabla\hat{V}\left|\frac{d_{(m)}\mathbf{F}_t^t(s)}{dt}\right|\right)$$

$$= \delta\hat{A}_{s=0}^{\infty}\left(\mathbf{F}_t^t(s), T, \nabla T, \hat{V}, \nabla\hat{V}\left|\left[\nabla\mathbf{v}(t - s) - \mathbf{F}_t^t(s) \cdot \nabla\mathbf{v}\right]\right.\right) \tag{5.3.1-41}$$

Remembering that this last expression is linear in

$$\nabla\mathbf{v}(t - s) - \mathbf{F}_t^t(s) \cdot \nabla\mathbf{v}$$

we let $\delta\mathbf{B}_{s=0}^{\infty}$ be a tensor-valued functional such that (Coleman 1964, p. 20; Johnson 1977, Eq. 8)

$$\mathrm{tr}\left(\delta\mathbf{B}_{s=0}^{\infty} \cdot \nabla\mathbf{v}\right) \equiv \rho\,\delta\hat{A}_{s=0}^{\infty}\left(\mathbf{F}_t^t(s), T, \nabla T, \hat{V}, \nabla\hat{V}\left|\left[-\mathbf{F}_t^t(s) \cdot \nabla\mathbf{v}\right]\right.\right) \tag{5.3.1-42}$$

and let

$$\delta\mathbf{C}_{s=0}^{\infty} \equiv \rho\,\delta\hat{A}_{s=0}^{\infty}\left(\mathbf{F}_t^t(s), T, \nabla T, \hat{V}, \nabla\hat{V}\left|\nabla\mathbf{v}(t - s)\right.\right). \tag{5.3.1-43}$$

Combine (5.3.1-40) through (5.3.1-43) with the entropy inequality (5.3.1-5) and argue that, for the result to be satisfied for all processes, we must have

$$\hat{A} = \hat{A}_{s=0}^{\infty}\left(\mathbf{F}_t^t(s), T, \hat{V}\right)$$

$$\hat{S} = \hat{S}_{s=0}^{\infty}\left(\mathbf{F}_t^t(s), T, \hat{V}\right)$$

$$= -\left(\frac{\partial \hat{A}}{\partial T}\right)_{\hat{V}, \mathbf{F}_t^t(s)}$$

(5.3.1-44)

and that the entropy inequality reduces to

$$\delta \mathbf{C}_{s=0}^{\infty} - \mathrm{tr}\left\{\left[\mathbf{T} - \left(\frac{\partial \hat{A}}{\partial \hat{V}}\right)_{T, \mathbf{F}_t^t(s)}\mathbf{I} - \delta \mathbf{B}_{s=0}^{\infty}\right] \cdot \nabla \mathbf{v}\right\}$$

$$+ \mathrm{div}(\mathbf{q} - \mathbf{e}) + \frac{1}{T}\mathbf{e} \cdot \nabla T \leq 0$$

(5.3.1-45)

v) Because of the inability of a fluid to support a shear stress at equilibrium, Truesdell (1977, p. 202) argues that

$$\mathbf{T} = -P\mathbf{I} + \mathbf{S}$$

where

$$P = P\left(\hat{V}, T\right)$$

and

$$\mathbf{S} = \mathbf{S}_{s=0}^{\infty}\left(\mathbf{F}_t^t(s), T, \nabla T, \hat{V}, \nabla \hat{V}\right)$$

Reason that, because (5.3.1-45) must be true for all processes,

$$P = P\left(\hat{V}, T\right)$$

$$= -\left(\frac{\partial \hat{A}}{\partial \hat{V}}\right)_{T, \mathbf{F}_t^t(s)}$$

(5.3.1-46)

and that the entropy inequality (5.3.1-45) can be written as

$$\delta \mathbf{C}_{s=0}^{\infty} - \mathrm{tr}\left\{\left[\mathbf{S} - \delta \mathbf{B}_{s=0}^{\infty}\right] \cdot \nabla \mathbf{v}\right\} + \mathrm{div}(\mathbf{q} - \mathbf{e}) + \frac{1}{T}\mathbf{e} \cdot \nabla T \leq 0$$

(5.3.1-47)

vi) Argue that (5.3.1-44) and (5.3.1-46) allow us to write

$$\hat{A} = \hat{A}_{(v)}\left(T, \hat{V}\right) + \hat{A}_{(e)}$$

(5.3.1-48)

where $\hat{A}_{(e)}$, a function of the deformation history and temperature, is the elastic component of $\hat{A}$. From (5.3.1-48), look ahead to the next section to conclude that *Euler's equation* for a viscoelastic fluid becomes

$$\hat{A} = -P\hat{V} + \mu + \hat{A}_{(e)}$$

(5.3.1-49)

in which we have defined

$$\mu(T, \rho) \equiv \left(\frac{\partial \check{A}_{(v)}}{\partial \rho}\right)_T$$

as the *chemical potential on a mass basis.*

vii) Finally, develop a discussion similar to the one given in the text to determine that

$$\mathbf{e} = \mathbf{q}$$

and that the entropy inequality (5.3.1-47) further reduces to

$$\delta \mathbf{C}_{s=0}^{\infty} - \operatorname{tr}\left\{\left[\mathbf{S} - \delta \mathbf{B}_{s=0}^{\infty}\right] \cdot \nabla \mathbf{v}\right\} + \frac{1}{T}\mathbf{q} \cdot \nabla T \leq 0 \tag{5.3.1-50}$$

This exercise was written with the help of P. K. Dhori. For a similar development, see Johnson (1977).

## 5.3.2   Restrictions on Caloric Equation of State

In the preceding section, we began with some broad statements about material behavior and concluded in (5.3.1-12) that, if the entropy inequality was to be obeyed,

$$\hat{A} = \hat{A}(\hat{V}, T) \tag{5.3.2-1}$$

or

$$\check{A} = \check{A}(\rho, T) \tag{5.3.2-2}$$

or

$$\tilde{A} = \tilde{A}(\tilde{V}, T) \tag{5.3.2-3}$$

or

$$\check{A} = \check{A}(c, T) \tag{5.3.2-4}$$

where $\hat{A}$, $\check{A}$, and $\tilde{A}$ are the *Helmholtz free energy per unit mass, per unit volume,* and *per unit mole,* respectively. We will refer to these statements as alternative forms of the *caloric equation of state.* At the same time, we found in (5.3.1-13) that

$$\hat{S} = -\left(\frac{\partial \hat{A}}{\partial T}\right)_{\hat{V}}$$

$$\check{S} = -\left(\frac{\partial \check{A}}{\partial T}\right)_{\rho}$$

$$\tilde{S} = -\left(\frac{\partial \tilde{A}}{\partial T}\right)_{\tilde{V}} \tag{5.3.2-5}$$

$$\check{S} = -\left(\frac{\partial \check{A}}{\partial T}\right)_{c}$$

In addition to the thermodynamic pressure introduced in (5.3.1-30), we will define the *chemical potential on a mass basis*[8]

$$\mu \equiv \left( \frac{\partial \check{A}}{\partial \rho} \right)_T \tag{5.3.2-6}$$

and the *chemical potential on a molar basis*

$$\mu^{(m)} \equiv \left( \frac{\partial \check{A}}{\partial c} \right)_T \tag{5.3.2-7}$$

The differentials of (5.3.2-1) through (5.3.2-4) may consequently be expressed as

$$d\hat{A} = -P \, d\hat{V} - \hat{S} \, dT \tag{5.3.2-8}$$

$$d\check{A} = -\check{S} \, dT + \mu \, d\rho \tag{5.3.2-9}$$

$$d\tilde{A} = -P \, d\tilde{V} - \tilde{S} \, dT \tag{5.3.2-10}$$

$$d\check{A} = -\check{S} \, dT + \mu^{(m)} \, dc \tag{5.3.2-11}$$

Equations (5.3.2-8) and (5.3.2-10) are two forms of the *Gibbs equation*.

Equations (5.3.2-9) and (5.3.2-11) may be rearranged to read

$$d \left( \frac{\hat{A}}{\hat{V}} \right) = -\frac{\hat{S}}{\hat{V}} dT + \mu \, d \left( \frac{1}{\hat{V}} \right)$$

$$d\hat{A} = \left( \frac{\hat{A}}{\hat{V}} - \frac{\mu}{\hat{V}} \right) d\hat{V} - \hat{S} \, dT \tag{5.3.2-12}$$

and

$$d \left( \frac{\tilde{A}}{\tilde{V}} \right) = -\frac{\tilde{S}}{\tilde{V}} dT + \mu^{(m)} \, d \left( \frac{1}{\tilde{V}} \right)$$

$$d\tilde{A} = \left( \frac{\tilde{A}}{\tilde{V}} - \frac{\mu^{(m)}}{\tilde{V}} \right) d\tilde{V} - \tilde{S} \, dT \tag{5.3.2-13}$$

Comparison of the coefficients in (5.3.2-8) with those in (5.3.2-12) and comparison of the coefficients in (5.3.2-10) with those in (5.3.2-13) give two forms of *Euler's equation*:

$$\hat{A} = -P\hat{V} + \mu$$

$$\tilde{A} = -P\tilde{V} + \mu^{(m)} \tag{5.3.2-14}$$

Two forms of the *Gibbs–Duhem equation* follow immediately by subtracting (5.3.2-12) and (5.3.2-13) from the differentials of (5.3.2-14):

$$\hat{S} \, dT - \hat{V} \, dP + d\mu = 0$$

$$\tilde{S} \, dT - \tilde{V} \, dP + d\mu^{(m)} = 0 \tag{5.3.2-15}$$

---

[8] These expressions for the chemical potential were suggested by G. M. Brown, Department of Chemical Engineering, Northwestern University, Evanston, Illinois 60201-3120.

We would like to emphasize that Euler's equation, the Gibbs equation, and the Gibbs–Duhem equation all apply to dynamic processes, so long as the statements about behavior made in Section 5.3.1 are applicable to the materials being considered.

**Exercise 5.3.2-1** *Alternative forms of the specific variables* Show that alternative expressions for temperature, thermodynamic pressure, chemical potential on a mass basis, and chemical potential on a molar basis are

$$T = \left( \frac{\partial \hat{U}}{\partial \hat{S}} \right)_{\hat{V}} = \left( \frac{\partial \check{U}}{\partial \check{S}} \right)_{\rho} = \left( \frac{\partial \tilde{U}}{\partial \tilde{S}} \right)_{\tilde{V}} = \left( \frac{\partial \check{U}}{\partial \check{S}} \right)_{c}$$

$$P = - \left( \frac{\partial \hat{U}}{\partial \hat{V}} \right)_{\hat{S}} = - \left( \frac{\partial \tilde{U}}{\partial \tilde{V}} \right)_{\tilde{S}}$$

$$\mu = \left( \frac{\partial \check{U}}{\partial \rho} \right)_{\check{S}}$$

$$\mu^{(m)} = \left( \frac{\partial \check{U}}{\partial c} \right)_{\check{S}}$$

**Exercise 5.3.2-2** *The Maxwell relations* Let us define

$$\hat{A} \equiv \hat{U} - T\hat{S}$$
$$\hat{H} \equiv \hat{U} + P\hat{V}$$
$$\hat{G} \equiv \hat{H} - T\hat{S}$$

We refer to $\hat{A}$ as *Helmholtz free energy* per unit mass, $\hat{S}$ as *enthalpy* per unit mass, and $\hat{G}$ as *Gibbs free energy* per unit mass. Determine that

i) $\left( \frac{\partial T}{\partial \hat{V}} \right)_{\hat{S}} = - \left( \frac{\partial P}{\partial \hat{S}} \right)_{\hat{V}}$

ii) $\left( \frac{\partial \hat{S}}{\partial \hat{V}} \right)_{T} = \left( \frac{\partial P}{\partial T} \right)_{\hat{V}}$

iii) $\left( \frac{\partial T}{\partial P} \right)_{\hat{S}} = \left( \frac{\partial \hat{V}}{\partial \hat{S}} \right)_{P}$

iv) $- \left( \frac{\partial \hat{S}}{\partial P} \right)_{T} = \left( \frac{\partial \hat{V}}{\partial T} \right)_{P}$

**Exercise 5.3.2-3** *More Maxwell relations* Following the definitions introduced in Exercise 8.4.2-2, we have that

$$\check{A} = \check{U} - T\check{S}$$
$$\check{H} = \check{U} + P\check{V}$$
$$\check{G} = \check{H} - T\check{S}$$

Determine that

i) $\left(\dfrac{\partial T}{\partial \rho}\right)_{\check{S}} = \left(\dfrac{\partial \mu}{\partial \check{S}}\right)_{\rho}$

ii) $-\left(\dfrac{\partial \check{S}}{\partial \rho}\right)_{T} = \left(\dfrac{\partial \mu}{\partial T}\right)_{\rho}$

**Exercise 5.3.2-4** *Heat capacities*  We define the heat capacity per unit mass at constant pressure $\hat{c}_P$ and the heat capacity per unit mass at constant specific volume $\hat{c}_V$ as

$$\hat{c}_P \equiv T \left(\dfrac{\partial \hat{S}}{\partial T}\right)_{P}$$

and

$$\hat{c}_V \equiv T \left(\dfrac{\partial \hat{S}}{\partial T}\right)_{\hat{V}}$$

i) Determine that

$$\hat{c}_P = \left(\dfrac{\partial \hat{H}}{\partial T}\right)_{P}$$

and

$$\hat{c}_V = \left(\dfrac{\partial \hat{U}}{\partial T}\right)_{\hat{V}}$$

ii) Prove that

$$\rho \hat{c}_P - \left(\dfrac{\partial P}{\partial T}\right)_{\hat{V}} \left(\dfrac{\partial \ln \hat{V}}{\partial \ln T}\right)_{P} = \rho \hat{c}_V$$

iii) For an ideal gas, conclude that

$$\hat{c}_P = \hat{c}_V + \dfrac{R}{M}$$

where $M$ is the molecular weight

## 5.3.3  Energy and Thermal Energy Flux Vectors

In (5.3.1-25), we found that

$$\mathbf{q} = \mathbf{e} \tag{5.3.3-1}$$

As a result, it is necessary only to investigate the behavior of the energy flux vector $\mathbf{q}$.

By (5.3.1-6), we restricted ourselves to a class of material behavior such that

$$\mathbf{q} = \mathbf{h}\left(\hat{V},\ T,\ \nabla \hat{V},\ \nabla T\right) \tag{5.3.3-2}$$

Any description of material behavior such as this must be consistent with four principles:

1) the principle of determinism (Section 2.3.1),
2) the principle of local action (Section 2.3.1),
3) the principle of frame indifference (Section 2.3.1), and
4) the differential entropy inequality (5.3.1-31).

The first two are satisfied identically by (5.3.3-2). We wish to explore here the implications of the other two.

The principle of frame indifference requires that

$$\mathbf{q}^* = \mathbf{Q} \cdot \mathbf{q}$$
$$= \mathbf{Q} \cdot \mathbf{h} \left( \hat{V}, \, T, \, \nabla \hat{V}, \, \nabla T \right)$$
$$= \mathbf{h} \left( \hat{V}, \, T, \, (\nabla \hat{V})^*, \, (\nabla T)^* \right)$$
$$= \mathbf{h} \left( \hat{V}, \, T, \, \mathbf{Q} \cdot \nabla \hat{V}, \, \mathbf{Q} \cdot \nabla T \right) \tag{5.3.3-3}$$

or $\mathbf{h}$ is a vector-valued, isotropic function of two vectors $\nabla \hat{V}$ and $\nabla T$:

$$\mathbf{Q} \cdot \mathbf{h} \left( \hat{V}, \, T, \, \nabla \hat{V}, \, \nabla T \right) = \mathbf{h} \left( \hat{V}, \, T, \, \mathbf{Q} \cdot \nabla \hat{V}, \, \mathbf{Q} \cdot \nabla T \right) \tag{5.3.3-4}$$

By the representation theorems of Spencer and Rivlin (1959, Sec. 7) and of Smith (1965), the most general polynomial vector function of two vectors is of the form

$$\mathbf{q} = \kappa_{(1)} \nabla T + \kappa_{(2)} \nabla \hat{V} + \kappa_{(3)} \nabla T \wedge \nabla \hat{V} \tag{5.3.3-5}$$

where $\kappa_{(1)}$, $\kappa_{(2)}$, and $\kappa_{(3)}$ are scalar-valued polynomials in $|\nabla T|$, $|\nabla \hat{V}|$, and $(\nabla T \cdot \nabla \hat{V})$. (In applying the theorem of Spencer and Rivlin, we identify a vector $\mathbf{b}$, which has rectangular Cartesian components $b_i$, with the skew-symmetric tensor that has rectangular Cartesian components $e_{ijk} b_i$.) Since

$$(\mathbf{Q} \cdot \nabla T) \wedge (\mathbf{Q} \cdot \nabla \hat{V}) = \det \mathbf{Q} [\mathbf{Q} \cdot (\nabla T \wedge \nabla \hat{V})] \tag{5.3.3-6}$$

then $(\nabla T \wedge \nabla \hat{V})$ is not a frame-indifferent vector (see Section 1.2.1). It follows that, for (5.3.3-4) to be satisfied, (5.3.3-5) must reduce to

$$\mathbf{q} = \kappa_{(1)} \nabla T + \kappa_{(2)} \nabla \hat{V} \tag{5.3.3-7}$$

To explore the implications of the differential entropy inequality, let us begin by restricting ourselves to a class of processes in which $\mathbf{v} = \mathbf{0}$ and $\nabla \hat{V} = \mathbf{0}$. With these restrictions, with the recognition that $T$ was introduced as a positive scalar field in (5.2.2-6), and in view of (5.3.3-1), Equation (5.3.1-31) reduces to

$$\mathbf{q} \cdot \nabla T \leq 0 \tag{5.3.3-8}$$

or, in view of (5.3.3-7),

$$\kappa_{(1)} \nabla T \cdot \nabla T \leq 0 \tag{5.3.3-9}$$

Since (5.3.3-7) is required to be valid for all processes, this implies that, in the limit $\nabla \hat{V} \to 0$,

$$k_{(1)} \equiv -\kappa_{(1)}$$
$$> 0 \tag{5.3.3-10}$$

Here we have replaced the $\geq$ by $>$, recognizing that the equality applies only at equilibrium.

Let us examine a more general set of processes, in which simply $\mathbf{v} = \mathbf{0}$. Under these restrictions, (5.3.3-7) and (5.3.3-8) imply that

$$\kappa_{(1)}\nabla T \cdot \nabla T + \kappa_{(2)}\nabla \hat{V} \cdot \nabla T \leq 0 \tag{5.3.3-11}$$

Recognizing that we can construct processes in which $\nabla \hat{V} \cdot \nabla T$ can be as small or as large as desired, we conclude in view of (5.3.3-10) that

$$\kappa_{(2)} = -k_{(2)}\nabla \hat{V} \cdot \nabla T \tag{5.3.3-12}$$

where

$$k_{(2)} \geq 0 \tag{5.3.3-13}$$

In view of (5.3.3-10) and (5.3.3-12), Equation (5.3.3-7) reduces to

$$\mathbf{q} = -k_{(1)}\nabla T - k_{(2)}\left(\nabla \hat{V} \cdot \nabla T\right)\nabla \hat{V} \tag{5.3.3-14}$$

where (5.3.3-10) and (5.3.3-13) apply.

Solids for which dynamic response in a process depends upon a direction (or a set of directions) intrinsically associated with the material, such as $\nabla \hat{V}$ in (5.3.3-14), are said to be *anisotropic*. This seems to be an unfortunate use of the word anisotropic, since we see in (5.3.3-4) that $\mathbf{q}$ is an *isotropic* function of the various independent variables. We prefer to say a material described by (5.3.3-11) and (5.3.3-14) is *oriented*. If $k$ were independent of $|\nabla \hat{V}|$ and $|\nabla T \cdot \nabla \hat{V}|$, we would say that the material was *nonoriented* rather than *isotropic*.

One should probably not apply (5.3.3-14) to wood or stratified limestone, in which $\hat{V}$ might be regarded as a function of position. These solids are probably better represented as porous media, as discussed in Section 7.3.

The most common special case of (5.3.3-14) is *Fourier's law*:

$$\mathbf{q} = \mathbf{e}$$
$$= -k\nabla T \tag{5.3.3-15}$$

where the *thermal conductivity*

$$k = k\left(T, \hat{V}\right) > 0 \tag{5.3.3-16}$$

Finally, under what conditions is it sufficient to describe the behavior of an apparently oriented material with Fourier's law and a spatially dependent thermal conductivity $k$, and when must one recognize local orientation of the material with (5.3.3-14)? To my knowledge, these are questions that have not been addressed in the literature.

### 5.3.4 Stress Tensor

In (5.3.1-6), we restricted ourselves to a class of stress-deformation behavior such that

$$\mathbf{T} = \mathbf{T}\left(\hat{V}, T, \nabla \hat{V}, \nabla T, \mathbf{D}\right) \tag{5.3.4-1}$$

As we noted in the preceding section, any description of material behavior must be consistent with

- the principle of determinism (Section 2.3.1),
- the principle of local action (Section 2.3.2),

- the principle of frame indifference (Section 2.3.2), and
- the differential entropy inequality (5.3.1-31).

The first two are satisfied identically by (5.3.4-1). It is necessary to explore the implications of only the other two.

In a manner similar to that used in the preceding section, we can examine the implications of the principle of frame indifference on (5.3.4-1). In particular, we can derive the most general tensor-valued polynomial function of $\nabla \hat{V}$, $\nabla T$, and the rate of deformation tensor $\mathbf{D}$ (Spencer and Rivlin 1959, Sec. 7; Smith 1965). The result is a function with too many independent parameters to be immediately useful. Since there is little or no experimental evidence available to guide us, we recommend the current practice in engineering, which is to use the constitutive equations for $\mathbf{T}$ developed for isothermal materials, recognizing that all parameters should be functions of the local thermodynamic state variables $T$ and $\hat{V}$.

As an example of how the differential entropy inequality can be used to place constraints on material behavior, let us consider the most general *linear* relation between the stress tensor and the rate of deformation tensor that is consistent with the principle of frame indifference (see Section 2.3.2):

$$\mathbf{T} = (\alpha + \lambda \operatorname{div} \mathbf{v})\mathbf{I} + 2\mu\mathbf{D} \tag{5.3.4-2}$$

We already know from Section 5.3.1 that $\alpha = -P$, the thermodynamic pressure, and (5.3.4-2) can be written as

$$\mathbf{S} \equiv \mathbf{T} + P\mathbf{I}$$
$$= \lambda \operatorname{div} \mathbf{v}\,\mathbf{I} + 2\mu\mathbf{D} \tag{5.3.4-3}$$

Since (5.3.4-3) is assumed to apply in every possible process, let us begin by considering an isothermal flow in which the nondiagonal components of $\mathbf{D}$ are equal to zero and

$$D_{11} = D_{22} = D_{33} \tag{5.3.4-4}$$

Under these conditions, (5.3.1-31) reduces to

$$(3\lambda + 2\mu)(D_{11})^2 \geq 0 \tag{5.3.4-5}$$

and we conclude that

$$3\lambda + 2\mu > 0 \tag{5.3.4-6}$$

Note that in (5.3.4-6) we have replaced the $\geq$ by $>$, recognizing that the equality applies only at equilibrium. It has been stated that *Stokes's relation*,

$$\kappa \equiv 3\lambda + 2\mu$$
$$= 0 \tag{5.3.4-7}$$

for the *bulk viscosity* $\kappa$ has been substantiated for low-density monatomic gases (Bird et al. 1960, p. 79). Truesdell (1952, Sec. 61A) argues that this result is implicitly assumed in that theory. To our knowledge, experimental measurements indicate that $\lambda$ is positive and that for many fluids it is orders of magnitude greater than $\mu$, which as we see below is also positive (Truesdell 1952, Sec. 61A; Karim and Rosenhead 1952).

Now consider an isothermal, isochoric motion:

$$\operatorname{div} \mathbf{v} = 0 \tag{5.3.4-8}$$

Then (5.3.1-31) reduces to

$$2\mu \, \text{tr}(\mathbf{D} \cdot \mathbf{D}) \geq 0 \tag{5.3.4-9}$$

or

$$\mu > 0 \tag{5.3.4-10}$$

In view of (5.3.4-6) and (5.3.4-10), Equation (5.3.4-3) reduces to the *Newtonian fluid* (see Section 2.3.2):

$$\mathbf{T} = (-P + \lambda \, \text{div} \, \mathbf{v})\mathbf{I} + 2\mu \mathbf{D} \tag{5.3.4-11}$$

**Exercise 5.3.4-1** *Generalized Newtonian fluid*

i) Assume that the viscous portion of the stress tensor of an incompressible fluid is described by (Section 2.3.3)

$$\mathbf{S} \equiv \mathbf{T} + P\mathbf{I}$$
$$= 2\eta(\gamma)\mathbf{D}$$

where

$$\gamma \equiv \sqrt{2 \, \text{tr} \, (\mathbf{D} \cdot \mathbf{D})}$$

and that the energy flux vector is represented by Fourier's law. Use the approach outlined in the text to prove that

$$\eta(\gamma) > 0$$

Note here that the definition for $\mathbf{S}$ differs from that introduced in (2.3.3-1). Here it has been appropriate to recognize that all fluids are compressible, even though the effect of compressibility may be negligible in a given set of experiments.

ii) Consider an incompressible fluid for which the energy flux vector is described by Fourier's law and for which the viscous portion of the stress tensor and the rate of deformation tensor are related by

$$2\mathbf{D} = \varphi(\tau)\mathbf{S}$$

where

$$\tau = \sqrt{\frac{1}{2} \, \text{tr} \, (\mathbf{S} \cdot \mathbf{S})}$$

Prove that

$$\varphi(\tau) > 0$$

## 5.4 Summary of Useful Equations

In Section 5.1.3, we derived the differential energy balance in terms of the internal energy per unit mass $\hat{U}$. We are usually more interested in determining the temperature distribution.

From the definition for $\hat{A}$ in terms of $\hat{U}$ given in Exercise 5.3.2-2 as well as Equation (5.3.2-8), we have that

$$d\hat{U} = T\,d\hat{S} - P\,d\hat{V}$$

$$= T\left(\frac{\partial \hat{S}}{\partial T}\right)_{\hat{V}} dT + \left[T\left(\frac{\partial \hat{S}}{\partial \hat{V}}\right)_{T} - P\right]d\hat{V}$$

$$= \hat{c}_V\,dT + \left[T\left(\frac{\partial P}{\partial T}\right)_{\hat{V}} - P\right]d\hat{V} \tag{5.4.0-1}$$

Here we introduce the heat capacity per unit mass at constant specific volume (see Exercise 5.3.2-4),

$$\hat{c}_V \equiv T\left(\frac{\partial \hat{S}}{\partial T}\right)_{\hat{V}} \tag{5.4.0-2}$$

The differential mass balance tells us that

$$\rho \frac{d_{(m)}\hat{V}}{dt} = -\frac{1}{\rho}\frac{d_{(m)}\rho}{dt}$$

$$= \operatorname{div} \mathbf{v} \tag{5.4.0-3}$$

Equations (5.4.0-1) and (5.4.0-3) allow us to rewrite (5.1.3-12) as

$$\rho\hat{c}_V \frac{d_{(m)}T}{dt} = -\operatorname{div}\mathbf{q} - T\left(\frac{\partial P}{\partial T}\right)_{\hat{V}} \operatorname{div}\mathbf{v} + \operatorname{tr}(\mathbf{S}\cdot\nabla\mathbf{v}) + \rho Q$$

$$\rho\hat{c}_V \frac{d_{(m)}T}{dt} = -\operatorname{div}\mathbf{q} - T\left(\frac{\partial P}{\partial T}\right)_{\hat{V}} \operatorname{div}\mathbf{v} + \operatorname{tr}(\mathbf{S}\cdot\mathbf{D}) + \rho Q \tag{5.4.0-4}$$

In Table 5.4.0-1, we present seven equivalent forms of the differential energy balance discussed in Section 5.3.2 that may be derived in a similar fashion. Often, one form will have a particular advantage in any given problem.

Commonly, the only external force to be considered is a uniform gravitational field and we may represent it as (see Section 2.4.1 and Exercise 2.4.1-1)

$$\mathbf{f} = -\nabla\phi. \tag{5.4.0-5}$$

We will hereafter refer to $\phi$ as *potential energy per unit mass*.

Equation (5.4.0-4) is shown for rectangular Cartesian, cylindrical, and spherical coordinates in Table 5.4.0-2. It should not be difficult to use these as guides in immediately writing the corresponding expressions for the other six forms of the differential energy balance given in Table 5.4.0-1. You will also find useful the rectangular Cartesian, cylindrical, and spherical components of Fourier's law shown in Table 5.4.0-3.

For an incompressible Newtonian fluid with constant viscosity and constant thermal conductivity, (5.4.0-4) reduces to

$$\rho\hat{c}\frac{d_{(m)}T}{dt} = k\operatorname{div}\nabla T + 2\mu\operatorname{tr}(\mathbf{D}\cdot\mathbf{D}) + \rho Q \tag{5.4.0-6}$$

**Table 5.4.0-1.** Various forms of the differential energy balance

$$\rho \frac{d_{(m)}}{dt} \left( \hat{U} + \frac{1}{2}v^2 + \phi \right) = -\text{div } \mathbf{q} + \text{div } (\mathbf{T} \cdot \mathbf{v}) + \rho Q$$

$$\rho \frac{d_{(m)}}{dt} \left( \hat{U} + \frac{1}{2}v^2 \right) = -\text{div } \mathbf{q} + \text{div } (\mathbf{T} \cdot \mathbf{v}) + \rho (\mathbf{v} \cdot \mathbf{f}) + \rho Q$$

$$\rho \frac{d_{(m)}\hat{U}}{dt} = -\text{div } \mathbf{q} - P \text{ div } \mathbf{v} + \text{tr} (\mathbf{S} \cdot \nabla \mathbf{v}) + \rho Q$$

$$\rho \frac{d_{(m)}\hat{H}}{dt} = -\text{div } \mathbf{q} + \frac{d_{(m)}P}{dt} + \text{tr} (\mathbf{S} \cdot \nabla \mathbf{v}) + \rho Q$$

$$\rho \hat{c}_V \frac{d_{(m)}T}{dt} = -\text{div } \mathbf{q} - T \left( \frac{\partial P}{\partial T} \right)_{\hat{V}} \text{div } \mathbf{v} + \text{tr} (\mathbf{S} \cdot \nabla \mathbf{v}) + \rho Q$$

$$\rho \hat{c}_P \frac{d_{(m)}T}{dt} = -\text{div } \mathbf{q} + \left( \frac{\partial \ln \hat{V}}{\partial \ln T} \right)_{P} \frac{d_{(m)}P}{dt} + \text{tr} (\mathbf{S} \cdot \nabla \mathbf{v}) + \rho Q$$

$$\rho \frac{d_{(m)}\hat{S}}{dt} = -\text{div } \left( \frac{\mathbf{q}}{T} \right) - \frac{1}{T^2} \mathbf{q} \cdot \nabla T + \frac{1}{T} \text{tr} (\mathbf{S} \cdot \nabla \mathbf{v}) + \frac{\rho Q}{T}$$

**Table 5.4.0-2.** The differential energy balance in several coordinate systems

*Rectangular Cartesian coordinates:*

$$\rho \hat{c}_V \left( \frac{\partial T}{\partial t} + v_1 \frac{\partial T}{\partial z_1} + v_2 \frac{\partial T}{\partial z_2} + v_3 \frac{\partial T}{\partial z_3} \right)$$

$$= -\left( \frac{\partial q_1}{\partial z_1} + \frac{\partial q_2}{\partial z_2} + \frac{\partial q_3}{\partial z_3} \right) - T \left( \frac{\partial P}{\partial T} \right)_{\hat{V}} \left( \frac{\partial v_1}{\partial z_1} + \frac{\partial v_2}{\partial z_2} + \frac{\partial v_3}{\partial z_3} \right) + S_{11} \frac{\partial v_1}{\partial z_1} + S_{22} \frac{\partial v_2}{\partial z_2} + S_{33} \frac{\partial v_3}{\partial z_3}$$

$$+ S_{12} \left( \frac{\partial v_1}{\partial z_2} + \frac{\partial v_2}{\partial z_1} \right) + S_{13} \left( \frac{\partial v_1}{\partial z_3} + \frac{\partial v_3}{\partial z_1} \right) + S_{23} \left( \frac{\partial v_2}{\partial z_3} + \frac{\partial v_3}{\partial z_2} \right) + \rho Q$$

*Cylindrical coordinates:*

$$\rho \hat{c}_V \left( \frac{\partial T}{\partial t} + v_r \frac{\partial T}{\partial r} + \frac{v_\theta}{r} \frac{\partial T}{\partial \theta} + v_z \frac{\partial T}{\partial z} \right)$$

$$= -\left[ \frac{1}{r} \frac{\partial}{\partial r} (r q_r) + \frac{1}{r} \frac{\partial q_\theta}{\partial \theta} + \frac{\partial q_z}{\partial z} \right] - T \left( \frac{\partial P}{\partial T} \right)_{\hat{V}} \left[ \frac{1}{r} \frac{\partial}{\partial r} (r v_r) + \frac{1}{r} \frac{\partial v_\theta}{\partial \theta} + \frac{\partial v_z}{\partial z} \right]$$

$$+ S_{rr} \frac{\partial v_r}{\partial r} + S_{\theta\theta} \frac{1}{r} \left( \frac{\partial v_\theta}{\partial \theta} + v_r \right) + S_{zz} \frac{\partial v_z}{\partial z} + S_{r\theta} \left[ r \frac{\partial}{\partial r} \left( \frac{v_\theta}{r} \right) + \frac{1}{r} \frac{\partial v_r}{\partial \theta} \right] + S_{rz} \left( \frac{\partial v_z}{\partial r} + \frac{\partial v_r}{\partial z} \right)$$

$$+ S_{\theta z} \left[ \frac{1}{r} \frac{\partial v_z}{\partial \theta} + \frac{\partial v_\theta}{\partial z} \right] + \rho Q$$

*(cont.)*

**Table 5.4.0-2.** (cont.)

*Spherical coordinates:*

$$\rho \hat{c}_V \left( \frac{\partial T}{\partial t} + v_r \frac{\partial T}{\partial r} + \frac{v_\theta}{r} \frac{\partial T}{\partial \theta} + \frac{v_\varphi}{r \sin \theta} \frac{\partial T}{\partial \varphi} \right)$$

$$= - \left[ \frac{1}{r^2} \frac{\partial}{\partial r}(r^2 q_r) + \frac{1}{r \sin \theta} \frac{\partial}{\partial \theta}(q_\theta \sin \theta) + \frac{1}{r \sin \theta} \frac{\partial q_\varphi}{\partial \varphi} \right]$$

$$- T \left( \frac{\partial P}{\partial T} \right)_{\hat{v}} \left[ \frac{1}{r^2} \frac{\partial}{\partial r}(r^2 v_r) + \frac{1}{r \sin \theta} \frac{\partial}{\partial \theta}(v_\theta \sin \theta) + \frac{1}{r \sin \theta} \frac{\partial v_\varphi}{\partial \varphi} \right] + S_{rr} \frac{\partial v_r}{\partial r} + S_{\theta\theta} \left( \frac{1}{r} \frac{\partial v_\theta}{\partial \theta} + \frac{v_r}{r} \right)$$

$$+ S_{\varphi\varphi} \left( \frac{1}{r \sin \theta} \frac{\partial v_\varphi}{\partial \varphi} + \frac{v_r}{r} + \frac{v_\theta \cot \theta}{r} \right) + S_{r\theta} \left( \frac{\partial v_\theta}{\partial r} + \frac{1}{r} \frac{\partial v_r}{\partial \theta} - \frac{v_\theta}{r} \right)$$

$$+ S_{r\varphi} \left( \frac{\partial v_\varphi}{\partial r} + \frac{1}{r \sin \theta} \frac{\partial v_r}{\partial \varphi} - \frac{v_\varphi}{r} \right) + S_{\theta\varphi} \left( \frac{1}{r} \frac{\partial v_\varphi}{\partial \theta} + \frac{1}{r \sin \theta} \frac{\partial v_\theta}{\partial \varphi} - \frac{\cot \theta}{r} v_\varphi \right) + \rho Q$$

**Table 5.4.0-3.** The differential energy balance for Newtonian fluids with constant $\rho$ and $k$

*Rectangular Cartesian coordinates:*

$$\rho \hat{c} \left( \frac{\partial T}{\partial t} + v_1 \frac{\partial T}{\partial z_1} + v_2 \frac{\partial T}{\partial z_2} + v_3 \frac{\partial T}{\partial z_3} \right)$$

$$= k \left( \frac{\partial^2 T}{\partial z_1^2} + \frac{\partial^2 T}{\partial z_2^2} + \frac{\partial^2 T}{\partial z_3^2} \right) + 2\mu \left[ \left( \frac{\partial v_1}{\partial z_1} \right)^2 + \left( \frac{\partial v_2}{\partial z_2} \right)^2 + \left( \frac{\partial v_3}{\partial z_3} \right)^2 \right]$$

$$+ \mu \left[ \left( \frac{\partial v_1}{\partial z_2} + \frac{\partial v_2}{\partial z_1} \right)^2 + \left( \frac{\partial v_1}{\partial z_3} + \frac{\partial v_3}{\partial z_1} \right)^2 + \left( \frac{\partial v_2}{\partial z_3} + \frac{\partial v_3}{\partial z_2} \right)^2 \right] + \rho Q$$

*Cylindrical coordinates:*

$$\rho \hat{c} \left( \frac{\partial T}{\partial t} + v_r \frac{\partial T}{\partial r} + \frac{v_\theta}{r} \frac{\partial T}{\partial \theta} + v_z \frac{\partial T}{\partial z} \right)$$

$$= k \left[ \frac{1}{r} \frac{\partial}{\partial r} \left( r \frac{\partial T}{\partial r} \right) + \frac{1}{r^2} \frac{\partial^2 T}{\partial \theta^2} + \frac{\partial^2 T}{\partial z^2} \right] + 2\mu \left\{ \left( \frac{\partial v_r}{\partial r} \right)^2 + \left[ \frac{1}{r} \left( \frac{\partial v_\theta}{\partial \theta} + v_r \right) \right]^2 + \left( \frac{\partial v_z}{\partial z} \right)^2 \right\}$$

$$+ \mu \left\{ \left( \frac{\partial v_\theta}{\partial z} + \frac{1}{r} \frac{\partial v_z}{\partial \theta} \right)^2 + \left( \frac{\partial v_z}{\partial r} + \frac{\partial v_r}{\partial z} \right)^2 + \left[ \frac{1}{r} \frac{\partial v_r}{\partial \theta} + r \frac{\partial}{\partial r} \left( \frac{v_\theta}{r} \right) \right]^2 \right\} + \rho Q$$

*Spherical coordinates:*

$$\rho \hat{c} \left( \frac{\partial T}{\partial t} + v_r \frac{\partial T}{\partial r} + \frac{v_\theta}{r} \frac{\partial T}{\partial \theta} + \frac{v_\varphi}{r \sin \theta} \frac{\partial T}{\partial \varphi} \right)$$

$$= k \left[ \frac{1}{r^2} \frac{\partial}{\partial r} \left( r^2 \frac{\partial T}{\partial r} \right) + \frac{1}{r^2 \sin \theta} \frac{\partial}{\partial \theta} \left( \sin \theta \frac{\partial T}{\partial \theta} \right) + \frac{1}{r^2 \sin^2 \theta} \frac{\partial^2 T}{\partial \varphi^2} \right] + 2\mu \left[ \left( \frac{\partial v_r}{\partial r} \right)^2 \right.$$

$$+ \left( \frac{1}{r} \frac{\partial v_\theta}{\partial \theta} + \frac{v_r}{r} \right)^2 + \left( \frac{1}{r \sin \theta} \frac{\partial v_\varphi}{\partial \varphi} + \frac{v_r}{r} + \frac{v_\theta \cot \theta}{r} \right)^2 \left. \right] + \mu \left\{ \left[ r \frac{\partial}{\partial r} \left( \frac{v_\theta}{r} \right) + \frac{1}{r} \frac{\partial v_r}{\partial \theta} \right]^2 \right.$$

$$+ \left[ \frac{1}{r \sin \theta} \frac{\partial v_r}{\partial \varphi} + r \frac{\partial}{\partial r} \left( \frac{v_\varphi}{r} \right) \right]^2 + \left[ \frac{\sin \theta}{r} \frac{\partial}{\partial \theta} \left( \frac{v_\varphi}{\sin \theta} \right) + \frac{1}{r \sin \theta} \frac{\partial v_\theta}{\partial \varphi} \right]^2 \left. \right\} + \rho Q$$

**Table 5.4.0-4.** Components of the energy flux vector as represented by Fourier's law

*Rectangular Cartesian coordinates:*

$$q_1 = -k \frac{\partial T}{\partial z_1}$$

$$q_2 = -k \frac{\partial T}{\partial z_2}$$

$$q_3 = -k \frac{\partial T}{\partial z_3}$$

*Cylindrical coordinates:*

$$q_r = -k \frac{\partial T}{\partial r}$$

$$q_\theta = -k \frac{1}{r} \frac{\partial T}{\partial \theta}$$

$$q_z = -k \frac{\partial T}{\partial z}$$

*Spherical coordinates:*

$$q_r = -k \frac{\partial T}{\partial r}$$

$$q_\theta = -k \frac{1}{r} \frac{\partial T}{\partial \theta}$$

$$q_\varphi = -k \frac{1}{r \sin \theta} \frac{\partial T}{\partial \varphi}$$

Here we make the identification for an incompressible fluid:

$$\hat{c} \equiv \hat{c}_V \equiv T \left( \frac{\partial \hat{S}}{\partial T} \right)_{\hat{V}} = \hat{c}_P \equiv T \left( \frac{\partial \hat{S}}{\partial T} \right)_P \tag{5.4.0-7}$$

The rectangular Cartesian, cylindrical, and spherical components of this equation are displayed in Table 5.4.0-3. The reader may use these as guides in immediately writing the corresponding expressions for the various forms of the differential energy balance given in Table 5.4.0-1.

In Section 2.4.1, we discussed the notation to be used for vector and tensor components in cylindrical and spherical coordinate systems. These comments continue to apply.

**Exercise 5.4.0-1** *Alternative forms of differential energy balance* Starting with (5.1.3-8) and (5.1.3-12), derive the other forms of the differential energy balance presented in Table 5.4.0-1.

# 6

---

# Differential Balances in Energy Transfer

I N THIS CHAPTER, we are primarily concerned with the formulation and solution of boundary-value problems involving the differential energy balance. You should look upon this chapter as paralleling Chapter 3, where we focused our attention upon solutions of the differential momentum balance.

There is one point that may be worth emphasizing. You will perhaps notice as you go through this chapter that none of the problems directly involve satisfying the differential entropy inequality. This inequality has been satisfied identically by placing certain restrictions upon the constitutive equations for the stress tensor and the energy flux vector. We saw, for example, in Sections 5.3.3 and 5.3.4 that, for a Newtonian fluid that obeys Fourier's law,

$$k > 0$$
$$\mu > 0$$

and

$$\lambda > -\frac{2}{3}\mu$$

This situation is entirely analogous to the way in which symmetry of the stress tensor was satisfied identically in Chapter 3.

---

## 6.1 Philosophy

There is an essential complication in the problems we consider here as compared with those we treated in Chapter 3. There we were concerned with simultaneous solutions of the differential mass and momentum balances for some assumed stress–deformation behavior. Here we analyze problems that require the simultaneous solution of the differential mass, momentum, and energy balances, given particular constitutive equations for both the stress tensor and the energy flux vector.

As we described in Section 3.1, the first step is to decide what the problem is. In part this means that constitutive equations for the stress tensor and energy flux vector must be chosen. Though a variety of representations for the stress tensor are used in this chapter, only Fourier's law is employed for the energy flux vector.

To complete the specification of a particular problem, we must describe the geometry of the material or the geometry through which the material moves, the forces that cause the material to move, and any energy transmission to the material. As in Chapter 3, every problem requires a statement of boundary conditions in its formulation. Beyond those indicated in Section 3.1, there are several common types of boundary conditions.

1. We shall assume temperature to be continuous across a phase interface. This is suggested by anticipating that local equilibrium is established at the phase boundary (Slattery 1990, p. 842). This is the same argument we used to justify continuity of the tangential components of velocity in Section 3.1.
2. The jump energy balance discussed in (5.1.3-9) must be satisfied at every phase interface.
3. We assume that temperatures and energy fluxes remain finite at all points in a material.

The advice we gave in Section 3.1 to launch the discussion of solutions for specific fluid mechanics problems is again applicable. We will not be willing to spend the time or money to solve every problem we formulate. Sometimes, it is more worthwhile to approximate a realistic but difficult problem by a somewhat simpler problem for which a solution is more readily available. This may be all that is needed. At worst, it should prove helpful as a limiting-case check on whatever further work is to be done.

The problems considered in this chapter are ones that can be solved easily. Concepts and principles can be introduced in this way in the minimum amount of time. Further, we feel that relatively simple problems must be understood before examining more difficult ones.

As in our discussion of solutions for the differential momentum balance, we do not say that the solutions we find here are unique, though some uniqueness and existence theorems for Laplace's equation are available (Kellogg 1929, pp. 211 and 277). We are more interested in finding a solution. Sometimes experimental evidence will suggest that the solutions we seek are unique, but this may not always be so clear.

## 6.2    Conduction

One of the simplest classes of problems involving energy transmission is conduction in a stationary solid. Since the velocity vector is identically zero, the differential energy balance reduces to

$$\rho \hat{c} \frac{\partial T}{\partial t} = \text{div}\,(k\,\nabla T) + \rho Q \qquad (6.2.0\text{-}1)$$

In writing this equation, we recognize that, for a solid,

$$\hat{c} \equiv \hat{c}_V = \hat{c}_P \qquad (6.2.0\text{-}2)$$

We often are not concerned with the stress distribution in the heated solid and, since the velocity distribution is known, the differential mass and momentum balances need not be considered.

In learning how to solve a problem, one often begins by assuming that $k$ is independent of temperature and, consequently, independent of position within the solid. For a steady-state temperature distribution with no external or mutual energy transmission ($Q = 0$), the differential energy balance reduces to Laplace's equation:

$$\text{div}\,(\nabla T) = 0 \qquad\qquad (6.2.0\text{-}3)$$

The most attractive feature of this limiting case is that considerable attention has been given to Laplace's equation in the literature (Kellogg 1929).

Remember that Laplace's equation occurred in our treatment of incompressible potential flows. The mathematical problems that we saw there occur here with a different physical significance. For example, a boundary condition in which the normal component of velocity is specified for a potential flow corresponds here to one in which the normal component of the energy flux vector is designated.

For a discussion of many more solutions of (6.2.0-1), see Carslaw and Jaeger (1959).

## 6.2.1  Cooling of a Semi-Infinite Slab: Constant Surface Temperature

Let us consider a quenching operation in which a hot body of metal with a uniform temperature $T_0$ is suddenly plunged into a cooling bath whose average temperature is controlled automatically by a refrigeration system. We wish to leave the body in the bath until the maximum temperature anywhere in the body is no greater than some specified value. To achieve the maximum rate of production with the available equipment, we wish to remove the body from the quenching bath as soon as possible. This means that we must determine the temperature distribution in the body as a function of time.

In principle, we should solve for the temperature distribution both in the metal and in the surrounding quenching oil. Energy cannot flow from the solid into the oil, unless there is a temperature gradient in the oil near the surface of the body. Our boundary conditions at the bounding surface of the metal should be that temperature is continuous across the phase interface and that the jump energy balance is satisfied. For the sake of simplicity, we shall say that the surface temperature of the metal is the same as the average temperature of the quenching oil, and we shall ignore the temperature distribution in the well-mixed oil.

In order that this problem be as simple as possible, let us also replace our finite body with a semi-infinite solid that occupies all space corresponding to $z_2 \geq 0$. The initial and boundary conditions become

$$\text{at } t = 0 \text{ for all } z_2 > 0 : \ T = T_0 \qquad\qquad (6.2.1\text{-}1)$$

and

$$\text{at } z_2 = 0 \text{ for all } t > 0 : \ T = T_1 \qquad\qquad (6.2.1\text{-}2)$$

Some authors feel that as an additional boundary condition one should take the temperature to be $T_0$ as $z_2$ approaches infinity for all values of time (Bird et al. 1960, p. 353; Schlichting 1979, p. 90). We take the view here that temperature is not directly constrained to be $T_0$ at infinity for times greater than zero (Carslaw and Jaeger 1959, p. 62); it is only necessary that temperature be finite as $z_2$ approaches infinity (Carslaw and Jaeger 1959, p. 71).

We shall further assume that the thermal conductivity of the metal may be taken to be a constant. Since there is no external or mutual energy transmission, (6.2.0-1) reduces to

$$\rho \hat{c} \frac{\partial T}{\partial t} = k \, \text{div}\,(\nabla T) \qquad\qquad (6.2.1\text{-}3)$$

Let us look for a solution of the form

$$T = T(t, z_2) \tag{6.2.1-4}$$

From Table 5.4.1-3, we see that (6.2.1-3) reduces to

$$\rho \hat{c} \frac{\partial T}{\partial t} = k \frac{\partial^2 T}{\partial z_2{}^2} \tag{6.2.1-5}$$

We seek a solution for this equation consistent with (6.2.1-1) and (6.2.1-2).

Our search can be simplified if we introduce a dimensionless temperature:

$$T^\star \equiv \frac{T - T_0}{T_1 - T_0} \tag{6.2.1-6}$$

Equation (6.2.1-5) becomes

$$\frac{\partial T^\star}{\partial t} = \alpha \frac{\partial^2 T^\star}{\partial z_2{}^2} \tag{6.2.1-7}$$

where

$$\alpha \equiv \frac{k}{\rho \hat{c}} \tag{6.2.1-8}$$

Equations (6.2.1-1) and (6.2.1-2) reduce to

$$\text{at } t = 0 \text{ for all } z_2 > 0 : \ T^\star = 0 \tag{6.2.1-9}$$

and

$$\text{at } z_2 = 0 \text{ for } t > 0 : \ T^\star = 1 \tag{6.2.1-10}$$

This problem is mathematically of the same form as the one that resulted when we analyzed the flow in a fluid bounded by a wall that is suddenly set in motion, in Section 3.2.4. Consequently, the solution is

$$T^\star \equiv \frac{T - T_0}{T_1 - T_0}$$

$$= 1 - \text{erf}\left( \frac{z_2}{\sqrt{4\alpha t}} \right) \tag{6.2.1-11}$$

Because $\text{erf}(2) = 0.995$, the temperature of the slab remains essentially unchanged outside a region of thickness

$$\delta_T \approx 4\sqrt{\alpha t} \tag{6.2.1-12}$$

and (6.2.1-11) may be used as an approximate temperature distribution for a slab whose thickness is large compared with $\delta_T$.

**Exercise 6.2.1-1** *An infinitely long cylinder* (Carslaw and Jaeger 1959, p. 199)   An infinitely long solid circular cylinder is initially at a uniform temperature $T_0$. For time greater than zero, the surface temperature is constrained to be $T_1$. Determine the temperature distribution in a rod whose radius is $R$.

*Answer:*

$$\frac{T - T_1}{T_0 - T_1} = \sum_{n=1}^{\infty} \frac{2J_0[\lambda_n(r/R)]}{\lambda_n J_1(\lambda_n)} \exp\left(\frac{-\lambda_n^2 kt}{\rho \hat{c} R^2}\right)$$

where the $\lambda_n$ ($n = 1, 2, \ldots$) are the positive roots of $J_0(\lambda_n) = 0$.

**Exercise 6.2.1-2** *Two large blocks brought into contact* A large block of metal $A$ at a uniform temperature $T_0^{(A)}$ is brought into contact with a large block of metal $B$ at a uniform temperature $T_0^{(B)}$ along two plane faces of the blocks. Estimate the temperature of the interface as a function of time.

*Answer:* At the phase interface,

$$T^\star = \left(1 + \sqrt{\frac{\rho^{(A)} \hat{c}^{(A)} k^{(A)}}{\rho^{(B)} \hat{c}^{(B)} k^{(B)}}}\right)^{-1}$$

*Hint:* Pose the problem in terms of

$$T^\star \equiv \frac{T - T_0^{(A)}}{T_0^{(B)} - T_0^{(A)}}$$

Take the Laplace transform and solve for the temperature distribution in each phase.

**Exercise 6.2.1-3** *Periodic surface temperature* (Carslaw and Jaeger 1959, p. 65) Conduction of energy in a solid with a periodic surface temperature is of considerable practical importance. Problems of this type arise in designing automatic temperature control systems, in estimating the periodic temperatures (and periodic thermal stresses) in the cylinder walls of internal combustion engines, and in studying the periodic heating of Earth's crust by the Sun (Carslaw and Jaeger 1959, p. 81).

Perhaps the simplest problem of this type is to consider a semi-infinite solid ($z_2 > 0$) whose surface is subjected to a periodic temperature variation:

at $z_2 = 0 :$ $T = T_0 + A \cos(\omega t - \epsilon)$

Determine that

$$T^\star \equiv \frac{T - T_0}{A}$$
$$= \exp(-K z_2) \cos(\omega t - K z_2 - \epsilon)$$

where

$$K \equiv \sqrt{\frac{\omega \rho \hat{c}}{2k}}$$

*Hint:* See Exercise 3.2.4-3.

**6.2.2** Cooling a Semi-Infinite Slab: Newton's "Law" of Cooling

To solve the complete quenching problem discussed in the introduction of Section 6.2.1, it would be necessary to determine the temperature distribution both in the metal and in the oil. At the metal–oil phase interface, temperature is continuous and the jump energy balance must be satisfied. For simplicity, we approximated the temperature of the metal surface as the average temperature of the oil, and we ignored the temperature distribution in the oil.

A somewhat better approximation would be to employ

**Newton's "law" of cooling**[1]    The energy flux across a fluid–solid phase interface is roughly proportional to the difference between the temperature of the surface and the temperature of the surrounding bulk fluid (which might be assumed to be well mixed):

$$\mathbf{q} \cdot \mathbf{n} = h(T_{\text{surface}} - T_{\text{surroundings}})$$

Our understanding is that $\mathbf{n}$ is the unit normal to the phase interface directed into the surroundings. The coefficient $h$ is referred to as the *heat-transfer coefficient*. Experimentally, $h$ is not a constant, although it is often assumed to be a constant in simple simulations. For another use of the heat-transfer coefficient concept, see Section 7.4.2.

Let us solve for the temperature distribution in the semi-infinite slab that was described in Section 6.2.1. We make only one change in the statement of the problem. Instead of requiring the temperature of the phase interface to be the average temperature of the oil, we specify that the energy transfer across the phase interface be described by Newton's "law" of cooling with a constant heat-transfer coefficient $h$,

$$\text{at } z_2 = 0 \text{ for all } t > 0 : \ k\frac{\partial T}{\partial z_2} = h\,(T - T_1) \tag{6.2.2-1}$$

With the change of variable

$$T^\star \equiv \frac{T - T_1}{T_0 - T_1} \tag{6.2.2-2}$$

(6.2.1-5) becomes

$$\frac{\partial T^\star}{\partial t} = \alpha\frac{\partial^2 T^\star}{\partial z_2{}^2} \tag{6.2.2-3}$$

The initial condition (6.2.1-1) and the boundary condition (6.2.2-1) may be written as

$$\text{at } t = 0 \text{ for all } z_2 > 0 : \ T^\star = 1 \tag{6.2.2-4}$$

and

$$\text{at } z_2 = 0 \text{ for all } t > 0 : \ \frac{\partial T^\star}{\partial z_2} = \frac{h}{k}T^\star \tag{6.2.2-5}$$

Let us define the function

$$A = A(t, z_2) \equiv T^\star - \frac{k}{h}\frac{\partial T^\star}{\partial z_2} \tag{6.2.2-6}$$

In terms of $A$, (6.2.2-3) through (6.2.2-5) become

$$\frac{\partial A}{\partial t} = \alpha\frac{\partial^2 A}{\partial z_2{}^2} \tag{6.2.2-7}$$

$$\text{at } t = 0 \text{ for all } z_2 > 0 : \ A = 1 \tag{6.2.2-8}$$

---

[1]  This should not be viewed as a law or postulate. It is nothing more than a definition for $h$.

and

at $z_2 = 0$ for all $t > 0$ : $A = 0$  (6.2.2-9)

Taking the approach used in Section 3.2.4, we find that

$$A = \text{erf}\left(\frac{z_2}{\sqrt{4\alpha t}}\right)$$  (6.2.2-10)

Equations (6.2.2-6) and (6.2.2-10) combine to yield a first-order ordinary differential equation for $T^\star$:

$$\frac{\partial T^\star}{\partial z_2} - \frac{h}{k}T^\star = -\frac{h}{k}\text{erf}\left(\frac{z_2}{\sqrt{4\alpha t}}\right)$$  (6.2.2-11)

We require that $T^\star$ remain finite as $z_2$ approaches infinity. The solution to (6.2.2-11) that satisfies this condition is

$$T^\star = -\frac{h}{k}\exp\left(\frac{hz_2}{k}\right)\int_\infty^{z_2}\text{erf}\left(\frac{\zeta}{\sqrt{4\alpha t}}\right)\exp\left(\frac{-h\zeta}{k}\right)d\zeta$$  (6.2.2-12)

A change of variable

$$\eta \equiv \zeta - z_2$$  (6.2.2-13)

allows this solution to be expressed as

$$T^\star = \frac{2h}{\sqrt{\pi}k}\int_0^\infty \exp\left(-\frac{h\eta}{k}\right)\left[\int_0^{(z_2+\eta)/\sqrt{4\alpha t}}\exp(-u^2)\,du\right]d\eta$$  (6.2.2-14)

Upon an integration by parts, we have

$$T^\star = \text{erf}\left(\frac{z_2}{\sqrt{4\alpha t}}\right)$$

$$+ \frac{1}{\sqrt{\pi\alpha t}}\int_0^\infty \exp\left(-\frac{h\eta}{k} - \frac{[z_2+\eta]^2}{4\alpha t}\right)d\eta$$  (6.2.2-15)

or

$$T^\star = \text{erf}\left(\frac{z_2}{\sqrt{4\alpha t}}\right) + \frac{1}{\sqrt{\pi\alpha t}}\exp\left(\frac{hz_2}{k} + \frac{h^2\alpha t}{k^2}\right)$$

$$\times \int_0^\infty \exp\left(-\frac{[z_2+\eta+2h\alpha t/k]^2}{4\alpha t}\right)d\eta$$  (6.2.2-16)

This may be simplified by defining

$$u \equiv \frac{z_2+\eta+2h\alpha t/k}{\sqrt{4\alpha t}}$$  (6.2.2-17)

with the result

$$T^\star = \text{erf}\left(\frac{z_2}{\sqrt{4\alpha t}}\right) + \frac{2}{\sqrt{\pi}}\exp\left(\frac{hz_2}{k} + \frac{h^2\alpha t}{k^2}\right)$$

$$\times \int_{(z_2+2h\alpha t/k)/\sqrt{4\alpha t}}^\infty \exp(-u^2)\,du$$  (6.2.2-18)

or

$$T^{\star} = \mathrm{erf}\left(\frac{z_2}{\sqrt{4\alpha t}}\right) + \exp\left(\frac{hz_2}{k} + \frac{h^2\alpha t}{k^2}\right)\left[1 - \mathrm{erf}\left(\frac{z_2}{\sqrt{4\alpha t}} + \frac{h}{k}\sqrt{\alpha t}\right)\right] \qquad (6.2.2\text{-}19)$$

At the surface

$$T^{\star} = \exp\frac{h^2\alpha t}{k^2}\left[1 - \mathrm{erf}\left(\frac{h}{k}\sqrt{\alpha t}\right)\right] \qquad (6.2.2\text{-}20)$$

This may be expressed as a power series (Carslaw and Jaeger 1959, p. 483)

$$T^{\star} = \frac{k}{h\sqrt{\pi\alpha t}}\left(1 - \frac{k^2}{2h^2\alpha t} + \frac{3k^4}{4h^4\alpha^2 t} - \cdots\right) \qquad (6.2.2\text{-}21)$$

After a long time, the surface temperature may be approximated by

$$T^{\star} \equiv \frac{T - T_1}{T_0 - T_1}$$

$$= \frac{k}{h\sqrt{\pi\alpha t}} \qquad (6.2.2\text{-}22)$$

The error involved is less than

$$\frac{k^3}{2h^3(\pi\alpha^3 t^3)^{1/2}}$$

**Exercise 6.2.2-1**

i) Starting with (6.2.2-6), derive (6.2.2-7).

ii) Starting with (6.2.2-7) through (6.2.2-9), determine the solution given by (6.2.2-10).

iii) Starting with (6.2.2-11) and the requirement that $T^{\star}$ remain finite as $z_2$ approaches infinity, determine the solution given by (6.2.2-12).

**Exercise 6.2.2-2** *Energy transfer from a pipe* (Carslaw and Jaeger 1959, p. 189)    A pipe is used to transport a fluid whose average temperature is $T_0$. The pipe is mounted in an airstream, the average temperature of which is $T_1$.

You may assume that the temperature of the interior surface of the pipe ($r = \kappa R$) is $T_0$. But assume that Newton's "law" of cooling applies on the exterior surface ($r = R$).

i) Show that the temperature distribution in the wall of the pipe is given by

$$\frac{T - T_1}{T_0 - T_1} = 1 - \frac{(hR/k)\,\ln(r/\kappa R)}{1 + (hR/k)\,\ln(1/\kappa)}$$

ii) Show that, if $\kappa Rh/k > 1$, the magnitude of the energy transfer between the exterior wall of the pipe and the surroundings continuously decreases as $R$ increases for a fixed value of $\kappa R$. Show that, if $\kappa Rh/k < 1$, the magnitude of this energy transfer is a maximum at $R = k/h$.

### 6.2.3   Cooling a Flat Sheet: Constant Surface Temperature

Let us replace the semi-infinite slab of Section 6.2.1 with a stationary infinite flat sheet that occupies all space between $z_2 = -a$ and $z_2 = +a$. The material initially has a uniform temperature

$$\text{at } t = 0 \text{ for} -a < z_2 < a : \quad T = T_0 \tag{6.2.3-1}$$

and we approximate the plunge of the hot sheet into a quenching bath by stating that

$$\text{at } z_2 = \pm a \text{ for } t > 0 : \quad T = T_1 \tag{6.2.3-2}$$

We again take thermal conductivity to be a constant.

Let us look for a solution of the form

$$T = T(t, z_2) \tag{6.2.3-3}$$

The search for a solution can be somewhat simplified if we introduce an appropriate set of dimensionless variables. Let

$$T^\star \equiv \frac{T - T_1}{T_0 - T_1} \tag{6.2.3-4}$$

$$t^\star \equiv \frac{tk}{\rho \hat{c} a^2} \tag{6.2.3-5}$$

and

$$z_2^\star \equiv \frac{z_2}{a} \tag{6.2.3-6}$$

The differential energy balance, (6.2.0-1), becomes

$$\frac{\partial T^\star}{\partial t^\star} = \frac{\partial^2 T^\star}{\partial z_2^{\star 2}} \tag{6.2.3-7}$$

with the boundary conditions

$$\text{at } t^\star = 0 \text{ for} -1 < z_2^\star < 1 : \quad T^\star = 1 \tag{6.2.3-8}$$

and

$$\text{at } z_2^\star = \pm 1 \text{ for } t^\star > 0 : \quad T^\star = 0 \tag{6.2.3-9}$$

Let us examine solutions to (6.2.3-7) that are of the form

$$T^\star = \mathcal{T}(t^\star)\mathcal{Z}(z_2^\star) \tag{6.2.3-10}$$

This implies that

$$\frac{\mathcal{T}'}{\mathcal{T}} = \frac{\mathcal{Z}''}{\mathcal{Z}} = -\lambda^2 \tag{6.2.3-11}$$

where $\lambda^2$ is a constant.

The solution for

$$\mathcal{T}' + \lambda^2 \mathcal{T} = 0 \tag{6.2.3-12}$$

is of the form

$$\mathcal{T} = A e^{-\lambda^2 t^\star} \tag{6.2.3-13}$$

The solution for

$$Z'' + \lambda^2 Z = 0 \tag{6.2.3-14}$$

is of the form

$$Z = B \sin(\lambda z_2^\star) + C \cos(\lambda z_2^\star) \tag{6.2.3-15}$$

The boundary conditions (6.2.3-9) imply that

$$\lambda = \frac{(2n+1)\pi}{2} \quad n = 0, 1, 2, \ldots \tag{6.2.3-16}$$

and

$$B = 0 \tag{6.2.3-17}$$

A linear combination of all solutions to (6.2.3-7) indicated by (6.2.3-10), (6.2.3-13), and (6.2.3-15) to (6.2.3-17) yields

$$T^\star = \sum_{n=0}^{\infty} D_n \exp\left(\frac{-[2n+1]^2\pi^2 t^\star}{4}\right) \cos\left(\frac{[2n+1]\pi z_2^\star}{2}\right) \tag{6.2.3-18}$$

Boundary condition (6.2.3-8) requires that

$$1 = \sum_{n=0}^{\infty} D_n \cos\left(\frac{[2n+1]\pi z_2^\star}{2}\right) \tag{6.2.3-19}$$

which implies that

$$\int_{-1}^{1} \cos\left(\frac{[2m+1]\pi\xi}{2}\right) d\xi$$

$$= \sum_{n=0}^{\infty} D_n \int_{-1}^{1} \cos\left(\frac{[2m+1]\pi\xi}{2}\right) \cos\left(\frac{[2n+1]\pi\xi}{2}\right) d\xi$$

$$= \begin{cases} 0 & \text{if } m \neq n \\ D_m & \text{if } m = n \end{cases} \tag{6.2.3-20}$$

or

$$D_n = \frac{4(-1)^n}{(2n+1)\pi} \tag{6.2.3-21}$$

The final result for temperature has the form (Carslaw and Jaeger 1959, p. 100)

$$T^\star = \sum_{n=0}^{\infty} \frac{4(-1)^n}{[2n+1]\pi} \exp\left(\frac{-[2n+1]^2\pi^2 t^\star}{4}\right) \cos\left(\frac{(2n+1)\pi z_2^\star}{2}\right) \tag{6.2.3-22}$$

or

$$\frac{T - T_1}{T_0 - T_1} = \sum_{n=0}^{\infty} \frac{4(-1)^n}{[2n+1]\pi} \exp\left(\frac{-[2n+1]^2\pi^2 kt}{4\rho\hat{c}a^2}\right) \cos\left(\frac{[2n+1]\pi z_2}{2a}\right) \tag{6.2.3-23}$$

**Exercise 6.2.3-1** *A flat sheet* (Carslaw and Jaeger 1959, p. 122)   Repeat the problem in the text. Now, assuming Newton's "law" of cooling (Section 6.2.2) with a constant heat transfer coefficient, $h$ can be used to describe the loss of energy from the sheet to a fluid with an average temperature $T_1$.

*Answer:*

$$T^\star = \sum_{n=1}^{\infty} \frac{2H \cos(\lambda_n z_2^\star)\sec \lambda_n}{H(H+1) + \lambda_n^2} \exp(-\lambda_n^2 t^\star)$$

where the $\lambda_n$ ($n = 1, 2, \ldots$) denote the positive roots of

$$\lambda \tan \lambda = H$$
$$\equiv \frac{ha}{k}$$

Carslaw and Jaeger (1959, p. 491) tabulate the first six roots.

**Exercise 6.2.3-2** *A bar* (Carslaw and Jaeger 1959, p. 173)   Consider the cooling of a semi-infinite bar of metal with a uniform cross section $-a \leq z_2 \leq a$, $-b \leq z_1 \leq b$. The initial temperature of the bar is $T_0$ and the temperature of the surface of the bar is maintained at $T_1$. If we denote the solution given in (6.2.3-23) as $\varphi(z_2, t, a)$, verify that the solution to this problem is

$$\frac{T - T_1}{T_0 - T_1} = \varphi(z_2, t, a)\varphi(z_1, t, b)$$

**Exercise 6.2.3-3** *More on a bar* (Carslaw and Jaeger 1959, p. 173)   Consider the problem of Exercise 6.2.3-2 but use Newton's "law" of cooling (Section 6.2.2) with a constant heat transfer coefficient to describe the loss of energy from the bar to a fluid whose average temperature is $T_1$. If we denote the solution given in Exercise 6.2.3-1 as $\psi(z_2, t, a, h)$, verify that the solution to this problem is

$$\frac{T - T_1}{T_0 - T_1} = \psi(z_2, t, a, h)\psi(z_1, t, b, h)$$

**Exercise 6.2.3-4** *A sphere* (Carslaw and Jaeger 1959, p. 233)   A sphere is initially at a uniform temperature $T_0$. For time greater than zero, the surface temperature is constrained to be $T_1$. Determine the temperature distribution in the sphere whose radius is $R$.

*Answer:*

$$\frac{T - T_1}{T_0 - T_1} = \frac{2R}{\pi} \sum_{n=1}^{\infty} \frac{(-1)^{n+1}}{nr} \sin\left(n\pi \frac{r}{R}\right) \exp\left(\frac{-n^2\pi^2 kt}{\rho\hat{c}R^2}\right)$$

*Hint:*   Introduce a transformation of the general form $u = rT$.

**Exercise 6.2.3-5** *More on a flat sheet* (Carslaw and Jaeger 1959, p. 100)   A stationary flat sheet occupies all space between $z_2 = 0$ and $z_2 = b$, and it initially has a uniform temperature $T_0$. For time greater than zero, the wall at $z_2 = b$ is held at a fixed temperature $T_1$ while the wall at $z_2 = 0$ is insulated. Determine the temperature distribution in this sheet as a function of time and position.

*Answer:*

$$\frac{T - T_1}{T_0 - T_1} = \sum_{m=0}^{\infty} \frac{4(-1)^m}{(2m+1)\pi} \cos\left(\frac{[2m+1]\pi}{2} \frac{z_2}{b}\right) \exp\left(\frac{-[2m+1]^2\pi^2 kt}{4\rho\hat{c}b^2}\right)$$

---

## 6.3    More Complete Solutions

In the succeeding sections, there are two general types of problems with which we will be concerned: Either the fluids will be required to be incompressible or the motion will be found to be isochoric.

For both of these cases, the differential energy balance from Table 5.4.1-1 reduces to

$$\rho\hat{c}_V \frac{d_{(m)}T}{dt} = -\text{div }\mathbf{q} + \text{tr}(\mathbf{S} \cdot \nabla\mathbf{v}) + \rho Q \qquad (6.3.0\text{-}1)$$

We will assume that Fourier's law applies. We will not be concerned here with external or mutual energy transmission ($Q = 0$), and for simplicity we will generally neglect the temperature dependence of thermal conductivity. Under these conditions, (6.3.0-1) further simplifies to

$$\rho\hat{c}_V \frac{d_{(m)}T}{dt} = k\,\text{div}(\nabla T) + \text{tr}(\mathbf{S} \cdot \nabla\mathbf{v}) \qquad (6.3.0\text{-}2)$$

In the sections that follow, we shall generally be concerned with simultaneous solutions of (6.3.0-2), the differential mass balance

$$\frac{d_{(m)}\rho}{dt} + \rho\,\text{div }\mathbf{v} = 0 \qquad (6.3.0\text{-}3)$$

the differential momentum balance

$$\rho\frac{d_{(m)}\mathbf{v}}{dt} = \text{div }\mathbf{T} + \rho\mathbf{f} \qquad (6.3.0\text{-}4)$$

and an appropriate constitutive equation for the stress tensor.

### 6.3.1    Couette Flow of a Compressible Newtonian Fluid

In Figure 6.3.1-1 a compressible Newtonian fluid is trapped between two parallel planes that are separated by a distance $b$. The lower plane is stationary and has a fixed uniform temperature:

at $z_2 = 0$:  $\mathbf{v} = \mathbf{0}$

$$T = T_0 \qquad (6.3.1\text{-}1)$$

The upper plane moves with a constant speed $V$ in the $z_1$ direction, and it has a uniform

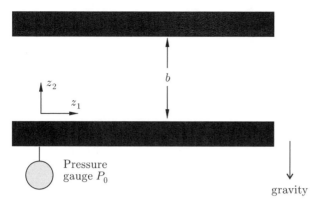

**Figure 6.3.1-1.** Couette flow.

temperature:

$$\text{at } z_2 = b : \quad v_1 = V$$
$$v_2 = v_3$$
$$= 0$$
$$T = T_1$$

(6.3.1-2)

A pressure gauge mounted on the lower plate reads

$$\text{at } z_2 = z_1 = z_3 = 0 : \quad -T_{22} = P_0$$

(6.3.1-3)

Let us determine the velocity and temperature distributions within the fluid.

For lack of further information we will assume that the planes are very large and that edge effects may be neglected. We shall also take all physical properties (other than density) to be constants.

The boundary conditions (6.3.1-1) and (6.3.1-2) suggest that we assume

$$v_1 = v_1(z_2)$$
$$v_2 = 0$$
$$v_3 = 0$$

(6.3.1-4)

This form of velocity distribution satisfies the condition for an isochoric motion:

$$\text{div } \mathbf{v} = 0$$

(6.3.1-5)

In view of (6.3.1-4) and (6.3.1-5), the differential mass balance tells us that

$$\frac{d_{(m)}\rho}{dt} = \frac{\partial \rho}{\partial z_1} v_1$$
$$= 0$$

(6.3.1-6)

or

$$\frac{\partial \rho}{\partial z_1} = 0$$

(6.3.1-7)

This implies that density may be a function of $z_2$ and $z_3$:

$$\rho = \rho(z_2, z_3)$$

(6.3.1-8)

The components of the differential momentum balance simplify to

$$0 = -\frac{\partial P}{\partial z_1} + \mu \frac{\partial^2 v_1}{\partial z_2^2} \qquad \text{. (6.3.1-9)}$$

$$0 = -\frac{\partial P}{\partial z_2} - \rho g \qquad \text{(6.3.1-10)}$$

$$0 = -\frac{\partial P}{\partial z_3} \qquad \text{(6.3.1-11)}$$

Equations (6.3.1-10) and (6.3.1-11) indicate that

$$\frac{\partial^2 P}{\partial z_3 \, \partial z_2} = -\frac{\partial \rho}{\partial z_3} g$$

$$= 0 \qquad \text{(6.3.1-12)}$$

This means that density is not a function of $z_3$, and (6.3.1-8) reduces to

$$\rho = \rho(z_2) \qquad \text{(6.3.1-13)}$$

From (6.3.1-10) and (6.3.1-11), we have that

$$P = -g \int_0^{z_2} \rho \, dz_2 + h(z_1) \qquad \text{(6.3.1-14)}$$

Equation (6.3.1-9) allows us to say

$$\frac{\partial P}{\partial z_1} = \frac{dh}{dz_1}$$

$$= \mu \frac{d^2 v_1}{dz_2^2}$$

$$= A = \text{a constant} \qquad \text{(6.3.1-15)}$$

or

$$P = -g \int_0^{z_2} \rho \, dz_2 + A z_1 + B \qquad \text{(6.3.1-16)}$$

In view of (6.3.1-3) and (6.3.1-4), we reason that

$$B = P_0 \qquad \text{(6.3.1-17)}$$

We must wait in order to determine the constant $A$.

Integration of (6.3.1-15) yields

$$v_1 = \frac{A}{2\mu} z_2^2 + C_1 z_2 + C_2 \qquad \text{(6.3.1-18)}$$

The boundary conditions on velocity given by (6.3.1-1) and (6.3.1-2) require

$$C_1 = \frac{V}{b} - \frac{A}{2\mu} b \qquad \text{(6.3.1-19)}$$

and

$$C_2 = 0 \qquad \text{(6.3.1-20)}$$

In view of the boundary conditions on temperature given by (6.3.1-1) and (6.3.1-2), it seems reasonable to look for a temperature distribution in the form of

$$T = T(z_2) \tag{6.3.1-21}$$

Let us assume that our compressible fluid is such that pressure can be expressed as a function of density and temperature:

$$P = P(\rho, T) \tag{6.3.1-22}$$

Equations (6.3.1-13), (6.3.1-21), and (6.3.1-22) require that in (6.3.1-16)

$$A = 0 \tag{6.3.1-23}$$

The differential energy balance from Table 5.4.1-2 reduces for this situation to

$$0 = k\frac{d^2T}{dz_2{}^2} + \mu \left(\frac{dv_1}{dz_2}\right)^2 \tag{6.3.1-24}$$

or, to be consistent with (6.3.1-18) through (6.3.1-20) and (6.3.1-23),

$$\frac{d^2T}{dz_2{}^2} = -\frac{\mu}{k}\left(\frac{V}{b}\right)^2 \tag{6.3.1-25}$$

It is helpful to write this in a dimensionless form as

$$\frac{d^2T^\star}{dz_2^{\star 2}} = -N_{Br} \tag{6.3.1-26}$$

where

$$T^\star \equiv \frac{T - T_0}{T_1 - T_0}$$
$$z_2^\star \equiv \frac{z_2}{b} \tag{6.3.1-27}$$

and the Brinkman number

$$N_{Br} \equiv \frac{\mu V^2}{k(T_1 - T_0)} \tag{6.3.1-28}$$

From (6.3.1-1) and (6.3.1-2), the boundary conditions that must be satisfied by (6.3.1-26) are

$$\text{at } z_2^\star = 0: \ T^\star = 0$$
$$\text{at } z_2^\star = 1: \ T^\star = 1 \tag{6.3.1-29}$$

Equation (6.3.1-26) is easily integrated to satisfy (6.3.1-29). We find

$$T^\star = -\frac{N_{Br}}{2}\left(z_2^{\star 2} - z_2^\star\right) + z_2^\star \tag{6.3.1-30}$$

In summary, the velocity, pressure, and temperature distributions in the fluid are

$$\frac{v_1}{V} = \frac{z_2}{b} \tag{6.3.1-31}$$

$$P = -g\int_0^{z_2} \rho \, dz_2 + P_0 \tag{6.3.1-32}$$

and

$$\frac{T - T_0}{T_1 - T_0} = \frac{N_{Br}}{2}\left[\frac{z_2}{b} - \left(\frac{z_2}{b}\right)^2\right] + \frac{z_2}{b} \tag{6.3.1-33}$$

Upon comparison of (6.3.1-26) with (6.3.1-24), it becomes obvious that the Brinkman number $N_{Br}$ is indicative of the rate at which energy is dissipated by viscous forces within the fluid. If we neglect this dissipation of energy, we see that the temperature distribution is linear:

$$\text{as } N_{Br} \to 0 : \quad \frac{T - T_0}{T_1 - T_0} \to \frac{z_2}{b} \tag{6.3.1-34}$$

The next section takes up a similar problem, where not only density but the other physical properties as well are allowed to depend upon temperature.

**Exercise 6.3.1-1**    How is the temperature distribution of the problem discussed in the text altered if the temperature of the upper wall is changed to $T_0$?

*Answer:*

$$\frac{T - T_0}{T_0} = \frac{1}{2}N_{Br}\left[\frac{z_2}{b} - \left(\frac{z_2}{b}\right)^2\right]$$

**Exercise 6.3.1-2**    Redo the problem of the text, now taking the lower wall to be insulated.

*Answer:*

$$\frac{T - T_1}{T_1} = \frac{1}{2}N_{Br}\left[1 - \left(\frac{z_2}{b}\right)^2\right]$$

**Exercise 6.3.1-3**    For the problem of the text, when is the upper wall cooled and when is it heated?

**Exercise 6.3.1-4**    How is the analysis of the text altered if the fluid is taken to be incompressible?

**Exercise 6.3.1-5**    What happens to the problem of the text when both walls are said to be insulated?

**Exercise 6.3.1-6** *Couette flow of an incompressible power-law fluid*    Rework the problem of the text, assuming that the gap between the planes is filled with an incompressible power-law fluid. Show that the temperature distribution has the same form with $N_{Br}$ replaced by

$$N_{Br(PL)} \equiv \frac{mV^{n+1}}{kb^{n-1}(T_1 - T_0)}$$

For this problem, use Figure 6.3.1-2 in place of Figure 6.3.1-1.

**Exercise 6.3.1-7** *Flow through a channel*    An incompressible Newtonian fluid flows through the channel of width $2b$ shown in Figure 3.2.1-5 (see Exercise 3.2.1-7). Both planes are maintained at a constant temperature $T_0$. Determine the temperature distribution in the fluid. The temperature dependence of viscosity and thermal conductivity may be neglected.

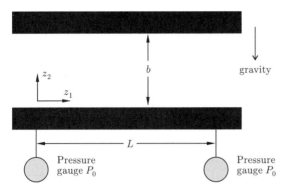

**Figure 6.3.1-2.** Couette flow of an incompressible fluid.

*Answer:*

$$T - T_0 = \frac{\mu(v_{1(max)})^2}{3k}\left[1 - \left(\frac{z_2}{b}\right)^4\right]$$

**Exercise 6.3.1-8** Repeat Exercise 6.3.1-7, assuming that the lower wall ($z_2 = -b$) is insulated.

**Exercise 6.3.1-9** *Tangential annular flow of an incompressible Newtonian fluid* As explained in Section 3.2.3, tangential annular flow is the basis for one common type of viscometer. As the result of viscous dissipation, the temperature distribution within the fluid can be appreciably different from the known temperature of the walls. Apparent non-Newtonian behavior in viscous materials may be attributable to the temperature dependence of a Newtonian viscosity.

Referring to Figure 3.2.3-1, let us assume that the inner wall is stationary, that the angular velocity $\Omega$ of the outer wall is specified, and that the temperature of both walls is a constant $T_0$. For simplicity, we will take the viscosity and thermal conductivity of the fluid to be independent of temperature.

Solve for the velocity and temperature distribution in the fluid and determine the position at which temperature is maximized. The results for four values of $\kappa$ are shown in Figure 6.3.1-3.

*Answer:*

$$T^\star \equiv \frac{T - T_0}{T_0}$$

$$= N\left[1 - \left(\frac{R}{r}\right)^2\right] + N\left(\frac{1}{\kappa^2} - 1\right)\frac{\ln(r/R)}{\ln \kappa},$$

where

$$N \equiv N_{Br}\frac{\kappa^4}{(1 - \kappa^2)^2}$$

$$N_{Br} \equiv \frac{\mu\Omega^2 R^2}{kT_0}$$

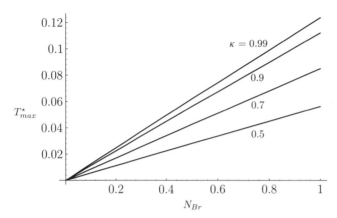

**Figure 6.3.1-3.** The maximum temperature $T^\star_{max} \equiv (T - T_0)/T_0$ as a function of $N_{Br} \equiv \mu\Omega^2 R^2/(kT_0)$ for selected values of $\kappa$ in the tangential annular flow of an incompressible Newtonian fluid.

**Exercise 6.3.1-10** *More on tangential annular flow of an incompressible Newtonian fluid*  Repeat Exercise 6.3.1-9, assuming that the inner wall is insulated.

*Answer:*

$$\frac{T - T_0}{T_0} = N\left[1 - \left(\frac{R}{r}\right)^2\right] - \frac{2N}{\kappa^2}\ln\left(\frac{r}{R}\right)$$

**Exercise 6.3.1-11** *Tangential annular flow of an incompressible power-law fluid*  Repeat Exercise 6.3.1-9 for an incompressible power-law fluid.

*Answer:*

$$\frac{T - T_0}{T_0} = \left(\frac{n}{2}\right)^2 N_{Br(PL)}\left[\frac{2\kappa^{2/n}}{n(1 - \kappa^{2/n})}\right]^{n+1}$$
$$\times \left[\frac{1}{\kappa^{2/n}} - \left(\frac{R}{r}\right)^{2/n} - \frac{\ln(r/\kappa R)}{\ln(1/\kappa)}\left(\frac{1}{\kappa^{2/n}} - 1\right)\right]$$

**Exercise 6.3.1-12**  Show that the result of Exercise 6.3.1-9 reduces to that for Exercise 6.3.1-1 as $\kappa$ approaches unity.

**Exercise 6.3.1-13** *Viscous heating in an oscillatory flow* (**Bird 1965b**)  For the flow described in Exercise 3.2.4-5, determine the temperature distribution in the fluid in the limit as $\omega \rightarrow \infty$. You may assume that both walls are maintained at a constant and uniform temperature $T_0$.

   In the limit as $\omega \rightarrow \infty$, it is not really the instantaneous temperature distribution that is of practical interest. We are more concerned with the distribution of the average temperature over one period of the oscillation:

$$\overline{T} \equiv \frac{\omega}{2\pi}\int_t^{t+2\pi/\omega} T\, dt'$$

It is reasonable to begin this problem by looking for a solution of the form

$$T = T(t, z_2)$$

for which the differential energy balance takes the form

$$\rho \hat{c} \frac{\partial T}{\partial t} = k \frac{\partial^2 T}{\partial z_2{}^2} + \mu \left(\frac{\partial v_1}{\partial z_2}\right)^2$$

It is the time average of this equation with which we shall be concerned:

$$\frac{\omega}{2\pi} \int_t^{t+2\pi/\omega} \left[ \rho \hat{c} \frac{\partial T}{\partial t} - k \frac{\partial^2 T}{\partial z_2{}^2} - \mu \left(\frac{\partial v_1}{\partial z_2}\right)^2 \right] dt' = 0$$

Referring to Section 4.1.1, we see that this reduces to

$$\rho \hat{c} \frac{\partial \overline{T}}{\partial t} = k \frac{\partial^2 \overline{T}}{\partial z_2{}^2} + \mu \overline{\left(\frac{\partial v_1}{\partial z_2}\right)^2}$$

But the time-averaged temperature defined here should be independent of time:

$$\overline{T} = \overline{T}(z_2)$$

and the time-averaged differential energy balance reduces to

$$0 = k \frac{d^2 \overline{T}}{d z_2{}^2} + \mu \overline{\left(\frac{d v_1}{d z_2}\right)^2}$$

i) Determine that

$$\frac{1}{a^2} \left(\frac{b}{v_0}\right)^2 \overline{\left(\frac{d v_1}{d z_2}\right)^2}$$

$$= \frac{1}{a^2} \overline{\left(\frac{d v_1^\star}{d z_2^\star}\right)^2}$$

$$= \left[\sinh^2 a \left(1 - z_2^\star\right) \sin^2 a \left(1 - z_2^\star\right) + \cosh^2 a \left(1 - z_2^\star\right) \cos^2 a \left(1 - z_2^\star\right)\right]$$

$$\times \left[\sinh^2 a \cos^2 a + \sin^2 a \cosh^2 a\right]^{-1}$$

where

$$a \equiv \sqrt{\frac{\omega \rho b^2}{2\mu}}$$

ii) We are concerned here with the limit as $\omega \to \infty$. Find that

$$\lim a \to \infty : \quad \frac{1}{a^2} \overline{\left(\frac{d v_1^\star}{d z_2^\star}\right)^2} = \exp\left(-2 a z_2^\star\right)$$

iii) Calculate the time-averaged temperature distribution in the fluid to be

$$\overline{T} - T_0 = \frac{\mu v_0{}^2}{4k} \left\{1 - \exp(-2 a z_2^\star) - \left[1 - \exp(-2a)\right] z_2^\star\right\}$$

Conclude that, in the limit as $a \to \infty$, the maximum temperature occurs at the lower wall:

$$\overline{T}_{\max} - T_0 = \frac{\mu v_0{}^2}{4k}$$

**Exercise 6.3.1-14** *Viscous heating in flow through a tube*    Assuming constant physical properties, determine the temperature distribution in laminar flow of an incompressible Newtonian fluid through a tube, assuming that the wall of the tube is held at a constant temperature $T_0$.

*Answer:*

$$T^\star \equiv \frac{T - T_0}{T_0}$$

$$= \frac{N_{Br}}{4}\left[1 - \left(\frac{r}{R}\right)^4\right]$$

Here

$$N_{Br} \equiv \frac{\mu v_{z(\max)}{}^2}{kT_0}$$

## 6.3.2   Couette Flow of a Compressible Newtonian Fluid with Variable Viscosity and Thermal Conductivity

Let us repeat the analysis of Exercise 6.3.1-1 for a compressible Newtonian fluid in laminar flow between the two planes shown in Figure 6.3.1-1. The principal changes from Section 6.3.1 are that now both planes are maintained at a constant temperature $T_0$ and that both viscosity and thermal conductivity are functions of temperature. For the moment, we shall leave the functional dependence of $\mu$ and $k$ unspecified. Our objective is to determine the pressure, velocity, and temperature distributions within the fluid.

The boundary conditions to be satisfied are

at $z_2 = 0$ :  $\mathbf{v} = \mathbf{0}$

$$T = T_0 \qquad\qquad\qquad\qquad (6.3.2\text{-}1)$$

at $z_2 = b$ :  $v_1 = V$
$$v_2 = v_3$$
$$= 0$$
$$T = T_0 \qquad\qquad\qquad\qquad (6.3.2\text{-}2)$$

at $z_2 = z_1 = z_3 = 0$ :  $-T_{22} = P_0 \qquad\qquad (6.3.2\text{-}3)$

We shall look for a velocity distribution of the form

$$v_1 = v_1(z_2), \quad v_2 = v_3 = 0 \qquad\qquad (6.3.2\text{-}4)$$

Use the differential mass balance to conclude that

$$\rho = \rho(z_2, z_3) \qquad\qquad\qquad\qquad (6.3.2\text{-}5)$$

The three components of the differential momentum balance take a somewhat different form here:

$$0 = -\frac{\partial P}{\partial z_1} + \frac{\partial}{\partial z_2}\left(\mu \frac{dv_1}{dz_2}\right) \tag{6.3.2-6}$$

$$0 = -\frac{\partial P}{\partial z_2} - \rho g \tag{6.3.2-7}$$

$$0 = -\frac{\partial P}{\partial z_3} \tag{6.3.2-8}$$

From (6.3.2-7) and (6.3.2-8), we are able to argue that

$$\frac{\partial^2 P}{\partial z_3 \partial z_2} = -\frac{\partial \rho}{\partial z_3} g = 0 \tag{6.3.2-9}$$

and

$$\rho = \rho(z_2) \tag{6.3.2-10}$$

In view of the boundary conditions on temperature given by (6.3.2-1) and (6.3.2-2), we shall look for a temperature distribution of the form

$$T = T(z_2) \tag{6.3.2-11}$$

Since viscosity is a function only of temperature, (6.3.2-11) allows us to rewrite (6.3.2-6) as

$$0 = -\frac{\partial P}{\partial z_1} + \frac{d}{dz_2}\left(\mu \frac{dv_1}{dz_2}\right) \tag{6.3.2-12}$$

Equations (6.3.2-7) and (6.3.2-12) indicate that the thermodynamic pressure distribution is of the form

$$P = -g \int_0^{z_2} \rho \, dz_2 + h(z_1) \tag{6.3.2-13}$$

From (6.3.2-12) and (6.3.2-13), we see that

$$\frac{\partial P}{\partial z_1} = \frac{dh}{dz_1} = \frac{d}{dz_2}\left(\mu \frac{dv_1}{dz_2}\right) = A = \text{a constant} \tag{6.3.2-14}$$

The pressure distribution reduces to

$$P = -g \int_0^{z_2} \rho \, dz_2 + A z_1 + P_0 \tag{6.3.2-15}$$

which satisfies boundary condition (6.3.2-3). If we assume that thermodynamic pressure is a function only of density and temperature,

$$P = P(\rho, T) \tag{6.3.2-16}$$

Equations (6.3.2-10) and (6.3.2-11) tell us that

$$A = 0 \tag{6.3.2-17}$$

At this point, we see from (6.3.2-14) that

$$\frac{d}{dz_2}\left(\mu\frac{dv_1}{dz_2}\right) = 0 \tag{6.3.2-18}$$

must be solved simultaneously with the differential energy balance,

$$\frac{d}{dz_2}\left(k\frac{dT}{dz_2}\right) + \mu\left(\frac{dv_1}{dz_2}\right)^2 = 0 \tag{6.3.2-19}$$

for the velocity and temperature distributions. Let us introduce as dimensionless variables

$$v_1^\star \equiv \frac{v_1}{V}$$

$$T^\star \equiv \frac{T - T_0}{T_0} \tag{6.3.2-20}$$

$$z_2^\star \equiv \frac{z_2}{b}$$

in terms of which (6.3.2-18) and (6.3.2-19) become

$$\frac{dv_1^\star}{dz_2^\star} = \frac{\mu_0}{\mu}C_1 \tag{6.3.2-21}$$

and

$$\frac{\mu_0}{\mu}\frac{d}{dz_2^\star}\left(\frac{k}{k_0}\frac{dT^\star}{dz_2^\star}\right) + N_{Br}\left(\frac{dv_1^\star}{dz_2^\star}\right)^2 = 0 \tag{6.3.2-22}$$

Here $\mu_0$ and $k_0$ are a characteristic viscosity and characteristic thermal conductivity; the Brinkman number is defined here as

$$N_{Br} \equiv \frac{\mu_0 V^2}{T_0 k_0} \tag{6.3.2-23}$$

The corresponding boundary conditions are

$$\text{at } z_2^\star = 0: \ v_1^\star = T^\star$$
$$= 0 \tag{6.3.2-24}$$

and

$$\text{at } z_2^\star = 1: \ v_1^\star = 1$$
$$T^\star = 0 \tag{6.3.2-25}$$

Equations (6.3.2-21) and (6.3.2-22) suggest that we expand $\mu_0/\mu$ and $k/k_0$ in Taylor series as functions of the dimensionless temperature $T^\star$:

$$\frac{k}{k_0} = 1 + \alpha_1 T^\star + \alpha_2 T^{\star 2} + \cdots \tag{6.3.2-26}$$

$$\frac{\mu_0}{\mu} = 1 + \beta_1 T^\star + \beta_2 T^{\star 2} + \cdots \tag{6.3.2-27}$$

In these series expansions we have identified $\mu_0$ and $k_0$ as the viscosity and the thermal conductivity of the fluid at $T = T_0$. We will assume that the parameters $\alpha_i$ and $\beta_i$ ($i = 1, 2, 3, \ldots$) are known from available experimental data.

With (6.3.2-26) and (6.3.2-27), we see that (6.3.2-21) and (6.3.2-22) are nonlinear equations that must be solved numerically. In such cases, one should ask whether the analysis of some limiting case might be almost as interesting. For example, here we can ask about the effects of viscous dissipation in the limit of very small values of the Brinkman number $N_{Br}$. If we did have a solution to the complete problem, we could visualize that $T^\star$, $v_1^\star$, and the constant of integration $C_1$ could all be expanded in Taylor series as functions of the Brinkman number:

$$T^\star = T^{\star(0)} + T^{\star(1)} N_{Br} + T^{\star(2)} N_{Br}^2 + \cdots \tag{6.3.2-28}$$

$$v_1^\star = v_1^{\star(0)} + v_1^{\star(1)} N_{Br} + v_1^{\star(2)} N_{Br}^2 + \cdots \tag{6.3.2-29}$$

$$C_1 = C_1^{(0)} + C_1^{(1)} N_{Br} + C_1^{(2)} N_{Br}^2 + \cdots \tag{6.3.2-30}$$

The quantities $T^{\star(0)}$, $v_1^{\star(0)}$, and $C_1^{(0)}$ are known, respectively, as the zeroth perturbations (with respect to $N_{Br} = 0$) of $T^\star$, $v_1^\star$, and $C_1$; $T^{\star(1)}$, $v_1^{\star(1)}$, and $C_1^{(1)}$ are the first perturbations (with respect to $N_{Br} = 0$) of these variables.

Substituting (6.3.2-27) through (6.3.2-30) into (6.3.2-21) we find

$$\frac{dv_1^{\star(0)}}{dz_2^\star} + N_{Br} \frac{dv_1^{\star(1)}}{dz_2^\star} + \cdots$$

$$= \left[ 1 + \beta_1 \left( T^{\star(0)} + T^{\star(1)} N_{Br} + \cdots \right) + \beta_2 \left( T^{\star(0)} + T^{\star(1)} N_{Br} + \cdots \right)^2 + \cdots \right]$$

$$\times \left[ C_1^{(0)} + C_1^{(1)} N_{Br} + \cdots \right] \tag{6.3.2-31}$$

$$\left[ \frac{dv_1^{\star(0)}}{dz_2^\star} - C_1^{(0)} \left( 1 + \beta_1 T^{\star(0)} + \beta_2 T^{\star(0)2} + \cdots \right) \right]$$

$$+ N_{Br} \left[ \frac{dv_1^{\star(1)}}{dz_2^\star} - C_1^{(0)} \left( \beta_1 T^{\star(1)} + 2\beta_2 T^{\star(0)} T^{\star(1)} + \cdots \right) \right.$$

$$\left. - C_1^{(1)} \left( 1 + \beta_1 T^{\star(0)} + \beta_2 T^{\star(0)2} + \cdots \right) \right] + \cdots = 0 \tag{6.3.2-32}$$

Since $N_{Br}^0$, $N_{Br}^1$, $N_{Br}^2$, ... are linearly independent, we conclude that the coefficients of these quantities in (6.3.2-32) must individually be zero. Looking at the first two coefficients in (6.3.2-32), we have

$$\frac{dv_1^{\star(0)}}{dz_2^\star} - C_1^{(0)} \left( 1 + \beta_1 T^{\star(0)} + \beta_2 T^{\star(0)2} + \cdots \right) = 0 \tag{6.3.2-33}$$

and

$$\frac{dv_1^{\star(1)}}{dz_2^\star} - C_1^{(0)} \left( \beta_1 T^{\star(1)} + 2\beta_2 T^{\star(0)} T^{\star(1)} + \cdots \right)$$

$$- C_1^{(1)} \left( 1 + \beta_1 T^{\star(0)} + \beta_2 T^{\star(0)2} + \cdots \right) = 0 \tag{6.3.2-34}$$

Going through the same argument with (6.3.2-22), we find that the coefficient of $N_{Br}{}^0$ is

$$\left(1 + \beta_1 T^{\star(0)} + \beta_2 T^{\star(0)2} + \cdots\right)$$
$$\frac{d}{dz_2^\star}\left[\left(1 + \alpha_1 T^{\star(0)} + \alpha_2 T^{\star(0)2} + \cdots\right)\frac{dT^{\star(0)}}{dz_2^\star}\right] = 0 \qquad (6.3.2\text{-}35)$$

The coefficients of $N_{Br}{}^0$ in (6.3.2-24) and (6.3.2-25) tell us that

$$\text{at } z_2^\star = 0: \ T^{\star(0)} = 0 \qquad (6.3.2\text{-}36)$$

and

$$\text{at } z_2^\star = 1: \ T^{\star(0)} = 0 \qquad (6.3.2\text{-}37)$$

Clearly, (6.3.2-35) through (6.3.2-37) are satisfied by

$$T^{\star(0)} = 0 \qquad (6.3.2\text{-}38)$$

Equation (6.3.2-33) simplifies to

$$\frac{dv_1^{\star(0)}}{dz_2^\star} - C_1^{(0)} = 0 \qquad (6.3.2\text{-}39)$$

The corresponding boundary conditions are found by looking at the coefficients $N_{Br}{}^0$ in (6.3.2-24) and (6.3.2-25):

$$\text{at } z_2^\star = 0: \ v_1^{\star(0)} = 0 \qquad (6.3.2\text{-}40)$$

$$\text{at } z_2^\star = 1: \ v_1^{\star(0)} = 1 \qquad (6.3.2\text{-}41)$$

Equation (6.3.2-39) can be integrated with boundary conditions (6.3.2-40) and (6.3.2-41) to find

$$v_1^{\star(0)} = z_2^\star \qquad (6.3.2\text{-}42)$$

and

$$C_1^{(0)} = 1 \qquad (6.3.2\text{-}43)$$

The coefficient of $N_{Br}$ in (6.3.2-22) is

$$\frac{d^2 T^{\star(1)}}{dz_2^{\star 2}} + 1 = 0 \qquad (6.3.2\text{-}44)$$

The boundary conditions for this equation are found from the coefficients of $N_{Br}$ in (6.3.2-24) and (6.3.2-25):

$$\text{at } z_2^\star = 0: \ T^{\star(1)} = 0 \qquad (6.3.2\text{-}45)$$

$$\text{at } z_2^\star = 1: \ T^{\star(1)} = 0 \qquad (6.3.2\text{-}46)$$

The solution to (6.3.2-44) that satisfies (6.3.2-45) and (6.3.2-46) is

$$T^{\star(1)} = \frac{1}{2}\left(z_2^\star - z_2^{\star 2}\right) \qquad (6.3.2\text{-}47)$$

From (6.3.2-34), (6.3.2-38), (6.3.2-43), and (6.3.2-47), the coefficient of $N_{Br}$ in (6.3.2-21) yields

$$\frac{dv_1^{\star(1)}}{dz_2^\star} - \frac{1}{2}\beta_1\left(z_2^\star - z_2^{\star(2)}\right) - C_1^{(1)} = 0 \tag{6.3.2-48}$$

The boundary conditions for this equation are determined by the coefficient $N_{Br}$ in (6.3.2-24) and (6.3.2-25):

$$\text{at } z_2^\star = 0: \quad v_1^{\star(1)} = 0 \tag{6.3.2-49}$$

$$\text{at } z_2^\star = 1: \quad v_1^{\star(1)} = 0 \tag{6.3.2-50}$$

Equations (6.3.2-48) to (6.3.2-50) are satisfied by

$$v_1^{\star(1)} = \frac{-\beta_1}{12}\left(z_2^\star - 3z_2^{\star 2} + 2z_2^{\star 3}\right) \tag{6.3.2-51}$$

and

$$C_1^{(1)} = -\frac{\beta_1}{12} \tag{6.3.2-52}$$

Looking at the coefficient of $N_{Br}^{\ 2}$ in (6.3.2-22), we have

$$\frac{d^2 T^{\star(2)}}{dz_2^{\star 2}} + \alpha_1\left(\frac{1}{4} - \frac{3}{2}z_2^\star + \frac{3}{2}z_2^{\star 2}\right) + \frac{\beta_1}{2}\left(-\frac{1}{3} + z_2^\star - z_2^{\star 2}\right) = 0 \tag{6.3.2-53}$$

The boundary conditions that the solutions to this equation must satisfy are found by examining the coefficients of $N_{Br}^{\ 2}$ in (6.3.2-24) and (6.3.2-25):

$$\text{at } z_2^\star = 0: \quad T^{\star(2)} = 0 \tag{6.3.2-54}$$

$$\text{at } z_2^\star = 1: \quad T^{\star(2)} = 0 \tag{6.3.2-55}$$

Equations (6.3.2-53) through (6.3.2-55) are satisfied by

$$T^{\star(2)} = -\frac{\alpha_1}{8}\left(z_2^{\star 2} - 2z_2^{\star 3} + z_2^{\star 4}\right) - \frac{\beta_1}{24}\left(z_2^\star - 2z_2^{\star 2} + 2z_2^{\star 3} - z_2^{\star 4}\right) \tag{6.3.2-56}$$

In summary, (6.3.2-28), (6.3.2-29), (6.3.2-38), (6.3.2-42), (6.3.2-47), (6.3.2-51), and (6.3.2-56) describe the dimensionless velocity and temperature profiles as

$$\begin{aligned}
\frac{v_1}{V} &= v_1^\star \\
&= z_2^\star - N_{Br}\frac{\beta_1}{12}\left(z_2^\star - 3z_2^{\star 2} + 2z_2^{\star 3}\right) + \cdots
\end{aligned} \tag{6.3.2-57}$$

and

$$\begin{aligned}
\frac{T - T_0}{T_0} &= T^\star \\
&= N_{Br}\frac{1}{2}\left(z_2^\star - z_2^{\star 2}\right) - N_{Br}^{\ 2}\frac{\alpha_1}{8}\left(z_2^{\star 2} - 2z_2^{\star 3} + z_2^{\star 4}\right) \\
&\quad - N_{Br}^{\ 2}\frac{\beta_1}{24}\left(z_2^\star - 2z_2^{\star 2} + 2z_2^{\star 3} - z_2^{\star 4}\right) + \cdots
\end{aligned} \tag{6.3.2-58}$$

This is an example of how one carries out a *perturbation solution* of one or more nonlinear differential equations. The *perturbation parameter* was chosen to be the Brinkman number $N_{Br}$. Equations (6.3.2-57) and (6.3.2-58) describe the velocity and temperature distributions for sufficiently small values of the perturbation parameter. There are two drawbacks to a perturbation solution:

1) One has no firm guarantee that the series solution developed converges.
2) Assuming that the series converges, there is no firm estimate of the error involved in truncating the series. The best that can be said is that the error should be of the order of the last term neglected.

The solution developed here is based upon a similar solution suggested by Bird et al. (1960, p. 306) for an incompressible Newtonian fluid. Other approaches to the same general problem have been suggested (Illingworth 1950; DeGroff 1956a,b; Morgan 1957).

A somewhat similar approach has been taken by Turian and Bird (1963) in their discussion of viscous heating in a cone–plate viscometer for an incompressible Newtonian fluid. Their comparison with experimental data is particularly helpful in indicating how important it is to consider viscous heating in that geometry.

**Exercise 6.3.2-1**    Fill in the missing steps in the text between (6.3.2-35) and (6.3.2-56).

**Exercise 6.3.2-2** *Tangential annular flow of an incompressible Newtonian fluid*    Repeat the problem described in the text for an incompressible Newtonian fluid undergoing the tangential annular flow described in Figure 3.2.3-1. The inner wall is stationary, the outer wall rotates at a constant angular velocity $\Omega$, and both walls are maintained at a fixed temperature $T_0$.

Conclude that results for the maximum temperature given in Figure 6.3.1-3 still apply. If

$$T_z^\star \equiv \frac{T_z}{2\pi L \mu_0 \Omega R^2}$$

$$= T_z^{\star(0)} + N_{Br} T_z^{\star(1)} + \cdots$$

compute the percent deviation in $T_z$ attributable to the first perturbation as a function of $N_{Br}$ as shown in Figure 6.3.2-1:

$$T_{z\,(\text{dev})}^\star \equiv N_{Br} \frac{T_z^{\star(1)}}{T_z^{\star(0)}} \tag{6.3.2-59}$$

**Exercise 6.3.2-3** *Couette flow of an incompressible power-law fluid*    Repeat the problem described in the text for an incompressible power-law fluid undergoing Couette flow described in Figure 3.2.2-2. Both walls are maintained at a fixed temperature $T_0$. Assume that in (2.3.3-6) $n$ is a constant, (6.3.2-26) applies, and

$$\frac{1}{m^\star} = 1 + \beta_1 T^\star + \beta_2 T^{\star 2} + \cdots$$

where $T^\star$ is defined by (6.3.2-20).

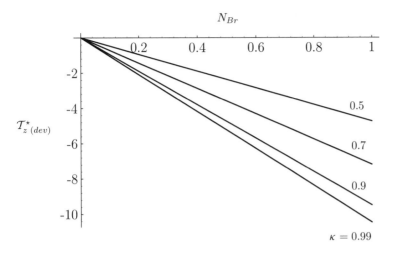

**Figure 6.3.2-1.** $\mathcal{T}^{\star}_{z\,(dev)}$ defined by (6.3.2-59) as a function of $N_{Br} \equiv \mu_0\Omega^2 R^2/(k_0T_0)$ for selected values of $\kappa$ in the tangential annular flow of $n$-hexane at $T_0 = 298$ K.

*Answer:*

$$\frac{v_1}{V} = v_1^{\star}$$

$$= z_2^{\star} - N_{Br}\frac{\beta_1}{12n}\left(z_2^{\star} - 3z_2^{\star 2} + 2z_2^{\star 3}\right) + \cdots$$

and

$$\frac{T - T_0}{T_0} = T^{\star}$$

$$= N_{Br}\frac{1}{2}\left(z_2^{\star} - z_2^{\star 2}\right) + \cdots$$

**Exercise 6.3.2-4** *Couette flow of an incompressible power-law fluid*   Repeat Exercise 6.3.2-3, now assuming that $n$ is a function of $T$ as well.

**6.3.3**   Rate at Which a Fluid Freezes (or a Solid Melts)

A semi-infinite, incompressible fluid (or solid) $A$ is bounded by a stationary wall at $z_2 = 0$. The system, which is at a uniform pressure, is initially at a uniform temperature $T_0 > T_M$ (or $T_0 < T_M$), where $T_M$ is the melting point:

at $t = 0$ for all $z_2 > 0$ :   $T^{(A)} = T_0$                            (6.3.3-1)

For time $t > 0$, the temperature of stationary wall is changed to $T_1 < T_M$ (or $T_1 > T_M$), the fluid begins to freeze (or the solid begins to melt), and a layer of solid (or liquid) $B$ forms:

at $z_2 = 0$ for all $t > 0$ :   $T^{(B)} = T_1$                            (6.3.3-2)

Our objective is to determine the rate at which the solid–liquid interface moves across the material.

In this analysis, we will seek temperature distributions in the fluid and solid of the form

$$T^{(A)} = T^{(A)}(z_2, t)$$
$$T^{(B)} = T^{(B)}(z_2, t)$$

(6.3.3-3)

We will assume that the densities of the fluid and solid differ and that the fluid moves as the solid either expands or contracts:

$$v_2^{(A)} = v_2^{(A)}(z_2, t)$$

$$v_1^{(A)} = v_3^{(A)}$$

$$= 0$$

$$v_1^{(B)} = v_2^{(B)}$$

$$= v_3^{(B)}$$

$$= 0$$

(6.3.3-4)

We will assume that the solid–liquid phase interface

$$z_2 = h(t)$$

(6.3.3-5)

is a plane having a speed of displacement

$$u_2 = u_2(t)$$

(6.3.3-6)

Under these circumstances, the differential mass balance is satisfied identically in the solid phase $B$, and for the fluid phase $A$ it demands that

$$\frac{\partial v_2^{(A)}}{\partial z_2} = 0$$

(6.3.3-7)

or

$$v_2^{(A)} = v_2^{(A)}(t)$$

(6.3.3-8)

The jump mass balance requires

$$v_2^{(A)} = \frac{\left(\rho^{(A)} - \rho^{(B)}\right)}{\rho^{(A)}} u_2$$

(6.3.3-9)

Neglecting the effects of inertia, we find that the differential momentum balance requires that modified pressure be independent of position in the fluid, phase $A$. Since we are not concerned with the stresses in the solid phase $B$, it is unnecessary to consider either the differential momentum balance for the solid phase $B$ or the jump momentum balance.

For the solid phase $B$, the differential energy balance requires

$$\frac{\partial T^{(B)\star}}{\partial t} = \alpha^{(B)} \frac{\partial^2 T^{(B)\star}}{\partial z_2^2}$$

(6.3.3-10)

where we have found it convenient to introduce the dimensionless temperature

$$T^\star \equiv \frac{T - T_1}{T_0 - T_1}$$

(6.3.3-11)

and the *thermal diffusivity*

$$\alpha \equiv \frac{k}{\rho \hat{c}} \tag{6.3.3-12}$$

With the change of variable

$$\eta^{(B)} \equiv \frac{z_2}{\sqrt{4\alpha^{(B)}t}} \tag{6.3.3-13}$$

Equation (6.3.3-10) takes the form

$$\frac{d^2 T^{(B)\star}}{d\eta^{(B)2}} + 2\eta^{(B)} \frac{dT^{(B)\star}}{d\eta^{(B)}} = 0 \tag{6.3.3-14}$$

From (6.3.3-2), Equation (6.3.3-14) must be solved with the boundary condition

$$\text{at } \eta^{(B)} = 0: \ T^{(B)\star} = 0 \tag{6.3.3-15}$$

The solution of (6.3.3-14) consistent with (6.3.3-15) is

$$T^{(B)\star} = D_1 \, \text{erf} \left( \frac{z_2}{\sqrt{4\alpha^{(B)}t}} \right) \tag{6.3.3-16}$$

As noted in Section 6.1, we will assume that temperature is continuous across the solid–liquid interface. Using the phase rule, we conclude that the temperature of the phase interface must be $T_M$. From (6.3.3-5) and (6.3.3-16), we conclude that

$$T_M^{\star} = D_1 \, \text{erf} \left( \frac{h}{\sqrt{4\alpha^{(B)}t}} \right) \tag{6.3.3-17}$$

or that

$$\frac{h}{\sqrt{4\alpha^{(B)}t}} = \lambda$$
$$= \text{a constant} \tag{6.3.3-18}$$

and

$$D_1 = \frac{T_M^{\star}}{\text{erf}(\lambda)} \tag{6.3.3-19}$$

Equation (6.3.3-18) implies that

$$u_2 = \frac{dh}{dt}$$
$$= \lambda \left( \frac{\alpha^{(B)}}{t} \right)^{1/2} \tag{6.3.3-20}$$

For the liquid phase $A$, the differential energy balance requires

$$\frac{\partial T^{(A)\star}}{\partial t} + v_2^{(A)} \frac{\partial T^{(A)\star}}{\partial z_2} = \alpha^{(A)} \frac{\partial^2 T^{(A)\star}}{\partial z_2^{2}} \tag{6.3.3-21}$$

or, in view of (6.3.3-9) and (6.3.3-20),

$$\frac{\partial T^{(A)\star}}{\partial t} + \lambda \frac{\left(\rho^{(A)} - \rho^{(B)}\right)}{\rho^{(A)}} \left(\frac{\alpha^{(B)}}{t}\right)^{1/2} \frac{\partial T^{(A)\star}}{\partial z_2}$$

$$= \alpha^{(A)} \frac{\partial^2 T^{(A)\star}}{\partial z_2{}^2} \tag{6.3.3-22}$$

With the change of variable

$$\eta^{(A)} \equiv \frac{z_2}{\sqrt{4\alpha^{(A)}t}} \tag{6.3.3-23}$$

Equation (6.3.3-22) takes the form

$$\frac{d^2 T^{(A)\star}}{d\eta^{(A)2}} + \left[2\lambda \left(\frac{\rho^{(B)} - \rho^{(A)}}{\rho^{(A)}}\right) \left(\frac{\alpha^{(B)}}{\alpha^{(A)}}\right)^{1/2} + 2\eta^{(A)}\right] \frac{d T^{(A)\star}}{d\eta^{(A)}}$$

$$= 0 \tag{6.3.3-24}$$

From (6.3.3-1), Equation (6.3.3-24) must be solved with the boundary condition

$$\text{as } \eta^{(A)} \to \infty : \ T^{(A)\star} \to 1 \tag{6.3.3-25}$$

The solution is

$$T^{(A)\star} = 1 + D_2 \, \text{erfc} \left[\frac{z_2}{\sqrt{4\alpha^{(A)}t}} + \lambda \left(\frac{\rho^{(B)} - \rho^{(A)}}{\rho^{(A)}}\right) \left(\frac{\alpha^{(B)}}{\alpha^{(A)}}\right)^{1/2}\right] \tag{6.3.3-26}$$

Since temperature must be continuous across the phase interface, we conclude that

$$D_2 = \left(T_M^\star - 1\right) \left\{\text{erfc}\left[\lambda \frac{\rho^{(B)}}{\rho^{(A)}} \left(\frac{\alpha^{(B)}}{\alpha^{(A)}}\right)^{1/2}\right]\right\}^{-1} \tag{6.3.3-27}$$

Because the system is at a uniform pressure, the jump energy balance (5.1.3-9) takes the form

$$\text{at } z_2 = h : k^{(B)} \frac{\partial T^{(B)}}{\partial z_2} - k^{(A)} \frac{\partial T^{(A)}}{\partial z_2}$$

$$= -\rho^{(A)} \hat{U}^{(A)} \left(v_2^{(A)} - u_2\right) - \rho^{(B)} \hat{U}^{(B)} u_2 + v_2^{(A)} T_{22}^{(A)}$$

$$= \left[\rho^{(B)} \left(\hat{U}^{(A)} - \hat{U}^{(B)}\right) + \frac{\left(\rho^{(B)} - \rho^{(A)}\right)}{\rho^{(A)}} P^{(A)}\right] u_2$$

$$= \rho^{(B)} \Delta \hat{H} u_2$$

$$= \lambda \rho^{(B)} \Delta \hat{H} \left(\frac{\alpha^{(B)}}{t}\right)^{1/2} \tag{6.3.3-28}$$

where

$$\Delta \hat{H} \equiv \hat{H}^{(A)} - \hat{H}^{(B)} \tag{6.3.3-29}$$

In arriving at (6.3.3-28), we have neglected the change in kinetic energy across the dividing surface with respect to $\Delta \hat{H}$. In view of (6.3.3-16), (6.3.3-19), (6.3.3-26), and (6.3.3-27), equation (6.3.3-28) gives us a transcendental equation which can be solved for $\lambda$:

$$\frac{\exp\left(-\lambda^2\right)}{\mathrm{erf}(\lambda)} - \frac{\left(1 - T_M^\star\right)}{T_M^\star} \left(\frac{k^{(A)}}{k^{(B)}}\right) \left(\frac{\alpha^{(B)}}{\alpha^{(A)}}\right)^{1/2}$$

$$\times \exp\left[-\lambda^2 \frac{\alpha^{(B)}}{\alpha^{(A)}} \left(\frac{\rho^{(B)}}{\rho^{(A)}}\right)^2\right] \left\{\mathrm{erfc}\left[\lambda \frac{\rho^{(B)}}{\rho^{(A)}} \left(\frac{\alpha^{(B)}}{\alpha^{(A)}}\right)^{1/2}\right]\right\}^{-1}$$

$$= \lambda \frac{\sqrt{\pi}\,\Delta \hat{H}}{\hat{c}^{(B)}\left(T_M - T_1\right)} \tag{6.3.3-30}$$

For water, $k^{(A)} = 0.6028$ J/(m s K), $k^{(B)} = 2.219$ J/(m s K), $\rho^{(A)} = 1,000$ kg/m$^3$, $\rho^{(B)} = 920$ kg/m$^3$, $\hat{c}^{(A)} = 4,187$ J/(kg K), $\hat{c}^{(B)} = 2,100$ J/(kg K), $\alpha^{(A)} = 1.440 \times 10^{-7}$, $\alpha^{(B)} = 1.148 \times 10^{-6}$, $T_M = 273$ K, and $\Delta H = 3.338 \times 10^5$ J/kg. We will consider the case in which $T_0 = 278$ K and $T_1 = 268$ K. With these assumptions, $\lambda = 0.1159$, and the corresponding temperature distribution at various times is shown in Figure 6.3.3-1. Note the discontinuity in the

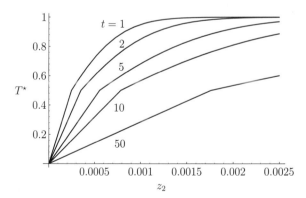

**Figure 6.3.3-1.** The dimensionless temperature $T^\star$ as a function of dimensional position $z_2$ measured in meters for water at several times: 1, 2, 5, 10, and 50 seconds. The initial temperature of the liquid water (phase $A$) is $T_0 = 278$ K; the temperature of the refrigerated wall at $z_2 = 0$ is $T_1 = 268$ K. The discontinuity in the first derivative occurs at the phase interface where $T = T_M = 273$ K and $T^\star = 0.5$.

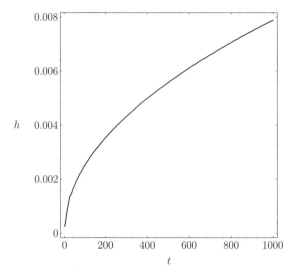

**Figure 6.3.3-2.** The position $h$ of the interface measured in meters as a function of time $t$ measured in seconds for the case shown in Figure 6.3.3-1.

derivative of temperature at the interface. The position of the interface as a function of time is shown in Figure 6.3.3-2.

It is interesting to note that for water, out to 10,000, the results for the case described above are indistinguishable from those corresponding to $\rho^{(B)} = \rho^{(A)} = 1,000 \text{ kg/m}^3$. The case in which the densities of the two phases are assumed to be equal is described by Carslaw and Jaeger (1959, p. 285).

Unfortunately, the one-dimensional analysis given here does not readily extend to a multi-dimensional problem. For a guide in attacking such a case, see Section 9.3.8.

This section was written with the help of R. L. Robinson.

## 6.4    When Viscous Dissipation Is Neglected

As we have mentioned previously, many problems are difficult to solve in full generality. Some ingenuity in the use of numerical analysis may be required as well as considerable effort on the part of the programmer, not to mention the cost of computer time. This dollar outlay may be justified, but before undertaking such an expenditure it is usually worthwhile to consider some pertinent limiting cases. It may turn out that a limiting case is all that a situation demands. At worst, the solution for the limiting case should provide a necessary check on whatever numerical work is required.

Limiting cases are most easily delineated in terms of dimensionless forms of the differential mass, momentum, and energy balances. To be specific, let the material be a compressible

Newtonian fluid with variable coefficients of viscosity, thermal conductivity, and heat capacity. We define the following dimensionless variables:

$$\rho^{\star} \equiv \frac{\rho}{\rho_0}, \qquad \mathbf{v}^{\star} \equiv \frac{\mathbf{v}}{v_0}, \ \ P^{\star} \equiv \frac{P}{P_0}$$

$$T^{\star} \equiv \frac{T}{T_0}, \qquad t^{\star} \equiv \frac{t}{t_0}, \ \ \mathbf{D}^{\star} \equiv \frac{L_0}{v_0}\mathbf{D}$$

$$\mathbf{S}^{\star} \equiv \frac{L_0}{\mu_0 v_0}\mathbf{S}, \ z_i^{\star} \equiv \frac{z_i}{L_0}, \ \mu^{\star} \equiv \frac{\mu}{\mu_0} \tag{6.4.0-1}$$

$$\lambda^{\star} \equiv \frac{\lambda}{\mu_0}, \qquad k^{\star} \equiv \frac{k}{k_0}, \ \hat{c}_V^{\star} \equiv \frac{\hat{c}_V}{\hat{c}_0}$$

$$\mathbf{f}^{\star} \equiv \frac{\mathbf{f}}{f_0}$$

Here the quantities distinguished by a subscript 0 are reference (or characteristic) quantities. For example, $v_0$ is a characteristic magnitude of velocity and $f_0$ might be the acceleration of gravity. The choice of the reference quantities in any particular problem is arbitrary; it is usually made so as to simplify the equations and boundary conditions as much as possible.

In terms of these dimensionless variables, the differential mass balance has the form

$$\frac{1}{N_{St}}\frac{\partial \rho^{\star}}{\partial t^{\star}} + \mathrm{div}\,(\rho^{\star}\mathbf{v}^{\star}) = 0 \tag{6.4.0-2}$$

The differential momentum balance for a compressible Newtonian fluid becomes

$$\frac{1}{N_{St}}\rho^{\star}\frac{\partial \mathbf{v}^{\star}}{\partial t^{\star}} + \rho^{\star}(\nabla \mathbf{v}^{\star}) \cdot \mathbf{v}^{\star} = -\frac{1}{N_{Ru}}\nabla P^{\star}$$

$$+ \frac{1}{N_{Re}}\mathrm{div}\,(2\mu^{\star}\mathbf{D}^{\star} + \lambda^{\star}\mathrm{div}\,\mathbf{v}^{\star}\mathbf{I}) + \frac{1}{N_{Fr}}\rho^{\star}\mathbf{f}^{\star} \tag{6.4.0-3}$$

In a similar way for the differential energy balance in Table 5.4.1-1, assuming there is no external or mutual energy transmission, we have

$$\frac{1}{N_{St}}\rho^{\star}\hat{c}_V^{\star}\frac{\partial T^{\star}}{\partial t^{\star}} + \rho^{\star}\hat{c}_V^{\star}(\nabla T^{\star}) \cdot \mathbf{v}^{\star} = \frac{1}{N_{Pe}}\mathrm{div}\,\left(k^{\star}\nabla T^{\star}\right)$$

$$- \frac{N_{Br}}{N_{Ru}N_{Pr}}T^{\star}\left(\frac{\partial P^{\star}}{\partial T^{\star}}\right)_{\hat{V}^{\star}}\mathrm{div}\,\mathbf{v}^{\star} + \frac{N_{Br}}{N_{Pe}}\mathrm{tr}\,\left(\mathbf{S}^{\star} \cdot \nabla \mathbf{v}^{\star}\right) \tag{6.4.0-4}$$

In these equations, we have defined the Strouhal, Ruark, Reynolds, Froude, Prandtl, Peclet, and Brinkman numbers as

$$N_{St} \equiv \frac{t_0 v_0}{L_0}, \qquad N_{Ru} \equiv \frac{\rho_0 v_0{}^2}{P_0}$$

$$N_{Re} \equiv \frac{\rho_0 v_0 L_0}{\mu_0}, \qquad N_{Fr} \equiv \frac{v_0{}^2}{f_0 L_0}$$

$$N_{Pr} \equiv \frac{\hat{c}_0 \mu_0}{k_0}, \qquad N_{Pe} \equiv N_{Re} N_{Pr} = \frac{\rho_0 \hat{c}_0 v_0 L_0}{k_0}$$

$$(6.4.0\text{-}5)$$

$$N_{Br} \equiv \frac{\mu_0 v_0{}^2}{k_0 T_0}$$

(The ratio $N_{Br}/N_{Pe}$ is sometimes referred to as the *Eckert number*.)

For sufficiently small values of $N_{Br}/N_{Pe}$, one should be able to neglect the effects of viscous dissipation with respect to convection in (6.4.0-4):

$$\frac{1}{N_{St}} \rho^\star \hat{c}_V^\star \frac{\partial T^\star}{\partial t^\star} + \rho^\star \hat{c}_V^\star (\nabla T^\star) \cdot \mathbf{v}^\star$$
$$= \frac{1}{N_{Pe}} \text{div}(k^\star \nabla T^\star) - \frac{N_{Br}}{N_{Ru} N_{Pr}} T^\star \left( \frac{\partial P^\star}{\partial T^\star} \right)_{\hat{V}^\star} \text{div } \mathbf{v}^\star \qquad (6.4.0\text{-}6)$$

Although this is an intuitive argument, it should not lead to any contradictions such as the one noted in the introduction to Section 3.4.

### 6.4.1   Natural Convection

By *natural convection*, we refer to a flow in a gravitational field resulting from a density gradient, which in turn is created either by a temperature gradient or a concentration gradient. For example, a common method of heating a home is to use a radiator through which either hot water or steam circulates. As the temperature of the air adjacent to the radiator rises, its density decreases, and it rises. In this way, a slow circulation of air past the radiator warms the rest of the room.

### Theory

We begin with a general treatment of natural convection, restricting ourselves to situations where density may be considered to be a function only of temperature and not of pressure. We shall assume that the natural convection arises as the result of a temperature difference. (Natural convection resulting from concentration gradients could be developed in the context of Chapter 9 in a very similar fashion.)

Before proceeding, we must specify the dependence of density upon temperature. Let us write $\rho$ as a Taylor series with respect to some reference temperature $T_0$:

$$\rho = \rho \big|_{T=T_0} + \left( \frac{d\rho}{dT} \right)_{T=T_0} (T - T_0) + \cdots \qquad (6.4.1\text{-}1)$$

We know that

$$\frac{d\rho}{dT} = \frac{d}{dT} \left( \frac{1}{\hat{V}} \right) = -\frac{1}{\hat{V}^2} \frac{d\hat{V}}{dT} = -\rho \beta \qquad (6.4.1\text{-}2)$$

where the quantity

$$\beta \equiv \frac{1}{\hat{V}} \frac{d\hat{V}}{dT} \tag{6.4.1-3}$$

is known as the *coefficient of volume expansion*. This suggests defining

$$\rho_0 \equiv \rho \big|_{T=T_0} \tag{6.4.1-4}$$

and

$$\beta_0 \equiv \beta \big|_{T=T_0} \tag{6.4.1-5}$$

and rearranging (6.4.1-1) as

$$\rho^\star \equiv \frac{\rho}{\rho_0}$$
$$= 1 - (\beta_0 \Delta T)T^\star + \cdots \tag{6.4.1-6}$$

Here we have introduced

$$T^\star \equiv \frac{T - T_0}{\Delta T} \tag{6.4.1-7}$$

where the temperature difference $\Delta T$ should be defined in the context of a particular problem.

In taking only the first two terms on the right of (6.4.1-6), our discussion is restricted to small values of $\beta_0 \Delta T$. For a wide range of liquids (Dean 1979, pp. 10–127), $\beta_0 \approx 10^{-3}\,^\circ\mathrm{C}^{-1}$. This means that $\Delta T$ probably could be as large as $10^2\,^\circ\mathrm{C}$ without seriously affecting the accuracy of this result.

Substituting (6.4.1-6) in (6.4.0-2), we find

$$\frac{1}{N_{St}} \frac{\partial}{\partial t^\star} \left(1 - \beta_0 \Delta T T^\star + \cdots\right) + \mathrm{div}\left(\left[1 - \beta_0 \Delta T T^\star + \cdots\right]\mathbf{v}^\star\right) = 0 \tag{6.4.1-8}$$

In the limit $\beta_0 \Delta T \ll 1$, the differential mass balance (6.4.1-8) reduces to

$$\mathrm{div}\,\mathbf{v}^\star = 0 \tag{6.4.1-9}$$

In this limit, the fluid behaves as though it were incompressible.

In a similar manner, we can substitute (6.4.1-6) in (6.4.0-3) to find

$$[1 - \beta_0 \Delta T T^\star + \cdots]\left[\frac{1}{N_{St}}\frac{\partial \mathbf{v}^\star}{\partial t^\star} + \nabla \mathbf{v}^\star \cdot \mathbf{v}^\star\right] = -\frac{1}{N_{Ru}}\nabla \mathcal{P}^\star$$

$$+ \frac{1}{N_{Re}}\,\mathrm{div}(2\mu^\star \mathbf{D}^\star + \lambda^\star[\mathrm{div}\,\mathbf{v}^\star]\mathbf{I}) - \frac{N_{Gr}}{N_{Re}^2}T^\star \mathbf{f}^\star + \cdots \tag{6.4.1-10}$$

where

$$\mathcal{P} \equiv P + \rho_0 \phi \tag{6.4.1-11}$$

and

$$\mathbf{f}^\star \equiv \frac{\mathbf{f}}{g} = -\frac{1}{g}\nabla \phi \tag{6.4.1-12}$$

Here $N_{Gr}$ is the Grashof number

$$N_{Gr} \equiv \frac{\beta_0 \Delta T \, g L_0^3 \rho_0^2}{\mu_0^2} \qquad (6.4.1\text{-}13)$$

and $g$ is the acceleration of gravity. In arriving at this result, we have recognized that

$$-\frac{1}{N_{Ru}} \nabla P^\star + \frac{1}{N_{Fr}} \rho^\star \mathbf{f}^\star$$

$$= -\frac{P_0}{\rho_0 v_0^2} \nabla P^\star + \left( \frac{L_0}{\rho_0 v_0^2} \right) \rho \mathbf{f}$$

$$= -\frac{P_0}{\rho_0 v_0^2} \nabla P^\star + \left( \frac{L_0}{\rho_0 v_0^2} \right) \left( \rho_0 - \rho_0 \beta_0 \Delta T \, T^\star \ldots \right) (-\nabla \phi)$$

$$= -\frac{1}{N_{Ru}} \nabla \mathcal{P}^\star - \left( \frac{L_0 g \beta_0 \Delta T}{v_0^2} \right) T^\star \mathbf{f}^\star$$

$$= -\frac{1}{N_{Ru}} \nabla \mathcal{P}^\star - \frac{N_{Gr}}{N_{Re}^2} T^\star \mathbf{f}^\star \qquad (6.4.1\text{-}14)$$

It will be important to recognize that $N_{Gr}$ may be large, even though $\beta_0 \Delta T \ll 1$.

Our discussion here will be confined to the limit $\beta_0 \Delta T \ll 1$. But before we can take this limit, we must settle two points. First, we must ensure that, in this limit, the last term on the right of (6.4.1-10) is retained, since this is the driving force for the flow. Second, we must define $v_0$, because a characteristic speed will normally not arise in the boundary conditions for a natural convection problem. Both of these difficulties are resolved if we define $v_0$ by requiring

$$N_{Re} = N_{Gr}^{1/2} \qquad (6.4.1\text{-}15)$$

in which case

$$v_0 \equiv (\beta_0 \Delta T \, g L_0)^{1/2} \qquad (6.4.1\text{-}16)$$

With this understanding, the differential momentum balance for a Newtonian fluid (6.4.1-10) reduces in the limit $\beta_0 \Delta T \ll 1$ to

$$\frac{1}{N_{St}} \frac{\partial \mathbf{v}^\star}{\partial t^\star} + \left( \nabla \mathbf{v}^\star \right) \cdot \mathbf{v}^\star = -\frac{1}{N_{Ru}} \nabla \mathcal{P}^\star + \frac{1}{N_{Gr}^{1/2}} \operatorname{div} \left( 2\mu^\star \mathbf{D}^\star \right) - T^\star \mathbf{f}^\star \qquad (6.4.1\text{-}17)$$

Similarly, in the limit $\beta_0 \Delta T \ll 1$, Equation (6.4.0-4) reduces to

$$\frac{1}{N_{St}} \hat{c}_V^\star \frac{\partial T^\star}{\partial t^\star} + \hat{c}_V^\star \nabla T^\star \cdot \mathbf{v}^\star = \frac{1}{N_{Pr} N_{Gr}^{1/2}} \operatorname{div}(k^\star \nabla T^\star) \qquad (6.4.1\text{-}18)$$

with the additional constraint that

$$\frac{N_{Br}}{N_{Pr} N_{Gr}^{1/2}} = \frac{\mu \beta_0^{1/2} g^{1/2}}{\rho_0 \hat{c}_0 (\Delta T)^{1/2} L_0^{1/2}}$$

$$\ll 1 \qquad (6.4.1\text{-}19)$$

In the limit of (6.4.1-19), we can neglect the effects of viscous dissipation with respect to the effects of convection, which in most natural convection problems would be an excellent approximation.

### Natural Convection Between Vertical Heated Plates

As an example of how this theory can be used, consider steady-state natural convection between the two vertical parallel plates shown in Figure 6.4.1-1. We do not consider the circulation patterns near the top and bottom of the channel formed by these plates. If we say that $T_2 > T_1$, we visualize the warmer fluid on the left rising and the cooler fluid on the right descending. We recognize that the circulation arises because the density of the warmer fluid is less than that of the cooler fluid. We must explicitly recognize the Newtonian fluid to be compressible. The coefficients of viscosity and the thermal conductivity are assumed to be constants.

We are given two boundary conditions on velocity:

$$\text{at } z_2^\star \equiv \frac{z_2}{b} = \pm 1: \ \mathbf{v}^\star = \mathbf{0} \tag{6.4.1-20}$$

and two boundary conditions on temperature:

$$\text{at } z_2 = b: \ T^\star = \frac{T_1 - T_0}{\Delta T} \tag{6.4.1-21}$$

$$\text{at } z_2 = -b: \ T^\star = \frac{T_2 - T_0}{\Delta T} \tag{6.4.1-22}$$

If we visualize this very long channel between the two plates to be closed by plates at either end, there is no net flow either up or down in the channel. This gives us a further condition to be met:

$$\text{for all } z_1^\star: \ \int_{-1}^{1} v_1^\star \, dz_2^\star = 0 \tag{6.4.1-23}$$

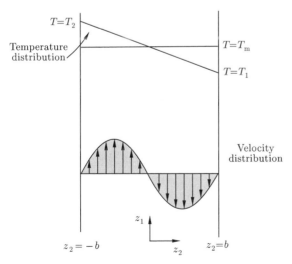

**Figure 6.4.1-1.** Natural convection between vertical heated plates.

The boundary conditions on temperature and the velocity distribution intuitively sketched in Figure 6.4.1-1 suggest that there is only one nonzero component of velocity, $v_1$, and that both $v_1$ and $T$ are functions only of $z_2$:

$$v_2^\star = v_3^\star$$
$$= 0$$
$$v_1^\star = v_1^\star \left( z_2^\star \right)$$
$$T^\star = T^\star \left( z_2^\star \right)$$

(6.4.1-24)

This form of a velocity distribution satisfies the differential mass balance (6.4.1-9) identically.

The three components of the differential momentum balance (6.4.1-17) reduce to

$$0 = -\frac{1}{N_{Ru}} \frac{\partial \mathcal{P}^\star}{\partial z_1^\star} + \frac{1}{N_{Gr}^{1/2}} \frac{d^2 v_1^\star}{dz_2^{\star 2}} + T^\star$$

(6.4.1-25)

and

$$\frac{\partial \mathcal{P}^\star}{\partial z_2^\star} = \frac{\partial \mathcal{P}^\star}{\partial z_3^\star} = 0$$

(6.4.1-26)

where we have specified the characteristic pressure $P_0$ by setting $N_{Ru} = 1$.

The differential energy balance (6.4.1-18) becomes

$$\frac{d^2 T^\star}{dz_2^{\star 2}} = 0$$

(6.4.1-27)

Integrating this twice consistent with (6.4.1-21) and (6.4.1-22) and defining

$$T_0 \equiv \frac{T_1 + T_2}{2}$$

(6.4.1-28)

$$\Delta T \equiv T_2 - T_1$$

(6.4.1-29)

we find

$$T^\star = -\frac{z_2^\star}{2}$$

(6.4.1-30)

From (6.4.1-25), (6.4.1-26), and (6.4.1-30), we have

$$\frac{1}{N_{Ru}} \frac{\partial \mathcal{P}^\star}{\partial z_1^\star} = \frac{1}{N_{Gr}^{1/2}} \frac{d^2 v_1^\star}{dz_2^{\star 2}} - \frac{z_2^\star}{2}$$
$$= C_1$$
$$= \text{a constant}$$

(6.4.1-31)

Integrating this twice consistent with (6.4.1-20), we have

$$\frac{v_1^\star}{N_{Gr}^{1/2}} = \frac{1}{12} \left( z_2^{\star 3} - z_2^\star \right) + \frac{1}{2} C_1 \left( z_2^{\star 2} - 1 \right)$$

(6.4.1-32)

In view of (6.4.1-23), we find

$$C_1 = 0$$

(6.4.1-33)

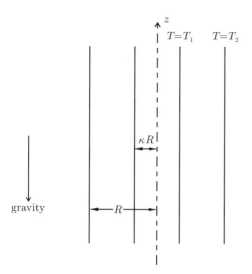

**Figure 6.4.1-2.** Natural convection between concentric vertical heated cylinders.

and

$$\frac{v_1^\star}{N_{Gr}^{1/2}} = \frac{1}{12}\left(z_2^{\star 3} - z_2^\star\right) \tag{6.4.1-34}$$

or (Bird et al. 1960, p. 300)

$$v_1 = \frac{\rho_0 \beta_0 g b^2 \Delta T}{12\mu_0}\left(z_2^{\star 3} - z_2^\star\right) \tag{6.4.1-35}$$

**Exercise 6.4.1-1** *Concentric vertical heated cylinders*    Repeat the problem discussed in the text for natural convection between concentric vertical cylinders as described in Figure 6.4.1-2. The temperature of the inner wall (at $r = \kappa R$) is $T_1$; the temperature of the outer wall (at $r = R$) is $T_2$. Again, assume that the fluid is a Newtonian liquid whose density is a function only of temperature.

*Answer:*

$$\frac{v_z \rho_0 R}{\mu} = \left[\left(\frac{r}{R}\right)^2 - 1\right](A - B) - \frac{\ln(r/R)}{\ln\kappa}\left[A\left(\kappa^2 - 1\right)\right.$$

$$\left. + B\left(\kappa^2 \ln\kappa - \kappa^2 + 1\right)\right] + B\left(\frac{r}{R}\right)^2 \ln\left(\frac{r}{R}\right)$$

where

$$A \equiv \frac{1}{4}\left(\frac{R^3\rho_0}{\mu^2}\frac{dP}{dz} + \frac{R^3\rho_0^2 g}{\mu^2}\right)$$

$$= B\frac{\left[\left(7\kappa^2/4 + 3/4\right)\left(\kappa^2 - 1\right) - \kappa^4\ln\kappa - \left(\kappa^2 - 1\right)^2/\ln\kappa\right]}{\left[\kappa^4 - 1 - \left(\kappa^2 - 1\right)^2/\ln\kappa\right]}$$

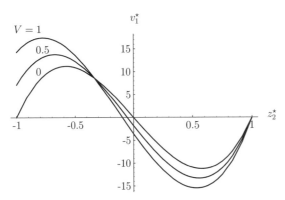

**Figure 6.4.1-3.** The dimensionless velocity component $v_1^\star$ as a function of $z_2^\star$ for water, $\beta_0 = 10^{-3}$ K$^{-1}$, $T_1 = 300$ K, $T_2 = 400$ K (and an appropriately elevated pressure), $L_0 \equiv b = 0.5$ cm, $v_0 = 0.07$ m/s, and $V = 0, 0.5, 1$ m/s.

and

$$B \equiv \frac{\beta_0(T_2 - T_1)R^3\rho_0^2 g}{4\mu^2 \ln \kappa}$$

The parameters $\rho_0$ and $\beta_0$ are evaluated at $T_2$.

**Exercise 6.4.1-2** *Relative effects of natural and forced convection in a channel* (with M. W. Vaughn) Let us again consider the flow shown in Figure 6.4.1-1. Adopt the same boundary conditions as in the text with the exception that

at $z_2 = -b$ : $v_1 = V$, $v_2 = v_3 = 0$

Neglecting the effects of viscous dissipation, determine the temperature and velocity distributions in the channel. The results for a particular case are shown in Figure 6.4.1-3.

**Exercise 6.4.1-3** *Natural convection, forced convection, and viscous dissipation* (with M. W. Vaughn) Reconsider Exercise 6.4.1-2, now constructing a perturbation analysis to take into account the effects of viscous dissipation on the velocity and temperature distributions.

Although the temperature distribution shown in Figure 6.4.1-4 is strongly affected by viscous dissipation, it has no discernible effect on the velocity distribution in Figure 6.4.1-5. Because $N_{Gr}$ is so small, there are negligible effects of the temperature distribution upon the velocity distribution. It appears that, for any flow in which the effects of viscous dissipation are important, the effects of natural convection can be neglected with respect to those of forced convection.

*Hint:* I encourage you to use one of the symbolic operator programs available to develop your solution.

**Exercise 6.4.1-4** *More about natural convection, forced convection, and viscous dissipation* (with M. W. Vaughn) Develop a full numerical solution for the problem described in Exercise 6.4.1-3 rather than a perturbation solution.

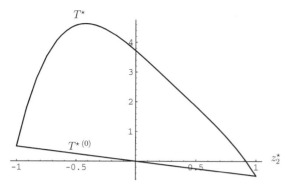

**Figure 6.4.1-4.** $T^\star$ correct to the zeroth and first perturbations in $N_{Br}$ (the latter cannot be distinguished from the full numerical solution) as functions of $z_2^\star$ for a very viscous water ($\mu = 1$ Pa s), $\beta_0 = 10^{-3}$ K$^{-1}$, $T_1 = 300$ K, $T_2 = 400$ K (and an appropriately elevated pressure), $L_0 \equiv b = 0.5$ cm, $V = 50$ m/s, $v_0 = 0.07$ m/s, and $N_{Br} = 2.33 \times 10^{-5}$.

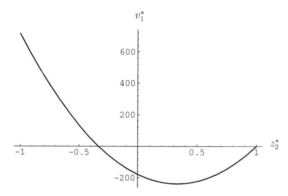

**Figure 6.4.1-5.** $v_1^\star$ correct to the zeroth perturbation in $N_{Br}$ (which cannot be distinguished from the full numerical solution) as a function of $z_2^\star$ for a very viscous water ($\mu = 1$ Pa s), $\beta_0 = 10^{-3}$ K$^{-1}$, $T_1 = 300$ K, $T_2 = 400$ K (and an appropriately elevated pressure), $L_0 \equiv b = 0.5$ cm, $V = 50$ m/s, $v_0 = 0.07$ m/s, and $N_{Br} = 2.33 \times 10^{-5}$.

Consider the results shown in Figures 6.4.1-4 and 6.4.1-5. Although viscous dissipation has a dramatic effect on the temperature profile, the velocity distribution is unaffected.

Note that, because the perturbation parameter $N_{Br} = 2.33 \times 10^{-5}$ is very small, the first-order perturbation solution for $T^\star$ in Figure 6.4.1-4 is indistinguishable from the full numerical solution.

**Exercise 6.4.1-5** *More about natural convection between vertical heated plates* In what way has natural convection enhanced the energy transfer between parallel plates described in the text?

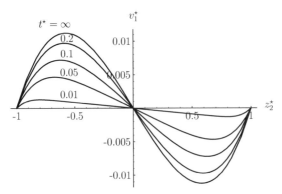

**Figure 6.4.1-6.** $v_1^\star$ as a function of $z_2^\star$ for $t^\star \equiv t\mu/\left(b^2\rho\right)$
$= 0.01, 0.05, 0.1, 0.2, \infty$.

**Exercise 6.4.1-6** *More about natural convection between vertical heated plates*    For time $t < 0$, the Newtonian fluid in the vertical channel shown in Figure 6.4.1-1 is stationary, and the temperature of the system is $T_1$. For $t > 0$, the temperature of the left wall is $T_2$. We would like to determine the temperature and velocity distributions as functions of time. Unfortunately, this is a relatively difficult problem. To simplify the computation, let us assume that the steady-state temperature distribution is achieved instantaneously and that all physical properties are constants. The resulting velocity profiles are shown in Figure 6.4.1-6.

*Hint:*    Review Exercise 3.2.4-6.

## 6.5    No Convection

Let us begin by restricting ourselves to incompressible fluids and in this way rule out the possibility of natural convection (Section 6.4.1).

Referring to the dimensionless form of the differential energy balance displayed in the introduction to Section 6.4,

$$\frac{1}{N_{St}}\hat{c}^\star\frac{\partial T^\star}{\partial t^\star} + \hat{c}^\star(\nabla T^\star)\cdot\mathbf{v}^\star = \frac{1}{N_{Pe}}\,\text{div}\,(k^\star\nabla T^\star) + \frac{N_{Br}}{N_{Pe}}\,\text{tr}(\mathbf{S}^\star\cdot\nabla\mathbf{v}^\star) \qquad (6.5.0\text{-}1)$$

we see that, in the limit as $N_{Pe} \ll 1$ for arbitrary values of $N_{St}$ and $N_{Br}/N_{Pe}$, it is intuitively appealing to neglect convection (as represented by the second term on the left) with respect to conduction (described by the first term on the right). For this limiting case, the differential energy balance reduces to

$$\frac{1}{N_{St}}\hat{c}^\star\frac{\partial T^\star}{\partial t^\star} = \frac{1}{N_{Pe}}\,\text{div}\,(k^\star\nabla T^\star) + \frac{N_{Br}}{N_{Pe}}\,\text{tr}(\mathbf{S}^\star\cdot\nabla\mathbf{v}^\star) \qquad (6.5.0\text{-}2)$$

When neglecting convection with respect to conduction, it is also common to neglect viscous dissipation. For the limiting case, where both $N_{Re} \ll 1$ and $N_{Br}/N_{Re} \ll 1$ for a fixed value of $N_{Pr}$, the energy balance reduces to

$$\frac{1}{N_{St}} \hat{c}^\star \frac{\partial T^\star}{\partial t^\star} = \frac{1}{N_{Pe}} \text{ div}(k^\star \nabla T^\star) \tag{6.5.0-3}$$

An interesting aspect of this limiting case is that the energy balance has the same form as that used to describe conduction in solids. Consequently, all of the problems discussed in the context of conduction in solids (Sections 6.2) can be understood to be applicable in this limit to fluids as well.

It is important to realize that our discussion here has a certain degree of artificiality associated with it. We began by assuming that we were dealing only with incompressible fluids. In effect this meant that we were ruling out the possibility that density might depend upon temperature. However, the densities of all real fluids are dependent upon temperature to some degree.

Let us think a little further about the type of problem discussed in Section 6.4.1. There is no forced convection and yet there is motion in the fluid (natural convection) as the result of the temperature dependence of the fluid's density. Since there is no forced convection, the characteristic velocity of the problem is chosen by setting $N_{Re} = (N_{Gr})^{1/2}$. Assuming that

$$\frac{N_{Br}}{N_{Pr}(N_{Gr})^{1/2}} \ll 1$$

that density is a function of temperature, and that $\beta_0 \Delta T \ll 1$, we can consider the limiting case as $N_{Pe} = N_{Pr}(N_{Gr})^{1/2} \ll 1$ for arbitrary values of $N_{St}$. Under these circumstances the dimensionless energy balance of Section 6.4.1 reduces to

$$\frac{1}{N_{St}} \hat{c} \frac{\partial T^\star}{\partial t^\star} = \frac{1}{N_{Pr}(N_{Gr})^{1/2}} \text{ div}(k^\star \nabla T^\star) \tag{6.5.0-4}$$

**Exercise 6.5.0-1** *Conduction from a sphere to a stagnant fluid* (Bird et al. 1960, p. 303)   A heated sphere of diameter $D$ is suspended in a large body of an incompressible fluid (the temperature dependence of density is neglected). The temperature of the sphere is maintained at $T_0$; the temperature of the fluid at the large distance from the sphere is known to be $T_\infty$. The thermal conductivity $k$ of the fluid may be assumed to be constant.

i) Determine the temperature distribution in the fluid.

ii) Use this temperature distribution to obtain an expression for the energy flux at the surface of the sphere. Equate this result to the expression for the energy flux written in terms of Newton's "law" of cooling (Section 6.2.2). Show that the Nusselt number

$$N_{Nu} \equiv \frac{hD}{k} = 2$$

This is a well-known result in agreement with experimental data in the limit $(N_{Gr})^{1/2} \ll 1$ for a fixed value of $N_{Pr}$ (Bird et al. 1960, p. 409).

**Exercise 6.5.0-2** *More about freezing (or melting) a layer of fluid (or solid): a limiting case*   An incompressible fluid $A$ is bounded by two stationary walls that are separated by a distance $2L$. The system, which is at a uniform pressure, is initially at the melting point $T_M$. For time $t > 0$, the temperature of these walls is changed to $T_1 < T_M$ (or $T_1 > T_M$), the fluid begins to freeze (or the solid begins to melt), and a layer of solid (or liquid) $B$ forms at both walls. For the sake of simplicity, neglect the difference between the fluid and solid densities. Consider the limiting case such that $N_{Pe} \ll 1$. For this limiting case, determine the time $t_f$ required to freeze the fluid.

*Hint:*   Write the differential energy balance for phase $B$ and the jump energy balance in dimensionless forms. Choose the characteristic time $t_0$ by requiring the Strouhal number $N_{St} = 1$, and define the characteristic speed $v_0$ in such a way as to simplify the form of the jump energy balance.

*Answer:*

$$t_f = \frac{\rho \Delta \hat{H} L^2}{2k^{(B)} (T_M - T_1)}$$

**Exercise 6.5.0-3** *Still more about freezing (or melting) a layer of fluid (or solid)*   Repeat the preceding exercise for a cylindrical tube of radius $R$ filled with an incompressible fluid $A$.

*Answer:*

$$t_f = \frac{\rho \Delta \hat{H} R^2}{4k^{(B)} (T_M - T_1)}$$

---

## 6.6    No Conduction

In Section 6.5.1, we considered the limit $N_{Pe} \ll 1$ or $N_{Re} \ll 1$ for a fixed value of $N_{Pr}$ (i.e., for a particular material) and for arbitrary values of $N_{St}$ and $N_{Br}/N_{Pe}$. Now let us go to the other extreme and study the limit $N_{Re} \gg 1$, again for a fixed value of $N_{Pr}$.

In terms of the dimensionless form of the differential energy balance presented in the introduction to Section 6.4,

$$\frac{1}{N_{St}} \rho^\star \hat{c}_V^\star \frac{\partial T^\star}{\partial t^\star} + \rho^\star \hat{c}_V^\star (\nabla T^\star) \cdot \mathbf{v}^\star = \frac{1}{N_{Pr} N_{Re}} \text{div} (k^\star \nabla T^\star)$$

$$- \frac{N_{Br}}{N_{Ru} N_{Pr}} T^\star \left( \frac{\partial P^\star}{\partial T^\star} \right)_{\hat{V}^\star} \text{div} \mathbf{v}^\star + \frac{N_{Br}}{N_{Pr} N_{Re}} \text{tr}(\mathbf{S}^\star \cdot \nabla \mathbf{v}^\star) \qquad (6.6.0\text{-}1)$$

This limit suggests that the first term on the right, representing conduction, can be neglected with respect to the second term on the left, representing convection. As a result of this intuitive argument, the energy balance is simplified to

$$\frac{1}{N_{St}} \rho^\star \hat{c}_V^\star \frac{\partial T^\star}{\partial t^\star} + \rho^\star \hat{c}_V^\star (\nabla T^\star) \cdot \mathbf{v}^\star$$

$$= - \frac{N_{Br}}{N_{Ru} N_{Pr}} T^\star \left( \frac{\partial P^\star}{\partial T^\star} \right)_{\hat{V}^\star} \text{div} \mathbf{v}^\star + \frac{N_{Br}}{N_{Pe}} \text{tr} (\mathbf{S}^\star \cdot \nabla \mathbf{v}^\star) \qquad (6.6.0\text{-}2)$$

This neglect of conduction with respect to convection should remind you of potential flow discussed in Section 3.4, where the viscous terms in the differential momentum balance were neglected with respect to the convection inertial terms. In the introduction to Section 3.4, we found that this limiting case was not capable of representing important aspects of real fluid behavior, at least in the neighborhood of boundary surfaces. This was not altogether surprising, since all the second-derivative terms representing viscous effects had been dropped with a resulting reduction in the order of the differential equations to be solved for the velocity distribution.

The situation is not so different here, where conduction is neglected with respect to convection to arrive at (6.6.0-2). All the second derivatives of temperature are dropped from the differential energy balance with a resulting reduction in the order of the differential equation. On the basis of our discussion in Section 3.4, we can anticipate that (6.6.0-2) will not allow for a realistic description of the temperature distribution in a fluid near its bounding surfaces. By analogy with the applications found for potential flow, we might expect (6.6.0-2) to be used to describe the temperature distribution in fluids outside the immediate neighborhood of their bounding surfaces.

## 6.6.1 Speed of Propagation of Sound Waves

For the moment let us define the speed of sound to be the speed of propagation of pressure waves resulting from a small-amplitude disturbance in a compressible fluid.

If $a$ is the dimensionless amplitude of the disturbance, we wish to determine the speed of the propagation of pressure waves in the limit $a \ll 1$. When $a = 0$, the temperature, pressure, and density assume uniform values throughout the fluid:

$$T = T^{(0)} = \text{a constant}$$

$$P = P^{(0)} = \text{a constant}$$

$$\rho = \rho^{(0)} = \text{a constant} \tag{6.6.1-1}$$

$$\mathbf{v} = \mathbf{0}$$

Our analysis must be based upon the differential mass, momentum, and energy balances. In terms of dimensionless variables, these three equations become

$$\frac{1}{N_{St}} \frac{\partial \rho^\star}{\partial t^\star} + \text{div} \, (\rho^\star \mathbf{v}^\star) = 0 \tag{6.6.1-2}$$

$$\frac{1}{N_{St}} \rho^\star \frac{\partial \mathbf{v}^\star}{\partial t^\star} + \rho^\star \nabla \mathbf{v}^\star \cdot \mathbf{v}^\star = -\frac{1}{N_{Ru}} \nabla P^\star + \frac{1}{N_{Re}} \text{div} \, \mathbf{S}^\star + \frac{1}{N_{Fr}} \rho^\star \mathbf{f}^\star \tag{6.6.1-3}$$

and

$$\frac{1}{N_{St}} \rho^\star \hat{c}_P^\star \frac{\partial T^\star}{\partial t^\star} + \rho^\star \hat{c}_P^\star \nabla T^\star \cdot \mathbf{v}^\star$$

$$= \frac{1}{N_{Pr} N_{Re}} \text{div}(k^\star \nabla T^\star) + \frac{N_{Br}}{N_{St} N_{Ru} N_{Pr}} \left( \frac{\partial \ln \hat{V}^\star}{\partial \ln T^\star} \right)_{P^\star} \frac{\partial P^\star}{\partial t^\star}$$

$$+ \frac{N_{Br}}{N_{Ru} N_{Pr}} \left( \frac{\partial \ln \hat{V}^\star}{\partial \ln T^\star} \right)_{P^\star} \nabla P^\star \cdot \mathbf{v}^\star + \frac{N_{Br}}{N_{Pe}} \text{tr}(\mathbf{S}^\star \cdot \nabla \mathbf{v}^\star) \tag{6.6.1-4}$$

If our characteristic length is chosen to be representative of the macroscopic system and if our characteristic velocity is the speed of sound, it seems reasonable to confine our attention to the limit $N_{Re} \gg 1$, $N_{Fr} \gg 1$, and $N_{Br}/N_{Re} \ll 1$ for arbitrary values of $N_{St}$, $N_{Ru}$, $N_{Fr}$, and

$N_{Br}$ and a fixed value of $N_{Pr}$ (a specified material). In this limit, (6.6.1-2) to (6.6.1-4) become in dimensional form

$$\frac{d_{(m)}\rho}{dt} + \rho \, \text{div} \, \mathbf{v} = 0 \tag{6.6.1-5}$$

$$\rho \frac{d_{(m)}\mathbf{v}}{dt} = -\nabla P \tag{6.6.1-6}$$

and

$$\rho \hat{c}_P \frac{d_{(m)}T}{dt} = \left(\frac{\partial \ln \hat{V}}{\partial \ln T}\right)_P \frac{d_{(m)}P}{dt} \tag{6.6.1-7}$$

We wish to obtain a solution to these equations valid in the limit $a \ll 1$. This suggests that we carry out a perturbation analysis as we did in Section 6.3.2. We begin by expressing $T, P, \rho$, and $\mathbf{v}$ as Taylor series:

$$T = T^{(0)} + aT^{(1)} + a^2 T^{(2)} + \cdots \tag{6.6.1-8}$$

$$P = P^{(0)} + aP^{(1)} + a^2 P^{(2)} + \cdots \tag{6.6.1-9}$$

$$\rho = \rho^{(0)} + a\rho^{(1)} + a^2 \rho^{(2)} + \cdots \tag{6.6.1-10}$$

and

$$\mathbf{v} = a\mathbf{v}^{(1)} + a^2 \mathbf{v}^{(2)} + \cdots \tag{6.6.1-11}$$

Substituting these series into (6.6.1-7) and looking only at the coefficient of $a$, we find

$$\rho^{(0)} \hat{c}_P^{(0)} \frac{\partial T^{(1)}}{\partial t} = \left(\frac{\partial \ln \hat{V}}{\partial \ln T}\right)_{P^{(0)}} \frac{\partial P^{(1)}}{\partial t} \tag{6.6.1-12}$$

where we define

$$\left(\frac{\partial \ln \hat{V}}{\partial \ln T}\right)_{P^{(0)}} \equiv \left(\frac{\partial \ln \hat{V}}{\partial \ln T}\right)_P \bigg|_{\substack{T=T^{(0)} \\ P=P^{(0)}}} \tag{6.6.1-13}$$

and

$$\hat{c}_P^{(0)} \equiv \hat{c}_P \big|_{\substack{T=T^{(0)} \\ P=P^{(0)}}} \tag{6.6.1-14}$$

We also know from (5.3.1-12) and (5.3.1-30) that

$$\frac{\partial P}{\partial t} = \left(\frac{\partial P}{\partial \rho}\right)_T \frac{\partial \rho}{\partial t} + \left(\frac{\partial P}{\partial T}\right)_\rho \frac{\partial T}{\partial t} \tag{6.6.1-15}$$

the first perturbation of which says

$$\frac{\partial P^{(1)}}{\partial t}\left[1 - \left(\frac{\partial P}{\partial T}\right)_{\rho^{(0)}} \frac{\partial T^{(1)}/\partial t}{\partial P^{(1)}/\partial t}\right] = \left(\frac{\partial P}{\partial \rho}\right)_{T^{(0)}} \frac{\partial \rho^{(1)}}{\partial t} \tag{6.6.1-16}$$

Equations (6.6.1-12) and (6.6.1-16) may now be combined to tell us

$$\frac{\partial P^{(1)}/\partial t}{\partial \rho^{(1)}/\partial t} = \left(\frac{\partial P}{\partial \rho}\right)_{T^{(0)}} \frac{\rho^{(0)}\hat{c}_P^{(0)}}{\rho^{(0)}\hat{c}_P^{(0)} - (\partial P/\partial T)_{\rho^{(0)}}(\partial \ln \hat{V}/\partial \ln T)_{P^{(0)}}} \tag{6.6.1-17}$$

Since (see Exercise 8.4.2-7)

$$\rho \hat{c}_P - \left(\frac{\partial P}{\partial T}\right)_\rho \left(\frac{\partial \ln \hat{V}}{\partial \ln T}\right)_P = \rho \hat{c}_V \tag{6.6.1-18}$$

Equation (6.6.1-17) may finally be rearranged in what will prove to be a more interesting form:

$$\frac{\partial P^{(1)}/\partial t}{\partial \rho^{(1)}/\partial t} = \gamma^{(0)} \left(\frac{\partial P}{\partial \rho}\right)_{T^{(0)}}$$

$$= (v_s)^2 \tag{6.6.1-19}$$

We have introduced as definitions here

$$\gamma_0 \equiv \frac{\hat{c}_P^{(0)}}{\hat{c}_V^{(0)}} \tag{6.6.1-20}$$

and

$$v_s \equiv \sqrt{\gamma^{(0)} \left(\frac{\partial P}{\partial \rho}\right)_{T^{(0)}}} \tag{6.6.1-21}$$

The first perturbation of (6.6.1-5) yields

$$\frac{\partial \rho^{(1)}}{\partial t} + \rho^{(0)} \operatorname{div} \mathbf{v}^{(1)} = 0 \tag{6.6.1-22}$$

We can use (6.6.1-19) to write this as

$$\frac{\partial P^{(1)}}{\partial t} + \rho^{(0)} {v_s}^2 \operatorname{div} \mathbf{v}^{(1)} = 0 \tag{6.6.1-23}$$

Upon differentiating with respect to time, we find that this becomes

$$\frac{\partial^2 P^{(1)}}{\partial t^2} + \rho^{(0)} {v_s}^2 \operatorname{div} \frac{\partial \mathbf{v}^{(1)}}{\partial t} = 0 \tag{6.6.1-24}$$

The second term in this last equation suggests that we look at the first perturbation of (6.6.1-6):

$$\rho^{(0)} \frac{\partial \mathbf{v}^{(1)}}{\partial t} = -\nabla P^{(1)} \tag{6.6.1-25}$$

From (6.6.1-24) and (6.6.1-25), we see that the first perturbation in pressure is a solution of the wave equation:

$$\frac{\partial^2 P^{(1)}}{\partial t^2} - {v_s}^2 \operatorname{div} \nabla P^{(1)} = 0 \tag{6.6.1-26}$$

The discussion at this point is clarified if we restrict ourselves to a one-dimensional pressure wave in rectangular Cartesian coordinates:

$$P^{(1)} = P^{(1)}(t, z_1) \tag{6.6.1-27}$$

For this case, (6.6.1-26) reduces to

$$\frac{\partial^2 P^{(1)}}{\partial t^2} - v_s^2 \frac{\partial^2 P^{(1)}}{\partial z_1^2} = 0 \tag{6.6.1-28}$$

If we introduce as changes of variable

$$\xi \equiv z_1 - v_s t \tag{6.6.1-29}$$

and

$$\eta \equiv z_1 + v_s t \tag{6.6.1-30}$$

then (6.6.1-28) takes the simpler form

$$\frac{\partial^2 P^{(1)}}{\partial \xi \, \partial \eta} = 0 \tag{6.6.1-31}$$

This equation can be integrated immediately to yield

$$P^{(1)} = F(\xi) + G(\eta) \tag{6.6.1-32}$$

where $F(\xi)$ and $G(\eta)$ are arbitrary functions.

The physical interpretation of this result becomes more obvious if we set $G(\eta) = 0$. In this event

$$\begin{aligned} P &= P^{(0)} + a P^{(1)} \\ &= P^{(0)} + a F(\xi) \end{aligned} \tag{6.6.1-33}$$

Taking the derivative of this equation with respect to time while holding $P$ constant, we find

$$0 = \left[ \left( \frac{\partial z_1}{\partial t} \right)_P - v_s \right] \frac{dF}{d\xi} \tag{6.6.1-34}$$

Since $F(\xi)$ is an arbitrary function, we conclude that the

$$\text{speed of sound}: \quad \left( \frac{\partial z_1}{\partial t} \right)_P = v_s \equiv \sqrt{\gamma^{(0)} \left( \frac{\partial P}{\partial \rho} \right)_{T^{(0)}}} \tag{6.6.1-35}$$

The function $F(\xi)$ describes a pressure wave traveling in the positive $z_1$ direction. The speed of propagation of a surface on which pressure is a constant is $v_s$, our definition for the speed of sound.

We can now recognize that $G(\eta)$ describes a pressure wave traveling in the negative $z_1$ direction with a speed $v_s$.

We often think in terms of ideal gases for which

$$\left( \frac{\partial P}{\partial \rho} \right)_{T^{(0)}} = \frac{P^{(0)}}{\rho^{(0)}} \tag{6.6.1-36}$$

and

$$\text{ideal gas}: \quad v_s = \sqrt{\gamma^{(0)} \frac{P^{(0)}}{\rho^{(0)}}} \tag{6.6.1-37}$$

The discussion that we have presented here is based upon those given by Serrin (1959, p. 179) and by Landau and Lifshitz (1987, p. 252). It rests upon a definition for the speed of sound that is not unequivocal when applied to an entirely arbitrary fluid motion. Serrin (1959, p. 212) presents a more satisfying treatment based upon the conception of sound waves as surfaces of discontinuity with respect to the pressure gradient.

**Exercise 6.6.1-1** *Neglecting conduction and viscous dissipation in an ideal gas*   In the limit $N_{Re} \gg 1$ and $N_{Br}/N_{Re} \ll 1$ for arbitrary values of $N_{St}$, $N_{Ru}$, and $N_{Br}$ and a fixed value of $N_{Pr}$ we have seen that the differential energy balance appears to simplify to (6.6.1-7). Conclude that for an ideal gas undergoing such a process

$$P^{(\gamma-1)/\gamma} T^{-1} = \text{a constant}$$

or

$$P\rho^{-\gamma} = \text{a constant}$$

**Exercise 6.6.1-2** *More about speed of sound*   Find the speed of sound in a power-law fluid.

---

## 6.7  Boundary-Layer Theory

This is the second step of our development of boundary-layer theory, in which we consider both momentum transfer and energy transfer. For the first step described in Section 3.5, we confined our attention to momentum transfer. For the third step described in Section 9.5, we will consider momentum, energy, and mass transfer.

### 6.7.1  Plane Flow Past a Flat Plate

In the introduction to Section 6.6, we explored the possibility of neglecting all the conduction terms of the energy balance in the limit $N_{Re} \gg 1$ for a fixed value of $N_{Pr}$ (i.e., for a particular material). We anticipated that elimination of all of the conduction terms would be inconsistent with a realistic description of the temperature distribution near its bounding surfaces.

This problem is really quite similar to the one that we encountered in our discussion of potential flow in Section 3.4. Our answer there was to develop boundary-layer theory in Section 3.5.1. In boundary-layer theory, we reasoned that a portion of the viscous terms might be neglected for $N_{Re} \gg 1$, with a resulting considerable simplification in the components of the differential momentum balance.

Our intention here is to extend the boundary-layer concept to the energy balance. Unfortunately, we are missing direct experimental evidence, such as the temperature distribution in the immediate neighborhood of a flat plate, which might be used to suggest our next step. We shall rely heavily upon our previous success with boundary-layer theory in Sections 3.5.1 to 3.5.6 and argue here by analogy.

Let us begin by considering in some detail the same class of flows that we used to introduce boundary-layer theory in Section 3.5.1: plane flow past a flat plate. With reference to Figure 6.7.1-1, the temperature of the fluid as it approaches the plate at $z_1 = 0$ is known to be $T_\infty$

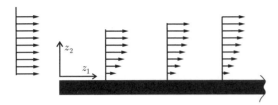

**Figure 6.7.1-1.** Flow past a flat plate.

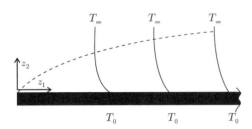

**Figure 6.7.1-2.** Temperature distribution in flow past an isothermal flat plate.

for all values of $z_2$. For the moment, we will say no more about the external flow and the thermal boundary condition at the plate.

We shall find it convenient to work in terms of the following dimensionless variables:

$$v_i^\star \equiv \frac{v_i}{v_0}, \qquad z_i^\star \equiv \frac{z_i}{L}$$
$$T^\star \equiv \frac{T - T_\infty}{T_0 - T_\infty}, \qquad t^\star \equiv \frac{t}{t_0} \tag{6.7.1-1}$$

Here $v_0$ is a magnitude of velocity characteristic of the plane nonviscous flow outside the boundary layer, $L$ is the length of the plate, $T_0$ is characteristic of the temperature distribution on the plate, and $t_0$ is a characteristic time. For this plane flow, it seems reasonable to assume that

$$T^\star = T^\star \left(z_1^\star, z_2^\star, t^\star\right) \tag{6.7.1-2}$$

For simplicity, we limit ourselves to an incompressible Newtonian fluid whose viscosity and thermal conductivity are constants independent of temperature. As a result, the velocity distribution appropriate to this flow is precisely that described in Section 3.5.1, since both density and viscosity are assumed to be independent of temperature. We can consequently concentrate our attention upon the temperature distribution.

If we were able to make very careful measurements of temperature in the fluid, we would intuitively expect to find the temperature to be $T_\infty$ nearly everywhere in the fluid, the only exception being a very thin region next to the plate in which the temperature undergoes a rapid change in order to satisfy whatever condition is required of temperature at the plate. As illustrated in Figure 6.7.1-2 for an isothermal plate, we would anticipate that the thickness of this region of nonuniform temperature increases as we go downstream (for increasing values of $z_1$). We should also expect to find that at any given value of $z_1$ the thickness of this region decreases as the velocity in the outer region increases.

If we go back to the argument given in the introduction to Section 6.6 and neglect conduction with respect to convection in high-speed flows, for this situation we predict the temperature to be a uniform $T_\infty$ outside the immediate neighborhood of the plate.

This picture suggests that conduction is just as important as convection in the energy balance for $N_{Re} \gg 1$, at least in the boundary layer.

The reasoning we presented in the introduction to Section 6.6 is not applicable to this region, since it says that conduction is negligibly small compared with convection; apparently, this treatment must be reserved for the temperature distribution outside the boundary layer. To use an approximation similar to that given in the introduction to Section 6.6 for the boundary layer, we must revise the definitions of our dimensionless variables so as to prevent the disappearance of all the conduction terms. If, as we indicated above, the thickness of the thermal boundary layer decreases as the Reynolds number increases, we should magnify the thickness of the boundary layer by introducing as we did in Section 3.5.1

$$z_2^{\star\star} \equiv \sqrt{N_{Re}}\, z_2^\star \tag{6.7.1-3}$$

where

$$N_{Re} \equiv \frac{v_0 L \rho}{\mu} \tag{6.7.1-4}$$

As explained in Section 3.5.1, the differential mass balance requires that we also introduce a magnified $z_2$ component of velocity:

$$v_2^{\star\star} \equiv \sqrt{N_{Re}}\, v_2^\star \tag{6.7.1-5}$$

In view of (6.7.1-2), the dimensionless energy balance given in the introduction to Section 6.4 reduces for this plane flow to

$$\frac{1}{N_{St}} \frac{\partial T^\star}{\partial t^\star} + \frac{\partial T^\star}{\partial z_1^\star} v_1^\star + \frac{\partial T^\star}{\partial z_2^\star} v_2^\star = \frac{1}{N_{Pr} N_{Re}} \left( \frac{\partial^2 T^\star}{\partial z_1^{\star 2}} + \frac{\partial^2 T^\star}{\partial z_2^{\star 2}} \right)$$
$$+ \frac{2N_{Br}}{N_{Pr} N_{Re}} \left[ \left( \frac{\partial v_1^\star}{\partial z_1^\star} \right)^2 + \left( \frac{\partial v_2^\star}{\partial z_2^\star} \right)^2 \right] + \frac{N_{Br}}{N_{Pr} N_{Re}} \left( \frac{\partial v_1^\star}{\partial z_2^\star} + \frac{\partial v_2^\star}{\partial z_1^\star} \right)^2 \tag{6.7.1-6}$$

or, in terms of $z_2^{\star\star}$ and $v_2^{\star\star}$,

$$\frac{1}{N_{St}} \frac{\partial T^\star}{\partial t^\star} + \frac{\partial T^\star}{\partial z_1^\star} v_1^\star + \frac{\partial T^\star}{\partial z_2^{\star\star}} v_2^{\star\star} = \frac{1}{N_{Pr}} \left( \frac{1}{N_{Re}} \frac{\partial^2 T^\star}{\partial z_1^{\star 2}} + \frac{\partial^2 T^\star}{\partial z_2^{\star\star 2}} \right)$$
$$+ \frac{2N_{Br}}{N_{Pr} N_{Re}} \left[ \left( \frac{\partial v_1^\star}{\partial z_1^\star} \right)^2 + \left( \frac{\partial v_2^{\star\star}}{\partial z_2^{\star\star}} \right)^2 \right] + \frac{N_{Br}}{N_{Pr}} \left( \frac{\partial v_1^\star}{\partial z_2^{\star\star}} + \frac{1}{N_{Re}} \frac{\partial v_2^{\star\star}}{\partial z_1^\star} \right)^2 \tag{6.7.1-7}$$

Here

$$N_{Pr} \equiv \frac{\hat{c}\mu}{k}, \qquad N_{Br} \equiv \frac{\mu v_0^2}{k(T_0 - T_\infty)}$$
$$N_{St} \equiv \frac{t_0 v_0}{L}, \qquad N_{Re} \equiv \frac{L v_0 \rho}{\mu} \tag{6.7.1-8}$$

Equation (6.7.1-7) suggests that, for $N_{Re} \gg 1$, a fixed value of $N_{Pr}$, and arbitrary values of $N_{St}$ and $N_{Br}$, the dimensionless differential energy balance may be simplified to

$$\frac{1}{N_{St}} \frac{\partial T^\star}{\partial t^\star} + \frac{\partial T^\star}{\partial z_1^\star} v_1^\star + \frac{\partial T^\star}{\partial z_2^{\star\star}} v_2^{\star\star} = \frac{1}{N_{Pr}} \frac{\partial^2 T^\star}{\partial z_2^{\star\star 2}} + \frac{N_{Br}}{N_{Pr}} \left( \frac{\partial v_1^\star}{\partial z_2^{\star\star}} \right)^2 \tag{6.7.1-9}$$

Since the velocity distribution for this flow may be presumed to be already known following the discussion of Section 3.5.1, Equation (6.7.1-9) represents the differential equation to be solved for the dimensionless temperature distribution.

Outside the boundary-layer region, conduction can be neglected with respect to convection in the differential energy balance. Let

$$T^{(e)\star} = T^{(e)\star}\left(z_1^\star, z_2^\star\right) = T^{(e)\star}\left(z_1^\star, \frac{z_2^{\star\star}}{\sqrt{N_{Re}}}\right) \tag{6.7.1-10}$$

denote the dimensionless temperature distribution for the nonconducting, nonviscous external flow. Within a region where both the boundary-layer solution and the external nonviscous, nonconducting flow are valid

$$\tilde{T}^\star \equiv \lim N_{Re} \gg 1 \text{ for } z_1^\star, z_2^{\star\star} \text{ fixed}:$$

$$T^{(e)\star}\left(z_1^\star, \frac{z_2^{\star\star}}{\sqrt{N_{Re}}}\right) = T^{(e)\star}\left(z_1^\star, 0\right) \tag{6.7.1-11}$$

For $z_2^{\star\star} \gg 1$, we must require that $T^\star$ from the boundary-layer solution must approach asymptotically the corresponding temperature from the nonconducting, nonviscous flow:

$$\text{for } z_2^{\star\star} \to \infty: \quad T^\star \to \tilde{T}^\star \tag{6.7.1-12}$$

**Exercise 6.7.1-1** *The boundary-layer equations for natural convection*   In Section 6.4.1, we derived the dimensionless forms of the differential mass balance, the differential momentum balance, and the differential energy balance appropriate to natural convection in a Newtonian fluid. Let us assume that $\hat{c}_V$, $k$, and $\mu$ are all constants.

i) Extend the discussion of Section 3.5.1 for the velocity boundary layer on a flat plate. Find that, for $N_{Gr} \gg 1$, the differential mass balance and the differential momentum balance imply

$$\frac{\partial v_1^\star}{\partial z_1^\star} + \frac{\partial v_2^{\star\star}}{\partial z_2^{\star\star}} = 0$$

and

$$\frac{1}{N_{St}} \frac{\partial v_1^\star}{\partial t^\star} + v_1^\star \frac{\partial v_1^\star}{\partial z_1^\star} + v_2^{\star\star} \frac{\partial v_1^\star}{\partial z_2^{\star\star}} = -\frac{1}{N_{Ru}} \frac{d\mathcal{P}^\star}{dz_1^\star} + \frac{\partial^2 v_1^\star}{\partial z_2^{\star\star 2}} - T^\star f_1^\star$$

where

$$-\frac{1}{N_{Ru}} \frac{d\mathcal{P}^\star}{dz_1^\star} = -\frac{1}{N_{Ru}} \frac{d\tilde{\mathcal{P}}^\star}{dz_1^\star} = \frac{1}{N_{St}} \frac{\partial \tilde{v}_1^\star}{\partial t^\star} + \tilde{v}_1^\star \frac{\partial \tilde{v}_1^\star}{\partial z_1^\star} + \tilde{T}^\star f_1^\star$$

Here $\tilde{\mathcal{P}}^\star$, $\tilde{v}_1^\star$, and $\tilde{T}^\star$ are the dimensionless modified pressure, velocity, and temperature distributions at the plate as determined by the nonviscous, nonconducting outer flow (outside the boundary layer).

ii) Determine that the corresponding differential energy balance appropriate to the inner problem or thermal boundary layer is

$$\frac{1}{N_{St}}\frac{\partial T^\star}{\partial t^\star} + \frac{\partial T^\star}{\partial z_1^\star}v_1^\star + \frac{\partial T^\star}{\partial z_2^{\star\star}}v_2^{\star\star} = \frac{1}{N_{Pr}}\frac{\partial^2 T^\star}{\partial z_2^{\star\star 2}}$$

## 6.7.2 More on Plane Flow Past a Flat Plate

Here we consider a particular case to which the theory in the preceding section is applicable: steady-state flow past a flat plate at zero incidence of an incompressible Newtonian fluid. The temperature of the plate is maintained constant at $T_0$; the temperature of the fluid has a uniform value $T_\infty$ on the plane $z_1 = 0$ in Figure 6.7.1-1. For simplicity, we take the viscosity and thermal conductivity of the fluid to be independent of temperature.

Let us define the characteristic magnitude of velocity of the fluid to be $\tilde{v}_1$, the $z_1$ component of velocity in the nonviscous, nonconducting fluid in the limit as the plate is approached; the characteristic length is $L$, the length of the plate; the characteristic temperature is taken to be the temperature difference $T_0 - T_\infty$. In terms of these characteristic quantities we define the dimensionless velocity components, coordinates, and temperature as

$$v_i^\star \equiv \frac{v_i}{\tilde{v}_1}, \qquad z_i^\star \equiv \frac{z_i}{L}$$
$$T^\star \equiv \frac{T - T_\infty}{T_0 - T_\infty} \tag{6.7.2-1}$$

The velocity distribution for this flow is unchanged from that found in Section 3.5.2, since both viscosity and density are taken to be independent of temperature. Just to review, we concluded there that

$$v_1^\star \equiv \frac{v_1}{\tilde{v}_1}$$
$$= f'$$
$$v_2^{\star\star} \equiv \sqrt{N_{Re}}\,\frac{v_2}{\tilde{v}_1} \tag{6.7.2-2}$$
$$= \frac{1}{2\sqrt{z_1^\star}}(\eta f' - f)$$

where $f = f(\eta)$ and

$$\eta \equiv \frac{z_2^{\star\star}}{\sqrt{z_1^\star}}$$
$$= \frac{\sqrt{N_{Re}}\,z_2^\star}{\sqrt{z_1^\star}} \tag{6.7.2-3}$$

The prime is used to denote differentiation with respect to $\eta$. The function $f$ is a solution of

$$ff'' + 2f''' = 0 \tag{6.7.2-4}$$

that satisfies the boundary conditions

$$\text{at } \eta = 0: \ f = f' = 0 \tag{6.7.2-5}$$

and

$$\text{as } \eta \to \infty : \ f' \to 1 \tag{6.7.2-6}$$

By the Reynolds number we mean here

$$N_{Re} \equiv \frac{\tilde{v}_1 L \rho}{\mu} \tag{6.7.2-7}$$

For the nonviscous, nonconducting flow at the outer edge of the boundary layer, we know that

$$\text{at } z_1^\star = 0 : \ T^{(e)\star} = 0 \tag{6.7.2-8}$$

Since

$$v_1^{(e)\star} = 1$$
$$v_2^{(e)\star} = v_3^{(e)\star}$$
$$= 0 \tag{6.7.2-9}$$

the differential energy balance applicable to this flow simplifies to

$$\frac{\partial T^{(e)\star}}{\partial z_1{}^\star} v_1^{(e)\star} = \frac{\partial T^{(e)\star}}{\partial z_1^\star}$$
$$= 0 \tag{6.7.2-10}$$

We conclude from (6.7.2-8) through (6.7.2-10) that

$$T^{(e)\star}\left(z_1{}^\star, z_2^\star\right) = 0 \tag{6.7.2-11}$$

and in particular

$$\tilde{T}^\star \equiv T^{(e)\star}\left(z_1{}^\star, 0\right)$$
$$= 0 \tag{6.7.2-12}$$

From Section 6.7.1, the differential energy balance applicable to the boundary layer is

$$\frac{\partial T^\star}{\partial z_1^\star} v_1^\star + \frac{\partial T^\star}{\partial z_2^{\star\star}} v_2^{\star\star} = \frac{1}{N_{Pr}} \frac{\partial^2 T^\star}{\partial z_2^{\star\star 2}} + \frac{N_{Br}}{N_{Pr}} \left(\frac{\partial v_1^\star}{\partial z_2^{\star\star}}\right)^2 \tag{6.7.2-13}$$

Here

$$N_{Pr} \equiv \frac{\hat{c}\mu}{k}$$
$$N_{Br} \equiv \frac{\mu \tilde{v}_1^2}{k\left(T_0 - T_\infty\right)} \tag{6.7.2-14}$$

Equation (6.7.2-13) can be regarded as the differential equation to be solved for the dimensionless temperature distribution in the boundary layer, since the velocity distribution in the boundary layer is already known from Section 3.5.2. The boundary conditions to be satisfied by the desired solution to (6.7.2-13) are that

$$\text{at } z_2^{**} = 0 : \ T^* = 1 \tag{6.7.2-15}$$

and

$$\text{as } z_2^{**} \to \infty : \ T^* \to \tilde{T}^* = 0 \tag{6.7.2-16}$$

By analogy with our analysis for the boundary-layer velocity distribution in Section 3.5.2, we anticipate that we might be able to find a solution to (6.7.2-13) by combining the two independent variables in such a way as to transform this equation into an ordinary differential equation. If we anticipate a solution of the form

$$T^* = T^*(\eta) \tag{6.7.2-17}$$

with the help of (6.7.2-2), Equation (6.7.2-13) becomes

$$T^{*\prime\prime} + \frac{1}{2} N_{Pr} f T^{*\prime} = -N_{Br} (f^{\prime\prime})^2 \tag{6.7.2-18}$$

where primes denote differentiation with respect to $\eta$.

The corresponding boundary conditions are

$$\text{at } \eta = 0 : \ T^* = 1 \tag{6.7.2-19}$$

and

$$\text{as } \eta \to \infty : \ T^* \to 0 \tag{6.7.2-20}$$

This boundary value problem has been solved numerically using Mathematica (1993). The results are shown in Figure 6.7.2-1 for $N_{Pr} = 0.7$ (air) and for both positive values of $N_{Br}$ $(T_0 - T_\infty > 0)$ and negative values of $N_{Br}$ $(T_0 - T_\infty < 0)$. In interpreting these results, it will be helpful to note that

$$\text{at } z_2 = 0 : \ q_2 = -k \frac{\partial T}{\partial z_2}$$

$$= \frac{-k(T_0 - T_\infty)\sqrt{N_{Re}}}{L\sqrt{z_1^*}} \frac{dT^*}{d\eta} \tag{6.7.2-21}$$

This means that, for $N_{Br} = 3$, energy is transferred to the wall. As the result of viscous dissipation, the temperature has been raised above $T_0$, a short distance away from the wall, even though $(T_0 - T_\infty > 0)$. For $N_{Br} = 2$, the energy transfer to the wall is approaching zero. As $N_{Br}$ decreases from 1 to 0, there is increasing energy transfer away from the wall. As $N_{Br}$ further decreases from 0 to $-3$, we see from (6.7.2-21) that the energy transfer is now to the wall.

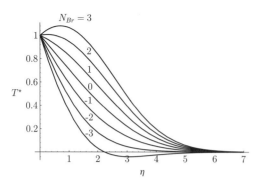

**Figure 6.7.2-1.** $T^\star$ as a function of $\eta$ for flow past an isothermal wall with $N_{Pr} = 0.7$ (air). From top to bottom, $N_{Br} = 3, 2, 1, 0, -1, -2, -3$. Positive values of $N_{Br}$ correspond to $T_0 - T_\infty > 0$; negative values to $T_0 - T_\infty < 0$.

Let us look at the total energy transfer per unit width of plate from the wall to the fluid:

$$Q \equiv \int_0^L q_2|_{z_2=0} \, dz_1$$

$$= L \int_0^1 q_2|_{z_2=0} \, dz_1^\star$$

$$= -k(T_0 - T_\infty)\sqrt{N_{Re}} \int_0^1 \frac{1}{\sqrt{z_1^\star}} \frac{dT^\star}{d\eta}\bigg|_{\eta=0} dz_1^\star$$

$$= -k(T_0 - T_\infty)\sqrt{N_{Re}} \frac{dT^\star}{d\eta}\bigg|_{\eta=0} \int_0^1 \frac{1}{\sqrt{z_1^\star}} dz_1^\star$$

$$= 2k(T_0 - T_\infty)\sqrt{N_{Re}} \left( -\frac{dT^\star}{d\eta}\bigg|_{\eta=0} \right) \tag{6.7.2-22}$$

It will be more convenient to rearrange this in terms of the Nusselt number

$$N_{Nu} \equiv \frac{Q}{k(T_0 - T_\infty)}$$

$$= 2\sqrt{N_{Re}} \left( -\frac{dT^\star}{d\eta}\bigg|_{\eta=0} \right) \tag{6.7.2-23}$$

Figure 6.7.2-2 summarizes in these terms the results shown in Figure 6.7.2-1.

Let us consider a specific example. From Figure 6.7.2-2, we can say that for $N_{Pr} = 0.7$ the wall will be cooled so long as (Schlichting 1979, p. 297)

$$0 < N_{Br} \equiv \frac{\mu \tilde{v}_1^2}{k(T_0 - T_\infty)} < 1.7 \tag{6.7.2-24}$$

For a stream of air flowing at $\tilde{v}_1 = 30$ m/s, with $N_{Pr} = 0.7$, $\mu = 21 \times 10^{-6}$ Pa s, and

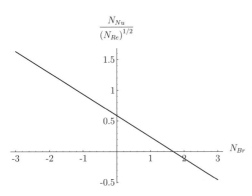

**Figure 6.7.2-2.** $N_{Nu}/(N_{Re})^{1/2}$ as a function of $N_{Br}$ for flow past an isothermal wall with $N_{Pr} = 0.7$ (air).

$k = 28 \times 10^{-3}$ W/(m·K), Equation (6.7.2-24) says that the wall will be cooled for

$$T_0 - T_\infty > 0.4°C \tag{6.7.2-25}$$

**Exercise 6.7.2-1** *Flow past an isothermal flat plate*   For flow past an isothermal flat plate, determine that the dimensionless velocity and temperature distributions have the same form when $N_{Pr} = 1$. We say that, under these circumstances, the velocity and temperature distributions are *similar*.

**Exercise 6.7.2-2** *Natural convection at a vertical wall* (Schlichting 1979, p. 315)   In Exercise 6.7.1-1, we introduced the boundary-layer equations appropriate to natural convection in the limit where the Grashof number $N_{Gr} \gg 1$. Let us apply these equations to analyze the heat transfer from a vertical hot plate.

   A vertical flat plate of uniform temperature $T_0$ is immersed in a Newtonian fluid that has a uniform temperature $T_\infty$ very far away from the plate. The coordinate $z_1$ is measured along the plate in the opposite direction from gravity, starting at the leading edge of the plate; $z_2$ is measured into the fluid from the plate.

   i) What are the boundary conditions to be satisfied by the simultaneous solution of the boundary-layer equations?
   ii) The differential mass balance can be identically satisfied by the introduction of a dimensionless stream function $\psi^\star$. The resulting two partial differential equations can be reduced to ordinary differential equations by looking for a combination-of-variables solution of the form

$$\psi^\star = \left(4z_1^\star\right)^{3/4} Z(\eta)$$

$$T^\star \equiv \frac{T - T_\infty}{T_0 - T_\infty}$$

$$= T^\star(\eta)$$

where

$$\eta \equiv \frac{z_2^{\star\star}}{\left(4z_1^\star\right)^{1/4}}$$

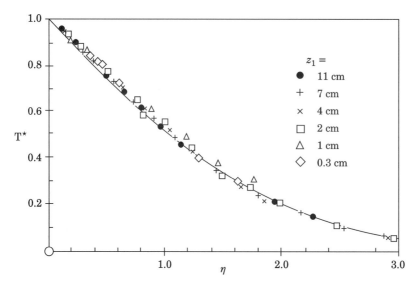

**Figure 6.7.2-3.** $T^{\star}$ as a function of $\eta$ in natural convection adjacent to an isothermal wall, with experimental data for air ($N_{Pr} = 0.73$) by Schmidt and Beckmann (1930). Taken from Schlichting (1979, Fig. 12.25).

Determine that the equation of motion and the differential energy balance reduce respectively to

$$Z''' + 3ZZ'' - 2Z'^2 + T^{\star} = 0$$

and

$$T^{\star''} + 3N_{Pr}ZT^{\star'} = 0$$

iii) Write down the boundary conditions that these equations must satisfy.

iv) Use, for example, Mathematica (1993) to solve this problem numerically. The results shown in Figures 6.7.2-3 and 6.7.2-4 for air ($N_{Pr} = 0.73$) are in excellent agreement with available experimental data.

v) Determine that

$$N_{Nu} \equiv 0.478(N_{Gr})^{1/4}$$

**Exercise 6.7.2-3** *Flow past an adiabatic wall*   Use, for example, Mathematica (1993) to solve the problem described in the text, assuming that the wall is adiabatic (insulated). In this case, it will be appropriate to define

$$T^{\star} \equiv \frac{T - T_{\infty}}{T_{\infty}}$$

$$N_{Br} \equiv \frac{\mu \tilde{v}_1^2}{k(T_{\infty})}$$

The solution is shown in Figure 6.7.2-5.

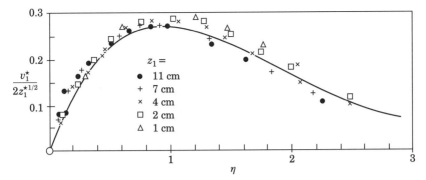

**Figure 6.7.2-4.** $Z' = v_1^\star/\left(2z_1^{\star 1/2}\right)$ as a function of $\eta$ in natural convection adjacent to an isothermal wall, with experimental data for air ($N_{Pr} = 0.73$) by Schmidt and Beckmann (1930). Taken from Schlichting (1979, Fig. 12.26).

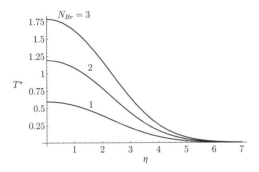

**Figure 6.7.2-5.** $T^\star$ as a function of $\eta$ for several values of $N_{Br}$ in flow past an adiabatic (insulated) wall.

**Exercise 6.7.2-4** *More on flow past an isothermal flat plate*   Show that the solution developed in the text can also be written as (Schlichting 1979, p. 293)

$$T^\star = AT_{N_{Br}=0}^\star + \frac{N_{Br}}{N_{Pr}}T_{adb}^\star \tag{6.7.2-26}$$

where $kT_{adb}^\star$ is the solution for flow past an adiabatic wall, developed in Exercise 6.7.2-3.

**Exercise 6.7.2-5** *Ice formation on an airfoil*   Construct an analysis to define the conditions under which icing will take place on an airfoil. I suggest that you may make the following assumptions:

- Since we are not yet ready to consider multicomponent systems, assume that the airfoil is moving through pure water vapor.
- Replace the airfoil by a flat plate.
- Assume that only ice forms on the airfoil; any liquid water is swept away.
- Assume that the plate is adiabatic.
- It is not necessary to follow the thickness of the ice as a function of time.

*Hint:*   The primary difference between this problem and the one described in the text is that the momentum and energy problems should be solved simultaneously. The velocity

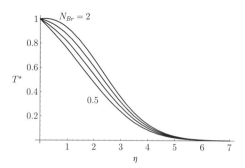

**Figure 6.7.2-6.** $T^\star$ as a function of $\eta$ for flow of water vapor past an insulated wall, the temperature of which has a constant value $T_0 = 273$ K as the result of ice formation. In these computations, we have assumed that $N_h = 324$ and $N_{Pr} = 0.785$, and we have shown the results for $N_{Br} = 0.5, 1, 1.5,$ and 2. The various values of $N_{Br}$ correspond to different values of $T_0 - T_\infty$ in (6.7.2-14).

distribution is no longer that found in Section 3.5.2, because mass transfer occurs at the wall. Use the jump energy balance to determine the normal component of velocity at the plate.

*Answer:*  The $z_1$ component of velocity is virtually unchanged from that found in Section 3.5.2. The vapor must be subcooled. The result will depend not only upon $N_{Br}$ but also upon

$$N_h \equiv \frac{\Delta \hat{H}}{\tilde{v}_1^2}$$

Figure 6.7.2-6 shows the results for $\tilde{v}_1 = 89.4$ m/s, the enthalpy change on condensation $\Delta \hat{H} = 2.59 \times 10^6$ J/Kgm, $N_h = 324$, and $N_{Pr} = 0.785$. The various values of $N_{Br}$ correspond to different values of $T_0 - T_\infty$ in (6.7.2-14). The computations suggest that ice will not form on an airfoil for $N_{Br} > 1.81$.

### 6.7.3  Plane Flow Past a Curved Wall

In Section 6.7.1 we discussed the temperature distribution in plane flow past a flat plate. In the limit $N_{Re} \gg 1$, we found that the differential energy balance may be considerably simplified for the fluid in a thin boundary layer next to the plate. This simplification involved neglecting some (though not all) of the conduction terms with respect to the convection terms.

In this section, we wish to consider the thermal boundary layer formed by an incompressible Newtonian fluid in plane flow past a curved wall. For simplicity, we take both viscosity and thermal conductivity to be constants, independent of temperature. We anticipate that the temperature distribution in a sufficiently thin boundary layer will be much the same whether the wall is curved or flat. Our object here is to show in what sense this intuitive feeling is correct.

Our approach follows closely that in Section 3.5.3, where we determined the forms of the differential mass and momentum balances appropriate to plane flow past a curved wall.

A portion of a typical curved wall is shown in Figure 3.5.3-1. With respect to the rectangular Cartesian coordinate system indicated, the equation of this surface is

$$z_2 = f(z_1) \tag{6.7.3-1}$$

To better compare flow past a curved wall with flow past a flat plate, let us view this problem in terms of an orthogonal curvilinear coordinate system such that

$x \equiv x^1$ is defined to be the arc length measured along the wall in a plane of constant $z$,
$y \equiv x^2$ is defined to be the arc length measured along straight lines that are normal to the wall, and
$z \equiv x^3 \equiv z_3$ is the coordinate normal to the plane of flow.

By plane flow, we mean here that

$$
\begin{aligned}
v_x &\equiv v_{(1)} \\
&= v_x(x, y, t) \\
v_y &\equiv v_{(2)} \\
&= v_y(x, y, t) \\
v_z &\equiv v_{(3)} \\
&= 0 \\
T &= T(x, y, t)
\end{aligned}
\tag{6.7.3-2}
$$

With these restrictions, the differential energy balance may be written as (see Exercise 6.7.3-1)

$$
\begin{aligned}
\rho \hat{c} \left( \frac{\partial T}{\partial t} + \frac{\partial T}{\partial x} \frac{v_x}{1 + \kappa y} + \frac{\partial T}{\partial y} v_y \right) &= k \left\{ \frac{1}{1 + \kappa y} \frac{\partial}{\partial x} \left( \frac{1}{1 + \kappa y} \frac{\partial T}{\partial x} \right) \right. \\
&\left. + \frac{1}{1 + \kappa y} \frac{\partial}{\partial y} \left[ (1 + \kappa y) \frac{\partial T}{\partial y} \right] \right\} + 2\mu \left[ \frac{1}{(1 + \kappa y)^2} \left( \frac{\partial v_x}{\partial x} + \kappa v_y \right)^2 \right. \\
&\left. + \left( \frac{\partial v_y}{\partial y} \right)^2 + \frac{1}{2} \left( \frac{\partial v_x}{\partial y} + \frac{1}{1 + \kappa y} \frac{\partial v_y}{\partial x} - \frac{\kappa v_x}{1 + \kappa y} \right)^2 \right]
\end{aligned}
\tag{6.7.3-3}
$$

where we define

$$\kappa \equiv \frac{-f''}{[1 + (f')^2]^{3/2}} = \kappa(x) \tag{6.7.3-4}$$

The primes here indicate differentiation with respect to $z_1$. We may think of $-\kappa$ as twice the mean curvature of the surface (Slattery 1990, p. 1118), the normal curvature (Slattery 1990, p. 1118) of the surface in the direction $x$, or the only nonzero principal curvature (Slattery 1990, p. 1119) of the surface.

In addition to the dimensionless velocity, dimensionless temperature, and dimensionless time suggested in the introduction to Section 6.4, let us define

$$x^\star \equiv \frac{x}{L}, \qquad y^\star \equiv \frac{y}{L}, \qquad \kappa^\star \equiv \kappa L \tag{6.7.3-5}$$

If we extend the arguments of Section 6.7.1 to this geometry, we are motivated to express our results in terms of

$$y^{\star\star} \equiv \sqrt{N_{Re}}\, y^{\star}$$
$$v_y^{\star\star} \equiv \sqrt{N_{Re}}\, v_y^{\star} \tag{6.7.3-6}$$

For $N_{Re} \gg 1$, Equation (6.7.3-3) reduces to

$$\frac{1}{N_{St}}\frac{\partial T^{\star}}{\partial t^{\star}} + \frac{\partial T^{\star}}{\partial x^{\star}}\frac{v_x^{\star}}{1 + \kappa^{\star\star}y^{\star\star}} + \frac{\partial T^{\star}}{\partial y^{\star\star}}v_y^{\star\star}$$

$$= \frac{1}{N_{Pr}}\frac{1}{1 + \kappa^{\star\star}y^{\star\star}}\frac{\partial}{\partial y^{\star\star}}\left[(1 + \kappa^{\star\star}y^{\star\star})\frac{\partial T^{\star}}{\partial y^{\star\star}}\right]$$

$$+ \frac{N_{Br}}{N_{Pr}}\left(\frac{\partial v_x^{\star}}{\partial y^{\star\star}} - \frac{\kappa^{\star\star}v_x^{\star}}{1 + \kappa^{\star\star}y^{\star\star}}\right)^2 \tag{6.7.3-7}$$

in which

$$\kappa^{\star\star} \equiv \sqrt{N_{Re}}\,\kappa^{\star} \tag{6.7.3-8}$$

For a fixed-wall configuration, $\kappa^{\star\star} \ll 1$ in the limit $N_{Re} \gg 1$. Equation (6.7.3-7) further simplifies under these conditions to

$$\frac{1}{N_{St}}\frac{\partial T^{\star}}{\partial t^{\star}} + \frac{\partial T^{\star}}{\partial x^{\star}}v_x^{\star} + \frac{\partial T^{\star}}{\partial y^{\star\star}}v_y^{\star\star} = \frac{1}{N_{Pr}}\frac{\partial^2 T^{\star}}{\partial y^{\star\star 2}} + \frac{N_{Br}}{N_{Pr}}\left(\frac{\partial v_x^{\star}}{\partial y^{\star\star}}\right)^2 \tag{6.7.3-9}$$

As we suggested in Section 6.7.1,

$$\text{as } y^{\star\star} \to \infty: \quad T^{\star} \to \tilde{T}^{\star} \tag{6.7.3-10}$$

where $\tilde{T}^{\star}$ is the dimensionless temperature distribution at the curved wall for the corresponding nonviscous, nonconducting flow:

$$\frac{1}{N_{St}}\frac{\partial \tilde{T}^{\star}}{\partial t^{\star}} + \frac{\partial \tilde{T}^{\star}}{\partial x^{\star}}\tilde{v}_x^{\star} = 0 \tag{6.7.3-11}$$

By $\tilde{v}_x^{\star}$, we mean the dimensionless $x$ component of velocity at the curved wall for the corresponding nonviscous flow.

Since we assume that $v_x^{\star}$ and $v_y^{\star\star}$ are known *a priori*, Equation (6.7.3-9) can be solved for the dimensionless temperature distribution in the thermal boundary layer. As our intuition suggested, the differential energy balance appropriate to the thermal boundary layer developed in plane flow past a curved wall has the same form as that appropriate to plane flow past a flat plate found in Section 6.7.1.

**Exercise 6.7.3-1** *Derivation of (6.7.3-3)*    Noting the results of Exercise 3.5.3-1, derive (6.7.3-3) starting from Table 5.4.1-1.

**Exercise 6.7.3-2** *Derivation of (6.7.3-7)*    Introduce in (6.7.3-3) the dimensionless velocity, dimensionless temperature, and dimensionless time defined in the introduction to Section 6.4

as well as those dimensionless variables defined in (6.7.3-5) and (6.7.3-6). Construct the reasoning that leads to (6.7.3-7).

### 6.7.4 Flow Past a Wedge

As an illustration of the development given in the preceding section, let us consider plane flow of an incompressible Newtonian fluid past the wedge shown in Figure 6.7.4-1. The temperature of the wedge's wall is maintained constant at $T_0$; the temperature of the gas stream at $x = 0$ is known to be $T_\infty$. Our object here is to determine the temperature distribution within the gas in the immediate neighborhood of the wedge, as well as the local rate of the energy transfer from the wedge to the gas. For simplicity, we shall assume that both the viscosity and thermal conductivity of the gas are independent of temperature and we shall neglect viscous dissipation within the gas.

The nonviscous, potential flow outside the boundary layer predicts that, at the surface of the wedge and in the immediate neighborhood of the apex (Schlichting 1979, p. 156),

$$\tilde{v}_x = u x^m \tag{6.7.4-1}$$

where the included angle of the wedge is

$$\pi \beta = \frac{2\pi m}{1 + m} \tag{6.7.4-2}$$

In Section 6.7.3 let $L$ be a characteristic length associated with the wedge, perhaps its length. Take the characteristic velocity to be $uL^m$ and the characteristic temperature to be $(T_0 - T_\infty)$. With this understanding, we define

$$
\begin{aligned}
v_x^\star &\equiv \frac{v_x}{uL^m} \\
v_y^{\star\star} &\equiv \sqrt{N_{Re}} \frac{v_y}{uL^m} \\
T^\star &\equiv \frac{T - T_\infty}{T_0 - T_\infty} \\
x^\star &\equiv \frac{x}{L} \\
y^{\star\star} &\equiv \sqrt{N_{Re}} \frac{y}{L}
\end{aligned}
\tag{6.7.4-3}
$$

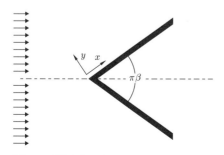

**Figure 6.7.4-1.** Flow past a wedge.

where

$$N_{Re} \equiv \frac{uL^{1+m}\rho}{\mu} \tag{6.7.4-4}$$

We have been vague in introducing the characteristic length $L$, since we will find it drops out of the final results.

We found in Section 6.7.3 that the dimensionless form of the differential energy balance appropriate to the boundary layer can be written as

$$\frac{\partial T^\star}{\partial x^\star}v_x^\star + \frac{\partial T^\star}{\partial y^{\star\star}}v_y^{\star\star} = \frac{1}{N_{Pr}}\frac{\partial^2 T^\star}{\partial y^{\star\star 2}} + \frac{N_{Br}}{N_{Pr}}\left(\frac{\partial v_x^\star}{\partial y^{\star\star}}\right)^2 \tag{6.7.4-5}$$

For this problem,

$$N_{Pr} \equiv \frac{\hat{c}\mu}{k}$$

$$N_{Br} \equiv \frac{\mu u^2 L^{2m}}{k(T_0 - T_\infty)} \tag{6.7.4-6}$$

In introducing this problem we said that we would neglect viscous effects. More precisely, we will restrict ourselves to the limit as $N_{Br} \ll 1$ for a fixed value of $N_{Pr}$, in which it appears reasonable to approximate (6.7.4-5) as

$$\frac{\partial T^\star}{\partial x^\star}v_x^\star + \frac{\partial T^\star}{\partial y^{\star\star}}v_y^{\star\star} = \frac{1}{N_{Pr}}\frac{\partial^2 T^\star}{\partial y^{\star\star 2}} \tag{6.7.4-7}$$

The velocity distribution for the boundary layer was discussed in Section 3.5.4, Case 1a. In view of our definitions for dimensionless variables in (6.7.4-3), Equation (6.7.4-1) may be written as

$$\tilde{v}_x^\star = x^{\star m} \tag{6.7.4-8}$$

This means that in Section 3.5.4 we are forced to define the as yet unspecified constant

$$K \equiv \left(\frac{1+m}{2}\right)^{m/1+m} \tag{6.7.4-9}$$

It follows immediately that

$$v_x^\star = \frac{\partial \psi}{\partial y^{\star\star}} = x^{\star m}\frac{df}{d\eta} \tag{6.7.4-10}$$

and

$$v_y^{\star\star} = -\frac{\partial \psi}{\partial x^\star}$$

$$= -\left(\frac{1+m}{2}\right)^{1/2}x^{\star(m-1)/2}f$$

$$- \left(\frac{2}{1+m}\right)^{1/2}\frac{m-1}{2}x^{\star(m-1)/2}\eta\frac{df}{d\eta} \tag{6.7.4-11}$$

Here $f$, a function of

$$\eta \equiv \frac{y^{\star\star}}{g}$$

$$= \frac{y^{\star\star}}{[2/(1+m)]^{1/2}x^{\star(1-m)/2}} \tag{6.7.4-12}$$

is a solution to

$$\frac{d^3 f}{d\eta^3} + f\frac{d^2 f}{d\eta^2} + \frac{2m}{1+m}\left[1 - \left(\frac{df}{d\eta}\right)^2\right] = 0 \tag{6.7.4-13}$$

consistent with the boundary conditions

$$\text{at } \eta = 0: \quad f = \frac{df}{d\eta} = 0 \tag{6.7.4-14}$$

and

$$\text{as } \eta \to \infty: \quad \frac{df}{d\eta} \to 1 \tag{6.7.4-15}$$

Mathematica (1993) was used to obtain the results shown in Figure 6.7.4-2 for $N_{Pr} = 0.72$ and several values of $m$ (the corresponding values of $360\beta$ can be found in Table 6.7.4-1).

If we anticipate a solution of the form

$$T^\star = T^\star(\eta) \tag{6.7.4-16}$$

then (6.7.4-7) becomes

$$\frac{d^2 T^\star}{d\eta^2} + N_{Pr} f\frac{dT^\star}{d\eta} = 0 \tag{6.7.4-17}$$

The appropriate boundary conditions are

$$\text{at } \eta = 0: \quad T^\star = 1 \tag{6.7.4-18}$$

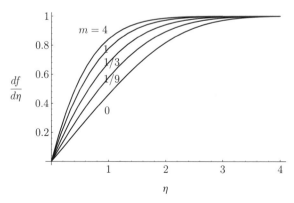

**Figure 6.7.4-2.** $df/d\eta$ for $N_{Pr} = 0.72$ and several values of $m$ (the corresponding values of $360\beta$ can be found in Table 6.7.4-1) in flow past an isothermal-walled wedge.

**Table 6.7.4-1.** $N_{Nu}x^*(x^{*m+1}N_{Re})^{-1/2}$ for $N_{Pr} = 0.72$ and several values of $m$ in flow past an isothermal-walled wedge

| $m$ | $360\beta$ | $N_{Nu}x^*(x^{*\ m+1}N_{Re})^{-1/2}$ |
|-----|------------|--------------------------------------|
| 4   | 288        | 0.845                                |
| 1   | 180        | 0.503                                |
| 1/3 | 90         | 0.390                                |
| 1/9 | 36         | 0.337                                |
| 0   | 0          | 0.298                                |

and

$$\text{as } \eta \to \infty : \ T^\star \to 0 \tag{6.7.4-19}$$

Equation (6.7.4-17) is easily integrated with (6.7.4-18) and (6.7.4-19) to find

$$T^\star = \left[ \int_\eta^\infty \exp\left(-N_{Pr} \int_0^\xi f\, d\tau\right) d\xi \right] \left[ \int_0^\infty \exp\left(-N_{Pr} \int_0^\xi f\, d\tau\right) d\xi \right]^{-1} \tag{6.7.4-20}$$

The $y$ component of the energy flux vector evaluated at the wall of the wedge is computed to be

$$\text{at } \eta = 0 : \ q_y = -k\frac{\partial T}{\partial y}$$

$$= -\frac{k(T_0 - T_\infty)}{L} N_{Re}^{1/2} \left(\frac{1+m}{2}\right)^{1/2} x^{\star(m-1)/2} T^{\star\prime}(0)$$

$$= \frac{k(T_0 - T_\infty)}{L} N_{Re}^{1/2} \left(\frac{1+m}{2}\right)^{1/2} x^{\star(m-1)/2}$$

$$\times \left[ \int_0^\infty \exp\left(-N_{Pr} \int_0^\xi f\, d\tau\right) d\xi \right]^{-1} \tag{6.7.4-21}$$

or

$$\frac{N_{Nu}x^\star}{(x^{\star m+1}N_{Re})^{1/2}} = \left(\frac{1+m}{2}\right)^{1/2} \left[ \int_0^\infty \exp\left(-N_{Pr} \int_0^\xi f\, d\tau\right) d\xi \right]^{-1} \tag{6.7.4-22}$$

where we define the Nusselt number as

$$N_{Nu} \equiv \frac{q_y(0)L}{k(T_0 - T_\infty)} \tag{6.7.4-23}$$

Results computed using Mathematica (1993) are shown in Table 6.7.4-1.

Notice that the characteristic length $L$ has dropped out of (6.7.4-20) and (6.7.4-22) ($\eta$ is independent of $L$). The results are not applicable to a semi-infinite wedge, since (6.7.4-1) is applicable only in the neighborhood of the apex.

### 6.7.5 Flow Past a Body of Revolution

In Sections 6.7.1 and 6.7.3 we discussed the temperature distribution in plane flow past a flat plate and plane flow past a curved wall. We found that the thermal boundary layer could be described by the same set of equations in both cases.

In what follows, we investigate the thermal boundary layer formed by an incompressible Newtonian fluid flowing past a body of revolution. We expect the temperature distribution in a sufficiently thin boundary layer on a body of revolution to be similar to that on a flat plate. We again take the viscosity and thermal conductivity to be constants, independent of temperature.

Our approach here follows closely that in Section 3.5.6, where we determined the forms of the differential mass and momentum balances appropriate to flow past a body of revolution.

A portion of a typical body of revolution is shown in Figure 3.5.6-1. With respect to the cylindrical coordinate system indicated, the equation of the axially symmetric surface is

$$r = f(z) \tag{6.7.5-1}$$

To better compare flow past a body of revolution with flow past a flat plate, let us view this problem in terms of an orthogonal curvilinear coordinate system such that

$x \equiv x^3$ is defined to be arc length measured along the wall (in the direction of flow) in a plane of constant $\theta$,

$y \equiv x^1$ is defined to be arc length measured along straight lines that are normal to the wall, and

$\theta \equiv x^2$ is the azimuthal cylindrical coordinate (measured around the axis of the body).

The shape of the wall suggests

$$v_x \equiv v_{(3)}$$
$$= v_x(x, y, t)$$
$$v_y \equiv v_{(1)}$$
$$= v_y(x, y, t) \tag{6.7.5-2}$$
$$v_\theta \equiv v_{(2)}$$
$$= 0$$
$$T = T(x, y, t)$$

With these restrictions, the differential energy balance may be written as (see Exercise 6.7.5-1)

$$\rho \hat{c} \left( \frac{\partial T}{\partial t} + \frac{\partial T}{\partial x} + \frac{v_x}{1 + \kappa y} + \frac{\partial T}{\partial y} v_y \right)$$

$$= k \left\{ \frac{1}{1 + \kappa y} \frac{\partial}{\partial x} \left( \frac{1}{1 + \kappa y} \frac{\partial T}{\partial x} \right) + \frac{1}{1 + \kappa y} \frac{\partial}{\partial y} \left[ (1 + \kappa y) \frac{\partial T}{\partial y} \right] \right.$$

$$\left. + \frac{f'}{g(1 + \kappa y)} \frac{\partial T}{\partial x} + \frac{1}{g} \frac{\partial T}{\partial y} \right\} + 2\mu \left\{ \frac{1}{(1 + \kappa y)^2} \left( \frac{\partial v_x}{\partial x} + \kappa v_y \right)^2 + \left( \frac{\partial v_y}{\partial y} \right)^2 + \left( \frac{v_y}{g} \right)^2 \right.$$

$$\left. + \frac{1}{2} \left[ (1 + \kappa y) \frac{\partial}{\partial y} \left( \frac{v_x}{1 + \kappa y} \right) + \frac{1}{1 + \kappa y} \frac{\partial v_y}{\partial x} \right]^2 \right\} \tag{6.7.5-3}$$

where we define

$$\kappa = \kappa(x) \equiv \frac{-f''}{[1 + (f')^2]^{3/2}} \tag{6.7.5-4}$$

and

$$g \equiv f[1 + (f')^2]^{1/2} + y \tag{6.7.5-5}$$

The primes here indicate differentiation with respect to the cylindrical coordinate $z$ measured along the axis of revolution. We may think of $-\kappa$ as the normal curvature (McConnell 1957, p. 210) of the surface in the direction $x$; it is also one of the principal curvatures (McConnell 1957, p. 211) of the surface.

In addition to the dimensionless velocity, dimensionless temperature, and dimensionless time suggested in the introduction to Section 6.4, let us define

$$x^\star \equiv \frac{x}{L}, \qquad y^\star \equiv \frac{y}{L}$$
$$f^\star \equiv \frac{f}{L}, \qquad \kappa^\star \equiv \kappa L \tag{6.7.5-6}$$

If we extend the arguments of Section 6.7.1 to this geometry, we are motivated to express our results in terms of

$$y^{\star\star} \equiv \sqrt{N_{Re}}\, y^\star$$
$$v_y^{\star\star} \equiv \sqrt{N_{Re}}\, v_y^\star \tag{6.7.5-7}$$

For $N_{Re} \gg 1$, Equation (6.7.5-3) appears to reduce to

$$\frac{1}{N_{St}} \frac{\partial T^\star}{\partial t^\star} + \frac{\partial T^\star}{\partial x^\star} \frac{v_x^\star}{1 + \kappa^{\star\star} y^{\star\star}} + \frac{\partial T^\star}{\partial y^{\star\star}} v_y^{\star\star}$$
$$= \frac{1}{N_{Pr}} \left\{ \frac{1}{1 + \kappa^{\star\star} y^{\star\star}} \frac{\partial}{\partial y^{\star\star}} \left[ (1 + \kappa^{\star\star} y^{\star\star}) \frac{\partial T^\star}{\partial y^{\star\star}} \right] \right\}$$
$$+ \frac{N_{Br}}{N_{Pr}} \left[ (1 + \kappa^{\star\star} y^{\star\star}) \frac{\partial}{\partial y^{\star\star}} \left( \frac{v_x^\star}{1 + \kappa^{\star\star} y^{\star\star}} \right) \right]^2 \tag{6.7.5-8}$$

in which

$$\kappa^{\star\star} \equiv N_{Re}^{-1/2} \kappa^\star \tag{6.7.5-9}$$

For a fixed-wall configuration, $\kappa^{\star\star} \to 0$ in the limit $N_{Re} \gg 1$, and (6.7.5-8) further simplifies to

$$\frac{1}{N_{St}} \frac{\partial T^\star}{\partial t^\star} + \frac{\partial T^\star}{\partial x^\star} v_x^\star + \frac{\partial T^\star}{\partial y^{\star\star}} v_y^{\star\star} = \frac{1}{N_{Pr}} \frac{\partial^2 T^\star}{\partial y^{\star\star 2}} + \frac{N_{Br}}{N_{Pr}} \left( \frac{\partial v_x^\star}{\partial y^{\star\star}} \right)^2 \tag{6.7.5-10}$$

Equation (6.7.5-10) indicates that the differential energy balance for the thermal boundary layer has the same form for plane flow past flat plates, plane flow past curved walls, and axisymmetric flow past bodies of revolution. But if the differential energy balance must be solved simultaneously with the differential mass balance and the form of the differential momentum balance appropriate to the boundary layer, we see from Section 3.5.6 that the differential mass balance appropriate to the boundary layer of a body of revolution has a somewhat different form from that for a flat plate. In other words, the total boundary-value problem describing simultaneous momentum and energy transfer within a boundary layer on a body of revolution differs from the similar boundary-value problem for a flat plate.

In Section 3.5.6, we discussed the transformation suggested by Mangler (Schlichting 1979, p. 245; see also Exercises 3.5.6-3 and 3.5.6-4) by means of which the boundary-layer equations appropriate to a body of revolution can be transformed into those for a flat plate. Mangler suggests the introduction of the following variables:

$$\overline{x}^\star \equiv \int_0^{x^\star} f^{\star 2} dx^\star \tag{6.7.5-11}$$

$$\overline{y}^{\star\star} \equiv f^\star y^{\star\star} \tag{6.7.5-12}$$

$$\overline{t}^\star \equiv f^{\star 2} t^\star \tag{6.7.5-13}$$

$$\overline{v}_y^{\star\star} \equiv \frac{1}{f^\star} v_y^{\star\star} + \frac{f^{\star\prime}}{f^{\star 2}[1 + (f^{\star\prime})^2]^{1/2}} y^{\star\star} v_x^\star \tag{6.7.5-14}$$

With this change of variables, (6.7.5-10) becomes

$$\frac{1}{N_{St}} \frac{\partial T^\star}{\partial \overline{t}^\star} + \frac{\partial T^\star}{\partial \overline{x}^\star} v_x^\star + \frac{\partial T^\star}{\partial \overline{y}^{\star\star}} \overline{v}_y^{\star\star} = \frac{1}{N_{Pr}} \frac{\partial^2 T^\star}{\partial \overline{y}^{\star\star 2}} + \frac{N_{Br}}{N_{Pr}} \left( \frac{\partial v_x^\star}{\partial \overline{y}^{\star\star}} \right)^2 \tag{6.7.5-15}$$

This confirms our original intuitive feelings. The mathematical problems that describe boundary-layer flow past a body of revolution and plane flow past a flat plate can have the same form.

**Exercise 6.7.5-1** *Derivation of (6.7.5-3)*   Noting the results of Exercise 3.5.6-1, derive (6.7.5-3) starting from Table 5.4.1-1.

**Exercise 6.7.5-2** *Derivation of (6.7.5-8)*   Introduce in (6.7.5-3) the dimensionless velocity, dimensionless temperature, and dimensionless time defined in the introduction to Section 6.4, as well as those dimensionless variables defined in (6.7.5-6) and (6.7.5-7). Construct the reasoning that leads to (6.7.5-8).

**Exercise 6.7.5-3** *Mangler's transformation* (Schlichting 1979, p. 245)   Starting with (6.7.5-10), use (6.7.5-11) through (6.7.5-14) to change variables and arrive at (6.7.5-15). The results of Exercise 3.5.6-3 are helpful.

### 6.7.6   Energy Transfer in the Entrance of a Heated Section of a Tube

An incompressible Newtonian fluid with constant viscosity and thermal conductivity flows through a tube of radius $R$. For $z < 0$, the wall of the tube is insulated:

$$\text{at } r = R \text{ for } z < 0: \quad \frac{\partial T}{\partial r} = 0 \tag{6.7.6-1}$$

For $z > 0$, the temperature of the wall is maintained constant at $T_1$:

$$\text{at } r = R \text{ for } z > 0: \quad T = T_1 \tag{6.7.6-2}$$

Very far upstream of the entrance to this heated section, the fluid is known to be at a uniform temperature $T_0$:

$$\text{as } z \to -\infty \text{ for } r \leq R : \ T \to T_0 \qquad (6.7.6\text{-}3)$$

We wish to determine the rate of energy transfer to the fluid in the heated portion of the tube. For the moment we will focus our attention on the entrance to this heated section.

Since the viscosity and density of this fluid are taken to be constants independent of temperature, the velocity distribution is that found in Section 3.2.1:

$$v_r = v_\theta = 0$$

$$v_z = v_{z(\text{max})} \left[ 1 - \left( \frac{r}{R} \right)^2 \right] \qquad (6.7.6\text{-}4)$$

where $v_{z(\text{max})}$ is the $z$ component of velocity along the centerline.

The boundary conditions on temperature and the known velocity distribution suggest that the temperature distribution in the heated portion of the tube is axisymmetric:

$$T = T(r, z) \qquad (6.7.6\text{-}5)$$

From Table 5.4.1-3, we see that the differential energy balance for this situation is

$$\rho \hat{c} v_z \frac{\partial T}{\partial z} = k \left[ \frac{1}{r} \frac{\partial}{\partial r} \left( r \frac{\partial T}{\partial r} \right) + \frac{\partial^2 T}{\partial z^2} \right] + \mu \left( \frac{\partial v_z}{\partial r} \right)^2 \qquad (6.7.6\text{-}6)$$

or

$$\left( 1 - r^{\star 2} \right) \frac{\partial T^\star}{\partial z^\star} = \frac{1}{N_{Pe}} \left[ \frac{1}{r^\star} \frac{\partial}{\partial r^\star} \left( r^\star \frac{\partial T^\star}{\partial r^\star} \right) + \frac{\partial^2 T^\star}{\partial z^{\star 2}} \right] + \frac{4 N_{Br}}{N_{Pe}} r^{\star 2} \qquad (6.7.6\text{-}7)$$

Here the dimensionless temperature $T^\star$, dimensionless radial coordinate $r^\star$, and dimensionless axial coordinate $z^\star$ are defined as

$$T^\star \equiv \frac{T - T_0}{T_1 - T_0}$$

$$r^\star \equiv \frac{r}{R} \qquad (6.7.6\text{-}8)$$

$$z^\star \equiv \frac{z}{R}$$

The Peclet and Brinkman numbers are

$$N_{Pe} \equiv \frac{\hat{c} R v_{z(\text{max})} \rho}{k}$$

$$N_{Br} \equiv \frac{\mu v_{z(\text{max})}^2}{k(T_1 - T_0)} \qquad (6.7.6\text{-}9)$$

In the initial statement of this problem, we limit ourselves to the entrance of the heated portion of this tube. For $N_{Re} \gg 1$ for a specified value of $N_{Pr}$ (for $N_{Pe} \gg 1$), we anticipate a thermal boundary layer developing along the wall of the tube. It is this thermal boundary layer with which we are primarily concerned in this section.

Our approach in describing this thermal boundary layer is very similar to that taken in the preceding sections, where we were concerned with thermal boundary layers on submerged bodies. But there is one important change. In the preceding sections, the velocity distribution was obtained by making the boundary-layer approximations in the differential momentum balance. Here we know the velocity distribution throughout the fluid *a priori*. This will require a different definition for the expanded coordinate to be used in describing the boundary layer.

Since we are primarily concerned with the thermal boundary layer along the wall of the tube, let us introduce

$$s^\star \equiv 1 - r^\star \tag{6.7.6-10}$$

a dimensionless distance measured from the wall. In terms of $s^\star$, Equation (6.7.6-7) becomes

$$\left(2s^\star - s^{\star 2}\right) \frac{\partial T^\star}{\partial z^\star}$$

$$= \frac{1}{N_{Pe}} \left( \frac{\partial^2 T^\star}{\partial s^{\star 2}} - \frac{1}{1 - s^\star} \frac{\partial T^\star}{\partial s^\star} + \frac{\partial^2 T^\star}{\partial z^{\star 2}} \right) + \frac{4 N_{Br}}{N_{Pe}} (1 - s^\star)^2 \tag{6.7.6-11}$$

We are concerned with a very thin boundary layer that we intuitively feel must get thinner at a fixed value of $z^\star$ for $N_{Re} \gg 1$ and a fixed value of $N_{Pr}$. The discussion in Sections 3.5.1 and 6.7.1 motivates us to introduce as an expanded coordinate

$$s^{\star\star} \equiv (N_{Pe})^a s^\star \tag{6.7.6-12}$$

There is a slight advantage in working in terms of $s^{\star\star}$, in that $N_{Pr}$ is eliminated from the final result. In terms of $s^{\star\star}$, (6.7.6-11) becomes

$$\left(2s^{\star\star} - N_{Pe}^{-a} s^{\star\star 2}\right) \frac{\partial T^\star}{\partial z^\star}$$

$$= N_{Pe}^{3a-1} \frac{\partial^2 T^\star}{\partial s^{\star\star 2}} - \frac{N_{Pe}^{2a-1}}{1 - N_{Pe}^{-a} s^{\star\star}} \frac{\partial T^\star}{\partial s^{\star\star}} + N_{Pe}^{a-1} \frac{\partial^2 T^\star}{\partial z^{\star 2}}$$

$$+ 4 N_{Br} N_{Pe}^{a-1} (1 - N_{Pe}^{-a} s^{\star\star})^2 \tag{6.7.6-13}$$

In order that some of the conduction terms survive for $N_{Pe} \gg 1$ and finite values of $N_{Br}$, we conclude that

$$a = \frac{1}{3} \tag{6.7.6-14}$$

The form of the differential energy balance appropriate to the thermal boundary layer in this problem is

$$\text{for } N_{Pe} \gg 1 : \quad 2s^{\star\star} \frac{\partial T^\star}{\partial z^\star} = \frac{\partial^2 T^\star}{\partial s^{\star\star 2}} \tag{6.7.6-15}$$

In view of (6.7.6-14), the dependence upon $N_{Re}$ in the expanded coordinate for the thermal boundary layer differs from that ($a = 1/2$) introduced in Section 6.7.1. The reason is that in Sections 6.7.1 to 6.7.5 the velocity distribution is determined by a boundary-layer analysis. In this problem we have an exact solution for the velocity distribution throughout the entire flow, both within and outside of the thermal boundary layer.

Since we are neglecting both axial conduction and viscous dissipation in this analysis, we may rewrite boundary condition (6.7.6-3) as

$$\text{at } z = 0 \text{ for all } r < R : \quad T = T_0 \tag{6.7.6-16}$$

If $\tilde{T}$ represents the temperature distribution for the nonviscous, nonconducting external flow evaluated at the wall, the differential energy balance requires

$$\tilde{v}_z \frac{\partial \tilde{T}}{\partial z} = 0 \tag{6.7.6-17}$$

We conclude that

$$\text{for } s^{\star\star} \to \infty \text{ and all } z^\star > 0 : \ T^\star \to \tilde{T}^\star = 0 \tag{6.7.6-18}$$

This, together with (6.7.6-2) in the form of

$$\text{for } s^{\star\star} = 0 \text{ and all } z^\star > 0 : \ T^\star = 1 \tag{6.7.6-19}$$

gives the boundary conditions that must be satisfied in solving (6.7.6-15).
    With the change of variable

$$\eta \equiv \frac{s^{\star\star}}{\sqrt[3]{9z^\star/2}} \tag{6.7.6-20}$$

Equation (6.7.6-15) is reduced to an ordinary differential equation:

$$-3\eta^2 \frac{dT^\star}{d\eta} = \frac{d^2 T^\star}{d\eta^2} \tag{6.7.6-21}$$

Boundary conditions (6.7.6-18) and (6.7.6-19) are transformed into

$$\text{as } \eta \to \infty : \ T^\star \to 0 \tag{6.7.6-22}$$

and

$$\text{at } \eta = 0 : \ T^\star = 1 \tag{6.7.6-23}$$

A solution consistent with (6.7.6-21) through (6.7.6-23) is readily found to be

$$T^\star = \frac{\int_\eta^\infty \exp(-\eta^3)\,d\eta}{\int_0^\infty \exp(-\eta^3)\,d\eta} \tag{6.7.6-24}$$

In terms of the gamma function,

$$\Gamma(n) \equiv \int_0^\infty x^{n-1} e^{-x}\,dx \tag{6.7.6-25}$$

Equation (6.7.6-24) can be written somewhat more conveniently as

$$T^\star = \frac{1}{\Gamma\left(\frac{4}{3}\right)} \int_\eta^\infty \exp(-\eta^3)\,d\eta \tag{6.7.6-26}$$

Since

$$-q_r = k \frac{\partial T}{\partial r}$$

$$= -\frac{k(T_1 - T_0)}{R} N_{Pe}^{1/3} \frac{1}{\sqrt[3]{9z^\star/2}} \frac{dT^\star}{d\eta}$$

$$= \frac{k(T_1 - T_0)}{R} N_{Pe}^{1/3} \frac{1}{\sqrt[3]{9z^\star/2}} \frac{1}{\Gamma\left(\frac{4}{3}\right)} \exp(-\eta^3) \tag{6.7.6-27}$$

we can readily calculate the average energy flux from the wall to the fluid in a heated portion of tube of length $L$ to be

$$(-q_r|_{r=R})_{av} \equiv \frac{R}{L} \int_0^{L/R} -q_r|_{r=R} \, dz^\star$$

$$= \left(\frac{9}{2}\right)^{2/3} \frac{k(T_1 - T_0)}{R} N_{Pe}{}^{1/3} \frac{1}{\Gamma\left(\frac{1}{3}\right)} \left(\frac{R}{L}\right)^{1/3} \tag{6.7.6-28}$$

In terms of the Nusselt number,

$$N_{Nu} \equiv 2 \frac{(-q_r|_{r=R})_{av} R}{(T_1 - T_0)k} \tag{6.7.6-29}$$

this is somewhat more conveniently written as

$$N_{Nu} = \left(\frac{9}{2}\right)^{2/3} \frac{2}{\Gamma\left(\frac{1}{3}\right)} N_{Pe}{}^{1/3} \left(\frac{R}{L}\right)^{1/3} \tag{6.7.6-30}$$

With the abrupt change in boundary conditions described by (6.7.6-1) and (6.7.6-2), we can anticipate that axial conduction neglected in the above analysis must in fact be significant in a small region near $z = 0$ and $r = R$. This is confirmed by Newman's (1969) numerical solution for this region. Not unexpectedly, he finds that this region where axial conduction cannot be neglected becomes smaller as $N_{Pe}$ increases.

For more about this and similar problems, see the exercises that follow as well as Sections 6.8.1 and 6.8.2.

**Exercise 6.7.6-1** *Derivation of (6.7.6-26)* Integrate (6.7.6-21) consistent with (6.7.6-22) and (6.7.6-23) to arrive at (6.7.6-26).

**Exercise 6.7.6-2** *Forced convection in a tube with constant energy flux at the wall* Let us repeat the problem of the text, replacing boundary conditions (6.7.6-2) by

$$\text{at } r = R \text{ for } z > 0 : \quad k\frac{\partial T}{\partial r} = q$$
$$= \text{a constant}$$

i) Introduce the dimensionless temperature

$$T^\star = \frac{T - T_0}{T_1}$$

where $T_1$ is a characteristic temperature that will be defined in such a way as to make the boundary-value problem as simple as possible. Repeat the discussion in the text to conclude that, for the thermal boundary layer in the entrance of the heated portion of the tube, $T^\star$ must satisfy

$$2s^{\star\star} \frac{\partial T^\star}{\partial z^\star} = \frac{\partial^2 T^\star}{\partial s^{\star\star 2}}$$

for

$$N_{Pe} \equiv \frac{\hat{c} R v_{z(\max)} \rho}{k}$$
$$\gg 1$$

consistent with the boundary conditions

$$\text{as } s^{**} \to \infty \text{ for all } z^* > 0 : \ T^* \to 0$$

and

$$\text{at } s^{**} = 0 \text{ for all } z^* > 0 : \ \frac{\partial T^*}{\partial s^{**}} = -1$$

In arriving at this form of the problem, you will find it necessary to define

$$T_1 \equiv \frac{Rq}{kN_{Pe}^{1/3}}$$

ii) Determine that

$$u \equiv \frac{\partial T^*}{\partial s^{**}}$$

must satisfy

$$2\frac{\partial u}{\partial z^*} = \frac{\partial}{\partial s^{**}} \left( \frac{1}{s^{**}} \frac{\partial u}{\partial s^{**}} \right)$$

consistent with the boundary conditions

$$\text{as } s^{**} \to \infty \text{ for all } z^* > 0 : \ u \to 0$$

and

$$\text{at } s^{**} = 0 : \ u = -1$$

iii) Anticipate a solution to the problem posed in (ii) of the form

$$u = u(\eta)$$

where

$$\eta \equiv \frac{s^{**}}{\sqrt[3]{9z^*/2}}$$

Conclude that

$$\frac{\partial T^*}{\partial s^{**}} = u$$

$$= -\frac{\int_\eta^\infty \eta \exp(-\eta^3)\, d\eta}{\int_0^\infty \eta \exp(-\eta^3)\, d\eta}$$

$$= \frac{-3}{\Gamma\left(\frac{2}{3}\right)} \int_\eta^\infty \eta \exp(-\eta^3)\, d\eta \tag{6.7.6-31}$$

iv) It is clear that

$$T^* = T^*(\eta, s^{**})$$

or

$$T^* = T^*(\eta, z^*)$$

Let us take this later point of view and write (6.7.6-31) in the form

$$\frac{\partial T^\star}{\partial \eta} \frac{1}{\sqrt[3]{9z^\star/2}} = \frac{-3}{\Gamma\left(\frac{2}{3}\right)} \int_\eta^\infty \eta \exp(-\eta^3)\, d\eta \tag{6.7.6-32}$$

The last boundary condition to be satisfied is that

$$\text{as } \eta \to \infty \text{ for all } z^\star > 0 : \quad T^\star \to 0 \tag{6.7.6-33}$$

We consequently may integrate (6.7.6-32) consistent with (6.7.6-33) to find[2]

$$\frac{T - T_0}{Rq/k} = \sqrt[3]{\frac{9z^\star/2}{N_{Pe}}} \left\{ \eta \left[ \frac{\Gamma\left(\frac{2}{3}; \eta^3\right)}{\Gamma\left(\frac{2}{3}\right)} - 1 \right] + \frac{\exp(-\eta^3)}{\Gamma\left(\frac{2}{3}\right)} \right\}$$

We have introduced here the incomplete gamma function,

$$\Gamma\left(\frac{2}{3}; \eta^3\right) \equiv \int_0^{\eta^3} x^{\frac{2}{3}-1} \exp(-x)\, dx$$

v) If we introduce the Nusselt number,

$$N_{Nu} \equiv \frac{2qR}{[(T|_{r=R})_{av} - T_0]k}$$

conclude that

$$N_{Nu} = \frac{8\Gamma\left(\frac{2}{3}\right)}{3(9/2)^{1/3}} N_{Pe}{}^{1/3} \left(\frac{R}{L}\right)^{1/3}$$

**Exercise 6.7.6-3** *Heat transfer from an isothermal wall to a falling film*   An incompressible Newtonian fluid flows down an inclined plane as shown in Figure 3.2.5-4 (see also Exercise 3.2.5-5). The wall is insulated for $z_1 < 0$:

$$\text{at } z_2 = 0 \text{ for } z_1 < 0 : \quad \frac{\partial T}{\partial z_2} = 0$$

For $z_1 > 0$, the wall is maintained at a constant temperature $T_1$:

$$\text{at } z_2 = 0 \text{ for } z_1 > 0 : \quad T = T_1$$

Very far upstream from the entrance to the heated portion of the wall, the fluid has a uniform temperature $T_0$:

$$\text{as } z_1 \to -\infty \text{ for } 0 \le z_2 \le \delta : \quad T \to T_0$$

Following the general outline of the discussion of the text, determine that the temperature distribution in the fluid near the entrance to the heated section has the form

$$T^\star \equiv \frac{T - T_0}{T_1 - T_0}$$

$$= \frac{1}{\Gamma\left(\frac{4}{3}\right)} \int_\eta^\infty \exp(-\eta^3)\, d\eta$$

---

[2] In the Bird et al. (1960, p. 309) solution to this problem, $v_0 = 2v_{z(max)}$.

in the limit as the Peclet number

$$N_{Pe} \equiv \frac{\hat{c}\delta v_{1(\max)}\rho}{k}$$
$$\gg 1$$

where

$$\eta \equiv \frac{z_2^{\star\star}}{\sqrt[3]{9z_1^{\star}/2}}$$

and

$$z_1^{\star} \equiv \frac{z_1}{\delta}$$

$$z_2^{\star\star} \equiv (N_{Pe})^{1/3}\frac{z_2}{\delta}$$

Here $v_{1(\max)}$ is the maximum velocity of the fluid in the film:

$$v_{1(\max)} \equiv \frac{\delta^2 \rho g \cos\alpha}{2\mu}$$

**Exercise 6.7.6-4** *Heat transfer from a wall to a falling film with constant energy flux at the wall*    Repeat Exercise 6.7.6-3, replacing the isothermal boundary condition with

$$\text{at } z_2 = 0 \text{ for } z_1 > 0 : \ k\frac{\partial T}{\partial z_2} = q$$

Conclude that, for $N_{Pe} \gg 1$, the temperature distribution in the thermal boundary layer, near the entrance of the heated portion of the wall, is exactly the same as that found in Exercise 6.7.6-2 when $R$ is replaced by $\delta$ and $v_{z(\max)}$ by $v_{1(\max)}$ defined in Exercise 3.2.5-3.

**Exercise 6.7.6-5** *Heat transfer from a gas stream to a falling film*    An incompressible Newtonian fluid flows down an inclined plane as shown in Figure 3.2.5-4 (see Exercise 3.2.5-3). Let us assume that there is no energy transfer from the gas stream to the falling film for $z_1 < 0$:

$$\text{at } z_2 = \delta \text{ for } z_1 < 0 : \ \frac{\partial T}{\partial z_2} = 0$$

Outside the immediate neighborhood of the liquid film, the gas stream has a uniform temperature $T_1$. To simplify the problem somewhat, we will assume that for $z_1 > 0$ the temperature of the gas–liquid phase interface is $T_1$:

$$\text{at } z_2 = \delta \text{ for } z_1 > 0 : \ T = T_1$$

Very far upstream, the temperature of the fluid is uniform at $T_0$:

$$\text{as } z_1 \to -\infty \text{ for } 0 \leq z_2 \leq \delta : \ T \to T_0$$

Determine the temperature distribution in the thermal boundary layer near the entrance to the heated portion of the film for $N_{Pe} \gg 1$.

*Answer:*

$$T^{\star} \equiv \frac{T - T_0}{T_1 - T_0} = 1 - \operatorname{erf}\left(\frac{1 - z_2/\delta}{\sqrt{4kz_1/(\hat{c}\delta^2 v_{1(\max)}\rho)}}\right)$$

*Hint:* This problem is similar to the one discussed in the text, but the expanded $z_2$ coordinate is defined differently.

**Exercise 6.7.6-6** *More on heat transfer from a gas stream to a falling film* Let us repeat Exercise 6.7.6-5 and attempt to describe the boundary condition at the gas–liquid phase interface more realistically. Rather than saying that the phase interface is in equilibrium with the gas very far away from it, let us describe the energy transfer in terms of Newton's "law" of cooling (see Section 6.2.2).

*Answer:*

$$T^\star \equiv \frac{T - T_1}{T_0 - T_1}$$

$$= \mathrm{erf}\left(\frac{s^{\star\star}}{\sqrt{4z_1^\star}}\right) + \exp\left(\frac{s^{\star\star}}{B} + \frac{z_1^\star}{B^2}\right)\left[1 - \mathrm{erf}\left(\frac{s^{\star\star}}{\sqrt{4z_1^\star}} + \frac{\sqrt{z_1^\star}}{B}\right)\right]$$

$$s^{\star\star} \equiv \sqrt{N_{Pe}}\left(1 - \frac{z_2}{\delta}\right)$$

$$z_1^\star \equiv \frac{z_1}{\delta}$$

$$B \equiv \frac{(N_{Pe})^{1/2}}{N_{Nu}}$$

$$N_{Pe} \equiv \frac{\hat{c}\,\delta v_{1(\max)}\rho}{k}$$

$$N_{Nu} \equiv \frac{h\,\delta}{k}$$

*Hint:* See Section 6.2.2 as well as the hint for Exercise 6.7.6-5.

**Exercise 6.7.6-7** *Energy transfer in the entrance of a heated section of a tube for a power-law fluid* Repeat the problem discussed in the text for a power-law fluid to conclude that the temperature distribution is again given by (6.7.6-26), where

$$\eta \equiv \frac{s^{\star\star}}{\sqrt[3]{9nz^\star/(1+n)}}$$

This means that the $N_{Nu}$ as defined by (6.7.6-29) becomes

$$N_{Nu} = 9\left(\frac{1+n}{9n}\right)^{1/3}\frac{1}{\Gamma\left(\frac{1}{3}\right)}(N_{Pe})^{1/3}\left(\frac{R}{L}\right)^{1/3}$$

## 6.7.7 More on Melt Spinning

In our discussion of melt spinning in Section 3.3.4, we assumed that the temperature of the polymer–gas interface was independent of axial position. I would like to address that issue here.

The static gas outside the boundary layer, formed adjacent to the polymer–gas interface, has a uniform temperature $T_\infty^{(g)}$. In what follows, we will confine our attention to the momentum and energy transfer within the boundary layer. Our objective is to determine within a constant the temperature at the polymer–gas interface.

We will follow the suggestion of Section 6.7.5, working in the orthogonal coordinate system recommended in Section 3.5.6 for bodies of revolution. In view of (3.3.4-22) and (3.3.4-29),

$$\frac{dR^\star/dz^\star}{R^\star\left[1+(dR^\star/dz^\star)^2\right]^{1/2}} = -\frac{a^\star}{2\left(a^\star z^\star + 1\right)^2\left[1+a^\star\left(a^\star z^\star + 1\right)^{-3/2}\right]^{1/2}}$$

$$\ll 1 \tag{6.7.7-1}$$

In this limit, the differential mass and momentum balances for the boundary layer take the same forms as they do for a boundary layer on a flat plate (Section 3.5.1).

### Surrounding Air: Momentum Transfer

With constant physical properties, the reasoning of Sakiadis (1961a,b; see also Exercise 3.5.2-3), applies directly. We conclude that

$$v_x^{(a)\star} \equiv \frac{v_x^{(g)}}{V}$$

$$= \frac{\partial\psi}{\partial y^{\star\star}}$$

$$= f'$$

$$v_y^{(a)\star\star} \equiv \sqrt{N_{Re}^{(g)}}\,\frac{v_y^{(g)}}{V}$$

$$= -\frac{\partial\psi}{\partial x^\star}$$

$$= \frac{1}{2\sqrt{x^\star}}(\eta f' - f) \tag{6.7.7-2}$$

where

$$\psi = \sqrt{x^\star}f(\eta) \tag{6.7.7-3}$$

and

$$\eta \equiv \frac{y^{\star\star}}{\sqrt{x^\star}}$$

$$= \frac{\sqrt{N_{Re}^{(g)}}\,y^\star}{\sqrt{x^\star}} \tag{6.7.7-4}$$

The function $f(\eta)$ is the solution of

$$ff'' + 2f''' = 0 \tag{6.7.7-5}$$

consistent with the boundary conditions

$$\text{at } \eta = 0: \quad f = 0, \quad f' = 1 \tag{6.7.7-6}$$

and

$$\text{as } \eta \to \infty : \quad f' \to 0 \tag{6.7.7-7}$$

Here the primes denote differentiation with respect to $\eta$.

## Surrounding Air: Energy Transfer

Let us define

$$T^\star \equiv \frac{T - T_\infty^{(g)}}{T_0^{(p)} - T_\infty^{(g)}} \tag{6.7.7-8}$$

In view of (6.7.7-2), Equation (6.7.1-9) can be written in terms of $\eta$ as

$$\frac{f}{2} \frac{dT^{(a)\star}}{d\eta} + \frac{1}{N_{Pr}^{(g)}} \frac{d^2 T^{(a)\star}}{d\eta^2} + \frac{N_{Br}^{(g)}}{N_{Pr}^{(g)}} f''^2 = 0 \tag{6.7.7-9}$$

in which the prime here denotes differentiation with respect to $\eta$ and

$$N_{Br}^{(g)} \equiv \frac{V^2 \mu^{(g)}}{k^{(g)} \left( T_{\text{initial}}^{(p)} - T_\infty^{(l)} \right)} \tag{6.7.7-10}$$

This equation is to be solved consistent with (6.7.1-12) in the form

$$\text{as } \eta \to \infty : \quad T^{(a)\star} \to 0 \tag{6.7.7-11}$$

as well as continuity of temperature and the jump energy balance at the polymer–gas interface. This requires solving simultaneously for the temperature distribution in the polymer phase, which we have not done.

For present purposes it is sufficient to recognize that, at the polymer–gas phase interface $\eta = 0$,

$$T^\star = T^\star(0)$$

$$= \text{a constant} \tag{6.7.7-12}$$

Of course, while this is a solution, it is not necessarily unique.

This section was prepared with the assistance of P. K. Dhori.

---

## 6.8 More About Energy Transfer in a Heated Section of a Tube

Let us go back and take another look at the problem discussed in Section 6.7.6. We will consider two cases: a constant wall temperature, which might correspond to a steam-jacketed pipe, and a constant energy flux at the wall, which might be thought of as a pipe wrapped with electrical tape.

### 6.8.1    Constant Temperature at Wall

An incompressible Newtonian fluid with constant viscosity and thermal conductivity flows through a tube of radius $R$. For $z < 0$, the wall of the tube is insulated:

$$\text{at } r = R \text{ for } z < 0 : \quad \frac{\partial T}{\partial r} = 0 \tag{6.8.1-1}$$

For $z > 0$, the temperature of the wall is maintained constant at $T_1$:

$$\text{at } r = R \text{ for } z > 0 : \quad T = T_1 \tag{6.8.1-2}$$

Very far upstream from the entrance to this heated section, the fluid is known to be at a uniform temperature $T_0$:

$$\text{as } z \to -\infty \text{ for } r \leq R : \quad T \to T_0 \tag{6.8.1-3}$$

In Section 6.7.6, we examined the temperature distribution and the rate of energy transfer to the fluid near the entrance to this heated portion of the tube for very large values of $N_{Pe}$. In what follows, we give our attention to the temperature distribution in the fluid somewhat downstream from the entrance. We will still restrict ourselves to the limit $N_{Pe} \gg 1$.

We continue to assume that the temperature distribution is axisymmetric:

$$T = T(r, z) \tag{6.8.1-4}$$

Applying the velocity distribution found in Section 3.2.1, we conclude that the dimensionless differential energy balance has a form similar to that found in Section 6.7.6:

$$\left(1 - r^{\star 2}\right) \frac{\partial T^\star}{\partial z^\star} = \frac{1}{N_{Pe}} \left[ \frac{1}{r^\star} \frac{\partial}{\partial r^\star} \left( r^\star \frac{\partial T^\star}{\partial r^\star} \right) + \frac{\partial^2 T^\star}{\partial z^{\star 2}} \right] + \frac{4 N_{Br}}{N_{Pe}} r^{\star 2} \tag{6.8.1-5}$$

where

$$T^\star \equiv \frac{T - T_1}{T_0 - T_1}$$
$$r^\star \equiv \frac{r}{R} \tag{6.8.1-6}$$
$$z^\star \equiv \frac{z}{R}$$

and

$$N_{Pe} \equiv \frac{\hat{c} R v_{z(\max)} \rho}{k}$$
$$N_{Br} \equiv \frac{\mu v_{z(\max)}^2}{k (T_0 - T_1)} \tag{6.8.1-7}$$

Intuitively, we expect that, sufficiently far downstream from the entrance, effects attributable to the curvature of the tube wall should become important, although axial conduction should continue to be a negligible effect. Since we are primarily concerned with what is happening at relatively large values of $z^\star$, we are motivated to introduce a contracted dimensionless axial coordinate

$$z^{\star\star} \equiv \frac{z^\star}{(N_{Pe})^b} \tag{6.8.1-8}$$

In terms of $z^{**}$, Equation (6.8.1-5) may be written as

$$\left(1 - r^{*2}\right) \frac{\partial T^*}{\partial z^{**}} = \frac{(N_{Pe})^{b-1}}{r^*} \frac{\partial}{\partial r^*} \left(r^* \frac{\partial T^*}{\partial r^*}\right) + (N_{Pe})^{-1-b} \frac{\partial^2 T^*}{\partial z^{**2}} + 4N_{Br} (N_{Pe})^{b-1} r^{*2}$$

(6.8.1-9)

In order that the convection and radial conduction terms be of the same order of magnitude for $N_{Pe} \gg 1$, we choose

$$b \equiv 1 \tag{6.8.1-10}$$

$$z^{**} \equiv \frac{z^*}{(N_{Pe})} \tag{6.8.1-11}$$

Then the differential energy balance (6.8.1-9) reduces to

$$\left(1 - r^{*2}\right) \frac{\partial T^*}{\partial z^{**}} = \frac{1}{r^*} \frac{\partial}{\partial r^*} \left(r^* \frac{\partial T^*}{\partial r^*}\right) + 4N_{Br} r^{*2} \tag{6.8.1-12}$$

For $N_{Pe} \gg 1$, this is the differential equation to be solved for the temperature distribution in the fluid relatively far downstream from the entrance to the heated section. The appropriate boundary conditions are

$$\text{at } r^* = 1 \text{ for } z^{**} < 0: \quad \frac{\partial T^*}{\partial r^*} = 0 \tag{6.8.1-13}$$

$$\text{at } r^* = 1 \text{ for } z^{**} > 0: \quad T^* = 0 \tag{6.8.1-14}$$

and

$$\text{as } z^{**} \to -\infty \text{ for } r^* \le 1: \quad T^* \to 1 \tag{6.8.1-15}$$

For sufficiently small values of the Brinkman number $N_{Br}$, we can neglect viscous dissipation and write (6.8.1-12) as

$$\left(1 - r^{*2}\right) \frac{\partial T^*}{\partial z^{**}} = \frac{1}{r^*} \frac{\partial}{\partial r^*} \left(r^* \frac{\partial T^*}{\partial r^*}\right) \tag{6.8.1-16}$$

Having neglected both axial conduction and viscous dissipation, we see that boundary conditions (6.8.1-13) and (6.8.1-15) imply

$$\text{at } z^{**} = 0 \text{ for } r^* < 1: \quad T^* = 1 \tag{6.8.1-17}$$

Our objective here is to solve (6.8.1-16) consistent with (6.8.1-14) and (6.8.1-17). This special case is known as the *Graetz* (1883, 1885) problem. In what follows, we are guided by the excellent summaries of this problem given by Jakob (1949, p. 451) and Brown (1960).

Seeking a separable solution to (6.8.1-16) of the form

$$T^* = X \left(z^{**}\right) Y \left(r^*\right) \tag{6.8.1-18}$$

we find

$$\frac{1}{X} \frac{dX}{dz^{**}} = \frac{1}{Y} \frac{1}{r^* \left(1 - r^{*2}\right)} \frac{d}{dr^*} \left(r^* \frac{dY}{dr^*}\right)$$

$$= -\lambda^2 \tag{6.8.1-19}$$

**Table 6.8.1-1.** Roots of (6.8.1-22) as computed by Peng (personal communication, 1997)

| $n$ | $\lambda_n$ |
|-----|-------------|
| 1 | 2.7043644199 |
| 2 | 6.6790314493 |
| 3 | 10.67337953817 |
| 4 | 14.6710784626 |
| 5 | 18.6698718682 |
| 6 | 22.6691434231 |
| 7 | 26.6686619960 |
| 8 | 30.6683231499 |
| 9 | 34.6680738224 |
| 10 | 38.6678798118 |
| 11 | 42.6676199803 |
| 12 | 46.6767891942 |

where $\lambda$ is a constant to be determined. We quickly come to the conclusion that

$$T^\star = C \exp\left(-\lambda^2 z^{\star\star}\right) Y\left(r^\star\right) \tag{6.8.1-20}$$

in which the function $Y\left(\lambda, r^\star\right)$ satisfies

$$\frac{1}{r^\star} \frac{d}{dr^\star}\left(r^\star \frac{dY}{dr^\star}\right) + \lambda^2\left(1 - r^{\star 2}\right) Y = 0 \tag{6.8.1-21}$$

A particular solution to (6.8.1-21) takes the form

$$Y\left(\lambda, r^\star\right) = \sum_{i=0}^{\infty} a_i r^{\star i} \tag{6.8.1-22}$$

in which

$$a_i = 0 \quad \text{if } i < 0$$

$$a_i = 1 \quad \text{if } i = 0 \tag{6.8.1-23}$$

$$a_i = -\left(\frac{\lambda}{i}\right)^2 (a_{i-2} - a_{i-4}) \quad \text{if } i > 0$$

In view of (6.8.1-14), we must require

$$Y\left(\lambda, 1\right) = 0 \tag{6.8.1-24}$$

The $n$ roots of this equation will be designated as $\lambda_n$ ($n = 1, 2, 3, \ldots$). The first eleven roots have been given by Brown (1960); they are shown in Table 6.8.1-1.

We now recognize that (6.8.1-20) should be written as a superposition of all possible solutions:

$$T^\star = \sum_{n=1}^{\infty} C_n \exp\left(-\lambda_n^2 z^{\star\star}\right) Y_n\left(r^\star\right) \tag{6.8.1-25}$$

Here the function $Y_n\left(\lambda, r^\star\right)$ is a solution to

$$\frac{1}{r^\star}\frac{d}{dr^\star}\left(r^\star \frac{dY_n}{dr^\star}\right) + \lambda_n^2\left(1 - r^{\star 2}\right) Y_n = 0 \tag{6.8.1-26}$$

or

$$Y_n\left(\lambda, r^\star\right) = \sum_{i=0}^{\infty} a_{n,i} r^{\star i} \tag{6.8.1-27}$$

in which

$$a_{n,i} = 0 \qquad \text{if } i < 0$$

$$a_{n,i} = 1 \qquad \text{if } i = 0 \tag{6.8.1-28}$$

$$a_{n,i} = -\left(\frac{\lambda_n}{i}\right)^2 \left(a_{n,i-2} - a_{n,i-4}\right) \qquad \text{if } i > 0$$

Graetz proved that if $m \neq n$ (Jakob 1949, p. 453)

$$\int_0^1 Y_n Y_m r^\star \left(1 - r^{\star 2}\right) dr^\star = 0 \tag{6.8.1-29}$$

and otherwise

$$\int_0^1 Y_n{}^2 r^\star \left(1 - r^{\star 2}\right) dr^\star = \frac{1}{2\lambda_n}\left[\frac{\partial Y_n}{\partial \lambda_n}\frac{\partial Y_n}{\partial r^\star}\right]_{r^\star = 1} \tag{6.8.1-30}$$

In view of (6.8.1-17), Equation (6.8.1-23) requires

$$1 = \sum_{n=1}^{\infty} C_n Y_n\left(r^\star\right) \tag{6.8.1-31}$$

Multiplying both sides of (6.8.1-29) by $r^\star(1 - r^{\star 2})Y_n$ and integrating, we can use (6.8.1-29) and (6.8.1-30) to find

$$\int_0^1 Y_n r^\star \left(1 - r^{\star 2}\right) dr^\star = C_n \int_0^1 Y_n{}^2 r^\star \left(1 - r^{\star 2}\right) dr^\star$$

$$= C_n \frac{1}{2\lambda_n}\left[\frac{\partial Y_n}{\partial \lambda_n}\frac{\partial Y_n}{\partial r^\star}\right]_{r^\star = 1} \tag{6.8.1-32}$$

Employing (6.8.1-26), we can express

$$\int_0^1 Y_n r^\star \left(1 - r^{\star 2}\right) dr^\star = -\frac{1}{\lambda_n{}^2}\left(r^\star \frac{\partial Y_n}{\partial r^\star}\right)_{r^\star = 1} \tag{6.8.1-33}$$

This permits us to conclude from (6.8.1-32) that

$$C_n = -\frac{2}{\lambda_n(\partial Y_n/\partial \lambda_n)_{r^\star = 1}} \tag{6.8.1-34}$$

Let us assume that our primary interest is in the film coefficient for heat transfer $h$, where

$$-q_r|_{r=R} = h(T_1 - T_m) \tag{6.8.1-35}$$

and $T_m$ is the mean or *cup-mixing* temperature:

$$T_m^* \equiv \int_0^1 T^* \left(1 - r^{*2}\right) r^* \, dr^* \left[\int_0^1 \left(1 - r^{*2}\right) r^* \, dr^*\right]^{-1}$$

$$= 4 \int_0^1 T^* \left(1 - r^{*2}\right) r^* \, dr^* \tag{6.8.1-36}$$

Alternatively, we can write

$$-q_r|_{r=R} = k \left.\frac{\partial T}{\partial r}\right|_{r=R}$$

$$= \frac{k(T_0 - T_1)}{R} \left.\frac{\partial T^*}{\partial r^*}\right|_{r^*=1}$$

$$= \frac{k(T_0 - T_1)}{R} \sum_{n=1}^{\infty} \left[C_n \exp\left(-\lambda_n^2 z^{\star\star}\right) \sum_{i=2}^{\infty} i a_{n,i}\right] \tag{6.8.1-37}$$

Using (6.8.1-35) and (6.8.1-37), we can express the Nusselt number as

$$N_{Nu} \equiv \frac{hR}{k}$$

$$= -\frac{1}{T_m^*} \sum_{n=1}^{\infty} \left[C_n \exp\left(-\lambda_n^2 z^{\star\star}\right) \sum_{i=2}^{\infty} i a_{n,i}\right] \tag{6.8.1-38}$$

or the mean Nusselt number for a length of the heated section as

$$N_{Nu,m} \equiv \frac{1}{L^{**}} \int_0^{L^{**}} N_{Nu} \, dz^{**}$$

$$= -\sum_{n=1}^{\infty} \left\{C_n \frac{1}{L^{**}} \int_0^{L^{**}} \left[\frac{1}{T_m^*} \exp\left(-\lambda_n^2 z^{\star\star}\right)\right] dz^{**} \sum_{i=2}^{\infty} i a_{n,i}\right\} \tag{6.8.1-39}$$

Figure 6.8.1-1 compares the predictions of (6.8.1-39) with data presented by Kays (1955, Fig. 7). Except for the first point on the left, the data appear to approach the prediction in the limit as $1/z^{**} \to 0$, as required.

This section was prepared with the help of P. K. Dhori.

**Exercise 6.8.1-1**    Formalize the argument that takes boundary conditions (6.8.1-13) and (6.8.1-15) into (6.8.1-17).

### 6.8.2    Constant Energy Flux at Wall

Instead of specifying the temperature of the wall as in Section 6.8.1, let us instead specify a constant energy flux.

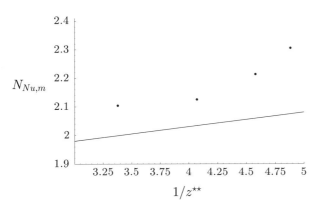

$N_{Nu,m}$

$1/z^{\star\star}$

**Figure 6.8.1-1.** $N_{Nu,m}$ as a function of $1/z^{\star\star}$ for a constant wall temperature. The prediction of (6.8.1-39) is compared with data presented by Kays (1955, Fig. 7).

To be specific, we again have an incompressible Newtonian fluid with constant viscosity and thermal conductivity flowing through a tube of radius $R$. For $z < 0$, the wall of the tube is insulated:

$$\text{at } r = R \text{ for } z < 0: \quad \frac{\partial T}{\partial r} = 0 \tag{6.8.2-1}$$

But for $z > 0$, the radial component of the energy flux vector is a constant $q$:

$$\text{at } r = R \text{ for } z > 0: \quad k\frac{\partial T}{\partial r} = q \tag{6.8.2-2}$$

Very far upstream of the entrance to this heated section, the fluid is known to be at a uniform temperature $T_0$:

$$\text{as } z \to -\infty \text{ for } r \leq R: \quad T \to T_0 \tag{6.8.2-3}$$

We will focus our attention upon the temperature distribution in the fluid very far downstream from the entrance to the heated section. As in Section 6.8.1, we will still restrict ourselves to the limit $N_{Pe} \gg 1$.

Repeating the argument in Section 6.8.1, we find that for $N_{Pe} \gg 1$ the dimensionless form of the differential energy balance appropriate somewhat downstream from the entrance of the tube is

$$\left(1 - r^{\star 2}\right)\frac{\partial T^{\star}}{\partial z^{\star\star}} = \frac{1}{r^{\star}}\frac{\partial}{\partial r^{\star}}\left(r^{\star}\frac{\partial T^{\star}}{\partial r^{\star}}\right) + 4N_{Br}r^{\star 2} \tag{6.8.2-4}$$

where

$$T^{\star} \equiv \frac{T - T_0}{qR/k}$$

$$r^{\star} \equiv \frac{r}{R} \tag{6.8.2-5}$$

$$z^{\star\star} \equiv \frac{z}{RN_{Pe}}$$

and

$$N_{Pe} \equiv \frac{\hat{c} R v_{z(\max)} \rho}{k}$$

$$N_{Br} \equiv \frac{\mu v_{z(\max)}^2}{q R} \tag{6.8.2-6}$$

Equation (6.8.2-4) must be solved consistent with the boundary conditions

at $r^\star = 1$ for $z^{\star\star} < 0$ : $\dfrac{\partial T^\star}{\partial r^\star} = 0$ $\tag{6.8.2-7}$

at $r^\star = 1$ for $z^{\star\star} > 0$ : $\dfrac{\partial T^\star}{\partial r^\star} = 1$ $\tag{6.8.2-8}$

and

as $z^{\star\star} \to -\infty$ for $r^\star \le 1$ : $T^\star \to 0$ $\tag{6.8.2-9}$

For $N_{Br} \ll 1$, it seems reasonable to neglect the effects of viscous dissipation in (6.8.2-4):

$$\left(1 - r^{\star 2}\right) \frac{\partial T^\star}{\partial z^{\star\star}} = \frac{1}{r^\star} \frac{\partial}{\partial r^\star} \left(r^\star \frac{\partial T^\star}{\partial r^\star}\right) \tag{6.8.2-10}$$

Since both axial conduction and viscous dissipation are neglected, boundary conditions (6.8.2-7) and (6.8.2-9) imply

at $z^{\star\star} = 0$ for $r^\star < 1$ : $T^\star = 0$ $\tag{6.8.2-11}$

Equation (6.8.2-10) is to be solved using (6.8.2-8) and (6.8.2-11) as boundary conditions.

Such a solution has been presented by Siegel, Sparrow, and Hallman (1958). But before describing their analysis, let us first construct an approximate solution that we will require.

## An Approximate Solution

Let us begin by examining the temperature distribution in the fluid very far downstream from the entrance to the heated section, where we might expect that, sufficiently far downstream from the entrance to the heated section, the temperature of the fluid should be a linear function of axial position:

$$T^\star_{\text{approx.}} = z^{\star\star} \Theta_1(r^\star) + \Theta_2(r^\star) \tag{6.8.2-12}$$

With this assumption, (6.8.2-10) becomes

$$\left(1 - r^{\star 2}\right) \Theta_1 = \frac{z^{\star\star}}{r^\star} \frac{d}{dr^\star} \left(r^\star \frac{d\Theta_1}{dr^\star}\right) + \frac{1}{r^\star} \frac{d}{dr^\star} \left(r^\star \frac{d\Theta_2}{dr^\star}\right) \tag{6.8.2-13}$$

But if $\Theta_2$ is to be a function only of $r^\star$, we must require

$$\frac{d}{dr^\star} \left(r^\star \frac{d\Theta_1}{dr^\star}\right) = 0 \tag{6.8.2-14}$$

or

$$\Theta_1 = C_1 \ln r^\star + C_2 \tag{6.8.2-15}$$

Since $T^\star$ must be finite at $r^\star = 0$, we must set $C_1 = 0$ and conclude that

$$\Theta_1 = C_2$$

$$= \text{a constant} \tag{6.8.2-16}$$

and that (6.8.2-13) reduces to

$$\frac{1}{r^\star} \frac{d}{dr^\star} \left( r^\star \frac{d\Theta_2}{dr^\star} \right) = C_2 \left( 1 - r^{\star 2} \right) \tag{6.8.2-17}$$

A solution of the form (6.8.2-12) and (6.8.2-16) is inconsistent with boundary conditions (6.8.2-8) and (6.8.2-11). As an alternative to boundary condition (6.8.2-11), we can observe that the energy entering with the fluid at cross section $z^{\star\star} = 0$ must leave either through the cross section at $z^{\star\star}$ or through the bounding walls of the tube:

$$\int_0^{2\pi} \int_0^R \rho \hat{c}(T - T_0) v_z r \, dr \, d\theta - 2\pi R z q = 0 \tag{6.8.2-18}$$

Using the velocity distribution developed in Section 3.2.1, we can write (6.8.2-18) in terms of our dimensionless variables as

$$\int_0^1 T^\star \left( 1 - r^{\star 2} \right) r^\star \, dr^\star - z^{\star\star} = 0 \tag{6.8.2-19}$$

In view of (6.8.2-12) and (6.8.2-16), Equation (6.8.2-19) requires

$$C_2 = 4 \tag{6.8.2-20}$$

and

$$\int_0^1 \Theta_2 \left( 1 - r^{\star 2} \right) r^\star \, dr^\star = 0 \tag{6.8.2-21}$$

This and the requirement that

$$\text{at } r^\star = 1 : \quad \frac{d\Theta_2}{dr^\star} = 1 \tag{6.8.2-22}$$

are the two boundary conditions to be satisfied by the required solution to (6.8.2-17).

We may integrate (6.8.2-17) to find

$$\int_0^{r^\star (d\Theta_2/dr^\star)} d\left( r^\star \frac{d\Theta_2}{dr^\star} \right) = 4 \int_0^{r^\star} \left( r^\star - r^{\star 3} \right) dr^\star$$

$$\frac{d\Theta_2}{dr^\star} = 2r^\star - r^{\star 3} \tag{6.8.2-23}$$

Notice that boundary condition (6.8.2-22) is automatically satisfied. Carrying out another integration, we find

$$\Theta_2 = r^{\star 2} - \frac{r^{\star 4}}{4} + C_3 \tag{6.8.2-24}$$

In view of (6.8.2-21), we must require

$$C_3 = -\frac{7}{24} \tag{6.8.2-25}$$

In summary, Equations (6.8.2-12), (6.8.2-16), (6.8.2-24), and (6.8.2-25) tell us that the temperature distribution very far downstream from the entrance to the heated section should be approximately (Goldstein 1938, p. 622; Bird et al. 1960, p. 296)

$$T^\star_{\text{approx.}} = 4z^{\star\star} + r^{\star 2} - \frac{1}{4}r^{\star 4} - \frac{7}{24} \tag{6.8.2-26}$$

### Complete Solution

In seeking a complete solution, it will be convenient to make the change of variable

$$T^{\star\star} \equiv T^\star - T^\star_{\text{approx.}} \tag{6.8.2-27}$$

In terms of $T^{\star\star}$, (6.8.2-10) becomes

$$\left(1 - r^{\star 2}\right)\frac{\partial T^{\star\star}}{\partial z^{\star\star}} = \frac{1}{r^\star}\frac{\partial}{\partial r^\star}\left(r^\star\frac{\partial T^{\star\star}}{\partial r^\star}\right) \tag{6.8.2-28}$$

This must be solved consistent with (6.8.2-8) and (6.8.2-11) in the forms

$$\text{at } r^\star = 1 \text{ for } z^{\star\star} > 0 : \quad \frac{\partial T^{\star\star}}{\partial r^\star} = 0 \tag{6.8.2-29}$$

and

$$\text{at } z^{\star\star} = 0 \text{ for } r^\star < 1 : \quad T^{\star\star} = -\left(r^{\star 2} - \frac{1}{4}r^{\star 4} - \frac{7}{24}\right) \tag{6.8.2-30}$$

Seeking a separable solution to (6.8.2-28) of the form

$$T^{\star\star} = X\left(z^{\star\star}\right) Y\left(r^\star\right) \tag{6.8.2-31}$$

we find

$$\frac{1}{X}\frac{dX}{dz^{\star\star}} = \frac{1}{Y}\frac{1}{r^\star\left(1 - r^{\star 2}\right)}\frac{d}{dr^\star}\left(r^\star\frac{dY}{dr^\star}\right)$$

$$= -\lambda^2 \tag{6.8.2-32}$$

where $\lambda$ is a constant to be determined. We quickly come to the conclusion that

$$T^{\star\star} = C \exp\left(-\lambda^2 z^{\star\star}\right) Y\left(r^\star\right) \tag{6.8.2-33}$$

in which the function $Y\left(\lambda, r^\star\right)$ satisfies

$$\frac{1}{r^\star}\frac{d}{dr^\star}\left(r^\star\frac{dY}{dr^\star}\right) + \lambda^2\left(1 - r^{\star 2}\right) Y = 0 \tag{6.8.2-34}$$

A particular solution to (6.8.2-34) takes the form

$$Y\left(\lambda, r^\star\right) = \sum_{i=0}^{\infty} a_i r^{\star i} \tag{6.8.2-35}$$

in which

$$a_i = 0 \qquad \text{if } i < 0$$
$$a_i = 1 \qquad \text{if } i = 0 \tag{6.8.2-36}$$
$$a_i = -\left(\frac{\lambda}{i}\right)^2 (a_{i-2} - a_{i-4}) \qquad \text{if } i > 0$$

**Table 6.8.2-1.** Roots of (6.8.2-37) as given by Siegel et al. (1958)

| $n$ | $\lambda_n$ |
|---|---|
| 1 | 5.0675 |
| 2 | 9.1576 |
| 3 | 13.1972 |
| 4 | 17.2202 |
| 5 | 21.2355 |
| 6 | 25.2465 |
| 7 | 29.2549 |

In view of (6.8.2-29), we must require

$$\text{at } r^\star = 1: \quad \frac{dY}{dr^\star} = 0 \tag{6.8.2-37}$$

The $n$ roots of this equation will be designated as $\lambda_n$ ($n = 1, 2, 3, \ldots$). The first seven roots have been given by Siegel et al. (1958); they are shown in Table 6.8.2-1.

We now recognize that (6.8.2-33) should be written as a superposition of all possible solutions:

$$T^{\star\star} = \sum_{n=1}^{\infty} C_n \exp\left(-\lambda_n^2 z^{\star\star}\right) Y_n\left(r^\star\right) \tag{6.8.2-38}$$

Here the function $Y_n\left(\lambda, r^\star\right)$ is a solution to

$$\frac{1}{r^\star} \frac{d}{dr^\star} \left(r^\star \frac{dY_n}{dr^\star}\right) + \lambda_n^2 \left(1 - r^{\star 2}\right) Y_n = 0 \tag{6.8.2-39}$$

or

$$Y_n\left(\lambda, r^\star\right) = \sum_{i=0}^{\infty} a_{n,i} r^{\star i} \tag{6.8.2-40}$$

in which

$$a_{n,i} = 0 \quad \text{if } i < 0$$
$$a_{n,i} = 1 \quad \text{if } i = 0 \tag{6.8.2-41}$$
$$a_{n,i} = -\left(\frac{\lambda_n}{i}\right)^2 \left(a_{n,i-2} - a_{n,i-4}\right) \quad \text{if } i > 0$$

Equations (6.8.2-30) and (6.8.2-38) require

$$-\left(r^{\star 2} - \frac{1}{4} r^{\star 4} - \frac{7}{24}\right) = \sum_{n=1}^{\infty} C_n Y_n\left(r^\star\right) \tag{6.8.2-42}$$

Multiplying both sides of (6.8.2-42) by $r^\star(1 - r^{\star2})Y_n$ and integrating, we find in view of (6.8.1-29)

$$-\int_0^1 r^\star \left(1 - r^{\star2}\right)\left(r^{\star2} - \frac{1}{4}r^{\star4}\right) Y_n \, dr^\star = C_n \int_0^1 r^\star \left(1 - r^{\star2}\right) Y_n^{\,2} \, dr^\star \qquad (6.8.2\text{-}43)$$

or

$$C_n = -\int_0^1 r^\star \left(1 - r^{\star2}\right)\left(r^{\star2} - \frac{1}{4}r^{\star4}\right) Y_n \, dr^\star \left[\int_0^1 r^\star \left(1 - r^{\star2}\right) Y_n^{\,2} \, dr^\star\right]^{-1}$$

$$(6.8.2\text{-}44)$$

Let us assume that our primary interest is in the film coefficient for heat transfer $h$, where

$$-q_r|_{r=R} = q$$

$$= h(T_{\text{wall}} - T_m) \qquad (6.8.2\text{-}45)$$

Here $T_{\text{wall}}$ is the temperature at $r^\star = 1$:

$$T_{\text{wall}}^\star = 4z^{\star\star} + \frac{11}{24} + T_{\text{wall}}^{\star\star}$$

$$= 4z^{\star\star} + \frac{11}{24} + \sum_{n=1}^{\infty}\sum_{i=0}^{\infty} C_n \exp\left(-\lambda_n^{\,2}z^{\star\star}\right) a_{n,i} \qquad (6.8.2\text{-}46)$$

and from (6.8.1-36)

$$T_m^\star = 4\int_0^1 T^\star \left(1 - r^{\star2}\right) r^\star \, dr^\star$$

$$= 4\int_0^1 \left(T^{\star\star} + T_{\text{approx}}^\star\right)\left(1 - r^{\star2}\right) r^\star \, dr^\star$$

$$= 4z^{\star\star} \qquad (6.8.2\text{-}47)$$

The Nusselt number can be expressed as

$$N_{Nu} \equiv \frac{hR}{k}$$

$$= \frac{qR}{k\left(T_{\text{wall}} - T_m\right)}$$

$$= \frac{1}{T_{\text{wall}}^\star - T_m^\star}$$

$$= \frac{1}{\frac{11}{24} + \sum_{n=1}^{\infty}\sum_{i=0}^{\infty} C_n \exp\left(-\lambda_n^{\,2}z^{\star\star}\right) a_{n,i}} \qquad (6.8.2\text{-}48)$$

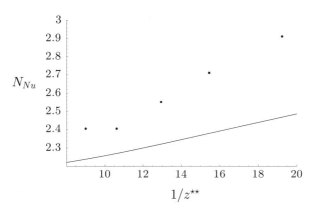

**Figure 6.8.2-1.** $N_{Nu}$ as a function of $1/z^{\star\star}$ for a constant energy flux at the wall. The prediction of (6.8.1-39) is compared with data presented by Kays (1955, Fig. 7).

or the mean Nusselt number for a length of the heated section as

$$N_{Nu,m} \equiv \frac{1}{L^{\star\star}} \int_0^{L^{\star\star}} N_{Nu}\, dz^{\star\star} \tag{6.8.2-49}$$

Figure 6.8.2-1 compares the predictions of (6.8.2-48) with data presented by Kays (1955, Fig. 8). Except for the first point on the left, the data appear to approach the prediction in the limit as $1/z^{\star\star} \to 0$, as required.

This section was prepared with the help of P. K. Dhori.

# 7

# Integral Averaging in Energy Transfer

THE DISCUSSION OF INTEGRAL AVERAGING techniques begun in Chapter 4 is continued here, but the emphasis is upon energy transfer. If you have not already done so, I recommend that you read Chapter 4 before beginning this material. The derivations are more detailed there and much more is said to motivate the discussion.

## 7.1 Time Averaging

*Turbulent energy transfer* implies that at least one of the materials involved in the energy transfer process is in turbulent flow. Everything that we said in Section 4.1 is equally applicable here. The only additional complication is that, like the velocity and pressure distributions, the temperature distribution varies randomly with time in all possible frames of reference.

In the next few sections we shall be concerned with the time average of the differential energy balance and its solutions.

### 7.1.1 The Time-Averaged Differential Energy Balance

As noted in Section 4.1.1, for simplicity we shall limit ourselves to incompressible fluids. We shall also assume that there is no external or mutual energy transmission, that the ratio of the Brinkman number to the Peclet number is sufficiently small for viscous dissipation to be neglected (see the introduction to Section 6.4), and that the heat capacity per unit mass may be treated as a constant. Under these circumstances, the differential energy balance of Table 5.4.1-1 reduces to

$$\rho \hat{c} \left[ \frac{\partial T}{\partial t} + \operatorname{div}(T\mathbf{v}) \right] + \operatorname{div} \mathbf{q} = 0 \tag{7.1.1-1}$$

Using the definition introduced in Section 4.1.1, let us take the time average of this equation:

$$\frac{1}{\Delta t} \int_t^{t+\Delta t} \left[ \rho \hat{c} \frac{\partial T}{\partial t'} + \operatorname{div}(\rho \hat{c} T\mathbf{v} + \mathbf{q}) \right] dt' = 0 \tag{7.1.1-2}$$

The time-averaging operation commutes with partial differentiation with respect to time (see Section 4.1.1) and the divergence operation:

$$\rho \hat{c} \frac{\partial \overline{T}}{\partial t} + \mathrm{div}\left(\rho \hat{c} \overline{T \mathbf{v}} + \overline{\mathbf{q}}\right) = 0 \tag{7.1.1-3}$$

It is more common to write this result as

$$\rho \hat{c} \left( \frac{\partial \overline{T}}{\partial t} + \nabla \overline{T} \cdot \overline{\mathbf{v}} \right) = -\mathrm{div}\left(\overline{\mathbf{q}} + \mathbf{q}^{(t)}\right) \tag{7.1.1-4}$$

where we have introduced the *turbulent energy flux*

$$\mathbf{q}^{(t)} \equiv \rho \hat{c} \left( \overline{T \mathbf{v}} - \overline{T}\, \overline{\mathbf{v}} \right) \tag{7.1.1-5}$$

When we recognize Fourier's law as representing the energy flux, (7.1.1-4) takes the form

$$\rho \hat{c} \left( \frac{\partial \overline{T}}{\partial t} + \nabla \overline{T} \cdot \overline{\mathbf{v}} \right) = \mathrm{div}\left(k \nabla \overline{T} - \mathbf{q}^{(t)}\right) \tag{7.1.1-6}$$

Our problem here is very similar to that encountered in Section 4.1.1. Just as there we had to stop and propose empirical data correlations for the Reynolds stress tensor $\mathbf{S}^{(t)}$, here we must stop and consider empirical representations for the turbulent energy flux $\mathbf{q}^{(t)}$.

### 7.1.2 Empirical Correlation for the Turbulent Energy Flux

Three examples are used here to illustrate how empirical data correlations for the turbulent energy flux $\mathbf{q}^{(t)}$ can be prepared. Three ideas are used in this discussion.

1) For changes of frames such that

$$\overline{\mathbf{Q}} \doteq \mathbf{Q} \tag{7.1.2-1}$$

we may use the result of Section 4.1.2 to find that $\mathbf{q}^{(t)}$ is frame indifferent:

$$\begin{aligned}
\mathbf{q}^{(t)*} &\equiv \rho \hat{c} \left( \overline{T^* \mathbf{v}^*} - \overline{T}^* \overline{\mathbf{v}}^* \right) \\
&= \rho \hat{c} \left[ \overline{T^* (\mathbf{v}^* - \overline{\mathbf{v}}^*)} \right] \\
&= \rho \hat{c} \mathbf{Q} \cdot \left[ \overline{T(\mathbf{v} - \overline{\mathbf{v}})} \right] \\
&= \mathbf{Q} \cdot \mathbf{q}^{(t)}
\end{aligned} \tag{7.1.2-2}$$

Here $\mathbf{Q}$ is a (possibly) time-dependent orthogonal second-order tensor. In arriving at this result, we make use of the fact that a velocity difference is frame indifferent (see Exercise 1.2.2-1).

2) We shall assume that the principle of frame indifference discussed in Section 2.3.1 applies to any empirical correlations developed for $\mathbf{q}^{(t)}$, so long as the changes of frame considered satisfy (7.1.2-1).

3) We shall use the Buckingham–Pi theorem (Brand 1957) to further limit the form of any expression for $\mathbf{q}^{(t)}$.

## Example 1: Prandtl's Mixing-Length Theory

Let us attempt to develop an empirical correlation for $\mathbf{q}^{(t)}$ appropriate to the fully developed flow regime in wall turbulence.

Influenced by the manner in which we approached empirical correlations for the Reynolds stress tensor in Section 4.1.3, we propose that the turbulent energy flux vector be regarded as a function of the density of the fluid, its heat capacity per unit mass $\hat{c}$, the distance $l$ from the wall, $\overline{\mathbf{D}}$, and $\nabla \overline{T}$:

$$\mathbf{q}^{(t)} = \mathbf{q}^{(t)}(\rho, \hat{c}, l, \overline{\mathbf{D}}, \nabla \overline{T}) \tag{7.1.2-3}$$

We specifically do not include thermal conductivity or viscosity as independent variables, because we are considering the fully developed turbulent flow regime. The most general expression of this form that satisfies our restricted form of the principle of frame indifference is rather lengthy (see Exercise 7.1.2-1). It is perhaps sufficient to say that the literature has been primarily concerned with the special case

$$\mathbf{q}^{(t)} = -\kappa \nabla \overline{T} \tag{7.1.2-4}$$

where

$$\kappa = \kappa \left( \rho, \hat{c}, l, \sqrt{2 \operatorname{tr} \left( \overline{\mathbf{D}} \cdot \overline{\mathbf{D}} \right)} \right) \tag{7.1.2-5}$$

The Buckingham–Pi theorem (Brand 1957) requires that (7.1.2-5) be of the form

$$\kappa = \kappa^{\star} \rho \hat{c} l^2 \sqrt{2 \operatorname{tr} \left( \overline{\mathbf{D}} \cdot \overline{\mathbf{D}} \right)} \tag{7.1.2-6}$$

and (7.1.2-4) becomes

$$\mathbf{q}^{(t)} = -\kappa^{\star} \rho \hat{c} l^2 \sqrt{2 \operatorname{tr} \left( \overline{\mathbf{D}} \cdot \overline{\mathbf{D}} \right)} \nabla \overline{T} \tag{7.1.2-7}$$

Here $\kappa^{\star}$ is a dimensionless constant. Equation (7.1.2-7) should be viewed as the tensorial form of *Prandtl's mixing-length theory* for energy transfer (Goldstein 1938, p. 648).

I would like to emphasize that we do not expect the Prandtl mixing-length theory to be appropriate to the laminar sublayer or buffer zone. We assumed at the beginning that we were constructing a representation for the turbulent energy flux in the fully developed turbulent flow regime.

## Example 2: Deissler's Expression for the Region near the Wall

Let us consider that portion of the turbulent flow of an incompressible Newtonian fluid in the immediate vicinity of a bounding wall: the laminar sublayer and the buffer zone. Looking back at our discussion of Example 2 in Section 4.1.3, it seems reasonable to propose

$$\mathbf{q}^{(t)} = \mathbf{q}^{(t)} \left( \rho, \hat{c}, \mu, l, \overline{\mathbf{v}} - \mathbf{v}^{(s)}, \nabla \overline{T} \right) \tag{7.1.2-8}$$

Remember that $\mathbf{v}^{(s)}$ indicates the velocity of the bounding wall.

Let us focus our attention in this relationship on the dependence of $\mathbf{q}^{(t)}$ upon the two vectors:

$$\mathbf{q}^{(t)} = \hat{\mathbf{q}}^{(t)} \left( \overline{\mathbf{v}} - \mathbf{v}^{(s)}, \nabla \overline{T} \right) \tag{7.1.2-9}$$

By our restricted form of the principle of frame indifference, the functional relationship between these variables should be the same in every frame of reference. This means that

$$
\begin{aligned}
\mathbf{q}^{(t)*} &= \mathbf{Q} \cdot \mathbf{q}^{(t)} \\
&= \mathbf{Q} \cdot \hat{\mathbf{q}}^{(t)} \left( \overline{\mathbf{v}} - \mathbf{v}^{(s)}, \nabla \overline{T} \right) \\
&= \hat{\mathbf{q}}^{(t)} \left( \mathbf{Q} \cdot [\overline{\mathbf{v}} - \mathbf{v}^{(s)}], \mathbf{Q} \cdot \nabla \overline{T} \right)
\end{aligned}
\tag{7.1.2-10}
$$

or $\hat{\mathbf{q}}^{(t)}$ is an isotropic function (Truesdell and Noll 1965, p. 22):

$$
\hat{\mathbf{q}}^{(t)} \left( \overline{\mathbf{v}} - \mathbf{v}^{(s)}, \nabla \overline{T} \right) = \mathbf{Q}^T \cdot \hat{\mathbf{q}}^{(t)} \left( \mathbf{Q} \cdot \left[ \overline{\mathbf{v}} - \mathbf{v}^{(s)} \right], \mathbf{Q} \cdot \nabla \overline{T} \right)
\tag{7.1.2-11}
$$

By representation theorems of Spencer and Rivlin (1959, Sec. 7) and of Smith (1965), the most general polynomial isotropic vector function of two vectors has the form

$$
\mathbf{q}^{(t)} = \kappa_{(1)} \nabla \overline{T} + \kappa_{(2)} \left( \overline{\mathbf{v}} - \mathbf{v}^{(s)} \right)
\tag{7.1.2-12}
$$

Here $\kappa_{(1)}$ and $\kappa_{(2)}$ are scalar-valued polynomials of the general form[1]

$$
\kappa_{(i)} = \kappa_{(i)} \left( \rho, \hat{c}, \mu, l, |\nabla \overline{T}|, |\overline{\mathbf{v}} - \mathbf{v}^{(s)}|, \left[ \overline{\mathbf{v}} - \mathbf{v}^{(s)} \right] \cdot \nabla \overline{T} \right)
\tag{7.1.2-13}
$$

An application of the Buckingham–Pi theorem (Brand 1957) tells us that

$$
\frac{\kappa_{(1)}}{\rho \hat{c} l \left| \overline{\mathbf{v}} - \mathbf{v}^{(s)} \right|} = \kappa_{(1)}^{\star}
\tag{7.1.2-14}
$$

and

$$
\frac{\kappa_{(2)}}{\rho \hat{c} l \left| \nabla \overline{T} \right|} = \kappa_{(2)}^{\star}
\tag{7.1.2-15}
$$

Here

$$
\kappa_{(i)}^{\star} = \kappa_{(i)}^{\star} \left( N, \frac{\hat{c} \left| \nabla \overline{T} \right| l}{\left| \overline{\mathbf{v}} - \mathbf{v}^{(s)} \right|^2}, \frac{\left[ \overline{\mathbf{v}} - \mathbf{v}^{(s)} \right] \cdot \nabla \overline{T}}{\left| \overline{\mathbf{v}} - \mathbf{v}^{(s)} \right| \left| \nabla \overline{T} \right|} \right)
\tag{7.1.2-16}
$$

and

$$
N \equiv \frac{\rho l \left| \overline{\mathbf{v}} - \mathbf{v}^{(s)} \right|}{\mu}
\tag{7.1.2-17}
$$

Deissler (1955) has proposed on empirical grounds that

$$
\mathbf{q}^{(t)} = -n^2 \rho \hat{c} l \left| \overline{\mathbf{v}} - \mathbf{v}^{(s)} \right| [1 - \exp(-n^2 N)] \nabla \overline{T}
\tag{7.1.2-18}
$$

The $n$ appearing here is meant to be the same as that used in (4.1.3-21) and evaluated in Section 4.1.4.

---

[1] In applying the theorem of Spencer and Rivlin, we identify the vector $\mathbf{b}$ that has covariant components $b_i$ with the skew-symmetric tensor that has contravariant components $\epsilon^{ijk} b_i$. Their theorem requires an additional term in (7.1.2-12) proportional to the vector product $(\overline{\mathbf{v}} - \mathbf{v}^{(s)}) \cdot \nabla \overline{T}$. This term is not consistent with the requirement that $\mathbf{q}^{(t)}$ be isotropic (Truesdell and Noll 1965, p. 24) and consequently is dropped.

### Example 3: Eddy Conductivity in Free Turbulence

If we move very far away from any walls into a region of free turbulence, it is tempting to look at (7.1.2-3) and assume instead

$$\mathbf{q}^{(t)} = \mathbf{q}^{(t)}(\rho, \hat{c}, l, \nabla \overline{T}) \tag{7.1.2-19}$$

By analogy with the development in Section 5.3.3, we can say that the principle of frame indifference requires

$$\mathbf{q}^{(t)} = -\kappa \left(\rho, \hat{c}, l, \left|\nabla \overline{T}\right|\right) \nabla \overline{T} \tag{7.1.2-20}$$

But the Buckingham–Pi theorem (Brand 1957) tells us that $\kappa$ must in fact be a constant scalar under the assumptions made:

$$\mathbf{q}^{(t)} = -\kappa \nabla \overline{T} \tag{7.1.2-21}$$

The scalar $\kappa$ is usually known as the *eddy conductivity*.

Notice that the type of theoretical objection raised against the use of an eddy viscosity in Section 4.1.3 does not apply here.

**Exercise 7.1.2-1**    Use a representation theorem due to Noll (Truesdell and Noll 1965, p. 35) to conclude that the most general expression of the form of (7.1.2-3) that satisfies our restricted form of the principle of frame indifference is

$$\mathbf{q}^{(t)} = \left(\varphi_{(0)}\mathbf{I} + \varphi_{(1)}\overline{\mathbf{D}} + \varphi_{(2)}\overline{\mathbf{D}} \cdot \overline{\mathbf{D}}\right) \cdot \nabla \overline{T}$$

where

$$\varphi_{(k)} = \varphi_{(k)} \left(\rho, \hat{c}, l, \operatorname{div}\overline{\mathbf{v}}, \operatorname{tr}\left(\overline{\mathbf{D}} \cdot \overline{\mathbf{D}}\right), \det \overline{\mathbf{D}}, \left|\nabla \overline{T}\right|\right.$$
$$\left.\nabla \overline{T} \cdot [\overline{\mathbf{D}} \cdot \nabla \overline{T}], \nabla \overline{T} \cdot \overline{\mathbf{D}} \cdot \overline{\mathbf{D}} \cdot \nabla \overline{T}\right)$$

### 7.1.3    Turbulent Energy Transfer in a Heated Section of a Tube

An incompressible Newtonian fluid with constant viscosity and thermal conductivity is in turbulent flow through a tube of radius $R$. For $z < 0$, the wall of the tube is insulated:

$$\text{at } r = R, \text{ for } z < 0 : \quad \frac{\partial \overline{T}}{\partial r} = 0 \tag{7.1.3-1}$$

But for $z > 0$, the radial component of the energy flux is constant:

$$\text{at } r = R, \text{ for } z > 0 : \quad k\frac{\partial \overline{T}}{\partial r} = q \tag{7.1.3-2}$$

Very far upstream from the entrance to this heated section, the fluid has a uniform temperature $T_0$:

$$\text{as } z \to -\infty, \text{ for } r \leq R : \quad T \to T_0 \tag{7.1.3-3}$$

For the moment, let us direct our attention to the temperature distribution in the laminar sublayer and buffer zone.

In Section 4.1.4, we examined the velocity distribution that develops in turbulent flow through a tube. We found that there is only one nonzero component of the time-averaged velocity vector:

$$\bar{v}_r = \bar{v}_\theta$$
$$= 0 \tag{7.1.3-4}$$
$$\bar{v}_z = \bar{v}_z(r)$$

In the laminar sublayer and buffer zone where

$$s^\star \equiv \frac{s}{R}$$
$$\ll 1 \tag{7.1.3-5}$$

the velocity distribution can be obtained by integrating

$$\left\{1 + n^2 v^\star s^{\star\star}\left[1 - \exp\left(-n^2 v^\star s^{\star\star}\right)\right]\right\}\frac{dv^\star}{ds^{\star\star}} = 1 \tag{7.1.3-6}$$

consistent with the boundary condition

$$\text{at } r = R: \ v^\star = 0 \tag{7.1.3-7}$$

Here

$$v^\star \equiv \frac{\bar{v}_z}{v_0}$$
$$= \sqrt{\frac{\rho}{S_0}}\,\bar{v}_z \tag{7.1.3-8}$$
$$s^{\star\star} \equiv \frac{s}{R}N^{(t)} \tag{7.1.3-9}$$

and

$$N^{(t)} \equiv \frac{\rho v_0 R}{\mu}$$
$$= \frac{\sqrt{\rho S_0}R}{\mu} \tag{7.1.3-10}$$

The result of this integration for $n = 0.124$ is shown in Figure 4.1.4-3. Since we are concerned here with an incompressible fluid of constant viscosity, the velocity distribution is unaffected by the temperature distribution. We will consequently assume that (7.1.3-4) and (7.1.3-6) again apply.

Let us assume that the time-averaged temperature distribution in the laminar sublayer and buffer zone has the form

$$\bar{T} = \bar{T}(r, z) \tag{7.1.3-11}$$

As indicated in Example 2 of Section 7.1.2, Deissler's expression for the turbulent energy flux is currently recommended for the region next to the wall. Consequently, the time-averaged differential energy balance for this region described by inequality (7.1.3-5) becomes

$$N^{(t)-1} v^\star \frac{\partial T^\star}{\partial z^\star}$$

$$= \frac{\partial}{\partial s^{\star\star}} \left( \left\{ \frac{1}{N_{Pr}} + n^2 s^{\star\star} v^\star [1 - \exp(-n^2 s^{\star\star} v^\star)] \right\} \frac{\partial T^\star}{\partial s^{\star\star}} \right)$$

$$+ \frac{1}{N^{(t)2}} \frac{\partial}{\partial z^\star} \left( \left\{ \frac{1}{N_{Pr}} + n^2 s^{\star\star} v^\star [1 - \exp(-n^2 s^{\star\star} v^\star)] \right\} \frac{\partial T^\star}{\partial z^\star} \right) \tag{7.1.3-12}$$

where

$$T^\star \equiv \frac{\overline{T} - T_0}{qR/k}$$

$$z^\star \equiv \frac{z}{R} \tag{7.1.3-13}$$

Since we are interested in turbulent flow, it is not unreasonable to confine our attention to the limit $N^{(t)} \to \infty$, in which case (7.1.3-12) reduces to

$$\frac{\partial}{\partial s^{\star\star}} \left( \left\{ \frac{1}{N_{Pr}} + n^2 s^{\star\star} v^\star [1 - \exp(-n^2 s^{\star\star} v^\star)] \right\} \frac{\partial T^\star}{\partial s^{\star\star}} \right) = 0 \tag{7.1.3-14}$$

Within the heated portion of the tube, this equation must be integrated consistent with (7.1.3-2), which takes the form

$$\text{at } s^{\star\star} = 0 : \quad \frac{\partial T^\star}{\partial s^{\star\star}} = -\frac{1}{N^{(t)}} \tag{7.1.3-15}$$

The result is

$$\left\{ \frac{1}{N_{Pr}} + n^2 s^{\star\star} v^\star [1 - \exp(-n^2 s^{\star\star} v^\star)] \right\} \frac{\partial T^\star}{\partial s^{\star\star}} = -\frac{1}{N_{Pr} N^{(t)}} \tag{7.1.3-16}$$

Either the next integration may be carried out using a matching condition with the fully developed turbulent flow at the edge of the buffer zone or the integration may be expressed in terms of the temperature at the wall of the tube.

This is most easily illustrated with the limiting case

for $N_{Pr} = 1$ :
$$\left\{ 1 + n^2 s^{\star\star} v^\star [1 - \exp(-n^2 s^{\star\star} v^\star)] \right\} \frac{\partial T^\star}{\partial s^{\star\star}} = -\frac{1}{N^{(t)}} \tag{7.1.3-17}$$

Comparing this with (7.1.3-6), we find

$$\frac{dT^\star}{dv^\star} = -\frac{1}{N^{(t)}} \tag{7.1.3-18}$$

This is easily integrated consistent with boundary condition (7.1.3-7) and

$$\text{at } s^{\star\star} = 0 : \quad T^\star = T_w^\star \tag{7.1.3-19}$$

to find that, within the laminar sublayer and buffer zone,

$$T^\star - T_w^\star = -\frac{v^\star}{N^{(t)}} \tag{7.1.3-20}$$

This prompts us to say that, for $N_{Pr} = 1$, the velocity and temperature distributions in the laminar sublayer and buffer zone are similar.

Note that the axial dependence of $T^\star$ has not been considered; no attempt has been made to satisfy boundary conditions (7.1.3-1) and (7.1.3-3).

**Exercise 7.1.3-1** *The temperature distribution in fully developed turbulent flow*   Continue the discussion in the text, focusing your attention upon the fully developed portion of the turbulent flow.

i) Use the Prandtl mixing-length theory discussed in Example 1 of Section 7.1.2 to determine that in the fully developed portion of the flow near the wall (although outside the buffer zone)

as $N^{(t)} \to \infty$ :

$$\frac{\partial}{\partial s^{\star\star}} \left[ \left( \frac{1}{N_{Pr}} + \kappa^\star s^{\star\star 2} \frac{dv^\star}{ds^{\star\star}} \right) \frac{\partial T^\star}{\partial s^{\star\star}} \right] = 0 \qquad (7.1.3\text{-}21)$$

ii) Formulate the appropriate boundary condition at the edge of the buffer zone $s^{\star\star} = s_1^{\star\star}$, and integrate (7.1.3-21) to find

$$\left( \frac{1}{N_{Pr}} + \kappa^\star s^{\star\star 2} \frac{dv^\star}{ds^{\star\star}} \right) \frac{\partial T^\star}{\partial s^{\star\star}} = -\frac{1}{N_{Pr} N^{(t)}} \qquad (7.1.3\text{-}22)$$

iii) Prompted by a similar statement in Section 4.1.4, assume that

$$\kappa^\star s^{\star\star 2} \frac{dv^\star}{ds^{\star\star}} \gg \frac{1}{N_{Pr}} \qquad (7.1.3\text{-}23)$$

Integrate (7.1.3-22) with the boundary condition

at $s^{\star\star} = s_1^{\star\star} : \; T^\star = T_1^\star$ \qquad (7.1.3-24)

to find

$$T^\star - T_1^\star = -\frac{1}{N_{Pr} N^{(t)}} \frac{\sqrt{\eta_1^\star}}{\kappa^\star} \ln \frac{s^{\star\star}}{s_1^{\star\star}} \qquad (7.1.3\text{-}25)$$

iv) We are again prompted by Section 4.1.4 to assume that

$$s_1^{\star\star} = 26 \qquad (7.1.3\text{-}26)$$

and

$$\kappa^\star = n_1^\star$$
$$= (0.36)^2 \qquad (7.1.3\text{-}27)$$

Under these conditions, verify inequality (7.1.3-23).

v) Assuming (7.1.3-27), calculate that

$$\text{for } N_{Pr} = 1 : \; \frac{T^\star - T_1^\star}{v^\star - v_1^\star} = -\frac{1}{N^{(t)}} \qquad (7.1.3\text{-}28)$$

Using the result of this section for the laminar sublayer and buffer zone, calculate that

for $N_{Pr} = 1$ :
$$T^\star - T_w^\star = -\frac{v^\star}{N^{(t)}}$$

$$= -\frac{1}{N^{(t)}} \left( \frac{1}{0.36} \ln s^{\star\star} + 3.8 \right) \tag{7.1.3-29}$$

**Exercise 7.1.3-2** *The Reynolds analogy for $N_{Pr} = 1$*   Multiply (7.1.3-29) of Exercise 7.1.3-1 by $v^\star$ and integrate over the cross section of the tube to find

$$\langle v^\star T^\star \rangle - \langle v^\star \rangle T_w^\star = -\frac{\langle v^{\star 2} \rangle}{N^{(t)}} \tag{7.1.3-30}$$

where we define for any scalar $\varphi$

$$\langle \varphi \rangle \equiv \frac{1}{\pi R^2} \int_0^{2\pi} \int_0^R \varphi r \, dr \, d\theta \tag{7.1.3-31}$$

Rearrange (7.1.3-30) to conclude that

$$\frac{N_{Nu}}{N_{Re}} = \frac{c}{2} \tag{7.1.3-32}$$

Here

$$N_{Nu} \equiv \frac{2hR}{k}$$
$$N_{Re} \equiv \frac{2\rho \langle \bar{v}_z \rangle R}{\mu} \tag{7.1.3-33}$$

and the drag coefficient $c$ is defined as

$$c \equiv \frac{S_0}{\frac{1}{2}\rho \langle \bar{v}_z \rangle^2} \tag{7.1.3-34}$$

The film coefficient $h$ has been defined in terms of the bulk mixing temperature $\langle \bar{v}_z \bar{T} \rangle / \langle \bar{v}_z \rangle$:

$$h \equiv \frac{q}{\bar{T}_w - \langle \bar{v}_z \bar{T} \rangle / \langle \bar{v}_z \rangle} \tag{7.1.3-35}$$

Equation (7.1.3-32) represents a special case of an empirical analogy between the film coefficient for heat transfer and the drag coefficient first pointed out by Colburn (1933).

---

## 7.2   Area Averaging

The concept of area averaging introduced in Section 4.2 is generalized here to apply to energy transfer. The essence of this approach is to average the differential energy balance over a cross section normal to the macroscopic energy transfer.

When we are primarily concerned with evaluating the time or the position dependence of an area-averaged variable, we will normally be required to make some statement about the energy flux at a bonding phase interface, perhaps employing Newton's law of cooling. This class of problems is illustrated in Section 7.2.1.

Because of the similarity of the material, I encourage you to review Section 4.2 before proceeding.

### 7.2.1  A Straight Cooling Fin of Rectangular Profile

This material is taken from Jacob (1949, p. 221).

Let us determine the efficiency of the cooling fin pictured in Figure 7.2.1-1. We will assume that we know the temperature of the wall upon which the fin is mounted:

$$\text{at } z_1 = 0 : \quad T = T_w \tag{7.2.1-1}$$

We will further neglect the energy lost through the end of the fin:

$$\text{at } z_1 = L : \quad \frac{\partial T}{\partial z_1} = 0 \tag{7.2.1-2}$$

Our first problem is to determine the temperature distribution in the fin. It seems reasonable to recognize that the temperature will be a function of all three coordinates:

$$T = T(z_1, z_2, z_3) \tag{7.2.1-3}$$

The differential energy balance to be solved consequently takes the form

$$\frac{\partial^2 T}{\partial z_1{}^2} + \frac{\partial^2 T}{\partial z_2{}^2} + \frac{\partial^2 T}{\partial z_3{}^2} = 0 \tag{7.2.1-4}$$

This seems to be a situation where we are not primarily concerned with the details of the temperature distribution in the fin. It will be sufficient to determine the dependence of the

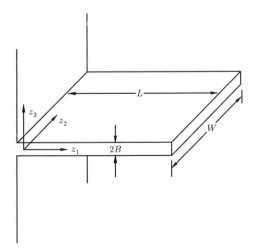

**Figure 7.2.1-1.** A straight cooling fin of rect-angular profile.

average temperature,

$$\overline{T} \equiv \frac{1}{2BW} \int_{-B}^{B} \int_{0}^{W} T \, dz_2 \, dz_3 \tag{7.2.1-5}$$

upon $z_1$. With this definition for the area average, we find from (7.2.1-4)

$$\frac{\partial^2 \overline{T}}{\partial z_1{}^2} + \frac{1}{2BW} \int_{-B}^{B} \left( \frac{\partial T}{\partial z_2}\Big|_{z_2=W} - \frac{\partial T}{\partial z_2}\Big|_{z_2=0} \right) dz_3$$

$$+ \frac{1}{2BW} \int_{0}^{W} \left( \frac{\partial T}{\partial z_3}\Big|_{z_3=B} - \frac{\partial T}{\partial z_3}\Big|_{z_3=-B} \right) dz_2 = 0 \tag{7.2.1-6}$$

Let us apply Newton's law of cooling in the form

at a surface: $\mathbf{q} \cdot \mathbf{n} = h(\overline{T} - T_a)$ (7.2.1-7)

where $T_a$ is the ambient temperature of the surrounding air. This implies that

at $z_2 = 0$: $k\dfrac{\partial T}{\partial z_2} = h(\overline{T} - T_a)$ (7.2.1-8)

at $z_2 = W$: $-k\dfrac{\partial T}{\partial z_2} = h(\overline{T} - T_a)$ (7.2.1-9)

at $z_3 = -B$: $k\dfrac{\partial T}{\partial z_3} = h(\overline{T} - T_a)$ (7.2.1-10)

and

at $z_3 = B$: $-k\dfrac{\partial T}{\partial z_3} = h(\overline{T} - T_a)$ (7.2.1-11)

In view of these relationships, (7.2.1-6) reduces to

$$\frac{\partial^2 \overline{T}}{\partial z_1{}^2} = \frac{h}{k} \left( \frac{2}{W} + \frac{1}{B} \right) (\overline{T} - T_a) \tag{7.2.1-12}$$

We find it convenient to introduce as dimensionless variables

$$\overline{T}^\star \equiv \frac{\overline{T} - T_a}{T_w - T_a} \tag{7.2.1-13}$$

and

$$z_1^\star \equiv \frac{z_1}{L} \tag{7.2.1-14}$$

This allows us to write (7.2.1-12) in the considerably simpler form

$$\frac{\partial^2 \overline{T}^\star}{\partial z_1{}^2} = N^2 \overline{T}^\star \tag{7.2.1-15}$$

where

$$N^2 \equiv \frac{hL^2}{k} \left( \frac{2}{W} + \frac{1}{B} \right) \tag{7.2.1-16}$$

From (7.2.1-1) and (7.2.1-2), the appropriate boundary conditions are

$$\text{at } z_1^\star = 0 : \ \overline{T}^\star = 1 \tag{7.2.1-17}$$

and

$$\text{at } z_1^\star = 1 : \ \frac{d\overline{T}^\star}{dz_1^\star} = 0 \tag{7.2.1-18}$$

The solution to (7.2.1-15) consistent with boundary conditions (7.2.1-17) and (7.2.1-18) can easily be determined to be

$$\overline{T}^\star = \frac{\cosh(N[1 - z_1^\star])}{\cosh N} \tag{7.2.1-19}$$

The effectiveness of a fin has been defined to be (Jacob 1949, p. 235)

$$\eta \equiv \frac{\text{heat actually dissipated by fin}}{\text{heat dissipated if fin temperature were at } T_w} \tag{7.2.1-20}$$

From (7.2.1-19), we can readily calculate

$$
\begin{aligned}
\eta &= \frac{\int_0^W \int_0^L h(\overline{T} - T_a)dz_1\, dz_2 + \int_{-B}^B \int_0^L h(\overline{T} - T_a)dz_1\, dz_3}{\int_0^W \int_0^L h(T_w - T_a)dz_1\, dz_2 + \int_{-B}^B \int_0^L h(T_w - T_a)dz_1\, dz_3} \\
&= \int_0^L \overline{T}^\star \, dz_1^\star \\
&= \frac{\tanh(N)}{N} \tag{7.2.1-21}
\end{aligned}
$$

which is shown in Figure 7.2.1-2.

For a further discussion of fins, see Jacob (1949) and Eckert and Drake (1959).

**Exercise 7.2.1-1** *A circular cooling fin of rectangular profile ( Jakob 1949, p. 232 )* Let us consider a circular cooling fin of width $2B$ and radius $R$ mounted on a pipe, the outside radius of which

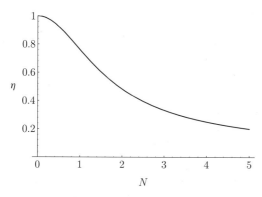

**Figure 7.2.1-2.** Reduction of evaporation rate by transpiration.

is $\kappa R$. The temperature at the base of the fin is known:

$$\text{at } r = \kappa R: \quad T = T_w$$

As in the problem discussed in the text, we will neglect the energy lost through the outer edge:

$$\text{at } r = R: \quad \frac{\partial T}{\partial r} = 0$$

Follow the discussion in the text as a model in determining that the area-averaged temperature distribution in the fin is given by

$$\frac{\overline{T} - T_a}{T_w - T_a} = \frac{I_1(N)K_0(Nr/R) + I_0(Nr/R)K_1(N)}{I_1(N)K_0(N\kappa) + I_0(N\kappa)K_1(N)}$$

where

$$\overline{T} \equiv \frac{1}{4\pi B} \int_0^{2\pi} \int_{-B}^{B} T \, dz \, d\theta$$

and

$$N^2 \equiv \frac{hR^2}{kB}$$

By $I_n(x)$ and $K_n(x)$ we mean respectively the $n$th-order modified Bessel functions of the first and second kinds. Determine also that the efficiency of a circular fin of rectangular profile is

$$\eta = \frac{2\kappa}{N^2(1 - \kappa^2)} \left[ \frac{I_1(N)K_1(N\kappa) - I_1(N\kappa)K_1(N)}{I_1(N)K_0(N\kappa) + I_0(N\kappa)K_1(N)} \right]$$

## 7.3    Local Volume Averaging

We are commonly concerned with exothermic chemical reactions in beds of porous pellets impregnated with a catalyst. Both in the western portion of the United States and in Canada, there are very large deposits in sandstone formations of nearly solid asphalts; it has been suggested that controlled underground combustion might be used to thermally decompose these asphalts and distill the products in place. Various schemes have been proposed and are currently being used to produce energy from the naturally occurring, relatively shallow, hot rock that is found in many parts of the world. All of these processes have one thing in common. To analyze them, one must understand energy transfer in porous media.

Perhaps the most common approach to energy transfer in porous media is to view the porous solid and whatever gases and liquids it contains as a continuum and simply to employ the usual differential energy balance discussed in Section 5.1.3 (Bird et al. 1960, p. 328).[2] No attempt is made to distinguish between energy transfer in the fluid and energy transfer in

[2] The discussion of transpiration cooling given by Jacob (1957, p. 399) is a notable exception. He does distinguish between heat transfer in the solid and heat transfer in the fluid; however, he does not define his temperatures and velocities as averages of the actual velocity and temperature distributions in the solid and fluid phases.

the solid. A more serious objection is that the velocity and temperature used are not defined in terms of the actual velocity and temperature distributions in the solid and fluid phases.

Our successful discussion of momentum transfer in Section 4.3 suggests that we take the same point of view here in studying energy transfer. This means that we should begin by developing the local volume average of the differential energy balance.

For simplicity, we restrict this discussion to a single incompressible fluid flowing through a stationary, rigid porous medium.

### 7.3.1 The Local Volume Average of the Differential Energy Balance

In Section 4.3, we developed the local volume averages of the differential mass and momentum balances. To be more specific, these were local volume averages of the differential mass and momentum balances for the fluid flowing through a porous structure. It was not important there to examine the implications of conservation of mass and of the momentum balance for the solid. But here it is essential that we account for energy transfer in both the solid and the fluid. Let us begin by taking the local volume average of the differential energy balance for each phase.

I said that we would limit ourselves to the flow of an incompressible fluid. Let us also neglect the effect of viscous dissipation in the differential energy balance (in principle, we should argue that $N_{Br}/N_{Pr}N_{Re} \rightarrow 0$; see the introduction to Section 6.4), and let us explicitly rule out the possibility of mutual or external energy transmission (radiation). With these restrictions, the local volume average of the differential energy balance of Table 5.4.1-1 says for the fluid that

$$\frac{1}{V} \int_{R^{(f)}} \left( \rho^{(f)} \hat{c}^{(f)} \frac{\partial T}{\partial t} + \rho^{(f)} \hat{c}^{(f)} \operatorname{div}(T\mathbf{v}) + \operatorname{div} \mathbf{q} \right) dV = 0 \tag{7.3.1-1}$$

The operations of volume integration and differentiation with respect to time may be interchanged in the first term on the left:

$$\frac{1}{V} \int_{R^{(f)}} \rho^{(f)} \hat{c}^{(f)} \frac{\partial T}{\partial t} dV = \rho^{(f)} \hat{c}^{(f)} \frac{\partial \overline{T}^{(f)}}{\partial t} \tag{7.3.1-2}$$

Here we take the heat capacity per unit mass $\hat{c}^{(f)}$ for the fluid to be a constant. The theorem of Section 4.3.2 can be used to express the second and third terms on the left of (7.3.1-1) as

$$\frac{1}{V} \int_{R^{(f)}} \rho^{(f)} \hat{c}^{(f)} \operatorname{div}(T\mathbf{v}) dV = \rho^{(f)} \hat{c}^{(f)} \operatorname{div}\left( \overline{T\mathbf{v}}^{(f)} \right) \tag{7.3.1-3}$$

and

$$\frac{1}{V} \int_{R^{(f)}} \operatorname{div} \mathbf{q} dV = \operatorname{div} \overline{\mathbf{q}}^{(f)} + \frac{1}{V} \int_{S_w} \mathbf{q}^{(f)} \cdot \mathbf{n} dA \tag{7.3.1-4}$$

We define $\mathbf{n}$ to be the unit normal to $S_w$ directed from the fluid phase into the solid. In arriving at (7.3.1-3), we have observed that the velocity vector must be zero at the fluid–solid phase interface $S_w$. We will assume that Fourier's law describes the thermal behavior of both the fluid and the solid. Further, we shall take the thermal conductivities for both phases to be constants. Applying the theorem from Section 4.3.2, we learn that

$$
\begin{aligned}
\overline{\mathbf{q}}^{(f)} &\equiv \frac{1}{V} \int_{R^{(f)}} \mathbf{q}^{(f)} \, dV \\
&= -\frac{k^{(f)}}{V} \int_{R^{(f)}} \nabla T^{(f)} \, dV \\
&= -k^{(f)} \nabla \overline{T}^{(f)} - \frac{k^{(f)}}{V} \int_{S_w} T \mathbf{n} \, dA
\end{aligned}
\tag{7.3.1-5}
$$

In view of (7.3.1-2) and (7.3.1-3) through (7.3.1-5), Equation (7.3.1-1) becomes

$$
\rho^{(f)} \hat{c}^{(f)} \frac{\partial \overline{T}^{(f)}}{\partial t} + \rho^{(f)} \hat{c}^{(f)} \, \mathrm{div} \left( \overline{T \mathbf{v}}^{(f)} \right) - k^{(f)} \, \mathrm{div} \left( \nabla \overline{T}^{(f)} \right)
$$
$$
- k^{(f)} \, \mathrm{div} \left( \frac{1}{V} \int_{S_w} T \mathbf{n} \, dA \right) + \frac{1}{V} \int_{S_w} \mathbf{q}^{(f)} \cdot \mathbf{n} \, dA = 0
\tag{7.3.1-6}
$$

Turning our attention to the stationary solid phase, essentially the same development yields as the local volume average of the differential energy balance:

$$
\rho^{(s)} \hat{c}^{(s)} \frac{\partial \overline{T}^{(s)}}{\partial t} - k^{(s)} \, \mathrm{div} \left( \nabla \overline{T}^{(s)} \right)
$$
$$
+ k^{(s)} \, \mathrm{div} \left( \frac{1}{V} \int_{S_w} T \mathbf{n} \, dA \right) - \frac{1}{V} \int_{S_w} \mathbf{q}^{(s)} \cdot \mathbf{n} \, dA = 0
\tag{7.3.1-7}
$$

Comparing this with (7.3.1-6), we notice that the last two terms have different signs. This is the result of our definition of $\mathbf{n}$ as the unit normal to $S_w$ directed from the fluid phase into the solid phase.

We shall not concern ourselves here with whatever small temperature differences there are between the solid and fluid phases, and we will assume that Jacob (1957, p. 399) made a somewhat similar assumption in his discussion of transpiration cooling:

$$
\begin{aligned}
\langle T \rangle^{(f)} &= \langle T \rangle^{(s)} \\
&= \langle T \rangle
\end{aligned}
\tag{7.3.1-8}
$$

where we introduce the total volume average (4.3.1-7) of temperature,

$$
\langle T \rangle \equiv \frac{1}{V} \int_R T \, dV
\tag{7.3.1-9}
$$

The sum of (7.3.1-6) and (7.3.1-7) requires

$$
\left[ \Psi \rho^{(f)} \hat{c}^{(f)} + (1 - \Psi) \rho^{(s)} \hat{c}^{(s)} \right] \frac{\partial \langle T \rangle}{\partial t} + \rho^{(f)} \hat{c}^{(f)} \, \mathrm{div} \left( \langle T \rangle \overline{\mathbf{v}}^{(f)} \right)
$$
$$
= k^{(f)} \, \mathrm{div}[\nabla (\Psi \langle T \rangle)] + k^{(s)} \, \mathrm{div}\{\nabla [(1 - \Psi)\langle T \rangle]\} + \mathrm{div} \, \mathbf{h}
\tag{7.3.1-10}
$$

Here we define the *thermal tortuosity vector*,

$$
\mathbf{h} \equiv \rho^{(f)} \hat{c}^{(f)} \left( \langle T \rangle \overline{\mathbf{v}}^{(f)} - \overline{T \mathbf{v}}^{(f)} \right) + \left( k^{(f)} - k^{(s)} \right) \frac{1}{V} \int_{S_w} T \mathbf{n} \, dA
\tag{7.3.1-11}
$$

In arriving at (7.3.1-10), we have satisfied the jump energy balance of Section 5.1.3 at the fluid–solid phase interface $S_w$.

The physical meaning of the thermal tortuosity vector is clarified by noting that, if both $T$ and $\Psi$ are independent of position,

$$\mathbf{h} = (k^{(f)} - k^{(s)})\frac{T}{V}\int_{S_w} \mathbf{n}\, dA$$
$$= 0 \tag{7.3.1-12}$$

In reaching this conclusion, we have used the theorem of Section 4.3.2 applied to a constant. It may also be helpful to express (7.3.1-10) as

$$\left[\Psi\rho^{(f)}\hat{c}^{(f)} + (1-\Psi)\rho^{(s)}\hat{c}^{(s)}\right]\frac{\partial\langle T\rangle}{\partial t} + \rho^{(f)}\hat{c}^{(f)}\,\mathrm{div}\left(\langle T\rangle\overline{\mathbf{v}}^{(f)}\right) = -\mathrm{div}\,\mathbf{q}^{(e)} \tag{7.3.1-13}$$

where

$$\mathbf{q}^{(e)} \equiv -k^{(f)}\nabla\left(\Psi\langle T\rangle\right) - k^{(s)}\nabla\left[(1-\Psi)\langle T\rangle\right] - \mathbf{h} \tag{7.3.1-14}$$

may be thought of as an *effective energy flux*. If $\Psi$ is independent of position,

$$\mathbf{q}^{(e)} = -\left[\Psi k^{(f)} + (1-\Psi)k^{(s)}\right]\nabla\langle T\rangle - \mathbf{h} \tag{7.3.1-15}$$

Because of the simplifications shown in (7.3.1-12) and (7.3.1-15), we shall direct our attention to structures of uniform porosity in the sections that immediately follow.

In the next section, I discuss the form that I might expect empirical correlations for $\mathbf{h}$ to take.

### 7.3.2 Empirical Correlations for $\mathbf{h}$

In this section, we give three examples of how experimental data can be used to prepare correlations for $\mathbf{h}$, introduced in Section 7.3.1. Four points form the foundation for this discussion.

1) The thermal tortuosity vector $\mathbf{h}$ is frame indifferent:

$$\mathbf{h}^* \equiv \rho^{(f)}\hat{c}^{(f)}\left(\psi^{-1}\overline{T^*}^{(f)}\overline{\mathbf{v}^*}^{(f)} - \overline{T^*\mathbf{v}^*}^{(f)}\right) + (k^{(f)} - k^{(s)})\frac{1}{V}\int_{S_w} T^*\mathbf{n}^*\, dA$$

$$= \rho^{(f)}\hat{c}^{(f)}\overline{\left[T^*(\psi^{-1}\overline{\mathbf{v}^*}^{(f)} - \mathbf{v}^*)\right]}^{(f)} + (k^{(f)} - k^{(s)})\frac{1}{V}\int_{S_w} T^*\mathbf{n}^*\, dA$$

$$= \rho^{(f)}\hat{c}^{(f)}\overline{\left[T\mathbf{Q}\cdot\left(\psi^{-1}\overline{\mathbf{v}}^{(f)} - \mathbf{v}\right)\right]}^{(f)} + (k^{(f)} - k^{(s)})\frac{1}{V}\int_{S_w} T\mathbf{Q}\cdot\mathbf{n}\, dA$$

$$= \mathbf{Q}\cdot\mathbf{h} \tag{7.3.2-1}$$

In the second line we observe that the superficial volume average of a superficial volume average is simply the superficial volume average (see Exercise 4.3.7-1); in the third line we employ the frame indifference of temperature and of the velocity difference. Here $\mathbf{Q}$ is a (possibly) time-dependent, orthogonal, second-order tensor.

2) We assume that the principle of frame indifference introduced in Section 2.3.1 applies to any empirical correlation developed for $\mathbf{h}$.

3) The Buckingham–Pi theorem (Brand 1957) serves to further restrict the form of any expression for **h**.

4) The averaging surface $S$ is so large that **h** may be assumed not to be an explicit function of position in the porous structure, though it very well may be an implicit function of position as a result of its dependence upon other variables.

### Example 1: Nonoriented Porous Solids Filled with a Stagnant Fluid

By a stagnant fluid, we mean that there is no macroscopic motion of the fluid:

$$\overline{\mathbf{v}}^{(f)} = 0 \tag{7.3.2-2}$$

This suggests that we might neglect the first term on the right of (7.3.1-11) and write

$$\mathbf{h} \doteq (k^{(f)} - k^{(s)}) \frac{1}{V} \int_{S_w} T \mathbf{n} \, dA \tag{7.3.2-3}$$

For geometrically similar nonoriented porous media, **h** might be thought of as a function of the local particle diameter $l_0$, the thermal conductivities $k^{(f)}$ and $k^{(s)}$, the porosity $\Psi$, as well as some measure of the local temperature distribution such as $\nabla \langle T \rangle$:

$$\mathbf{h} = \mathbf{h}\left(l_0, k^{(f)}, k^{(s)}, \Psi, \nabla \langle T \rangle\right) \tag{7.3.2-4}$$

For the moment, let us fix our attention on the dependence of **h** upon $\nabla \langle T \rangle$:

$$\mathbf{h} = \hat{\mathbf{h}}(\nabla \langle T \rangle) \tag{7.3.2-5}$$

By the principle of frame indifference, the functional relationship between these two variables should be the same in every frame of reference. This means that

$$\begin{aligned}
\mathbf{h}^* &= \mathbf{Q} \cdot \mathbf{h} \\
&= \mathbf{Q} \cdot \hat{\mathbf{h}}(\nabla \langle T \rangle) \\
&= \hat{\mathbf{h}}(\mathbf{Q} \cdot \nabla \langle T \rangle)
\end{aligned} \tag{7.3.2-6}$$

or $\hat{\mathbf{h}}$ is an isotropic function (Truesdell and Noll 1965, p. 22):

$$\hat{\mathbf{h}}(\nabla \langle T \rangle) = \mathbf{Q}^T \cdot \hat{\mathbf{h}}(\mathbf{Q} \cdot \nabla \langle T \rangle) \tag{7.3.2-7}$$

By a representation theorem for a vector-valued isotropic function of one vector (Truesdell and Noll 1965, p. 35), we may write

$$\begin{aligned}
\mathbf{h} &= \hat{\mathbf{h}}(\nabla \langle T \rangle) \\
&= H \nabla \langle T \rangle
\end{aligned} \tag{7.3.2-8}$$

where

$$H = \hat{H}(|\nabla \langle T \rangle|) \tag{7.3.2-9}$$

Comparing (7.3.2-8) and (7.3.2-9) with (7.3.2-4), we see that

$$H = H\left(l_0, k^{(f)}, k^{(s)}, \Psi, |\nabla \langle T \rangle|\right) \tag{7.3.2-10}$$

An application of the Buckingham–Pi theorem (Brand 1957) allows us to conclude that

$$H = k^{(s)} K^\star \tag{7.3.2-11}$$

Here

$$K^\star = K^\star \left( \frac{k^{(f)}}{k^{(s)}}, \Psi \right) \tag{7.3.2-12}$$

In summary, Equations (7.3.2-8), (7.3.2-11), and (7.3.2-12) represent possibly the simplest form that empirical correlations for the thermal tortuosity vector **h** can take in a nonoriented porous medium.

### Example 2: Nonoriented Porous Solids Filled with a Flowing Fluid

For geometrically similar, nonoriented porous media through which a fluid is flowing, **h** might be thought of as a function of the local particle diameter $l_0$, the thermal conductivities $k^{(f)}$ and $k^{(s)}$, the porosity $\Psi$, the density $\rho^{(f)}$ of the fluid, the heat capacity $\hat{c}^{(f)}$ per unit mass of the fluid, the local volume-averaged velocity of the fluid with respect to the local-averaged velocity of the solid $\overline{\mathbf{v}}^{(f)} - \overline{\mathbf{v}}^{(s)}$, as well as some measure of the local temperature distribution such as $\nabla\langle T\rangle$:

$$\mathbf{h} = \mathbf{h}\left( l_0, k^{(f)}, k^{(s)}, \Psi, \rho^{(f)}, \hat{c}^{(f)}, \overline{\mathbf{v}}^{(f)} - \overline{\mathbf{v}}^{(s)}, \nabla\langle T\rangle \right) \tag{7.3.2-13}$$

Let us first examine the dependence of **h** upon the two vectors:

$$\mathbf{h} = \hat{\mathbf{h}}\left( \overline{\mathbf{v}}^{(f)} - \overline{\mathbf{v}}^{(s)}, \nabla\langle T\rangle \right) \tag{7.3.2-14}$$

By the principle of frame indifference, the functional relationship between these variables should be the same in every frame of reference. This means that

$$\begin{aligned}
\mathbf{h}^* &= \mathbf{Q} \cdot \mathbf{h} \\
&= \mathbf{Q} \cdot \hat{\mathbf{h}}\left( \overline{\mathbf{v}}^{(f)} - \overline{\mathbf{v}}^{(s)}, \nabla\langle T\rangle \right) \\
&= \hat{\mathbf{h}}\left( \mathbf{Q} \cdot \left[ \overline{\mathbf{v}}^{(f)} - \overline{\mathbf{v}}^{(s)} \right], \mathbf{Q} \cdot \nabla\langle T\rangle \right)
\end{aligned} \tag{7.3.2-15}$$

or $\hat{\mathbf{h}}$ is an isotropic function (Truesdell and Noll 1965, p. 22):

$$\begin{aligned}
&\hat{\mathbf{h}}\left( \overline{\mathbf{v}}^{(f)} - \overline{\mathbf{v}}^{(s)}, \nabla\langle T\rangle \right) \\
&\quad = \mathbf{Q}^T \cdot \hat{\mathbf{h}}\left( \mathbf{Q} \cdot \left[ \overline{\mathbf{v}}^{(f)} - \overline{\mathbf{v}}^{(s)} \right], \mathbf{Q} \cdot \nabla\langle T\rangle \right)
\end{aligned} \tag{7.3.2-16}$$

By representation theorems of Spencer and Rivlin (1959, Sec. 7) and of Smith (1965), the most general polynomial isotropic vector function of two vectors has the form

$$\mathbf{h} = H_{(1)}\nabla\langle T\rangle + H_{(2)}\left( \overline{\mathbf{v}}^{(f)} - \overline{\mathbf{v}}^{(s)} \right) \tag{7.3.2-17}$$

Here $H_{(1)}$ and $H_{(2)}$ are scalar-valued polynomials[3]

$$\begin{aligned}
H_{(i)} &= H_{(i)}\left( \left| \overline{\mathbf{v}}^{(f)} - \overline{\mathbf{v}}^{(s)} \right|, \left| \nabla\langle T\rangle \right|, \left[ \overline{\mathbf{v}}^{(f)} - \overline{\mathbf{v}}^{(s)} \right] \right. \\
&\qquad \left. \cdot \nabla\langle T\rangle, l_0, k^{(f)}, k^{(s)}, \Psi, \rho^{(f)}, \hat{c}^{(f)} \right)
\end{aligned} \tag{7.3.2-18}$$

An application of the Buckingham–Pi theorem (Brand 1957) allows us to conclude that

$$H_{(1)} = k^{(s)}H_{(1)}^\star \tag{7.3.2-19}$$

---

[3] See footnote 1 of Section 4.3.5.

and

$$H_{(2)} = \rho^{(f)} \hat{c}^{(f)} l_0 \, |\nabla \langle T \rangle| \, H_{(2)}^{\star} \tag{7.3.2-20}$$

Here

$$H_{(i)}^{\star} = H_{(i)}^{\star} \left( \frac{k^{(f)}}{k^{(s)}}, N_{Pe}, \frac{\left[ \overline{\mathbf{v}}^{(f)} - \overline{\mathbf{v}}^{(s)} \right] \cdot \nabla \langle T \rangle}{\left| \overline{\mathbf{v}}^{(f)} - \overline{\mathbf{v}}^{(s)} \right| \, |\nabla \langle T \rangle|}, \Psi \right) \tag{7.3.2-21}$$

and

$$N_{Pe} \equiv \frac{\rho^{(f)} \hat{c}^{(f)} \left| \overline{\mathbf{v}}^{(f)} - \overline{\mathbf{v}}^{(s)} \right| l_0}{k^{(s)}} \tag{7.3.2-22}$$

As we might expect, $H_{(2)} = 0$ for $|\nabla \langle T \rangle| = 0$, with the result that $\mathbf{h} = 0$ in this limit.

In summary, Equations (7.3.2-17) and (7.3.2-19) through (7.3.2-21) represent perhaps the simplest form that empirical correlations for the thermal tortuosity vector $\mathbf{h}$ can take when a fluid flows through a nonoriented porous medium.

### Example 3: Oriented Porous Solids Filled with a Stagnant Fluid

We should not expect (7.3.2-8), (7.3.2-11), and (7.3.2-12) to represent the thermal tortuosity vector for a porous structure in which particle diameter $l$ is a function of position. For such a structure, (7.3.2-4) must be altered to include a dependence upon additional vectors and possibly tensors. For example, one might postulate a dependence of $\mathbf{h}$ upon a local gradient of particle diameter as well as $\nabla \langle T \rangle$:

$$\mathbf{h} = \mathbf{h} \left( l, k^{(f)}, k^{(s)}, \Psi, \nabla \langle T \rangle, \nabla l \right) \tag{7.3.2-23}$$

The principle of frame indifference and the Buckingham–Pi theorem (Brand 1957) may be employed as in Example 2 to conclude that

$$\mathbf{h} = K_{(1)} \nabla \langle T \rangle + K_{(2)} \nabla l \tag{7.3.2-24}$$

where

$$K_{(1)} = k^{(s)} K_{(1)}^{\star} \tag{7.3.2-25}$$

$$K_{(2)} = k^{(s)} \, |\nabla \langle T \rangle| \, K_{(2)}^{\star} \tag{7.3.2-26}$$

and

$$K_{(i)}^{\star} = K_{(i)}^{\star} \left( \frac{k^{(f)}}{k^{(s)}}, \frac{\nabla l \cdot \nabla \langle T \rangle}{|\nabla l| \, |\nabla \langle T \rangle|}, |\nabla l|, \Psi \right) \tag{7.3.2-27}$$

Equations (7.3.2-24) through (7.3.2-27) represent perhaps the simplest form that empirical correlations for thermal tortuosity vector $\mathbf{h}$ can take in a porous structure whose orientation can be described by the local gradient in particle diameter.

### 7.3.3  Summary of Results for a Nonoriented, Uniform-Porosity Structure

In Section 7.3.1 we found that, when the porosity $\Psi$ is independent of position in the porous medium, the local volume average of the differential energy balance requires

$$\left[ \Psi \rho^{(f)} \hat{c}^{(f)} + (1 - \Psi) \rho^{(s)} \hat{c}^{(s)} \right] \frac{\partial \langle T \rangle}{\partial t}$$
$$+ \rho^{(f)} \hat{c}^{(f)} \operatorname{div} \left( \langle T \rangle \overline{\mathbf{v}}^{(f)} \right) = -\operatorname{div} \mathbf{q}^{(e)} \tag{7.3.3-1}$$

Here

$$\mathbf{q}^{(e)} \equiv - \left[ \Psi k^{(f)} + (1 - \Psi) k^{(s)} \right] \nabla \langle T \rangle - \mathbf{h} \tag{7.3.3-2}$$

should be thought of as the *effective energy flux*. In arriving at this result, we have neglected any effects attributable to viscous dissipation and radiation.

For a porous medium filled with a stagnant fluid, we suggested in Section 7.3.2 that the thermal tortuosity vector $\mathbf{h}$ can be described by (7.3.2-8) and (7.3.2-11). This result suggests that we may write (7.3.3-2) in terms of an *effective thermal conductivity* $k^{(e)}$:

$$\mathbf{q}^{(e)} = -k^{(e)} \nabla \langle T \rangle \tag{7.3.3-3}$$

Here

$$k^{(e)} \equiv \Psi k^{(f)} + (1 - \Psi) k^{(s)} - k^{(s)} K^{\star} \tag{7.3.3-4}$$

and

$$K^{\star} = K^{\star} \left( \frac{k^{(f)}}{k^{(s)}}, \Psi \right) \tag{7.3.3-5}$$

Experimental studies (Kunii and Smith 1960, 1961b; Mischke and Smith 1962; Masamune and Smith 1963a,b; Huang and Smith 1963) for the stagnant-fluid case confirm the general form of this expression for the effective thermal conductivity $k^{(e)}$.

When there is simultaneous flow through the porous structure, we suggested in Section 7.3.2 that the thermal tortuosity vector $\mathbf{h}$ might be represented by (7.3.2-17), (7.3.2-19), and (7.3.2-20). In these terms, the effective energy flux can be expressed as

$$\mathbf{q}^{(e)} = - \left[ \Psi k^{(f)} + (1 - \Psi) k^{(s)} - k^{(s)} K^{\star}_{(1)} \right] \nabla \langle T \rangle + \rho^{(f)} \hat{c}^{(f)} l_0 |\nabla \langle T \rangle| K^{\star}_{(2)} \overline{\mathbf{v}}^{(f)} \tag{7.3.3-6}$$

where

$$K^{\star}_{(i)} = K^{\star}_{(i)} \left( \frac{k^{(f)}}{k^{(s)}}, N_{Pe}, \frac{\overline{\mathbf{v}}^{(f)} \cdot \nabla \langle T \rangle}{|\overline{\mathbf{v}}^{(f)}| \, |\nabla \langle T \rangle|}, \Psi \right) \tag{7.3.3-7}$$

and

$$N_{Pe} \equiv \frac{\rho^{(f)} \hat{c}^{(f)} |\overline{\mathbf{v}}^{(f)}| \, l_0}{k^{(f)}} \tag{7.3.3-8}$$

In arriving at this expression, we have assumed that the porous medium is stationary. Sometimes it may be more convenient to think of the effective energy flux $\mathbf{q}^{(e)}$ in terms of an effective thermal conductivity tensor $\mathbf{K}^{(e)}$,

$$\mathbf{q}^{(e)} = -\mathbf{K}^{(e)} \cdot \nabla \langle T \rangle \tag{7.3.3-9}$$

in which

$$\mathbf{K}^{(e)} \equiv \left[ \Psi k^{(f)} + (1 - \Psi) k^{(s)} - k^{(s)} K^{\star}_{(1)} \right] \mathbf{I} + \frac{\rho^{(f)} \hat{c}^{(f)} l_0 |\nabla \langle T \rangle| K^{\star}_{(2)}}{\overline{\mathbf{v}}^{(f)} \cdot \nabla \langle T \rangle} \overline{\mathbf{v}}^{(f)} \overline{\mathbf{v}}^{(f)} \tag{7.3.3-10}$$

The results of experimental studies (Kunii and Smith 1961a; Willhite, Kunii, and Smith 1962; Willhite, Dranoff, and Smith 1963; Adivarahan, Kunii, and Smith 1962) may be thought of in terms of the components of this tensor.

For the sake of simplicity, in what follows we take

$$K^\star_{(i)} = K^\star_{(i)} \left( \frac{k^{(f)}}{k^{(s)}}, \Psi \right) \tag{7.3.3-11}$$

and write (7.3.3-6) as

$$\mathbf{q}^{(e)} = -\alpha \nabla \langle T \rangle + \beta \, |\nabla \langle T \rangle| \, \overline{\mathbf{v}}^{(f)} \tag{7.3.3-12}$$

where $\alpha$ and $\beta$ are independent of position in any particular situation.

## 7.3.4 Transpiration Cooling

By *transpiration*, we mean that there is simultaneous flow and energy transfer in a porous structure. Transpiration sometimes can be used to reduce the rate of heat transfer or to decrease the amount of insulation needed for a particular application. The following problem illustrates this idea.

It has been proposed (Bird et al. 1960, p. 345) that the rate of evaporation of liquefied oxygen in small containers might be reduced by taking advantage of transpiration. The liquid could be stored in a spherical container surrounded by a spherical shell of porous insulating material like that shown in Figure 7.3.4-1. A small gap is to be left between the container and the insulation, and the opening through the insulation is to be plugged. In operation we can visualize the evaporating oxygen leaving the spherical flask, moving through the gap between the flask and insulation, and flowing uniformly out through the porous structure. Let us say that we have set as our design criterion that the rate of energy transfer to the oxygen flask should be no more than $\mathcal{Q}$. Oxygen enters the insulation at $r = \kappa R$ at approximately the boiling point $T_\kappa$; the temperature of the oxygen leaving the insulation at $r = R$ is estimated to be $T_1$. The inner radius $\kappa R$ of the insulation shell is fixed by the diameter of the oxygen flask. We wish to determine the outer radius $R$ and in this way the thickness of insulation required.

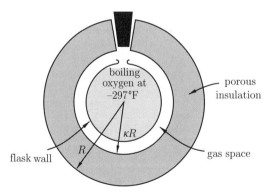

**Figure 7.3.4-1.** Reduction of evaporation rate by transpiration.

It seems reasonable to begin by assuming in spherical coordinates

$$\bar{v}_r^{(f)} = \bar{v}_r^{(f)}(r)$$

$$\bar{v}_\theta^{(f)} = \bar{v}_\varphi^{(f)} = 0 \tag{7.3.4-1}$$

$$\langle T \rangle = \langle T \rangle (r)$$

We estimate the pressure gradient through the porous insulation to be so small that the oxygen may be treated as an incompressible fluid. Under these circumstances, the local volume-averaged differential mass balance in the form of (4.3.3-3) requires

$$\frac{1}{r^2} \frac{d}{dr} \left( r^2 \bar{v}_r^{(f)} \right) = 0 \tag{7.3.4-2}$$

or

$$\bar{v}_r^{(f)} = \frac{Q}{4\pi \lambda_{(\kappa)} \rho_{(\kappa)} r^2} \tag{7.3.4-3}$$

Here $\lambda_{(\kappa)}$ and $\rho_{(\kappa)}$ are the heat of vaporization of oxygen and the density of oxygen evaluated at the temperature of the oxygen flask $T_\kappa$.

We will further assume that everywhere

$$\left| \frac{d\langle T \rangle}{dr} \right| = \frac{d\langle T \rangle}{dr} \tag{7.3.4-4}$$

In view of (7.3.4-1) and (7.3.4-4), we see that (7.3.3-12) requires

$$q_r^{(e)} = -\alpha \frac{d\langle T \rangle}{dr} + \beta \frac{d\langle T \rangle}{dr} \bar{v}_r^{(f)}$$

$$q_\theta^{(e)} = q_\varphi^{(e)}$$

$$= 0 \tag{7.3.4-5}$$

As suggested in Section 7.3.3, for the sake of simplicity, we will take both $\alpha$ and $\beta$ to be independent of position in the porous insulation.

The local volume-averaged differential energy balance in the form of (7.3.1-13) reduces under these circumstances to

$$\rho^{(f)} \hat{c}^{(f)} \bar{v}_r^{(f)} \frac{d\langle T \rangle}{dr} = -\frac{1}{r^2} \frac{d}{dr} \left( r^2 q_r^{(e)} \right) \tag{7.3.4-6}$$

Because of (7.3.4-3) and (7.3.4-5), this becomes

$$\frac{d}{dr} \left[ \frac{\hat{c}^{(f)} Q}{4\pi \lambda_{(\kappa)}} \langle T \rangle - r^2 \left( \alpha \frac{d\langle T \rangle}{dr} - \beta \frac{d\langle T \rangle}{dr} \frac{Q}{4\pi \lambda_{(\kappa)} \rho_{(\kappa)} r^2} \right) \right] = 0 \tag{7.3.4-7}$$

This is more simply expressed in terms of dimensionless variables as

$$\frac{d}{d\xi} \left[ \Theta - (\xi^2 - B) \frac{d\Theta}{d\xi} \right] = 0 \tag{7.3.4-8}$$

where

$$\Theta \equiv \frac{\langle T \rangle - T_1}{T_\kappa - T_1} \tag{7.3.4-9}$$

$$\xi \equiv \frac{r}{R_0} \tag{7.3.4-10}$$

$$R_0 \equiv \frac{\hat{c}^{(f)} \mathcal{Q}}{4\pi \alpha \lambda_{(\kappa)}} \tag{7.3.4-11}$$

and

$$B \equiv \frac{\beta \mathcal{Q}}{4\pi \alpha \lambda_{(\kappa)} \rho_{(\kappa)} R_0^2} \tag{7.3.4-12}$$

The corresponding boundary conditions describe the temperatures that exist in the gas phases surrounding the spherical shell of insulation:

$$\text{at } \xi = \xi_\kappa \equiv \frac{\kappa R - \epsilon}{R_0}: \quad \Theta = 1 \tag{7.3.4-13}$$

and

$$\text{at } \xi = \xi_1 \equiv \frac{R + \epsilon}{R_0}: \quad \Theta = 0 \tag{7.3.4-14}$$

By $\epsilon$, I mean the diameter of the averaging surface $S$.

Integrating once, we have

$$\Theta - (\xi^2 - B)\frac{d\Theta}{d\xi} = C_1 \tag{7.3.4-15}$$

in which $C_1$ is a constant. Integrating again and applying boundary conditions (7.3.4-13) and (7.3.4-14), we learn

$$\Theta = \frac{f(\xi) - f(\xi_1)}{f(\xi_\kappa) - f(\xi_1)} \tag{7.3.4-16}$$

Here we define

$$\text{for } B \neq 0: \quad f(\xi) \equiv \left| \frac{\sqrt{B} - \xi}{\sqrt{B} + \xi} \right|^{1/(2\sqrt{B})} \tag{7.3.4-17}$$

and

$$\text{for } B = 0: \quad f(\xi) \equiv \exp\left(-\frac{1}{\xi}\right) \tag{7.3.4-18}$$

The rate of heat transfer to the oxygen flask is specified as

$$\begin{aligned} \mathcal{Q} &= -4\pi (\kappa R - \epsilon)^2 \, q_r^{(e)} \big|_{r = \kappa R - \epsilon} \\ &= 4\pi \alpha R_0 (T_\kappa - T_1) \left( \xi_\kappa^2 - B \right) \frac{d\Theta}{d\xi} \bigg|_{\xi - \xi_\kappa} \\ &= 4\pi \alpha R_0 (T_1 - T_\kappa) \frac{f(\xi_\kappa)}{f(\xi_1) - f(\xi_\kappa)} \end{aligned} \tag{7.3.4-19}$$

This is more conveniently expressed as

$$Q^\star \equiv \frac{Q}{4\pi\alpha R_0 (T_1 - T_\kappa)}$$

$$= \frac{f(\xi_\kappa)}{f(\xi_1) - f(\xi_\kappa)} \tag{7.3.4-20}$$

Equation (7.3.4-20) can in turn be solved for the unknown $\xi_1$ to find

$$\text{for } B \neq 0: \ \xi_1 = \sqrt{B} \left( \frac{1 - A}{1 + A} \right) \tag{7.3.4-21}$$

and

$$\text{for } B = 0: \ \xi_1 = \left( \frac{1}{\xi_\kappa} + \ln \frac{Q^\star}{1 + Q^\star} \right)^{-1} \tag{7.3.4-22}$$

where

$$A \equiv \left( \frac{Q^\star}{1 + Q^\star} \right)^{-2\sqrt{B}} \left( \frac{\sqrt{B} + \xi_\kappa}{\sqrt{B} - \xi_\kappa} \right) \tag{7.3.4-23}$$

As a basis for comparison, let us ask what happens when there is no transpiration. We require the inner radius of the insulation $\kappa R$ to be the same, but we allow the outer radius of the insulation $R_{(wot)}$ to take a different value in order to compensate for the lack of transpiration. Equation (7.3.4-7) reduces to

$$\frac{d}{dr} \left( r^2 \frac{d\langle T \rangle}{dr} \right) = 0 \tag{7.3.4-24}$$

This is solved consistent with the boundary conditions

$$\text{at } r = \kappa R - \epsilon: \ \langle T \rangle = T_\kappa \tag{7.3.4-25}$$

and

$$\text{at } r = R_{(wot)} + \epsilon: \ \langle T \rangle = T_1 \tag{7.3.4-26}$$

to find

$$\Theta = \frac{1 - \xi_{1(wot)}/\xi}{1 - \xi_{1(wot)}/\xi_\kappa} \tag{7.3.4-27}$$

This allows us to predict the dimensionless rate of energy transfer to the oxygen flask as

$$Q^\star = \frac{\xi_{1(wot)}\xi_\kappa}{\xi_{1(wot)} - \xi_\kappa} \tag{7.3.4-28}$$

Here we define

$$\xi_{1(wot)} \equiv \frac{R_{(wot)} + \epsilon}{R_0} \tag{7.3.4-29}$$

Equation (7.3.4-28) gives us the relation for

$$\xi_{1(wot)} = \frac{\xi_\kappa Q^\star}{Q^\star - \xi_\kappa} \tag{7.3.4-30}$$

which can be compared with (7.3.4-21) and (7.3.4-22).

Perhaps the effect of transpiration is best appreciated by considering a specific example:

$$\begin{aligned}
\mathcal{Q} &= 60\,\text{Btu/h} \\
\kappa R &= 0.5\,\text{ft} \\
T_\kappa &= -297°\text{F} \\
T_1 &= 30°\text{F} \\
\alpha &= 0.02\,\text{Btu/h\,ft\,°F} \\
\beta &= 0 \\
\hat{c}^{(f)} &= 0.22\,\text{Btu/lb}_\text{m}\,°\text{F} \\
\lambda_{(\kappa)} &= 91.7\,\text{Btu/lb}_\text{m} \\
\Psi &= 0.7 \\
\epsilon &\doteq 0\,\text{ft}
\end{aligned}$$

From (7.3.4-22) and (7.3.4-30), we find the thickness of insulation.

with transpiration: $R - \kappa R = 6$ in.
without transpiration: $R_{(wot)} - \kappa R = 13$ in.

Clearly, transpiration can be of practical importance.

To help us further to evaluate this effect, let us ask what happens when the exterior of the insulation is sealed (perhaps with paint) to prevent transpiration. Without transpiration, we learn by analogy with (7.3.4-28) that the dimensionless rate of energy transfer to the oxygen flask is

$$\begin{aligned}
\mathcal{Q}^\star_{(wot)} &\equiv \frac{\mathcal{Q}_{(wot)}}{4\pi\alpha R_0\,(T_1 - T_\kappa)} \\
&= \frac{\xi_1\xi_\kappa}{\xi_1 - \xi_\kappa}
\end{aligned} \tag{7.3.4-31}$$

If we define the effectiveness $E$ of the transpiration as

$$E \equiv \frac{\mathcal{Q}_{(wot)} - \mathcal{Q}}{\mathcal{Q}_{(wot)}} \tag{7.3.4-32}$$

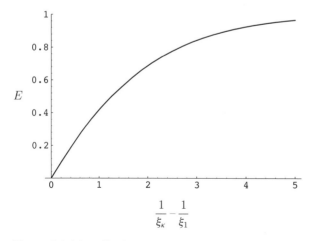

**Figure 7.3.4-2.** Effectiveness $E$ for $B = 0$.

we can compute from (7.3.4-20) and (7.3.4-31) that

$$E = 1 - \left( \frac{1}{\xi_\kappa} - \frac{1}{\xi_1} \right) \frac{f(\xi_\kappa)}{f(\xi_1) - f(\xi_\kappa)} \tag{7.3.4-33}$$

This takes a particularly simple form

$$\text{for } B = 0: \quad E = 1 - \frac{1/\xi_\kappa - 1/\xi_1}{\exp(1/\xi_\kappa - 1/\xi_1) - 1} \tag{7.3.4-34}$$

Equation (7.3.4-34) is shown in Figure 7.3.4-2.

---

## 7.4 More on Integral Balances

In the usual undergraduate course on thermodynamics for engineers, the problems do not center around determining velocity, temperature, and concentration distributions in materials. Rather, the student is asked to determine macroscopic heat-transfer rates and power requirements, or he or she may be asked to comment on whether a given process is feasible. The student is given as tools the integral mass and energy balances, the mechanical energy balance, and the integral entropy inequality. My purpose here is to indicate how the integral energy balance and the integral entropy inequality arise in the context of the preceding discussion. At the same time I shall indicate how one may obtain two additional limiting cases of the mechanical energy balance.

The discussion in the next few sections is closely related to that of Section 4.4. It might be worthwhile at this point for the reader to review this section or at least to reread the introduction, which discusses the place of integral balances in engineering.

As I pointed out in that introduction, by an integral balance I mean an equation that describes an accumulation of any quantity in terms of influx and outflow. The implication is that the system over which the balance is made need not be a collection of material particles. The system might be the fluid on the shell side of a heat exchanger aboard a space capsule that is moving along some arbitrary trajectory through space. The system in this example is not a collection of material particles, since fluid is continuously entering and leaving the heat exchanger.

In the sections that follow, I will assume that the system has a volume $V^{(s)}$ and a closed bounding surface $S^{(s)}$. I will denote the velocity of the closed bounding surface as $\mathbf{v}^{(s)}$; in general, this velocity may be a function of position on the surface. I shall refer to that portion of $S^{(s)}$ across which mass moves as being entrances and exits of the system; I shall denote the collection of entrances and exits by $S_{(\text{ent ex})}$.

Many of the ideas associated with integral balances presented here originate with Bird (1957).

### 7.4.1 The Integral Energy Balance

We wish to develop here an energy balance that describes the time rate of change of the energy associated with an arbitrary system. We will take the same approach that we used in arriving at the integral mass and momentum balances in Sections 4.4.1 and 4.4.3.

By means of the differential mass balance, the differential energy balance (see Table 5.4.1-1) may be written as

$$\frac{\partial}{\partial t}\left[\rho\left(\hat{U}+\frac{1}{2}v^2\right)\right] + \text{div}\left[\rho\left(\hat{U}+\frac{1}{2}v^2\right)\mathbf{v}\right]$$
$$= -\text{div}\,\mathbf{q} + \text{div}(\mathbf{T}\cdot\mathbf{v}) + \rho(\mathbf{v}\cdot\mathbf{f}) + \rho Q \qquad (7.4.1\text{-}1)$$

Integrating this over the volume of our arbitrary system, we have

$$\int_{R^{(s)}}\left\{\frac{\partial}{\partial t}\left[\rho\left(\hat{U}+\frac{1}{2}v^2\right)\right] + \text{div}\left[\rho\left(\hat{U}+\frac{1}{2}v^2\right)\mathbf{v}\right]\right.$$
$$\left. + \text{div}\,\mathbf{q} - \text{div}(\mathbf{T}\cdot\mathbf{v}) - \rho(\mathbf{v}\cdot\mathbf{f}) - \rho Q\right\}dV = 0 \qquad (7.4.1\text{-}2)$$

The first integral on the left of (7.4.1-2) may be expressed more conveniently using the generalized transport theorem of Section 1.3.2:

$$\frac{d}{dt}\int_{R^{(s)}}\rho\left(\hat{U}+\frac{1}{2}v^2\right)dV$$
$$= \int_{R^{(s)}}\frac{\partial}{\partial t}\left[\rho\left(\hat{U}+\frac{1}{2}v^2\right)\right]dV + \int_{S^{(s)}}\rho\left(\hat{U}+\frac{1}{2}v^2\right)(\mathbf{v}^{(s)}\cdot\mathbf{n})\,dA \qquad (7.4.1\text{-}3)$$

The physical meaning of the second, third, and fourth terms is clearer after an application of Green's transformation:

$$\int_{R^{(s)}}\left\{\text{div}\left(\rho\left[\hat{U}+\frac{1}{2}v^2\right]\mathbf{v}\right) + \text{div}\,\mathbf{q} - \text{div}(\mathbf{T}\cdot\mathbf{v})\right\}dV$$
$$= \int_{S^{(s)}}\left[\rho\left(\hat{U}+\frac{1}{2}v^2\right)\mathbf{v}\cdot\mathbf{n} + \mathbf{q}\cdot\mathbf{n} - \mathbf{v}\cdot(\mathbf{T}^T\cdot\mathbf{n})\right]dA \qquad (7.4.1\text{-}4)$$

Equations (7.4.1-3) and (7.4.1-4) allow us to express (7.4.1-2) as

$$\frac{d}{dt}\int_{R^{(s)}}\rho\left(\hat{U}+\frac{1}{2}v^2\right)dV$$
$$= \int_{S^{(s)}}\rho\left(\hat{U}+\frac{1}{2}v^2\right)(\mathbf{v}-\mathbf{v}^{(s)})\cdot(-\mathbf{n})\,dA$$
$$+ \mathcal{Q} - \int_{S^{(s)}}\mathbf{v}\cdot[\mathbf{T}\cdot(-\mathbf{n})]\,dA + \int_{R^{(s)}}\rho(\mathbf{v}\cdot\mathbf{f} + Q)\,dV \qquad (7.4.1\text{-}5)$$

The first term on the right is the net rate at which internal and kinetic energy is brought into the system with whatever material is crossing the boundaries. By $\mathcal{Q}$ we mean the rate of (contact) energy transfer to the system across the bounding surfaces of the system:

$$\mathcal{Q} \equiv \int_{S^{(s)}}\mathbf{q}\cdot(-\mathbf{n})\,dA \qquad (7.4.1\text{-}6)$$

The last term in (7.4.1-5) describes the rate at which work is done on the system by the external force, usually gravity, and the rate of external energy transmission to the system, usually in the form of radiation.

The third term on the right of (7.4.1-5) describes the rate at which work is done by the system upon its bounding surfaces or the rate at which work is done by the system upon its surroundings. This is the total work done by the contact forces upon the boundary, whereas we usually speak in terms of the work done beyond that attributable to a constant ambient pressure $p_0$. To correct for the effect of $p_0$, we may begin by noting that

$$\int_{S^{(s)}} \mathbf{v} \cdot [\mathbf{T} \cdot (-\mathbf{n})] \, dA = \int_{S^{(s)}} \mathbf{v} \cdot [(\mathbf{T} + p_0 \mathbf{I}) \cdot (-\mathbf{n})] \, dA + \int_{S^{(s)}} p_0 (\mathbf{v} \cdot \mathbf{n}) \, dA \quad (7.4.1\text{-}7)$$

Since $p_0$ is independent of time, we may use the generalized transport theorem to write the second term on the right of (7.4.1-7) as

$$\int_{S^{(s)}} p_0 (\mathbf{v} \cdot \mathbf{n}) \, dA = \int_{R^{(s)}} \frac{\partial p_0}{\partial t} \, dV + \int_{S^{(s)}} p_0 (\mathbf{v} \cdot \mathbf{n}) \, dA$$

$$= \frac{d}{dt} \int_{R^{(s)}} p_0 \, dV - \int_{S^{(s)}} p_0 \left( \mathbf{v} - \mathbf{v}^{(s)} \right) \cdot (-\mathbf{n}) \, dA \quad (7.4.1\text{-}8)$$

Using (7.4.1-7) and (7.4.1-8), we may replace (7.4.1-5) with

$$\frac{d}{dt} \int_{R^{(s)}} \rho \left( \hat{U} + \frac{1}{2} v^2 + \frac{p_0}{\rho} \right) dV$$

$$= \int_{S^{(s)}} \rho \left( \hat{U} + \frac{1}{2} v^2 + \frac{p_0}{\rho} \right) (\mathbf{v} - \mathbf{v}^{(s)}) \cdot (-\mathbf{n}) \, dA + \mathcal{Q}$$

$$- \int_{S^{(s)}} \mathbf{v} \cdot [(\mathbf{T} + p_0 \mathbf{I}) \cdot (-\mathbf{n})] \, dA + \int_{R^{(s)}} \rho (\mathbf{v} \cdot \mathbf{f} + Q) \, dV \quad (7.4.1\text{-}9)$$

or

$$\frac{d}{dt} \int_{R^{(s)}} \rho \left( \hat{U} + \frac{1}{2} v^2 + \frac{p_0}{\rho} \right) dV$$

$$= \int_{S_{(\text{ent ex})}} \rho \left( \hat{H} + \frac{1}{2} v^2 \right) (\mathbf{v} - \mathbf{v}^{(s)}) \cdot (-\mathbf{n}) \, dA$$

$$+ \mathcal{Q} - \mathcal{W} + \int_{R^{(s)}} \rho (\mathbf{v} \cdot \mathbf{f} + Q) \, dV$$

$$+ \int_{S_{(\text{ent ex})}} \left[ -(P - p_0) \left( \mathbf{v}^{(s)} \cdot \mathbf{n} \right) + \mathbf{v} \cdot (\mathbf{S} \cdot \mathbf{n}) \right] dA \quad (7.4.1\text{-}10)$$

Introduced in Section 4.4.5,

$$\mathcal{W} \equiv \int_{S^{(s)} - S_{(\text{ent ex})}} \mathbf{v} \cdot [(\mathbf{T} + p_0 \mathbf{I}) \cdot (-\mathbf{n})] \, dA \quad (7.4.1\text{-}11)$$

is the work done by the system on the surroundings at the impermeable surfaces of the system. Equation (7.4.1-10) will be referred to as a *general form of the integral energy balance*.

This is only one form of the integral energy balance for single-phase systems. If we had started with a different expression for the differential energy balance in Table 5.4.1-1, our final result would be somewhat different. Various forms of the integral energy balance appropriate to single-phase systems are presented in Table 7.4.1-1.

**Table 7.4.1-1.** General forms of the integral energy balance applicable to a single-phase system

$$\frac{d}{dt} \int_{R^{(s)}} \rho \left( \hat{U} + \frac{1}{2} v^2 + \varphi + \frac{p_0}{\rho} \right) dV$$

$$= \int_{S_{(ent\ ex)}} \rho \left( \hat{H} + \frac{1}{2} v^2 + \varphi \right) \left( \mathbf{v} - \mathbf{v}^{(s)} \right) \cdot (-\mathbf{n})\ dA$$

$$+ Q - W + \int_{R^{(s)}} \rho Q\ dV + \int_{S_{(ent\ ex)}} \left[ -(P - p_0)\left( \mathbf{v}^{(s)} \cdot \mathbf{n} \right) + \mathbf{v} \cdot (\mathbf{S} \cdot \mathbf{n}) \right] dA^a$$

$$\frac{d}{dt} \int_{R^{(s)}} \rho \left( \hat{U} + \frac{1}{2} v^2 + \frac{p_0}{\rho} \right) dV$$

$$= \int_{S_{(ent\ ex)}} \rho \left( \hat{H} + \frac{1}{2} v^2 \right) \left( \mathbf{v} - \mathbf{v}^{(s)} \right) \cdot (-\mathbf{n})\ dA$$

$$+ Q - W + \int_{R^{(s)}} \rho (\mathbf{v} \cdot \mathbf{f} + Q)\ dV$$

$$+ \int_{S_{(ent\ ex)}} \left[ -(P - p_0)\left( \mathbf{v}^{(s)} \cdot \mathbf{n} \right) + \mathbf{v} \cdot (\mathbf{S} \cdot \mathbf{n}) \right] dA$$

$$\frac{d}{dt} \int_{R^{(s)}} \rho \hat{S}\ dV = \int_{S_{(ent\ ex)}} \rho \hat{S} \left( \mathbf{v} - \mathbf{v}^{(s)} \right) \cdot (-\mathbf{n})\ dA + \int_{S^{(s)}} \frac{1}{T} \mathbf{q} \cdot (-\mathbf{n})\ dA$$

$$+ \int_{R^{(s)}} \left[ -\frac{1}{T^2} \mathbf{q} \cdot \nabla T + \frac{1}{T} \operatorname{tr}(\mathbf{S} \cdot \nabla \mathbf{v}) + \frac{1}{T} \rho Q \right] dV$$

$$\frac{d}{dt} \int_{R^{(s)}} \rho \left( \hat{U} + \frac{p_0}{\rho} \right) dV$$

$$= \int_{S_{(ent\ ex)}} \rho \left( \hat{U} + \frac{p_0}{\rho} \right) \left( \mathbf{v} - \mathbf{v}^{(s)} \right) \cdot (-\mathbf{n})\ dA$$

$$+ Q + \int_{R^{(s)}} [-(P - p_0)\operatorname{div} \mathbf{v} + \operatorname{tr}(\mathbf{S} \cdot \nabla \mathbf{v}) + \rho Q]\ dV$$

*For an incompressible fluid:*

$$\frac{d}{dt} \int_{R^{(s)}} \rho \hat{U}\ dV = \int_{S_{(ent\ ex)}} \rho \hat{U} \left( \mathbf{v} - \mathbf{v}^{(s)} \right) \cdot (-\mathbf{n})\ dA + Q$$

$$+ \int_{R^{(s)}} [\operatorname{tr}(\mathbf{S} \cdot \nabla \mathbf{v}) + \rho Q]\ dV$$

*For an isothermal fluid:*

$$\frac{d}{dt} \int_{R^{(s)}} \rho \hat{S}\ dV = \int_{S_{(ent\ ex)}} \rho \hat{S} \left( \mathbf{v} - \mathbf{v}^{(s)} \right) \cdot (-\mathbf{n})\ dA + \frac{Q}{T}$$

$$+ \frac{1}{T} \int_{R^{(s)}} [\operatorname{tr}(\mathbf{S} \cdot \nabla \mathbf{v}) + \rho Q]\ dV$$

*For an isentropic fluid:*

$$\mathcal{Q} + \int_{R^{(s)}} [\mathrm{tr}(\mathbf{S} \cdot \nabla \mathbf{v}) + \rho Q]\, dV = 0$$

*For an isobaric fluid:*

$$\frac{d}{dt} \int_{R^{(s)}} \rho \hat{H}\, dV = \int_{S_{(\mathrm{ent\ ex})}} \rho \hat{H}\, (\mathbf{v} - \mathbf{v}^{(s)}) \cdot (-\mathbf{n})\, dA + \mathcal{Q} + \int_{R^{(s)}} [\mathrm{tr}(\mathbf{S} \cdot \nabla \mathbf{v}) + \rho Q]\, dV$$

---

$^a$We assume here that $\partial \varphi / \partial t = 0$.

As we pointed out in Section 4.4.1, we are more commonly concerned with multiphase systems. Using the approach and notation of Section 4.4.1 and making no additional assumptions, we find that the parallel of (7.4.1-10) for multiphase systems is

$$\frac{d}{dt} \int_{R^{(s)}} \rho \left( \hat{U} + \frac{1}{2} v^2 + \frac{p_0}{\rho} \right) dV$$

$$= \int_{S_{(\mathrm{ent\ ex})}} \rho \left( \hat{H} + \frac{1}{2} v^2 \right) (\mathbf{v} - \mathbf{v}^{(s)}) \cdot (-\mathbf{n})\, dA$$

$$+ \mathcal{Q} - \mathcal{W} + \int_{R^{(s)}} \rho (\mathbf{v} \cdot \mathbf{f} + Q)\, dV$$

$$+ \int_{S_{(\mathrm{ent\ ex})}} [-(P - p_0)\,(\mathbf{v}^{(s)} \cdot \mathbf{n}) - \mathbf{v} \cdot (\mathbf{S} \cdot \mathbf{n})]\, dA$$

$$+ \int_{\Sigma} \left[ \rho \left( \hat{U} + \frac{1}{2} v^2 \right) (\mathbf{v} - \mathbf{u}) \cdot \boldsymbol{\xi} + \mathbf{q} \cdot \boldsymbol{\xi} - \mathbf{v} \cdot (\mathbf{T} \cdot \boldsymbol{\xi}) \right] dA \qquad (7.4.1\text{-}12)$$

If we assume that the jump energy balance of Section 5.1.3 applies, this reduces to the equivalent result for single-phase systems, (7.4.1-10).

As before, if we start with a different form of the differential energy balance from Table 5.4.1-1, we will find a somewhat different form for the integral energy balance applicable to a multiphase system. The various possibilities are shown in Table 7.4.1-2.

Normally a number of assumptions are made in the course of analyzing a particular physical situation with the help of the integral energy balance. The most commonly invoked assumptions are these:

1) No mass transfer occurs across internal phase interfaces.
2) The jump mass, momentum, and energy balances of Sections 1.3.6, 2.2.3, and 5.1.3 apply.
3) No mutual or external energy transmission occurs.
4) Entrances and exits are fixed in space.
5) Work done by viscous forces (as described by the extra-stress tensor) may be neglected at entrances and exits.

**Table 7.4.1-2.** General forms of the integral energy balance applicable to a multiphase system where the jump energy balance (5.1.3-9) applies

$$\frac{d}{dt} \int_{R^{(s)}} \rho \left( \hat{U} + \frac{1}{2}v^2 + \varphi + \frac{p_0}{\rho} \right) dV$$

$$= \int_{S_{(ent\ ex)}} \rho \left( \hat{H} + \frac{1}{2}v^2 + \varphi \right) (\mathbf{v} - \mathbf{v}^{(s)}) \cdot (-\mathbf{n}) \, dA + \mathcal{Q} - \mathcal{W} + \int_{R^{(s)}} \rho Q \, dV$$

$$+ \int_{S_{(ent\ ex)}} \left[ -(P - p_0) (\mathbf{v}^{(s)} \cdot \mathbf{n}) + \mathbf{v} \cdot (\mathbf{S} \cdot \mathbf{n}) \right] dA$$

$$+ \int_{\Sigma} \left[ \rho \varphi (\mathbf{v} - \mathbf{u}) \cdot \boldsymbol{\xi} \right] dA^a$$

$$\frac{d}{dt} \int_{R^{(s)}} \rho \left( \hat{U} + \frac{1}{2}v^2 + \frac{p_0}{\rho} \right) dV$$

$$= \int_{S_{(ent\ ex)}} \rho \left( \hat{H} + \frac{1}{2}v^2 \right) (\mathbf{v} - \mathbf{v}^{(s)}) \cdot (-\mathbf{n}) \, dA + \mathcal{Q} - \mathcal{W} + \int_{R^{(s)}} \rho (\mathbf{v} \cdot \mathbf{f} + Q) \, dV$$

$$+ \int_{S_{(ent\ ex)}} \left[ -(P - p_0) (\mathbf{v}^{(s)} \cdot \mathbf{n}) + \mathbf{v} \cdot (\mathbf{S} \cdot \mathbf{n}) \right] dA$$

$$\frac{d}{dt} \int_{R^{(s)}} \rho \hat{S} \, dV = \int_{S_{(ent\ ex)}} \rho \hat{S} (\mathbf{v} - \mathbf{v}^{(s)}) \cdot (-\mathbf{n}) \, dA + \int_{S^{(s)}} \frac{1}{T} \mathbf{q} \cdot (-\mathbf{n}) \, dA$$

$$+ \int_{R^{(s)}} \left[ -\frac{1}{T^2} \mathbf{q} \cdot \nabla T + \frac{1}{T} \operatorname{tr}(\mathbf{S} \cdot \nabla \mathbf{v}) + \frac{1}{T} \rho Q \right] dV$$

$$+ \int_{\Sigma} \left[ \rho \hat{S} (\mathbf{v} - \mathbf{u}) \cdot \boldsymbol{\xi} + \frac{1}{T} \mathbf{q} \cdot \boldsymbol{\xi} \right] dA$$

$$\frac{d}{dt} \int_{R^{(s)}} \rho \left( \hat{U} + \frac{p_0}{\rho} \right) dV$$

$$= \int_{S_{(ent\ ex)}} \rho \left( \hat{U} + \frac{p_0}{\rho} \right) (\mathbf{v} - \mathbf{v}^{(s)}) \cdot (-\mathbf{n}) \, dA$$

$$+ \mathcal{Q} + \int_{R^{(s)}} \left[ -(P - p_0) \operatorname{div} \mathbf{v} + \operatorname{tr}(\mathbf{S} \cdot \nabla \mathbf{v}) + \rho Q \right] dV$$

$$+ \int_{\Sigma} \left[ \rho \left( \hat{U} + \frac{p_0}{\rho} \right) (\mathbf{v} - \mathbf{u}) \cdot \boldsymbol{\xi} + \mathbf{q} \cdot \boldsymbol{\xi} \right] dA$$

*For incompressible fluids:*

$$\frac{d}{dt} \int_{R^{(s)}} \rho \hat{U} \, dV = \int_{S_{(ent\ ex)}} \rho \hat{U} (\mathbf{v} - \mathbf{v}^{(s)}) \cdot (-\mathbf{n}) \, dA + \mathcal{Q}$$

$$+ \int_{R^{(s)}} \left[ \operatorname{tr}(\mathbf{S} \cdot \nabla \mathbf{v}) + \rho Q \right] dV + \int_{\Sigma} \left[ \rho \hat{U} (\mathbf{v} - \mathbf{u}) \cdot \boldsymbol{\xi} + \mathbf{q} \cdot \boldsymbol{\xi} \right] dA$$

*For an isothermal system:*

$$\frac{d}{dt} \int_{R^{(s)}} \rho \hat{S} \, dV = \int_{S_{(\text{ent ex})}} \rho \hat{S} \left( \mathbf{v} - \mathbf{v}^{(s)} \right) \cdot (-\mathbf{n}) \, dA + \frac{Q}{T}$$

$$+ \frac{1}{T} \int_{R^{(s)}} [\text{tr}(\mathbf{S} \cdot \nabla \mathbf{v}) + \rho Q] \, dV + \int_{\Sigma} \left[ \rho \hat{S}(\mathbf{v} - \mathbf{u}) \cdot \boldsymbol{\xi} + \frac{1}{T} \mathbf{q} \cdot \boldsymbol{\xi} \right] dA$$

*For an isentropic system:*

$$Q + \int_{R^{(s)}} [\text{tr}(\mathbf{S} \cdot \nabla \mathbf{v}) + \rho Q] \, dV + \int_{\Sigma} \left[ \mathbf{q} \cdot \boldsymbol{\xi} \right] dA = 0$$

*For an isobaric system:*

$$\frac{d}{dt} \int_{R^{(s)}} \rho \hat{H} \, dV = \int_{S_{(\text{ent ex})}} \rho \hat{H} \left( \mathbf{v} - \mathbf{v}^{(s)} \right) \cdot (-\mathbf{n}) \, dA + Q$$

$$+ \int_{R^{(s)}} [\text{tr}(\mathbf{S} \cdot \nabla \mathbf{v}) + \rho Q] \, dV + \int_{\Sigma} \left[ \rho \hat{H}(\mathbf{v} - \mathbf{u}) \cdot \boldsymbol{\xi} + \mathbf{q} \cdot \boldsymbol{\xi} \right] dA$$

---

[a] We assume here that $\partial \varphi / \partial t = 0$.

With assumptions 2 through 5, Equation (7.4.1-12) simplifies to

$$\frac{d}{dt} \int_{R^{(s)}} \rho \left( \hat{U} + \frac{1}{2} v^2 + \frac{p_0}{\rho} \right) dV = \int_{S_{(\text{ent ex})}} \rho \left( \hat{H} + \frac{1}{2} v^2 \right) (-\mathbf{v} \cdot \mathbf{n}) \, dA$$

$$+ Q - W + \int_{R^{(s)}} \rho \mathbf{v} \cdot \mathbf{f} \, dV \qquad (7.4.1\text{-}13)$$

Table 7.4.1-3 indicates a number of other possibilities.

There are three common types of problems in which the integral energy balance is applied: The rate of energy transfer $Q$ may be neglected, it may be the unknown to be determined, or it may be known from previous experimental data. In this last case, one employs an empirical correlation of data for $Q$. In Section 7.4.2, we discuss the form that these empirical correlations should take.

**Exercise 7.4.1-1** *An isentropic fluid*  Let us define an isentropic fluid to be one in which specific entropy is independent of time and position. Prove that sufficient conditions for a fluid to be isentropic are that its specific internal energy and thermodynamic pressure are independent of time and position.

**Exercise 7.4.1-2** *More about an isentropic fluid*

i) If entropy is independent of time and position, prove that

$$\nabla P = \rho \nabla \hat{H}$$

**Table 7.4.1-3.** Restricted forms of the integral energy balance applicable to a multiphase system. These forms are applicable in the context of assumptions 1 through 5 in the text

$$\frac{d}{dt} \int_{R^{(s)}} \rho \left( \hat{U} + \frac{1}{2}v^2 + \varphi + \frac{p_0}{\rho} \right) dV$$

$$= \int_{S_{(ent\ ex)}} \rho \left( \hat{H} + \frac{1}{2}v^2 + \varphi \right) (-\mathbf{v} \cdot \mathbf{n})\, dA + \mathcal{Q} - \mathcal{W}^a$$

$$\frac{d}{dt} \int_{R^{(s)}} \rho \left( \hat{U} + \frac{1}{2}v^2 + \frac{p_0}{\rho} \right) dV$$

$$= \int_{S_{(ent\ ex)}} \rho \left( \hat{H} + \frac{1}{2}v^2 \right) (-\mathbf{v} \cdot \mathbf{n})\, dA + \mathcal{Q} - \mathcal{W} + \int_{R^{(s)}} \rho \mathbf{v} \cdot \mathbf{f}\, dV$$

$$\frac{d}{dt} \int_{R^{(s)}} \rho \hat{S}\, dV$$

$$= \int_{S_{(ent\ ex)}} \rho \hat{S}(-\mathbf{v} \cdot \mathbf{n})\, dA + \int_{S^{(s)}} \frac{1}{T} \mathbf{q} \cdot (-\mathbf{n})\, dA$$

$$+ \int_{R^{(s)}} \left[ -\frac{1}{T^2} \mathbf{q} \cdot \nabla T + \frac{1}{T} \operatorname{tr}(\mathbf{S} \cdot \nabla \mathbf{v}) \right] dV$$

$$\frac{d}{dt} \int_{R^{(s)}} \rho \left( \hat{U} + \frac{p_0}{\rho} \right) dV$$

$$= \int_{S_{(ent\ ex)}} \rho \left( \hat{U} + \frac{p_0}{\rho} \right) (-\mathbf{v} \cdot \mathbf{n})\, dA$$

$$+ \mathcal{Q} + \int_{R^{(s)}} \left[ -(P - p_0) \operatorname{div} \mathbf{v} + \operatorname{tr}(\mathbf{S} \cdot \nabla \mathbf{v}) \right] dV$$

*For incompressible fluids:*

$$\frac{d}{dt} \int_{R^{(s)}} \rho \hat{U}\, dV = \int_{S_{(ent\ ex)}} \rho \hat{U}(-\mathbf{v} \cdot \mathbf{n})\, dA + \mathcal{Q} + \int_{R^{(s)}} \operatorname{tr}(\mathbf{S} \cdot \nabla \mathbf{v})\, dV$$

*For an isothermal system:*

$$\frac{d}{dt} \int_{R^{(s)}} \rho \hat{S}\, dV = \int_{S_{(ent\ ex)}} \rho \hat{S}(-\mathbf{v} \cdot \mathbf{n})\, dA + \frac{\mathcal{Q}}{T} + \frac{1}{T} \int_{R^{(s)}} \operatorname{tr}(\mathbf{S} \cdot \nabla \mathbf{v})\, dV$$

*For an isentropic system:*

$$\mathcal{Q} + \int_{R^{(s)}} \operatorname{tr}(\mathbf{S} \cdot \nabla \mathbf{v})\, dV = 0$$

*For an isobaric system:*

$$\frac{d}{dt} \int_{R^{(s)}} \rho \hat{H}\, dV = \int_{S_{(ent\ ex)}} \rho \hat{H}(-\mathbf{v} \cdot \mathbf{n})\, dA + \mathcal{Q} + \int_{R^{(s)}} \operatorname{tr}(\mathbf{S} \cdot \nabla \mathbf{v})\, dA$$

[a] We assume here that $\partial \varphi / \partial t = 0$.

and

$$\frac{\partial P}{\partial t} = \rho \frac{\partial \hat{H}}{\partial t}$$

ii) Use Green's transformation to prove that

$$\int_{R^{(s)}} -(P - p_0) \operatorname{div} \mathbf{v} \, dV = \int_{S^{(s)}} (P - p_0) \left( \mathbf{v} - \mathbf{v}^{(s)} \right) \cdot (-\mathbf{n}) \, dA$$
$$- \int_{S^{(s)}} (P - p_0) \mathbf{v}^{(s)} \cdot \mathbf{n} \, dA + \int_{R^{(s)}} \mathbf{v} \cdot \nabla P \, dV$$

iii) Use Green's transformation and the differential mass balance to prove that

$$\int_{R^{(s)}} \mathbf{v} \cdot \nabla P \, dV = \int_{S^{(s)}} \rho \hat{H} \mathbf{v} \cdot \mathbf{n} \, dA$$
$$+ \int_{R^{(s)}} \frac{\partial (\rho \hat{H})}{\partial t} \, dV - \int_{R^{(s)}} \rho \frac{\partial \hat{H}}{\partial t} \, dV$$

iv) Use the generalized transport theorem to find that

$$\int_{R^{(s)}} \mathbf{v} \cdot \nabla P \, dV = \frac{d}{dt} \int_{R^{(s)}} \rho \left( \hat{U} + \frac{p_0}{\rho} \right) \, dV - \int_{S_{(\text{ent ex})}} \rho \hat{H} \left( \mathbf{v} - \mathbf{v}^{(s)} \right)$$
$$\cdot (-\mathbf{n}) \, dA + \int_{S^{(s)}} (P - p_0) \mathbf{v}^{(s)} \cdot \mathbf{n} \, dA$$

v) Deduce that

$$\int_{R^{(s)}} -(P - p_0) \operatorname{div} \mathbf{v} \, dV = \frac{d}{dt} \int_{R^{(s)}} \rho \left( \hat{U} + \frac{p_0}{\rho} \right) \, dV$$
$$- \int_{S_{(\text{ent ex})}} \rho \left( \hat{U} + \frac{p_0}{\rho} \right) \left( \mathbf{v} - \mathbf{v}^{(s)} \right) \cdot (-\mathbf{n}) \, dA$$

and that the result for an isentropic fluid in Table 7.4.1-1 follows.

Similar arguments can be given leading to the results for isentropic fluids in Tables 7.4.1-2 and 7.4.1-3.

**Exercise 7.4.1-3** *The integral energy balance for turbulent flows* I recommend following Section 4.4.2 in developing forms of the integral energy balance appropriate to turbulent flows. Because of the relative complexity of the equations, it generally seems to be more practical to time average the various forms of the integral energy balance found in Tables 7.4.1-1 through 7.4.1-3.

## 7.4.2 Empirical Correlations for $Q$

By means of two illustrations, we indicate here how empirical data correlations for $Q$ ($\overline{Q}$ when dealing with turbulent flows), introduced in Section 7.4.1, can be constructed. There are three principal ideas to be considered in this discussion.

1) The total rate of contact energy transmission to the system is frame indifferent:

$$
\begin{aligned}
\mathcal{Q}^{\star} &= \int_{S^{(s)}} \mathbf{q}^{\star} \cdot (-\mathbf{n}^{\star}) \, dA \\
&= \int_{S^{(s)}} \mathbf{q} \cdot (-\mathbf{n}) \, dA \\
&= \mathcal{Q}
\end{aligned}
\tag{7.4.2-1}
$$

2) We assume that the principle of frame indifference, introduced in Section 2.3.1, applies to any empirical correlation developed for $\mathcal{Q}$.

3) The form of any expression for $\mathcal{Q}$ must satisfy the Buckingham–Pi theorem (Brand 1957).

### Example 1: Forced Convection in Plane Flow Past a Cylindrical Body

An infinitely long cylindrical body is submerged in a large mass of an incompressible Newtonian fluid. The surface temperature of the solid body is $T_0$; the fluid has a nearly uniform temperature $T_\infty$ outside the immediate neighborhood of the body. In a frame of reference that is fixed with respect to the earth, the cylindrical body translates without rotation at a constant velocity $\mathbf{v}_0$; the fluid at a very large distance from the body moves with a uniform velocity $\mathbf{v}_\infty$. The vectors $\mathbf{v}_0$ and $\mathbf{v}_\infty$ are normal to the axis of the cylinder so that we may expect that the fluid moves in a plane flow. One unit vector $\boldsymbol{\alpha}$ is sufficient to describe the orientation of the cylinder with respect to $\mathbf{v}_0$ and $\mathbf{v}_\infty$.

It seems reasonable to say that $\mathcal{Q}$ should be a function of the fluid density $\rho$, the fluid viscosity $\mu$, the fluid heat capacity per unit mass $\hat{c}$, the fluid's thermal conductivity $k$, a length $L$ that is characteristic of the cylinder's cross section, $\mathbf{v}_\infty - \mathbf{v}_0$, $\boldsymbol{\alpha}$, and $\Delta T \equiv T_\infty - T_0$:

$$
\mathcal{Q} = f(\rho, \mu, \hat{c}, k, L, \mathbf{v}_\infty - \mathbf{v}_0, \boldsymbol{\alpha}, \Delta T)
\tag{7.4.2-2}
$$

Let us concentrate our attention upon the independent variables $\mathbf{v}_\infty - \mathbf{v}_0$ and $\boldsymbol{\alpha}$:

$$
\mathcal{Q} = \tilde{f}(\mathbf{v}_\infty - \mathbf{v}_0, \boldsymbol{\alpha})
\tag{7.4.2-3}
$$

By the principle of frame indifference, we conclude that $\tilde{f}$ is a scalar-valued isotropic function of two vectors:

$$
\tilde{f}(\mathbf{v}_\infty - \mathbf{v}_0, \boldsymbol{\alpha}) = \tilde{f}(\mathbf{Q} \cdot [\mathbf{v}_\infty - \mathbf{v}_0], \mathbf{Q} \cdot \boldsymbol{\alpha})
\tag{7.4.2-4}
$$

Here $\mathbf{Q}$ is an orthogonal second-order tensor that describes in part a change of frame. A representation theorem due to Cauchy (Truesdell and Noll 1965, p. 29) tells us that the most general isotropic scalar-valued function of two vectors has the form

$$
\tilde{f}(\mathbf{v}_\infty - \mathbf{v}_0, \boldsymbol{\alpha}) = \tilde{F}\left(|\mathbf{v}_\infty - \mathbf{v}_0|, [\mathbf{v}_\infty - \mathbf{v}_0] \cdot \boldsymbol{\alpha}\right)
\tag{7.4.2-5}
$$

This allows us to express (7.4.2-2) as

$$
\mathcal{Q} = F(\rho, \mu, \hat{c}, k, L, |\mathbf{v}_\infty - \mathbf{v}_0|, [\mathbf{v}_\infty - \mathbf{v}_0] \cdot \boldsymbol{\alpha}, \Delta T)
\tag{7.4.2-6}
$$

But the Buckingham–Pi theorem (Brand 1957) requires that this last expression be of the form

$$
N_{Nu} = N_{Nu}\left(N_{Re}, N_{Pr}, N_{Br}, \frac{\mathbf{v}_\infty - \mathbf{v}_0}{|\mathbf{v}_\infty - \mathbf{v}_0|} \cdot \boldsymbol{\alpha}\right)
\tag{7.4.2-7}
$$

where the Nusselt, Reynolds, Prandtl, and Brinkman numbers are defined as

$$
N_{Nu} \equiv \frac{Q}{kL\,\Delta T}, \quad N_{Re} \equiv \frac{L\rho|\mathbf{v}_\infty - \mathbf{v}_0|}{\mu}
$$
$$
N_{Pr} \equiv \frac{\hat{c}\mu}{k}, \quad N_{Br} \equiv \frac{\mu|\mathbf{v}_\infty - \mathbf{v}_0|^2}{k\,\Delta T}
\tag{7.4.2-8}
$$

It is traditional to define a heat-transfer coefficient $h$ as

$$
h \equiv \frac{Q}{A\,\Delta T}
\tag{7.4.2-9}
$$

where $A$ is proportional to $L^2$ and denotes the area available for contact energy transfer. The Nusselt number is in turn expressed in terms of this heat-transfer coefficient:

$$
N_{Nu} = \frac{hL}{k}
\tag{7.4.2-10}
$$

One computes the rate of contact energy transfer to a system as

$$
Q = hA\,\Delta T
\tag{7.4.2-11}
$$

estimating the heat-transfer coefficient $h$ from an empirical data correlation of the form of (7.4.2-7).

Most empirical correlations for the Nusselt number are not as general as (7.4.2-7) indicates. Commonly, the Brinkman number $N_{Br}$ is quite small, suggesting that viscous dissipation may be neglected. Further, most studies are for a single orientation of a body (or a set of bodies such as a tube bundle) with respect to a fluid stream. Under these conditions, (7.4.2-7) assumes a simpler form (Kays and London 1964; Bird et al. 1960, p. 408):

$$
N_{Nu} = N_{Nu}(N_{Re}, N_{Pr})
\tag{7.4.2-12}
$$

### Example 2: Natural Convection from a Submerged Sphere

Consider a sphere of radius $a$ and surface temperature $T_0$ that is submerged in a large body of a Newtonian fluid. Outside the immediate neighborhood of the sphere, the temperature of the fluid has a constant value $T_\infty$. No relative motion between the sphere and the fluid is imposed, although a circulation pattern is set up in the fluid as the result of natural convection.

In addition to saying that $Q$ is a function of $a$, the fluid viscosity $\mu$, the fluid heat capacity per unit mass $\hat{c}$, and the local magnitude of the acceleration of gravity $g$, we must account for the temperature dependence of the fluid density, since this is the primary cause of natural convection. We can do this by saying that $Q$ must also be a function of the fluid density $\rho$ and the coefficient of volume expansion of the fluid

$$
\beta \equiv -\frac{1}{\rho}\left(\frac{\partial \rho}{\partial T}\right)_P
\tag{7.4.2-13}
$$

evaluated as some temperature characteristic of the experiment. This characteristic temperature is usually chosen to be the *film temperature* $T_f = (T_0 - T_\infty)/2$. In summary, we postulate that

$$
Q = f(\mu, \hat{c}, a, \Delta T, \rho, \beta)
\tag{7.4.2-14}
$$

This relationship automatically satisfies the principle of frame indifference.

The Buckingham–Pi theorem requires that (7.4.2-14) assume the general form

$$N_{Nu} = N_{Nu}(\beta\Delta T, N_{Fr}, N_{Pr}, N_{Br})  \tag{7.4.2-15}$$

where the Nusselt and Froude numbers are defined as

$$N_{Nu} \equiv \frac{Q}{ka\Delta T}, \qquad N_{Fr} \equiv \frac{\mu^2}{\rho^2 a^3 g}  \tag{7.4.2-16}$$

The Prandtl and Brinkman numbers are again defined by (7.4.2-8). (This definition for the Froude number is consistent with the more common definition $N_{Fr} = v^2/ga$, if we define the characteristic speed $v$ to be such that $N_{Re} = \rho a v/\mu = 1$.)

As suggested in the discussion of Example 1, it is traditional in the literature to define a heat-transfer coefficient $h$ by (7.4.2-9) or in this case

$$h \equiv \frac{Q}{4\pi a^2 \Delta T}  \tag{7.4.2-17}$$

The rate of contact energy transfer to the sphere is consequently to be calculated by setting

$$Q = h4\pi a^2 \Delta T  \tag{7.4.2-18}$$

where $h$ is to be determined from the empirical correlation of data of the form of (7.4.2-15).

As one might expect, the Brinkman number $N_{Br}$ is so small for most situations as to suggest that viscous dissipation may be neglected and that (7.4.2-15) may be approximated by

$$N_{Nu} = N_{Nu}(\beta\Delta T, N_{Fr}, N_{Pr})  \tag{7.4.2-19}$$

Ranz and Marshall (Ranz and Marshall 1952a,b) found that a data correlation of the form

$$N_{Nu} \equiv \frac{ha}{k}$$
$$= 2 + 0.60\left(\frac{\beta\Delta T}{N_{Fr}}\right)^{1/4}\left(N_{Pr}\right)^{1/3}  \tag{7.4.2-20}$$

agrees well with available experimental data for $(\beta\Delta T/N_{Fr})^{1/4}\left(N_{Pr}\right)^{1/3} < 200$.

### 7.4.3   More About the Mechanical Energy Balance

In Section 4.4.5, we derived one of the general forms of the mechanical energy balance for single-phase systems:

$$\frac{d}{dt}\int_{R^{(s)}} \rho\left(\frac{1}{2}v^2 + \varphi\right) dV$$

$$= \int_{S_{(\text{ent ex})}} \rho\left(\frac{1}{2}v^2 + \varphi + \frac{P - p_0}{\rho}\right)\left(\mathbf{v} - \mathbf{v}^{(s)}\right) \cdot (-\mathbf{n})\, dA$$

$$+ \int_{R^{(s)}} (P - p_0)\, \text{div } \mathbf{v}\, dV - \mathcal{W} - \mathcal{E}$$

$$+ \int_{S_{(\text{ent ex})}} \left[-(P - p_0)\mathbf{v}^{(s)} \cdot \mathbf{n} + \mathbf{v} \cdot (\mathbf{S} \cdot \mathbf{n})\right] dA  \tag{7.4.3-1}$$

This is not one of the more useful forms of the mechanical energy balance in the sense that the value of the second integral on the right will not be immediately obvious for most situations. We got around this difficulty in Section 4.4.5 by restricting ourselves to incompressible fluids, in which case

$$\int_{R^{(s)}} (P - p_0) \operatorname{div} \mathbf{v} \, dV = 0 \tag{7.4.3-2}$$

Our object here is to indicate that there are other useful forms of the mechanical energy balance that are not restricted to incompressible fluids. We illustrate this point by devoting the bulk of this section to isothermal fluids.

For the moment, let us concentrate on rearranging the second integral on the right of (7.4.3-1) for the case of an isothermal fluid. We can begin by using Green's transformation to find that

$$\int_{R^{(s)}} (P - p_0) \operatorname{div} \mathbf{v} \, dV$$

$$= \int_{R^{(s)}} \operatorname{div} \left( [P - p_0] \mathbf{v} \right) dV - \int_{R^{(s)}} \mathbf{v} \cdot \nabla P \, dV$$

$$= - \int_{S^{(s)}} (P - p_0) \left( \mathbf{v} - \mathbf{v}^{(s)} \right) \cdot (-\mathbf{n}) \, dA$$

$$\quad + \int_{S^{(s)}} (P - p_0) \mathbf{v}^{(s)} \cdot \mathbf{n} \, dA - \int_{R^{(s)}} \mathbf{v} \cdot \nabla P \, dV \tag{7.4.3-3}$$

Since

$$\nabla P = \rho \nabla \hat{G} \tag{7.4.3-4}$$

we can use another application of Green's transformation and the differential mass balance to arrive at

$$\int_{R^{(s)}} \mathbf{v} \cdot \nabla P \, dV$$

$$= \int_{R^{(s)}} \rho \mathbf{v} \cdot \nabla \hat{G} \, dV$$

$$= \int_{R^{(s)}} [\operatorname{div}(\rho \hat{G} \mathbf{v}) - \hat{G} \operatorname{div}(\rho \mathbf{v})] \, dV$$

$$= \int_{S^{(s)}} \rho \hat{G} \mathbf{v} \cdot \mathbf{n} \, dA + \int_{R^{(s)}} \hat{G} \frac{\partial \rho}{\partial t} \, dV$$

$$= \int_{S^{(s)}} \rho \hat{G} \mathbf{v} \cdot \mathbf{n} \, dA + \int_{R^{(s)}} \frac{\partial(\rho \hat{G})}{\partial t} \, dV$$

$$\quad - \int_{R^{(s)}} \rho \frac{\partial \hat{G}}{\partial t} \, dV \tag{7.4.3-5}$$

Noting that

$$\rho \frac{\partial \hat{G}}{\partial t} = \frac{\partial P}{\partial t}$$

$$\quad = \frac{\partial (P - p_0)}{\partial t} \tag{7.4.3-6}$$

we may use the generalized transport theorem to express (7.4.3-5) as

$$
\int_{R^{(s)}} \mathbf{v} \cdot \nabla P \, dV = \frac{d}{dt} \int_{R^{(s)}} \rho \left( \hat{A} + \frac{p_0}{\rho} \right) dV
$$

$$
- \int_{S_{(ent\ ex)}} \rho \hat{G} \left( \mathbf{v} - \mathbf{v}^{(s)} \right) \cdot (-\mathbf{n}) \, dA
$$

$$
+ \int_{S^{(s)}} (P - p_0) \mathbf{v}^{(s)} \cdot \mathbf{n} \, dA \tag{7.4.3-7}
$$

Substituting this into (7.4.3-3), we have

$$
\int_{R^{(s)}} (P - p_0) \, \mathrm{div}\, \mathbf{v} \, dV = -\frac{d}{dt} \int_{R^{(s)}} \rho \left( \hat{A} + \frac{p_0}{\rho} \right) dV
$$

$$
+ \int_{S_{(ent\ ex)}} \rho \left( \hat{A} + \frac{p_0}{\rho} \right) \left( \mathbf{v} - \mathbf{v}^{(s)} \right) \cdot (-\mathbf{n}) \, dA \tag{7.4.3-8}
$$

This last allows us to express (7.4.3-1) as

$$
\frac{d}{dt} \int_{R^{(s)}} \rho \left( \hat{A} + \frac{1}{2} v^2 + \varphi + \frac{p_0}{\rho} \right) dV
$$

$$
= \int_{S_{(ent\ ex)}} \rho \left( \hat{G} + \frac{1}{2} v^2 + \varphi \right) \left( \mathbf{v} - \mathbf{v}^{(s)} \right) \cdot (-\mathbf{n}) \, dA - \mathcal{W} - \mathcal{E}
$$

$$
+ \int_{S_{(ent\ ex)}} \left[ -(P - p_0) \mathbf{v}^{(s)} \cdot \mathbf{n} + \mathbf{v} \cdot (\mathbf{S} \cdot \mathbf{n}) \right] dA \tag{7.4.3-9}
$$

which is a general form of the mechanical energy balance appropriate to single-phase systems composed of isothermal fluids.

Other general forms of the integral mechanical energy balance that can be derived are presented in Table 7.4.3-1.

The generalization of these relations to multiphase systems closely follows the discussion of Section 4.4.5. The results are given in Table 7.4.3-2.

The most common applications of the integral mechanical energy balances are to systems such that the following restrictions are reasonable:

1) There is no mass transfer across internal phase interfaces.
2) The jump mass and momentum balances of Sections 1.3.6 and 2.2.3 apply.
3) Entrances and exits are fixed in space.
4) Work done by viscous forces (as described by **S**) may be neglected at entrances and exits.

The forms of the integral mechanical energy balance applicable under these restrictions are shown in Table 7.4.3-3.

The extension of this discussion to turbulent flows follows along the lines of Exercise 4.4.5-4.

The remarks made concerning $\mathcal{E}$ at the conclusion of Section 4.4.5 are still applicable. Empirical data correlations for $\mathcal{E}$ are often useful. The approach recommended in preparing these correlations is outlined in Section 4.4.6.

Finally, I wish to call particular attention to R. B. Bird's (1957) discussion of the mechanical energy balance. With only minor extensions, I have adopted his viewpoint.

**Table 7.4.3-1.** General forms of the integral mechanical energy balance applicable to a single-phase system

$$\frac{d}{dt} \int_{R^{(s)}} \rho \left( \frac{1}{2} v^2 + \varphi \right) dV$$

$$= \int_{S_{(\text{ent ex})}} \rho \left( \frac{1}{2} v^2 + \varphi + \frac{P - p_0}{\rho} \right) (\mathbf{v} - \mathbf{v}^{(s)}) \cdot (-\mathbf{n}) \, dA$$

$$+ \int_{R^{(s)}} (P - p_0) \operatorname{div} \mathbf{v} \, dV - \mathcal{W} - \mathcal{E}$$

$$+ \int_{S_{(\text{ent ex})}} \left[ -(P - p_0)\mathbf{v}^{(s)} \cdot \mathbf{n} + \mathbf{v} \cdot (\mathbf{S} \cdot \mathbf{n}) \right] dA^a$$

$$\frac{d}{dt} \int_{R^{(s)}} \frac{1}{2} \rho v^2 \, dV$$

$$= \int_{S_{(\text{ent ex})}} \rho \left( \frac{1}{2} v^2 + \frac{P - p_0}{\rho} \right) (\mathbf{v} - \mathbf{v}^{(s)}) \cdot (-\mathbf{n}) \, dA$$

$$+ \int_{R^{(s)}} (P - p_0) \operatorname{div} \mathbf{v} \, dV - \mathcal{W} - \mathcal{E} + \int_{R^{(s)}} \mathbf{v} \cdot \rho \mathbf{f} \, dV$$

$$+ \int_{S_{(\text{ent ex})}} \left[ -(P - p_0)\mathbf{v}^{(s)} \cdot \mathbf{n} + \mathbf{v} \cdot (\mathbf{S} \cdot \mathbf{n}) \right] dA$$

*For an incompressible fluid:*

$$\frac{d}{dt} \int_{R^{(s)}} \rho \left( \frac{1}{2} v^2 + \varphi \right) dV$$

$$= \int_{S_{(\text{ent ex})}} \rho \left( \frac{1}{2} v^2 + \varphi + \frac{p - p_0}{\rho} \right) (\mathbf{v} - \mathbf{v}^{(s)}) \cdot (-\mathbf{n}) \, dA - \mathcal{W}$$

$$- \mathcal{E} + \int_{S_{(\text{ent ex})}} \left[ -(p - p_0)\mathbf{v}^{(s)} \cdot \mathbf{n} + \mathbf{v} \cdot (\mathbf{S} \cdot \mathbf{n}) \right] dA^a$$

*For an isothermal fluid:*

$$\frac{d}{dt} \int_{R^{(s)}} \rho \left( \hat{A} + \frac{1}{2} v^2 + \varphi + \frac{p_0}{\rho} \right) dV$$

$$= \int_{S_{(\text{ent ex})}} \rho \left( \hat{G} + \frac{1}{2} v^2 + \varphi \right) (\mathbf{v} - \mathbf{v}^{(s)}) \cdot (-\mathbf{n}) \, dA - \mathcal{W}$$

$$- \mathcal{E} + \int_{S_{(\text{ent ex})}} \left[ -(P - p_0)\mathbf{v}^{(s)} \cdot \mathbf{n} + \mathbf{v} \cdot (\mathbf{S} \cdot \mathbf{n}) \right] dA^{a,b}$$

*For an isentropic fluid:*

$$\frac{d}{dt} \int_{R^{(s)}} \rho \left( \hat{U} + \frac{1}{2} v^2 + \varphi + \frac{p_0}{\rho} \right) dV$$

$$= \int_{S_{(\text{ent ex})}} \rho \left( \hat{H} + \frac{1}{2} v^2 + \varphi \right) (\mathbf{v} - \mathbf{v}^{(s)}) \cdot (-\mathbf{n}) \, dA - \mathcal{W}$$

$$- \mathcal{E} + \int_{S_{(\text{ent ex})}} \left[ -(P - p_0)\mathbf{v}^{(s)} \cdot \mathbf{n} + \mathbf{v} \cdot (\mathbf{S} \cdot \mathbf{n}) \right] dA^{a,b}$$

*(cont.)*

**Table 7.4.3-1.** (cont.)

---

*For an isobaric fluid:*

$$\frac{d}{dt} \int_{R^{(s)}} \rho \left( \frac{1}{2}v^2 + \varphi \right) dV$$

$$= \int_{S_{(ent\ ex)}} \rho \left( \frac{1}{2}v^2 + \varphi \right) (\mathbf{v} - \mathbf{v}^{(s)}) \cdot (-\mathbf{n})\, dA - \mathcal{W} - \mathcal{E}$$

$$+ \int_{S^{(s)} - S_{(ent\ ex)}} -(P - p_0)\mathbf{v} \cdot \mathbf{n}\ dA$$

$$+ \int_{S_{(ent\ ex)}} \mathbf{v} \cdot (\mathbf{S} \cdot \mathbf{n})\, dA^a$$

---

[a] We assume that $\partial \varphi / \partial t = 0$.
[b] Applicable to systems where composition is independent of time and position.

**Table 7.4.3-2.** General forms of the integral mechanical energy balance applicable to a multiphase system

---

$$\frac{d}{dt} \int_{R^{(s)}} \rho \left( \frac{1}{2}v^2 + \varphi \right) dV$$

$$= \int_{S_{(ent\ ex)}} \rho \left( \frac{1}{2}v^2 + \varphi + \frac{P - p_0}{\rho} \right) (\mathbf{v} - \mathbf{v}^{(s)}) \cdot (-\mathbf{n})\, dA$$

$$+ \int_{R^{(s)}} (P - p_0) \operatorname{div} \mathbf{v}\, dV - \mathcal{W} - \mathcal{E}$$

$$+ \int_{S_{(ent\ ex)}} \left[ -(P - p_0)\mathbf{v}^{(s)} \cdot \mathbf{n} + \mathbf{v} \cdot (\mathbf{S} \cdot \mathbf{n}) \right] dA$$

$$+ \int_{\Sigma} \left[ \rho \left( \frac{1}{2}v^2 + \varphi \right) (\mathbf{v} - \mathbf{u}) \cdot \boldsymbol{\xi} - \mathbf{v} \cdot (\mathbf{T} + p_0\mathbf{I}) \cdot \boldsymbol{\xi} \right] dA^a$$

$$\frac{d}{dt} \int_{R^{(s)}} \frac{1}{2}\rho v^2\, dV$$

$$= \int_{S_{(ent\ ex)}} \rho \left( \frac{1}{2}v^2 + \frac{P - p_0}{\rho} \right) (\mathbf{v} - \mathbf{v}^{(s)}) \cdot (-\mathbf{n})\, dA$$

$$+ \int_{R^{(s)}} (P - p_0) \operatorname{div} \mathbf{v}\, dV - \mathcal{W} - \mathcal{E} + \int_{R^{(s)}} \mathbf{v} \cdot \rho \mathbf{f}\, dV$$

$$+ \int_{S_{(ent\ ex)}} \left[ -(P - p_0)\mathbf{v}^{(s)} \cdot \mathbf{n} + \mathbf{v} \cdot (\mathbf{S} \cdot \mathbf{n}) \right] dA$$

$$+ \int_{\Sigma} \left[ \frac{1}{2}\rho v^2 (\mathbf{v} - \mathbf{u}) \cdot \boldsymbol{\xi} - \mathbf{v} \cdot (\mathbf{T} + p_0\mathbf{I}) \cdot \boldsymbol{\xi} \right] dA$$

*For incompressible fluids:*

$$\frac{d}{dt} \int_{R^{(s)}} \rho \left( \frac{1}{2} v^2 + \varphi \right) dV$$

$$= \int_{S_{(ent\ ex)}} \rho \left( \frac{1}{2} v^2 + \varphi + \frac{p - p_0}{\rho} \right) (\mathbf{v} - \mathbf{v}^{(s)}) \cdot (-\mathbf{n}) \, dA$$

$$- \mathcal{W} - \mathcal{E} + \int_{S_{(ent\ ex)}} \left[ -(p - p_0)\mathbf{v}^{(s)} \cdot \mathbf{n} + \mathbf{v} \cdot (\mathbf{S} \cdot \mathbf{n}) \right] dA$$

$$+ \int_{\Sigma} \left[ \rho \left( \frac{1}{2} v^2 + \varphi \right) (\mathbf{v} - \mathbf{u}) \cdot \boldsymbol{\xi} - \mathbf{v} \cdot (\mathbf{T} + p_0 \mathbf{I}) \cdot \boldsymbol{\xi} \right] dA^a$$

*For an isothermal system:*

$$\frac{d}{dt} \int_{R^{(s)}} \rho \left( \hat{A} + \frac{1}{2} v^2 + \varphi + \frac{p_0}{\rho} \right) dV$$

$$= \int_{S_{(ent\ ex)}} \rho \left( \hat{G} + \frac{1}{2} v^2 + \varphi \right) (\mathbf{v} - \mathbf{v}^{(s)}) \cdot (-\mathbf{n}) \, dA - \mathcal{W}$$

$$- \mathcal{E} + \int_{S_{(ent\ ex)}} \left[ -(P - p_0)\mathbf{v}^{(s)} \cdot \mathbf{n} + \mathbf{v} \cdot (\mathbf{S} \cdot \mathbf{n}) \right] dA$$

$$+ \int_{\Sigma} \left[ \rho \left( \hat{A} + \frac{1}{2} v^2 + \varphi \right) (\mathbf{v} - \mathbf{u}) \cdot \boldsymbol{\xi} - \mathbf{v} \cdot (\mathbf{T} + p_0 \mathbf{I}) \cdot \boldsymbol{\xi} \right] dA^{a,b}$$

*For an isentropic system:*

$$\frac{d}{dt} \int_{R^{(s)}} \rho \left( \hat{U} + \frac{1}{2} v^2 + \varphi + \frac{p_0}{\rho} \right) dV$$

$$= \int_{S_{(ent\ ex)}} \rho \left( \hat{H} + \frac{1}{2} v^2 + \varphi \right) (\mathbf{v} - \mathbf{v}^{(s)}) \cdot (-\mathbf{n}) \, dA - \mathcal{W}$$

$$- \mathcal{E} + \int_{S_{(ent\ ex)}} \left[ -(P - p_0)\mathbf{v}^{(s)} \cdot \mathbf{n} + \mathbf{v} \cdot (\mathbf{S} \cdot \mathbf{n}) \right] dA$$

$$+ \int_{\Sigma} \left[ \rho \left( \hat{U} + \frac{1}{2} v^2 + \varphi \right) (\mathbf{v} - \mathbf{u}) \cdot \boldsymbol{\xi} - \mathbf{v} \cdot (\mathbf{T} + p_0 \mathbf{I}) \cdot \boldsymbol{\xi} \right] dA^{a,b}$$

*For an isobaric system:*

$$\frac{d}{dt} \int_{R^{(s)}} \rho \left( \frac{1}{2} v^2 + \varphi \right) dV$$

$$= \int_{S_{(ent\ ex)}} \rho \left( \frac{1}{2} v^2 + \varphi \right) (\mathbf{v} - \mathbf{v}^{(s)}) \cdot (-\mathbf{n}) \, dA - \mathcal{W} - \mathcal{E}$$

$$+ \int_{S^{(s)} - S_{(ent\ ex)}} -(P - p_0)\mathbf{v}^{(s)} \cdot \mathbf{n} + \int_{S_{(ent\ ex)}} \mathbf{v} \cdot (\mathbf{S} \cdot \mathbf{n}) \, dA$$

$$+ \int_{\Sigma} \left[ \rho \left( \frac{1}{2} v^2 + \varphi \right) (\mathbf{v} - \mathbf{u}) \cdot \boldsymbol{\xi} - \mathbf{v} \cdot (\mathbf{T} + p_0 \mathbf{I}) \cdot \boldsymbol{\xi} \right] dA^a$$

---

[a] We assume that $\partial \varphi / \partial t = 0$.

[b] Applicable to systems where composition is independent of time and position.

**Table 7.4.3-3.** Restricted forms of the integral mechanical energy balance applicable to a multiphase system. These forms are applicable following assumptions 1 through 4 given in the text

$$\frac{d}{dt} \int_{R^{(s)}} \rho \left( \frac{1}{2} v^2 + \varphi \right) dV$$

$$= \int_{S_{(ent\ ex)}} \rho \left( \frac{1}{2} v^2 + \varphi + \frac{P - p_0}{\rho} \right) (-\mathbf{v} \cdot \mathbf{n}) \, dA$$

$$+ \int_{R^{(s)}} (P - p_0) \operatorname{div} \mathbf{v} \, dV - \mathcal{W} - \mathcal{E}^a$$

$$\frac{d}{dt} \int_{R^{(s)}} \frac{1}{2} \rho v^2 \, dV$$

$$= \int_{S_{(ent\ ex)}} \rho \left( \frac{1}{2} v^2 + \frac{P - p_0}{\rho} \right) (-\mathbf{v} \cdot \mathbf{n}) \, dA$$

$$+ \int_{R^{(s)}} (P - p_0) \operatorname{div} \mathbf{v} \, dV - \mathcal{W} - \mathcal{E} + \int_{R^{(s)}} \mathbf{v} \cdot \rho \mathbf{f} \, dV$$

*For incompressible fluids:*

$$\frac{d}{dt} \int_{R^{(s)}} \rho \left( \frac{1}{2} v^2 + \varphi \right) dV$$

$$= \int_{S_{(ent\ ex)}} \rho \left( \frac{1}{2} v^2 + \varphi + \frac{p - p_0}{\rho} \right) (-\mathbf{v} \cdot \mathbf{n}) \, dA - \mathcal{W} - \mathcal{E}^a$$

*For an isothermal system:*

$$\frac{d}{dt} \int_{R^{(s)}} \rho \left( \hat{A} + \frac{1}{2} v^2 + \varphi + \frac{p_0}{\rho} \right) dV$$

$$= \int_{S_{(ent\ ex)}} \rho \left( \hat{G} + \frac{1}{2} v^2 + \varphi \right) (-\mathbf{v} \cdot \mathbf{n}) \, dA - \mathcal{W} - \mathcal{E}^{a,b}$$

*For an isentropic system:*

$$\frac{d}{dt} \int_{R^{(s)}} \rho \left( \hat{U} + \frac{1}{2} v^2 + \varphi + \frac{p_0}{\rho} \right) dV$$

$$= \int_{S_{(ent\ ex)}} \rho \left( \hat{H} + \frac{1}{2} v^2 + \varphi \right) (-\mathbf{v} \cdot \mathbf{n}) \, dA - \mathcal{W} - \mathcal{E}^{a,b}$$

*For an isobaric system:*

$$\frac{d}{dt} \int_{R^{(s)}} \rho \left( \frac{1}{2} v^2 + \varphi \right) dV$$

$$= \int_{S_{(ent\ ex)}} \rho \left( \frac{1}{2} v^2 + \varphi \right) (-\mathbf{v} \cdot \mathbf{n}) \, dA - \mathcal{W} - \mathcal{E} + \int_{S_{(ent\ ex)}} (P - p_0) \mathbf{v} \cdot \mathbf{n} \, dA^a$$

[a] We assume that $\partial \varphi / \partial t = 0$.
[b] Applicable to systems where composition is independent of time and position.

**7.4.4**  The Integral Entropy Inequality

The entropy inequality is the only postulate for which we have not as yet derived the corresponding integral relationship.

Starting with (5.2.3-8) and (5.3.1-25), we can write the differential entropy inequality as

$$\frac{\partial(\rho\hat{S})}{\partial t} + \text{div}(\rho\hat{S}\mathbf{v}) + \text{div}\left(\frac{1}{T}\mathbf{q}\right) - \rho\frac{Q}{T} \geq 0 \tag{7.4.4-1}$$

Integrating (7.4.4-1) over an arbitrary system, we have

$$\int_{R^{(s)}} \left[\frac{\partial(\rho\hat{S})}{\partial t} + \text{div}(\rho\hat{S} \cdot \mathbf{v}) + \text{div}\left(\frac{1}{T}\mathbf{q}\right) - \rho\frac{Q}{T}\right] dV \geq 0 \tag{7.4.4-2}$$

The generalized transport theorem can be used to express the first term on the left as

$$\int_{R^{(s)}} \frac{\partial(\rho\hat{S})}{\partial t}\,dV = \frac{d}{dt}\int_{R^{(s)}} \rho\hat{S}\,dV - \int_{S^{(s)}} \rho\hat{S}\mathbf{v}^{(s)} \cdot \mathbf{n}\,dA \tag{7.4.4-3}$$

After an application of Green's transformation, the second and third terms on the left of (7.4.4-2) become

$$\int_{R^{(s)}} \left[\text{div}(\rho\hat{S}\mathbf{v}) + \text{div}\left(\frac{1}{T}\mathbf{q}\right)\right] dV$$

$$= \int_{S^{(s)}} \left[\rho\hat{S}\mathbf{v} + \frac{1}{T}\mathbf{q}\right] \cdot \mathbf{n}\,dA \tag{7.4.4-4}$$

By means of (7.4.4-3) and (7.4.4-4), we are able to write (7.4.4-2) as

$$\frac{d}{dt}\int_{R^{(s)}} \rho\hat{S}\,dA \geq \int_{S_{(ent\,ex)}} \rho\hat{S}\left(\mathbf{v} - \mathbf{v}^{(s)}\right) \cdot (-\mathbf{n})\,dA$$

$$+ \int_{S^{(s)}} \frac{1}{T}\mathbf{q} \cdot (-\mathbf{n})\,dA + \int_{R^{(s)}} \frac{\rho Q}{T}\,dV \tag{7.4.4-5}$$

This says that the time rate change of the entropy associated with an arbitrary system is greater than or equal to the net rate at which entropy is brought into the system with whatever material flows across the boundaries of this system, the net rate at which entropy is transferred to the system as a result of contact energy transfer to the system, and the net rate at which entropy is produced in the system as a result of mutual and external energy transmission to the system. We will refer to (7.4.4-5) as a form of the *integral entropy inequality* appropriate to single-phase systems.

For multiphase systems, we can take the approach of Section 4.4.1 and immediately arrive at

$$\frac{d}{dt}\int_{R^{(s)}} \rho\hat{S}\,dV \geq \int_{S_{(ent\,ex)}} \rho\hat{S}\left(\mathbf{v} - \mathbf{v}^{(s)}\right) \cdot (-\mathbf{n})\,dA$$

$$+ \int_{S^{(s)}} \frac{1}{T}\mathbf{q} \cdot (-\mathbf{n})\,dA + \int_{R^{(s)}} \frac{\rho Q}{T}\,dV$$

$$+ \int_{\Sigma} \left[\rho\hat{S}(\mathbf{v} - \mathbf{u}) \cdot \boldsymbol{\xi} + \frac{1}{T}\mathbf{q} \cdot \boldsymbol{\xi}\right] dA \tag{7.4.4-6}$$

This is another form of the *general integral entropy inequality* appropriate to multiphase systems.

We usually will be willing to say that the jump entropy inequality (5.2.3-9) together with (5.3.1-25) applies, in which case (7.4.4-6) reduces to (7.4.4-5).

**Exercise 7.4.4-1** *The integral entropy inequality for turbulent flows*    I recommend following Section 4.4.2 in developing forms of the integral entropy inequality appropriate for turbulent flows. It generally seems more straightforward to simply time average the integral entropy inequality derived in the text.

### 7.4.5    Integral Entropy Inequality for Turbulent Flows

Our approach here is basically the same as that which we have taken in arriving at forms of the other integral balances appropriate to turbulent flows. See Section 4.4.2.

We could repeat the analysis of Section 7.4.4 using time averages of the differential entropy inequalities derived in Section 5.2.3.

It seems much more straightforward to time average the integral entropy inequality of Section 7.4.4 to find for any single-phase or multiphase system

$$
\frac{d}{dt} \overline{\int_{R^{(s)}} \rho \hat{S} \, dV} \geq \overline{\int_{(ent\ ex)} \rho \hat{S} \left( \mathbf{v} - \mathbf{v}^{(s)} \right) \cdot (-\mathbf{n}) \, dA}
$$

$$
+ \overline{\int_{S^{(s)}} \frac{1}{T} \mathbf{q} \cdot (-\mathbf{n}) \, dA} + \overline{\int_{R^{(s)}} \rho \frac{Q}{T} \, dV}
\tag{7.4.5-1}
$$

This is the *integral entropy inequality for turbulent flows*. The only assumption that we have made in arriving at this result is that the jump entropy inequalities (5.2.3-9) and (5.3.1-25) are applicable at all phase interfaces involved.

A somewhat simpler result can be obtained for single-phase or multiphase systems that do not involve fluid–fluid phase interfaces. Under these circumstances, we can use (4.4.2-7) through (4.4.2-9) to show that (7.4.5-1) reduces to

$$
\frac{d}{dt} \int_{R^{(s)}} \overline{\rho \hat{S}} \, dV \geq \int_{S_{(ent\ ex)}} \left( \overline{\rho \hat{S} \mathbf{v}} - \overline{\rho \hat{S} \mathbf{v}^{(s)}} \right) \cdot (-\mathbf{n}) \, dA
$$

$$
+ \int_{S^{(s)}} \overline{\frac{1}{T} \mathbf{q}} \cdot (-\mathbf{n}) \, dA + \int_{R^{(s)}} \overline{\frac{\rho Q}{T}} \, dV
\tag{7.4.5-2}
$$

### 7.4.6    Example

An insulated, evacuated tank is connected through a valved pipe to a constant-pressure line containing an ideal diatomic gas maintained at a constant pressure $P_0$ and a constant temperature $T_0$. We may assume that the constant-pressure heat capacity per unit mass is

$$
\hat{c}_P = \frac{7}{2} \frac{R}{M}
\tag{7.4.6-1}
$$

where $R$ is the gas-law constant and $M$ is the molecular weight of the gas. The volume of the tank $\mathcal{V}$ is known.

The valve between the tank and the line is suddenly opened, admitting the gas to the tank. We wish to compute the amount and the temperature of the gas in the tank, when the pressure in the tank is $P_{(final)}$.

Let us choose our system to be the gas in the tank. This system has only one entrance, through the pipeline, and no exits. The boundary of the system is fixed in space.

For simplicity, we shall neglect the effects of turbulence.

From the integral mass balance,

$$\frac{d\mathcal{M}}{dt} = -\int_{S_{(ent)}} \rho \mathbf{v} \cdot \mathbf{n} \, dA \tag{7.4.6-2}$$

Here $\mathcal{M}$ indicates the mass of the gas in the tank:

$$\mathcal{M} \equiv \int_{R^{(s)}} \rho \, dV \tag{7.4.6-3}$$

which is a function of time. By $S_{(ent)}$ we refer to the tank's entrance.

If we neglect the changes in kinetic energy and potential energy and if we make the assumptions noted in Section 7.4.1, the integral energy balance of Table 7.4.1-3 requires

$$\frac{d\mathcal{U}}{dt} = -\int_{S_{(ent)}} \rho \hat{H} \mathbf{v} \cdot \mathbf{n} \, dA \tag{7.4.6-4}$$

where

$$\mathcal{U} \equiv \int_{R^{(s)}} \rho \hat{U} \, dV \tag{7.4.6-5}$$

is the internal energy associated with the system.

The specific enthalpy should be very nearly a constant with respect to position in the entrance, so that (7.4.6-4) may be combined with (7.4.6-2) to obtain

$$\frac{d\mathcal{U}}{dt} = -\hat{H}_{(ent)} \int_{S_{(ent)}} \rho \mathbf{v} \cdot \mathbf{n} \, dA$$

$$= \hat{H}_{(ent)} \frac{d\mathcal{M}}{dt} \tag{7.4.6-6}$$

Furthermore, the specific enthalpy of the incoming gas should be nearly a constant as a function of time. With this assumption, (7.4.6-6) may be integrated to find that at any particular time

$$\mathcal{U}_{(final)} = \hat{H}_{(ent)} \mathcal{M}_{(final)} \tag{7.4.6-7}$$

If the gas is well mixed in the tank, this last expression becomes

$$\hat{U}_{(final)} = \hat{H}_{(ent)} \tag{7.4.6-8}$$

It is easily shown that

$$\hat{c}_P \equiv T \left( \frac{\partial \hat{S}}{\partial T} \right)_P$$

$$= \left( \frac{\partial \hat{H}}{\partial T} \right)_P$$

$$= \left( \frac{\partial \hat{U}}{\partial T} \right)_P + P \left( \frac{\partial \hat{V}}{\partial T} \right)_P \tag{7.4.6-9}$$

For an ideal gas,

$$\hat{U} = \hat{U}(T) \tag{7.4.6-10}$$

and

$$p\hat{V} = \frac{RT}{M} \tag{7.4.6-11}$$

Consequently, (7.4.6-9) may be rearranged to read

$$\frac{d\hat{U}}{dT} = \hat{c}_p - \frac{R}{M} \tag{7.4.6-12}$$

In view of (7.4.6-1), this last equation may be integrated to find that

$$\hat{U}_{(\text{final})} - \hat{U}_{(\text{ent})} = \frac{5}{2}\frac{R}{M}\left(T_{(\text{final})} - T_{(\text{ent})}\right) \tag{7.4.6-13}$$

By means of (7.4.6-8), (7.4.6-11), and (7.4.6-13), we conclude that

$$\begin{aligned}
\hat{U}_{(\text{final})} - \hat{U}_{(\text{ent})} &= \hat{H}_{(\text{ent})} - \hat{U}_{(\text{ent})} \\
&= P_{(\text{ent})}\hat{V}_{(\text{ent})} \\
&= \frac{R}{M}T_{(\text{ent})} \\
&= \frac{5}{2}\frac{R}{M}\left(T_{(\text{final})} - T_{(\text{ent})}\right)
\end{aligned} \tag{7.4.6-14}$$

or

$$\begin{aligned}
T_{(\text{final})} &= \frac{7}{5}T_{(\text{ent})} \\
&= \frac{7}{5}T_0
\end{aligned} \tag{7.4.6-15}$$

It follows immediately that the mass of gas in the tank at the end of the process is

$$\mathcal{M}_{(\text{final})} = \frac{5}{7}\frac{M}{R}\frac{P_{(\text{final})}}{T_0}\mathcal{V} \tag{7.4.6-16}$$

This example was suggested by Prof. G. M. Brown, Northwestern University.

**Exercise 7.4.6-1**[4]　An insulated, evacuated vessel of 3-m³ capacity is connected to a steam line that transports 1.4 MPa saturated steam. The valve between the steam line and the vessel is opened to admit steam to the vessel. Compute the amount of steam in the vessel and its temperature, if the valve is closed when the pressure in the vessel reaches 0.34 MPa.

**Exercise 7.4.6-2**　A well-insulated tank of 3-m³ capacity is connected to a valved line containing an unlimited supply of saturated steam at 1.4 MPa. Initially the tank is filled with saturated steam at 0.1 MPa. At a given time, the valve is opened. The valve is closed again when the tank pressure reaches 0.7 MPa. How much steam flows into the tank? What is the final temperature of the steam in the tank?

---

[4] Exercises 7.4.6-1 through 7.4.6-8 were suggested by Prof. W. W. Graessley, Department of Chemical Engineering, Northwestern University.

**Exercise 7.4.6-3**   A valve connects two identical insulated vessels, each with a volume of $2.8 \times 10^{-2}$ m³. The valve is initially closed. One vessel contains steam at 316°C and 6.9 MPa; the other vessel is evacuated.

   i) The valve is opened and flow occurs, until the pressures are identical. The valve is then closed. Determine the final temperature in each vessel.
   ii) If the valve is left open so that thermal equilibrium is eventually attained between the two vessels, calculate the final temperature.

**Exercise 7.4.6-4**   An exit high-pressure line from a chemical reactor contains almost pure Freon-12 at 6.9 MPa and 138°C, according to the instruments. However, an operator opens a small valve in the side of the line and claims the recorded temperature must be wrong, since the gas issuing from the line feels cold. Resolve this question if possible by appropriate calculations.

**Exercise 7.4.6-5**   Freon-12 at 68.9 kPa gauge and $-17.8$°C enters our plant at the rate of 454 kg$_m$/h. An adiabatic compressor raises the pressure to 1.4 MPa gauge, at which point a thermometer in the line reads 116°C. A heat exchanger cools the stream to 60°C, while the pressure remains constant. Calculate the power input to the compressor.

**Exercise 7.4.6-6**   Steam flows in a large uninsulated pipeline at the rate of 4.5 kg/s. At the first station, temperature and pressure gauges indicate 316°C and 6.9 MPa. Downstream at the second station the pressure is 5.2 MPa, and the quality is 0.85 (vapor fraction). What is the temperature at the second station, and what is the rate at which heat is transferred to the pipeline?

**Exercise 7.4.6-7**   Oxygen passes through an adiabatic steady-flow compressor at the rate of 454 kg/h, entering as a saturated vapor at 252 kPa, and emerging at 1.8 MPa and 175 K. Determine the shaft power per unit of mass of $O_2$ and the required motor horsepower.

**Exercise 7.4.6-8**   Carbon dioxide passes through an adiabatic steady-state flow turbine at the rate of 3.8 kg/s. In enters at 2.8 MPa and 37.8°C, and emerges as a saturated vapor at 0.7 MPa. What is the shaft-work output per unit mass of carbon dioxide, and what is the power delivered by the turbine?

**Exercise 7.4.6-9** *Stagnation temperature*   A *total-temperature* probe illustrated in Figure 7.4.6-1 can be used to measure the temperature $T_1$ of an ideal gas that moves with a speed $v_1$. A portion of the gas enters the open end of the probe and decelerates to nearly zero velocity

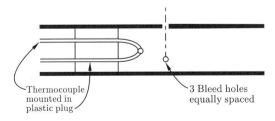

Thermocouple
mounted in
plastic plug

3 Bleed holes
equally spaced

**Figure 7.4.6-1.**  A total-temperature probe.

before slowly leaking out of the bleed holes. You may assume that the tubing surrounding the thermocouple has a small thermal conductivity.

Determine that the temperature $T_2$ measured by the thermocouple is

$$T_2 = T_1 + \frac{(v_1)^2}{2\hat{c}_P}$$

*Hint:* Choose your system so as to include a portion of the gas stream that is undisturbed by the presence of the probe. What other assumptions must be made?

### Exercise 7.4.6-10

i) Repeat the analysis of this section, assuming that the engine operates in a periodic cycle.
ii) Repeat the analysis of this section, assuming that the engine involves a turbulent flow.

### Exercise 7.4.6-11    Smith and Van Ness (1959, p. 176) state that

No apparatus can operate in such a way that its only effect (in system and surroundings) is to convert the heat taken in completely into work.

Prove this statement, starting with the integral energy balance and the integral entropy inequality.

*Hint:* In carrying out this proof, one must interpret what is meant by the authors' statement.

### Exercise 7.4.6-12    An inventor has devised a process that makes energy continuously available at an elevated temperature. Saturated steam at 220°F is the only source of energy. There is an abundance of cooling water at 75°F. What is the maximum amount of energy that could be made available at 400°F/Btu of energy given up by the steam? (Smith and Van Ness 1959, p. 338)

### Exercise 7.4.6-13    Smith and Van Ness (1959, p. 176) state that

Any process which consists solely in the transfer of heat from one temperature to a higher one is impossible.

Use the integral entropy inequality to prove this statement.

# 8

# Foundations for Mass Transfer

THIS IS AN EXCELLENT TIME for review. In fact, a review is practically forced upon you in this chapter.

A body composed of a single species is a limiting case of multicomponent materials. In what follows, we reformulate the fundamental postulates made in Chapters 1, 2, and 5, to enable their application to multicomponent bodies. With a few relatively minor modifications, all we have said about single-component systems can be applied to multicomponent ones.

## 8.1 Viewpoint

Up to this point we have been primarily concerned with single-component materials or materials of uniform composition. Hereafter, we shall be treating a material consisting of $N$ species or constituents that is undergoing an arbitrary number of homogeneous and heterogeneous chemical reactions. Not only are we interested in the velocity and temperature distributions in such a material, but we wish to follow its composition as a function of time and position. We may ask, for example, how rapidly a particular species formed by a catalytic reaction at an adjacent surface will distribute itself throughout a material; or we may wish to determine the rate at which a liquid droplet will evaporate into a surrounding gas stream.

Our first task is to choose a continuum model for an $N$-component material. We will wish to follow each species individually as the $N$-component mixture goes through some operation, possibly involving deformation, flow, and chemical reactions. We will view each species as a continuous medium with a variable mass-density field. The model for the $N$-component mixture is a superposition of the $N$ continuous media.

One feature of this representation initially may seem confusing. At any point in space occupied by the mixture, $N$ material particles (one from each of the continuous media representing individual species) coexist. The confusion usually arises from a dangerous and incorrect identification of a material particle in a continuous medium of species $A$ with a molecule of species $A$. This superposition of $N$ constituent media is consistent with our usual practice of identifying compositions with each point in a multicomponent mixture.

### 8.1.1    Body, Motion, and Material Coordinates

The ideas we introduced in Section 1.1 for a single-component material may be extended easily to a particular species $A$ in an $N$-constituent mixture.

A *body of species A* is a set, the elements $\zeta_{(A)}$ of which are called particles of species $A$ or *material particles of species A*. A one-to-one continuous mapping of this set onto a region of three-dimensional euclidean point space exists and is called a *configuration* of the body of constituent $A$:

$$\mathbf{z} = \mathbf{\chi}_{(A)}\left(\zeta_{(A)}\right) \tag{8.1.1-1}$$

or

$$\zeta_{(A)} = \chi_{(A)}^{-1}(\mathbf{z}) \tag{8.1.1-2}$$

Here $\chi_{(A)}^{-1}$ indicates the inverse mapping of $\mathbf{\chi}_{(A)}$. The point $\mathbf{z} = \mathbf{\chi}_{(A)}(\zeta_{(A)})$ is called the place occupied by the particle $\zeta_{(A)}$, and $\zeta_{(A)} = \chi_{(A)}^{-1}(\mathbf{z})$ is called the particle of species $A$ whose place is $\mathbf{z}$.

For the moment, we will not concern ourselves directly with a body of an $N$-component mixture, but rather with the $N$ constituent bodies. It is the *superposition* of these $N$ constituent bodies that forms the model for the $N$-component mixture.

A *motion* of a body of species $A$ is a one-parameter family of configurations; the real parameter $t$ is time. We write

$$\mathbf{z} = \mathbf{\chi}_{(A)}\left(\zeta_{(A)}, t\right) \tag{8.1.1-3}$$

and

$$\zeta_{(A)} = \chi_{(A)}^{-1}(\mathbf{z}, t) \tag{8.1.1-4}$$

Although we do not wish to confuse a body of species $A$ with any of its spatial configurations, we must recognize that it is available to us for observation and study only in those configurations. It is often convenient to take advantage of this by using positions in some particular configuration as a means of identifying the particles of species $A$ that form the body. This reference configuration may be, but need not be, one actually occupied by the body in the course of its motion. The place of a particle of species $A$ in the reference configuration $\kappa_{(A)}$ will be denoted by

$$\mathbf{z}_{\kappa(A)} = \kappa_{(A)}\left(\zeta_{(A)}\right) \tag{8.1.1-5}$$

The particle at the place $\mathbf{z}_{\kappa(A)}$ in the configuration $\kappa_{(A)}$ may be expressed as

$$\zeta_{(A)} = \kappa_{(A)}^{-1}\left(\mathbf{z}_{\kappa(A)}\right) \tag{8.1.1-6}$$

If $\mathbf{\chi}_{(A)}$ is a motion of the body of species $A$,

$$\begin{aligned}
\mathbf{z} &= \mathbf{\chi}_{(A)}\left(\zeta_{(A)}, t\right) \\
&= \mathbf{\chi}_{(A)}\left(\kappa_{(A)}^{-1}\left(\mathbf{z}_{\kappa(A)}\right), t\right) \\
&= \mathbf{\chi}_{\kappa(A)}\left(\mathbf{z}_{\kappa(A)}, t\right)
\end{aligned} \tag{8.1.1-7}$$

This defines a family of *deformations* from the reference configuration of constituent $A$. The subscript $\kappa$ is to remind one that the form of

$$\chi_{\kappa(A)}\left(\mathbf{z}_{\kappa(A)}, t\right)$$

depends upon the choice of reference configuration $\kappa_{(A)}$.

The coordinates $\mathbf{z}_{\kappa(A)i}$ identify the place $\mathbf{z}_{\kappa(A)}$ in an arbitrary coordinate system that is stationary as a function of time with respect to the reference configuration $\kappa_{(A)}$; these coordinates are referred to as the *material coordinates* of species $A$.

Let $B$ be any quantity, scalar or tensor. We shall wish to talk about the time derivative of $B$ following the motion of a particle of species $A$ or the *material derivative* following species $A$. We define

$$\frac{d_{(A)}B}{dt} \equiv \left(\frac{\partial B}{\partial t}\right)_{\mathbf{z}_{\kappa(A)}}$$

$$\equiv \left(\frac{\partial B}{\partial t}\right)_{\mathbf{z}_{\kappa(A)1}, \mathbf{z}_{\kappa(A)2}, \mathbf{z}_{\kappa(A)3}} \tag{8.1.1-8}$$

As an example, the velocity vector of species $A$,

$$\mathbf{v}_{(A)} \equiv \frac{d_{(A)}\mathbf{z}}{dt}$$

$$= \left[\frac{\partial \chi_{\kappa(A)}\left(\mathbf{z}_{\kappa(A)}, t\right)}{\partial t}\right]_{\mathbf{z}_{\kappa(A)}} \tag{8.1.1-9}$$

represents the time rate of change of position of a material particle of species $A$.

**Exercise 8.1.1-1** Let $B$ be any real scalar field, spatial vector field, or second-order tensor field. Show that

$$\frac{d_{(A)}B}{dt} = \frac{\partial B}{\partial t} + (\nabla B) \cdot \mathbf{v}_{(A)}$$

**Exercise 8.1.1-2** Show that the contravariant components of the velocity vector of species $A$ are

$$v^i_{(A)} = \frac{d_{(A)}x^i}{dt}$$

**Exercise 8.1.1-3** Let $\mathbf{a} = \mathbf{a}(\mathbf{z}, t)$ be some vector field that is a function both of position and time.

i) Show that

$$\frac{d_{(A)}\mathbf{a}}{dt} = \frac{\delta_{(A)}a^n}{\delta t}\mathbf{g}_n$$

in which we have defined

$$\frac{\delta_{(A)}a^n}{\delta t} \equiv \frac{\partial a^n}{\partial t} + a^n_{\,,i}v^i_{(A)}$$

ii) Show that

$$\frac{d_{(A)}\mathbf{a}}{dt} = \frac{\delta_{(A)}a_n}{\delta t}\mathbf{g}^n$$

where

$$\frac{\delta_{(A)}a_n}{\delta t} \equiv \frac{\partial a_n}{\partial t} + a_{n,i}v_{(A)}^i$$

The quantity $\delta_{(A)}a^n/\delta t$ may be referred to as the *intrinsic derivative* following species $A$ of the contravariant vector-field component $a^n$; it should be viewed as the contravariant component of the vector field $d_{(A)}\mathbf{a}/dt$.

**Exercise 8.1.1-4**   The intrinsic derivative following species $A$ introduced in Exercise 8.1.1-3 may be extended readily to higher-order tensor fields. Consider the second-order tensor field $\mathbf{T} = \mathbf{T}(\mathbf{z}, t)$.

i) Show that

$$\frac{d_{(A)}\mathbf{T}}{dt} = \frac{\delta_{(A)}T_{ij}}{\delta t}\mathbf{g}^i\mathbf{g}^j$$

where

$$\frac{\delta_{(A)}T_{ij}}{\delta t} \equiv \frac{\partial T_{ij}}{\partial t} + T_{ij,k}v_{(A)}^k$$

ii) Show that

$$\frac{d_{(A)}\mathbf{T}}{dt} = \frac{\delta_{(A)}T_{i.}^{\ j}}{\delta t}\mathbf{g}^i\mathbf{g}_j$$

where

$$\frac{\delta_{(A)}T_{i.}^{\ j}}{\delta t} \equiv \frac{\partial T_{i.}^{\ j}}{\partial t} + T_{i.\ ,k}^{\ j}v_{(A)}^k$$

The quantity $\delta_{(A)}T_{ij}/\delta t$ may be referred to as the *intrinsic derivative* following species $A$ of the doubly covariant tensor-field component $T_{ij}$; it should be thought of as the doubly covariant component of the tensor field $d_{(A)}\mathbf{T}/dt$. The quantity $\delta_{(A)}T_{i.}^{\ j}/\delta t$ may be referred to as the *intrinsic derivative* following species $A$ of the mixed tensor-field component $T_{i.}^{\ j}$; it is a mixed component of the tensor field $d_{(A)}\mathbf{T}/dt$.

---

## 8.2   Species Mass Balance

### 8.2.1   Differential and Jump Balances

To discuss the movements of the various constituents in a multicomponent mixture, we require as a sixth postulate the

**Mass balance for an individual species**   The time rate of change of the mass of a body of species $A$ is equal to the rate at which the mass of $A$ is produced by chemical reactions.

Let $R_{(A)}$ denote the region occupied by a body of species $A$ (a set of particles of species $A$), $\rho_{(A)}$ denote the *mass density* of species $A$, $r_{(A)}$ denote the rate of production of species $A$ per unit volume by homogeneous chemical reactions, and $r_{(A)}^{(\sigma)}$ denote the rate of production of species $A$ per unit area by heterogeneous chemical reactions (on the phase interfaces). This postulate states that

$$\frac{d}{dt}\int_{R_{(A)}}\rho_{(A)}\,dV = \int_{R_{(A)}}r_{(A)}\,dV + \int_{\Sigma}r_{(A)}^{(\sigma)}\,dA \tag{8.2.1-1}$$

where the limits on these integrations are functions of time.

Applying the transport theorem of Exercise 8.2.1-2 to the left side of (8.2.1-1), we find

$$\int_{R_{(A)}}\left(\frac{d_{(A)}\rho_{(A)}}{dt} + \rho_{(A)}\,\mathrm{div}\,\mathbf{v}_{(A)} - r_{(A)}\right)dV$$

$$+ \int_{\Sigma}\left\{\left[\rho_{(A)}\left(\mathbf{v}_{(A)} - \mathbf{u}\right)\cdot\boldsymbol{\xi}\right] - r_{(A)}^{(\sigma)}\right\}dA = 0 \tag{8.2.1-2}$$

This implies that at each point in each phase the *differential mass balance for species A*,

$$\frac{d_{(A)}\rho_{(A)}}{dt} + \rho_{(A)}\,\mathrm{div}\,\mathbf{v}_{(A)} = r_{(A)} \tag{8.2.1-3}$$

must be satisfied and that at each point on each phase interface the *jump mass balance for species A*,

$$\left[\rho_{(A)}\left(\mathbf{v}_{(A)} - \mathbf{u}\right)\cdot\boldsymbol{\xi}\right] = r_{(A)}^{(\sigma)} \tag{8.2.1-4}$$

must be obeyed.

A commonly used alternative form of (8.2.1-3) is

$$\frac{\partial\rho_{(A)}}{\partial t} + \mathrm{div}\left(\rho_{(A)}\mathbf{v}_{(A)}\right) = r_{(A)} \tag{8.2.1-5}$$

Sometimes it is more convenient to work in terms of *molar density*

$$c_{(A)} \equiv \frac{\rho_{(A)}}{M_{(A)}} \tag{8.2.1-6}$$

where $M_{(A)}$ is the *molecular weight* of species $A$. In this case, (8.2.1-5) may be rewritten as

$$\frac{\partial c_{(A)}}{\partial t} + \mathrm{div}\left(c_{(A)}\mathbf{v}_{(A)}\right) = \frac{r_{(A)}}{M_{(A)}} \tag{8.2.1-7}$$

**Exercise 8.2.1-1** *Transport theorem for species A*    Let $\Phi$ be any scalar, vector, or tensor field. Show that

$$\frac{d}{dt}\int_{R_{(A)}}\Phi\,dV = \int_{R_{(A)}}\left(\frac{d_{(A)}\Phi}{dt} + \Phi\,\mathrm{div}\,\mathbf{v}_{(A)}\right)dV$$

This is a form of the *transport theorem* applicable to a single-phase body of species $A$.

*Hint:*   Review Section 1.3.2.

**Exercise 8.2.1-2** *Transport theorem for a discontinuous body of species A*    Let $\Phi$ be any scalar, vector, or tensor field. Derive the following *transport theorem* for a body of species $A$ that contains a dividing surface $\Sigma$:

$$\frac{d}{dt}\int_{R_{(A)}}\Phi\,dV = \int_{R_{(A)}}\left(\frac{d_{(A)}\Phi}{dt} + \Phi\,\mathrm{div}\,\mathbf{v}_{(A)}\right)dV + \int_{\Sigma}\left[\Phi\left(\mathbf{v}_{(A)} - \mathbf{u}\right)\cdot\boldsymbol{\xi}\right]dA$$

The notation has the same meaning as in Section 1.3.5 with the understanding that we are restricting ourselves to a phase interface that occurs in a body of species $A$.

**Exercise 8.2.1-3** *Overall differential mass balance*   Let us define the density $\rho$ of a multicomponent mixture in the usual manner:

$$\rho \equiv \sum_{A=1}^{N} \rho_{(A)}$$

Let us also define the *mass-averaged velocity* of a multicomponent mixture

$$\mathbf{v} \equiv \frac{1}{\rho} \sum_{A=1}^{N} \rho_{(A)} \mathbf{v}_{(A)}$$

i) Show that the sum of (8.2.1-5) over all $N$ species may be written as

$$\frac{\partial \rho}{\partial t} + \text{div}\,(\rho \mathbf{v}) = \sum_{A=1}^{N} r_{(A)}$$

Our conception of conservation of mass, formalized in Section 8.3.1, tells us that

$$\sum_{A=1}^{N} r_{(A)} = 0$$

Consequently, we have that

$$\frac{\partial \rho}{\partial t} + \text{div}(\rho \mathbf{v}) = 0$$

This equation, known as the *overall differential mass balance*, is formally identical with the differential mass balance derived in Section 1.3.3 for a single-component material.

ii) Show that the above result may be expressed as

$$\frac{d_{(v)}\rho}{dt} + \rho\,\text{div}\,\mathbf{v} = \sum_{A=1}^{N} r_{(A)} = 0$$

where

$$\frac{d_{(v)}\rho}{dt} \equiv \frac{\partial \rho}{\partial t} + (\nabla \rho) \cdot \mathbf{v}$$

**Exercise 8.2.1-4**   Derive (8.2.1-5).

**Exercise 8.2.1-5** *Frame indifference of differential mass balance*   Prove that the differential mass balance for species $A$ takes the same form in every frame of reference.

**Exercise 8.2.1-6** *Frame indifference of jump mass balance for species A*   Prove that the jump mass balance for species $A$ takes the same form in every frame of reference.

## 8.2.2   Concentration, Velocities, and Mass Fluxes

One of the most confusing aspects of mass-transfer problems is that there are several sets of terminology in common use.

In discussing the concentration of species $A$ in a multicomponent mixture, one may refer to the mass density $\rho_{(A)}$, the molar density $c_{(A)}$, the mass fraction $\omega_{(A)}$, or the mole fraction $x_{(A)}$. The relations between these quantities are explored in Tables 8.5.2-1 and 8.5.1-1.

Experimentalists and theoreticians are not content to use only $\mathbf{v}_{(A)}$. Frequently, one refers to the velocity of species $A$ with respect to the *mass-averaged velocity*

$$\mathbf{u}_{(A)} \equiv \mathbf{v}_{(A)} - \mathbf{v} \tag{8.2.2-1}$$

where the mass-averaged velocity is defined by

$$\mathbf{v} \equiv \sum_{A=1}^{N} \omega_{(A)} \mathbf{v}_{(A)} \tag{8.2.2-2}$$

In Tables 8.5.1-3 and 8.5.1-4, we see that we may also work with the velocity of species $A$ with respect to the molar-averaged velocity

$$\mathbf{u}_{(A)}^{\diamond} \equiv \mathbf{v}_{(A)} - \mathbf{v}^{\diamond} \tag{8.2.2-3}$$

where we have introduced the *molar-averaged velocity*

$$\mathbf{v}^{\diamond} \equiv \sum_{A=1}^{N} x_{(A)} \mathbf{v}_{(A)} \tag{8.2.2-4}$$

In Section 8.2.1, we derived the differential mass balance for species $A$,

$$\frac{\partial \rho_{(A)}}{\partial t} + \text{div } \mathbf{n}_{(A)} = r_{(A)} \tag{8.2.2-5}$$

where we have introduced

$$\mathbf{n}_{(A)} \equiv \rho_{(A)} \mathbf{v}_{(A)} \tag{8.2.2-6}$$

as the *mass flux of species A with respect to the fixed frame of reference* (the inertial frame of reference or the laboratory walls). The alternative ways we have of looking at concentrations and velocities suggest the variety of expressions for the mass fluxes outlined in Table 8.5.1-4. For example, the mass flux of species $A$ *with respect to the mass-averaged velocity* is defined as

$$\mathbf{j}_{(A)} \equiv \rho_{(A)} \left( \mathbf{v}_{(A)} - \mathbf{v} \right) \tag{8.2.2-7}$$

which permits us to write (8.2.2-5) as

$$\frac{\partial \rho_{(A)}}{\partial t} + \text{div}\left(\rho_{(A)} \mathbf{v}\right) + \text{div } \mathbf{j}_{(A)} = r_{(A)} \tag{8.2.2-8}$$

From the overall differential mass balance (Exercise 8.2.1-3 and Section 8.3.1):

$$\begin{aligned}
\frac{\partial \rho_{(A)}}{\partial t} + \text{div } \left(\rho_{(A)} \mathbf{v}\right) &= \frac{\partial \rho_{(A)}}{\partial t} + \nabla \rho_{(A)} \cdot \mathbf{v} + \rho_{(A)}(\text{div } \mathbf{v}) \\
&= \frac{d_{(v)}\rho_{(A)}}{dt} - \omega_{(A)} \frac{d_{(v)}\rho}{dt} \\
&= \rho \frac{d_{(v)}\omega_{(A)}}{dt}
\end{aligned} \tag{8.2.2-9}$$

in which

$$\frac{d_{(v)}B}{dt} \equiv \frac{\partial B}{\partial t} + \nabla B \cdot \mathbf{v} \tag{8.2.2-10}$$

denotes the derivative of any quantity $B$ (scalar or tensor) with respect to time following a fictitious particle that moves with the local mass-averaged velocity of the multicomponent mixture. Equations (8.2.2-8) and (8.2.2-9) allow us to conclude that

$$\rho \frac{d_{(v)}\omega_{(A)}}{dt} + \text{div } \mathbf{j}_{(A)} = r_{(A)} \tag{8.2.2-11}$$

Various alternative forms of the differential mass balance for species $A$ are presented in Table 8.5.1-5.

To assist the reader in formulating problems, the first form of the differential mass balance for species $A$ in Table 8.5.1-5 is presented for three specific coordinate systems in Table 8.5.1-6.

### Exercise 8.2.2-1

i) Given Table 8.5.1-1, derive the relations given in Table 8.5.1-2.

ii) Given Tables 8.5.1-1 and 8.5.1-3, derive the relations given in Table 8.5.1-4.

iii) Starting with the first form of the differential mass balance for species $A$ given in Table 8.5.1-5, derive the rest.

**Exercise 8.2.2-2** *More about the mass flux vector*   Show that the mass flux vector $\mathbf{j}_{(A)}$ is frame indifferent.

---

## 8.3    Revised Postulates

In Section 8.2.1, we began by postulating a mass balance for each individual species. The corresponding differential mass balance for a species was expressed in terms of the velocity of that species. By analogy with our discussion in Chapters 1 and 2, in order to determine the velocities of the various species present, we would expect to have to postulate forms of the momentum balance and the moment-of-momentum balance for each species (as well as an energy balance and an entropy inequality) together with descriptions of stress–deformation behavior for each species (Truesdell and Toupin 1960, pp. 469, 567, 612, and 645; Truesdell 1962; Bowen 1967; Müller 1968). Put aside for the moment the complexity of such a theory. A more important difficulty is that there has not been proposed even one experiment for measuring the stress–deformation behavior of an individual species! This is not a practical approach to the description of mass transfer, and it is not the approach taken in the vast majority of the literature in this area.

Instead of attempting to directly determine the velocities of the $N$ species present, we will determine $N - 1$ of the mass flux vectors $\mathbf{n}_{(A)}$ together with the mass-averaged velocity

$$\mathbf{v} \equiv \sum_{A=1}^{N} \omega_{(A)}\mathbf{v}_{(A)} \tag{8.3.0-1}$$

Rather than describing the stress–deformation behavior of the $N$ species, we will write constitutive equations for $N - 1$ of the mass flux vectors $\mathbf{n}_{(A)}$ (Sections 8.4.4 through 8.4.6) as well as for the stress deformation of the multicomponent material (Section 8.4.7). Rather than postulating forms of the momentum balance, of the moment-of-momentum balances, of the energy balance, and of the entropy inequality for each species, we will write these postulates only for the multicomponent material. Not only is the theory simplified with this approach, but the mass fluxes for the individual species are more directly observable than are the stresses.

Our discussions of momentum and energy transfer in single-component systems (Chapters 1 to 7) are based upon five postulates: conservation of mass (Section 1.3.1), the momentum and moment-of-momentum balances (Section 2.2.1), the energy balance (Section 5.1.1), and the entropy inequality (Section 5.2.1).

We wish to preserve these postulates for our work with multicomponent systems, but it is clear that some changes must be made. We have not defined for a multicomponent mixture what we might mean by a material particle or by a body. Exercise 8.2.1-3 suggests how we may proceed. We define a *multicomponent material particle* of a multicomponent mixture to be a (artificial) particle that moves with the mass-averaged velocity of the mixture (see Table 8.5.1-3). A *multicomponent body* is defined to be a set, the elements of which are multicomponent material particles.

Like a phase interface in a single-component body, a phase interface in a multicomponent body is a three-dimensional region on a molecular scale. By analogy with the philosophy developed in Section 1.3.4, we will represent multicomponent phase interfaces as dividing surfaces. For a more precise definition of multicomponent phase interfaces, see Slattery (1990, p. 698).

In what follows, the notation $d_{(v)}/dt$ indicates a derivative with respect to time following a multicomponent material particle. By $R_{(v)}$ we mean the region of space occupied by a set of multicomponent material particles; $S_{(v)}$ denotes the closed bounding surface of $R_{(v)}$; $\Sigma$ is the collection of dividing surfaces within $R_{(v)}$.

**Exercise 8.3.0-1** *Transport theorem*  Let $\Phi$ be any scalar, vector, or tensor field. Working by analogy with the discussion in Section 1.3.2, determine that

$$\frac{d}{dt} \int_{R_{(v)}} \Phi \, dV = \int_{R_{(v)}} \left( \frac{d_{(v)}\Phi}{dt} + \Phi \operatorname{div} \mathbf{v} \right) dV$$

This is the *transport theorem for a multicomponent body*.

**Exercise 8.3.0-2** *More about transport theorem*  Let $\Phi$ be any scalar, vector, or tensor field. Derive the following transport theorem for a multicomponent body that contains a singular surface $\Sigma$:

$$\frac{d}{dt} \int_{R_{(v)}} \Phi \, dV = \int_{R_{(v)}} \left( \frac{d_{(v)}\Phi}{dt} + \Phi \operatorname{div} \mathbf{v} \right) dV + \int_{\Sigma} \left[\!\left[ \Phi(\mathbf{v} - \mathbf{u}) \cdot \boldsymbol{\xi} \right]\!\right] dA$$

The notation has the same meaning it had in Section 1.3.5, with the understanding that we are concerned with a dividing surface that occurs in a multicomponent body.

### 8.3.1   Conservation of Mass

We will restate our first postulate as

**Conservation of mass**   The mass of a multicomponent body is independent of time.

We define the *mass $M$* of the multicomponent body as

$$\mathcal{M} \equiv \int_{R_{(v)}} \rho \, dV \qquad (8.3.1\text{-}1)$$

where

$$\rho \equiv \sum_{A=1}^{N} \rho_{(A)} \qquad (8.3.1\text{-}2)$$

Our postulate says that

$$\frac{d}{dt} \int_{R_{(v)}} \rho \, dV = 0 \qquad (8.3.1\text{-}3)$$

The transport theorem of Exercise 8.3.0-2 allows us to rewrite (8.3.1-3) as

$$\int_{R_{(v)}} \left( \frac{d_{(v)}\rho}{dt} + \rho \operatorname{div} \mathbf{v} \right) dV + \int_{\Sigma} \llbracket \rho(\mathbf{v} - \mathbf{u}) \cdot \boldsymbol{\xi} \rrbracket \, dA = 0 \qquad (8.3.1\text{-}4)$$

Since this must be true for an arbitrary multicomponent body, we conclude that at each point in each phase the *overall differential mass balance* for a multicomponent mixture,

$$\frac{d_{(v)}\rho}{dt} + \rho \operatorname{div} \mathbf{v} = 0 \qquad (8.3.1\text{-}5)$$

must be observed and at each point on each phase interface the *overall jump mass balance* at a multicomponent phase interface,

$$\llbracket \rho(\mathbf{v} - \mathbf{u}) \cdot \boldsymbol{\xi} \rrbracket = 0 \qquad (8.3.1\text{-}6)$$

must be satisfied. Not surprisingly, (8.3.1-5) has the same form as the differential mass balance for a single-component material developed in Section 1.3.3. It confirms the intuitive statement expressed in Exercise 8.2.1-3 that

$$\sum_{A=1}^{N} r_{(A)} = 0 \qquad (8.3.1\text{-}7)$$

and

$$\sum_{A=1}^{N} r_{(A)}^{\sigma} = 0 \qquad (8.3.1\text{-}8)$$

Additional commonly used forms of the overall differential mass balance and the overall jump mass balance are presented in Table 8.5.1-10 and Exercise 8.3.1-1.

**Exercise 8.3.1-1** *More on the overall jump mass balance*   Starting with the jump mass balance for an individual species (8.2.1-4), derive as an alternative form of the *overall jump mass balance* at a phase interface,

$$\left[\!\left[ c \left( \mathbf{v}^\diamond - \mathbf{u} \right) \cdot \boldsymbol{\xi} \right]\!\right] = \sum_{A=1}^{N} \frac{r_{(A)}^{(\sigma)}}{M_{(A)}}$$

**Exercise 8.3.1-2** *Alternative form of the transport theorem for a region containing a dividing surface*   Show that an alternative form of the transport theorem of Exercise 8.3.0-2 is

$$\frac{d}{dt} \int_{R_{(v)}} \rho \Phi \, dV = \int_{R_{(v)}} \rho \frac{d_{(v)} \Phi}{dt} \, dV + \int_{\Sigma} \left[\!\left[ \rho \Phi (\mathbf{v} - \mathbf{u}) \cdot \boldsymbol{\xi} \right]\!\right] dA$$

It is to be again understood that $\Phi$ is any scalar, vector, or tensor field.

**Exercise 8.3.1-3** *Alternative form of overall differential mass balance*   Derive the alternative form of the overall differential mass balance in Table 8.5.1-10 starting from the differential mass balance for an individual species.

**Exercise 8.3.1-4** *Frame indifference of overall differential mass balance*   Prove that the overall differential mass balance takes the same form in every frame of reference.

**Exercise 8.3.1-5** *Frame indifference of overall jump mass balance*   Prove that the overall jump mass balance takes the same form in every frame of reference.

### 8.3.2  Momentum Balance

We can restate our second postulate as the

**Momentum balance**   The time rate of change of the momentum of a multicomponent body relative to an inertial frame of reference is equal to the sum of the forces acting on the body.

Our discussion of forces in the introduction to Section 2.1 continues to apply, but we must recognize that each constituent may be subject to different external forces. Consider, for example, a dilute aqueous solution of sodium chloride subjected to an electric field. The sodium chloride will be nearly fully dissociated, which means that we must consider three separate species: the sodium ions, the chloride ions, and water. The force of the electric field upon the sodium ions will be equal in magnitude but opposite in direction to the force it exerts on the chloride ions. The electric field does not act directly upon the water. Yet all three species are under the influence of gravity.

With this thought in mind, our statement of the momentum balance for a multicomponent, multiphase body may be written as

$$\frac{d}{dt} \int_{R_{(v)}} \rho \mathbf{v} \, dV = \int_{S_{(v)}} \mathbf{t} \, dA + \int_{R_{(v)}} \sum_{A=1}^{N} \rho_{(A)} \mathbf{f}_{(A)} \, dV \qquad (8.3.2\text{-}1)$$

Here, $\mathbf{f}_{(A)}$ denotes the external force per unit mass acting upon species $A$.

If we introduce the stress tensor $\mathbf{T}$ in the usual manner, we may use the transport theorem of Exercise 8.3.1-2 to conclude that at each point within each phase the *overall differential momentum balance,*

$$\rho \frac{d_{(v)}\mathbf{v}}{dt} = \text{div } \mathbf{T} + \sum_{A=1}^{N} \rho_{(A)}\mathbf{f}_{(A)} \qquad (8.3.2\text{-}2)$$

must be satisfied and that at each point on each phase interface the *overall jump momentum balance,*

$$\left[\rho\mathbf{v}(\mathbf{v} - \mathbf{u}) \cdot \boldsymbol{\xi} - \mathbf{T} \cdot \boldsymbol{\xi}\right] = 0 \qquad (8.3.2\text{-}3)$$

must be obeyed.

If we define the *mass-averaged external force* per unit mass as

$$\mathbf{f} = \frac{1}{\rho} \sum_{A=1}^{N} \rho_{(A)}\mathbf{f}_{(A)} \qquad (8.3.2\text{-}4)$$

then (8.3.2-2) has the same form as the differential momentum balance for a single-component material:

$$\rho \frac{d_{(v)}\mathbf{v}}{dt} = \text{div } \mathbf{T} + \rho\mathbf{f} \qquad (8.3.2\text{-}5)$$

### 8.3.3  Moment-of-Momentum Balance

Let us restate our third postulate as

**Moment-of-momentum balance**   The time rate of change of the moment of momentum of a multicomponent body relative to an inertial frame of reference is equal to the sum of the moments of all forces acting on the body.

In writing the moment-of-momentum balance in this manner, we confine our attention to the so-called nonpolar case (i.e., we assume that all torques acting on the body are the result of forces acting on the body). Deviations from this case are thought to be unusual in the context of continuum mechanics. See Section 2.2.1 for more details.

The moment-of-momentum balance can be expressed as

$$\frac{d}{dt} \int_{R_{(v)}} \rho(\mathbf{p} \wedge \mathbf{v}) \, dV = \int_{S_{(v)}} \mathbf{p} \wedge \mathbf{t} \, dA + \int_{R_{(v)}} \rho(\mathbf{p} \wedge \mathbf{f}) \, dV \qquad (8.3.3\text{-}1)$$

Note that we have introduced here the mass-averaged external force per unit mass $\mathbf{f}$ defined by (8.3.2-4).

If we introduce the stress tensor $\mathbf{T}$, we may parallel the discussion in Section 2.2.4 in using the transport theorem of Exercise 8.3.1-2 as well as the overall differential momentum balance (8.3.2-2) and the overall jump momentum balance (8.3.2-3) to conclude that

$$\mathbf{T} = \mathbf{T}^{T} \qquad (8.3.3\text{-}2)$$

which is known as the *differential momentum balance.*

### 8.3.4 Energy Balance

Our fourth postulate can be restated as the

**Energy balance** In an inertial frame of reference, the time rate of change of the internal and kinetic energy of a multicomponent body is equal to the rate at which work is done on the body by the system of contact, external, and mutual forces acting upon it plus the rate of energy transmission to the body.

With the assumption that the concepts of contact energy transmission, external energy transmission, and mutual energy transmission developed in Section 5.1.2 are extended to multicomponent bodies, the energy balance in an inertial frame of reference says

$$\frac{d}{dt} \int_{R_{(v)}} \rho \left( \hat{U} + \frac{1}{2}v^2 \right) dV = \int_{S_{(v)}} \mathbf{v} \cdot (\mathbf{T} \cdot \mathbf{n}) \, dA$$

$$+ \int_{R_{(v)}} \sum_{A=1}^{N} \rho_{(A)} \left( \mathbf{v}_{(A)} \cdot \mathbf{f}_{(A)} \right) dV + \int_{S_{(v)}} h \, dA + \int_{R_{(v)}} \rho Q \, dV \qquad (8.3.4\text{-}1)$$

where the notation has the same meaning as used in Section 5.1.3.

Let us limit ourselves to a multicomponent body in which all quantities are continuous and differentiable as many times as desired. Following Exercise 5.1.3-2, we may express the contact energy flux in terms of the energy flux vector $\mathbf{q}$:

$$h = h(\mathbf{z}, \mathbf{n})$$

$$= -\mathbf{q} \cdot \mathbf{n} \qquad (8.3.4\text{-}2)$$

Arguing as we did in Section 5.1.3 and using the transport theorem of Exercise 8.3.1-2, we obtain the *overall differential energy balance* for a multicomponent mixture:

$$\rho \frac{d_{(v)}}{dt} \left( \hat{U} + \frac{1}{2}v^2 \right) = -\text{div} \, \mathbf{q} + \text{div} \, (\mathbf{T} \cdot \mathbf{v}) + \sum_{A=1}^{N} \rho_{(A)} \left( \mathbf{v}_{(A)} \cdot \mathbf{f}_{(A)} \right) + \rho Q \qquad (8.3.4\text{-}3)$$

or

$$\rho \frac{d_{(v)} \hat{U}}{dt} = -\text{div} \, \mathbf{q} + \text{tr} \, (\mathbf{T} \cdot \nabla \mathbf{v}) + \sum_{A=1}^{N} \mathbf{j}_{(A)} \cdot \mathbf{f}_{(A)} + \rho Q \qquad (8.3.4\text{-}4)$$

as well as the *overall jump energy balance*:

$$\left[ \rho \left( \hat{U} + \frac{1}{2}v^2 \right) (\mathbf{v} - \mathbf{u}) \cdot \boldsymbol{\xi} + \mathbf{q} \cdot \boldsymbol{\xi} - \mathbf{v} \cdot (\mathbf{T} \cdot \boldsymbol{\xi}) \right] = 0 \qquad (8.3.4\text{-}5)$$

Here $\mathbf{j}_{(A)}$ is the mass flux of species $A$ with respect to the mass-averaged velocity $\mathbf{v}$ (see Table 8.5.1-4).

Several different forms of the overall differential energy balance that are in common use are given in Table 8.5.2-1.

### 8.3.5 Entropy Inequality

Our fifth postulate can be revised as to read:

**Entropy inequality** The minimum time rate of change of the entropy of a multicomponent body is equal to the rate of entropy transmission to the body.

With the assumption that the concepts of contact entropy transmission, external entropy transmission, mutual entropy transmission, and temperature developed in Section 5.2.2 are extended to multicomponent bodies, the entropy inequality requires

$$\text{minimum } \frac{d}{dt} \int_{R_{(v)}} \rho \hat{S} \, dV = \int_{S_{(v)}} \eta \, dA + \int_{R_{(v)}} \rho \frac{Q}{T} \, dV \tag{8.3.5-1}$$

or

$$\frac{d}{dt} \int_{R_{(v)}} \rho \hat{S} \, dV \geq \int_{S_{(v)}} \eta \, dA + \int_{R_{(v)}} \rho \frac{Q}{T} \, dV \tag{8.3.5-2}$$

where $\hat{S}$ denotes entropy per unit mass and

$$Q \equiv Q_e + Q_m \tag{8.3.5-3}$$

is the scalar field that represents the sum of the external and mutual radiant energy transmission rates per unit mass. The notation used here has the same meaning as that employed in Section 5.2.3.

Let us limit ourselves to a multicomponent body in which all quantities are continuous and differentiable as many times as desired. By analogy with Exercise 5.2.3-1, we may express the contact entropy flux in terms of the thermal energy flux vector $\mathbf{e}$ and the positive scalar field *temperature $T$*:

$$\eta = \eta(\mathbf{z}, \mathbf{n})$$

$$= -\frac{1}{T} \mathbf{e} \cdot \mathbf{n} \tag{8.3.5-4}$$

Arguing as we did in Section 5.2.3-7 and using the transport theorem of Exercise 8.3.1-2, we find that the *overall differential entropy inequality*,

$$\rho \frac{d_{(v)} \hat{S}}{dt} \geq -\text{div} \left( \frac{\mathbf{e}}{T} \right) + \rho \frac{Q}{T} \tag{8.3.5-5}$$

must be satisfied at each point within each phase and the *overall jump entropy inequality*,

$$\left[ \rho \hat{S} (\mathbf{v} - \mathbf{u}) \cdot \boldsymbol{\xi} + \frac{1}{T} \mathbf{e} \cdot \boldsymbol{\xi} \right] \geq 0 \tag{8.3.5-6}$$

must be obeyed at each point on every interface.

**Exercise 8.3.5-1** *Still more about overall jump entropy inequality*   As explained in Exercise 5.2.3-3, because I have placed little emphasis on interfacial effects in this text, we will have no further use for the jump entropy inequality. To see its role in placing constraints upon interfacial behavior, refer to Slattery (1990).

---

## 8.4   Behavior

We have been concerned in the preceding sections with the form and implications of postulates stated for all materials. We know of course that all materials do not exhibit the same behavior.

As we did in Chapter 5, we wish to recognize that any material should be capable of undergoing all processes that are consistent with our fundamental axioms. In particular, we shall use the differential entropy inequality to restrict the form of descriptions from material behavior.

### 8.4.1 Implications of the Differential Entropy Inequality

We shall begin by investigating the restrictions that the overall differential entropy inequality (8.3.5-5) places upon the form of descriptions for bulk material behavior. The approach is similar to that suggested by Gurtin and Vargas (1971) and outlined for single-component materials in Section 5.3.1.

Multiplying (8.3.5-5) by $T$ and subtracting it from (8.3.4-4), we have

$$\rho\frac{d_{(v)}\hat{U}}{dt} - \rho T\frac{d_{(v)}\hat{S}}{dt} - \text{tr}(\mathbf{T} \cdot \nabla\mathbf{v}) + \text{div}(\mathbf{q} - \mathbf{e})$$
$$+ \frac{1}{T}\mathbf{e} \cdot \nabla T - \sum_{A=1}^{N}\mathbf{j}_{(A)} \cdot \mathbf{f}_{(A)} \leq 0 \tag{8.4.1-1}$$

We will find it more convenient to work in terms of the Helmholtz free energy per unit mass,

$$\hat{A} \equiv \hat{U} - T\hat{S} \tag{8.4.1-2}$$

in terms of which (8.4.1-1) becomes

$$\rho\frac{d_{(v)}\hat{A}}{dt} + \rho\hat{S}\frac{d_{(v)}T}{dt} - \text{tr}(\mathbf{T} \cdot \nabla\mathbf{v}) + \text{div}(\mathbf{q} - \mathbf{e})$$
$$+ \frac{1}{T}\mathbf{e} \cdot \nabla T - \sum_{A=1}^{N}\mathbf{j}_{(A)} \cdot \mathbf{f}_{(A)} \leq 0 \tag{8.4.1-3}$$

Let us restrict ourselves to a class of material behavior or to a class of constitutive equations that includes the form of equation of state most often discussed in thermodynamics (8.4.1-12), the predictions of kinetic theory for $\mathbf{q}$ and $\mathbf{e}$ (Section 8.4.3), the predictions of kinetic theory for $\mathbf{j}_{(A)}$ (Section 8.4.4), common descriptions for the rates of reactions (Section 8.4.8), and the common descriptions of generalized Newtonian fluids (Section 2.3.3). In particular, we will assume that

$$\hat{A} = \hat{A}(\mathbf{\Lambda}, \mathbf{D})$$

$$\mathbf{q} = \mathbf{q}(\mathbf{\Lambda})$$

$$\mathbf{e} = \mathbf{e}(\mathbf{\Lambda})$$

$$\mathbf{j}_{(A)} = \mathbf{j}_{(A)}(\mathbf{\Lambda}) \tag{8.4.1-4}$$

$$r_{(A)} = r_{(A)}(\mathbf{\Lambda})$$

$$\mathbf{T} = \mathbf{T}(\mathbf{\Lambda}, \mathbf{D})$$

where the set of variables

$$\mathbf{\Lambda} \equiv \big(\hat{V}, T, \omega_{(1)}, \omega_{(2)}, \dots, \omega_{(N-1)}, \nabla\hat{V}, \nabla T,$$
$$\nabla\omega_{(1)}, \nabla\omega_{(2)}, \dots, \nabla\omega_{(N-1)}, \mathbf{f}_{(1)}, \mathbf{f}_{(2)}, \dots, \mathbf{f}_{(N)}\big) \tag{8.4.1-5}$$

is a set of independent variables common to all of these constitutive equations and $\mathbf{D}$ is the rate-of-deformation tensor.

Using the chain rule, we can say from (8.4.1-4) that (see footnotes in Section 5.3.1)

$$
\begin{aligned}
\frac{d_{(v)}\hat{A}}{dt} &= \frac{\partial \hat{A}}{\partial \hat{V}}\frac{d_{(v)}\hat{V}}{dt} + \frac{\partial \hat{A}}{\partial T}\frac{d_{(v)}T}{dt} + \sum_{B=1}^{N-1}\frac{\partial \hat{A}}{\partial \omega_{(B)}}\frac{d_{(v)}\omega_{(B)}}{dt} + \frac{\partial \hat{A}}{\partial \nabla \hat{V}} \cdot \frac{d_{(v)}\nabla \hat{V}}{dt} \\
&\quad + \frac{\partial \hat{A}}{\partial \nabla T} \cdot \frac{d_{(v)}\nabla T}{dt} + \sum_{B=1}^{N-1}\frac{\partial \hat{A}}{\partial \nabla \omega_{(B)}} \cdot \frac{d_{(v)}\nabla \omega_{(B)}}{dt} \\
&\quad + \sum_{B=1}^{N}\frac{\partial \hat{A}}{\partial \mathbf{f}_{(B)}} \cdot \frac{d_{(v)}\mathbf{f}_{(B)}}{dt} + \mathrm{tr}\left(\frac{\partial \hat{A}}{\partial \mathbf{D}} \cdot \frac{d_{(v)}\mathbf{D}^{\mathrm{T}}}{dt}\right)
\end{aligned}
\tag{8.4.1-6}
$$

This together with the overall differential mass balance (Section 8.3.1),

$$
\rho\frac{d_{(v)}\hat{V}}{dt} = \mathrm{div}\,\mathbf{v}
\tag{8.4.1-7}
$$

and the differential mass balance for species $B$ (Section 8.5.1-5),

$$
\rho\frac{d_{(v)}\omega_{(B)}}{dt} = -\mathrm{div}\,\mathbf{j}_{(B)} + r_{(B)}
\tag{8.4.1-8}
$$

allow us to rearrange (8.4.1-3) in the form

$$
\begin{aligned}
&\rho\left[\left(\frac{\partial \hat{A}}{\partial T} + \hat{S}\right)\frac{d_{(v)}T}{dt} + \frac{\partial \hat{A}}{\partial \nabla \hat{V}} \cdot \frac{d_{(v)}\nabla \hat{V}}{dt} + \frac{\partial \hat{A}}{\partial \nabla T} \cdot \frac{d_{(v)}\nabla T}{dt} + \sum_{B=1}^{N-1}\frac{\partial \hat{A}}{\partial \nabla \omega_{(B)}} \cdot \frac{d_{(v)}\nabla \omega_{(B)}}{dt}\right. \\
&\quad \left.+ \sum_{B=1}^{N}\frac{\partial \hat{A}}{\partial \mathbf{f}_{(B)}} \cdot \frac{d_{(v)}\mathbf{f}_{(B)}}{dt}\right] + \mathrm{tr}\left(\frac{\partial \hat{A}}{\partial \mathbf{D}} \cdot \frac{d_{(v)}\mathbf{D}^{\mathrm{T}}}{dt}\right) - \mathrm{tr}\left[\left(\mathbf{T} - \frac{\partial \hat{A}}{\partial \hat{V}}\mathbf{I}\right) \cdot \nabla \mathbf{v}\right] \\
&\quad + \mathrm{div}\left(\mathbf{q} - \mathbf{e} - \sum_{B=1}^{N-1}\frac{\partial \hat{A}}{\partial \omega_{(B)}}\mathbf{j}_{(B)}\right) + \sum_{B=1}^{N-1}\mathbf{j}_{(B)} \cdot \nabla\left(\frac{\partial \hat{A}}{\partial \omega_{(B)}}\right) + \sum_{B=1}^{N-1}\frac{\partial \hat{A}}{\partial \omega_{(B)}}r_{(B)} \\
&\quad + \frac{1}{T}\mathbf{e} \cdot \nabla T - \sum_{B=1}^{N}\mathbf{j}_{(B)} \cdot \mathbf{f}_{(B)} \\
&\leq 0
\end{aligned}
\tag{8.4.1-9}
$$

It is a simple matter to construct $T$, $\hat{V}$, $\omega_{(B)}$, $\nabla T$, $\nabla \hat{V}$, $\nabla \omega_{(B)}$, and $\mathbf{f}_{(B)}$ fields such that at any given point within a phase at any specified time

$$
\frac{d_{(v)}T}{dt}, \quad \frac{d_{(v)}\nabla \hat{V}}{dt}, \quad \frac{d_{(v)}\nabla T}{dt}, \quad \frac{d_{(v)}\nabla \omega_{(B)}}{dt}, \quad \frac{d_{(v)}\mathbf{f}_{(B)}}{dt}, \quad \frac{d_{(v)}\mathbf{D}^{\mathrm{T}}}{dt}
\tag{8.4.1-10}
$$

take arbitrary values. We conclude that

$$
\hat{S} = -\left(\frac{\partial \hat{A}}{\partial T}\right)_{\hat{V},\,\omega_{(B)}}
\tag{8.4.1-11}
$$

$$
\hat{A} = \hat{A}\left(\hat{V}, T, \omega_{(1)}, \omega_{(2)}, \ldots, \omega_{(N-1)}\right)
\tag{8.4.1-12}
$$

and

$$-\text{tr}\left[\left(\mathbf{T} - \frac{\partial\hat{A}}{\partial\hat{V}}\mathbf{I}\right)\cdot\nabla\mathbf{v}\right] + \text{div}\left(\mathbf{q} - \mathbf{e} - \sum_{B=1}^{N-1}\frac{\partial\hat{A}}{\partial\omega_{(B)}}\mathbf{j}_{(B)}\right) + \sum_{B=1}^{N-1}\mathbf{j}_{(B)}\cdot\nabla\left(\frac{\partial\hat{A}}{\partial\omega_{(B)}}\right)$$

$$+ \sum_{B=1}^{N-1}\frac{\partial\hat{A}}{\partial\omega_{(B)}}r_{(B)} + \frac{1}{T}\mathbf{e}\cdot\nabla T - \sum_{B=1}^{N}\mathbf{j}_{(B)}\cdot\mathbf{f}_{(B)} \leq 0 \tag{8.4.1-13}$$

For simplicity, let us introduce

$$\mathbf{k} \equiv \mathbf{q} - \mathbf{e} - \sum_{B=1}^{N-1}\left(\frac{\partial\hat{A}}{\partial\omega_{(B)}}\right)_{\hat{V},\,\omega_{(C)}\,(C\neq B,\,N)}\mathbf{j}_{(B)} \tag{8.4.1-14}$$

and write inequality (8.4.1-13) as

$$\text{div}\,\mathbf{k} + f(\mathbf{\Lambda},\,\mathbf{D}) \leq 0 \tag{8.4.1-15}$$

The vector $\mathbf{k}$ is frame indifferent, because $\mathbf{q}$ and $\mathbf{e}$ are frame indifferent (see Exercises 5.1.3-3 and 5.2.3-2):

$$\mathbf{k}^* = \mathbf{Q}\cdot\mathbf{k} \tag{8.4.1-16}$$

From (8.4.1-4), we see that it is a function only of $\mathbf{\Lambda}$:

$$\mathbf{k} = \mathbf{k}(\mathbf{\Lambda}) \tag{8.4.1-17}$$

By the principle of frame indifference, this same form of relationship must hold in every frame of reference,

$$\mathbf{k}^* = \mathbf{k}(\mathbf{Q}\cdot\mathbf{\Lambda}) \tag{8.4.1-18}$$

where the set of variables is

$$\mathbf{Q}\cdot\mathbf{\Lambda} \equiv \big(\hat{V},\,T,\,\omega_{(1)},\,\omega_{(2)},\,\ldots,\,\omega_{(N-1)},\,\mathbf{Q}\cdot\nabla\hat{V},\,\mathbf{Q}\cdot\nabla T,\mathbf{Q}\cdot\nabla\omega_{(1)},$$

$$\mathbf{Q}\cdot\nabla\omega_{(2)},\,\ldots,\,\mathbf{Q}\cdot\nabla\omega_{(N-1)},\,\mathbf{Q}\cdot\mathbf{f}_{(1)},\,\mathbf{Q}\cdot\mathbf{f}_{(2)},\,\ldots,\,\mathbf{Q}\cdot\mathbf{f}_{(N)}\big) \tag{8.4.1-19}$$

Equations (8.4.1-16) through (8.4.1-18) imply that $\mathbf{k}(\mathbf{\Lambda})$ is an isotropic function:

$$\mathbf{k}(\mathbf{Q}\cdot\mathbf{\Lambda}) = \mathbf{Q}\cdot\mathbf{k}(\mathbf{\Lambda}) \tag{8.4.1-20}$$

Applying the chain rule to (8.4.1-17), we have

$$\text{div}\,\mathbf{k} = \text{tr}\left(\frac{\partial\mathbf{k}}{\partial\nabla\hat{V}}\cdot\nabla\nabla\hat{V}\right) + \text{tr}\left(\frac{\partial\mathbf{k}}{\partial\nabla T}\cdot\nabla\nabla T\right) + \sum_{B=1}^{N-1}\text{tr}\left(\frac{\partial\mathbf{k}}{\partial\nabla\omega_{(B)}}\cdot\nabla\nabla\omega_{(B)}\right)$$

$$+ \sum_{B=1}^{N}\text{tr}\left(\frac{\partial\mathbf{k}}{\partial\mathbf{f}_{(B)}}\cdot\nabla\mathbf{f}_{(B)}\right) + \frac{\partial\mathbf{k}}{\partial T}\cdot\nabla T$$

$$+ \frac{\partial\mathbf{k}}{\partial\hat{V}}\cdot\nabla\hat{V} + \sum_{B=1}^{N-1}\frac{\partial\mathbf{k}}{\partial\omega_{(B)}}\cdot\nabla\omega_{(B)} \tag{8.4.1-21}$$

We can construct compatible $\hat{V}$ and $T$ fields such that, at any given point within a phase at any specified time,

$$\nabla \hat{V}, \; \nabla T, \; \nabla \omega_{(B)}, \; \nabla \nabla \hat{V}, \; \nabla \nabla T, \; \nabla \nabla \omega_{(B)}$$

take arbitrary values (Gurtin and Vargas 1971). In view of (8.4.1-15) and (8.4.1-21),

$$\operatorname{tr}\left(\frac{\partial \mathbf{k}}{\partial \nabla \hat{V}} \cdot \nabla \nabla \hat{V}\right) = \operatorname{tr}\left(\frac{\partial \mathbf{k}}{\partial \nabla T} \cdot \nabla \nabla T\right)$$

$$= \operatorname{tr}\left(\frac{\partial \mathbf{k}}{\partial \nabla \omega_{(B)}} \cdot \nabla \nabla \omega_{(B)}\right)$$

$$= \sum_{B=1}^{N} \operatorname{tr}\left(\frac{\partial \mathbf{k}}{\partial \mathbf{f}_{(B)}} \cdot \nabla \mathbf{f}_{(B)}\right)$$

$$= \frac{\partial \mathbf{k}}{\partial T} \cdot \nabla T$$

$$= \frac{\partial \mathbf{k}}{\partial \hat{V}} \cdot \nabla \hat{V}$$

$$= \frac{\partial \mathbf{k}}{\partial \omega_{(B)}} \cdot \nabla \omega_{(B)}$$

$$= 0 \tag{8.4.1-22}$$

which implies that the symmetric parts of

$$\frac{\partial \mathbf{k}}{\partial \nabla \hat{V}}, \; \frac{\partial \mathbf{k}}{\partial \nabla T}, \; \frac{\partial \mathbf{k}}{\partial \nabla \omega_{(B)}}$$

are all zero as well as

$$\frac{\partial \mathbf{k}}{\partial \hat{V}} = \frac{\partial \mathbf{k}}{\partial T}$$

$$= \frac{\partial \mathbf{k}}{\partial \omega_{(B)}}$$

$$= \frac{\partial \mathbf{k}}{\partial \mathbf{f}_{(B)}}$$

$$= 0$$

Using the principle of frame indifference, Gurtin (1971, Lemma 6.2) and Gurtin and Vargas (1971, Lemma 10.2) have proved that, when the symmetric portions of the derivatives of an isotropic vector function $\mathbf{k}(\Lambda)$ with respect to each of the independent vectors are all zero, the function itself is zero. In this case we conclude that

$$\mathbf{k} = \mathbf{0} \tag{8.4.1-23}$$

or

$$\mathbf{e} = \mathbf{q} - \sum_{B=1}^{N-1}\left(\frac{\partial \hat{A}}{\partial \omega_{(B)}}\right)_{\hat{V}, T, \omega_{(C)} (C \neq B, N)} \mathbf{j}_{(B)} \tag{8.4.1-24}$$

This in turn implies that (8.4.1-13) reduces to

$$
-\mathrm{tr}\left\{\left[\mathbf{T}-\left(\frac{\partial\hat{A}}{\partial\hat{V}}\right)_{T,\,\omega_{(B)}}\mathbf{I}\right]\cdot\nabla\mathbf{v}\right\}+\sum_{B=1}^{N-1}\mathbf{j}_{(B)}\cdot\nabla\left(\frac{\partial\hat{A}}{\partial\omega_{(B)}}\right)_{\hat{V},\,T,\,\omega_{(C)}\,(C\neq B,\,N)}
$$

$$
+\sum_{B=1}^{N-1}\left(\frac{\partial\hat{A}}{\partial\omega_{(B)}}\right)_{\hat{V},\,T,\,\omega_{(C)}\,(C\neq B,\,N)}r_{(B)}
$$

$$
+\frac{1}{T}\mathbf{e}\cdot\nabla T-\sum_{B=1}^{N-1}\mathbf{j}_{(B)}\cdot\left(\mathbf{f}_{(B)}-\mathbf{f}_{(N)}\right)\leq 0 \tag{8.4.1-25}
$$

In what follows, we will find it convenient to define the *extra stress tensor* (or viscous portion of the stress tensor) as

$$
\mathbf{S}\equiv\mathbf{T}+P\mathbf{I} \tag{8.4.1-26}
$$

and to write (8.4.1-25) as

$$
-\mathrm{tr}\left\{\left[\mathbf{S}-\left\{P+\left(\frac{\partial\hat{A}}{\partial\hat{V}}\right)_{T,\,\omega_{(B)}}\right\}\mathbf{I}\right]\cdot\nabla\mathbf{v}\right\}+\sum_{B=1}^{N-1}\mathbf{j}_{(B)}\cdot\nabla\left(\frac{\partial\hat{A}}{\partial\omega_{(B)}}\right)_{\hat{V},\,T,\,\omega_{(C)}\,(C\neq B,\,N)}
$$

$$
+\sum_{B=1}^{N-1}\left(\frac{\partial\hat{A}}{\partial\omega_{(B)}}\right)_{\hat{V},\,T,\,\omega_{(C)}\,(C\neq B,\,N)}r_{(B)}
$$

$$
+\frac{1}{T}\mathbf{e}\cdot\nabla T-\sum_{B=1}^{N-1}\mathbf{j}_{(B)}\cdot\left(\mathbf{f}_{(B)}-\mathbf{f}_{(N)}\right)\leq 0 \tag{8.4.1-27}
$$

This prompts us to define the *thermodynamic pressure* as

$$
P\equiv-\left(\frac{\partial\hat{A}}{\partial\hat{V}}\right)_{T,\,\omega_{(B)}} \tag{8.4.1-28}
$$

and to write (8.4.1-25) as

$$
-\mathrm{tr}\left\{\mathbf{S}\cdot\nabla\mathbf{v}\right\}+\sum_{B=1}^{N-1}\mathbf{j}_{(B)}\cdot\nabla\left(\frac{\partial\hat{A}}{\partial\omega_{(B)}}\right)_{\hat{V},\,T,\,\omega_{(C)}\,(C\neq B,\,N)}
$$

$$
+\sum_{B=1}^{N-1}\left(\frac{\partial\hat{A}}{\partial\omega_{(B)}}\right)_{\hat{V},\,T,\,\omega_{(C)}\,(C\neq B,\,N)}r_{(B)}
$$

$$
+\frac{1}{T}\mathbf{e}\cdot\nabla T-\sum_{B=1}^{N-1}\mathbf{j}_{(B)}\cdot\left(\mathbf{f}_{(B)}-\mathbf{f}_{(N)}\right)\leq 0 \tag{8.4.1-29}
$$

Before we examine the implications of the entropy inequality, let us determine the consequences of (8.4.1-12).

### 8.4.2 Restrictions on Caloric Equation of State

We started with some broad statements about material behavior in the preceding section, and we concluded in (8.4.1-12) that, if the entropy inequality was to be obeyed,

$$\hat{A} = \hat{A}\left(\hat{V},\ T,\ \omega_{(1)},\ \omega_{(2)},\ \ldots,\ \omega_{(N-1)}\right) \tag{8.4.2-1}$$

or

$$\check{A} = \check{A}\left(T,\ \rho_{(1)},\ \rho_{(2)},\ \ldots,\ \rho_{(N)}\right) \tag{8.4.2-2}$$

or

$$\tilde{A} = \tilde{A}\left(\tilde{V},\ T,\ x_{(1)},\ x_{(2)},\ \ldots,\ x_{(N-1)}\right) \tag{8.4.2-3}$$

or

$$\check{A} = \check{A}\left(T,\ c_{(1)},\ c_{(2)},\ \ldots,\ c_{(N)}\right) \tag{8.4.2-4}$$

Here $\hat{A}$, $\check{A}$, and $\tilde{A}$ are the *Helmholtz free energy per unit mass, per unit volume,* and *per unit mole,* respectively. We will refer to these statements as alternative forms of the *caloric equation of state.* At the same time, we found in (8.4.1-11)

$$
\begin{aligned}
\hat{S} &= -\left(\frac{\partial \hat{A}}{\partial T}\right)_{\hat{V},\,\omega_{(B)}} \\
\check{S} &= -\left(\frac{\partial \check{A}}{\partial T}\right)_{\rho_{(B)}} \\
\tilde{S} &= -\left(\frac{\partial \tilde{A}}{\partial T}\right)_{\tilde{V},\,x_{(B)}} \\
\check{S} &= -\left(\frac{\partial \check{A}}{\partial T}\right)_{c_{(B)}}
\end{aligned}
\tag{8.4.2-5}
$$

In addition to the thermodynamic pressure introduced in (8.4.1-28), we will define the *chemical potential for species A on a mass basis:*[1]

$$\mu_{(A)} \equiv \left(\frac{\partial \check{A}}{\partial \rho_{(A)}}\right)_{T,\,\rho_{(C)}\,(C \neq A)} \tag{8.4.2-6}$$

and the *chemical potential for species A on a molar basis:*

$$\mu_{(A)}^{(m)} \equiv \left(\frac{\partial \check{A}}{\partial c_{(A)}}\right)_{T,\,c_{(C)}\,(C \neq A)} \tag{8.4.2-7}$$

---

[1] These expressions for the chemical potential were suggested by Prof. G. M. Brown, Department of Chemical Engineering, Northwestern University, Evanston, Illinois 60201-3120.

The differentials of (8.4.2-1) through (8.4.2-4) may consequently be expressed as

$$d\hat{A} = -P\,d\hat{V} - \hat{S}\,dT + \sum_{B=1}^{N-1} \left( \frac{\partial \hat{A}}{\partial \omega_{(B)}} \right)_{\hat{V},T,\omega_{(C)}\,(C \neq B,\,N)} d\omega_{(B)} \tag{8.4.2-8}$$

$$d\check{A} = -\check{S}\,dT + \sum_{B=1}^{N} \mu_{(B)}\,d\rho_{(B)} \tag{8.4.2-9}$$

$$d\tilde{A} = -P\,d\tilde{V} - \tilde{S}\,dT + \sum_{B=1}^{N-1} \left( \frac{\partial \tilde{A}}{\partial x_{(B)}} \right)_{\tilde{V},T,x_{(C)}\,(C \neq B,\,N)} dx_{(B)} \tag{8.4.2-10}$$

$$d\check{A} = -\check{S}\,dT + \sum_{B=1}^{N} \mu_{(B)}^{(m)}\,dc_{(B)} \tag{8.4.2-11}$$

Equations (8.4.2-9) and (8.4.2-11) may be rearranged to read

$$d\left( \frac{\hat{A}}{\hat{V}} \right) = -\frac{\hat{S}}{\hat{V}}\,dT + \sum_{B=1}^{N} \mu_{(B)}\,d\left( \frac{\omega_{(B)}}{\hat{V}} \right)$$

$$d\hat{A} = \left( \frac{\hat{A}}{\hat{V}} - \sum_{B=1}^{N} \mu_{(B)}\rho_{(B)} \right) d\hat{V} - \hat{S}\,dT + \sum_{B=1}^{N-1} \left( \mu_{(B)} - \mu_{(N)} \right) d\omega_{(B)} \tag{8.4.2-12}$$

and

$$d\left( \frac{\tilde{A}}{\tilde{V}} \right) = -\frac{\tilde{S}}{\tilde{V}}\,dT + \sum_{B=1}^{N} \mu_{(B)}^{(B)}\,d\left( \frac{x_{(B)}}{\tilde{V}} \right)$$

$$d\tilde{A} = \left( \frac{\tilde{A}}{\tilde{V}} - \sum_{B=1}^{N} \mu_{(B)}^{(m)}c_{(B)} \right) d\tilde{V} - \tilde{S}\,dT + \sum_{B=1}^{N-1} \left( \mu_{(B)}^{(B)} - \mu_{(N)}^{(m)} \right) dx_{(B)} \tag{8.4.2-13}$$

Comparison of the coefficients in (8.4.2-8) and (8.4.2-12) and comparison of the coefficients in (8.4.2-10) and (8.4.2-13) give

$$\left( \frac{\partial \hat{A}}{\partial \omega_{(B)}} \right)_{\hat{V},T,\omega_{(C)}\,(C \neq B,\,N)} = \mu_{(B)} - \mu_{(N)}$$

$$\left( \frac{\partial \tilde{A}}{\partial x_{(B)}} \right)_{\tilde{V},T,x_{(C)}\,(C \neq B,\,N)} = \mu_{(B)}^{(m)} - \mu_{(N)}^{(m)} \tag{8.4.2-14}$$

as well as two forms of *Euler's equation*:

$$\hat{A} = -P\hat{V} + \sum_{B=1}^{N} \mu_{(B)}\omega_{(B)}$$

$$\tilde{A} = -P\tilde{V} + \sum_{B=1}^{N} \mu_{(B)}^{(m)}x_{(B)} \tag{8.4.2-15}$$

Equations (8.4.2-8), (8.4.2-10), and (8.4.2-14) yield two forms of the *Gibbs equation*:

$$d\hat{A} = -P\,d\hat{V} - \hat{S}\,dT + \sum_{B=1}^{N-1} \left( \mu_{(B)} - \mu_{(N)} \right) d\omega_{(B)}$$

$$d\tilde{A} = -P\,d\tilde{V} - \tilde{S}\,dT + \sum_{B=1}^{N-1} \left( \mu_{(B)}^{(m)} - \mu_{(N)}^{(m)} \right) dx_{(B)}$$

(8.4.2-16)

Two forms of the *Gibbs–Duhem equation* follow immediately by subtracting (8.4.2-16) from the differentials of (8.4.2-15):

$$\hat{S}\,dT - \hat{V}\,dP + \sum_{B=1}^{N} \omega_{(B)}\,d\mu_{(B)} = 0$$

$$\tilde{S}\,dT - \tilde{V}\,dP + \sum_{B=1}^{N} x_{(B)}\,d\mu_{(B)}^{(m)} = 0$$

(8.4.2-17)

We would like to emphasize that Euler's equation, the Gibbs equation, and the Gibbs–Duhem equation all apply to dynamic processes, so long as the statements about behavior made in Section 8.4.1 are applicable to the materials being considered.

**Exercise 8.4.2-1** *Specific variables per unit mole*   Let $c_{(A)}$ denote moles of species $A$ per unit volume. Denote by $\tilde{\ }$ that we are dealing with a quantity per unit mole.

Show that alternative expressions for temperature and thermodynamic pressure are

$$T = \left( \frac{\partial \hat{U}}{\partial \hat{S}} \right)_{\hat{V},\,\omega_{(A)}} = \left( \frac{\partial \check{U}}{\partial \check{S}} \right)_{\rho_{(A)}} = \left( \frac{\partial \tilde{U}}{\partial \tilde{S}} \right)_{\tilde{V},\,x_{(A)}} = \left( \frac{\partial \check{U}}{\partial \check{S}} \right)_{c_{(A)}}$$

$$P = -\left( \frac{\partial \hat{U}}{\partial \hat{V}} \right)_{\hat{S},\,\omega_{(A)}} = -\left( \frac{\partial \tilde{U}}{\partial \tilde{V}} \right)_{\tilde{S},\,x_{(A)}}$$

$$\mu_{(A)} = \left( \frac{\partial \check{U}}{\partial \rho_{(A)}} \right)_{\check{S},\,\rho_{(B)}\,(B \neq A)}$$

$$\mu_{(A)}^{(m)} = \left( \frac{\partial \check{U}}{\partial c_{(A)}} \right)_{\check{S},\,c_{(B)}\,(B \neq A)}$$

**Exercise 8.4.2-2** *The Maxwell relations*   Let us define

$$\hat{A} \equiv \hat{U} - T\hat{S}$$
$$\hat{H} \equiv \hat{U} + P\hat{V}$$
$$\hat{G} \equiv \hat{H} - T\hat{S}$$

We refer to $\hat{A}$ as *Helmholtz free energy* per unit mass, $\hat{S}$ as *enthalpy* per unit mass, and $\hat{G}$

as *Gibbs free energy* per unit mass. Determine that

i)
$$\left(\frac{\partial T}{\partial \hat{V}}\right)_{\hat{S},\,\omega_{(B)}} = -\left(\frac{\partial P}{\partial \hat{S}}\right)_{\hat{V},\,\omega_{(B)}}$$

$$\left(\frac{\partial T}{\partial \omega_{(A)}}\right)_{\hat{S},\,\hat{V},\,\omega_{(B)}\,(B\neq A,\,N)} = \left(\frac{\partial \left(\mu_{(A)} - \mu_{(N)}\right)}{\partial \hat{S}}\right)_{\hat{V},\,\omega_{(B)}}$$

$$-\left(\frac{\partial P}{\partial \omega_{(A)}}\right)_{\hat{S},\,\hat{V},\,\omega_{(B)}\,(B\neq A,\,N)} = \left(\frac{\partial \left(\mu_{(A)} - \mu_{(N)}\right)}{\partial \hat{V}}\right)_{\hat{S},\,\omega_{(B)}}$$

$$\left(\frac{\partial(\mu_{(A)} - \mu_{(N)})}{\partial \omega_{(B)}}\right)_{\hat{S},\,\hat{V},\,\omega_{(c)}\,(C\neq B,\,N)} = \left(\frac{\partial \left(\mu_{(B)} - \mu_{(N)}\right)}{\partial \omega_{(A)}}\right)_{\hat{S},\hat{V},\omega_{(C)}(C\neq A,\,N)}$$

ii)
$$\left(\frac{\partial \hat{S}}{\partial \hat{V}}\right)_{T,\,\omega_{(B)}} = \left(\frac{\partial P}{\partial T}\right)_{\hat{V},\,\omega_{(B)}}$$

$$-\left(\frac{\partial \hat{S}}{\partial \omega_{(A)}}\right)_{T,\,\hat{V},\,\omega_{(B)}(B\neq A,\,N)} = \left(\frac{\partial \left(\mu_{(A)} - \mu_{(N)}\right)}{\partial T}\right)_{\hat{V},\,\omega_{(B)}}$$

$$-\left(\frac{\partial P}{\partial \omega_{(A)}}\right)_{T,\,\hat{V},\,\omega_{(B)}\,(B\neq A,\,N)} = \left(\frac{\partial \left(\mu_{(A)} - \mu_{(N)}\right)}{\partial \hat{V}}\right)_{T,\,\omega_{(B)}}$$

$$\left(\frac{\partial(\mu_{(A)} - \mu_{(N)})}{\partial \omega_{(B)}}\right)_{T,\,\hat{V},\,\omega_{(C)}\,(C\neq B,\,N)} = \left(\frac{\partial \left(\mu_{(B)} - \mu_{(N)}\right)}{\partial \omega_{(A)}}\right)_{T,\,\hat{V},\,\omega_{(C)}\,(C\neq A,\,N)}$$

iii)
$$\left(\frac{\partial T}{\partial P}\right)_{\hat{S},\,\omega_{(B)}} = \left(\frac{\partial \hat{V}}{\partial \hat{S}}\right)_{P,\,\omega_{(B)}}$$

$$\left(\frac{\partial T}{\partial \omega_{(A)}}\right)_{\hat{S},\,P,\,\omega_{(B)}\,(B\neq A,\,N)} = \left(\frac{\partial \left(\mu_{(A)} - \mu_{(N)}\right)}{\partial \hat{S}}\right)_{P,\,\omega_{(B)}}$$

$$\left(\frac{\partial \hat{V}}{\partial \omega_{(A)}}\right)_{\hat{S},\,P,\,\omega_{(B)}\,(B\neq A,\,N)} = \left(\frac{\partial \left(\mu_{(A)} - \mu_{(N)}\right)}{\partial P}\right)_{\hat{S},\,\omega_{(B)}}$$

$$\left(\frac{\partial(\mu_{(A)} - \mu_{(N)})}{\partial \omega_{(B)}}\right)_{\hat{S},\,P,\,\omega_{(C)}\,(C\neq B,\,N)} = \left(\frac{\partial \left(\mu_{(B)} - \mu_{(N)}\right)}{\partial \omega_{(A)}}\right)_{\hat{S},\,P,\,\omega_{(C)}\,(C\neq A,\,N)}$$

iv)
$$-\left(\frac{\partial \hat{S}}{\partial P}\right)_{T, \omega_{(B)}} = \left(\frac{\partial \hat{V}}{\partial T}\right)_{P, \omega_{(B)}}$$

$$-\left(\frac{\partial \hat{S}}{\partial \omega_{(A)}}\right)_{T, P, \omega_{(B)}(B \neq A, N)} = \left(\frac{\partial \left(\mu_{(A)} - \mu_{(N)}\right)}{\partial T}\right)_{P, \omega_{(B)}}$$

$$\left(\frac{\partial \hat{V}}{\partial \omega_{(A)}}\right)_{T, P, \omega_{(B)}(B \neq A, N)} = \left(\frac{\partial \left(\mu_{(A)} - \mu_{(N)}\right)}{\partial P}\right)_{T, \omega_{(B)}}$$

$$\left(\frac{\partial (\mu_{(A)} - \mu_{(N)})}{\partial \omega_{(B)}}\right)_{T, P, \omega_{(C)}(C \neq B, N)} = \left(\frac{\partial \left(\mu_{(B)} - \mu_{(N)}\right)}{\partial \omega_{(A)}}\right)_{T, P, \omega_{(C)}(C \neq A, N)}$$

**Exercise 8.4.2-3** *More Maxwell relations*   Following the definitions introduced in Exercise 8.4.2-2, we have that

$$\breve{A} = \breve{U} - T\breve{S}$$

$$\breve{H} = \breve{U} + P$$

$$\breve{G} = \breve{H} - T\breve{S}$$

Determine that

i)
$$\left(\frac{\partial T}{\partial \rho_{(A)}}\right)_{\breve{S}, \rho_{(B)}(B \neq A)} = \left(\frac{\partial \mu_{(A)}}{\partial \breve{S}}\right)_{\rho_{(B)}}$$

$$\left(\frac{\partial \mu_{(A)}}{\partial \rho_{(B)}}\right)_{\breve{S}, \rho_{(C)}(C \neq B)} = \left(\frac{\partial \mu_{(B)}}{\partial \rho_{(A)}}\right)_{\breve{S}, \rho_{(C)}(C \neq A)}$$

ii)
$$-\left(\frac{\partial \breve{S}}{\partial \rho_{(A)}}\right)_{T, \rho_{(B)}(B \neq A)} = \left(\frac{\partial \mu_{(A)}}{\partial T}\right)_{\rho_{(B)}}$$

$$\left(\frac{\partial \mu_{(A)}}{\partial \rho_{(B)}}\right)_{T, \rho_{(C)}(C \neq B)} = \left(\frac{\partial \mu_{(B)}}{\partial \rho_{(A)}}\right)_{T, \rho_{(C)}(C \neq A)}$$

**Exercise 8.4.2-4** *Partial mass variables*

i) In the context of an equilibrium system, let $\Phi$ be any extensive variable

$$\Phi = \Phi\left(T, P, m_{(1)}, m_{(2)}, \ldots, m_{(N)}\right)$$

in which $m_{(A)}$ is the mass of species $A$. Let us define

$$\hat{\Phi} \equiv \frac{\Phi}{m}$$

where

$$m \equiv \sum_{A=1}^{N} m_{(A)}$$

and

$$\hat{\Phi} = \hat{\Phi}\left(T, P, \omega_{(1)}, \omega_{(2)}, \ldots, \omega_{(N-1)}\right)$$

We define the *partial mass variable for an equilibrium system* as

$$\overline{\Phi}_{(A)} \equiv \left(\frac{\partial \Phi}{\partial m_{(A)}}\right)_{T, P, m_{(B)} (B \neq A)}$$

Taking roughly the same approach as was used in deriving (8.4.2-14) and (8.4.2-15), determine that

$$\overline{\Phi}_{(A)} - \overline{\Phi}_{(N)} = \left(\frac{\partial \hat{\Phi}}{\partial \omega_{(A)}}\right)_{T, P, \omega_{(B)} (B \neq A, N)}$$

and

$$\hat{\Phi} = \sum_{A=1}^{N} \overline{\Phi}_{(A)} \omega_{(A)}$$

ii) The results of (i) suggest that, for a *nonequilibrium* system, we define[2]

$$\overline{\Phi}_{(A)} - \overline{\Phi}_{(N)} \equiv \left(\frac{\partial \hat{\Phi}}{\partial \omega_{(A)}}\right)_{T, P, \omega_{(B)} (B \neq A, N)}$$

and

$$\hat{\Phi} = \sum_{A=1}^{N} \overline{\Phi}_{(A)} \omega_{(A)}$$

In this way it is more obvious that partial mass variables are intensive variables that have meaning in a discussion of nonequilibrium thermodynamics. Show from these definitions that

$$\sum_{A=1}^{N} \left(\frac{\partial \overline{\Phi}_{(A)}}{\partial \omega_{(B)}}\right)_{T, P, \omega_{(C)} (C \neq B, N)} , \omega_{(A)} = 0$$

**Exercise 8.4.2-5** *Partial molar variables*   Let $\Phi$ be any extensive variable such as internal energy, enthalpy, or volume. The *partial molar* variable $\overline{\Phi}_{(A)}^{(m)}$ is defined for a system at equilibrium as

$$\overline{\Phi}_{(A)}^{(m)} \equiv \left(\frac{\partial \Phi}{\partial n_{(A)}}\right)_{T, P, n_{(B)} (B \neq A)}$$

where $n_{(A)}$ denotes the number of moles of species $A$. Following a similar argument to that used in Exercise 8.4.2-4, conclude that partial mass and partial molar variables are simply related by

$$\overline{\Phi}_{(A)}^{(m)} = M_{(A)} \overline{\Phi}_{(A)}$$

---

[2] These expressions for the partial mass and partial molar variables (see next exercise) were suggested by Prof. G. M. Brown, Department of Chemical Engineering, Northwestern University, Evanston, Illinois 60201-3120.

and that, for a nonequilibrium system, we should define partial molar variables by

$$\overline{\Phi}_{(A)}^{(m)} - \overline{\Phi}_{(N)}^{(m)} \equiv \left(\frac{\partial \tilde{\Phi}}{\partial x_{(A)}}\right)_{T,\,P,\,x_{(B)}\,(B \neq A,\,N)}$$

and

$$\tilde{\Phi} = \sum_{A=1}^{N} \overline{\Phi}_{(A)}^{(m)} x_{(A)}$$

**Exercise 8.4.2-6**   Determine

i)    $\overline{S}_{(A)} = -\left(\dfrac{\partial \mu_{(A)}}{\partial T}\right)_{P,\,\omega_{(B)}}$

ii)   $\overline{V}_{(A)} = \left(\dfrac{\partial \mu_{(A)}}{\partial P}\right)_{T,\,\omega_{(B)}}$

*Hint:*   See the Gibbs–Duhem equation and Exercise 8.4.2-2.

**Exercise 8.4.2-7** *Heat capacities*   We define the heat capacity per unit mass at constant pressure $\hat{c}_P$ and the heat capacity per unit mass at constant specific volume $\hat{c}_V$ as

$$\hat{c}_P \equiv T \left(\frac{\partial \hat{S}}{\partial T}\right)_{P,\,\omega_{(B)}}$$

and

$$\hat{c}_V \equiv T \left(\frac{\partial \hat{S}}{\partial T}\right)_{\hat{V},\,\omega_{(B)}}$$

i) Determine that

$$\hat{c}_P = \left(\frac{\partial \hat{H}}{\partial T}\right)_{P,\,\omega_{(B)}}$$

and

$$\hat{c}_V = \left(\frac{\partial \hat{U}}{\partial T}\right)_{\hat{V},\,\omega_{(B)}}$$

ii) Prove that

$$\rho \hat{c}_P - \left(\frac{\partial P}{\partial T}\right)_{\hat{V},\,\omega_{(B)}} \left(\frac{\partial \ln \hat{V}}{\partial \ln T}\right)_{P,\,\omega_{(B)}} = \rho \hat{c}_V$$

iii) For an ideal gas, conclude that

$$\hat{c}_P = \hat{c}_V + \frac{R}{M}$$

where $M$ is the average molecular weight

$$M \equiv \frac{\rho}{c}$$

Exercise 8.4.2-8  Determine

i)  $\overline{U}_{(A)} - \overline{U}_{(N)} + \left[ P - T \left( \dfrac{\partial P}{\partial T} \right)_{\hat{V}, \omega_{(B)}} \right] \left( \overline{V}_{(A)} - \overline{V}_{(N)} \right)$

$$= \mu_{(A)} - \mu_{(N)} - T \left[ \frac{\partial \left( \mu_{(A)} - \mu_{(N)} \right)}{\partial T} \right]_{\hat{V}, \omega_{(B)}}$$

ii)  $\overline{H}_{(A)} - \overline{H}_{(N)} = \mu_{(A)} - \mu_{(N)} - T \left[ \dfrac{\partial \left( \mu_{(A)} - \mu_{(N)} \right)}{\partial T} \right]_{P, \omega_{(B)}}$

## 8.4.3  Energy Flux Vectors

From (8.4.1-24) and (8.4.2-14), we see that

$$\mathbf{e} = \mathbf{q} - \sum_{B=1}^{N-1} \left( \mu_{(B)} - \mu_{(N)} \right) \mathbf{j}_{(B)}$$

$$= \mathbf{q} - \sum_{B=1}^{N} \mu_{(B)} \mathbf{j}_{(B)} \tag{8.4.3-1}$$

There is only one constitutive equation for $\mathbf{q}$ and $\mathbf{e}$ that has received significant attention in the literature. Working in the context of the kinetic theory of dilute gases, Hirschfelder, Curtiss, and Bird (1954, Eq. 11.2-32) predict

$$\epsilon = -k \nabla T - cRT \sum_{A=1}^{N} D_{(A)}^T \frac{\mathbf{d}_{(A)}}{\rho_{(A)}}$$

$$= -k \nabla T - cRT \sum_{A=1}^{N-1} D_{(A)}^T \left( \frac{\mathbf{d}_{(A)}}{\rho_{(A)}} - \frac{\mathbf{d}_{(N)}}{\rho_{(N)}} \right) \tag{8.4.3-2}$$

in the last line in which we have recognized (8.4.4-3). Here

$$\epsilon \equiv \mathbf{q} - \sum_{A=1}^{N} \overline{H}_{(A)} \mathbf{j}_{(A)}$$

$$= \mathbf{q} - \sum_{A=1}^{N} \left[ \overline{G}_{(A)} - \overline{G}_{(N)} + \overline{H}_{(N)} + T \left( \overline{S}_{(A)} - \overline{S}_{(N)} \right) \right] \mathbf{j}_{(A)}$$

$$= \mathbf{q} - \sum_{A=1}^{N} \left[ \mu_{(A)} - \mu_{(N)} + \overline{H}_{(N)} + T \left( \overline{S}_{(A)} - \overline{S}_{(N)} \right) \right] \mathbf{j}_{(A)}$$

$$= \mathbf{q} - \sum_{A=1}^{N} \left( \mu_{(A)} + T \overline{S}_{(A)} \right) \mathbf{j}_{(A)}$$

$$= \mathbf{e} - \sum_{A=1}^{N} T \overline{S}_{(A)} \mathbf{j}_{(A)} \tag{8.4.3-3}$$

and, recognizing the results of Exercise 8.4.2-6 (see also Exercise 8.4.4-1)

$$\mathbf{d}_{(A)} \equiv \frac{\rho_{(A)}}{cRT} \left( \overline{S}_{(A)} \nabla T + \nabla \mu_{(A)} - \mathbf{f}_{(A)} - \frac{1}{\rho} \nabla P + \sum_{B=1}^{N} \omega_{(B)} \mathbf{f}_{(B)} \right)$$

$$= \frac{\rho_{(A)}}{cRT} \left[ \sum_{\substack{B=1 \\ B \neq A}}^{N} \left( \frac{\partial \mu_{(A)}}{\partial \omega_{(B)}} \right)_{T, P, \omega_{(C)} (C \neq A, B)} \nabla \omega_{(B)} \right.$$

$$\left. + \left( \overline{V}_{(A)} - \frac{1}{\rho} \right) \nabla P - \left( \mathbf{f}_{(A)} - \sum_{B=1}^{N} \omega_{(B)} \mathbf{f}_{(B)} \right) \right] \tag{8.4.3-4}$$

From the definition of partial mass variables given in Exercise 8.4.2-4 and the Gibbs–Duhem equation (8.4.2-17),

$$\sum_{A=1}^{N} \mathbf{d}_{(A)} = \frac{1}{cRT} \left( \sum_{A=1}^{N} \rho_{(A)} \overline{S}_{(A)} \nabla T + \sum_{A=1}^{N} \rho_{(A)} \nabla \mu_{(A)} - \sum_{A=1}^{N} \rho_{(A)} \mathbf{f}_{(A)} - \nabla P + \sum_{B=1}^{N} \rho_{(A)} \mathbf{f}_{(A)} \right)$$

$$= \frac{1}{cRT} \left( \rho \hat{S} \nabla T + \sum_{A=1}^{N} \rho_{(A)} \nabla \mu_{(A)} - \nabla P \right)$$

$$= 0 \tag{8.4.3-5}$$

The coefficients $k$ and $D_{(A)}^{T}$ are understood to be functions of the local thermodynamic state variables. The direct dependence of $\epsilon$ upon concentration gradient, pressure gradient, and the external forces through the $\mathbf{d}_{(A)}$ is usually referred to as the *Dufour effect*. It is generally believed to be small (Hirschfelder et al. 1954, p. 717) and will often be neglected in what follows. Equation (8.4.3-2) can be regarded as the extension of *Fourier's law* to multicomponent materials.

Equation (8.4.3-2) automatically satisfies the principles of determinism and of local action (Section 2.3.1). It is easily shown that it also satisfies the principle of frame indifference (Section 2.3.1). To explore the implications of the differential entropy inequality, let us begin by rearranging (8.4.1-25) in terms of the thermodynamic pressure and chemical potentials introduced in Sections 8.4.1 and 8.4.2. Recognizing (8.4.2-14), we find

$$-\mathrm{tr}\left[ (\mathbf{T} + P\mathbf{I}) \cdot \nabla \mathbf{v} \right] + \sum_{B=1}^{N-1} \mathbf{j}_{(B)} \cdot \nabla \left( \mu_{(B)} - \mu_{(N)} \right)$$

$$+ \sum_{B=1}^{N-1} \left( \mu_{(B)} - \mu_{(N)} \right) r_{(B)} + \frac{1}{T} \mathbf{e} \cdot \nabla T - \sum_{B=1}^{N-1} \mathbf{j}_{(B)} \cdot \left( \mathbf{f}_{(B)} - \mathbf{f}_{(N)} \right) \leq 0 \tag{8.4.3-6}$$

Using (8.4.3-3) and (8.4.3-4), we can further rearrange this as

$$-\mathrm{tr}\left[ (\mathbf{T} + P\mathbf{I}) \cdot \nabla \mathbf{v} \right] + \sum_{B=1}^{N-1} \mathbf{j}_{(B)} \cdot \left[ \nabla \left( \mu_{(B)} - \mu_{(N)} \right) + \left( \overline{S}_{(B)} - \overline{S}_{(N)} \right) \nabla T \right.$$

$$\left. - \left( \mathbf{f}_{(B)} - \mathbf{f}_{(N)} \right) \right] + \sum_{B=1}^{N-1} \left( \mu_{(B)} - \mu_{(N)} \right) r_{(B)} + \frac{1}{T} \epsilon \cdot \nabla T \leq 0 \tag{8.4.3-7}$$

or

$$-\mathrm{tr}\left[(\mathbf{T} - P\mathbf{I}) \cdot \nabla\mathbf{v}\right] + cRT \sum_{B=1}^{N-1} \mathbf{j}_{(B)} \cdot \left(\frac{\mathbf{d}_{(B)}}{\rho_{(B)}} - \frac{\mathbf{d}_{(N)}}{\rho_{(N)}}\right)$$

$$+ \sum_{B=1}^{N-1} \left(\mu_{(B)} - \mu_{(N)}\right) r_{(B)} + \frac{1}{T}\boldsymbol{\epsilon} \cdot \nabla T \leq 0 \tag{8.4.3-8}$$

If the entropy inequality is to be satisfied for every physical problem, we can examine a particular case such that $\mathbf{v} = \mathbf{0}, r_{(A)} = 0$, and $\mathbf{d}_{(A)} = 0$ for $A = 1, \ldots, N-1$, then (8.4.3-8) reduces to

$$\frac{1}{T}\boldsymbol{\epsilon} \cdot \nabla T = -\frac{k}{T}\nabla T \cdot \nabla T$$
$$< 0 \tag{8.4.3-9}$$

or

$$k > 0 \tag{8.4.3-10}$$

In the next section, we will investigate whether (8.4.3-2) automatically satisfies (8.4.3-8) more generally.

For dense gases, liquids, and solids, we recommend using an empirical extension of (8.4.3-2).

### 8.4.4 Mass Flux Vector

The most useful discussion of multicomponent diffusion has been given by Curtiss [1968; see also Condiff (1969)], who worked in the context of the Chapman–Enskog solution of the Boltzmann equation (the kinetic theory of dilute gases) to arrive at

$$\mathbf{j}_{(A)} = -D_{(A)}^T \nabla \ln T - \rho_{(A)} \sum_{B=1}^{N} \tilde{D}_{(AB)}\mathbf{d}_{(B)}$$

$$= -D_{(A)}^T \nabla \ln T - \rho_{(A)} \sum_{B=1}^{N-1} \left(\tilde{D}_{(AB)} - \tilde{D}_{(AN)}\right)\mathbf{d}_{(B)} \tag{8.4.4-1}$$

where the coefficients $\tilde{D}_{(AB)}$ are symmetric. Since

$$\sum_{A=1}^{N} \mathbf{j}_{(A)} = 0 \tag{8.4.4-2}$$

we find that

$$\sum_{A=1}^{N} D_{(A)}^T = 0 \tag{8.4.4-3}$$

and

$$\sum_{A=1}^{N} \omega_{(A)}\tilde{D}_{(AB)} = 0 \tag{8.4.4-4}$$

It is easily shown that (8.4.4-1) automatically satisfies the principles of determinism, of local action, and of frame indifference (Section 2.3.1). To examine the restrictions of the entropy inequality (8.4.3-8), it will be convenient to write (8.4.3-2) as

$$\epsilon = -k\nabla T - cRT \sum_{A=1}^{N-1} \alpha_{(0A)} \frac{\mathbf{d}_{(A)}}{\rho_{(A)}} \tag{8.4.4-5}$$

and (8.4.4-1) as

$$\mathbf{j}_{(A)} = -\alpha_{(A0)} \nabla \ln T - cRT \sum_{B=1}^{N-1} \alpha_{(AB)} \left( \frac{\mathbf{d}_{(B)}}{\rho_{(B)}} - \frac{\mathbf{d}_{(N)}}{\rho_{(N)}} \right) \tag{8.4.4-6}$$

Here

$$\alpha_{(0A)} = \alpha_{(A0)}$$
$$\equiv D_{(A)}^T$$
$$\alpha_{(AB)} \equiv \frac{1}{cRT} \rho_{(A)}\rho_{(B)} \tilde{D}_{(AB)} \tag{8.4.4-7}$$

If the entropy inequality is to be satisfied for every physical problem, we can examine a particular case such that $\mathbf{v} = \mathbf{0}$ and $r_{(A)} = 0$. For behavior described by (8.4.3-2) and (8.4.4-1), the entropy inequality (8.4.3-8) reduces to

$$\frac{\alpha_{(00)}}{T^2} \nabla T \cdot \nabla T + cR \sum_{A=1}^{N-1} \left( \alpha_{(0A)} + \alpha_{(A0)} \right) \left( \frac{\mathbf{d}_{(A)}}{\rho_{(A)}} - \frac{\mathbf{d}_{(N)}}{\rho_{(N)}} \right) \cdot \nabla T$$
$$+ (cRT)^2 \sum_{A=1}^{N-1} \sum_{B=1}^{N-1} \alpha_{(AB)} \left( \frac{\mathbf{d}_{(A)}}{\rho_{(A)}} - \frac{\mathbf{d}_{(N)}}{\rho_{(N)}} \right) \cdot \left( \frac{\mathbf{d}_{(B)}}{\rho_{(B)}} - \frac{\mathbf{d}_{(N)}}{\rho_{(N)}} \right)$$
$$> 0 \tag{8.4.4-8}$$

A necessary condition that the entropy inequality (8.4.3-8) be satisfied automatically is that, in addition to (8.4.3-10), the symmetric coefficients $\alpha_{(AB)}$ ($A, B = 0, \ldots, N - 1$) must form a positive matrix (Hoffman and Kunze 1961, p. 251). This in turn means that $\alpha_{(AB)}$ and $\tilde{D}_{(AB)}$ ($A, B = 1, \ldots, N - 1$) must form positive matrices as well. If these matrices are not positive, the entropy inequality places constraints on the processes that are permitted, and the entropy inequality must be checked for each process. As has been our practice previously, we have written strict inequalities here, since we are particularly concerned with dynamic problems.

### Generalized Maxwell–Stefan Equation

The kinetic theory of dilute gases has also been used to derive the *generalized Maxwell–Stefan* equations (Curtiss and Hirschfelder 1949; Curtiss 1968)

$$\mathbf{d}_{(A)} - \sum_{\substack{B=1 \\ B \neq A}}^{N} \frac{x_{(A)}x_{(B)}}{\mathcal{D}_{(AB)}} \left( \frac{D_{(B)}^T}{\rho_{(B)}} - \frac{D_{(A)}^T}{\rho_{(A)}} \right) \nabla \ln T$$

$$= \sum_{\substack{B=1 \\ B \neq A}}^{N} \frac{x_{(A)}x_{(B)}}{\mathcal{D}_{(AB)}} \left( \mathbf{v}_{(B)} - \mathbf{v}_{(A)} \right)$$

$$= \sum_{\substack{B=1 \\ B \neq A}}^{N} \frac{x_{(A)}x_{(B)}}{\mathcal{D}_{(AB)}} \left( \frac{\mathbf{n}_{(B)}}{\rho_{(B)}} - \frac{\mathbf{n}_{(A)}}{\rho_{(A)}} \right)$$

$$= \sum_{\substack{B=1 \\ B \neq A}}^{N} \frac{x_{(A)}x_{(B)}}{\mathcal{D}_{(AB)}} \left( \frac{\mathbf{j}_{(B)}}{\rho_{(B)}} - \frac{\mathbf{j}_{(A)}}{\rho_{(A)}} \right)$$

$$= \sum_{\substack{B=1 \\ B \neq A}}^{N} \frac{1}{c\mathcal{D}_{(AB)}} \left( x_{(A)}\mathbf{N}_{(B)} - x_{(B)}\mathbf{N}_{(A)} \right)$$

$$= \sum_{\substack{B=1 \\ B \neq A}}^{N} \frac{1}{c\mathcal{D}_{(AB)}} \left( x_{(A)}\mathbf{J}^{\diamond}_{(B)} - x_{(B)}\mathbf{J}^{\diamond}_{(A)} \right) \tag{8.4.4-9}$$

For dilute gases, the $\mathcal{D}_{(AB)}$ are the *binary diffusion coefficients* (see Section 8.4.6). More generally, for dense gases, liquids and solids, they are empirical coefficients, referred to as the *Maxwell–Stefan diffusion coefficients*.

Equation (8.4.4-9) can also be derived directly from (8.4.4-1). Here I follow Curtiss and Bird (1998), although Merk (1959) gave a very similar development starting with (8.4.4-1) in terms of the Curtiss and Hirschfelder (1949) and Hirschfelder et al. (1954, Eq. 11.2-33) diffusion coefficients.

Let us begin by defining

$$\mathbf{v}'_{(A)} \equiv \mathbf{v}_{(A)} + \frac{D^{T}_{(A)}}{\rho_{(A)}} \nabla \ln T \tag{8.4.4-10}$$

which permits us to express (8.4.4-1) as

$$\mathbf{v}'_{(A)} = - \sum_{B=1}^{N} \tilde{D}_{(AB)}\mathbf{d}_{(B)} \tag{8.4.4-11}$$

This permits us to write

$$x_{(A)}x_{(C)}\mathbf{v}'_{(A)} = -x_{(A)}x_{(C)}\mathbf{v}_{(A)} \sum_{B=1}^{N} \tilde{D}_{(AB)}\mathbf{d}_{(B)} \tag{8.4.4-12}$$

as well as

$$x_{(A)}x_{(C)}\mathbf{v}'_{(C)} = -x_{(A)}x_{(C)}\mathbf{v}_{(A)} \sum_{B=1}^{N} \tilde{D}_{(CB)}\mathbf{d}_{(B)} \tag{8.4.4-13}$$

If we multiply both equations by $1/\mathcal{D}_{(AC)}$ and subtract one from the other, we arrive at

$$\frac{x_{(A)}x_{(C)}}{\mathcal{D}_{(AC)}} \left( \mathbf{v}'_{(C)} - \mathbf{v}'_{(A)} \right) = - \sum_{B=1}^{N} \frac{x_{(A)}x_{(C)}}{\mathcal{D}_{(AC)}} \left( \tilde{D}_{(CB)} - \tilde{D}_{(AB)} \right) \mathbf{d}_{(B)} \tag{8.4.4-14}$$

or

$$\sum_{\substack{C=1 \\ C \neq A}}^{N} \frac{x_{(A)}x_{(C)}}{\mathcal{D}_{(AC)}} \left( \mathbf{v}'_{(C)} - \mathbf{v}'_{(A)} \right) = - \sum_{B=1}^{N} \left[ \sum_{\substack{C=1 \\ C \neq A}}^{N} \frac{x_{(A)}x_{(C)}}{\mathcal{D}_{(AC)}} \left( \tilde{D}_{(CB)} - \tilde{D}_{(AB)} \right) \right] \mathbf{d}_{(B)} \tag{8.4.4-15}$$

If we define the $\mathcal{D}_{(AC)}$ to be such that

$$\sum_{\substack{C=1 \\ C \neq A}}^{N} \frac{x_{(A)}x_{(C)}}{\mathcal{D}_{(AC)}} \left( \tilde{D}_{(CB)} - \tilde{D}_{(AB)} \right) = -\delta_{(AB)} + \omega_{(A)} \tag{8.4.4-16}$$

Equation (8.4.4-15) becomes, in view of (8.4.3-5),

$$\sum_{\substack{C=1 \\ C \neq A}}^{N} \frac{x_{(A)}x_{(C)}}{\mathcal{D}_{(AC)}} \left( \mathbf{v}'_{(C)} - \mathbf{v}'_{(A)} \right) = \mathbf{d}_{(A)} \tag{8.4.4-17}$$

or the generalized Maxwell–Stefan equations (8.4.4-9).

### Symmetry of the Maxwell–Stefan Coefficients

The Maxwell–Stefan coefficients $\mathcal{D}_{(AB)}$ are symmetric. To prove this, we will require several relations.

We will define

$$\hat{D}_{(AB)} \equiv -\omega_{(A)} \left( \tilde{D}_{(AB)} - \tilde{D}_{(AA)} \right) \tag{8.4.4-18}$$

which implies

$$\omega_{(A)} \tilde{D}_{(AB)} = -\hat{D}_{(AB)} + \sum_{\substack{C=1 \\ C \neq A}}^{N} \omega_{(C)} \hat{D}_{(AC)} \tag{8.4.4-19}$$

Because $\tilde{D}_{(AB)}$ is nonsingular, we can introduce its inverse, $\hat{D}_{(AB)}^{-1}$:

$$\sum_{B=1}^{N} \hat{D}_{(CB)}^{-1} \hat{D}_{(BD)} = \delta_{(CD)} \tag{8.4.4-20}$$

We will define $\mathcal{D}_{(AA)}$ by requiring

$$\sum_{C=1}^{N} \frac{x_{(A)}x_{(C)}}{\mathcal{D}_{(AC)}} = 0 \tag{8.4.4-21}$$

Equation (8.4.4-22) allows us to write (8.4.4-16) as

$$\sum_{C=1}^{N} \frac{x_{(A)}x_{(C)}}{\mathcal{D}_{(AC)}} \tilde{D}_{(CB)} = -\delta_{(AB)} + \omega_{(A)} \tag{8.4.4-22}$$

For simplicity, we will introduce

$$\tilde{C}_{(AB)} \equiv \frac{x_{(A)}x_{(B)}}{\mathcal{D}_{(AB)}} \tag{8.4.4-23}$$

and write (8.4.4-21) and (8.4.4-22) as

$$\sum_{C=1}^{N} \tilde{C}_{(AC)} = 0 \tag{8.4.4-24}$$

and

$$\sum_{C=1}^{N} \tilde{C}_{(AC)} \tilde{D}_{(CB)} = -\delta_{(AB)} + \omega_{(A)} \qquad (8.4.4\text{-}25)$$

Multiplying both sides of (8.4.4-25) by $\omega_{(B)}$ and using (8.4.4-19) and (8.4.4-24), we find

$$\sum_{C=1}^{N} \tilde{C}_{(AC)} \left( -\hat{D}_{(BC)} + \sum_{\substack{D=1 \\ D \neq B}}^{N} \omega_{(D)} \hat{D}_{(BD)} \right)$$

$$= -\delta_{(AB)}\omega_{(B)} + \omega_{(A)}\omega_{(B)} - \sum_{C=1}^{N} \tilde{C}_{(AC)} \hat{D}_{(BC)} = \delta_{(AB)}\omega_{(B)} + \omega_{(A)}\omega_{(B)} \qquad (8.4.4\text{-}26)$$

Multiplying this by $\hat{D}_{(DB)}^{-1}$ and, summing over $B$, we have

$$-\sum_{B=1}^{N} \hat{D}_{(DB)}^{-1} \sum_{C=1}^{N} \tilde{C}_{(AC)} \tilde{D}_{(BC)} = -\sum_{B=1}^{N} \hat{D}_{(DB)}^{-1} \delta_{(AB)}\omega_{(B)} + \sum_{B=1}^{N} \hat{D}_{(DB)}^{-1} \omega_{(A)}\omega_{(B)}$$

$$-\tilde{C}_{(AD)} = -\omega_{(A)} \hat{D}_{(DA)}^{-1} + \omega_{(A)} \sum_{B=1}^{N} \omega_{(B)} \hat{D}_{(DB)}^{-1} \qquad (8.4.4\text{-}27)$$

$$\tilde{C}_{(AD)} = \omega_{(A)} \left( \hat{D}_{(DA)}^{-1} - \sum_{B=1}^{N} \omega_{(B)} \hat{D}_{(DB)}^{-1} \right)$$

Since $\tilde{D}_{(AB)}$ is symmetric,

$$\omega_{(A)}\omega_{(B)} \tilde{D}_{(AB)} = \omega_{(B)}\omega_{(A)} \tilde{D}_{(BA)} \qquad (8.4.4\text{-}28)$$

Using (8.4.4-19), we can express this as

$$\omega_{(B)} \left( -\hat{D}_{(AB)} + \sum_{\substack{C=1 \\ C \neq A}}^{N} \omega_{(C)} \hat{D}_{(AC)} \right) = \omega_{(A)} \left( -\hat{D}_{(BA)} + \sum_{\substack{C=1 \\ C \neq B}}^{N} \omega_{(C)} \hat{D}_{(BC)} \right) \qquad (8.4.4\text{-}29)$$

Multiplying this by $\hat{D}_{(DB)}^{-1}$ and summing over $B$, we discover that

$$\sum_{B=1}^{N} \left( -\omega_{(B)} \hat{D}_{(DB)}^{-1} \hat{D}_{(AB)} + \omega_{(B)} \hat{D}_{(DB)}^{-1} \sum_{\substack{C=1 \\ C \neq A}}^{N} \omega_{(C)} \hat{D}_{(AC)} \right)$$

$$= \sum_{B=1}^{N} \left( -\omega_{(A)} \hat{D}_{(DB)}^{-1} \hat{D}_{(BA)} + \omega_{(A)} \hat{D}_{(CB)}^{-1} \sum_{\substack{C=1 \\ C \neq B}}^{N} \omega_{(C)} \hat{D}_{(BC)} \right)$$

$$= -\omega_{(A)} \delta_{(DA)} + \omega_{(A)} \sum_{B=1}^{N} \hat{D}_{(DB)}^{-1} \sum_{\substack{C=1 \\ C \neq B}}^{N} \omega_{(C)} \hat{D}_{(BC)}$$

$$= -\omega_{(A)} \delta_{(DA)} + \omega_{(A)} \omega_{(D)} \qquad (8.4.4\text{-}30)$$

Multiplying this last by $\hat{D}_{(EA)}^{-1}$ and summing over $A$, we have

$$\sum_{A=1}^{N} \hat{D}_{(EA)}^{-1} \sum_{B=1}^{N} \omega_{(B)} \hat{D}_{(DB)}^{-1} \left( -\hat{D}_{(AB)} + \sum_{\substack{C=1 \\ C \neq A}}^{N} \omega_{(C)} \hat{D}_{(AC)} \right)$$

$$= \sum_{A=1}^{N} \omega_{(A)} \left( -\delta_{(DA)} + \omega_{(D)} \right) \hat{D}_{(EA)}^{-1} \qquad (8.4.4\text{-}31)$$

or

$$-\omega_{(E)} \hat{D}_{(DE)}^{-1} + \sum_{B=1}^{N} \omega_{(B)} \omega_{(E)} \hat{D}_{(DB)}^{-1} = -\omega_{(D)} \hat{D}_{(ED)}^{-1} + \omega_{(D)} \sum_{A=1}^{N} \omega_{(A)} \omega_{(D)} \hat{D}_{(EA)}^{-1} \qquad (8.4.4\text{-}32)$$

Recognizing (8.4.4-27), we conclude that

$$\tilde{C}_{(ED)} = \tilde{C}_{(DE)}$$

$$\mathcal{D}_{(ED)} = \mathcal{D}_{(DE)} \qquad (8.4.4\text{-}33)$$

This proof is a slight variation on one given by Curtiss and Bird (1998).

## Relating $\tilde{C}_{(AB)}$ and $\tilde{D}_{(AB)}$

Depending upon the application, for an $N$-component system, we must be able to express either the $\tilde{D}_{(AB)}$ as functions of the $\tilde{C}_{(AB)}$ or the $\tilde{C}_{(AB)}$ as functions of $\tilde{D}_{(AB)}$. Let's look at each case separately.

In view of symmetry, for a $N$-component system, there are $N + N(N-1)/2$ independent components of $\tilde{D}_{(AB)}$. These are found in terms of the $\tilde{C}_{(AB)}$ by solving simultaneously the $N$ equations (8.4.4-4) and the $N(N-1)/2$ equations (8.4.4-25) for $A, B = 1, \ldots, N$ and $B > A$. The results can be simplified using (8.4.4-24) to eliminate the diagonal components of $\tilde{C}_{(AB)}$. For $N \geq 3$, I recommend that you use Mathematica (1993) in solving and simplifying these equations. In this way, we find that, $N = 2$, for

$$\tilde{D}_{(AA)} = \frac{\omega_{(B)}^2}{\tilde{C}_{(AB)}}$$

$$\tilde{D}_{(BB)} = \frac{\omega_{(A)}^2}{\tilde{C}_{(AB)}} \qquad (8.4.4\text{-}34)$$

$$\tilde{D}_{(AB)} = \tilde{D}_{(BA)}$$

$$= -\frac{\omega_{(A)} \omega_{(B)}}{\tilde{C}_{(AB)}}$$

and for $N = 3$,

$$\tilde{D}_{(AA)} = \frac{1}{\Delta} \left[ \omega_{(C)}^2 \tilde{C}_{(AB)} + \omega_{(B)}^2 \tilde{C}_{(AC)} + \left( \omega_{(B)} + \omega_{(C)} \right)^2 \tilde{C}_{(BC)} \right]$$

$$\tilde{D}_{(AB)} = \frac{1}{\Delta} \left[ \omega_{(C)}^2 \tilde{C}_{(AB)} - \omega_{(B)} \left( \omega_{(A)} + \omega_{(C)} \right) \tilde{C}_{(AC)} \right.$$

$$\left. - \omega_{(A)} \left( \omega_{(B)} + \omega_{(C)} \right) \tilde{C}_{(BC)} \right] \qquad (8.4.4\text{-}35)$$

$$\Delta \equiv \tilde{C}_{(AB)} \tilde{C}_{(AC)} + \tilde{C}_{(AB)} \tilde{C}_{(BC)} + \tilde{C}_{(AC)} \tilde{C}_{(BC)}$$

The other coefficients for the ternary system can be obtained by an even permutation of $ABC$. These results are in agreement with those presented by Curtiss (1968) and Curtiss and Bird (1998).

In a similar manner, for an $N$-component system, there are $N + N(N - 1)/2$ independent components of $\tilde{C}_{(AB)}$. These are found in terms of the $\tilde{D}_{(AB)}$ by solving simultaneously the $N$ equations (8.4.4-24) and the $N(N - 1)/2$ equations (8.4.4-25) for $A$, $B = 1, \ldots, N$ and $B > A$. The results can be simplified using (8.4.4-4) to eliminate the diagonal components of $\tilde{D}_{(AB)}$. For $N \geq 3$, I recommend that you use Mathematica (1993) in solving and simplifying these equations. In this way, we find that, $N = 2$, for

$$
\begin{aligned}
\tilde{C}_{AB} &= \frac{\omega_{(B)}^2}{\tilde{D}_{(AA)}} \\
&= \frac{\omega_{(A)}^2}{\tilde{D}_{(BB)}} \\
&= -\frac{\omega_{(A)}\omega_{(B)}}{\tilde{D}_{(AB)}}
\end{aligned}
\tag{8.4.4-36}
$$

and for $N = 3$,

$$
\tilde{C}_{(AB)} = -\frac{\omega_{(A)}\omega_{(B)}\left(\tilde{D}_{(AB)} + \tilde{D}_{(CC)} - \tilde{D}_{(AC)} - \tilde{D}_{(BC)}\right)}{\tilde{D}_{(AB)}\tilde{D}_{(CC)} - \tilde{D}_{(AC)}\tilde{D}_{(BC)}}
\tag{8.4.4-37}
$$

The other coefficients for the ternary system can be obtained by an even permutation of $ABC$. These results are in agreement with those derived by Curtiss and Bird (1998).

From (8.4.4-34), we see that for binary mixtures the Maxwell–Stefan coefficients $\mathcal{D}_{(AB)}$ can be referred to as binary diffusion coefficients. However, for a multicomponent mixture, one should not expect the Maxwell–Stefan coefficients appearing in (8.4.4-9) to be the same coefficients observed in binary diffusion (see Section 8.4.5), unless the mixture forms a dilute gas (Curtiss and Hirschfelder 1949, Curtiss 1968).

## Discussion

There are several important points to keep in mind as you begin examining multicomponent diffusion problems in Chapter 9.

- The Maxwell–Stefan coefficients $\mathcal{D}_{(AB)}$ should be interpreted as binary diffusion coefficients only for binary solutions or multicomponent dilute gases.
- The Curtiss diffusion coefficients $\tilde{D}_{(AB)}$ (perhaps expressed in terms of the $\mathcal{D}_{(AB)}$) should not be expected to automatically satisfy the entropy inequality (8.4.4-8) except for dilute gas mixtures. The expectation is that, for dilute gas mixtures, the kinetic-theory arguments ensure that the entropy inequality is satisfied. For multicomponent diffusion in dense gases, liquids, and solids, one should expect to check the entropy inequality (8.4.4-8) for each process.
- In solving problems, the Curtiss equation (8.4.4-1) will generally be the preferred description of behavior. The literature has placed more emphasis on the Maxwell–Stefan equations (8.4.4-9) because the binary diffusivities $\mathcal{D}_{(AB)}$ are relatively easy to estimate for dilute gases.
- Alternative forms of the Curtiss equation (8.4.4-1) in terms of molar fluxes are perhaps most easily derived by inverting the Maxwell–Stefan equations as suggested in Exercise 8.4.4-2).

**Exercise 8.4.4-1** *More about the generalized Maxwell–Stefan equations*    Rearrange (8.4.3-4) as

$$
\mathbf{d}_{(A)} = x_{(A)} \sum_{\substack{B=1 \\ B \neq A}}^{N} \left( \frac{\partial \ln a_{(A)}^{(m)}}{\partial x_{(B)}} \right)_{T, P, x_{(C)} \, (C \neq A, B)} \nabla x_{(B)}
$$

$$
+ \frac{x_{(A)} M_{(A)}}{RT} \left[ \left( \overline{V}_{(A)} - \frac{1}{\rho} \right) \nabla P - \mathbf{f}_{(A)} + \sum_{B=1}^{N} \omega_{(B)} \mathbf{f}_{(B)} \right]
$$

$$
= -x_{(A)} \sum_{\substack{B=1 \\ B \neq A}}^{N} \left( \frac{\partial \ln a_{(A)}^{(m)}}{\partial x_{(A)}} \right)_{T, P, x_{(C)} \, (C \neq A, B)} \nabla x_{(B)}
$$

$$
+ \frac{x_{(A)} M_{(A)}}{RT} \left[ \left( \overline{V}_{(A)} - \frac{1}{\rho} \right) \nabla P - \mathbf{f}_{(A)} + \sum_{B=1}^{N} \omega_{(B)} \mathbf{f}_{(B)} \right]
$$

$$
= - \sum_{\substack{B=1 \\ B \neq A}}^{N} \left( \frac{\partial \ln a_{(A)}^{(m)}}{\partial \ln x_{(A)}} \right)_{T, P, x_{(C)} \, (C \neq A, B)} \nabla x_{(B)}
$$

$$
+ \frac{x_{(A)} M_{(A)}}{RT} \left[ \left( \overline{V}_{(A)} - \frac{1}{\rho} \right) \nabla P - \mathbf{f}_{(A)} + \sum_{B=1}^{N} \omega_{(B)} \mathbf{f}_{(B)} \right]
$$

where we have defined the *relative activity* (*on a molar basis*) as

$$
a_{(A)}^{(m)} \equiv \exp \frac{\mu_{(A)}^{(m)} - \mu_{(A)}^{(m)\circ}}{RT}
$$

By $\mu_{(A)}^{(m)\circ}$, we mean the chemical potential (on a molar basis) for pure species $A$ at the same temperature and pressure.

For ideal solutions,

$$
\left( \frac{\partial \ln a_{(A)}^{(m)}}{\partial \ln x_{(A)}} \right)_{T, P, x_{(C)} \, (C \neq A, B)} = 1
$$

and the result above reduces to

$$
\mathbf{d}_{(A)} = \nabla x_{(A)} + \frac{x_{(A)} M_{(A)}}{RT} \left[ \left( \overline{V}_{(A)} - \frac{1}{\rho} \right) \nabla P - \mathbf{f}_{(A)} + \sum_{B=1}^{N} \omega_{(B)} \mathbf{f}_{(B)} \right]
$$

This means that, for ideal solutions, the generalized Maxwell–Stefan equation (8.4.4-9) becomes

$$
\nabla x_{(A)} + \frac{x_{(A)} M_{(A)}}{RT} \left[ \left( \overline{V}_{(A)} - \frac{1}{\rho} \right) \nabla P - \mathbf{f}_{(A)} + \sum_{B=1}^{N} \omega_{(B)} \mathbf{f}_{(B)} \right]
$$

$$
= \sum_{B=1}^{N} \frac{1}{c \mathcal{D}_{(AB)}} \left( x_{(A)} \mathbf{N}_{(B)} - x_{(B)} \mathbf{N}_{(A)} \right) + \sum_{B=1}^{N} \frac{x_{(A)} x_{(B)}}{\mathcal{D}_{(AB)}} \left( \frac{D_{(B)}^{T}}{\rho_{(B)}} - \frac{D_{(A)}^{T}}{\rho_{(A)}} \right) \nabla \ln T
$$

When we further neglect any effect attributable to thermal, pressure, and forced diffusion, we have what are commonly referred to as the *Maxwell–Stefan* equations:

$$\nabla x_{(A)} = \sum_{B=1}^{N} \frac{1}{c\mathcal{D}_{(AB)}} \left( x_{(A)}\mathbf{N}_{(B)} - x_{(B)}\mathbf{N}_{(A)} \right)$$

**Exercise 8.4.4-2** *Alternative forms for the Curtiss (1968) equation (8.4.4-1)*    In the text we have worked with one form of the Curtiss (1968) equation (8.4.4-1). There are two ways in which alternative forms can be derived. The most obvious approach is to use the relations in Table 8.5.1-4 to arrange (8.4.4-1) in the desired form. However, sometimes it may be more convenient to invert the Maxwell–Stefan equation (8.4.4-9). In what follows, we will demonstrate the latter approach for a three-component system. The same approach can be used to invert these equations for an arbitrary number of species.

Let us begin with the Maxwell–Stefan equation (8.4.4-9) in the form

$$\nabla x_{(A)} = \sum_{\substack{B=1 \\ B \neq A}}^{N} \frac{1}{c\mathcal{D}_{(AB)}} \left( x_{(A)}\mathbf{J}_{(B)}^{\diamond} - x_{(B)}\mathbf{J}_{(A)}^{\diamond} \right)$$

For a three-component system consisting of species $A$, $B$, and $C$, write this in the form (Toor 1964a)

$$\mathbf{J}_{(A)}^{\diamond} = -c \left( \overline{D}_{(AA)} \nabla x_{(A)} + \overline{D}_{(AB)} \nabla x_{(B)} \right)$$

$$\mathbf{J}_{(B)}^{\diamond} = -c \left( \overline{D}_{(BA)} \nabla x_{(A)} + \overline{D}_{(BB)} \nabla x_{(B)} \right)$$

where

$$\overline{D}_{AA} \equiv \frac{\mathcal{D}_{(AC)} \left[ x_{(A)}\mathcal{D}_{(BC)} + \left( 1 - x_{(A)} \right) \mathcal{D}_{(AB)} \right]}{S}$$

$$\overline{D}_{AB} \equiv \frac{x_{(A)}\mathcal{D}_{(BC)} \left( \mathcal{D}_{(AC)} - \mathcal{D}_{(AB)} \right)}{S}$$

$$\overline{D}_{BA} \equiv \frac{x_{(B)}\mathcal{D}_{(AC)} \left( \mathcal{D}_{(BC)} - \mathcal{D}_{(AB)} \right)}{S}$$

$$\overline{D}_{BB} \equiv \frac{\mathcal{D}_{(BC)} \left[ x_{(B)}\mathcal{D}_{(AC)} + \left( 1 - x_{(B)} \right) \mathcal{D}_{(AB)} \right]}{S}$$

$$S \equiv x_{(A)}\mathcal{D}_{(BC)} + x_{(B)}\mathcal{D}_{(AC)} + x_{(C)}\mathcal{D}_{(AB)}$$

In executing this inversion, you may wish to use (A.2.1-12).

### 8.4.5  Mass Flux Vector in Binary Solutions

For a mixture consisting of two components, (8.4.4-1) reduces in view of (8.4.4-2) and (8.4.4-34) to

$$\mathbf{j}_{(A)} = -\mathbf{j}_{(A)}$$

$$= -\rho_{(A)} \left( \tilde{D}_{(AA)} - \tilde{D}_{(AB)} \right) \mathbf{d}_{(A)} - D_{(A)}^{T} \nabla \ln T$$

$$= -\frac{c^2}{\rho} M_{(A)} M_{(B)} \mathcal{D}_{(AB)} \mathbf{d}_{(A)} - D_{(A)}^{T} \nabla \ln T \tag{8.4.5-1}$$

From (8.4.3-4), we find that this may be put in the somewhat more useful form

$$
\mathbf{j}_{(A)} = - \left( \frac{c}{RT} \right) M_{(A)} M_{(B)} \mathcal{D}_{(AB)} \omega_{(A)} \left[ \left( \frac{\partial \mu_{(A)}}{\partial \omega_{(B)}} \right)_{T,\,P} \nabla \omega_{(B)} \right.
$$
$$
\left. + \left( \overline{V}_{(A)} - \frac{1}{\rho} \right) \nabla P - \omega_{(B)} (\mathbf{f}_{(A)} - \mathbf{f}_{(B)}) \right] - D_{(A)}^T \nabla \ln T \tag{8.4.5-2}
$$

This may also be written as

$$
\mathbf{j}_{(A)} = - c M_{(A)} M_{(B)} \mathcal{D}_{(AB)} \left[ \left( \frac{\partial \ln a_{(A)}}{\partial \ln \omega_{(A)}} \right)_{T,\,P} \nabla \omega_{(A)} + \frac{\omega_{(A)}}{RT} \left( \overline{V}_{(A)} - \frac{1}{\rho} \right) \nabla P \right.
$$
$$
\left. - \frac{\omega_{(A)} \omega_{(B)}}{RT} (\mathbf{f}_{(A)} - \mathbf{f}_{(B)}) \right] - D_{(A)}^T \nabla \ln T \tag{8.4.5-3}
$$

where we introduce the *relative activity* (*on a mass basis*) defined as

$$
a_{(A)} \equiv \exp \left( \frac{\mu_{(A)} - \mu_{(A)}^\circ}{RT} \right) \tag{8.4.5-4}
$$

By $\mu^\circ{}_{(A)}$, we mean the chemical potential (on a mass basis) for pure species $A$ at the same temperature and pressure.

Instead of (8.4.5-3), it is more common to express (8.4.5-2) as

$$
\mathbf{j}_{(A)} = - \left( \frac{c^2}{\rho} \right) M_{(A)} M_{(B)} \mathcal{D}_{(AB)} \left[ \left( \frac{\partial \ln a_{(A)}^{(m)}}{\partial \ln x_{(A)}} \right)_{T,P} \nabla x_{(A)} \right.
$$
$$
+ \frac{M_{(A)} x_{(A)}}{RT} \left( \frac{1}{M_{(A)}} \overline{V}_{(A)}^{(m)} - \frac{1}{\rho} \right) \nabla P
$$
$$
\left. - \frac{M_{(A)} x_{(A)} \omega_{(B)}}{RT} (\mathbf{f}_{(A)} - \mathbf{f}_{(B)}) \right] - D_{(A)}^T \nabla \ln T \tag{8.4.5-5}
$$

where we define the *relative activity* (*on a molar basis*) as

$$
a_{(A)}^{(m)} \equiv \exp \left( \frac{\mu_{(A)}^{(m)} - \mu_{(A)}^{(m)\circ}}{RT} \right) \tag{8.4.5-6}
$$

By $\mu_{(A)}^{(m)\circ}$, we mean the chemical potential (on a molar basis) for pure species $A$ at the same temperature and pressure. The principal advantage of working in terms of $a_{(A)}^{(m)}$ is that

$$
\text{for ideal solutions:} \quad \left( \frac{\partial \ln a_{(A)}^{(m)}}{\partial \ln x_{(A)}} \right)_{T,P} = 1 \tag{8.4.5-7}
$$

It is helpful to think of (8.4.5-5) as the sum of four terms:

$$
\mathbf{j}_{(A)} = \mathbf{j}_{(A)}^{(o)} + \mathbf{j}_{(A)}^{(P)} + \mathbf{j}_{(A)}^{(f)} + \mathbf{j}_{(A)}^{(T)} \tag{8.4.5-8}
$$

*Ordinary diffusion* is one of the most commonly discussed limiting cases of (8.4.5-5):

$$\mathbf{j}_{(A)}^{(o)} \equiv -\left(\frac{c^2}{\rho}\right) M_{(A)} M_{(B)} \mathcal{D}_{(AB)} \left(\frac{\partial \ln a_{(A)}^{(m)}}{\partial \ln x_{(A)}}\right)_{T,P} \nabla x_{(A)} \tag{8.4.5-9}$$

For an *ideal solution*, (8.4.5-7) indicates that this reduces to

$$\mathbf{j}_{(A)}^{(o)} = -\left(\frac{c^2}{\rho}\right) M_{(A)} M_{(B)} \mathcal{D}_{(AB)} \nabla x_{(A)} \tag{8.4.5-10}$$

For nonideal mixtures, it is common practice in the literature to write (8.4.5-9) as

$$\mathbf{j}_{(A)}^{(o)} = -\left(\frac{c^2}{\rho}\right) M_{(A)} M_{(B)} \mathcal{D}_{(AB)}^0 \nabla x_{(A)} \tag{8.4.5-11}$$

where we introduce

$$\mathcal{D}_{(AB)}^0 \equiv \left(\frac{\partial \ln a_{(A)}^{(m)}}{\partial \ln x_{(A)}}\right)_{T,P} \mathcal{D}_{(AB)} \tag{8.4.5-12}$$

Equation (8.4.5-11) is commonly referred to as *Fick's first law of binary diffusion.* Various equivalent forms of Fick's first law are presented in Table 8.5.1. Since much of the work that follows assumes Fick's first law, we give in Table 8.5.1 some important forms of the differential mass balance for species $A$ (Table 8.5.1-5) that are consistent with it. For the limiting case of constant $\rho$ and constant $\mathcal{D}_{(AB)}^0$, Table 8.5.1 shows the differential mass balance for species $A$ in the three principal coordinate systems.

*Pressure diffusion*

$$\mathbf{j}_{(A)}^{(P)} \equiv \frac{M_{(A)} x_{(A)}}{RT} \left(\frac{1}{M_{(A)}} \overline{V}_{(A)}^{(m)} - \frac{1}{\rho}\right) \nabla P \tag{8.4.5-13}$$

is significant in systems where there are very large pressure gradients. See Section 9.3.5.

*Forced diffusion*

$$\mathbf{j}_{(A)}^{(f)} \equiv -\frac{M_{(A)} x_{(A)} \omega_{(B)}}{RT} \left(\mathbf{f}_{(A)} - \mathbf{f}_{(B)}\right) \tag{8.4.5-14}$$

is always a consideration in aqueous electrolytes. See Section 9.3.6.

*Thermal diffusion*

$$\mathbf{j}_{(A)}^{(T)} \equiv -D_{(A)}^T \nabla \ln T \tag{8.4.5-15}$$

is not seen frequently, since processes in which there are large temperature gradients are less common. See Exercises 9.3.5-2 and 9.3.5-3.

**Exercise 8.4.5-1** Show that the various forms of Fick's first law presented in Table 8.5.1-7 are equivalent.

**Exercise 8.4.5-2** *An alternative form of the differential mass balance* Starting with the first equation of Table 8.5.1-8, determine that

$$c \left(\frac{\partial \ln M}{\partial t} + \nabla \ln M \cdot \mathbf{v}\right) = \text{div}(c \mathcal{D}_{(AB)}^0 \nabla \ln M) + \frac{r_{(A)}}{M_{(A)}} + \frac{r_{(B)}}{M_{(B)}}$$

where $M$ is the molar-averaged molecular weight defined in Table 8.5.1-1. A special case of this result has been suggested by Bedingfield and Drew (1950).

### 8.4.6    Mass Flux Vector: Limiting Cases in Ideal Solutions

The simplicity of the relationships developed in Section 8.4.5 explains the attention given in the literature to mass transfer in binary solutions. Our intuition suggests that there may be some limiting cases for which (8.4.5-6) may be extended for use in multicomponent ideal solutions in the form

$$
\mathbf{N}_{(A)} = -c\mathcal{D}_{(Am)} \left\{ \nabla x_{(A)} + \frac{M_{(A)}x_{(A)}}{RT} \left[ \left( \overline{V}_{(A)} - \frac{1}{\rho} \right) \nabla P - f_{(A)} + \sum_{B=1}^{N} \omega_{(B)}\mathbf{f}_{(B)} \right] \right\}
$$

$$
+ x_{(A)} \sum_{B=1}^{N} \mathbf{N}_{(B)} - c \sum_{B=1}^{N} x_{(A)}x_{(B)} \left( \frac{D_{(A)}^T}{\rho_{(A)}} - \frac{D_{(B)}^T}{\rho_{(B)}} \right) \nabla \ln T \tag{8.4.6-1}
$$

Here $\mathcal{D}_{(Am)}$ should be thought of as the diffusion coefficient for species $A$ in the multicomponent mixture. Our purpose in this section is to establish some conditions under which we are justified in writing this constitutive equation as well as expressions for $\mathcal{D}_{(Am)}$ in terms of the appropriate binary diffusion coefficients.

From Exercise 8.4.4-3, the generalized Stefan–Maxwell equation for ideal solutions is

$$
\nabla x_{(A)} + \frac{x_{(A)}M_{(A)}}{RT} \left[ \left( \overline{V}_{(A)} - \frac{1}{\rho} \right) \nabla P - \mathbf{f}_{(A)} + \sum_{B=1}^{N} \omega_{(B)}\mathbf{f}_{(B)} \right]
$$

$$
= \sum_{B=1}^{N} \frac{1}{c\mathcal{D}_{(AB)}} \left( x_{(A)}\mathbf{N}_{(B)} - x_{(B)}\mathbf{N}_{(A)} \right) + \sum_{B=1}^{N} \frac{x_{(A)}x_{(B)}}{\mathcal{D}_{(AB)}} \left( \frac{D_{(B)}^T}{\rho_{(B)}} - \frac{D_{(A)}^T}{\rho_{(A)}} \right) \nabla \ln T \tag{8.4.6-2}
$$

This suggests that we arrange (8.4.6-1) in the form

$$
\nabla x_{(A)} + \frac{M_{(A)}x_{(A)}}{RT} \left[ \left( \overline{V}_{(A)} - \frac{1}{\rho} \right) \nabla P - \mathbf{f}_{(A)} + \sum_{B=1}^{N} \omega_{(B)}\mathbf{f}_{(B)} \right]
$$

$$
= \frac{1}{c\mathcal{D}_{(Am)}} \left( -\mathbf{N}_{(A)} + x_{(A)} \sum_{B=1}^{N} \mathbf{N}_{(B)} \right)
$$

$$
- \frac{1}{\mathcal{D}_{(Am)}} \sum_{B=1}^{N} x_{(A)}x_{(B)} \left( \frac{D_{(A)}^T}{\rho_{(A)}} - \frac{D_{(B)}^T}{\rho_{(B)}} \right) \nabla \ln T \tag{8.4.6-3}
$$

From (8.4.6-2) and (8.4.6-3), it follows that

$$
\frac{1}{c\mathcal{D}_{(Am)}} \left( \mathbf{N}_{(A)} - x_{(A)} \sum_{B=1}^{N} \mathbf{N}_{(B)} \right) + \frac{1}{\mathcal{D}_{(Am)}} \sum_{B=1}^{N} x_{(A)}x_{(B)} \left( \frac{D_{(A)}^T}{\rho_{(A)}} - \frac{D_{(B)}^T}{\rho_{(B)}} \right) \nabla \ln T
$$

$$
= \sum_{B=1}^{N} \frac{1}{c\mathcal{D}_{(AB)}} \left( x_{(B)}\mathbf{N}_{(A)} - x_{(A)}\mathbf{N}_{(B)} \right) + \sum_{B=1}^{N} \frac{x_{(A)}x_{(B)}}{\mathcal{D}_{(AB)}} \left( \frac{D_{(A)}^T}{\rho_{(A)}} - \frac{D_{(B)}^T}{\rho_{(B)}} \right) \nabla \ln T \tag{8.4.6-4}
$$

Now let us examine some special cases.

For trace components $2, 3, \ldots, N$ in nearly pure species 1, Equation (8.4.6-4) simplifies to

$$
\frac{\mathbf{N}_{(A)} + \left( 1/M_{(A)} \right) D_{(A)}^T \nabla \ln T}{c\mathcal{D}_{(Am)}} \approx \frac{\mathbf{N}_{(A)} + \left( 1/M_{(A)} \right) D_{(A)}^T \nabla \ln T}{c\mathcal{D}_{(A1)}} \tag{8.4.6-5}
$$

In other words, (8.4.6-1) describes the mass flux vector for species $A$ in nearly pure component 1 when we interpret

$$\mathcal{D}_{(Am)} \approx \mathcal{D}_{(A1)} \tag{8.4.6-6}$$

This is a well-known result for ordinary diffusion (Bird et al. 1960, p. 571).

For ideal solutions in which all the binary diffusion coefficients are the same, (8.4.6-4) requires

$$\mathcal{D}_{(Am)} = \mathcal{D}_{(AB)} \tag{8.4.6-7}$$

This is also a well-known result for ordinary diffusion (Bird et al. 1960, p. 571).

For ideal solutions in which species $2, 3, \ldots, N$ all move with the same velocity (or are stationary) and in which thermal diffusion may be neglected, (8.4.6-4) may be arranged as follows:

$$\frac{x_{(1)}\mathbf{v}_{(1)} - x_{(1)}\sum_{B=1}^{N} x_{(B)}\mathbf{v}_{(B)}}{\mathcal{D}_{(1m)}} = \sum_{B=1}^{N} \frac{x_{(1)}x_{(B)}}{\mathcal{D}_{(1B)}} \left(\mathbf{v}_{(1)} - \mathbf{v}_{(B)}\right)$$

$$\frac{x_{(1)}\left(1 - x_{(1)}\right)\mathbf{v}_{(1)} - x_{(1)}\mathbf{v}_{(2)}\sum_{B=2}^{N} x_{(B)}}{\mathcal{D}_{(1m)}} = \left(\mathbf{v}_{(1)} - \mathbf{v}_{(2)}\right)\sum_{B=2}^{N} \frac{x_{(1)}x_{(B)}}{\mathcal{D}_{(1B)}}$$

$$\frac{\left(1 - x_{(1)}\right)\mathbf{v}_{(1)} - \mathbf{v}_{(2)}\sum_{B=2}^{N} x_{(B)}}{\mathcal{D}_{(1m)}} = \left(\mathbf{v}_{(1)} - \mathbf{v}_{(2)}\right)\sum_{B=2}^{N} \frac{x_{(B)}}{\mathcal{D}_{(1B)}} \tag{8.4.6-8}$$

$$\frac{\left(1 - x_{(1)}\right)\left(\mathbf{v}_{(1)} - \mathbf{v}_{(2)}\right)}{\mathcal{D}_{(1m)}} = \left(\mathbf{v}_{(1)} - \mathbf{v}_{(2)}\right)\sum_{B=2}^{N} \frac{x_{(B)}}{\mathcal{D}_{(1B)}}$$

$$\frac{1 - x_{(1)}}{\mathcal{D}_{(1m)}} = \sum_{B=2}^{N} \frac{x_{(B)}}{\mathcal{D}_{(1B)}}$$

It is clear that (8.4.6-1) again applies with the interpretation

$$\mathcal{D}_{(1m)} = \frac{1 - x_{(1)}}{\sum_{B=2}^{N} x_{(B)}/\mathcal{D}_{(1B)}} \tag{8.4.6-9}$$

This represents an extension of Wilke's (1950) result for ordinary diffusion.

Of these three limiting cases, the first involving the use of (8.4.6-1) and (8.4.6-6) to describe the diffusion of a trace contaminant in nearly pure species 1 is without question the most important. It is the basis for most of the work in the literature involving multicomponent solutions and will be used several times in this text.

For sufficiently dilute solutions,

$$\mathbf{v} \approx \mathbf{v}^{\circ} \tag{8.4.6-10}$$

In this limit, we can express (8.4.6-1) also as

$$\mathbf{n}_{(A)} = -\rho\mathcal{D}_{(Am)} \left\{\nabla\omega_{(A)} + \frac{M_{(A)}\omega_{(A)}}{RT}\left[\left(\overline{V}_{(A)} - \frac{1}{\rho}\right)\nabla P - \mathbf{f}_{(A)} + \sum_{B=1}^{N}\omega_{(B)}\mathbf{f}_{(B)}\right]\right\}$$

$$+ \omega_{(A)}\sum_{B=1}^{N}\mathbf{n}_{(B)} - \rho\sum_{B=1}^{N}\omega_{(A)}x_{(B)}\left(\frac{D_{(A)}^{T}}{\rho_{(A)}} - \frac{D_{(B)}^{T}}{\rho_{(B)}}\right)\nabla\ln T \tag{8.4.6-11}$$

### 8.4.7  Constitutive Equations for the Stress Tensor

In general, we should expect the stress tensor in a multicomponent material to be a function of the motions of all the species present in the material as well as the temperature distribution (Brown 1967, Müller 1968). Since there is little or no experimental evidence to guide us, we recommend current practice in engineering, which is to use the constitutive equations for $\mathbf{T}$ discussed in Sections 2.3.2 to 2.3.4, recognizing that all parameters should be functions of the local thermodynamic state variables: $T$, $\rho_{(1)}, \ldots, \rho_{(N)}$.

### 8.4.8  Rates of Reactions

To fully specify a system, we need constitutive equations for $r_{(A)}$ (Section 8.2.1), the rate at which mass of species $A$ is produced by homogeneous chemical reactions per unit volume ($A = 1, 2, \ldots, N$), and for $r_{(A)}^{(\sigma)}$, the rate at which mass of species $A$ is produced by heterogeneous chemical reactions per unit area ($A = 1, 2, \ldots, N$). Although these constitutive equations are the focus of an extensive kinetics and catalysis literature, less attention has been given to the implications of the entropy inequality.

In the preceding sections, our emphasis has been on constitutive equations that automatically satisfy the entropy inequality (8.4.3-9). This means that, with appropriate constitutive equations for $\epsilon$, $\mathbf{j}_{(B)}$, and $\mathbf{T}$, we need only ensure that

$$\sum_{B=1}^{N-1} \left( \mu_{(B)} - \mu_{(N)} \right) r_{(B)} \leq 0 \tag{8.4.8-1}$$

Let us assume that there are $K$ chemical reactions proceeding in a system, either simultaneously or in sequence:

$$\sum_{j=1}^{K} \sum_{B=1}^{N} \mu_{(B)} r_{(B,j)} \leq 0 \tag{8.4.8-2}$$

If we assume that the same species $N$ appears as either a reactant or a product in each reaction, then for each reaction $j$, the rates of consumption (or production) of species $A$ and $B$ are related by

$$\frac{r_{(A,j)}}{M_{(A)} \nu_{(A,j)}} = \frac{r_{(B,j)}}{M_{(B)} \nu_{(B,j)}} \tag{8.4.8-3}$$

and (8.4.8-2) can be rearranged as

$$\sum_{j=1}^{K} \sum_{B=1}^{N} \mu_{(B)} \frac{M_{(B)} \nu_{(B,j)}}{M_{(N)} \nu_{(N,j)}} r_{(N,j)} \leq 0 \tag{8.4.8-4}$$

or

$$\sum_{j=1}^{K} \sum_{B=1}^{N} \mu_{(B)}^{(m)} \frac{\nu_{(B,j)}}{M_{(N)} \nu_{(N,j)}} r_{(N,j)} \leq 0 \tag{8.4.8-5}$$

Here $\nu_{(A,j)}$ is the stoichiometric coefficient for species $A$ in chemical reaction $j$. The stoichiometric coefficient is taken to be a positive number for a species produced in the chemical reaction.

Let us define the *relative activity (on a molar basis)* of species $A$ (Prausnitz 1969, p. 20) as

$$\ln a_{(A)} \equiv \frac{\mu_{(A)}^{(m)} - \mu_{(A)}^{(m)o}}{RT} \tag{8.4.8-6}$$

where $\mu_{(A)}^{(m)o}$ is the chemical potential in the corresponding standard state. In terms of activities, we can express

$$\sum_{B=1}^{N} \nu_{(B,j)} \mu_{(B)}^{(m)} = \sum_{B=1}^{N} \nu_{(B,j)} \mu_{(B)}^{(m)o} + RT \sum_{B=1}^{N} \nu_{(B,j)} \ln a_{(B)}$$

$$= -RT \ln K^o + RT \sum_{B=1}^{N} \ln \left( a_{(B)}^{\nu_{(B,j)}} \right)$$

$$= RT \sum_{B=1}^{N} \ln \left( \frac{1}{K^o} \left[ \gamma_{(B)} x_{(B)} \right]^{\nu_{(B,j)}} \right) \tag{8.4.8-7}$$

in which we have introduced the *reaction equilibrium constant* $K^o$ (Smith and Van Ness 1987, p. 504),

$$\ln K^o \equiv -\frac{1}{RT} \sum_{B=1}^{N} \nu_{(B,j)} \mu_{(B)}^{(m)o} \tag{8.4.8-8}$$

and the *activity coefficient* (Prausnitz 1969, p. 186),

$$\gamma_{(B)} \equiv \frac{a_{(B)}}{x_{(B)}} \tag{8.4.8-9}$$

In view of (8.4.8-7), we can express (8.4.8-5) alternatively as

$$RT \sum_{j=1}^{K} \sum_{B=1}^{N} \ln \left( \frac{1}{K^o} \left[ \gamma_{(B)} x_{(B)} \right]^{\nu_{(B,j)}} \right) \frac{r_{(N,j)}}{M_{(N)} \nu_{(N,j)}} \leq 0 \tag{8.4.8-10}$$

There are two ways in which (8.4.8-4), (8.4.8-5), or (8.4.8-10) might be used.

Our preceding discussions suggest that we describe the dependence of the rates of reactions $r_{(N,j)}$ in such a way that (8.4.8-4), (8.4.8-5), or (8.4.8-10) is automatically defined. We explore this possibility in Exercise 8.4.8-1. Unfortunately, this is not the way the literature has developed.

Instead, empirical expressions for $r_{(N,j)}$ are employed that do not automatically satisfy (8.4.8-4), (8.4.8-5), or (8.4.8-10). In this context, we explore several examples below.

## A Single Reaction

For the reaction

$$A + B \rightarrow 2C \tag{8.4.8-11}$$

Equations (8.4.8-5) and (8.4.8-10) require

$$-\mu_{(A)}^{(m)} - \mu_{(B)}^{(m)} + 2\mu_{(C)}^{(m)} < 0$$

$$RT \ln \left( \frac{1}{K^o} \frac{\gamma_{(C)}^2}{\gamma_{(A)} \gamma_{(B)}} \frac{x_{(C)}^2}{x_{(A)} x_{(B)}} \right) < 0$$

or

$$\frac{\gamma_{(C)}^2}{\gamma_{(A)}\gamma_{(B)}} \frac{x_{(C)}^2}{x_{(A)}x_{(B)}} < K^o \tag{8.4.8-12}$$

Denbigh (1963, p. 136) has explored the implications of (8.4.8-12) for gas-phase reactions.

## A System of Biological Reactions: Case I

Let us consider the system of biological reactions[3]

$$PEP + ADP \rightarrow PYR + ATP \tag{8.4.8-13}$$

$$PYR + ATP \rightarrow PEP + AMP + P_i \tag{8.4.8-14}$$

$$2ADP \rightleftharpoons ATP + AMP \tag{8.4.8-15}$$

Here ATP, ADP, and AMP stand for adenosine 5′-triphosphate, adenosine 5′-diphosphate, and adenosine 5′-monophosphate; PEP and PYR are phosphoenolpyruvate and pyruvate; $P_i$ is inorganic phosphate. The compounds ATP, ADP, and AMP are known as the *energy currency*; they are the major compounds involved in energy transduction in a living cell. The energy released in catabolism (breakdown of nutrients) is often stored in ATP by forming an additional phosphate bond between $P_i$ and ADP, which yields ATP. The chemical energy stored in ATP can be released in biosynthesis by breaking down the phosphate bonds in ATP to form ADP or AMP. For this reason, the phosphorylation and hydrolysis of ATP are often coupled with other chemical transformations in a cell. In general, catabolic reactions yield ATP, which is used to drive biosynthesis, maintaining cell structure and some undefined functions.

Equation (8.4.8-13), which we will refer to as *reaction 1*, is catalyzed by an enzyme called pyruvate kinase, and it is a common reaction in carbohydrate metabolism in almost all living cells. The energy released through the conversion from PEP to PYR is stored by forming one additional phosphate bond in ADP.

Equation (8.4.8-14), which we will refer to as *reaction 2*, drives PYR back to PEP by spending two phosphate bonds in ATP. This reaction is catalyzed by an enzyme called PEP synthetase, which exists only in bacteria and some plants. Animals and humans have similar reactions but more complicated reaction paths. The simultaneous operation of these two reactions creates a cycle, whose net reaction is the hydrolysis of ADP to AMP and $P_i$. This

---

[3] In this discussion of biological reactions, we follow common practice in appearing to ignore the roles of water and protons. As will become evident as you read further, this practice is justified by the assumption that one is concerned with dilute aqueous solutions whose pH = 7 and by an arbitrary choice of standard states.

The standard state of a solute is a 1 M binary aqueous solution at the ambient temperature and pressure having pH = 7 (Lewis et al. 1961, p. 246). The solution is sufficiently dilute that for any solute $A$

$$\frac{a_{(A)}}{c_{(A)}} = \gamma_{(A)}$$
$$= 1$$

where $a_{(A)}$ is the activity, $c_{(A)}$ the molality, and $\gamma_{(A)}$ the corresponding activity coefficient.

The standard state of water is pure water at the existing temperature and pressure. Because we are concerned with dilute solutions, the activity of water in solution is approximately 1.

The standard state of the protons is water having pH = 7. Because we are concerned with dilute aqueous solutions whose pH = 7, the activity of the protons in solution is approximately 1.

kind of cycle is termed a *futile cycle*, because originally no apparent biological purposes were known. Cells have evolved sophisticated mechanisms to control these cycles, so that chemical energy is not wasted.

Equation (8.4.8-15), which we will refer to as *reaction 3*, is catalyzed by an enzyme called adenylate kinase. The reaction is thought to be fast, and the mass action ratio is close to equilibrium.

We will make two simplifying assumptions that appear to be consistent with experimental observation. As reactions (8.4.8-13) through (8.4.8-15) proceed, we will assume that (within the timescale of interest, the biosynthesis of adenosine is relatively slow):

$$c_{(ATP)} + c_{(ADP)} + c_{(AMP)} = C_1$$

$$= \text{a constant} \tag{8.4.8-16}$$

In addition, the concentration of $P_i$ is usually well buffered in the cell:

$$c_{(P_i)} = C_2$$

$$= \text{a constant} \tag{8.4.8-17}$$

Our objective here is to examine the constraint imposed by (8.4.8-5).

Noting that (8.4.8-13) will be referred to as reaction 1 and (8.4.8-14) as reaction 2 and taking species $N$ as ATP, we see from (8.4.8-5) and (8.4.8-10) that

$$\left(-\mu_{(PEP)}^{(m)} - \mu_{(ADP)}^{(m)} + \mu_{(PYR)}^{(m)} + \mu_{(ATP)}^{(m)}\right) \frac{r_{(ATP,1)}}{M_{(ATP)} \nu_{(ATP),1}}$$

$$+ \left(-\mu_{(PYR)}^{(m)} - \mu_{(ATP)}^{(m)} + \mu_{(PEP)}^{(m)} + \mu_{(AMP)}^{(m)} + \mu_{(P_i)}^{(m)}\right) \frac{r_{(ATP,2)}}{M_{(ATP)} \nu_{(ATP),2}}$$

$$< 0 \tag{8.4.8-18}$$

or

$$RT f_1 \frac{r_{(ATP,1)}}{M_{(ATP)} \nu_{(ATP),1}} + RT f_2 \frac{r_{(ATP,2)}}{M_{(ATP)} \nu_{(ATP),2}} < 0 \tag{8.4.8-19}$$

or

$$f_1 \frac{r_{(ATP,1)}}{M_{(ATP)} \nu_{(ATP),1}} + f_2 \frac{r_{(ATP,2)}}{M_{(ATP)} \nu_{(ATP),2}} < 0 \tag{8.4.8-20}$$

Here

$$f_1 \equiv \ln\left(\frac{\gamma_{(PYR)} \gamma_{(ATP)}}{K_1^o \gamma_{(PEP)} \gamma_{(ADP)}} \frac{c_{(PYR)} c_{(ATP)}}{c_{(PEP)} c_{(ADP)}}\right)$$

$$f_2 \equiv \ln\left(\frac{\gamma_{(PEP)} \gamma_{(AMP)} \gamma_{(P_i)}}{K_2^o \gamma_{(PYR)} \gamma_{(ATP)}} \frac{c_{(PEP)} c_{(AMP)} c_{(P_i)}}{c_{(PYR)} c_{(ATP)}}\right) \tag{8.4.8-21}$$

and

$$K_1^o \equiv \exp\left(-\frac{-\mu_{(PEP)}^{(m)o} - \mu_{(ADP)}^{(m)o} + \mu_{(PYR)}^{(m)o} + \mu_{(ATP)}^{(m)o}}{RT}\right)$$

$$K_2^o \equiv \exp\left[-\left(-\mu_{(PYR)}^{(m)o} - \mu_{(ATP)}^{(m)o} + \mu_{(PEP)}^{(m)o} + \mu_{(AMP)}^{(m)o} + \mu_{(P_i)}^{(m)o}\right)/(RT)\right] \tag{8.4.8-22}$$

We also have recognized that, because reaction 3 (8.4.8-15) is very fast,

$$-2\mu_{(ADP)}^{(m)} + \mu_{(ATP)}^{(m)} + \mu_{(AMP)}^{(m)} = 0 \tag{8.4.8-23}$$

Because ATP is being produced in reaction 1 (8.4.8-13) and consumed in reaction 2 (8.4.8-14),

$$\frac{r_{(ATP,1)}}{\nu_{(ATP,1)}} > 0$$
$$\frac{r_{(ATP,2)}}{\nu_{(ATP,2)}} > 0 \tag{8.4.8-24}$$

Since the relative magnitudes of these terms are unknown, a sufficient but not necessary condition that (8.4.8-20) be satisfied is that their coefficients be negative:

$$f_1 < 0$$
$$f_2 < 0 \tag{8.4.8-25}$$

In a similar fashion, (8.4.8-23) can be rewritten as

$$\ln\left(\frac{1}{K_3^o} \frac{\gamma_{(ATP)}\gamma_{(AMP)}}{\gamma_{(ADP)}^2} \frac{c_{(ATP)}c_{(AMP)}}{c_{(ADP)}^2}\right) = 0$$

or

$$\frac{\gamma_{(ATP)}\gamma_{(AMP)}}{\gamma_{(ADP)}^2} \frac{c_{(ATP)}c_{(AMP)}}{c_{(ADP)}^2} = K_3^o$$

or

$$\frac{c_{(ATP)}c_{(AMP)}}{c_{(ADP)}^2} = D_3 \tag{8.4.8-26}$$

where

$$K_3^o \equiv \exp\left(-\frac{-2\mu_{(ADP)}^{(m)o} + \mu_{(ATP)}^{(m)o} + \mu_{(AMP)}^{(m)o}}{RT}\right) \tag{8.4.8-27}$$

and

$$D_3 \equiv \frac{K_3^o \gamma_{(ADP)}^2}{\gamma_{(ATP)}\gamma_{(AMP)}} \tag{8.4.8-28}$$

Equations (8.4.8-16) and (8.4.8-26) can be solved simultaneously for $c_{(ADP)}$ and $c_{(AMP)}$ as functions of $c_{(ATP)}$:

$$c_{(ADP)} = \frac{1}{2}\left\{-D_3 c_{(ATP)} \mp \left[4C_1 D_3 c_{(ATP)} + \left(D_3^2 - 4D_3\right)c_{(ATP)}^2\right]^{1/2}\right\} \tag{8.4.8-29}$$

$$c_{(AMP)} = \frac{1}{2}\left\{2C_1 - 2c_{(ATP)} + D_3 c_{(ATP)} \pm \left[\left(-2C_1 + 2c_{(ATP)} - D_3 c_{(ATP)}\right)^2\right.\right.$$
$$\left.\left. - 4\left(C_1^2 - 2C_1 c_{(ATP)} + c_{(ATP)}^2\right)\right]^{1/2}\right\} \tag{8.4.8-30}$$

To illustrate these results, let us assign the following values to the parameters:

$$C_1 = c_{(ATP)} + c_{(ADP)} + c_{(AMP)}$$
$$\approx 2 \text{ mM}$$

$$C_2 = c_{(P_i)}$$
$$\approx 15 \text{ mM}$$

$$E_1 \equiv \frac{\gamma_{(PYR)}\gamma_{(ATP)}}{K_1^o \gamma_{(PEP)}\gamma_{(ADP)}}$$
$$\approx \frac{1}{K_1^o}$$
$$\approx 3 \times 10^{-6} \tag{8.4.8-31}$$

$$E_2 \equiv \frac{\gamma_{(PEP)}\gamma_{(AMP)}\gamma_{(Pi)}}{K_2^o \gamma_{(PYR)}\gamma_{(ATP)}}$$
$$\approx \frac{1}{K_2^o}$$
$$\approx 0.7$$

$$D_3 \approx K_3^o$$
$$\approx 1$$

$$\frac{c_{(PEP)}}{c_{(PYR)}} = 1$$

Only the second roots shown in (8.4.8-29) and (8.4.8-30) are physically meaningful (positive) for

$$c_{(ATP)} < C_1 \tag{8.4.8-32}$$

or

$$c_{(ATP)} < 2 \text{ mM} \tag{8.4.8-33}$$

Figures 8.4.8-1 and 8.4.8-2 show $c_{(ADP)}$ and $c_{(AMP)}$ as functions of $c_{(ATP)}$ with this limitation.

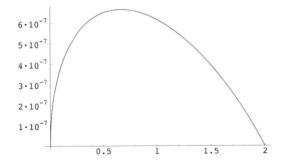

**Figure 8.4.8-1.** $c_{(ADP)}$ (mM) as a function of $c_{(ATP)}$ (mM) for the second root given in (8.4.8-29) and for the parameters shown in (8.4.8-31).

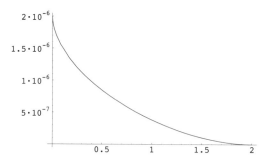

**Figure 8.4.8-2.** $c_{(AMP)}$ (mM) as a function of $c_{(ATP)}$ (mM) for the second root given in (8.4.8-30) and for the parameters shown in (8.4.8-31).

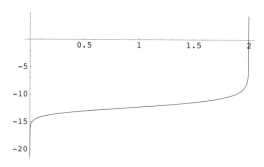

**Figure 8.4.8-3.** $f_1$ defined by (8.4.8-21) as a function of $c_{(ATP)}$ (mM) for the second set of roots in (8.4.8-29) and (8.4.8-30) and for the parameters shown in (8.4.8-31).

Figures 8.4.8-3 through 8.4.8-5 show $f_1$ and $f_2$ defined by (8.4.8-21) as functions of $c_{(ATP)}$ (mM) for the second set of roots in (8.4.8-29) and (8.4.8-30) and for the parameters shown in (8.4.8-31). For these parameters, inequality (8.4.8-20) is satisfied for $c_{(ATP)} > 0.018$ mM for all values of $r_{(ATP,1)}/v_{(ATP,1)}$ and $r_{(ATP,2)}/v_{(ATP,2)}$. For $c_{(ATP)} < 0.018$ mM, the reactions (8.4.8-13) through (8.4.8-15) are likely to proceed, so long as $r_{(ATP,2)}/v_{(ATP,2)}$ is not much larger than $r_{(ATP,1)}/v_{(ATP,1)}$ or the rate of reaction (8.4.8-14) is not much larger than the rate of reaction (8.4.8-13).

## A System of Biological Reactions: Case 2

Let us assume that the reactions (8.4.8-13) and (8.4.8-14) are reversed:

$$PEP + ADP \leftarrow PYR + ATP \tag{8.4.8-34}$$

$$PYR + ATP \leftarrow PEP + AMP + P_i \tag{8.4.8-35}$$

The discussion given for Case 1, above, can be repeated to conclude that sufficient but not necessary conditions for these reactions to proceed are

$$f_1 > 0$$
$$f_2 > 0 \tag{8.4.8-36}$$

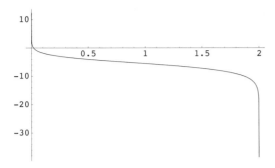

**Figure 8.4.8-4.** $f_2$ defined by (8.4.8-21) as a function of $c_{(ATP)}$ (mM) for the second set of roots in (8.4.8-29) and (8.4.8-30) and for the parameters shown in (8.4.8-31).

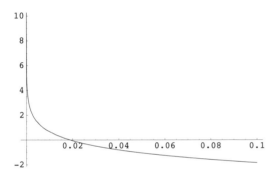

**Figure 8.4.8-5.** A further refinement of $f_2$ defined by (8.4.8-21) as a function of $c_{(ATP)}$ (mM) for the second set of roots in (8.4.8-29) and (8.4.8-30) and for the parameters shown in (8.4.8-31).

For the parameters shown in (8.4.8-31), inequality (8.4.8-20) cannot be satisfied for $c_{(ATP)} > 0.018$ mM. For $c_{(ATP)} < 0.018$ mM, the reactions (8.4.8-13) through (8.4.8-15) could proceed, if $r_{(ATP,2)}/v_{(ATP,2)}$ were sufficiently larger than $r_{(ATP,1)}/v_{(ATP,1)}$ or the rate of reaction (8.4.8-14) were sufficiently larger than the rate of reaction (8.4.8-13).

### A System of Biological Reactions: Case 3

Assume that only reaction (8.4.8-13) is reversed:

$$PEP + ADP \leftarrow PYR + ATP \tag{8.4.8-37}$$

$$PYR + ATP \rightarrow PEP + AMP + P_i \tag{8.4.8-38}$$

The discussion given for Case 1, above, can be repeated to conclude that sufficient but not necessary conditions for these reactions to proceed are

$$f_1 > 0$$
$$\tag{8.4.8-39}$$
$$f_2 < 0$$

For the parameters shown in (8.4.8-31), inequality (8.4.8-20) cannot be satisfied for $c_{(ATP)} < 0.018$ mM. For $c_{(ATP)} > 0.018$ mM, the reactions (8.4.8-13) through (8.4.8-15) could proceed, if $r_{(ATP,2)}/v_{(ATP,2)}$ were sufficiently larger than $r_{(ATP,1)}/v_{(ATP,1)}$ or the rate of reaction (8.4.8-14) were sufficiently larger than the rate of reaction (8.4.8-13).

### A System of Biological Reactions: Case 4

Now assume that only reaction (8.4.8-14) is reversed:

$$PEP + ADP \rightarrow PYR + ATP \tag{8.4.8-40}$$

$$PYR + ATP \leftarrow PEP + AMP + P_i \tag{8.4.8-41}$$

The discussion given for Case 1, above, can be repeated to conclude that sufficient but not necessary conditions for these reactions to proceed are

$$f_1 < 0$$
$$\tag{8.4.8-42}$$
$$f_2 > 0$$

For the parameters shown in (8.4.8-31), inequality (8.4.8-20) can be satisfied for $c_{(ATP)} <$ 0.018 mM, independent of the rates of reactions $r_{(ATP,1)}/v_{(ATP,1)}$ and $r_{(ATP,2)}/v_{(ATP,2)}$. For $c_{(ATP)} > 0.018$ mM, the reactions (8.4.8-13) through (8.4.8-15) could proceed, only if $r_{(ATP,1)}/v_{(ATP,1)}$ were sufficiently larger than $r_{(ATP,2)}/v_{(ATP,2)}$ or only if the rate of reaction (8.4.8-13) were sufficiently larger than the rate of reaction (8.4.8-14).

### Acknowledgment

This discussion of biological reactions was inspired by and written with the help of J. C. Liao. We are both grateful for the many questions answered by P. T. Eubank.

**Exercise 8.4.8-1** *Expressions for rates of reactions that automatically satisfy the entropy inequality*   Conclude that, if we could represent

$$\frac{r_{(N,j)}}{M_{(N)}v_{(N,j)}} = -k_{(N,j)} \sum_{B=1}^{N} \mu_{(B)}^{(m)} v_{(B,j)}$$

$$= -k_{(N,j)} RT \sum_{B=1}^{N} \ln \left( \frac{1}{K^o} \left[ \gamma_{(B)} x_{(B)} \right]^{v_{(B,j)}} \right)$$

the entropy inequality (8.4.8-5) or (8.4.8-10) would be automatically satisfied. Here $k_{(N,j)}$ is a positive scalar.

---

## 8.5   Summary

### 8.5.1   Summary of Useful Equations for Mass Transfer

In Tables 8.5.1-1 and 8.5.1-2, we explore the relations between the mass density $\rho_{(A)}$, the molar density $c_{(A)}$, the mass fraction $\omega_{(A)}$, and the mole fraction $x_{(A)}$.

In Tables 8.5.1-3 and 8.5.1-4, we summarize the variety of ways in which we can describe velocities and mass fluxes.

In Table 8.5.1-5, we present the various forms of the differential mass balance for species $A$. As a guide, the first equation in this table is presented for three specific coordinate systems in Table 8.5.1-6.

**Table 8.5.1-1.** Notation for concentrations

$$\rho \equiv \sum_{A=1}^{N} \rho_{(A)} = \text{mass density of solution}$$

$$\omega_{(A)} \equiv \frac{\rho_{(A)}}{\rho} = \text{mass fraction of component } A$$

$$c_{(A)} = \frac{\rho_{(A)}}{M_{(A)}} = \text{molar density of component } A$$

$$c \equiv \sum_{A=1}^{N} c_{(A)} = \text{molar density of solution}$$

$$x_{(A)} = \frac{c_{(A)}}{c} = \text{mole fraction of component } A$$

$$M = \frac{\rho}{c} = \sum_{A=1}^{N} x_{(A)} M_{(A)}$$
$$= \text{molar-averaged molecular weight of mixture}$$

$$\frac{1}{M} = \frac{c}{\rho} = \sum_{A=1}^{N} \frac{\omega_{(A)}}{M_{(A)}}$$

**Table 8.5.1-2.** Relations between mass and mole fractions

$$x_{(A)} = \frac{\omega_{(A)}/M_{(A)}}{\sum_{B=1}^{N} \omega_{(B)}/M_{(B)}}$$

$$\omega_{(A)} = \frac{x_{(A)} M_{(A)}}{\sum_{B=1}^{N} x_{(B)} M_{(B)}}$$

*For a binary system*:

$$dx_{(A)} = \frac{d\omega_{(A)}}{M_{(A)} M_{(B)} (\omega_{(A)}/M_{(A)} + \omega_{(B)}/M_{(B)})^2}$$

$$= \left(\frac{\rho}{c}\right)^2 \frac{d\omega_{(A)}}{M_{(A)} M_{(B)}}$$

$$d\omega_{(A)} = \frac{M_{(A)} M_{(B)} dx_{(A)}}{(x_{(A)} M_{(A)} + x_{(B)} M_{(B)})^2}$$

$$= \left(\frac{c}{\rho}\right)^2 M_{(A)} M_{(B)} dx_{(A)}$$

Various equivalent forms of Fick's first law for binary diffusion are shown in Table 8.5.1-7. Table 8.5.1-8 presents some important forms of the differential mass balance for species $A$ (see Table 8.5.1-5) that are consistent with it. For the limiting case of constant $\rho$ and constant $\mathcal{D}^0_{(AB)}$, Table 8.5.1-9 shows the differential mass balance for species $A$ in the three principal coordinate systems.

Two commonly used forms of the overall differential mass balance are given in Table 8.5.1-10.

**Table 8.5.1-3.** Various velocities and relations among them

$\mathbf{v}_{(A)} = $ Velocity of species $A$ with respect to the frame of reference

$$\mathbf{v} \equiv \sum_{A=1}^{N} \omega_{(A)} \mathbf{v}_{(A)}$$

$= $ mass-averaged velocity

$$\mathbf{v}^{\circ} \equiv \sum_{A=1}^{N} x_{(A)} \mathbf{v}_{(A)}$$

$= $ molar-averaged velocity

$\mathbf{u}_{(A)} \equiv \mathbf{v}_{(A)} - \mathbf{v}$

$= $ velocity of species $A$ relative to $\mathbf{v}$

$\mathbf{u}_{(A)}^{\circ} \equiv \mathbf{v}_{(A)} - \mathbf{v}^{\circ}$

$= $ velocity of species $A$ relative to $\mathbf{v}^{\circ}$

$$\mathbf{v} - \mathbf{v}^{\circ} = \sum_{A=1}^{N} \omega_{(A)} \left( \mathbf{v}_{(A)} - \mathbf{v}^{\circ} \right)$$

$$\mathbf{v}^{\circ} - \mathbf{v} = \sum_{A=1}^{N} x_{(A)} \left( \mathbf{v}_{(A)} - \mathbf{v} \right)$$

**Table 8.5.1-4.** Mass and molar fluxes

| Quantity | With respect to fixed frame | With respect to $\mathbf{v}$ | With respect to $\mathbf{v}^{\circ}$ |
|---|---|---|---|
| Velocity of species $A$ | $\mathbf{v}_{(A)}$ | $\mathbf{u}_{(A)} = \mathbf{v}_{(A)} - \mathbf{v}$ | $\mathbf{u}_{(A)}^{\circ} = \mathbf{v}_{(A)} - \mathbf{v}^{\circ}$ |
| Mass flux of species $A$ | $\mathbf{n}_{(A)} = \rho_{(A)} \mathbf{v}_{(A)}$ | $\mathbf{j}_{(A)} = \rho_{(A)} \left( \mathbf{v}_{(A)} - \mathbf{v} \right)$ | $\mathbf{j}_{(A)}^{\circ} = \rho_{(A)} \left( \mathbf{v}_{(A)} - \mathbf{v}^{\circ} \right)$ |
| Molar flux of species $A$ | $\mathbf{N}_{(A)} = c_{(A)} \mathbf{v}_{(A)}$ | $\mathbf{J}_{(A)} = c_{(A)} \left( \mathbf{v}_{(A)} - \mathbf{v} \right)$ | $\mathbf{J}_{(A)}^{\circ} = c_{(A)} \left( \mathbf{v}_{(A)} - \mathbf{v}^{\circ} \right)$ |
| Sum of mass fluxes | $\displaystyle\sum_{A=1}^{N} \mathbf{n}_{(A)} = \rho \mathbf{v}$ | $\displaystyle\sum_{A=1}^{N} \mathbf{j}_{(A)} = \mathbf{0}$ | $\displaystyle\sum_{A=1}^{N} \mathbf{j}_{(A)}^{\circ} = \rho(\mathbf{v} - \mathbf{v}^{\circ})$ |
| Sum of molar fluxes | $\displaystyle\sum_{A=1}^{N} \mathbf{N}_{(A)} = c\mathbf{v}^{\circ}$ | $\displaystyle\sum_{A=1}^{N} \mathbf{J}_{(A)} = c(\mathbf{v}^{\circ} - \mathbf{v})$ | $\displaystyle\sum_{A=1}^{N} \mathbf{J}_{(A)}^{\circ} = 0$ |
| Fluxes in terms of $\mathbf{n}_{(A)}$ | $\mathbf{N}_{(A)} = \dfrac{\mathbf{n}_{(A)}}{M_{(A)}}$ | $\mathbf{j}_{(A)} = \mathbf{n}_{(A)}$ $- \omega_{(A)} \displaystyle\sum_{B=1}^{N} \mathbf{n}_{(B)}$ | $\mathbf{j}_{(A)}^{\circ} = \mathbf{n}_{(A)}$ $- M_{(A)} x_{(A)} \displaystyle\sum_{B=1}^{N} \dfrac{\mathbf{n}_{(B)}}{M_{(B)}}$ |

Fluxes in terms of $\mathbf{N}_{(A)}$ $\quad \mathbf{n}_{(A)} = M_{(A)}\mathbf{N}_{(A)}$ $\qquad \mathbf{J}_{(A)} = \mathbf{N}_{(A)}$ $\qquad\qquad \mathbf{J}_{(A)}^{\diamond} = \mathbf{N}_{(A)}$

$$-\frac{\omega_{(A)}}{M_{(A)}}\sum_{B=1}^{N} M_{(B)}\mathbf{N}_{(B)} \qquad -x_{(A)}\sum_{B=1}^{N}\mathbf{N}_{(B)}$$

Fluxes in terms of $\mathbf{j}_{(A)}$ $\quad \mathbf{n}_{(A)} = \mathbf{j}_{(A)} + \rho_{(A)}\mathbf{v}$ $\qquad \mathbf{J}_{(A)} = \dfrac{\mathbf{j}_{(A)}}{M_{(A)}}$ $\qquad \mathbf{j}_{(A)}^{\diamond} = \mathbf{j}_{(A)}$

$$-\omega_{(A)}M\sum_{B=1}^{N}\frac{\mathbf{j}_{(B)}}{M_{(B)}}$$

for a binary system:

$$\mathbf{j}_{(A)}^{\diamond} = \frac{M}{M_{(B)}}\mathbf{j}_{(A)}$$

Fluxes in terms of $\mathbf{J}_{(A)}^{\diamond}$ $\quad \mathbf{N}_{(A)} = \mathbf{J}_{(A)}^{\diamond} + c_{(A)}\mathbf{v}^{\diamond}$ $\qquad \mathbf{J}_{(A)} = \mathbf{J}_{(A)}^{\diamond}$ $\qquad \mathbf{j}_{(A)}^{\diamond} = M_{(A)}\mathbf{J}_{(A)}^{\diamond}$

$$-\frac{x_{(A)}}{M}\sum_{B=1}^{N}M_{(B)}\mathbf{J}_{(B)}^{\diamond}$$

for a binary system:

$$\mathbf{J}_{(A)} = \frac{M_{(B)}}{M}\mathbf{J}_{(A)}^{\diamond}$$

**Table 8.5.1-5.** Forms of the differential mass balance for species $A$

$$\frac{\partial \rho_{(A)}}{\partial t} + \operatorname{div}\mathbf{n}_{(A)} = r_{(A)}$$

$$\frac{\partial c_{(A)}}{\partial t} + \operatorname{div}\mathbf{N}_{(A)} = \frac{r_{(A)}}{M_{(A)}}$$

$$\rho\frac{d_{(v)}\omega_{(A)}}{dt} + \operatorname{div}\mathbf{j}_{(A)} = r_{(A)}$$

$$\rho\frac{d_{(v)}}{dt}\left(\frac{c_{(A)}}{\rho}\right) + \operatorname{div}\mathbf{J}_{(A)} = \frac{r_{(A)}}{M_{(A)}}$$

$$\frac{d_{(v^{\diamond})}}{dt}\left(\frac{\rho_{(A)}}{c}\right) + \operatorname{div}\mathbf{j}_{(A)}^{\diamond} = r_{(A)} - M_{(A)}x_{(A)}\sum_{B=1}^{N}\frac{r_{(B)}}{M_{(B)}}$$

$$c\frac{d_{(v^{\diamond})}x_{(A)}}{dt} + \operatorname{div}\mathbf{J}_{(A)}^{\diamond} = \frac{r_{(A)}}{M_{(A)}} - x_{(A)}\sum_{B=1}^{N}\frac{r_{(B)}}{M_{(B)}}$$

$$\frac{d_{(v)}\Psi}{dt} \equiv \frac{\partial \Psi}{\partial t} + \nabla\Psi \cdot \mathbf{v}$$

$$\frac{d_{(v^{\diamond})}\Psi}{dt} \equiv \frac{\partial \Psi}{\partial t} + \nabla\Psi \cdot \mathbf{v}^{\diamond}$$

**Table 8.5.1-6.** The differential mass balance for species $A$ in several coordinate systems

---

*Rectangular Cartesian coordinates:*

$$\frac{\partial \rho_{(A)}}{\partial t} + \frac{\partial n_{(A)1}}{\partial z_1} + \frac{\partial n_{(A)2}}{\partial z_2} + \frac{\partial n_{(A)3}}{\partial z_3} = r_{(A)}$$

*Cylindrical coordinates:*

$$\frac{\partial \rho_{(A)}}{\partial t} + \frac{1}{r}\frac{\partial}{\partial r}\left(r n_{(A)r}\right) + \frac{1}{r}\frac{\partial n_{(A)\theta}}{\partial \theta} + \frac{\partial n_{(A)z}}{\partial z} = r_{(A)}$$

*Spherical coordinates:*

$$\frac{\partial \rho_{(A)}}{\partial t} + \frac{1}{r^2}\frac{\partial}{\partial r}\left(r^2 n_{(A)r}\right) + \frac{1}{r \sin\theta}\frac{\partial}{\partial \theta}\left(n_{(A)\theta} \sin\theta\right) + \frac{1}{r \sin\theta}\frac{\partial n_{(A)\phi}}{\partial \phi} = r_{(A)}$$

---

**Table 8.5.1-7.** Equivalent forms of Fick's first law of binary diffusion[a]

---

$$\mathbf{n}_{(A)} = \omega_{(A)}(\mathbf{n}_{(A)} + \mathbf{n}_{(B)}) - \rho \mathcal{D}^0_{(AB)} \nabla \omega_{(A)}$$

$$\mathbf{j}_{(A)} = -\rho \mathcal{D}^0_{(AB)} \nabla \omega_{(A)}$$

$$\mathbf{j}_{(A)} = -\left(\frac{c^2}{\rho}\right) M_{(A)} M_{(B)} \mathcal{D}^0_{(AB)} \nabla x_{(A)}$$

$$\mathbf{N}_{(A)} = x_{(A)}(\mathbf{N}_{(A)} + \mathbf{N}_{(B)}) - c \mathcal{D}^0_{(AB)} \nabla x_{(A)}$$

$$\mathbf{J}^\diamond_{(A)} = -c \mathcal{D}^0_{(AB)} \nabla x_{(A)}$$

$$\mathbf{J}^\diamond_{(A)} = -\left(\frac{\rho^2}{c M_{(A)} M_{(B)}}\right) \mathcal{D}^0_{(AB)} \nabla \omega_{(A)}$$

---

[a] Here, $\mathcal{D}^0_{(AB)} \equiv \left(\partial \ln a^{(m)}_{(A)}/\partial \ln x_{(A)}\right)_{T,P} \mathcal{D}_{(AB)}$.

---

**Table 8.5.1-8.** Some important forms of the differential mass balance for species $A$ consistent with Fick's first law of binary diffusion[a]

---

$$\rho \left(\frac{\partial \omega_{(A)}}{\partial t} + \nabla \omega_{(A)} \cdot \mathbf{v}\right) = \mathrm{div}\left(\rho \mathcal{D}^0_{(AB)} \nabla \omega_{(A)}\right) + r_{(A)}$$

$$c \left(\frac{\partial x_{(A)}}{\partial t} + \nabla x_{(A)} \cdot \mathbf{v}^\diamond\right) = \mathrm{div}\left(c \mathcal{D}^0_{(AB)} \nabla x_{(A)}\right) + x_{(B)} \frac{r_{(A)}}{M_{(A)}} - x_{(A)} \frac{r_{(B)}}{M_{(B)}}$$

*For constant $\rho$ and $\mathcal{D}^0_{(AB)}$:*

$$\frac{\partial \rho_{(A)}}{\partial t} + \nabla \rho_{(A)} \cdot \mathbf{v} = \mathcal{D}^0_{(AB)} \mathrm{div} \nabla \rho_{(A)} + r_{(A)}$$

*For constant c and $\mathcal{D}^0_{(AB)}$:*

$$\frac{\partial c_{(A)}}{\partial t} + \nabla c_{(A)} \cdot \mathbf{v}^\circ = \mathcal{D}^0_{(AB)} \, \text{div} \nabla c_{(A)} + x_{(B)} \frac{r_{(A)}}{M_{(A)}} - x_{(A)} \frac{r_{(B)}}{M_{(B)}}$$

*Either for constant $\rho$, constant $\mathcal{D}^0_{(AB)}$, $\mathbf{v} = \mathbf{0}$, and no chemical reactions
or for constant c, constant $\mathcal{D}^0_{(AB)}$, $\mathbf{v}^\circ = \mathbf{0}$, and no chemical reactions
(Fick's second law of diffusion):*

$$\frac{\partial \rho_{(A)}}{\partial t} = \mathcal{D}^0_{(AB)} \text{div} \nabla \rho_{(A)}$$

---

[a]Here, $\mathcal{D}^0_{(AB)} \equiv \left( \partial \ln a^{(m)}_{(A)} / \partial \ln x_{(A)} \right)_{T,P} \mathcal{D}_{(AB)}$.

**Table 8.5.1-9.** The differential mass balance for species $A$ for constant $\rho$ and $\mathcal{D}^0_{(AB)}$ [a]

*Rectangular Cartesian coordinates:*

$$\frac{\partial \rho_{(A)}}{\partial t} + \frac{\partial \rho_{(A)}}{\partial z_1} v_1 + \frac{\partial \rho_{(A)}}{\partial z_2} v_2 + \frac{\partial \rho_{(A)}}{\partial z_3} v_3 = \mathcal{D}^0_{(AB)} \left( \frac{\partial^2 \rho_{(A)}}{\partial z_1^2} + \frac{\partial^2 \rho_{(A)}}{\partial z_2^2} + \frac{\partial^2 \rho_{(A)}}{\partial z_3^2} \right) + r_{(A)}$$

*Cylindrical coordinates:*

$$\frac{\partial \rho_{(A)}}{\partial t} + \frac{\partial \rho_{(A)}}{\partial r} v_r + \frac{\partial \rho_{(A)}}{\partial \theta} \frac{v_\theta}{r} + \frac{\partial \rho_{(A)}}{\partial z} v_z = \mathcal{D}^0_{(AB)} \left[ \frac{1}{r} \frac{\partial}{\partial r} \left( r \frac{\partial \rho_{(A)}}{\partial r} \right) + \frac{1}{r^2} \frac{\partial^2 \rho_{(A)}}{\partial \theta^2} + \frac{\partial^2 \rho_{(A)}}{\partial z^2} \right] + r_{(A)}$$

*Spherical coordinates:*

$$\frac{\partial \rho_{(A)}}{\partial t} + \frac{\partial \rho_{(A)}}{\partial r} v_r + \frac{\partial \rho_{(A)}}{\partial \theta} \frac{v_\theta}{r} + \frac{\partial \rho_{(A)}}{\partial \varphi} \frac{v_\varphi}{r \sin \theta}$$

$$= \mathcal{D}^0_{(AB)} \left[ \frac{1}{r^2} \frac{\partial}{\partial r} \left( r^2 \frac{\partial \rho_{(A)}}{\partial r} \right) + \frac{1}{r^2 \sin \theta} \frac{\partial}{\partial \theta} \left( \sin \theta \frac{\partial \rho_{(A)}}{\partial \theta} \right) + \frac{1}{r^2 \sin^2 \theta} \frac{\partial^2 \rho_{(A)}}{\partial \varphi^2} \right] + r_{(A)}$$

---

[a]Here, $\mathcal{D}^0_{(AB)} \equiv \left( \partial \ln a^{(m)}_{(A)} / \partial \ln x_{(A)} \right)_{T,P} \mathcal{D}_{(AB)}$.

**Table 8.5.1-10.** Forms of the
overall differential mass balance

$$\frac{\partial \rho}{\partial t} + \text{div}\,(\rho \mathbf{v}) = 0$$

$$\frac{\partial c}{\partial t} + \text{div}\,(c \mathbf{v}^\circ) = \sum_{A=1}^{N} \frac{r_{(A)}}{M_{(A)}}$$

## 8.5.2 Summary of Useful Equations for Energy Transfer

In Section 8.3.4, we derived the differential energy balance in terms of the internal energy per unit mass $\hat{U}$. We are usually more interested in determining the temperature distribution within a material than the internal energy distribution. From the definition for $\hat{A}$ in terms of

$\hat{U}$ given in Exercise 8.4.2-2 as well as (8.4.2-16), we have that

$$d\hat{U} = d\hat{A} + d(T\hat{S})$$

$$= T\,d\hat{S} - P\,d\hat{V} + \sum_{A=1}^{N-1}\left(\mu_{(A)} - \mu_{(N)}\right)d\omega_{(A)}$$

$$= T\left(\frac{\partial\hat{S}}{\partial T}\right)_{\hat{V},\,\omega_{(B)}}dT + \left[T\left(\frac{\partial\hat{S}}{\partial\hat{V}}\right)_{T,\,\omega_{(B)}} - P\right]d\hat{V}$$

$$+ \sum_{A=1}^{N-1}\left[T\left(\frac{\partial\hat{S}}{\partial\omega_{(A)}}\right)_{T,\,\hat{V},\,\omega_{(B)}(B\neq A,N)} + \mu_{(A)} - \mu_{(N)}\right]d\omega_{(A)}$$

$$= \hat{c}_V\,dT + \left[T\left(\frac{\partial P}{\partial T}\right)_{\hat{V},\,\omega_{(B)}} - P\right]d\hat{V}$$

$$+ \sum_{A=1}^{N-1}\left\{-T\left[\frac{\partial\left(\mu_{(A)} - \mu_{(N)}\right)}{\partial T}\right]_{\hat{V},\,\omega_{(B)}} + \mu_{(A)} - \mu_{(N)}\right\}d\omega_{(A)}$$

$$= \hat{c}_V\,dT + \left[T\left(\frac{\partial P}{\partial T}\right)_{\hat{V},\,\omega_{(B)}} - P\right]d\hat{V}$$

$$+ \sum_{A=1}^{N}\left[\mu_{(A)} - T\left(\frac{\partial\mu_{(A)}}{\partial T}\right)_{\hat{V},\,\omega_{(B)}}\right]d\omega_{(A)} \tag{8.5.2-1}$$

Here we introduce the heat capacity per unit mass at constant specific volume (see Exercise 8.4.2-7),

$$\hat{c}_V \equiv T\left(\frac{\partial\hat{S}}{\partial T}\right)_{\hat{V},\,\omega_{(B)}} \tag{8.5.2-2}$$

and we employ one of the Maxwell relations (see Exercise 8.4.2-2):

$$\left(\frac{\partial\hat{S}}{\partial\omega_{(A)}}\right)_{T,\,\hat{V},\,\omega_{(B)}(B\neq A,N)} = -\left[\frac{\partial\left(\mu_{(A)} - \mu_{(N)}\right)}{\partial T}\right]_{\hat{V},\,\omega_{(B)}} \tag{8.5.2-3}$$

The overall differential mass balance tells us that

$$\rho\frac{d_{(v)}\hat{V}}{dt} = -\frac{1}{\rho}\frac{d_{(v)}\rho}{dt} = \operatorname{div}\mathbf{v} \tag{8.5.2-4}$$

and the differential mass balance for species $A$ gives (see Table 8.5.1-5)

$$\rho\frac{d_{(v)}\omega_{(A)}}{dt} = -\operatorname{div}\mathbf{j}_{(A)} + r_{(A)} \tag{8.5.2-5}$$

Equations (8.5.2-1), (8.5.2-4), and (8.5.2-5) allow us to rewrite (8.3.4-4) as

$$\rho \hat{c}_V \frac{d_{(v)}T}{dt} = -\operatorname{div} \mathbf{q} - T\left(\frac{\partial P}{\partial T}\right)_{\hat{V},\,\omega_{(B)}} \operatorname{div} \mathbf{v} + \operatorname{tr}(\mathbf{S} \cdot \nabla \mathbf{v}) + \rho Q + \sum_{A=1}^{N} \mathbf{j}_{(A)} \cdot \mathbf{f}_{(A)}$$

$$+ \sum_{A=1}^{N} \left[ \mu_{(A)} - T\left(\frac{\partial \mu_{(A)}}{\partial T}\right)_{\hat{V},\,\omega_{(B)}} \right] \left(\operatorname{div} \mathbf{j}_{(A)} - r_{(A)}\right) \tag{8.5.2-6}$$

In view of the extension of *Fourier's law* to multicomponent materials (Section 8.4.3), it is more convenient to write this last equation as

$$\rho \hat{c}_V \frac{d_{(v)}T}{dt} = -\operatorname{div}\left(\boldsymbol{\epsilon} + \sum_{A=1}^{N} \overline{H}_{(A)}\mathbf{j}_{(A)}\right) - T\left(\frac{\partial P}{\partial T}\right)_{\hat{V},\,\omega_{(B)}} \operatorname{div} \mathbf{v}$$

$$+ \operatorname{tr}(\mathbf{S} \cdot \nabla \mathbf{v}) + \rho Q + \sum_{A=1}^{N} \mathbf{j}_{(A)} \cdot \mathbf{f}_{(A)}$$

$$+ \sum_{A=1}^{N} \left[ \mu_{(A)} - T\left(\frac{\partial \mu_{(A)}}{\partial T}\right)_{\hat{V},\,\omega_{(B)}} \right] \left(\operatorname{div} \mathbf{j}_{(A)} - r_{(A)}\right) \tag{8.5.2-7}$$

where

$$\boldsymbol{\epsilon} \equiv \mathbf{q} - \sum_{A=1}^{N} \overline{H}_{(A)}\mathbf{j}_{(A)} \tag{8.5.2-8}$$

Using Exercise 8.4.2-8, the overall differential mass balance, and the differential mass balance for species $A$, we can express (8.5.2-7) as

$$\rho \hat{c}_V \frac{d_{(v)}T}{dt} = -\operatorname{div}\left(\boldsymbol{\epsilon} + \sum_{A=1}^{N} \overline{H}_{(A)}\mathbf{j}_{(A)}\right) - T\left(\frac{\partial P}{\partial T}\right)_{\hat{V},\,\omega_{(B)}} \operatorname{div} \mathbf{v} + \operatorname{tr}(\mathbf{S} \cdot \nabla \mathbf{v})$$

$$+ \rho Q + \sum_{A=1}^{N} \mathbf{j}_{(A)} \cdot \mathbf{f}_{(A)} + \sum_{A=1}^{N} \left[ \overline{H}_{(A)} - \overline{V}_{(A)} T \left(\frac{\partial P}{\partial T}\right)_{\hat{V},\,\omega_{(B)}} \right] \left(\operatorname{div} \mathbf{j}_{(A)} - r_{(A)}\right)$$

$$= -\operatorname{div} \boldsymbol{\epsilon} - \sum_{A=1}^{N} \nabla \overline{H}_{(A)} \cdot \mathbf{j}_{(A)} + \operatorname{tr}(\mathbf{S} \cdot \nabla \mathbf{v}) + \rho Q + \sum_{A=1}^{N} \mathbf{j}_{(A)} \cdot \mathbf{f}_{(A)} - \sum_{A=1}^{N} \overline{H}_{(A)}r_{(A)}$$

$$+ \sum_{A=1}^{N} T\left(\frac{\partial P}{\partial T}\right)_{\hat{V},\,\omega_{(B)}} \left(\rho \overline{V}_{(A)} \frac{d_{(v)}\omega_{(A)}}{dt} + \frac{1}{\rho}\frac{d_{(v)}\rho}{dt}\right)$$

$$= -\operatorname{div} \boldsymbol{\epsilon} - \sum_{A=1}^{N} \nabla \overline{H}_{(A)} \cdot \mathbf{j}_{(A)} + \operatorname{tr}(\mathbf{S} \cdot \nabla \mathbf{v})$$

$$+ \rho Q + \sum_{A=1}^{N} \mathbf{j}_{(A)} \cdot \mathbf{f}_{(A)} - \sum_{A=1}^{N} \overline{H}_{(A)}r_{(A)} - \sum_{A=1}^{N} \rho T\left(\frac{\partial P}{\partial T}\right)_{\hat{V},\,\omega_{(B)}} \omega_{(A)} \frac{d_{(v)}\overline{V}_{(A)}}{dt} \tag{8.5.2-9}$$

in which we have recognized that

$$\frac{1}{\rho} = \sum_{(A)}^{N} \overline{V}_{(A)}\omega_{(A)} \tag{8.5.2-10}$$

**Table 8.5.2-1.** Various forms of the overall differential energy balance

$$\rho \frac{d_{(v)}}{dt}\left(\hat{U} + \frac{1}{2}v^2 + \varphi\right) = -\mathrm{div}\left(\boldsymbol{\epsilon} + \sum_{A=1}^{N}\overline{H}_{(A)}\mathbf{j}_{(A)}\right) + \mathrm{div}(\mathbf{T}\cdot\mathbf{v}) + \rho Q + \sum_{A=1}^{N}\mathbf{j}_{(A)}\cdot\mathbf{f}_{(A)}$$

$$\rho \frac{d_{(v)}}{dt}\left(\hat{U} + \frac{1}{2}v^2\right) = -\mathrm{div}\left(\boldsymbol{\epsilon} + \sum_{A=1}^{N}\overline{H}_{(A)}\mathbf{j}_{(A)}\right) + \mathrm{div}(\mathbf{T}\cdot\mathbf{v}) + \sum_{A=1}^{N}\mathbf{n}_{(A)}\cdot\mathbf{f}_{(A)} + \rho Q$$

$$\rho \frac{d_{(v)}\hat{U}}{dt} = -\mathrm{div}\left(\boldsymbol{\epsilon} + \sum_{A=1}^{N}\overline{H}_{(A)}\mathbf{j}_{(A)}\right) - P\,\mathrm{div}\,\mathbf{v} + \mathrm{tr}\,(\mathbf{S}\cdot\nabla\mathbf{v}) + \rho Q + \sum_{A=1}^{N}\mathbf{j}_{(A)}\cdot\mathbf{f}_{(A)}$$

$$\rho \frac{d_{(v)}\hat{H}}{dt} = -\mathrm{div}\left(\boldsymbol{\epsilon} + \sum_{A=1}^{N}\overline{H}_{(A)}\mathbf{j}_{(A)}\right) + \frac{d_{(v)}P}{dt} + \mathrm{tr}\,(\mathbf{S}\cdot\nabla\mathbf{v}) + \rho Q + \sum_{A=1}^{N}\mathbf{j}_{(A)}\cdot\mathbf{f}_{(A)}$$

$$\frac{\partial\left(\rho\hat{H}\right)}{\partial t} + \mathrm{div}\left(\sum_{A=1}^{N}\overline{H}_{(A)}\mathbf{n}_{(A)}\right) = -\mathrm{div}\,\boldsymbol{\epsilon} + \frac{d_{(v)}P}{dt} + \mathrm{tr}\,(\mathbf{S}\cdot\nabla\mathbf{v}) + \rho Q + \sum_{A=1}^{N}\mathbf{j}_{(A)}\cdot\mathbf{f}_{(A)}$$

$$\frac{\partial\left(\rho\hat{H}\right)}{\partial t} + \mathrm{div}\left(\sum_{A=1}^{N}\overline{H}_{(A)}^{(m)}\mathbf{N}_{(A)}\right) = -\mathrm{div}\,\boldsymbol{\epsilon} + \frac{d_{(v)}P}{dt} + \mathrm{tr}\,(\mathbf{S}\cdot\nabla\mathbf{v}) + \rho Q + \sum_{A=1}^{N}\mathbf{j}_{(A)}\cdot\mathbf{f}_{(A)}$$

$$\rho\hat{c}_V\frac{d_{(v)}T}{dt} = -\mathrm{div}\left(\boldsymbol{\epsilon} + \sum_{A=1}^{N}\overline{H}_{(A)}\mathbf{j}_{(A)}\right) - T\left(\frac{\partial P}{\partial T}\right)_{\hat{V},\,\omega_{(B)}}\mathrm{div}\,\mathbf{v} + \mathrm{tr}(\mathbf{S}\cdot\nabla\mathbf{v}) + \rho Q$$
$$+ \sum_{A=1}^{N}\mathbf{j}_{(A)}\cdot\mathbf{f}_{(A)} + \sum_{A=1}^{N}\left[\mu_{(A)} - T\left(\frac{\partial\mu_{(A)}}{\partial T}\right)_{\hat{V},\,\omega_{(B)}}\right]\left(\mathrm{div}\,\mathbf{j}_{(A)} - r_{(A)}\right)$$

$$\rho\hat{c}_V\frac{d_{(v)}T}{dt} = -\mathrm{div}\left(\boldsymbol{\epsilon} + \sum_{A=1}^{N}\overline{H}_{(A)}\mathbf{j}_{(A)}\right) - T\left(\frac{\partial P}{\partial T}\right)_{\hat{V},\,\omega_{(B)}}\mathrm{div}\,\mathbf{v} + \mathrm{tr}(\mathbf{S}\cdot\nabla\mathbf{v}) + \rho Q$$
$$+ \sum_{A=1}^{N}\mathbf{j}_{(A)}\cdot\mathbf{f}_{(A)} + \sum_{A=1}^{N}\left[\overline{H}_{(A)} - \overline{V}_{(A)}T\left(\frac{\partial P}{\partial T}\right)_{\hat{V},\,\omega_{(B)}}\right]\left(\mathrm{div}\,\mathbf{j}_{(A)} - r_{(A)}\right)$$

$$\rho\hat{c}_V\frac{d_{(v)}T}{dt} = -\mathrm{div}\,\boldsymbol{\epsilon} - \sum_{A=1}^{N}\nabla\overline{H}_{(A)}\cdot\mathbf{j}_{(A)} + \mathrm{tr}(\mathbf{S}\cdot\nabla\mathbf{v}) + \rho Q + \sum_{A=1}^{N}\mathbf{j}_{(A)}\cdot\mathbf{f}_{(A)}$$
$$- \sum_{A=1}^{N}\overline{H}_{(A)}r_{(A)} - \sum_{A=1}^{N}\rho T\left(\frac{\partial P}{\partial T}\right)_{\hat{V},\,\omega_{(B)}}\omega_{(A)}\frac{d_{(v)}\overline{V}_{(A)}}{dt}$$

$$\rho\hat{c}_P\frac{d_{(v)}T}{dt} = -\mathrm{div}\left(\boldsymbol{\epsilon} + \sum_{A=1}^{N}\overline{H}_{(A)}\mathbf{j}_{(A)}\right) + \left(\frac{\partial\ln\hat{V}}{\partial\ln T}\right)_{P,\,\omega_{(B)}}\frac{d_{(v)}P}{dt} + \mathrm{tr}(\mathbf{S}\cdot\nabla\mathbf{v}) + \rho Q$$
$$+ \sum_{A=1}^{N}\mathbf{j}_{(A)}\cdot\mathbf{f}_{(A)} + \sum_{A=1}^{N}\left[\mu_{(A)} - T\left(\frac{\partial\mu_{(A)}}{\partial T}\right)_{P,\,\omega_{(B)}}\right]\left(\mathrm{div}\,\mathbf{j}_{(A)} - r_{(A)}\right)$$

$$\rho \hat{c}_P \frac{d_{(v)}T}{dt} = -\text{div}\left(\boldsymbol{\epsilon} + \sum_{A=1}^{N} \overline{H}_{(A)}\mathbf{j}_{(A)}\right) + \left(\frac{\partial \ln \hat{V}}{\partial \ln T}\right)_{P,\,\omega_{(B)}} \frac{d_{(v)}P}{dt} + \text{tr}(\mathbf{S} \cdot \nabla \mathbf{v}) + \rho Q$$

$$+ \sum_{A=1}^{N} \mathbf{j}_{(A)} \cdot \mathbf{f}_{(A)} + \sum_{A=1}^{N} \overline{H}_{(A)}\left(\text{div } \mathbf{j}_{(A)} - r_{(A)}\right)$$

$$\rho \hat{c}_P \frac{d_{(v)}T}{dt} = -\text{div } \boldsymbol{\epsilon} - \sum_{A=1}^{N} \nabla \overline{H}_{(A)} \cdot \mathbf{j}_{(A)} + \left(\frac{\partial \ln \hat{V}}{\partial \ln T}\right)_{P,\,\omega_{(B)}} \frac{d_{(v)}P}{dt}$$

$$+ \text{tr}(\mathbf{S} \cdot \nabla \mathbf{v}) + \rho Q + \sum_{A=1}^{N} \mathbf{j}_{(A)} \cdot \mathbf{f}_{(A)} - \sum_{A=1}^{N} \overline{H}_{(A)}r_{(A)}$$

These and other forms of the overall differential energy balance that may be derived in a similar fashion are presented in Table 8.5.2-1. Often one form will have a particular advantage in any given problem.

In deriving some of the forms of the differential energy balance shown in Table 8.5.2-1, we have assumed that (see also Section 2.4.1 and Exercise 2.4.1-1)

$$\mathbf{f} \equiv \sum_{A=1}^{N} \omega_{(A)}\mathbf{f}_{(A)}$$

$$= -\nabla \phi \qquad\qquad\qquad (8.5.2\text{-}11)$$

where the *potential energy* per unit mass $\phi$ is not a function of time.

The forms of differential energy balance in rectangular Cartesian, cylindrical, and spherical coordinates may be written as required by analogy with Tables 5.4.1-2 and 5.4.1-3.

**Exercise 8.5.2-1**   Derive the additional forms of the overall differential energy balance given in Table 8.5.2-1.

# 9

---

# Differential Balances in Mass Transfer

THIS CHAPTER PARALLELS Chapters 3 and 6. Although the applications are different, the approach is the same. One is better prepared to begin here after studying these earlier chapters.

There are three central questions that we will attempt to address.

- When is there a complete analogy between energy and mass transfer?
- When do energy and mass transfer problems take identical forms?
- When is diffusion-induced convection important?

In the absence of forced convection or of natural convection resulting from a density gradient in a gravitational field, the convective terms are often neglected in the differential mass balance for species $A$ with no explanation. Sometimes the qualitative argument is made that, since diffusion is a very slow process, the resulting diffusion-induced convection must certainly be negligible with respect to it. We hope that you will understand as a result of reading this chapter that this argument is too simplistic. Under certain circumstances, diffusion-induced convection is a major feature of the problem. Sometimes we are able to show that diffusion-induced convection is identically zero. More generally, diffusion-induced convection can be neglected with respect to diffusion as equilibrium is approached in the limit of dilute solutions.

When must we describe diffusion in multicomponent systems as multicomponent diffusion? We already know the answer from Section 8.4.6: in concentrated solutions. In this chapter, we will try to get a better feeling for just how concentrated the solution must be before the approach using binary diffusion fails.

---

## 9.1  Philosophy of Solving Mass Transfer Problems

In principle, the problems we are about to take up are considerably more complex than those we dealt with in either Chapter 3 or Chapter 6. We should begin to think in terms of simultaneous momentum, energy, and mass transfer, requiring a simultaneous solution of the differential mass balance for each species present, the differential momentum balance, and the differential energy balance, with particular constitutive equations for the mass flux vectors, the stress tensor, and the energy flux vector.

Much can be learned about problems of engineering importance by making simplifying assumptions. In the problems that follow we shall often neglect thermal effects and treat the fluids involved as though they were isothermal, ignoring the energy balance. In addition, although in every problem we will be quite concerned with the velocity distribution, we often will say nothing about the pressure distribution. The implication is that the differential momentum balance is approximately satisfied.

As we said in Section 3.1, the first step in analyzing a physical situation is to decide just exactly what the problem is. In part, this means that constitutive equations for the mass flux vectors, the stress tensor, and the energy flux vector must be chosen. With respect to the mass and energy flux vectors, the literature to date gives us very little choice beyond those constitutive equations described in Sections 8.4.3 and 8.4.4 (or the special cases taken up in Sections 8.4.5 and 8.4.6). Though Section 8.4.7 suggests a variety of constitutive equations for the stress tensor, they are basically all of the same form as those introduced in Sections 2.3.2 to 2.3.4. Unfortunately, mass transfer in viscoelastic fluids has been largely neglected until now.

To complete the specification of a particular problem, we must describe the geometry of the material or the geometry through which the material moves, the homogeneous and heterogeneous chemical reactions (see Section 8.2.1), the forces that cause the material to move, and any energy transfer to the material. Just as in Chapters 3 and 6, every problem requires a statement of boundary conditions in its formulation. Beyond those indicated in Sections 3.1 and 6.1, there are several common types of boundary conditions for which one should look in an unfamiliar physical situation.

1) We shall assume that at an interface the phases are in equilibrium. It might be somewhat more natural to say that the chemical potentials of all species present are continuous across the phase boundary. This is suggested by anticipating that, in a sense, local equilibrium is established at the phase boundary (Slattery 1990, p. 842). The use of chemical potentials in describing a physical problem is not generally recommended, because experimental data for chemical potential as a function of solution concentration are scarce.
2) The jump mass balance (8.2.1-4) must be satisfied for every species at every phase interface.
3) We assume that concentrations and mass fluxes remain finite at all points in the material.

The advice we gave in Section 3.1 is still applicable. Sometimes it will be relatively simple to formulate a problem, but either impossible to come up with an analytic solution or very expensive to execute a numerical solution. It is often worthwhile to approximate a realistic, difficult problem by one that is somewhat easier to handle. This may be all that is needed in some cases, or perhaps it can serve as a useful check on whatever numerical work is being done.

As in our discussions of solutions for momentum and energy balances, the results determined here are not unique. We are simply interested in finding a solution. Sometimes experiments will suggest that the solutions obtained are unique, but often this evidence is not available.

## 9.2 Energy and Mass Transfer Analogy

In beginning our study of mass transfer problems, it is important to understand when there are analogous energy transfer problems. Under what circumstances do energy and mass transfer problems take the same forms? Under what circumstances can we replace our mass transfer problem with an energy transfer problem whose solution may already be known? Energy and mass transfer problems take the same forms when the differential equations and boundary conditions describing the systems have the same forms. Let us begin by comparing the differential energy balance (Table 5.4.1-1),

$$\rho \hat{c}_V \frac{d_{(m)} T}{dt} = -\text{div } \mathbf{q} - T \left( \frac{\partial P}{\partial T} \right)_{\hat{V}} \text{div } \mathbf{v} + \text{tr} (\mathbf{S} \cdot \nabla \mathbf{v}) + \rho Q \tag{9.2.0-1}$$

with the differential mass balance for species $A$ (Table 8.5.1-5),

$$\rho \frac{d_{(v)} \omega_{(A)}}{dt} + \text{div } \mathbf{j}_{(A)} = r_{(A)} \tag{9.2.0-2}$$

We can simplify Equation (9.2.1-1) with the following assumptions:

1) The system has a uniform composition.
2) The system consists of a single phase, so that it is unnecessary to consider the jump energy balance.
3) The phase is incompressible.
4) Viscous dissipation can be neglected.
5) Radiation can be neglected.
6) Fourier's law (5.3.3-15)

$$\mathbf{q} = -k \nabla T \tag{9.2.0-3}$$

   is appropriate.
7) All physical properties are constants.

Equation (9.2.0-1) then reduces to

$$\frac{1}{N_{St}} \frac{\partial T^\star}{\partial t^\star} + \nabla T^\star \cdot \mathbf{v}^\star = \frac{1}{N_{Pe}} \text{div}(\nabla T^\star) \tag{9.2.0-4}$$

Here we have defined the Strouhal, Peclet, Prandtl, and Reynolds numbers as

$$N_{St} \equiv \frac{t_0 v_0}{L_0}, \quad N_{Pe} \equiv N_{Pr} N_{Re} = \frac{\rho_0 \hat{c}_0 v_0 L_0}{k_0}$$

$$N_{Pr} \equiv \frac{\hat{c}_0 \mu_0}{k_0}, \quad N_{Re} \equiv \frac{\rho_0 v_0 L_0}{\mu_0} \tag{9.2.0-5}$$

Simplification of Equation (9.2.0-2) can be accomplished with these assumptions:

1) The system is isothermal; all viscous dissipation and radiation can be neglected (to ensure that the differential and jump energy balances are satisfied).
2) The system consists of a single phase, so that the jump mass balance for species $A$ need not be considered.
3) The phase is incompressible (to ensure that the overall differential mass balances have the same forms).
4) The system is composed of only two components, or the multicomponent solution is sufficiently dilute that diffusion can be regarded as binary (Section 8.4.6). Under these circumstances, Fick's first law (Table 8.5.1-7) applies:

$$\mathbf{j}_{(A)} = -\rho \mathcal{D}^0_{(AB)} \nabla \omega_{(A)} \tag{9.2.0-6}$$

5) There are no homogeneous or heterogeneous chemical reactions.
6) Any mass transfer at phase boundaries is so slow that the normal component of **v** can be considered to be zero.
7) Effects attributable to thermal, pressure, and forced diffusion can be neglected.
8) All physical properties are constants.

Equation (9.2.0-2) then reduces to

$$\frac{1}{N_{St}} \frac{\partial \omega_{(A)}}{\partial t^\star} + \nabla \omega_{(A)} \cdot \mathbf{v}^\star = \frac{1}{N_{Pe,m}} \text{div}(\nabla \omega_{(A)}) \tag{9.2.0-7}$$

which has the same mathematical form as (9.2.0-4). Here the Peclet number for mass transfer and the Schmidt number are defined as

$$N_{Pe,m} \equiv N_{Sc} N_{Re} = \frac{v_0 L_0}{\mathcal{D}^0_{(AB)}}, \qquad N_{Sc} \equiv \frac{\mu_0}{\rho_0 \mathcal{D}^0_{(AB)}} \tag{9.2.0-8}$$

Note that energy and mass transfer problems take the same form under the conditions noted above, only when it is the mass-averaged velocity **v** that appears in the differential mass balance. If one chooses to work in terms of mole fractions $x_{(A)}$ rather than mass fractions $\omega_{(A)}$, the system must be so dilute that $\mathbf{v} \approx \mathbf{v}^\circ$.

In conclusion, there will be occasions where we will be satisfied to describe a mass transfer problem by an energy transfer problem whose boundary conditions have the same form and whose solution is available to us. However, these problems will not be the ones most frequently encountered.

### 9.2.1  Film Theory

A common situation in which we take advantage of the analogy between energy and mass transfer is in the construction of correlations for mass transfer coefficients.

If we are unwilling or unable to derive the temperature distribution in one of the phases adjoining an interface, we may approximate the energy flux at a *stationary* interface using Newton's "law" of cooling (Section 6.2.2):

$$\text{at an interface}: \ \mathbf{q} \cdot \boldsymbol{\xi} = h \ (T - T_\infty)$$
$$-\nabla T^\star = N_{Nu} \ (T^\star - T^\star_\infty) \tag{9.2.1-1}$$

where we have assumed that

$$\text{at an interface}: \ \mathbf{v} \cdot \boldsymbol{\xi} = 0 \tag{9.2.1-2}$$

Here $h$ is the *film coefficient for energy transfer in the limit of no mass transfer* and

$$N_{Nu} \equiv \frac{h L_0}{k} \tag{9.2.1-3}$$

is the *Nusselt number*. It is clear from (9.2.0-4) and (9.2.1-1) that an analysis or experimental study would show

$$N_{Nu} = N_{Nu} \ (N_{St}, N_{Re}, N_{Pr}) \tag{9.2.1-4}$$

(The separate dependence upon $N_{Re}$ would enter from the differential momentum balance.)

Just as we employed Newton's "law" of cooling above, we will find it helpful to introduce the empirical observation

**Newton's "law" of mass transfer**[1]    The mass flux across a fluid–solid phase interface is roughly proportional to the difference between the composition of the fluid adjacent to the interface (which could be assumed to be in equilibrium with the interface) and the composition of the surrounding bulk fluid (which might be assumed to be well mixed).

We write

$$\text{at an interface}: \ \mathbf{n}_{(A)} \cdot \boldsymbol{\xi} = k^{\bullet}_{(A)\omega} \left( \omega_{(A)} - \omega_{(A)\infty} \right) \tag{9.2.1-5}$$

or

$$\text{at an interface}: \ \mathbf{N}_{(A)} \cdot \boldsymbol{\xi} = k^{\bullet}_{(A)x} \left( x_{(A)} - x_{(A)\infty} \right) \tag{9.2.1-6}$$

The understanding here is that $\boldsymbol{\xi}$ is the unit normal to the phase interface that is directed into these surroundings. The coefficients $k^{\bullet}_{(A)\omega}$ and $k^{\bullet}_{(A)x}$ are usually referred to as *mass-transfer coefficients*. In the limit of sufficiently dilute solutions and no heterogeneous chemical reactions, for reasons that will become obvious below, we commonly write

$$\text{at an interface}: \ \mathbf{n}_{(A)} \cdot \boldsymbol{\xi} = k_{\omega} \left( \omega_{(A)} - \omega_{(A)\infty} \right) \tag{9.2.1-7}$$

or

$$\text{at an interface}: \ \mathbf{N}_{(A)} \cdot \boldsymbol{\xi} = k_x \left( x_{(A)} - x_{(A)\infty} \right) \tag{9.2.1-8}$$

We will refer to the coefficients $k_{\omega}$ and $k_x$ as *film coefficients for mass transfer in the limit of no mass transfer*.

In view of (9.2.1-2), we may approximate the mass flux of species $A$ at a stationary boundary by (9.2.1-7):

$$\text{at an interface}: \ \mathbf{n}_{(A)} \cdot \boldsymbol{\xi} = k_{\omega} \left( \omega_{(A)} - \omega_{(A)\infty} \right)$$
$$-\nabla \omega_{(A)} = N_{Nu,m} \left( \omega_{(A)} - \omega_{(A)\infty} \right) \tag{9.2.1-9}$$

and

$$N_{Nu,m} \equiv \frac{k_{\omega} L_0}{\rho \mathcal{D}_{(AB)}} \tag{9.2.1-10}$$

is known as the *Nusselt number for mass transfer*. From (9.2.0-7) and (9.2.1-9), we see that any analysis or experimental study would show

$$N_{Nu,m} = N_{Nu,m} \left( N_{St}, N_{Re}, N_{Sc} \right) \tag{9.2.1-11}$$

(Again, the separate dependence upon $N_{Re}$ would enter from the differential momentum balance.)

---

[1] I have adopted this name to stress the analogy with Newton's "law" of cooling introduced in Section 6.2.2. The relationship was originally suggested by A. N. Shchukarev and W. Nernst (Levich 1962, p. 41).

For a gas so dilute that $\mathbf{v} \approx \mathbf{v}^\circ$ (see above), we see from (9.2.1-8)

at an interface : $\mathbf{N}_{(A)} \cdot \boldsymbol{\xi} = k_x \left( x_{(A)} - x_{(A)\infty} \right)$

$$\mathbf{n}_{(A)} \cdot \boldsymbol{\xi} = M_{(A)} k_x \left( x_{(A)} - x_{(A)\infty} \right)$$

$$= M k_x \left( \omega_{(A)} - \omega_{(A)\infty} \right) \tag{9.2.1-12}$$

$$-\nabla \omega_{(A)} = \frac{M k_x}{\rho \mathcal{D}_{(AB)}} \left( \omega_{(A)} - \omega_{(A)\infty} \right)$$

$$= \frac{k_x}{c \mathcal{D}_{(AB)}} \left( \omega_{(A)} - \omega_{(A)\infty} \right)$$

Comparing this last expression with (9.2.1-9), we conclude that

$$k_x = \frac{1}{M} k_\omega$$

$$= \frac{c}{\rho} k_\omega \tag{9.2.1-13}$$

and as an alternative to (9.2.1-10)

$$N_{Nu,m} \equiv \frac{k_x L_0}{c \mathcal{D}_{(AB)}} \tag{9.2.1-14}$$

Our conclusion is that (9.2.1-4) and (9.2.1-11) could have the same functional forms. As long as the conditions for an analogy are met, a correlation for $k_\omega$ or $k_x$ may be constructed by relabeling the analogous correlation for $h$. Most correlations for $k_\omega$ or $k_x$ have been constructed in this way, since $h$ is generally easier to measure.

We often wish to use a film coefficient for energy transfer under conditions where the analogy between energy and mass transfer fails. Instead of (9.2.1-1), the overall jump energy balance (8.3.4-5) suggests that, at a *stationary* interface, the total energy transfer can often be expressed as

at an interface : $\left( \mathbf{q} + \rho \hat{H} \mathbf{v} \right) \cdot \boldsymbol{\xi} = h^\bullet (T - T_\infty)$

$$= h \mathcal{F}_{\text{energy}} (T - T_\infty) \tag{9.2.1-15}$$

in which

$$\mathcal{F}_{\text{energy}} \equiv \frac{\left( \mathbf{q} + \rho \hat{H} \mathbf{v} \right) \cdot \boldsymbol{\xi}}{(\mathbf{q} \cdot \boldsymbol{\xi})|_{\mathbf{v} \cdot \boldsymbol{\xi} = 0}} \tag{9.2.1-16}$$

and

$$h^\bullet \equiv h \mathcal{F}_{\text{energy}} \tag{9.2.1-17}$$

In the same way, for sufficiently dilute systems we can write instead of (9.2.1-5)

at an interface : $\mathbf{n}_{(A)} \cdot \boldsymbol{\xi} = k^\bullet_{(A)\omega} \left( \omega_{(A)} - \omega_{(A)\infty} \right)$

$$= k_\omega \mathcal{F}_{(A)} \left( \omega_{(A)} - \omega_{(A)\infty} \right)$$

$$= M_{(A)} k^\bullet_{(A)x} \left( x_{(A)} - x_{(A)\infty} \right)$$

$$= M_{(A)} k_x \mathcal{F}_{(A)} \left( x_{(A)} - x_{(A)\infty} \right) \tag{9.2.1-18}$$

where

$$\mathcal{F}_{(A)} \equiv \frac{\mathbf{n}_{(A)} \cdot \boldsymbol{\xi}}{\left(\mathbf{n}_{(A)} \cdot \boldsymbol{\xi}\right)\big|_{\mathbf{v} \cdot \boldsymbol{\xi}=0}}$$

$$= \frac{\mathbf{N}_{(A)} \cdot \boldsymbol{\xi}}{\left(\mathbf{N}_{(A)} \cdot \boldsymbol{\xi}\right)\big|_{\mathbf{v} \cdot \boldsymbol{\xi}=0}} \qquad (9.2.1\text{-}19)$$

and

$$k_{(A)\omega}^{\bullet} \equiv k_{\omega} \mathcal{F}_{(A)}$$
$$k_{(A)x}^{\bullet} \equiv k_{x} \mathcal{F}_{(A)} \qquad (9.2.1\text{-}20)$$

We face the task of constructing approximations for the correction coefficients $\mathcal{F}_{\text{energy}}$ and $\mathcal{F}_{(A)}$. Three classes of corrections have been offered in the literature (Bird et al. 1960, pp. 658–676): film theory, boundary-layer theory, and penetration theory.

Film theory assumes that there is a *stagnant* film of some thickness $L$ adjacent to an interface. It is stagnant in the sense that there is no lateral motion in the film, only mass transfer in the direction normal to the phase interface. There, of course, is no such thing as a stagnant film, normally. The stagnant film is a simplistic device that allows us to readily compute $\mathcal{F}_{\text{energy}}$ and $\mathcal{F}_{(A)}$.

Film theory can be useful to the extent that the ratios $\mathcal{F}_{\text{energy}}$ and $\mathcal{F}_{(A)}$ are more accurate than the component estimates for either the numerators or denominators. But it will be useful *only if the fictitious film thickness L drops out of the estimates for $\mathcal{F}_{energy}$ and $\mathcal{F}_{(A)}$*. Although our intention here is not to develop the full scope of this problem, we will explore film theory in this chapter as a means of investigating the magnitude of the effects of diffusion-induced and reaction-induced mass transfer.

This section as well as some of the notation was inspired by the discussion of film theory given by Bird et al. (1960).

## 9.3    Complete Solutions for Binary Systems

The discussions that follow deviate in two ways from our treatment of momentum and energy transfer in Chapters 1 through 6.

1) We never attempt to satisfy the differential momentum balance. The role of the differential momentum balance in these problems is to define the pressure gradient in the system. Because the pressure gradients are normally so small as to be undetectable with common instrumentation, there is no practical reason to determine the pressure distributions.

2) We do not employ the usual constraint that the tangential components of velocity are continuous across a phase interface. Although we will find that the effects of convection induced by diffusion are significant when compared with the effects of diffusion, the velocities induced by diffusion are small. The effects of violations of the *no-slip* boundary condition have not been detected experimentally. For more on this point, see Exercises 9.3.1-4 and 9.3.1-5.

### 9.3.1 Unsteady-State Evaporation

Evaporation of a volatile liquid from a partially filled, open container often is referred to as the *Stefan diffusion problem* (Stefan 1889, Rubinstein 1971). This problem has not been analyzed in detail previously, due to the complications introduced by the moving liquid–vapor interface. Arnold (1944) [see also Wilke (1950), Lee and Wilke (1954), and Bird et al. (1960, p. 594)] assumed a stationary phase interface, although he did account for diffusion-induced convection. Prata and Sparrow (1985) considered nonisothermal evaporation in an adiabatic tube with a stationary interface, but they presented no comparisons with experimental data. In what follows, we will consider a vertical tube, partially filled with a pure liquid $A$. For time $t < 0$, this liquid is isolated from the remainder of the tube, which is filled with a gas mixture of $A$ and $B$, by a closed diaphragm. The entire apparatus is maintained at a constant temperature and pressure (neglecting the very small hydrostatic effect). At time $t = 0$, the diaphragm is carefully opened, and the evaporation of $A$ commences. We wish to determine the concentration distribution of $A$ in the gas phase as well as the position of the liquid–gas phase interface as functions of time.

We will consider two cases, beginning with an experiment in which the phase interface is stationary. We will conclude by examining the relationship of the Arnold (1944) analysis to that developed here.

#### A Very Long Tube with a Stationary Interface

In this first case, let us assume that the tube is very long and that, with an appropriate arrangement of the apparatus, the liquid–gas phase interface remains fixed in space as the evaporation takes place.

Let us assume that $A$ and $B$ form an ideal-gas mixture. This allows us to say that the molar density $c$ is a constant throughout the gas phase.[2]

For simplicity, let us replace the finite gas phase with a semi-infinite gas that occupies all space corresponding to $z_2 > 0$. The initial and boundary conditions become

$$\text{at } t = 0 \text{ for all } z_2 > 0 : \quad x_{(A)} = x_{(A)0} \tag{9.3.1-1}$$

and

$$\text{at } z_2 = 0 \text{ for all } t > 0 : \quad x_{(A)} = x_{(A)\text{eq}} \tag{9.3.1-2}$$

By $x_{(A)\text{eq}}$ we mean the mole fraction of species $A$ in the $AB$ gas mixture that is in equilibrium with pure liquid $A$ at the existing temperature and pressure.

Equations (9.3.1-1) and (9.3.1-2) suggest that we seek a solution to this problem of the form

$$
\begin{aligned}
v_1^\diamond &= v_3^\diamond \\
&= 0 \\
v_2^\diamond &= v_2^\diamond(t, z_2) \\
x_{(A)} &= x_{(A)}(t, z_2)
\end{aligned}
\tag{9.3.1-3}
$$

---

[2] It is common in the analysis of this problem to assume that the liquid phase is saturated with species $B$ (Bird et al. 1960, p. 594). We will avoid this assumption through our use of the jump mass balances at the liquid–gas phase interface.

Since $c$ can be taken to be a constant and since there is no homogeneous chemical reaction, the overall differential mass balance from Table 8.5.1-10 requires

$$\frac{\partial v_2^\diamond}{\partial z_2} = 0 \tag{9.3.1-4}$$

This implies

$$v_2^\diamond = v_2^\diamond(t) \tag{9.3.1-5}$$

Let us assume that $\mathcal{D}_{(AB)}$ may be taken to be a constant. In view of (9.3.1-3), using Table 8.5.1-8 we may write the differential mass balance for species $A$ consistent with Fick's first law as

$$\frac{\partial x_{(A)}}{\partial t} + \frac{\partial x_{(A)}}{\partial z_2} v_2^\diamond - \mathcal{D}_{(AB)} \frac{\partial^2 x_{(A)}}{\partial z_2^2} = 0 \tag{9.3.1-6}$$

Before we can solve this problem, we must determine $v_2^\diamond = v_2^\diamond(t)$. We will do this without solving the differential momentum balance. As discussed in Section 9.1.1, the differential momentum balance could be used to determine the pressure distribution, but any effect beyond the hydrostatic effect would be too small to measure.

The overall jump mass balance (see Exercise 8.3.1-1) requires that

$$\text{at } z_2 = 0 : \ c^{(l)} v_2^{\diamond(l)} = c v_2^\diamond \tag{9.3.1-7}$$

The jump mass balance for species $A$ (8.2.1-4) demands that

$$\text{at } z_2 = 0 : \ c^{(l)} v_2^{\diamond(l)} = N_{(A)2} \tag{9.3.1-8}$$

This means that

$$\text{at } z_2 = 0 : \ N_{(A)2} = c v_2^\diamond \tag{9.3.1-9}$$

Using (9.3.1-9), the definition for the molar-averaged velocity $\mathbf{v}^\diamond$, and Fick's first law from Table 8.5.1-7, we can reason that

$$\text{at } z_2 = 0 \text{ for } t > 0 : \ v_2^\diamond = x_{(A)} v_2^\diamond - \mathcal{D}_{(AB)} \frac{\partial x_{(A)}}{\partial z_2}$$

$$= -\frac{\mathcal{D}_{(AB)}}{1 - x_{(A)}} \frac{\partial x_{(A)}}{\partial z_2} \tag{9.3.1-10}$$

Equations (9.3.1-2), (9.3.1-5), and (9.3.1-10) allow us to say that

everywhere for $t > 0$ and $z_2 > 0$ :

$$v_2^\diamond = -\frac{\mathcal{D}_{(AB)}}{1 - x_{(A)eq}} \frac{\partial x_{(A)}}{\partial z_2} \bigg|_{z_2=0} \tag{9.3.1-11}$$

Note that $(\partial x_{(A)}/\partial z_2)_{z_2=0}$ is a function of time.

Let us assume that $\mathcal{D}_{(AB)}$ may be taken to be a constant. In view of (9.3.1-11), Equation (9.3.1-6) becomes

$$\frac{\partial x_{(A)}}{\partial t} - \frac{\mathcal{D}_{(AB)}}{1 - x_{(A)eq}} \frac{\partial x_{(A)}}{\partial z_2} \bigg|_{z_2=0} \frac{\partial x_{(A)}}{\partial z_2} - \mathcal{D}_{(AB)} \frac{\partial^2 x_{(A)}}{\partial z_2^2} = 0 \tag{9.3.1-12}$$

We seek a solution to this equation consistent with boundary conditions (9.3.1-1) and (9.3.1-2).

Let us look for a solution by first transforming (9.3.1-12) into an ordinary differential equation. An earlier experience (see Section 3.2.4) suggests writing (9.3.1-12) in terms of a new independent variable

$$\eta \equiv \frac{z_2}{\sqrt{4\mathcal{D}_{(AB)}t}} \tag{9.3.1-13}$$

In terms of this variable, Equation (9.3.1-12) may be expressed as

$$\frac{d^2 x_{(A)}}{d\eta^2} + (2\eta + \varphi)\frac{dx_{(A)}}{d\eta} = 0 \tag{9.3.1-14}$$

where

$$\varphi \equiv \frac{1}{1 - x_{(A)\text{eq}}}\frac{dx_{(A)}}{d\eta}\bigg|_{\eta=0} \tag{9.3.1-15}$$

The appropriate boundary conditions for (9.3.1-14) are

$$\text{at } \eta = 0: \ x_{(A)} = x_{(A)\text{eq}} \tag{9.3.1-16}$$

and

$$\text{as } \eta \to \infty: \ x_{(A)} \to x_{(A)0} \tag{9.3.1-17}$$

Integrating (9.3.1-14) once, we find

$$\frac{dx_{(A)}}{d\eta} = C_1 \exp\left(-\left[\eta + \frac{\varphi}{2}\right]^2\right) \tag{9.3.1-18}$$

Here $C_1$ is a constant to be determined. Carrying out a second integration, consistent with (9.3.1-16), we learn

$$x_{(A)} - x_{(A)\text{eq}} = C_1 \int_{\varphi/2}^{\eta+\varphi/2} e^{-x^2}\,dx \tag{9.3.1-19}$$

Boundary condition (9.3.1-17) requires

$$x_{(A)0} - x_{(A)\text{eq}} = C_1 \int_{\varphi/2}^{\infty} e^{-x^2}\,dx$$

$$= C_1 \left(\int_0^{\infty} e^{-x^2}\,dx - \int_0^{\varphi/2} e^{-x^2}\,dx\right)$$

$$= C_1 \frac{\sqrt{\pi}}{2}\left[1 - \text{erf}\left(\frac{\varphi}{2}\right)\right] \tag{9.3.1-20}$$

We have as a final result that (Arnold 1944; Bird et al. 1960, p. 594)

$$\frac{x_{(A)} - x_{(A)\text{eq}}}{x_{(A)0} - x_{(A)\text{eq}}} = \frac{2/\sqrt{\pi}}{1 - \text{erf}(\varphi/2)}\int_{\varphi/2}^{\eta+\varphi/2} e^{-x^2}\,dx$$

$$= \frac{\text{erf}(\eta + \varphi/2) - \text{erf}(\varphi/2)}{1 - \text{erf}(\varphi/2)} \tag{9.3.1-21}$$

From (9.3.1-15) and (9.3.1-21), we determine

$$\varphi = \frac{-2\left(x_{(A)\text{eq}} - x_{(A)0}\right)\exp\left[-(\varphi/2)^2\right]}{\sqrt{\pi}\left(1 - x_{(A)\text{eq}}\right)\left[1 - \text{erf}(\varphi/2)\right]} \tag{9.3.1-22}$$

Starting (9.3.1-9) and (9.3.1-10), we can calculate the rate of evaporation of $A$ as

$$N_{(A)2}\big|_{z_2=0} = \frac{-c\mathcal{D}_{(AB)}}{1 - x_{(A)\text{eq}}}\frac{\partial x_{(A)}}{\partial z_2}\bigg|_{z_2=0}$$

$$= \frac{-c}{1 - x_{(A)\text{eq}}}\sqrt{\frac{\mathcal{D}_{(AB)}}{4t}}\frac{dx_{(A)}}{d\eta}\bigg|_{\eta=0}$$

$$= -c\varphi\sqrt{\frac{\mathcal{D}_{(AB)}}{4t}}$$

$$= \left[c\left(x_{(A)\text{eq}} - x_{(A)0}\right)\sqrt{\frac{\mathcal{D}_{(AB)}}{\pi t}}\right]\left[-\frac{\sqrt{\pi}\,\varphi}{2\left(x_{(A)\text{eq}} - x_{(A)0}\right)}\right] \tag{9.3.1-23}$$

It is natural to ask about the effect of convection in the differential mass balance for species $A$. We have the mental picture that diffusion takes place slowly and that $v_2^\circ$ is small. It is always dangerous to refer to a dimensional quantity as being small, since its magnitude depends upon the system of units chosen. If we arbitrarily set $v_2^\circ = 0$, we see from (9.3.1-11) and (9.3.1-15) that this has the effect of setting $\varphi = 0$ in the solution obtained above:

$$\text{no convection}: \quad \frac{x_{(A)} - x_{(A)\text{eq}}}{x_{(A)0} - x_{(A)\text{eq}}} = \text{erf}(\eta) \tag{9.3.1-24}$$

and

$$\text{no convection}: \quad N_{(A)2}\big|_{z_2=0} = -c\mathcal{D}_{(AB)}\frac{\partial x_{(A)}}{\partial z_2}\bigg|_{z_2=0}$$

$$= -c\sqrt{\frac{\mathcal{D}_{(AB)}}{4t}}\frac{dx_{(A)}}{d\eta}\bigg|_{\eta=0}$$

$$= c\left(x_{(A)\text{eq}} - x_{(A)0}\right)\sqrt{\frac{\mathcal{D}_{(AB)}}{\pi t}} \tag{9.3.1-25}$$

Upon comparison of (9.3.1-23) with this last expression, we see that

$$C_{\text{correction}} \equiv -\frac{\sqrt{\pi}\,\varphi}{2\left(x_{(A)\text{eq}} - x_{(A)0}\right)} \tag{9.3.1-26}$$

may be regarded as a correction to the evaporation rate accounting for diffusion-induced convection. Knowing

$$\frac{x_{(A)\text{eq}} - x_{(A)0}}{1 - x_{(A)\text{eq}}}$$

we can compute $\varphi$ from (9.3.1-22) as well as $\left(1 - x_{(A)\text{eq}}\right)C_{\text{correction}}$ from (9.3.1-26). Figure 9.3.1-1 shows us that $C_{\text{correction}} \to 1$, only in the limit $x_{(A)0} \to x_{(A)\text{eq}} \to 0$.

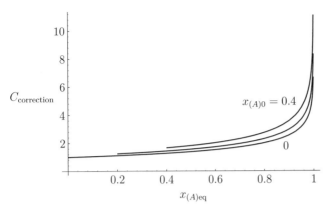

**Figure 9.3.1-1.** $C_{\text{correction}}$ as a function of $x_{(A)\text{eq}}$ for $x_{(A)0} = 0$ (bottom curve), for $x_{(A)0} = 0.2$ (middle curve), and for $x_{(A)0} = 0.4$ (top curve).

### A Very Long Tube with a Falling Interface

Let us once again consider a very long tube that is fixed in a laboratory frame of reference. We will make no special arrangements to maintain a stationary interface; the interface falls as evaporation takes place.

We will assume that the liquid–gas phase interface is a moving plane

$$z_2 = h(t) \tag{9.3.1-27}$$

and that, in place of (9.3.1-2), we have

$$\text{at } z_2 = h \text{ for all } t > 0: \ x_{(A)} = x_{(A)\text{eq}} \tag{9.3.1-28}$$

The overall jump mass balance (see Exercise 8.3.1-1) requires that

$$\text{at } z_2 = h: \ -c^{(l)}u_2 = c\left(v_2^\diamond - u_2\right) \tag{9.3.1-29}$$

where $u_2$ is the $z_2$ component of the speed of displacement of the interface. The jump mass balance for species $A$ (8.2.1-4) demands that

$$\text{at } z_2 = h: \ -c^{(l)}u_2 = N_{(A)2} - cx_{(A)}u_2 \tag{9.3.1-30}$$

This means that[3]

$$\text{at } z_2 = h: \ u_2 = -\frac{c}{c^{(l)} - c}v_2^\diamond \tag{9.3.1-31}$$

and

$$\text{at } z_2 = h: \ N_{(A)2} = \frac{c\left(cx_{(A)} - c^{(l)}\right)}{c - c^{(l)}}v_2^\diamond \tag{9.3.1-32}$$

---

[3] Since the speed of the interface is always finite, it is interesting to note in (9.3.1-31) that, as $c^{(l)} - c \to 0$, the molar-averaged velocity $v_2^\diamond \to 0$. In this limit, the effect of convection disappears, whether it is convection attributable to the moving interface or to diffusion-induced convection.

From Fick's first law of binary diffusion (Table 8.5.1-7),

$$N_{(A)2} = cx_{(A)}v_2^\diamond - cD_{(AB)}\frac{\partial x_{(A)}}{\partial z_2} \tag{9.3.1-33}$$

This together with (9.3.1-32) permits us to say that

$$\text{at } z_2 = h: \quad v_2^\diamond = \frac{\left(c - c^{(l)}\right)}{c^{(l)}\left(1 - x_{(A)eq}\right)}D_{(AB)}\frac{\partial x_{(A)}}{\partial z_2} \tag{9.3.1-34}$$

In view of (9.3.1-5), we conclude that

$$\text{everywhere}: \quad v_2^\diamond = -\frac{\left(c^{(l)} - c\right)}{c^{(l)}\left(1 - x_{(A)eq}\right)}D_{(AB)}\frac{\partial x_{(A)}}{\partial z_2}\bigg|_{z_2=h} \tag{9.3.1-35}$$

For the gas phase, the differential mass balance for species $A$ requires

$$\frac{\partial x_{(A)}}{\partial t} + v_2^\diamond\frac{\partial x_{(A)}}{\partial z_2} = D_{(AB)}\frac{\partial^2 x_{(A)}}{\partial z_2^2} \tag{9.3.1-36}$$

or in view of (9.3.1-35)

$$\frac{\partial x_{(A)}}{\partial t} - \frac{\left(c^{(l)} - c\right)}{c^{(l)}\left(1 - x_{(A)eq}\right)}D_{(AB)}\frac{\partial x_{(A)}}{\partial z_2}\bigg|_{z_2=h}\frac{\partial x_{(A)}}{\partial z_2} - D_{(AB)}\frac{\partial^2 x_{(A)}}{\partial z_2^2} = 0 \tag{9.3.1-37}$$

This must be solved consistent with (9.3.1-1) and (9.3.1-28).

With the transformations

$$\eta \equiv \frac{z_2}{\sqrt{4D_{(AB)}t}} \tag{9.3.1-38}$$

and

$$\lambda \equiv \frac{h}{\sqrt{4D_{(AB)}t}} \tag{9.3.1-39}$$

Equation (9.3.1-37) becomes

$$\frac{d^2 x_{(A)}}{d\eta^2} + (2\eta + \varphi)\frac{dx_{(A)}}{d\eta} = 0 \tag{9.3.1-40}$$

where

$$\varphi \equiv \frac{c^{(l)} - c}{c^{(l)}\left(1 - x_{(A)eq}\right)}\frac{dx_{(A)}}{d\eta}\bigg|_{\eta=\lambda}$$

$$= -2v_2^\diamond\sqrt{\frac{t}{D_{(AB)}}} \tag{9.3.1-41}$$

This last line follows directly from (9.3.1-35). From (9.3.1-1) and (9.3.1-28), we see that the appropriate boundary conditions for (9.3.1-40) are

$$\text{as } \eta \to \infty: \quad x_{(A)} \to x_{(A)0} \tag{9.3.1-42}$$

and

$$\text{at } \eta = \lambda: \quad x_{(A)} = x_{(A)eq} \tag{9.3.1-43}$$

with the recognition that we must require

$$\lambda = \text{a constant} \tag{9.3.1-44}$$

The solution for (9.3.1-40) consistent with (9.3.1-42) and (9.3.1-43) is

$$\frac{x_{(A)} - x_{(A)\text{eq}}}{x_{(A)0} - x_{(A)\text{eq}}} = \frac{\text{erf}(\eta + \varphi/2) - \text{erf}(\lambda + \varphi/2)}{1 - \text{erf}(\lambda + \varphi/2)} \tag{9.3.1-45}$$

From (9.3.1-41) and (9.3.1-45), we see that $\varphi$ is a solution of

$$\varphi = \frac{2 \left(x_{(A)0} - x_{(A)\text{eq}}\right)}{\sqrt{\pi} \left(1 - x_{(A)\text{eq}}\right)} \frac{\left(c^{(l)} - c\right)}{c^{(l)}} \frac{\exp\left[-(\lambda + \varphi/2)^2\right]}{\left[1 - \text{erf}(\lambda + \varphi/2)\right]} \tag{9.3.1-46}$$

Let us characterize the rate of evaporation by the position of the phase interface $z_2 = h(t)$. From (9.3.1-39) and (9.3.1-44), it follows that

$$\frac{dh}{dt} = u_2$$

$$= \lambda \sqrt{\frac{\mathcal{D}_{(AB)}}{t}} \tag{9.3.1-47}$$

From (9.3.1-31), (9.3.1-41), and (9.3.1-47), we have

$$\varphi = 2 \left(c^{(l)} - c\right) \frac{\lambda}{c} \tag{9.3.1-48}$$

For a given physical system, this together with (9.3.1-46) can be solved simultaneously for $\varphi$ and $\lambda$ using Mathematica (1993).

Let us conclude by examining the effect of neglecting convection in the liquid. If one simply says that $v_2^\diamond = 0$ and uses Fick's second law (Table 8.5.1-8) even though the overall jump mass balance (9.3.1-29) suggests that this is unreasonable, we find that

$$\lambda = \frac{\left(x_{(A)0} - x_{(A)\text{eq}}\right)}{\sqrt{\pi} \left(c^{(l)}/c - x_{(A)\text{eq}}\right)} \frac{\exp\left(-\lambda^2\right)}{\left[1 - \text{erf}(\lambda)\right]} \tag{9.3.1-49}$$

In the context of a particular physical system, this can be solved for $\lambda$ using Mathematica (1993).

Slattery and Mhetar (1996) observed the evaporation of a small amount of liquid from the bottom of a vertical tube that was 70 cm high and open at the top. A video camera in a previously calibrated configuration was used to record the position of the liquid–vapor interface. Changes in the position of the interface as small as 2 $\mu$m could be detected. As the evaporation proceeded, energy was transferred to the liquid–vapor interface. As the result of the small resistance to the flow of energy from the surrounding air, through the glass tube, to the liquid, the liquid temperature is nearly equal to the ambient temperature (Lee and Wilke 1954).

From (9.3.1-46), we see that, in the limit $x_{(A)\text{eq}} \to x_{(A)0}$, the dimensionless molar averaged velocity $\varphi \to 0$, and the effects of diffusion-induced convection can be neglected. We have chosen two liquids to emphasize this effect. Methanol has a relatively low vapor pressure at room temperature, and we anticipate that the effects of diffusion-induced convection will be small. Methyl formate has a larger vapor pressure, and the effects of diffusion-induced convection can be anticipated to be larger.

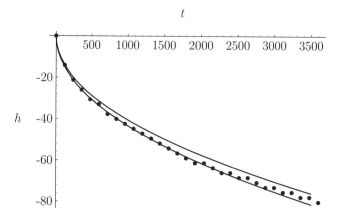

**Figure 9.3.1-2.** The lower curve gives the position of the phase interface $h$ ($\mu$m) as a function of $t$ (s) for evaporation of methanol into air at $T = 25.4°$C and $p = 1.006 \times 10^5$ Pa. The upper curve is the same case derived by arbitrarily neglecting convection.

For the evaporation of methanol, $T = 25.4°$C, $p = 1.006 \times 10^5$ Pa, $x_{(methanol)eq} = 0.172$, $x_{(methanol)0} = 0$, $\mathcal{D}_{(methanol,air)} = 1.558 \times 10^{-5}$ m$^2$/s [corrected from $1.325 \times 10^{-5}$ m$^2$/s at $0°$C and 1 atmosphere (Washburn 1929, p. 62) using a popular empirical correlation (Reid, Prausnitz, and Poling 1987; Fuller et al. on p. 587)], $c^{(l)} = 24.6$ kg mole/m$^3$ (Dean 1979, pp. 7-271 and 10-89), and $c = 0.0411$ kg mole/m$^3$ [estimated for air (Dean 1979, p. 10-92)]. Solving (9.3.1-46) and (9.3.1-48) simultaneously, we find $\lambda = -1.74 \times 10^{-4}$ and $\varphi = -0.208$. From Figure 9.3.1-2, it can be seen that the predicted height of the phase interface follows the experimental data closely up to 2,500 s. As the concentration front begins to approach the top of the tube, we would expect the rate of evaporation to be reduced. Note that neglecting diffusion-induced convection results in an underprediction of the rate of evaporation.

For the evaporation of methyl formate, $T = 25.4°$C, $p = 1.011 \times 10^5$ Pa, $x_{(mformate)eq} = 0.784$, $x_{(mformate)0} = 0$, $\mathcal{D}_{(mformate,air)} = 1.020 \times 10^{-5}$ m$^2$/s [corrected from $0.872 \times 10^{-5}$ m$^2$/s at $0°$C and 1 atmosphere (Washburn 1929, p. 62) using a popular empirical correlation (Reid, Prausnitz, and Poling 1987; Fuller et al., p. 587)], $c^{(l)} = 16.1$ kg mole/m$^3$ (Dean 1979, pp. 7-277 and 10-89), and $c = 0.0411$ kg mole/m$^3$ [estimated for air (Dean 1979, p. 10-92)]. Solving (9.3.1-46) and (9.3.1-48) simultaneously, we find $\lambda = -1.84 \times 10^{-3}$ and $\varphi = -1.44$. From Figure 9.3.1-3, it can be seen that the predicted height of the phase interface follows the experimental data closely over the entire range of observation. Once again, neglecting diffusion-induced convection results in an underprediction of the rate of evaporation.

The Stefan tube is ideally suited for measuring the diffusion coefficient of a volatile species in air. It is necessary only to do a least-square-error fit of the theoretical result to the experimental data. A comparison of the results obtained for the two experiments described above is shown in Table 9.3.1-1. Also shown are the predictions of a popular empirical correlation (Reid, Prausnitz, and Poling 1987; Fuller et al., p. 587).

### Relation to Analysis by Arnold (1944)

Arnold (1944) assumed that in a laboratory frame of reference the evaporating liquid–gas interface was stationary. Only species $A$ moved in the gas phase; species $B$ was stationary.

**Table 9.3.1-1.** Diffusion coefficients for methanol and methyl formate in air from a least-square-error fit of the experimental data, as reported in the literature (Washburn 1929) with appropriate corrections for temperature and pressure (Reid, Prausnitz, and Poling 1987; Fuller et al., p. 587), and from a popular empirical correlation (Reid, Prausnitz, and Poling 1987; Fuller et al., p. 587)

| | $\mathcal{D}_{\text{(methanol,air)}}$ (m$^2$/s) | $\mathcal{D}_{\text{(methylformate,air)}}$ (m$^2$/s) |
|---|---|---|
| Fit | $1.553 \times 10^{-5}$ | $1.022 \times 10^{-5}$ |
| Reported | $1.558 \times 10^{-5}$ | $1.020 \times 10^{-5}$ |
| Empirical | $1.662 \times 10^{-5}$ | $1.163 \times 10^{-5}$ |

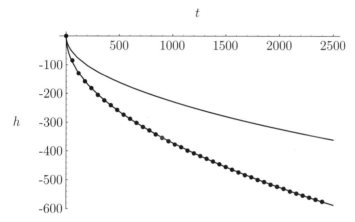

**Figure 9.3.1-3.** The lower curve gives the position of the phase interface $h$ ($\mu$m) as a function of $t$ (s) for evaporation of methyl formate into air at $T = 25.4°$C and $p = 1.011 \times 10^5$ Pa. The upper curve is the same case derived by arbitrarily neglecting convection.

An experiment to test this theory would be designed to have the liquid move to the interface as it evaporates, in order to maintain a stationary phase interface.

To determine the concentration profile in the gas phase, both we and Arnold (1944) solved (9.3.1-36) consistent with (9.3.1-1) and (9.3.1-28). The only difference between our solutions was that we represented $v_2^\diamond$ by (9.3.1-35), whereas Arnold (1944) used

$$\text{everywhere:} \quad v_2^\diamond = -\frac{\mathcal{D}_{(AB)}}{1 - x_{(A)\text{eq}}} \left. \frac{\partial x_{(A)}}{\partial z_2} \right|_{z_2 = h} \tag{9.3.1-50}$$

Let us consider the falling interface problem with a moving frame of reference, in which the phase interface is stationary. If we were to apply the Arnold (1944) analysis to this problem, we would not arrive at the correct concentration distribution. Although the Arnold (1944) analysis is correct for the stationary interface in a laboratory frame of reference, it does not completely account for gas-phase convection in the overall jump mass balance and

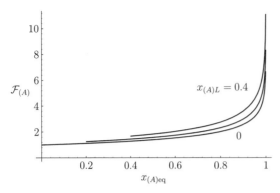

**Figure 9.3.1-4.** The evaporation of $A$ into a mixture of $A$ and $B$: $\mathcal{F}_{(A)}$ as a function of $x_{(A)\mathrm{eq}}$ for $x_{(A)\infty} = 0$ (bottom curve), for $x_{(A)\infty} = 0.2$ (middle curve), and for $x_{(A)\infty} = 0.4$ (top curve).

in the jump mass balance for species $A$ for the stationary interface in the moving frame of reference.

Whereas the Arnold (1944) analysis does not give the correct concentration distribution for the stationary interface in the moving frame of reference, we see by comparing (9.3.1-35) and (9.3.1-50) that the error will be very small, since normally $c \ll c^{(l)}$. Using the Arnold (1944) concentration distribution together with (9.3.1-31), (9.3.1-47), and (9.3.1-50), we arrive at results that are virtually indistinguishable from those shown in Figures 9.3.1-2 and 9.3.1-3.

**Exercise 9.3.1-1** *Film theory: evaporation*   Compute the film theory correction for evaporation of $A$ into a mixture of $A$ and $B$. The result,

$$\mathcal{F}_{(A)} = \frac{\ln\left[\left(1 - x_{(A)\infty}\right) / \left(1 - x_{(A)\mathrm{eq}}\right)\right]}{x_{(A)\mathrm{eq}} - x_{(A)\infty}}$$

is shown in Figure 9.3.1-4 for three values of $x_{(A)\infty}$.

**Exercise 9.3.1-2** *Mass transfer within a solid sphere (constant surface composition)*   A solid sphere of species $B$ contains a uniformly distributed trace of species $A$; the mass fraction of $A$ is $\omega_{(A)0}$. The radius of the sphere is $R$. At time $t = 0$, this sphere is placed in a large, well-stirred container of species $A$ (either vapor or liquid) containing a trace of species $B$. If such a solid were at equilibrium with this fluid, its composition would be $\omega_{(A)\mathrm{eq}}$. Determine the composition distribution in the sphere as a function of time.

In analyzing this problem, assume that the density $\rho$ of the sphere and the diffusion coefficient $\mathcal{D}^0_{(AB)}$ are constants independent of composition. This should be nearly true in the limit as $\omega_{(A)0} \to \omega_{(A)\mathrm{eq}}$. Do *not* assume $\mathbf{v} = \mathbf{0}$ merely because we are concerned with diffusion in a solid. If you think that this is true, prove it.

For a complete solution of this problem, we would have to solve for the concentration distributions in the fluid and the solid simultaneously. In carrying out such a solution, we would assume that the chemical potentials of both species are continuous across the phase interface and that the jump mass balances for both species must be satisfied at the phase interface. This would be a difficult problem, similar to the one that we encountered in Section 6.2.1 where we studied the temperature distribution in a semi-infinite slab. I suggest that you make the same kind of simplifying approximation that we made there. Assume that for the solid phase

at $r = R$ for $t > 0$ : $\omega_{(A)} = \omega_{(A)eq}$

*Answer:*

$$\frac{\omega_{(A)} - \omega_{(A)eq}}{\omega_{(A)0} - \omega_{(A)eq}} = \frac{2R}{\pi} \sum_{n=1}^{\infty} \frac{(-1)^{n+1}}{nr} \sin\left(n\pi \frac{r}{R}\right) \exp\left(\frac{-n^2\pi^2 \mathcal{D}_{(AB)}^0 t}{R^2}\right)$$

*Hint:* See Exercise 6.2.3-4.

**Exercise 9.3.1-3** *Mass transfer within a solid sphere* In Exercise 9.3.1-2, we assume that the surface of the solid sphere is in equilibrium with the fluid very far away from it. As suggested by our treatment of a somewhat similar heat-transfer problem in Section 6.2.2, there is a preferred approach.

Repeat Exercise 9.3.1-2 using Newton's "law" of mass transfer discussed in Section 9.2.1.

i) Having made the change of variables suggested in Exercise 6.2.3-4, look for a solution by the method of separation of variables. Satisfy all but the initial condition, and determine that the concentration distribution has the form

$$r^* \omega_{(A)}^* = \sum_{n=1}^{\infty} E_n \sin\left(\lambda_n r^*\right) \exp\left(-\lambda_n^2 t^*\right)$$

where

$$\omega_{(A)}^* = \frac{\omega_{(A)} - \omega_{(A)eq}}{\omega_{(A)0} - \omega_{(A)eq}}, \quad r^* \equiv \frac{r}{R}$$

$$t^* \equiv \frac{t \mathcal{D}_{(AB)}^0}{R^2}$$

and the $\lambda_n$ ($n = 1, 2, \ldots$) are the roots of

$$\lambda_n \cot(\lambda_n) = 1 - A \tag{9.3.1-51}$$

Here

$$A \equiv \frac{R k_{(A)\omega}}{\rho \mathcal{D}_{(AB)}^0}$$

The roots of (9.3.1-51) have been tabulated by Carslaw and Jaeger (1959, p. 492).

ii) Take essentially the same approach as we did in parts (iii) and (iv) of Exercise 6.2.3-1 to show that

$$\text{for } n \neq m : \int_0^1 \sin\left(\lambda_m r^*\right) \sin\left(\lambda_n r^*\right) dr^* = 0$$

and

$$\int_0^1 \sin^2\left(\lambda_n r^\star\right) dr^\star = \frac{\lambda_n^2 + A(A-1)}{2\left[\lambda_n^2 + (1-A)^2\right]}$$

iii) Use the results of (ii) in determining the coefficients $E_n$. Determine that the final expression for the concentration distribution is (Carslaw and Jaeger 1959, p. 238)

$$r^\star \omega_{(A)}^\star = \sum_{n=1}^\infty \frac{2A\left[\lambda_n^2 + (1-A)^2\right]\sin\lambda_n}{\lambda_n^2\left[\lambda_n^2 + A(A-1)\right]}\sin\left(\lambda_n r^\star\right)\exp\left(-\lambda_n^2 t^\star\right)$$

**Exercise 9.3.1-4** *Binary diffusion in a stagnant gas*  (Stevenson 1968)

i) Assume that species $B$ is stagnant:

$$\mathbf{N}_{(B)} = \mathbf{0}$$

Determine that

$$\mathbf{v}^\diamond = -\frac{\mathcal{D}_{(AB)}^0}{1 - x_{(A)}}\nabla x_{(A)}$$

ii) Let us limit ourselves to steady-state diffusion with no chemical reactions under conditions such that the total molar density $c$ and diffusion coefficient $\mathcal{D}_{(AB)}^0$ are constants. Conclude that the differential mass balance for species $A$ reduces to

$$\nabla x_{(B)} \cdot \nabla x_{(B)} = x_{(B)}\text{div}\left(\nabla x_{(B)}\right)$$

iii) Introduce as a new dependent variable

$$\alpha \equiv -\ln x_{(B)}$$

Prove that, in terms of $\alpha$, the differential mass balance for species $A$ becomes

$$\text{div}(\nabla \alpha) = 0$$

**Exercise 9.3.1-5** *More on steady-state diffusion through a stagnant gas film*    Reexamine the problem described in Exercise 9.3.1-1, assuming that the diffusion takes place in a cylindrical tube of radius $R$. Use the approach suggested in Exercise 9.3.1-4.

Conclude that we must relax the requirement that the tangential components of velocity must be zero at $r = R$, if species $B$ is stagnant (Whitaker 1967a).

**Exercise 9.3.1-6** *Constant evaporating mixture* (Bird et al. 1960, p. 587)    Consider a situation similar to that discussed in Exercise 9.3.1-1. A mixture of ethanol and toluene evaporates into an ideal-gas mixture of ethanol, toluene, and nitrogen. The apparatus is arranged in such a manner that the liquid–gas phase interface remains fixed in space as the evaporation takes place. Nitrogen is taken to be insoluble in the evaporating liquid. At the top of the column, the gas is maintained as essentially pure nitrogen. The entire system is maintained at $60°C$ and constant pressure. We have used the method of Fuller et al. (Reid, Prausnitz, and Poling 1987, p. 587) to estimate $\mathcal{D}_{(EN_2)} = 1.53 \times 10^{-5}$ m$^2$/s and $\mathcal{D}_{(TN_2)} = 9.42 \times 10^{-6}$ m$^2$/s.

**Table 9.3.1-2.** Vapor–liquid equilibrium data for the
ethanol–toluene system at 60°C[a]

| Mole fraction | | | | | |
|---|---|---|---|---|---|
| toluene in liquid | 0.096 | 0.155 | 0.233 | 0.274 | 0.375 |
| Mole fraction | | | | | |
| toluene in vapor | 0.147 | 0.198 | 0.242 | 0.256 | 0.277 |
| Total pressure, | | | | | |
| mmHg | 388 | 397 | 397 | 395 | 390 |

[a]From Wright (1933).

The vapor–liquid equilibrium data for the ethanol–toluene system at 60°C are given in Table 9.3.1-2.

i) Use jump mass balances for ethanol and toluene for the stationary phase interface to prove that

$$\frac{N_{(E)2}}{N_{(T)2}} = \frac{x_{(E)\text{liquid}}}{x_{(T)\text{liquid}}}$$

ii) Use Fick's first law in the gas phase to determine that for a mixture whose composition does not change as evaporation takes place (a constant evaporating mixture)

$$\frac{\mathcal{D}_{(Em)}}{\mathcal{D}_{(Tm)}} = \frac{\ln\left(1 - x_{(T)\text{eq}}/x_{(T)\text{liquid}}\right)}{\ln\left(1 - x_{(E)\text{eq}}/x_{(E)\text{liquid}}\right)}$$

Here $\mathcal{D}_{(Em)}$ and $\mathcal{D}_{(Tm)}$ are the diffusion coefficients for ethanol and toluene in the gas mixture as discussed in Section 8.4.6; $x_{(E)\text{liquid}}$ and $x_{(T)\text{liquid}}$ are the mole fractions of ethanol and toluene in the liquid phase; $x_{(E)\text{eq}}$ and $x_{(T)\text{eq}}$ are the mole fractions of ethanol and toluene in the gas at the phase interface.

iii) Determine that at high pressures

$$\frac{\mathcal{D}_{(EN_2)}}{\mathcal{D}_{(TN_2)}} = \frac{x_{(E)\text{liquid}}x_{(T)\text{eq}}}{x_{(T)\text{liquid}}x_{(E)\text{eq}}}$$

Use this relationship to estimate that at high pressures the composition of the constant evaporating mixture is $x_{(T)\text{liquid}} = 0.098$.

iv) Use the relationship developed in (ii) to *estimate* that at 760 mmHg the composition of the constant evaporating mixture is $x_{(T)\text{liquid}} = 0.15$. Robinson, Wright, and Bennett (1932) experimentally obtained $x_{(T)\text{liquid}} = 0.20$.

For a more complete discussion of this problem in the context of ternary diffusion as well as a better comparison with the experimental data, see Exercise 9.4.1-3.

### 9.3.2  Rate of Isothermal Crystallization

Our objective here is to determine how the rate of crystallization is affected by convection induced both by diffusion and by the density difference between the solid crystal and the adjacent liquid.

Most crystallizations take place under conditions such that forced convection is important. To construct a simulation for such operations, one requires empirical correlations for the energy and mass transfer coefficients. Typical empirical correlations apply under conditions such that induced convection is not important. For this reason, most analyses of crystallization have been done in the context of film theory (Bird et al. 1960, p. 658), although none of these analyses [see, e.g., Wilcox (1969)] has been used to construct corrections for the energy and mass transfer coefficients measured in the absence of induced convection.

Because the unknown film thickness can be expected to depend on the degree of induced convection, film theory is not well suited to investigate these effects. For this reason, we will examine them for crystallization from a semi-infinite adjacent liquid. This problem has received relatively little previous attention. Smith, Tiller, and Rutter (1955) considered isothermal crystallization, assuming that the speed of displacement of the interface was a constant. Tiller (1991, p. 183) reported without derivation the result for isothermal crystallization, assuming equal densities for the fluid and solid phases.

We wish to examine the complete problem, beginning with isothermal crystallization in this section and considering nonisothermal crystallization in the next.

The semi-infinite, incompressible liquid composed of species $A$ and $B$ shown in Figure 9.3.2-1 is subjected to a uniform pressure and temperature such that it is supersaturated with respect to $A$:

$$\text{at } t = 0 \text{ for all } z_2 > 0: \quad \omega_{(A)} = \omega_{(A)0} \tag{9.3.2-1}$$

For time $t > 0$, heterogeneous crystallization of pure species $A$ begins at the wall. We will assume that the rate of crystallization is controlled by diffusion, that all physical properties are constants, that the solid–liquid phase interface is a plane

$$z_2 = h(t) \tag{9.3.2-2}$$

and that

$$\text{at } z_2 = h \text{ for all } t > 0: \quad \omega_{(A)} = \omega_{(A)\text{eq}} \tag{9.3.2-3}$$

where $\omega_{(A)\text{eq}}$ denotes the solubility of $A$ at the imposed temperature and pressure. Our objective is to determine the rate at which the solid–liquid interface moves across the material.

In this analysis, we will seek a concentration distribution of the form

$$\omega_{(A)} = \omega_{(A)}(z_2, t) \tag{9.3.2-4}$$

Because it grows on a stationary wall, the velocity of the solid phase is zero. We will assume that, because the densities and concentrations of the fluid and solid differ, the fluid moves as the solid $A$ grows:

$$v_2^{(l)} = v_2^{(l)}(z_2, t)$$
$$v_1^{(l)} = v_3^{(l)}$$
$$\quad = 0 \tag{9.3.2-5}$$

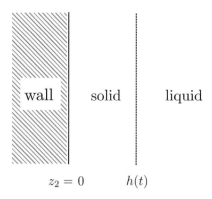

**Figure 9.3.2-1.** Moving solid–liquid phase interface $z_2 = h(t)$ during crystallization.

The overall differential mass balance (Section 8.3.1) requires that

$$\frac{\partial v_2^{(l)}}{\partial z_2} = 0 \tag{9.3.2-6}$$

and

$$v_2^{(l)} = v_2^{(l)}(t) \tag{9.3.2-7}$$

If we neglect inertial effects and recognize that the system is maintained at a uniform pressure, both the overall differential momentum balance for the liquid as well as the overall jump momentum balance are satisfied identically.

The overall jump mass balance (8.3.1-6) requires that

$$\text{at } z_2 = h: \quad -\rho^{(s)} u_2 = \rho^{(l)} \left( v_2^{(l)} - u_2 \right) \tag{9.3.2-8}$$

where $u_2$ is the $z_2$ component of the speed of displacement of the interface. The jump mass balance for species $A$ (8.2.1-4) demands that

$$\text{at } z_2 = h: \quad -\rho^{(s)} u_2 = n_{(A)2}^{(l)} - \rho^{(l)} \omega_{(A)} u_2 \tag{9.3.2-9}$$

This means that

$$\text{at } z_2 = h: \quad n_{(A)2}^{(l)} = \frac{\rho^{(l)} \left( \rho^{(l)} \omega_{(A)} - \rho^{(s)} \right)}{\rho^{(l)} - \rho^{(s)}} v_2^{(l)} \tag{9.3.2-10}$$

and

$$\text{at } z_2 = h: \quad u_2 = \frac{\rho^{(l)}}{\rho^{(l)} - \rho^{(s)}} v_2^{(l)} \tag{9.3.2-11}$$

From Fick's first law of binary diffusion (Section 8.4.5),

$$n_{(A)2}^{(l)} = \rho^{(l)} \omega_{(A)} v_2^{(l)} - \rho^{(l)} \mathcal{D}_{(AB)} \frac{\partial \omega_{(A)}}{\partial z_2} \tag{9.3.2-12}$$

This, together with (9.3.2-10), permits us to say that

$$\text{at } z_2 = h: \quad v_2^{(l)} = \frac{\left(\rho^{(l)} - \rho^{(s)}\right)}{\rho^{(s)} \left(1 - \omega_{(A)\text{eq}}\right)} \mathcal{D}_{(AB)} \frac{\partial \omega_{(A)}}{\partial z_2} \tag{9.3.2-13}$$

In view of (9.3.2-7), we conclude that

$$\text{everywhere}: \quad v_2^{(l)} = -\frac{\left(\rho^{(s)} - \rho^{(l)}\right)}{\rho^{(s)} \left(1 - \omega_{(A)\text{eq}}\right)} \mathcal{D}_{(AB)} \left.\frac{\partial \omega_{(A)}}{\partial z_2}\right|_{z_2 = h} \tag{9.3.2-14}$$

For the liquid phase, the differential mass balance for species $A$ requires

$$\frac{\partial \omega_{(A)}}{\partial t} + v_2^{(l)} \frac{\partial \omega_{(A)}}{\partial z_2} = \mathcal{D}_{(AB)} \frac{\partial^2 \omega_{(A)}}{\partial z_2{}^2} \tag{9.3.2-15}$$

or in view of (9.3.2-14)

$$\frac{\partial \omega_{(A)}}{\partial t} - \frac{\left(\rho^{(s)} - \rho^{(l)}\right)}{\rho^{(s)} \left(1 - \omega_{(A)\text{eq}}\right)} \mathcal{D}_{(AB)} \left.\frac{\partial \omega_{(A)}}{\partial z_2}\right|_{z_2=h} \frac{\partial \omega_{(A)}}{\partial z_2} - \mathcal{D}_{(AB)} \frac{\partial^2 \omega_{(A)}}{\partial z_2{}^2} = 0 \tag{9.3.2-16}$$

This must be solved consistent with the boundary conditions (9.3.2-1) and (9.3.2-3).

With the transformations

$$\eta \equiv \frac{z_2}{\sqrt{4\mathcal{D}_{(AB)}t}} \tag{9.3.2-17}$$

and

$$\lambda \equiv \frac{h}{\sqrt{4\mathcal{D}_{(AB)}t}} \tag{9.3.2-18}$$

Equation (9.3.2-16) becomes

$$\frac{d^2 \omega_{(A)}}{d\eta^2} + (2\eta + \varphi) \frac{d\omega_{(A)}}{d\eta} = 0 \tag{9.3.2-19}$$

where

$$\varphi \equiv \frac{\rho^{(s)} - \rho^{(l)}}{\rho^{(s)} \left(1 - \omega_{(A)\text{eq}}\right)} \left.\frac{d\omega_{(A)}}{d\eta}\right|_{\eta=\lambda} \tag{9.3.2-20}$$

From (9.3.2-1) and (9.3.2-3), we see that the appropriate boundary conditions for (9.3.2-19) are

$$\text{as } \eta \to \infty: \quad \omega_{(A)} \to \omega_{(A)0} \tag{9.3.2-21}$$

and

$$\text{at } \eta = \lambda: \quad \omega_{(A)} = \omega_{(A)\text{eq}} \tag{9.3.2-22}$$

with the recognition that we must require

$$\lambda = \text{a constant} \tag{9.3.2-23}$$

The solution for (9.3.2-19) consistent with (9.3.2-21) and (9.3.2-22) is

$$\frac{\omega_{(A)} - \omega_{(A)\text{eq}}}{\omega_{(A)0} - \omega_{(A)\text{eq}}} = \frac{\text{erf}(\eta + \varphi/2) - \text{erf}(\lambda + \varphi/2)}{1 - \text{erf}(\lambda + \varphi/2)} \tag{9.3.2-24}$$

From (9.3.2-20) and (9.3.2-24), we see that $\varphi$ is a solution of

$$\varphi = \frac{2\left(\omega_{(A)0} - \omega_{(A)eq}\right)}{\sqrt{\pi}\left(1 - \omega_{(A)eq}\right)} \frac{\left(\rho^{(s)} - \rho^{(l)}\right)}{\rho^{(s)}} \frac{\exp\left[-(\lambda + \varphi/2)^2\right]}{[1 - \mathrm{erf}(\lambda + \varphi/2)]} \tag{9.3.2-25}$$

Let us characterize the rate of crystallization by the position of the phase interface $z_2 = h(t)$. From (9.3.2-18) and (9.3.2-23), it follows that

$$\frac{dh}{dt} = u_2$$

$$= \lambda \sqrt{\frac{\mathcal{D}_{(AB)}}{t}} \tag{9.3.2-26}$$

From (9.3.2-11), (9.3.2-14), (9.3.2-17), and (9.3.2-24), we have

$$\frac{dh}{dt} = u_2$$

$$= \frac{\rho^{(l)}\mathcal{D}_{(AB)}}{\rho^{(s)}\left(1 - \omega_{(A)eq}\right)} \left.\frac{\partial\omega_{(A)}}{\partial z_2}\right|_{z_2=h}$$

$$= \frac{\rho^{(l)}}{2\rho^{(s)}\left(1 - \omega_{(A)eq}\right)} \left.\frac{d\omega_{(A)}}{d\eta}\right|_{\eta=\lambda} \sqrt{\frac{\mathcal{D}_{(AB)}}{t}}$$

$$= \frac{\rho^{(l)}\left(\omega_{(A)0} - \omega_{(A)eq}\right)}{\rho^{(s)}\left(1 - \omega_{(A)eq}\right)} \frac{\exp[-(\lambda + \varphi/2)^2]}{[1 - \mathrm{erf}(\lambda + \varphi/2)]} \sqrt{\frac{\mathcal{D}_{(AB)}}{\pi t}} \tag{9.3.2-27}$$

Comparing (9.3.2-26) and (9.3.2-27), we conclude that

$$\lambda = \frac{\rho^{(l)}\left(\omega_{(A)0} - \omega_{(A)eq}\right)}{\sqrt{\pi}\,\rho^{(s)}\left(1 - \omega_{(A)eq}\right)} \frac{\exp[-(\lambda + \varphi/2)^2]}{[1 - \mathrm{erf}(\lambda + \varphi/2)]} \tag{9.3.2-28}$$

From (9.3.2-25) and (9.3.2-28)

$$\varphi = 2\left(\rho^{(s)} - \rho^{(l)}\right)\frac{\lambda}{\rho^{(l)}} \tag{9.3.2-29}$$

and

$$\lambda + \frac{\varphi}{2} = \frac{\left(\omega_{(A)0} - \omega_{(A)eq}\right)}{\sqrt{\pi}\left(1 - \omega_{(A)eq}\right)} \frac{\exp[-(\lambda + \varphi/2)^2]}{[1 - \mathrm{erf}(\lambda + \varphi/2)]} \tag{9.3.2-30}$$

These equations must be solved simultaneously for $\varphi$ and $\lambda$.

Let us illustrate the predictions of (9.3.2-29) and (9.3.2-30) for isothermal crystallization of $n$-decane from a solution of $n$-decane in $n$-butane. We have assumed that $\rho^{(s)} = 903$ kg/m$^3$ (TRCTAMU), $\rho^{(l)} = 712$ kg/m$^3$ [a molar average of the pure component densities (Washburn 1928, p. 27)], $\omega_{(A)0} = 0.6$, and $\mathcal{D}_{(AB)} = 1.02 \times 10^{-9}$ m$^2$/s (Reid et al. 1987, pp. 598 and 611). In using (9.3.3-4) and (9.3.3-5), we have taken $a = 14.2$ and $b = 3.45 \times 10^3$ K (Reid et al. 1987, p. 373) and $T = 224$ K to conclude that $\omega_{(A)eq} = 0.501$. At this temperature, the solution is concentrated everywhere, including the region immediately adjacent to the interface.

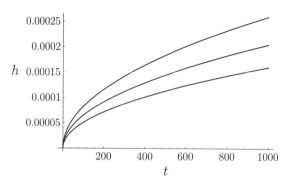

**Figure 9.3.2-2.** The middle curve gives the position of the phase interface $h$ ($\mu$m) as a function of $t$ (s) for isothermal crystallization of $n$-decane from a solution of $n$-decane in $n$-butane at $T = 224$ K, assuming $\omega_{(A)} = 0.6$. The top curve is the same case in which the densities of the two phases are assumed to be the same. The bottom curve is derived by arbitrarily neglecting convection.

Using Mathematica (1993) to solve (9.3.2-29) and (9.3.2-30) simultaneously for this case, we find $\varphi = 0.0544$ and $\lambda = 0.101$. The middle curve of Figure 9.3.2-2 shows the interface position $h$ as a function of $t$ as predicted by (9.3.2-18) for this value of $\lambda$.

Let us conclude by examining the effect of neglecting convection in the liquid. There are two ways in which this can be done.

If we simply say that $v_2^{(l)} = 0$ and use Fick's second law (Table 8.5.1-8) even though (assuming $\rho^{(s)} \neq \rho^{(l)}$) the overall jump mass balance (9.3.2-8) suggests that this is unreasonable, we find that

$$\lambda = \frac{\left(\omega_{(A)0} - \omega_{(A)\text{eq}}\right)}{\sqrt{\pi}\left(\rho^{(s)}/\rho^{(l)} - \omega_{(A)\text{eq}}\right)} \frac{\exp\left(-\lambda^2\right)}{[1 - \text{erf}(\lambda)]} \tag{9.3.2-31}$$

For the case described above, we find $\lambda = 0.0793$. The corresponding prediction of (9.3.2-18), shown as the bottom curve in Figure 9.3.2-2, is a significant underprediction of the rate of crystallization. This case involves a relatively concentrated solution. As the concentration of the bulk solution is reduced, the effects of convection become less important, and the complete solution approaches this limiting case.

If we assume that $\rho^{(s)} = \rho^{(l)}$ for the case described above, we see from the overall jump mass balance (9.3.2-8) together with (9.3.2-13) and (9.3.2-14) that $v_2^{(l)} = 0$ everywhere. Under these conditions, (9.3.2-29) and (9.3.2-30) reduce to

$$\varphi = 0 \tag{9.3.2-32}$$

and

$$\lambda = \frac{\left(\omega_{(A)0} - \omega_{(A)\text{eq}}\right)}{\sqrt{\pi}\left(1 - \omega_{(A)\text{eq}}\right)} \frac{\exp\left[-\lambda^2\right]}{[1 - \text{erf}(\lambda)]} \tag{9.3.2-33}$$

For the case described above, we find $\lambda = 0.129$. The corresponding prediction of (9.3.2-18), shown as the top curve in Figure 9.3.2-2, is a significant overprediction of the rate of crystallization. In effect, two separate errors or approximations have been made: $v_2^{(l)} = 0$

and $\rho^{(s)} = \rho^{(l)}$. It is for this reason that the complete solution does not approach this limit as the concentration of the bulk solution is reduced.

This section is taken from Slattery and Robinson (1996).

**Exercise 9.3.2-1** *Film theory: isothermal crystallization* Compute the film-theory correction for isothermal crystallization of $A$ from a solution of $A$ and $B$. By analogy with the discussion in Exercise 9.3.1-1, determine that

$$\mathcal{F}_{(A)} = \frac{\ln\left[\left(1 - \omega_{(A)\infty}\right) / \left(1 - \omega_{(A)\mathrm{eq}}\right)\right]}{\omega_{(A)\mathrm{eq}} - \omega_{(A)\infty}}$$

### 9.3.3 Rate of Nonisothermal Crystallization

Let us extend the discussion in the preceding section to account for energy transfer.

The system, which is at a uniform pressure, is initially at a uniform temperature and concentration:

$$\text{at } t = 0 \text{ for } z_2 > 0: \ T^{(l)} = T_0 > T_{\mathrm{eutectic}}$$

$$\omega_{(A)} = \omega_{(A)0} \tag{9.3.3-1}$$

where $T_{\mathrm{eutectic}}$ is the eutectic temperature, below which only a single, solid phase of mixed composition exists. For time $t > 0$, the temperature of stationary wall is changed,

$$\text{at } z_2 = 0 \text{ for } t > 0: \ T^{(s)} = T_1 \tag{9.3.3-2}$$

where

$$T_{\mathrm{eutectic}} < T_1 < T_0 \tag{9.3.3-3}$$

and heterogeneous crystallization of pure species $A$ begins at the wall. For the imposed pressure, we know that for an ideal solution (Reid et al. 1987, p. 373)

$$\text{at } z_2 = h(t) \text{ for all } t > 0: \ x_{(A)} = x_{(A)\mathrm{eq}}$$

$$= \exp\left(a - \frac{b}{T^{(l)}}\right) \tag{9.3.3-4}$$

or (Table 8.5.1-2)

$$\text{at } z_2 = h(t) \text{ for all } t > 0:$$

$$\omega_{(A)} = \omega_{(A)\mathrm{eq}}$$

$$= \frac{x_{(A)\mathrm{eq}} M_{(A)}}{x_{(A)\mathrm{eq}} M_{(A)} + \left(1 - x_{(A)\mathrm{eq}}\right) M_{(B)}} \tag{9.3.3-5}$$

Our objective is to determine the rate at which the solid–liquid interface moves across the material, assuming that the rate of crystallization is controlled both by the rate of diffusion and by the rate of energy transfer.

From Section 9.3.2, we have

$$\frac{\omega_{(A)} - \omega_{(A)\mathrm{eq}}}{\omega_{(A)0} - \omega_{(A)\mathrm{eq}}} = \frac{\mathrm{erf}(\eta + \varphi/2) - \mathrm{erf}(\lambda + \varphi/2)}{1 - \mathrm{erf}(\lambda + \varphi/2)} \tag{9.3.3-6}$$

where

$$\eta \equiv \frac{z_2}{\sqrt{4\mathcal{D}_{(AB)}t}} \tag{9.3.3-7}$$

and

$$\lambda \equiv \frac{h}{\sqrt{4\mathcal{D}_{(AB)}t}} \tag{9.3.3-8}$$

Equations (9.3.2-29) and (9.3.2-30) must be satisfied in determining $\varphi$ and $\lambda$. In what follows, we will find that the temperature of the interface is a constant and, as a result, that $\omega_{(A)\mathrm{eq}}$ and $\varphi$ are constants.

By analogy with (6.3.3-16), the solution of the differential energy balance for the solid phase (Table 8.5.2-1) consistent with (9.3.3-2) is

$$T^{(s)\star} = D_1 \operatorname{erf}\left(\sqrt{\frac{\mathcal{D}_{(AB)}}{\alpha^{(s)}}}\,\eta\right) \tag{9.3.3-9}$$

Here

$$T^{\star} \equiv \frac{T - T_1}{T_0 - T_1} \tag{9.3.3-10}$$

and

$$\alpha \equiv \frac{k}{\rho\hat{c}} \tag{9.3.3-11}$$

Since pressure is nearly independent of position in the liquid phase, the overall differential energy balance for a multicomponent system (Table 8.5.2-1) takes the form

$$\rho\hat{c}\left(\frac{\partial T^{(l)}}{\partial t} + \nabla T^{(l)} \cdot \mathbf{v}\right) = -\operatorname{div}\boldsymbol{\epsilon} - \sum_{C=1}^{N}\nabla\overline{H}_{(C)} \cdot \mathbf{j}_{(C)} \tag{9.3.3-12}$$

where, after neglecting the Dufour effect in (8.4.3-2),

$$\boldsymbol{\epsilon} = -k\nabla T \tag{9.3.3-13}$$

In view of (9.3.3-13), Equation (9.3.3-12) reduces to

$$\rho\hat{c}\left(\frac{\partial T^{(l)}}{\partial t} + \nabla T^{(l)} \cdot \mathbf{v}\right) = k\operatorname{div}\nabla T^{(l)} - \sum_{C=1}^{N}\nabla\overline{H}_{(C)} \cdot \mathbf{j}_{(C)} \tag{9.3.3-14}$$

In terms of the dimensionless variables

$$\overline{H}_{(A)}^{\star} \equiv \frac{\overline{H}_{(A)}}{\hat{c}\,(T_0 - T_1)}, \quad \mathbf{j}_{(A)}^{\star} \equiv \frac{L_0\mathbf{j}_{(A)}}{\rho^{(l)}\mathcal{D}_{AB}}$$

$$t^{\star} \equiv \frac{v_0 t}{L_0}, \qquad\qquad \mathbf{v}^{\star} \equiv \frac{\mathbf{v}}{v_0} \tag{9.3.3-15}$$

$$z_i^{\star} \equiv \frac{z_i}{L_0}$$

Equation (9.3.3-14) becomes

$$\frac{\partial T^{(l)\star}}{\partial t^{\star}} + \nabla T^{(l)\star} \cdot \mathbf{v}^{\star} = \frac{1}{N_{Re}N_{Pr}} \text{div} \nabla T^{(l)\star} - \frac{1}{N_{Re}N_{Sc}} \sum_{C=A}^{B} \nabla \overline{H}_{(C)}^{\star} \cdot \mathbf{j}_{(C)}^{\star} \qquad (9.3.3\text{-}16)$$

Here

$$N_{Pr} \equiv \frac{\hat{c}\mu}{k}$$

$$N_{Re} \equiv \frac{\rho v_0 L_0}{\mu} \qquad (9.3.3\text{-}17)$$

$$N_{Sc} \equiv \frac{\mu}{\rho \mathcal{D}_{AB}}$$

We limit our attention here to cases such that

$$N_{Sc} \gg N_{Pr} \qquad (9.3.3\text{-}18)$$

or

$$\mathcal{D}_{AB} \ll \alpha^{(l)} \equiv \frac{k^{(l)}}{\rho^{(l)}\hat{c}^{(l)}} \qquad (9.3.3\text{-}19)$$

Under these circumstances, (9.3.3-14) reduces to

$$\rho\hat{c} \left( \frac{\partial T^{(l)}}{\partial t} + \nabla T^{(l)} \cdot \mathbf{v} \right) = k \, \text{div} \, \nabla T^{(l)} \qquad (9.3.3\text{-}20)$$

For the one-dimensional problem with which we are concerned here, (9.3.3-20) reduces to

$$\frac{\partial T^{(l)}}{\partial t} + v_2^{(l)} \frac{\partial T^{(l)}}{\partial z_2} = \alpha^{(l)} \frac{\partial^2 T^{(l)}}{\partial z_2^2} \qquad (9.3.3\text{-}21)$$

From (9.3.2-7), (9.3.2-11), and (9.3.2-26), we have

$$v_2^{(l)} = \lambda \left( \frac{\rho^{(l)} - \rho^{(s)}}{\rho^{(l)}} \right) \sqrt{\frac{\mathcal{D}_{(AB)}}{t}} \qquad (9.3.3\text{-}22)$$

In terms of the dimensionless variables defined by (9.3.3-7) and (9.3.3-10), Equation (9.3.3-21) becomes

$$\frac{d^2 T^{(l)\star}}{d\eta^2} + 2\frac{\mathcal{D}_{(AB)}}{\alpha^{(l)}} \left[ \eta - \lambda \left( \frac{\rho^{(l)} - \rho^{(s)}}{\rho^{(l)}} \right) \right] \frac{dT^{(l)\star}}{d\eta} = 0 \qquad (9.3.3\text{-}23)$$

Integrating, we have

$$\frac{dT^{(l)\star}}{d\eta} = D_3 \exp \left\{ \frac{\mathcal{D}_{(AB)}}{\alpha^{(l)}} \left[ -\eta^2 + 2\lambda \left( \frac{\rho^{(l)} - \rho^{(s)}}{\rho^{(l)}} \right) \eta \right] \right\} \qquad (9.3.3\text{-}24)$$

Integrating again consistent with the condition

$$\text{as } \eta \to \infty: \ T^{(l)\star} \to 1 \qquad (9.3.3\text{-}25)$$

we conclude that [you may find it helpful to use Mathematica (1993)]

$$T^{(l)\star} = 1 + D_2 \operatorname{erfc}\left\{\sqrt{\frac{\mathcal{D}_{(AB)}}{\alpha^{(l)}}}\left[\eta + \lambda\left(\frac{\rho^{(s)} - \rho^{(l)}}{\rho^{(l)}}\right)\right]\right\} \tag{9.3.3-26}$$

In view of (8.4.3-2), (8.4.3-3), (9.3.2-11), (9.3.2-17), and (9.3.2-26), the jump energy balance (8.3.4-5) for a system at a uniform pressure takes the form

$$\text{at } z_2 = h: \ k^{(s)}(T_0 - T_1)\frac{\partial T^{(s)\star}}{\partial z_2} - k^{(l)}(T_0 - T_1)\frac{\partial T^{(l)\star}}{\partial z_2}$$

$$= -\rho^{(l)}\hat{U}^{(l)}\left(v_2^{(l)} - u_2\right) - \rho^{(s)}\hat{U}^{(s)}u_2 + v_2^{(l)}T_{22}^{(l)} - \sum_{C=A}^{B}\overline{H}_{(C)}^{(l)}j_{(C)2}^{(l)}$$

$$= \left[\rho^{(s)}\left(\hat{U}^{(l)} - \hat{U}^{(s)}\right) + \frac{\left(\rho^{(s)} - \rho^{(l)}\right)}{\rho^{(l)}}P^{(l)}\right]u_2 - \sum_{C=A}^{B}\overline{H}_{(C)}^{(l)}j_{(C)2}^{(l)}$$

$$= \rho^{(s)}\Delta\hat{H}\frac{dh}{dt} - \sum_{C=A}^{B}\overline{H}_{(C)}^{(l)}j_{(C)2}^{(l)} \tag{9.3.3-27}$$

or

$$\text{at } \eta = \lambda: \ \frac{dT^{(s)\star}}{d\eta} - \frac{k^{(l)}}{k^{(s)}}\frac{dT^{(l)\star}}{d\eta}$$

$$= \frac{2\rho^{(s)}\mathcal{D}_{(AB)}\Delta\hat{H}}{k^{(s)}(T_0 - T_1)}\lambda - \sum_{C=A}^{B}\frac{N_{Pr}^{(l)}}{N_{Sc}^{(l)}}\frac{k^{(l)}}{k^{(s)}}\overline{H}_{(C)}^{(l)\star}\frac{d\omega_{(C)}}{d\eta} \tag{9.3.3-28}$$

where

$$\Delta\hat{H} \equiv \hat{H}^{(l)} - \hat{H}^{(s)} \tag{9.3.3-29}$$

In the limit (9.3.3-18), Equation (9.3.3-28) reduces to

$$\text{at } \eta = \lambda: \ \frac{dT^{(s)\star}}{d\eta} - \frac{k^{(l)}}{k^{(s)}}\frac{dT^{(l)\star}}{d\eta}$$

$$= \frac{2\rho^{(s)}\mathcal{D}_{(AB)}\Delta\hat{H}}{k^{(s)}(T_0 - T_1)}\lambda \tag{9.3.3-30}$$

Finally, we observe that temperature is continuous across the phase boundary:

$$\text{at } \eta = \lambda: \ T^{(s)\star} = T^{(l)\star} \tag{9.3.3-31}$$

In summary, six equations, (9.3.3-4), (9.3.3-5), (9.3.2-29), (9.3.2-30), (9.3.3-30), and (9.3.3-31) must be solved simultaneously for six unknowns: $x_{(A)eq}$, $\omega_{(A)eq}$, $\varphi$, $\lambda$, $D_1$, and $D_2$.

The bottom curve in Figure 9.3.3-1 shows the position of the phase interface $h$ (m) as a function of $t$ (s) for nonisothermal crystallization of $n$-decane from a solution of $n$-decane in $n$-butane. We have estimated that $\rho^{(s)} = 903$ kg/m$^3$ (Marsh 1994), $\rho^{(l)} = 706$ kg/m$^3$ (Washburn 1929, p. 27), $\omega_{(A)0} = 0.6$, $\mathcal{D}_{(AB)} = 1.03 \times 10^{-9}$ m$^2$/s (Reid et al. 1987, pp. 598 and 611), $T_0 = 240$ K, $T_1 = 224$ K, $k^{(s)} = 0.135$ J m$^{-1}$ s$^{-1}$ K$^{-1}$, $k^{(l)} = 0.153$ J m$^{-1}$ s$^{-1}$ K$^{-1}$ (Jamieson 1975), $\hat{c}^{(s)} = 1.511 \times 10^3$ J kg$^{-1}$ K$^{-1}$ (Marsh 1994), $\hat{c}^{(l)} = 2.271 \times 10^3$ J kg$^{-1}$ K$^{-1}$ (Marsh 1994), and $\Delta H = 2.02 \times 10^5$ J/kg (Daubert and Danner 1989). In using (9.3.3-4) and (9.3.3-5), we have taken $a = 14.2$ and $b = 3.45 \times 10^3$ K (Reid et al. 1987, p. 373). Note

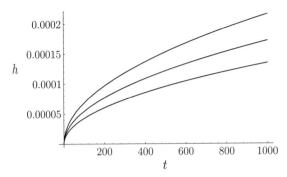

**Figure 9.3.3-1.** The middle curve shows the position of the phase interface $h$ ($\mu$m) as a function of $t$ (s) for nonisothermal crystallization of $n$-decane from a solution of $n$-decane in $n$-butane, assuming that the initial temperature is 240 K and that the temperature of the cooled wall is 224 K. The top curve is the same case in which the densities of the two phases are assumed to be the same. The bottom curve is derived by arbitrarily neglecting convection.

that, at the wall temperature $T_1 = 224$ K, the solution is concentrated everywhere, including the region immediately adjacent to the interface. Observe also that (9.3.3-19) is satisfied for this system.

A simultaneous solution of (9.3.3-4), (9.3.3-5), (9.3.2-29), (9.3.2-30), (9.3.3-30), and (9.3.3-31) gives $\omega_{(A)\text{eq}} = 0.518$, $\varphi = 0.0472$, $\lambda = 0.0847$, $D_1 = 1.28$, and $D_2 = -1.00$. The middle curve of Figure 9.3.3-1 shows $h$ as a function of $t$ as predicted by (9.3.3-8) for this value of $\lambda$.

Let us conclude by again examining the effect of neglecting convection in the liquid.

If we say that $v_2^{(l)} = 0$ even though (assuming $\rho^{(s)} \neq \rho^{(l)}$) the overall jump mass balance (9.3.2-8) suggests that this is unreasonable, we find that (9.3.2-31) is still valid and (9.3.3-26) becomes

$$T^{(l)\star} = 1 + D_2 \operatorname{erfc}\left(\sqrt{\frac{\mathcal{D}_{(AB)}}{\alpha^{(l)}}}\,\eta\right) \tag{9.3.3-32}$$

For the case described above, we find $\omega_{(A)\text{eq}} = 0.517$, $\lambda = 0.0665$, $D_1 = 1.25$, and $D_2 = -0.998$. The corresponding prediction of (9.3.3-8), shown as the bottom curve in Figure 9.3.3-1, is a significant underprediction of the rate of crystallization. This case involves a relatively concentrated solution. As the concentration of the bulk solution is reduced, the effects of convection become less important, and the complete solution approaches this limiting case.

If we assume that $\rho^{(s)} = \rho^{(l)}$ for the case described above, we see from the overall jump mass balance (9.3.2-8) together with (9.3.2-13) and (9.3.2-14) that $v_2^{(l)} = 0$ everywhere. Under these conditions, (9.3.2-29) and (9.3.2-30) reduce to (9.3.2-32) and (9.3.2-33). For the case described above, we find $\omega_{(A)\text{eq}} = 0.519$, $\lambda = 0.106$, $D_1 = 1.31$, and $D_2 = -0.996$.

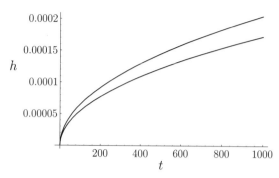

**Figure 9.3.3-2.** The top curve shows the position of the phase interface $h$ (m) as a function of $t$ (s) for isothermal crystallization of $n$-decane from a solution of $n$-decane in $n$-butane at 224 K. The bottom curve shows the interface position predicted for the nonisothermal case, assuming that the initial temperature is 240 K and that the temperature of the cooled wall is 224 K.

The corresponding prediction of (9.3.3-8), shown as the top curve in Figure 9.3.3-1, is a significant overprediction of the rate of crystallization.

The complete solution approaches the solution for the limiting case in which $v_2^{(l)} = 0$ as the concentration of the bulk solution is reduced. The complete solution does not approach the solution for the case in which $\rho^{(s)} = \rho^{(l)}$ (even though $v_2^{(l)} = 0$), because the densities are not equal in reality and such a solution does not describe a limiting case.

Figure 9.3.3-2 compares the complete isothermal solution with the complete nonisothermal solution from Figure 9.3.3-1, assuming that in both cases the wall temperature is 224 K. The rate of crystallization predicted by the isothermal analysis is larger than that predicted by the nonisothermal one. In the isothermal analysis, there is a resistance to mass transfer; in the nonisothermal analysis, there is an additional resistance to energy transfer.

For both the isothermal and nonisothermal analyses, we can summarize our results as follows:

1) Neglecting induced convection results in an underprediction of the rate of crystallization. In the limit of crystallization from a dilute solution, this difference disappears.
2) Neglecting the difference between the solid and liquid densities results in an overprediction of the rate of crystallization, even in the limit of crystallization from a dilute solution. In effect, two separate errors or approximations have been made: $v_2^{(l)} = 0$ and $\rho^{(s)} = \rho^{(l)}$. It is for this reason that the complete solution does not approach this limit as the concentration of the bulk solution is reduced.
3) An isothermal analysis results in an overprediction of the rate of crystallization, assuming that the temperature of the wall remains the same.

This section is taken from Slattery and Robinson (1996).

**Exercise 9.3.3-1** *Film theory: nonisothermal crystallization*    Redo Exercise 9.3.2-1 for the case of nonisothermal crystallization. Conclude that the result found there still holds, with the additional

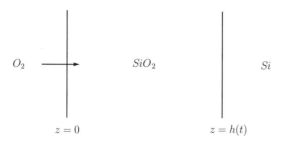

**Figure 9.3.4-1.** A film of silicon dioxide grows on silicon.

constraint that

$$\mathcal{F}_{\text{energy}} = \left\{ \frac{1}{[1 - \exp(-v_2 L/\alpha)]} - \frac{\hat{H}}{\hat{c}(T_\infty - T_{\text{eq}})} \right\} \frac{v_2 L}{\alpha}$$

Here

$$\frac{v_2 L}{\alpha} = \frac{N_{Pr}}{N_{Sc}} \ln\left( \frac{1 - \omega_{(A)\infty}}{1 - \omega_{(A)\text{eq}}} \right)$$

which means that $\mathcal{F}_{\text{energy}}$ is independent of $L$.

### 9.3.4 Silicon Oxidation

Deal and Grove (1965) proposed the original theory for the oxidation of silicon. They approximated this inherently unsteady-state process by a "steady-state" process, in which the concentration of molecular oxygen in silicon dioxide was a linear function of position. This "steady-state" assumption led them to make a further error in stating the mass balance for oxygen at the moving boundary. In what follows, I present the analysis presented by Peng, Wang, and Slattery (1996), which corrects these errors.

Referring to Figure 9.3.4-1, our objective here is to follow the formation of $SiO_2$ as a function of time on silicon subsequent to its initial exposure to $O_2$. We will make several assumptions:

1) The reaction at the $SiO_2$–Si interface is

$$Si + O_2 \rightarrow SiO_2 \tag{9.3.4-1}$$

This reaction is assumed to be instantaneous.

2) As suggested in Figure 9.3.4-1, we will work in a frame of reference in which the $O_2$–$SiO_2$ phase interface is stationary.

3) Molecular oxygen $O_2$ is the only diffusing component; $SiO_2$ is stationary in the oxide layer.

4) The molar density of silicon dioxide $c_{(SiO_2)}$ is independent of position and time in the oxide layer.

5) Equilibrium is established at the $SiO_2$–$O_2$ interface.

6) Temperature is independent of time and position. This means that the energy released by the oxidation reaction is dissipated rapidly, and the system remains in thermal equilibrium.

7) All physical parameters are considered to be constants.

Mass conservation for each species in the system must be satisfied. The differential mass balance equation for each species must be satisfied at each point in the $SiO_2$ phase. The jump mass balance equation for each species must be satisfied at each point on the $SiO_2$–Si interface.

The differential mass balance for $O_2$ requires (Table 8.5.1-5)

$$\frac{\partial c_{(O_2)}}{\partial t} + \frac{\partial N_{(O_2)z}}{\partial z} = 0 \tag{9.3.4-2}$$

The differential mass balance for $SiO_2$ is satisfied identically as the result of assumptions 9.3.4 and 9.3.4.

In view of assumption 9.3.4, jump mass balances for $O_2$ and $SiO_2$ require (Section 8.2.1)

$$\text{at } z = h: \quad -N_{(O_2)z} = \frac{r_{(O_2)}^{\sigma}}{M_{(O_2)}} \tag{9.3.4-3}$$

$$\text{at } z = h: c_{(SiO_2)}\frac{dh(t)}{dt} = -\frac{r_{(O_2)}^{\sigma}}{M_{(O_2)}} \tag{9.3.4-4}$$

Recognize here that $r_{(O_2)}^{(\sigma)}$ denotes the rate of production of $O_2$ at the phase interface. Since $O_2$ is actually consumed in reaction (9.3.4-1), the value of $r_{(O_2)}^{(\sigma)}$ will be a negative number. Adding (9.3.4-3) and (9.3.4-4), we find

$$\text{at } z = h: \quad \frac{dh}{dt} = \frac{N_{(O_2)z}}{c_{(SiO_2)}} \tag{9.3.4-5}$$

which can be used to replace either (9.3.4-3) or (9.3.4-4).

From Fick's first law (Table 8.5.1-7)

$$x_{(SiO_2)}N_{(O_2)z} = -c\mathcal{D}_{(O_2,SiO_2)}\frac{\partial x_{(O_2)}}{\partial z} \tag{9.3.4-6}$$

As the result of assumptions 9.3.4 and 9.3.4, we can write (9.3.4-2) as

$$\frac{\partial c_{(O_2)}}{\partial t} = \mathcal{D}_{(O_2,SiO_2)}\frac{\partial^2 c_{(O_2)}}{\partial z^2} \tag{9.3.4-7}$$

Note that, in arriving at this result, we have not assumed $c$ to be a constant. With reference to Figure 9.3.4-1, Equation (9.3.4-7) is to be solved consistent with the boundary conditions

$$\text{at } z = 0: \quad c_{(O_2)} = c_{(O_2)eq} \tag{9.3.4-8}$$

and, in view of assumption 9.3.4,

$$\text{at } z = h: \quad c_{(O_2)} = 0 \tag{9.3.4-9}$$

The initial condition will be implied by the form of the solution developed below.

With the change of variable

$$u \equiv \frac{z}{\sqrt{4\mathcal{D}_{(O_2,SiO_2)}t}} \tag{9.3.4-10}$$

(9.3.4-7) becomes

$$\frac{d^2 c_{(O_2)}}{du^2} + 2u\frac{dc_{(O_2)}}{du} = 0 \tag{9.3.4-11}$$

In view of (9.3.4-8) and (9.3.4-9), this is to be solved consistent with the boundary conditions:

$$\text{as } u = 0: \quad c_{(O_2)} = c_{(O_2)\text{eq}} \tag{9.3.4-12}$$

$$\text{at } u = \lambda: \quad c_{(O_2)} = 0 \tag{9.3.4-13}$$

where

$$\lambda \equiv \frac{h}{\sqrt{4\mathcal{D}_{(O_2, SiO_2)}t}}$$

$$= \text{a constant} \tag{9.3.4-14}$$

The solution of (9.3.4-11) consistent with (9.3.4-13) and (9.3.4-14) is

$$c_{(O_2)} = c_{(O_2)\text{eq}}\left(1 - \frac{\text{erf}(u)}{\text{erf}(\lambda)}\right) \tag{9.3.4-15}$$

From (9.3.4-5), (9.3.4-6), and (9.3.4-15), we have

$$\frac{dh}{dt} = \frac{c_{(O_2)\text{eq}}}{c_{(SiO_2)}} \sqrt{\frac{\mathcal{D}_{(O_2, SiO_2)}}{\pi t}} \frac{\exp\left(-\lambda^2\right)}{\text{erf}(\lambda)} \tag{9.3.4-16}$$

From (9.3.4-14), we also know that

$$\frac{dh}{dt} = \lambda\sqrt{\frac{\mathcal{D}_{(O_2, SiO_2)}}{t}} \tag{9.3.4-17}$$

Subtracting (9.3.4-17) from (9.3.4-16), we arrive at

$$\lambda = \frac{c_{(O_2)\text{eq}}}{\sqrt{\pi}\, c_{(SiO_2)}} \frac{\exp\left(-\lambda^2\right)}{\text{erf}(\lambda)} \tag{9.3.4-18}$$

which is used to specify $\lambda$.

Figure 9.3.4-2 compares the data of Lie, Razouk, and Deal (1982) for the growth of thick $SiO_2$ films at $20.3 \times 10^5$ Pa and 950°C with the predictions of (9.3.4-14) and (9.3.4-18). The diffusion coefficient $\mathcal{D}_{(O_2, SiO_2)}$ has been taken from the data correlation proposed by Peng et al. (1996). The equilibrium concentration $c_{(O_2)\text{eq}}$ at $P_0 = 1.01 \times 10^5$ Pa has been determined using the suggestion of Barrer (1951, p. 139) and the data of Norton (1961); its dependence upon pressure has been found using Henry's law in the form

$$c_{(O_2)\text{eq}} = \frac{P}{P_0} \times c_{(O_2)\text{eq}}(T, P_0) \tag{9.3.4-19}$$

A further comparison with experimental data is given by Peng et al. (1996), from which this work is taken.

### 9.3.5 Pressure Diffusion in a Natural Gas Well

A natural gas well of depth $L$ has been closed for some time. The mole fraction $x_{(A)0}$ of species $A$ and the pressure $P_0$ at the top of the well are known. We wish to determine the composition and pressure at the bottom of the well.[4]

---

[4] This problem was suggested by G. M. Brown, Department of Chemical Engineering, Northwestern University, Evanston, Illinois.

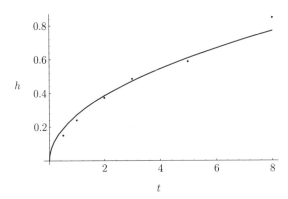

**Figure 9.3.4-2.** Thickness $h$ ($\mu$m) of the $SiO_2$ film as a function of time $t$ (h) from (9.3.4-14) and (9.3.4-18) compared with the data of Lie. et al. (1982) at 950°C and $20.3 \times 10^5$ Pa. In this calculation, we have taken $\mathcal{D}_{(O_2,SiO_2)} = 8.23 \times 10^{-13}$ m$^2$/s and $c_{(O_2)eq} = 4.58 \times 10^{-4}$ kg mole/m$^3$.

To simplify the computations, we will assume that we are dealing with a binary gas mixture of species $A$ and $B$, that temperature is uniform throughout the well, and that the mixture obeys the ideal-gas law and may be regarded as an ideal solution.

Since the system is closed and nothing is changing as a function of time, we postulate

$$\mathbf{v} = \mathbf{0}$$
$$x_{(A)} = x_{(A)}(z_3) \tag{9.3.5-1}$$
$$P = P(z_3)$$

We assume here that gravity acts in the $z_3$ direction. From the differential mass balance for species $A$ of Table 8.5.1-5,

$$\frac{\partial j_{(A)3}}{\partial z_3} = 0 \tag{9.3.5-2}$$

Because the system is closed, we may conclude that

$$j_{(A)3} = 0 \tag{9.3.5-3}$$

In view of (8.4.5-7), Equation (8.4.5-5) requires for an ideal solution

$$j_{(A)3} = -\left(\frac{c^2}{\rho}\right) M_{(A)} M_{(B)} \mathcal{D}_{(AB)} \left[ \frac{dx_{(A)}}{dz_3} + \frac{x_{(A)}}{RT} \left( \overline{V}_{(A)}^{(m)} - \frac{M_{(A)}}{\rho} \right) \frac{dP}{dz_3} \right]$$
$$= 0 \tag{9.3.5-4}$$

or

$$\frac{dx_{(A)}}{dz_3} = -\frac{x_{(A)}}{RT} \left( \overline{V}_{(A)}^{(m)} - \frac{M_{(A)}}{\rho} \right) \frac{dP}{dz_3} \tag{9.3.5-5}$$

From the ideal-gas law and the definitions introduced in Exercises 8.4.2-4 and 8.4.2-5,

$$\overline{V}_{(A)}^{(m)} = \overline{V}_{(B)}^{(m)}$$
$$= \tilde{V}$$
$$= \frac{RT}{P} \tag{9.3.5-6}$$

The differential momentum balance requires

$$\frac{dP}{dz_3} = \rho g \tag{9.3.5-7}$$

Equations (9.3.5-6) and (9.3.5-7) allow us to express (9.3.5-5) as

$$\frac{dx_{(A)}}{dz_3} = -\frac{gx_{(A)}}{RT}\left(\frac{RT\rho}{P} - M_{(A)}\right) \tag{9.3.5-8}$$

But the ideal-gas law further requires

$$\frac{RT\rho}{P} = x_{(A)}M_{(A)} + x_{(B)}M_{(B)} \tag{9.3.5-9}$$

so that we may eliminate $\rho$ and $P$ from (9.3.5-8) and say

$$\frac{dx_{(A)}}{dz_3} = \frac{gx_{(A)}}{RT}\left(M_{(A)} - M_{(B)}\right)\left(1 - x_{(A)}\right) \tag{9.3.5-10}$$

This is readily integrated:

$$\int_{x_{(A)0}}^{x_{(A)}} \frac{dx_{(A)}}{x_{(A)}\left(1 - x_{(A)}\right)} = \frac{g}{RT}\left(M_{(A)} - M_{(B)}\right)\int_0^{z_3} dz_3 \tag{9.3.5-11}$$

to find

$$\frac{1 - x_{(A)0}}{x_{(A)0}}\frac{x_{(A)}}{1 - x_{(A)}} = \exp\left(\frac{gz_3}{RT}[M_{(A)} - M_{(B)}]\right) \tag{9.3.5-12}$$

The composition at the bottom of the well is easily seen to be

$$\text{at } z = L : x_{(A)} = \frac{x_{(A)0}\exp\left([gL/(RT)][M_{(A)} - M_{(B)}]\right)}{1 - x_{(A)0} + x_{(A)0}\exp\left([gL/(RT)][M_{(A)} - M_{(B)}]\right)} \tag{9.3.5-13}$$

From (9.3.5-7) and (9.3.5-9), we see

$$\frac{dP}{dx_{(A)}}\frac{dx_{(A)}}{dz_3} = \frac{dP}{dz_3}$$

$$= \rho g$$

$$= \frac{gP}{RT}\left(M_{(A)}x_{(A)} + M_{(B)}x_{(B)}\right) \tag{9.3.5-14}$$

In view of (9.3.5-10), this last equation may be expressed as

$$\frac{dP}{dx_{(A)}} = \frac{\left[x_{(A)}\left(M_{(A)} - M_{(B)}\right) + M_{(B)}\right]P}{x_{(A)}\left(1 - x_{(A)}\right)\left(M_{(A)} - M_{(B)}\right)} \tag{9.3.5-15}$$

Upon integration, we learn

$$\frac{P}{P_0} = \left(\frac{x_{(A)}}{x_{(A)0}}\right)^{M_{(B)}/(M_{(A)}-M_{(B)})}\left(\frac{1 - x_{(A)0}}{1 - x_{(A)}}\right)^{M_{(A)}/(M_{(A)}-M_{(B)})} \tag{9.3.5-16}$$

In view of (9.3.5-12), we can say alternatively

$$\frac{x_{(A)}P}{x_{(A)0}P_0} = \exp\left(\frac{gz_3M_{(A)}}{RT}\right) \tag{9.3.5-17}$$

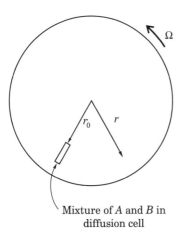

Mixture of $A$ and $B$ in
diffusion cell

**Figure 9.3.5-3.** A concentration
gradient is developed at steady state
in a centrifuge as the result of a
balance between pressure diffusion
and ordinary diffusion.

At the bottom of the well

$$\text{at } z = L : \quad \frac{P}{P_0} = \left\{ 1 - x_{(A)0} + x_{(A)0} \exp\left( \frac{gL}{RT} \left[ M_{(A)} - M_{(B)} \right] \right) \right\}$$

$$\times \left\{ \frac{\exp\left( gLM_{(A)}/RT \right)}{\exp\left( [gL/RT][M_{(A)} - M_{(B)}] \right)} \right\} \tag{9.3.5-18}$$

This same problem can be analyzed assuming that the gas has achieved a thermodynam-ically stable equilibrium (Slattery 1981, p. 492). It is reassuring that the same results are obtained using two apparently radically different approaches.

**Exercise 9.3.5-1** *The ultracentrifuge*    Figure 9.3.5-3 shows a binary liquid solution mounted in a cylindrical cell on a high-speed centrifuge. We wish to determine the concentration distri-bution of the two components $A$ and $B$ at steady state.

To somewhat simplify the analysis, we will assume that the two species form an ideal solution and that the partial molar volumes may be taken to be independent of composition.

i) Determine that

$$\overline{V}_{(B)}^{(m)} \frac{\partial \ln x_{(A)}}{\partial r} = \frac{r\Omega^2}{RT} \left( M_{(A)} \overline{V}_{(B)}^{(m)} - \rho \overline{V}_{(A)}^{(m)} \overline{V}_{(B)}^{(m)} \right)$$

and

$$\overline{V}_{(A)}^{(m)} \frac{\partial \ln x_{(B)}}{\partial r} = \frac{r\Omega^2}{RT} \left( M_{(B)} \overline{V}_{(A)}^{(m)} - \rho \overline{V}_{(A)}^{(m)} \overline{V}_{(B)}^{(m)} \right)$$

to conclude

$$\overline{V}_{(B)}^{(m)} \frac{\partial \ln x_{(A)}}{\partial r} - \overline{V}_{(A)}^{(m)} \frac{\partial \ln x_{(B)}}{\partial r} = \frac{r\Omega^2}{RT} \left( M_{(A)} \overline{V}_{(B)}^{(m)} - M_{(B)} \overline{V}_{(A)}^{(m)} \right)$$

ii) If

$$\text{at } r = r_0: \quad x_{(A)} = x_{(A)0}$$

$$x_{(B)} = x_{(B)0}$$

we may integrate the result of (i) to learn

$$\overline{V}_{(B)}^{(m)} \ln\left(\frac{x_{(A)}}{x_{(A)0}}\right) + \overline{V}_{(A)}^{(m)} \ln\left(\frac{x_{(B)0}}{x_{(B)}}\right) = \frac{\Omega^2}{2RT}\left(M_{(A)}\overline{V}_{(B)}^{(m)} - M_{(B)}\overline{V}_{(A)}^{(m)}\right)\left(r^2 - r_0^2\right)$$

iii) How does this simplify for the case in which the mole fraction of species $A$ is negligible?

**Exercise 9.3.5-2** *Thermal diffusion between vertical plates*  We discussed steady-state natural convection between vertical heated plates in Section 9.4.3. Now let us assume that we have a binary solution of ideal gases between the plates shown in Figure 6.4.1-1. Let us determine the steady-state concentration distribution, assuming that thermal diffusion (Section 8.4.4) is the dominant effect.

For gases, experimental data for the thermal diffusion coefficient $D_{(A)}^T$ are often presented in the form (Hirschfelder et al. 1954, p. 520)

$$\frac{\rho D_{(A)}^T}{c^2 M_{(A)} M_{(B)} D_{(AB)}} = \alpha x_{(A)} x_{(B)}$$

where the thermal diffusion factor $\alpha$ is nearly independent of concentration. Use the differential mass balance for species $A$ to argue that

$$\frac{\partial x_{(A)}}{\partial z_1} + \alpha x_{(A)} x_{(B)} \frac{\partial \ln T}{\partial z_1} = 0$$

which may be readily integrated to find

$$\int_{x_{(A)1}}^{x_{(A)}} \frac{dx_{(A)}}{x_{(A)}(1 - x_{(A)})} = -\alpha \int_{T_1}^{T} d \ln T$$

$$\left(\frac{x_{(A)1}}{x_{(A)}}\right)\left(\frac{1 - x_{(A)}}{1 - x_{(A)1}}\right) = \left(\frac{T}{T_1}\right)^\alpha$$

assuming $\alpha$ is a constant. Since $T$ is known as a function of $z_1$ from (6.4.1-30), this gives us $x_{(A)}$ as a function of $z_2$ and $x_{(A)2}$. To determine $x_{(A)1}$, argue by analogy with (6.4.1-23)

$$\text{for all } z_1^\star: \quad \int_{-1}^{1} x_{(A)} v_1^\star \, dz_2^\star = 0$$

in which $v_1^\star$ is given by (6.4.1-34) and $z_2^\star$ is defined by (6.4.1-20).

Whereas $\alpha$ is very nearly independent of concentration, its temperature dependence may be complex. It has been recommended (Brown 1940) that $\alpha$ in this result be evaluated at a mean temperature

$$T_m \equiv \frac{T_1 T_2}{T_1 - T_2} \ln\left(\frac{T_2}{T_1}\right)$$

**Exercise 9.3.5-3** *Natural convection between concentric vertical cylinders*   Assume that we have a binary solution of ideal gases undergoing natural convection between concentric vertical heated cylinders as described in Exercise 6.4.1-1. Extend the results of Exercise (9.3.5-2) to this problem as well.

The geometry described here is very similar to the Clusius–Dickel thermal diffusion column (Grew and Ibbs 1952, p. 91), which has been used successfully for the separation of isotopes. But there is one important difference. The Clusius–Dickel column has a finite length, whereas, in the problem described in Exercise 6.4.1-1, end effects are neglected. The reversal of flow at the top and bottom of the Clusius–Dickel column reinforces the separation, with the result that the primary concentration difference is not radial but axial. This particular aspect of the problem has been nicely explained by Grew and Ibbs (1952, p. 92).

### 9.3.6   Forced Diffusion in Electrochemical Systems

By forced diffusion, we refer to a situation in which the individual species in a solution are subjected to unequal external forces. As a result of these force differences, the various species are accelerated with respect to one another, and a separation occurs.

Perhaps the most common example of forced diffusion occurs when a salt solution is subjected to an electric field. When a salt such as $AgNO_3$ is dissolved in water, it dissociates. From our present viewpoint, we should almost certainly consider this to be a ternary rather than a binary solution; we should regard $Ag^+$ and $NO_3^-$ as individual species. The necessity for regarding these ions as individual species becomes more obvious when a solution is placed in an electric field. In this case, the force beyond gravity acting on $Ag^+$ is in the opposite direction to that acting on $NO_3^-$.

For simplicity, we shall neglect any effect attributable to pressure diffusion or thermal diffusion. We shall furthermore confine our attention to dilute solutions for which (8.4.6-1) is applicable and simplifies to

$$\mathbf{N}_{(A)} = -c\mathcal{D}_{(Am)} \left[ \nabla x_{(A)} + \frac{M_{(A)} x_{(A)}}{RT} \left( -\mathbf{f}_{(A)} + \sum_{B=1}^{N} \omega_{(B)} \mathbf{f}_{(B)} \right) \right] + c_{(A)} \mathbf{v}^\diamond \qquad (9.3.6\text{-}1)$$

We shall assume that this solution is subjected to an electric field for which the electrostatic potential is $\Phi$. Under these conditions, an ionic species $A$ is subjected to two external forces, gravity and that attributable to the electric field:

$$\mathbf{f}_{(A)} = \mathbf{g} - \frac{\epsilon_{(A)}}{\mathcal{M}_{(A)}} \nabla \Phi \qquad (9.3.6\text{-}2)$$

Here $\mathbf{g}$ is the acceleration of gravity, $\epsilon_{(A)}$ is the ionic charge, and $\mathcal{M}_{(A)}$ is the ionic mass.

In principle, the electrostatic potential $\Phi$ should be determined using Poisson's equation

$$\text{div}(\nabla \Phi) = \frac{F}{\epsilon_0 \epsilon} \sum_{A=1}^{N} a_{(A)} c_{(A)} \qquad (9.3.6\text{-}3)$$

Here, $F$ is the Faraday constant ($9.648 \times 10^4$ C/mol), $\epsilon_0$ the permittivity of vacuum ($8.854 \times 10^{-12}$ F/m), $\epsilon$ the (relative) dielectric constant (dimensionless), and $a_{(A)}$ the valence or charge number of species $A$.

For water at room temperature, $\epsilon \approx 80$. Because $F/(\epsilon_0 \epsilon)$ is so large, a seemingly negligible deviation from local electrical neutrality,

$$\sum_{A=1}^{N} a_{(A)} c_{(A)} = 0 \tag{9.3.6-4}$$

implies a large deviation from Laplace's equation

$$\text{div}(\nabla \Phi) = 0 \tag{9.3.6-5}$$

In practice, it is common to assume (9.3.6-4) or local electrical neutrality (Newman 1973, p. 231), which means that

$$\sum_{B=1}^{N} \omega_{(B)} \mathbf{f}_{(B)} = \sum_{B=1}^{N} \omega_{(B)} \mathbf{g} - \sum_{B=1}^{N} \omega_{(B)} \frac{\epsilon_{(B)}}{\mathcal{M}_{(B)}} \nabla \Phi$$

$$= \mathbf{g} - \frac{1}{\rho} \sum_{B=1}^{N} \rho_{(B)} \frac{\epsilon_{(B)} \tilde{N}}{M_{(B)}} \nabla \Phi$$

$$= \mathbf{g} - \frac{\tilde{N}}{\rho} \sum_{B=1}^{N} c_{(B)} \epsilon_{(B)} \nabla \Phi$$

$$= \mathbf{g} \tag{9.3.6-6}$$

where $\tilde{N}$ is Avogadro's number. This allows us to write (9.3.6-1) as

$$\mathbf{N}_{(A)} = -c \mathcal{D}_{(Am)} \left[ \nabla x_{(A)} + \frac{M_{(A)} x_{(A)}}{RT} \left( \frac{\epsilon_{(A)}}{\mathcal{M}_{(A)}} \nabla \Phi \right) \right] + c_{(A)} \mathbf{v}^{\diamond}$$

$$= -c \mathcal{D}_{(Am)} \left[ \nabla x_{(A)} + \frac{x_{(A)} \epsilon_{(A)}}{kT} \nabla \Phi \right] + c_{(A)} \mathbf{v}^{\diamond} \tag{9.3.6-7}$$

in which $k$ is the Boltzmann constant. In the limit of very dilute solutions, we will normally neglect the effects of convection to conclude that

$$\mathbf{N}_{(A)} = -c \mathcal{D}_{(Am)} \left[ \nabla x_{(A)} + \frac{x_{(A)} \epsilon_{(A)}}{kT} \nabla \Phi \right] \tag{9.3.6-8}$$

There are two very important points to note.

1) One cannot assume local electrical neutrality within the immediate neighborhood of a phase interface or electrode where appreciable charge separation has taken place to form an electric double layer. Normally a double layer is very thin, on the order of 1 to 10 nm. For a detailed treatment of the double layer, see Newman (1973, Ch. 7).
2) Local electrical neutrality (9.3.6-4) does NOT imply that Laplace's equation (9.3.6-5) can be used to determine $\Phi$. When we assume local electrical neutrality (9.3.6-4), we will not attempt to solve either (9.3.6-3) or (9.3.6-5). We will use (9.3.6-4) to eliminate $\nabla \Phi$, making no attempt to determine the magnitudes of its components.

Let me give you two examples of how (9.3.6-4) can be used to eliminate $\nabla \Phi$ with no attempt to determine the magnitudes of its components. For a more complete introduction to the transport processes in electrolytic solutions, with particular attention to the variety of possible boundary conditions, see Newman (1973) and Levich (1962).

## Dilute Binary Electrolyte: General Approach

A dilute binary electrolyte is the simplest case to handle. By a binary electrolyte, we mean the solution of a single salt composed of one kind of cation $(+)$ and one kind of anion $(-)$. Let $\nu_{(+)}$ and $\nu_{(-)}$ be the numbers of cations and anions produced by dissociation of one molecule of electrolyte. This suggests that, if we define

$$\kappa \equiv \frac{c_{(+)}}{\nu_{(+)}} = \frac{c_{(-)}}{\nu_{(-)}} \tag{9.3.6-9}$$

we can automatically satisfy the electroneutrality requirement (9.3.6-4). Assuming that there are no homogeneous chemical reactions, we can write the differential mass balances for the cation and anion as

$$\frac{\partial \kappa}{\partial t} + \operatorname{div}\left(\frac{\mathbf{N}_{(+)}}{\nu_{(+)}}\right) = 0 \tag{9.3.6-10}$$

and

$$\frac{\partial \kappa}{\partial t} + \operatorname{div}\left(\frac{\mathbf{N}_{(-)}}{\nu_{(-)}}\right) = 0 \tag{9.3.6-11}$$

Since we are dealing with a dilute solution, (9.3.6-1) requires for these two ions

$$\frac{\mathbf{N}_{(+)}}{\nu_{(+)}} = -\mathcal{D}_{(+m)}\left(\nabla \kappa + \frac{\kappa \epsilon_{(+)}}{kT} \nabla \Phi\right) + \kappa \mathbf{v}^{\diamond} \tag{9.3.6-12}$$

and

$$\frac{\mathbf{N}_{(-)}}{\nu_{(-)}} = -\mathcal{D}_{(-m)}\left(\nabla \kappa + \frac{\kappa \epsilon_{(-)}}{kT} \nabla \Phi\right) + \kappa \mathbf{v}^{\diamond} \tag{9.3.6-13}$$

Here we have introduced the Boltzmann constant $k = R/\tilde{N}$, where $\tilde{N}$ is Avogadro's number. By taking the difference between these last two equations and rearranging, we can find

$$\frac{-\mathcal{D}_{(+m)}\kappa \epsilon_{(+)}}{kT} \nabla \Phi = \frac{\epsilon_{(+)}\mathcal{D}_{(+m)}}{\epsilon_{(+)}\mathcal{D}_{(+m)} - \epsilon_{(-)}\mathcal{D}_{(-m)}}\left(\frac{\mathbf{N}_{(+)}}{\nu_{(+)}} - \frac{\mathbf{N}_{(-)}}{\nu_{(-)}}\right)$$
$$+ \frac{\epsilon_{(+)}\mathcal{D}_{(+m)}\left(\mathcal{D}_{(+m)} - \mathcal{D}_{(-m)}\right)}{\epsilon_{(+)}\mathcal{D}_{(+m)} - \epsilon_{(-)}\mathcal{D}_{(-m)}} \nabla \kappa \tag{9.3.6-14}$$

This allows us to eliminate the electrostatic potential $\Phi$ between (9.3.6-12) and (9.3.6-14) to find after some rearrangement

$$\frac{1}{a_{(+)}\mathcal{D}_{(+m)} - a_{(-)}\mathcal{D}_{(-m)}}\left(\frac{a_{(+)}\mathcal{D}_{(+m)}}{\nu_{(-)}}\mathbf{N}_{(-)} - \frac{a_{(-)}\mathcal{D}_{(-m)}}{\nu_{(+)}}\mathbf{N}_{(+)}\right) = -D \,\nabla \kappa + \kappa \mathbf{v}^{\diamond} \tag{9.3.6-15}$$

where

$$D \equiv \frac{\mathcal{D}_{(+m)}\mathcal{D}_{(-m)}\left(a_{(+)} - a_{(-)}\right)}{a_{(+)}\mathcal{D}_{(+m)} - a_{(-)}\mathcal{D}_{(-m)}} \tag{9.3.6-16}$$

Taking the divergence of (9.3.6-15) and employing (9.3.6-10) and (9.3.6-11), we can say finally that

$$\frac{\partial \kappa}{\partial t} + \operatorname{div}(\kappa \mathbf{v}^{\diamond}) = D \operatorname{div} \nabla \kappa \tag{9.3.6-17}$$

Equation (8.4.6-1) and the results here are applicable to dilute solutions as well as to the other cases explained in Section 8.4.6. For dilute solutions, it would appear that $c$ is nearly a constant and (9.3.6-12) can be further simplified to

$$\frac{\partial \kappa}{\partial t} + \nabla \kappa \cdot \mathbf{v}^\circ = D \text{ div } \nabla \kappa \qquad (9.3.6\text{-}18)$$

or

$$\frac{\partial \kappa}{\partial t} + \nabla \kappa \cdot \mathbf{v} = D \text{ div } \nabla \kappa \qquad (9.3.6\text{-}19)$$

For a dilute binary electrolyte, we must solve only (9.3.6-18) or (9.3.6-19) consistent with the overall differential mass balance and the appropriate boundary conditions. The resulting solution for $\kappa$ can be related to the desired concentration distributions through (9.3.6-9).

### Dilute Multiple Electrolyte: Specific Example

As an illustration of what can be done with a dilute multiple electrolyte, let us consider a simple case involving a ternary electrolyte.

Let us visualize that the cell shown in Figure 9.3.6-1 contains both $Ag^+NO_3^-$ at an average concentration $10^{-6} N$ and $K^+NO_3^-$ at an average concentration of $0.1 N$. A voltage is imposed upon the cell that is just sufficient to cause the silver ion concentration at the cathode to drop to essentially zero. We wish to determine the steady-state concentration distributions in the cell.

Because we are speaking about dissociated species, it will be convenient to work in molar terms. Since the solution is dilute, we will neglect the diffusion-induced convection, and the overall differential mass balance is satisfied identically. The differential mass balances for the $Ag^+$, $K^+$, and $NO_3^-$ ions require

$$N_{(Ag^+)} = \text{a constant}$$

$$N_{(K^+)} = 0 \qquad (9.3.6\text{-}20)$$

$$N_{(NO_3^-)} = 0$$

Here we have recognized that it is only the $Ag^+$ ion that is in motion.

Equation (9.3.6-8) requires

$$-\frac{N_{(Ag^+)}}{c\mathcal{D}_{(Ag^+m)}} = \frac{dx_{(Ag^+)}}{dz_2} + \frac{x_{(Ag^+)}\epsilon_{(Ag^+)}}{kT}\frac{d\Phi}{dz_2}$$

$$0 = \frac{dx_{(K^+)}}{dz_2} + \frac{x_{(K^+)}\epsilon_{(K^+)}}{kT}\frac{d\Phi}{dz_2} \qquad (9.3.6\text{-}21)$$

$$0 = \frac{dx_{(NO_3^-)}}{dz_2} + \frac{x_{(NO_3^-)}\epsilon_{(NO_3^-)}}{kT}\frac{d\Phi}{dz_2}$$

Adding these three equations and making use of local electrical neutrality (9.3.6-4) to say

$$x_{(Ag^+)} + x_{(K^+)} - x_{(NO_3^-)} = 0 \qquad (9.3.6\text{-}22)$$

we find

$$-\frac{N_{(Ag^+)}}{c\mathcal{D}_{(Ag^+m)}} = 2\frac{dx_{(NO_3^-)}}{dz_2} \qquad (9.3.6\text{-}23)$$

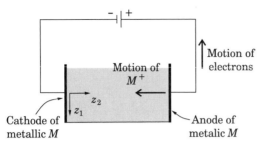

**Figure 9.3.6-1.** A simple cell filled with a binary electrolyte.

Integrating, we see that

$$x_{(NO_3^-)} - x_{(NO_3^-)\text{cathode}} = \frac{\alpha z_2}{L},$$
(9.3.6-24)

in which

$$\alpha \equiv \frac{-N_{(Ag^+)2}L}{2cD_{(Ag^+m)}}$$
(9.3.6-25)

In the same way, we reason from the second and third lines of (9.3.6-21)

$$x_{(K^+)} - x_{(K^+)\text{cath}} = \frac{-\alpha z_2}{L}$$

$$x_{(Ag^+)} = \frac{2\alpha z_2}{L}$$
(9.3.6-26)

This discussion was prompted by an exercise given by Bird et al. (1960, p. 588).

**Exercise 9.3.6-1** *The maximum current density in a simple cell*   The simple cell shown in Figure 9.3.6-1 is filled with a dilute binary electrolyte formed by dissolving a small amount of a salt $MX$ in water. We wish to relate the current density

$$\mathbf{i} \equiv F \sum_{A=1}^{N} a_{(A)}\mathbf{N}_{(A)}$$

to the concentration distributions for the ions $M^+$ and $X^-$. Here, $F$ is Faraday's constant. The cell may be assumed to be operating at steady state.

i)   Using the approach described in the text, determine that

$$\frac{i_2 z_2}{cDF} = \ln\left(\frac{D_{(-m)} - \left[D_{(+m)} + D_{(-m)}\right]x_{(+)}}{D_{(-m)} - \left[D_{(+m)} + D_{(-m)}\right]x_{(+)\text{cath}}}\right)$$

ii)   Conclude that for a very dilute solution we should be justified in approximating this expression by

$$\frac{-i_2 z_2}{2cD_{(+m)}F} = x_{(+)} - x_{(+)\text{cath}}$$

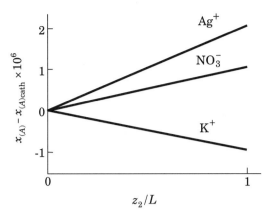

**Figure 9.3.6-2.** Ionic concentrations for very small values of $\alpha$.

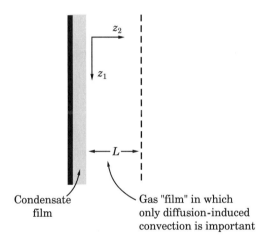

Condensate film

Gas "film" in which only diffusion-induced convection is important

**Figure 9.3.7-3.** Condensation of mixed vapors from a stagnant gas film.

iii) Determine that the maximum current density is

$$-i_{2(\text{max})} = 4c\mathcal{D}_{(+m)}\frac{F x_{(MX)\text{av}}}{L}$$

## 9.3.7 Film Theory: Condensation of Mixed Vapors

Chloroform and benzene condense continuously as shown in Figure 9.3.7-3 from an ideal-gas mixture of known composition and temperature at 1 atm. To achieve a specified composition of the condensate, what is the temperature of the condenser wall and the total molar rate of condensation (Bird et al. 1960, p. 586)? We will idealize the process by considering only a stagnant gas film in which the only convection considered is that induced by diffusion.

Let us begin by assuming

$$x_{(B)} = x_{(B)}(z_2) \tag{9.3.7-1}$$

and

$$v_1^\diamond = v_3^\diamond$$
$$= 0 \tag{9.3.7-2}$$
$$v_2^\diamond = v_2^\diamond(z_2)$$

In view of (9.3.7-1) and (9.3.7-2), Fick's first law (Table 8.5.1-7) requires that

$$N_{(B)1} = N_{(B)3}$$
$$= 0 \tag{9.3.7-3}$$
$$N_{(B)2} = N_{(B)2}(z_2)$$

From the differential mass balance for species $B$, we learn that

$$N_{(B)2} = \text{a constant} \tag{9.3.7-4}$$

From the overall differential mass balance (Table 8.5.1-10), we conclude that

$$c = \text{a constant} \tag{9.3.7-5}$$

You should be aware that this may ultimately contradict the temperature and concentration profiles that follow. I will say more about this at the end of this section.

From the jump mass balances for species $B$ and $C$,

$$\frac{N_{(B)2}}{N_{(C)2}} = \frac{x_{(B)\text{cond}}}{x_{(C)\text{cond}}} \tag{9.3.7-6}$$

where the subscript cond refers to the condensate. This allows us to rewrite the $z_2$ component of Fick's first law as

$$\frac{N_{(B)2}}{cD_{(BC)}} = -\frac{1}{1 - \left(x_{(B)}/x_{(B)\text{cond}}\right)} \frac{dx_{(B)}}{dz_2} \tag{9.3.7-7}$$

This and the equivalent for species $C$ can be integrated consistent with boundary conditions

$$\text{at } z_2 = 0 : \quad x_{(B)} = x_{(B)\text{eq}}$$
$$x_{(C)} = x_{(C)\text{eq}} \tag{9.3.7-8}$$

and

$$\text{at } z_2 = L : \quad x_{(B)} = x_{(B)\infty}$$
$$x_{(C)} = x_{(C)\infty} \tag{9.3.7-9}$$

to find

$$\frac{N_{(B)2}L}{cD_{(BC)}x_{(B)\text{cond}}} = \ln\left|\frac{x_{(B)\text{cond}} - x_{(B)\infty}}{x_{(B)\text{cond}} - x_{(B)\text{eq}}}\right| \tag{9.3.7-10}$$

and

$$\frac{N_{(C)2}L}{cD_{(BC)}x_{(C)\text{cond}}} = \ln\left|\frac{x_{(C)\text{cond}} - x_{(C)\infty}}{x_{(C)\text{cond}} - x_{(C)\text{eq}}}\right| \tag{9.3.7-11}$$

Here, the subscript $\infty$ refers to the bulk vapor stream outside the immediate neighborhood of the interface, and the subscript cond refers to the composition of the vapor in equilibrium with the liquid condensate.

If we are primarily interested in the total molar rate of condensation, from (9.3.7-6) it follows that

$$
v_2^\diamond = \frac{N_{(B)2}}{c x_{(B)\text{cond}}}
$$

$$
= \frac{N_{(C)2}}{c x_{(C)\text{cond}}}
$$

$$
= \frac{\mathcal{D}_{(BC)}}{L} \ln \left| \frac{x_{(B)\text{cond}} x_{(B)\infty}}{x_{(B)\text{cond}} - x_{(B)\text{eq}}} \right|
$$

$$
= \frac{\mathcal{D}_{(BC)}}{L} \ln \left| \frac{x_{(C)\text{cond}} - x_{(C)\infty}}{x_{(C)\text{cond}} - x_{(C)\text{eq}}} \right| \tag{9.3.7-12}
$$

The differential energy balance of Table 8.5.2-1 simplifies considerably for this case when we neglect viscous dissipation and the effect of pressure gradients:

$$
\text{div} \left( \sum_{A=1}^{N} \overline{H}_{(A)}^{(m)} \mathbf{N}_{(A)} + \boldsymbol{\epsilon} \right) = 0 \tag{9.3.7-13}
$$

From (8.4.3-2), when we neglect the Dufour effect, we may describe

$$
\boldsymbol{\epsilon} = -k \nabla T \tag{9.3.7-14}
$$

In view of (9.3.7-1) and (9.3.7-2), it seems reasonable to assume that

$$
T = T(z_2) \tag{9.3.7-15}
$$

This enables us to conclude from (9.3.7-13) and (9.3.7-14) that

$$
\sum_{A=1}^{N} \overline{H}_{(A)}^{(m)} N_{(A)2} - k \frac{dT}{dz_2} = C_1
$$

$$
= \text{a constant} \tag{9.3.7-16}
$$

For an ideal solution,

$$
\tilde{H} = x_{(B)} \tilde{H}_{(B)}^{(P)} + x_{(C)} \tilde{H}_{(C)}^{(P)}
$$

$$
= x_{(B)} \left( \tilde{H}_{(B)}^{(P)} - \tilde{H}_{(C)}^{(P)} \right) + \tilde{H}_{(C)}^{(P)} \tag{9.3.7-17}
$$

where $H_{(A)}^{(P)}$ is the enthalpy of the pure component. From the definition of partial molar quantities in Exercise 8.4.2-5,

$$
\overline{H}_{(B)}^{(m)} - \overline{H}_{(C)}^{(m)} = \tilde{H}_{(B)}^{(P)} - \tilde{H}_{(C)}^{(P)} \tag{9.3.7-18}
$$

and

$$
\tilde{H} = x_{(B)} \overline{H}_{(B)}^{(m)} + x_{(C)} \overline{H}_{(C)}^{(m)} \tag{9.3.7-19}
$$

Equations (9.3.7-17) through (9.3.7-19) can be solved simultaneously to conclude that

$$
\overline{H}_{(B)}^{(m)} = \tilde{H}_{(B)}^{(P)}
$$

$$
\overline{H}_{(C)}^{(m)} = \tilde{H}_{(C)}^{(P)} \tag{9.3.7-20}
$$

Let us measure the enthalpy of the pure components with respect to the pure liquid at its saturation temperature:

$$
\tilde{H}_{(B)}^{(P)} - \tilde{H}_{(B)}^{(P,R)} = \tilde{c}_{P(B)}^{(v)\circ}\left(T - T_{(B)}^{(R)}\right) + \tilde{\lambda}_{(B)}
$$
$$
\tilde{H}_{(C)}^{(P)} - \tilde{H}_{(C)}^{(P,R)} = \tilde{c}_{P(C)}^{(v)\circ}\left(T - T_{(C)}^{(R)}\right) + \tilde{\lambda}_{(B)}
$$
(9.3.7-21)

Here, $\tilde{H}_{(A)}^{(P,R)}$ is the absolute enthalpy of species $A$ in its reference state, $\tilde{c}_{P(A)}^{(v)\circ}$ the molar heat capacity of species $A$ at a very low pressure (ideal gas), $\tilde{\lambda}_{(A)}$ the molar heat of vaporization at its normal boiling point, and $T_{(A)}^{(R)}$ its reference temperature, its normal boiling point. Using (9.3.7-12), (9.3.7-20), and (9.3.7-21), we can write

$$
\sum_{A=1}^{N} \overline{\tilde{H}}_{(A)}^{(m)} N_{(A)2} = cv_2^{\diamond}\left(x_{(B)\mathrm{cond}}\tilde{H}_{(B)}^{(P)} + x_{(C)\mathrm{cond}}\tilde{H}_{(C)}^{(P)}\right)
$$
(9.3.7-22)

and rearrange (9.3.7-16) as

$$
k\frac{dT}{dz_2} = cv_2^{\diamond}\tilde{c}_{P,\mathrm{av}}^{(v)\circ}T + C_2
$$
(9.3.7-23)

where

$$
\tilde{c}_{P,\mathrm{av}}^{(v)\circ} \equiv x_{(B)\mathrm{cond}}\tilde{c}_{P(B)}^{(v)\circ} + x_{(C)\mathrm{cond}}\tilde{c}_{P(C)}^{(v)\circ}
$$
(9.3.7-24)

We can integrate (9.3.7-23) consistent with the boundary conditions

$$
\text{at } z_2 = 0: \ T = T_{\mathrm{eq}}
$$
(9.3.7-25)

and

$$
\text{at } z_2 = L: \ T = T_{\infty}
$$
(9.3.7-26)

to learn

$$
\frac{T - T_{\mathrm{eq}}}{T_{\infty} - T_{\mathrm{eq}}} = \frac{1 - \exp(-z_2 A/L)}{1 - \exp(-A)}
$$
(9.3.7-27)

where, in view of (9.3.7-12),

$$
A \equiv -\frac{1}{k}cv_2^{\diamond}\tilde{c}_{P,\mathrm{av}}^{(v)\circ}L
$$
$$
= -\frac{1}{k}c\mathcal{D}_{(BC)}\tilde{c}_{P,\mathrm{av}}^{(v)\circ}\ln\left|\frac{x_{(B)\mathrm{cond}} - x_{(B)\infty}}{x_{(B)\mathrm{cond}} - x_{(B)\mathrm{eq}}}\right|
$$
(9.3.7-28)

We know that

$$
\mathbf{q} + \rho\tilde{H}\mathbf{v} = \mathbf{q} + \sum_{A=1}^{N}\rho_{(A)}\overline{H}_{(A)}\mathbf{v}
$$
$$
= \mathbf{q} - \sum_{A=1}^{N}\overline{H}_{(A)}\mathbf{j}_{(A)} + \sum_{A=1}^{N}\overline{H}_{(A)}\mathbf{n}_{(A)}
$$
$$
= \boldsymbol{\epsilon} + \sum_{A=1}^{N}\overline{H}_{(A)}\mathbf{n}_{(A)}
$$
$$
= \boldsymbol{\epsilon} + \sum_{A=1}^{N}\overline{\tilde{H}}_{(A)}^{(m)}\mathbf{N}_{(A)}
$$
(9.3.7-29)

Equations (9.3.7-14), (9.3.7-22), and (9.3.7-29) permit us to compute the total energy flux to the condensate film as

$$\text{at } z_2 = 0: \quad -q_2 - \rho\hat{H}v_2 = k\frac{dT}{dz_2} - cv_2^\circ\left(x_{(B)\text{cond}}\tilde{H}_{(B)}^{(P)} + x_{(C)\text{cond}}\tilde{H}_{(C)}^{(P)}\right)$$

$$\approx k\frac{dT}{dz_2} - cv_2^\circ\tilde{\lambda}_{\text{av}}$$

$$= \frac{Ak\left(T_\infty - T_{\text{eq}}\right)}{L[1 - \exp(-A)]} - cv_2^\circ\tilde{\lambda}_{\text{av}}$$

$$= \frac{Ak\left(T_\infty - T_{\text{eq}}\right)}{L[1 - \exp(-A)]} - \frac{c\tilde{\lambda}_{\text{av}}\mathcal{D}_{(BC)}}{L}$$

$$\times \ln\left|\frac{x_{(B)\text{cond}} - x_{(B)\infty}}{x_{(B)\text{cond}} - x_{(B)\text{eq}}}\right| \tag{9.3.7-30}$$

in which

$$\tilde{\lambda}_{\text{av}} \equiv x_{(B)\text{cond}}\tilde{\lambda}_{(B)} + x_{(C)\text{cond}}\tilde{\lambda}_{(C)} \tag{9.3.7-31}$$

In the second step of (9.3.7-30), Equation (9.3.7-21) has been used; at the third step, (9.3.7-27); in the last step, (9.3.7-12).

In the absence of any net mass transfer, it is easy to see that the total energy flux to the condensate film is

$$\text{at } z_2 = 0: \quad -q_2 = \frac{k}{L}\left(T_\infty - T_{\text{eq}}\right) \tag{9.3.7-32}$$

and that the mass flux of species $A$ to the condensate film is

$$\text{at } z_2 = 0: \quad -N_{(B)2} = \frac{c\mathcal{D}_{(BC)}}{L}\left(x_{(B)\infty} - x_{(B)\text{eq}}\right) \tag{9.3.7-33}$$

Equations (9.3.7-10) and (9.3.7-30) through (9.3.7-33) permit us to compute the film-theory correction factor for energy and mass transfer described in Section 9.2.1:

$$\mathcal{F}_{\text{energy}} \equiv \frac{(\mathbf{q} + \rho\hat{H}\mathbf{v}) \cdot \xi}{(\mathbf{q} \cdot \xi)|_{\mathbf{v} \cdot \xi = 0}}$$

$$\approx \frac{A}{[1 - \exp(-A)]} - \frac{c\tilde{\lambda}_{\text{av}}\mathcal{D}_{(BC)}}{k\left(T_\infty - T_{\text{eq}}\right)}\ln\left|\frac{x_{(B)\text{cond}} - x_{(B)\infty}}{x_{(B)\text{cond}} - x_{(B)\text{eq}}}\right| \tag{9.3.7-34}$$

$$\mathcal{F}_{(B)} \equiv \frac{\mathbf{N}_{(B)} \cdot \xi}{\left(\mathbf{N}_{(B)} \cdot \xi\right)_{\mathbf{v} \cdot \xi = 0}}$$

$$= \left(\frac{x_{(B)\text{cond}}}{x_{(B)\text{eq}} - x_{(B)\infty}}\right)\ln\left|\frac{x_{(B)\text{cond}} - x_{(B)\infty}}{x_{(B)\text{cond}} - x_{(B)\text{eq}}}\right| \tag{9.3.7-35}$$

As explained in Section 9.2.1, these expressions are likely to be more useful than (9.3.7-10) and (9.3.7-30), which involve the film thickness $L$.

To complete our analysis, let us represent the rate of energy transfer to the condenser wall from the condensate film by Newton's "law" of cooling (Section 9.2.1):

$$-q_2 = h_{\text{wall}}\left(T_{\text{eq}} - T_{\text{wall}}\right) \tag{9.3.7-36}$$

where $T_{wall}$ is the temperature of the condenser wall. Similarly, let us represent the rate of energy transfer from the external gas stream to the condensate film as

$$- q_2 - \rho \hat{H} v_2 = \mathcal{F}_{energy} h_{ext} \left( T_\infty - T_{eq} \right) \tag{9.3.7-37}$$

If the film coefficients $h_{wall}$ and $h_{ext}$ for the limiting case of no net mass transfer can be estimated using standard correlations, the temperature $T_{wall}$ of the condenser wall necessary to achieve a liquid condensate of a given composition $x_{(C)}$ can be determined by solving (9.3.7-36) and (9.3.7-37) simultaneously.

Similarly, we can describe the rate of condensation using Newton's "law" of mass transfer (Section 9.2.1):

$$-N_{(B)2} - N_{(C)2} = - \left( 1 + \frac{x_{(C)cond}}{x_{(B)cond}} \right) N_{(B)2}$$

$$= \left( 1 + \frac{x_{(C)cond}}{x_{(B)cond}} \right) k_x \mathcal{F}_{(B)} \left( x_{(B)\infty} - x_{(B)eq} \right) \tag{9.3.7-38}$$

The film coefficient for mass transfer in the limit of no mass transfer, $k_x$, can be estimated using the analogy between energy and mass transfer described in Section 9.2.1.

Now let us return to (9.3.7-5) where we argued that $c =$ a constant. This is likely to be contradicted by the temperature and concentration profiles determined above. Although it has intuitive appeal, our initial assumption that this was a one-dimensional problem appears to be incorrect. In spite of this error, my expectation is that the analysis presented here is useful, since the temperature gradients in the stagnant film are likely to be small.

Exercises 9.3.7-1 and 9.3.7-2 illustrate the use of these results. For an early discussion of the condensation of mixed vapors, see Colburn and Drew (1937).

**Exercise 9.3.7-1** *More on the condensation of mixed vapors* (Bird et al. 1960, p. 586)  Consider continuous condensation on a 1 m × 1 m cooled surface from a flow stream of benzene and chloroform. What is the temperature $T_{wall}$ of the cooled surface if the composition of the condensate is specified? What is the rate of condensation?

To estimate $h$, you may assume that the results of Section 6.7.2 apply and that

$$v_\infty = 1 \, \text{m/s}$$

$$x_{(B)\infty} = 0.5$$

$$T_\infty = 100°C$$

$$h_{wall} = 56.8 \, \text{J/} \left( \text{s m}^2 \, \text{K} \right)$$

$$\mathcal{D}_{(BC)} = 4.38 \times 10^{-6} \, \text{m}^2/\text{s}$$

$$\tilde{c}_{P(B)}^{(v)\circ} = 10.1 \times 10^4 \, \text{J/(kg mol K)}$$

$$\tilde{c}_{P(C)}^{(v)\circ} = 7.14 \times 10^4 \, \text{J/(kg mol K)}$$

$$k = 1.21 \times 10^{-2} \, \text{J/(s ft K)}$$

$$\tilde{\lambda}_{(B)} = 3.08 \times 10^7 \, \text{J/kg mol}$$

$$\tilde{\lambda}_{(C)} = 2.96 \times 10^7 \, \text{J/kg mol}$$

The vapor–liquid equilibrium data for this system are given in Table 9.3.7-1.

**Table 9.3.7-1.** Vapor–liquid equilibrium data for chloroform–benzene system at 1 atm[a]

| $x_{(C)}$ vapor | $x_{(C)}$ liquid | Saturation temperature (°C) |
|---|---|---|
| 0.00 | 0.00 | 80.6 |
| 0.10 | 0.08 | 79.8 |
| 0.20 | 0.15 | 79.0 |
| 0.30 | 0.22 | 78.2 |
| 0.40 | 0.29 | 77.3 |
| 0.50 | 0.36 | 76.4 |
| 0.60 | 0.44 | 75.3 |
| 0.70 | 0.54 | 74.0 |
| 0.80 | 0.66 | 71.9 |
| 0.90 | 0.79 | 68.9 |
| 1.00 | 1.00 | 61.4 |

[a]From Chu et al. (1950, p. 61).

**Exercise 9.3.7-2** *Wet- and dry-bulb psychrometer* (Bird et al. 1960, pp. 649, 667) A simple method for measuring the humidity of air is to use two thermometers, the bulb of one of which is covered by a cloth sleeve that is saturated with water. In the usual arrangement, either humid air flows past the two thermometers or the thermometers are rotated in the humid air (the *sling psychrometer*). There are two principal differences from the analysis given in the text and extended in Exercise 9.3.7-1:

1) Only one of the two species is condensible.
2) The composition of the gas (its humidity) is unknown.

  i) Repeat the analysis of the text, recognizing that air $A$ is noncondensable. Determine that

$$\mathcal{F}_{\text{energy}} = \frac{A}{[1 - \exp(-A)]} - \frac{c\tilde{\lambda}_{(C)}\mathcal{D}_{(AW)}}{k(T_\infty - T_{\text{eq}})} \ln \left| \frac{1 - x_{(W)\infty}}{1 - x_{(W)\text{eq}}} \right|$$

and

$$\mathcal{F}_{(W)} = \left( \frac{1}{x_{(W)\text{eq}} - x_{(W)\infty}} \right) \ln \left| \frac{1 - x_{(W)\infty}}{1 - x_{(W)\text{eq}}} \right|$$

in which

$$A \equiv \frac{-cv_2^\diamond \tilde{c}_{P,(W)}^{(v)\circ} L}{k}$$

$$= -\frac{c\mathcal{D}_{(AW)}\tilde{c}_{P,(W)}^{(v)\circ}}{k} \ln \left| \frac{1 - x_{(W)\infty}}{1 - x_{(W)\text{eq}}} \right|$$

  ii) Use the jump energy balance to conclude that

$$k_x \mathcal{F}_{(W)} \left( x_{(W)\text{eq}} - x_{(W)\infty} \right) \tilde{\lambda}_{(W)} = h\mathcal{F}_{\text{energy}} \left( T_\infty - T_{\text{eq}} \right)$$

  iii) Use the Chilton–Colburn analogy (Bird et al. 1960, p. 647),

$$\frac{N_{Nu}}{N_{Nu,m}} = \left( \frac{N_{Sc}}{N_{Pr}} \right)^{-1/3}$$

together with (9.2.1-3) and (9.2.1-14) to estimate

$$\frac{h}{k_x} = \tilde{c}_{P,(A)} \left(\frac{N_{Sc}}{N_{Pr}}\right)^{2/3}$$

iv) Consider wet-bulb and dry-bulb temperatures, $T_{eq}$ and $T_\infty$, such as you might encounter in air conditioning:

$$T_{eq} = 21.1°C$$

$$T_\infty = 26.6°C$$

Determine the composition $x_{(W)\infty}$ of the air. You may assume that

$$N_{Sc} = 0.60$$

$$N_{Pr} = 0.73$$

$$x_{(W)eq} = 0.0247$$

$$\tilde{c}_{P,(A)} = 2.92 \times 10^4 \text{ J/(kg mol K)}$$

$$\tilde{c}_{P,(W)}^{(v)\circ} = 3.37 \times 10^4 \text{ J/(kg mol K)}$$

$$\mathcal{D}_{(AW)} = 2.50 \times 10^{-5} \text{ m}^2/\text{s}$$

$$\tilde{\lambda}_{(W)} = 4.43 \times 10^7 \text{ J/kg mol}$$

v) How is $x_{(W)\infty}$ changed, if you neglect the effects of convection?

## 9.3.8  Two- and Three-Dimensional Problems

Up to this point, we have discussed only one-dimensional (in space) diffusion problems. Two- and three-dimensional problems require a somewhat different approach. This can probably be best understood by contrasting one-dimensional and three-dimensional problems.

Since it is often helpful to have a process in mind, consider a cube and a large planar slab of sugar dissolving in water. Our objective is to determine the rate at which the sugar dissolves in each case.

Let us begin by considering the cube of sugar, a three-dimensional problem. To keep things simple, we will assume that the two-component aqueous phase is incompressible and that it extends to infinity. The mass-averaged velocity of the aqueous solution is likely to be non-zero for two reasons. First, close to the sugar cube, the solution may be concentrated. Second, the motion of the solid-liquid interface will induce convection. For an incompressible solution, we have six unknowns: the mass fraction of one species (sugar, perhaps; the mass fraction of water is found by difference, since the sum of mass fractions must be one), the speed of displacement of the solid-liquid interface, the three components of the mass-averaged velocity, and pressure. The six equations that we must solve are the differential mass balances (species and overall), the three components of the differential momentum balances, and the species (sugar) jump mass balance. The overall jump mass balance provides a boundary condition for the mass-averaged velocity.

Now consider a large planar slab of sugar for which we will neglect all end effects. For a one-dimensional problem such as this, we have three unknowns: the mass fraction of one species, the speed of displacement of the solid-liquid interface, and a single component of the mass-averaged velocity. The three equations that must be satisfied are the two differential mass balances (species and overall) and the species (sugar) jump mass balance. The overall jump mass balance provides a boundary condition for the mass-averaged velocity. We could solve the differential momentum balance for the pressure distribution, but the pressure gradient in one-dimensional diffusion problems is exceedingly small and, for this reason, ignored.

In summary, the major difference between one-dimensional and three-dimensional problems involving is that the three-dimensional problem requires us to solve the differential momentum balance. In a three-dimensional problem, the differential momentum balance cannot be ignored as we do in one-dimensional problems by simply saying that the pressure gradients are small and of no concern. The pressure gradients induced in a three-dimensional problem may alter the convection.

Unfortunately, to my knowledge there are no examples currently available of two- or three-dimensional diffusion in concentrated solutions.

Note that one must take a similar approach in discussing multidimensional melting or freezing (Section 6.3.3) and multidimensional, multicomponent diffusion (Section 9.4).

This section was written with P. K. Dhori.

## 9.4  Complete Solutions for Multicomponent Systems

In the preceding section, we have confined our attention to binary solutions or to solutions sufficiently dilute that they could be considered to be binary (see Section 8.4.6). In what follows, we explicitly consider multicomponent solutions. Although the example problems that I have chosen involve ternary solutions, I believe that problems involving four or more components could be handled in a similar manner.

### 9.4.1  Film Theory: Steady-State Evaporation

Let us consider a system that is similar to that described in Exercise 9.3.1-1. Pure liquid $A$ continuously evaporates into an ideal-gas mixture of $A$, $E$, and $F$. The apparatus is arranged in such a manner that the liquid–gas phase interface remains fixed in space as the evaporation takes place. Species $E$ and $F$ are assumed to be insoluble in liquid $A$:

$$\text{at } z_2 = 0: \ N_{(E)2} = N_{(F)2}$$

$$= 0 \tag{9.4.1-1}$$

For the existing conditions, the equilibrium composition of the gas phase is $x_{(A)\text{eq}}$:

$$\text{at } z_2 = 0: \ x_{(A)} = x_{(A)\text{eq}} \tag{9.4.1-2}$$

The composition of the gas phase at the top of the column is maintained constant,

$$\text{at } z_2 = L: \ x_{(E)} = x_{(E)\infty}$$

$$x_{(F)} = x_{(F)\infty} \tag{9.4.1-3}$$

We wish to determine the rate of evaporation of $A$ from the surface. Let us begin by asking for the mole fraction distribution of each species in the gas phase.

If we assume that our ideal-gas mixture is at a constant temperature and pressure (neglecting any hydrostatic effect), the total molar density $c$ is a constant. This suggests that we look for a solution of the form

$$v_2^\circ = v_2^\circ(z_2)$$

$$v_1^\circ = v_3^\circ$$

$$= 0 \tag{9.4.1-4}$$

$$x_{(E)} = x_{(E)}(z_2)$$

$$x_{(F)} = x_{(F)}(z_2)$$

It follows immediately from (9.4.1-4) that only the $z_2$ components of $\mathbf{N}_{(A)}$, $\mathbf{N}_{(E)}$, and $\mathbf{N}_{(F)}$ are nonzero. From the differential mass balances for these three species as well as from (9.4.1-1), we conclude that

$$N_{(A)2} = \text{a constant}$$

$$N_{(E)2} = N_{(F)2}$$

$$= 0 \tag{9.4.1-5}$$

When we neglect any effects attributable to thermal, pressure, and forced diffusion, the generalized Stefan–Maxwell equation (8.4.4-31) reduces for an ideal gas to (Exercise 8.4.4-1)

$$\nabla x_{(C)} = \sum_{B=1}^{N} \frac{1}{c\mathcal{D}_{(CB)}} \left( x_{(C)}\mathbf{N}_{(B)} - x_{(B)}\mathbf{N}_{(C)} \right) \tag{9.4.1-6}$$

Because of (9.4.1-5), Equation (9.4.1-6) says that, for species $E$ and $F$,

$$\frac{dx_{(E)}}{dz_2} = \frac{N_{(A)2}}{c\mathcal{D}_{(AE)}} x_{(E)} \tag{9.4.1-7}$$

and

$$\frac{dx_{(F)}}{dz_2} = \frac{N_{(A)2}}{c\mathcal{D}_{(AF)}} x_{(F)} \tag{9.4.1-8}$$

Equations (9.4.1-7) and (9.4.1-8) may be integrated consistent with boundary conditions (9.4.1-3) to find

$$\frac{x_{(E)}}{x_{(E)\infty}} = \exp\left(-\alpha\left[1 - \frac{z_2}{L}\right]\right) \tag{9.4.1-9}$$

and

$$\frac{x_{(F)}}{x_{(F)\infty}} = \exp\left(-\alpha\beta\left[1 - \frac{z_2}{L}\right]\right) \tag{9.4.1-10}$$

Here

$$\alpha \equiv \frac{N_{(A)2}L}{c\mathcal{D}_{(AE)}} \tag{9.4.1-11}$$

and

$$\beta \equiv \frac{\mathcal{D}_{(AE)}}{\mathcal{D}_{(AF)}} \tag{9.4.1-12}$$

Equations (9.4.1-9) and (9.4.1-10) determine the mole fraction distributions in the gas phase in terms of $\alpha$ and $\beta$.

Although we can assume that $\beta$ is known, $\alpha$ must be determined. This may be accomplished by requiring (9.4.1-9) and (9.4.1-10) to satisfy (9.4.1-2):

$$x_{(A)\text{eq}} = 1 - x_{(E)\infty}\exp(-\alpha) - x_{(F)\infty}\exp(-\alpha\beta) \tag{9.4.1-13}$$

Given $\beta$, we may solve (9.4.1-13) for $\alpha$.

Our final objective is to calculate the film correction factor (see Section 9.2.1):

$$\mathcal{F}_{(A)} \equiv \frac{N_{(A)2}}{N_{(A)2}\big|_{v^\circ=0}} \tag{9.4.1-14}$$

In Exercise 9.3.1-1, we learn

$$N_{(A)2}\big|_{v^\circ=0} = \frac{c\mathcal{D}_{(Am)}}{L}\left(x_{(A)\text{eq}} - x_{(A)\infty}\right) \tag{9.4.1-15}$$

This permits us to express

$$\mathcal{F}_{(A)} = \frac{\alpha\mathcal{D}_{(AE)}}{\mathcal{D}_{(Am)}\left(x_{(A)\text{eq}} - x_{(A)\infty}\right)} \tag{9.4.1-16}$$

We see from (9.4.1-13) that $\alpha$ is not an explicit function of $L$ and that the film thickness $L$ has dropped out of the expression for $\mathcal{F}_{(A)}$. Remember that $\mathcal{D}_{(Am)}$ is the binary diffusion coefficient for $A$ with respect to a gas composed of $E$ and $F$. The most common example would be the diffusion coefficient for $A$ with respect to air (a mixture of oxygen and nitrogen), which would be readily available.

For further information on multicomponent ordinary diffusion, I suggest reading Cussler and Lightfoot (1963a,b); Toor (1964a,b); Toor, Seshadri, and Arnold (1965), and Arnold and Toor (1967).

**Exercise 9.4.1-1** *Film theory: slow catalytic reaction*    A gas consisting of a mixture of species $A$, $E$, and $F$ is brought into contact with a solid surface that acts as a catalyst for the isomerization reaction $A \to E$. Determine that a film-theory correction factor cannot be developed, in the sense that such a correction factor would depend upon the fictitious film thickness $L$ (Hsu and Bird 1960).

**Exercise 9.4.1-2** *Film theory: instantaneous catalytic reaction*    Redo the preceding exercise, assuming that the reaction is instantaneous. Once again, conclude that a film-theory correction factor cannot be developed, in the sense that such a correction factor would depend upon the fictitious film thickness $L$.

**Exercise 9.4.1-3** *Constant evaporating mixture*    Let us reconsider Exercise 9.3.1-6. A mixture of
ethanol and toluene evaporates into a mixture of ethanol, toluene, and nitrogen. The apparatus
is arranged in such a manner that the liquid–gas phase interface remains fixed in space
as the evaporation takes place. The liquid is assumed to be saturated with nitrogen. The
entire system is maintained at 60°C and constant pressure. Our objective is to determine the
composition of the liquid phase, which remains constant as a function of time.

In Exercise 9.3.1-6, we assumed that the gas phase was sufficiently dilute that we could
employ Fick's first law. Here we follow the analysis outlined by Slattery and Lin (1978) in
recognizing that the gas is a three-component mixture. We will employ the Stefan–Maxwell
equations as we do in the text.

   i) Use jump mass balances for ethanol and toluene for the stationary phase interface to
      prove that

$$\frac{N_{(E)2}}{N_{(T)2}} = \frac{x_{(E)\text{liquid}}}{x_{(T)\text{liquid}}}$$

  ii) Using reasoning similar to that developed in the text, conclude that

$$x_{(N_2)} = \exp\left(-\alpha\left[1 - z_2^\star\right]\right)$$

   Here

$$z_2^\star \equiv \frac{z_2}{L}$$

   and

$$\alpha \equiv \frac{N_{(E)2}L}{c\mathcal{D}_{(EN_2)}}\left[1 + \frac{\mathcal{D}_{(EN_2)}x_{(T)\text{liquid}}}{\mathcal{D}_{(TN_2)}x_{(E)\text{liquid}}}\right]$$

 iii) Reason that

$$\frac{dx_{(E)}}{dz_2^\star} = \alpha\beta x_{(E)} - \alpha\gamma - \alpha\delta\exp\left(-\alpha\left[1 - z_2^\star\right]\right)$$

   where

$$\beta \equiv \frac{\mathcal{D}_{(EN_2)}}{\mathcal{D}_{(ET)}}\left[\frac{1/x_{(E)\text{liquid}}}{1 + \left(\mathcal{D}_{(EN_2)}/\mathcal{D}_{(TN_2)}\right)\left(x_{(T)\text{liquid}}/x_{(E)\text{liquid}}\right)}\right]$$

$$\gamma \equiv \frac{\mathcal{D}_{(EN_2)}}{\mathcal{D}_{(ET)}}\left[\frac{1}{1 + \left(\mathcal{D}_{(EN_2)}/\mathcal{D}_{(TN_2)}\right)\left(x_{(T)\text{liquid}}/x_{(E)\text{liquid}}\right)}\right]$$

$$= \beta x_{(E)\text{liquid}}$$

$$\delta \equiv \left[\frac{1 - \mathcal{D}_{(EN_2)}/\mathcal{D}_{(ET)}}{1 + \left(\mathcal{D}_{(EN_2)}/\mathcal{D}_{(TN_2)}\right)\left(x_{(T)\text{liquid}}/x_{(E)\text{liquid}}\right)}\right]$$

  iv) Determine that

$$x_{(N)\text{eq}} = \exp(-\alpha)$$

and

$$x_{(E)eq} = x_{(E)liquid}\left(1 - x_{(N_2)eq}{}^{\beta}\right) + \frac{\delta}{\beta - 1}\left(x_{(N_2)eq} - x_{(N_2)eq}{}^{\beta}\right)$$

v) We have used the method of Fuller et al. (Reid, Prausnitz, and Poling 1987, p. 587) to estimate $\mathcal{D}_{(EN_2)} = 1.53 \times 10^{-5}\,\text{m}^2/\text{s}$, $\mathcal{D}_{(TN_2)} = 9.42 \times 10^{-6}\,\text{m}^2/\text{s}$, and $\mathcal{D}_{(ET)} = 6.05 \times 10^{-6}$ $\text{m}^2/\text{s}$. Using the data of Exercise 9.3.1-6, we can estimate

$$x_{(E)eq} = \frac{\left(1 - x^*_{(T)eq}\right)P^*}{760}$$

$$x_{(T)eq} = \frac{x^*_{(T)eq}P^*}{760}$$

$$x_{(N)eq} = 1 - x_{(E)eq} - x_{(T)eq}$$

where $x^*_{(T)eq}$ is the mole fraction of toluene in the two-component vapor and $P^*$ is the corresponding pressure of the vapor. Conclude that, to two significant figures, $x_{(T)liquid} = 0.18$. Robinson et al. (1932) obtained experimentally $x_{(T)liquid} = 0.20$. Note that our computed result is very sensitive to errors in the binary vapor–liquid equilibrium data or to errors in representing these data. In arriving at our result, we have fitted polynomials to the data presented in Exercise 9.3.1-6.

## 9.4.2 More on Unsteady-State Evaporation

This section is taken from Mhetar and Slattery (1997).

Evaporation of a pure volatile liquid from a partially filled open container has been analyzed in detail (Slattery and Mhetar 1996), but little attention has been paid to the unsteady evaporation of a liquid consisting of two or more components. Richardson (1959) analyzed the evaporation of a volatile liquid from its solution with virtually nonvolatile liquid. Carty and Schrodt (1975) considered the steady-state evaporation of a binary liquid mixture (acetone and methanol) into air from the Stefan tube with a stationary interface.

In what follows we consider a long vertical tube, partially filled with a two-component liquid mixture. Imagine that, for $t < 0$, this liquid mixture is isolated from the remainder of the tube by a closed diaphragm, which is filled with a gaseous mixture of $A$, $B$, and $C$. The entire apparatus is maintained at constant temperature and pressure. At time $t = 0$, we imagine that the diaphragm is carefully opened, and $A$ and $B$ are allowed to evaporate.

Here we wish to determine the concentration distribution of $A$ and $B$ in the gas as well as in liquid phase and the position of the liquid–gas interface as a function of time. Also, we propose to measure the binary liquid diffusion coefficient by following the position of the phase interface as a function of time, assuming that the binary diffusion coefficients in the gas phase can be estimated. We conclude by comparing the result of a new measurement of the binary diffusion coefficient for benzene and chloroform at 25°C with a value previously reported by Sanni and Hutchison (1973).

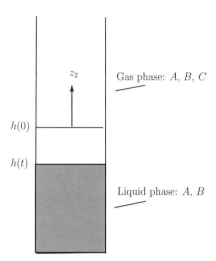

**Figure 9.4.2-1.** The gas–liquid interface $h(t)$ falls in a laboratory frame of reference as the liquid solution of $A$ and $B$ evaporates.

## Analysis

Let us consider a long tube that is fixed in a laboratory frame of reference as shown in Figure 9.4.2-1. In analyzing this problem, we shall make the following assumptions:

1) The temperature of the system is independent of time.
2) To describe ternary diffusion in the gas phase, we will use the alternative form of the Curtiss (1968) equation described in Exercise 8.4.4-2.
3) All binary diffusion coefficients, both in the liquid and the gas, are constants.
4) The molar density $c$ is a constant in both the liquid and gas phases.
5) Species $C$ is not soluble in the liquid phase.

For simplicity, let us replace the finite gas and liquid phases with semi-infinite phases. The initial and boundary conditions become

$$\text{at } t = 0 \text{ for all } z_2 > 0: \quad x_{(A)}^{(g)} = x_{(A)0}^{(g)}$$

$$x_{(B)}^{(g)} = x_{(B)0}^{(g)} \tag{9.4.2-1}$$

$$x_{(A)}^{(l)} = x_{(A)0}^{(l)}$$

and

$$\text{at } z_2 = h \text{ for all } t > 0: \quad x_{(A)}^{(g)} = x_{(A)\text{eq}}^{(g)}$$

$$x_{(B)}^{(g)} = x_{(B)\text{eq}}^{(g)} \tag{9.4.2-2}$$

By $x_{(A)\text{eq}}^{(g)}$ we mean the mole fraction of species $A$ in the gas mixture that is in equilibrium with the liquid adjacent to the phase interface at the existing temperature and pressure.

Assumption 9.4.2 together with Equations (9.4.2-1) and (9.4.2-2) suggest that we seek a solution to this problem of the form

$$v_1^{\diamond(g)} = v_3^{\diamond(g)}$$

$$= 0$$

$$v_2^{\diamond(g)} = v_2^{\diamond(g)}(t, z_2)$$

$$v_1^{\diamond(l)} = v_3^{\diamond(l)}$$

$$= 0 \qquad \text{(9.4.2-3)}$$

$$v_2^{\diamond(l)} = v_2^{\diamond(l)}(t, z_2)$$

$$x_{(A)}^{(g)} = x_{(A)}^{(g)}(t, z_2)$$

$$x_{(B)}^{(g)} = x_{(B)}^{(g)}(t, z_2)$$

$$x_{(A)}^{(l)} = x_{(A)}^{(l)}(t, z_2)$$

In short, we wish to analyze unsteady-state, one-dimensional evaporation of a binary liquid mixture. In view of assumption 9.4.2, the overall differential mass balance requires (Table 8.5.1-5)

$$\frac{\partial v_2^{\diamond(g)}}{\partial z_2} = 0$$

$$\frac{\partial v_2^{\diamond(l)}}{\partial z_2} = 0 \qquad \text{(9.4.2-4)}$$

This implies

$$v_2^{\diamond(g)} = v_2^{\diamond(g)}(t)$$

$$v_2^{\diamond(l)} = v_2^{\diamond(l)}(t) \qquad \text{(9.4.2-5)}$$

Since

$$\text{as } z_2 \to -\infty : \ v_2^{\diamond(l)} \to 0 \qquad \text{(9.4.2-6)}$$

we can observe that

$$\text{everywhere} : \ v_2^{\diamond(l)} = 0 \qquad \text{(9.4.2-7)}$$

With (9.4.2-7), the overall jump mass balance requires (Section 8.3.1)

$$u_2 = \frac{c^{(g)} v_2^{\diamond(g)}}{c^{(g)} - c^{(l)}} \qquad \text{(9.4.2-8)}$$

If we assume Fick's first law and recognize assumptions 9.4.2 and 9.4.2, we can write the differential mass balance for species $A$ in the liquid as (Table 8.5.1-6)

$$\frac{\partial x_{(A)}^{(l)}}{\partial t} = \mathcal{D}_{(AB)}^{(l)} \frac{\partial^2 x_{(A)}^{(l)}}{\partial z_2{}^2} \qquad \text{(9.4.2-9)}$$

With the change of variable

$$\eta \equiv \frac{z_2}{\sqrt{4\mathcal{D}^o t}} \tag{9.4.2-10}$$

Equation (9.4.2-9) can be rearranged as

$$\frac{d^2 x_{(A)}^{(l)}}{d\eta^2} + 2\eta \frac{\mathcal{D}^o}{\mathcal{D}_{(AB)}^{(l)}} \frac{d x_{(A)}^{(l)}}{d\eta} = 0 \tag{9.4.2-11}$$

Here $\mathcal{D}^o$ is a diffusion coefficient that for the moment will be left undefined. We can define $\mathcal{D}^o$ as $\mathcal{D}_{(AB)}^{(l)}$ or simply 1, whatever we find more convenient. Equation (9.4.2-11) can be solved consistent with the boundary conditions

$$\text{at } \eta = \lambda: \ x_{(A)}^{(l)} = x_{(A)\text{eq}}^{(l)} \tag{9.4.2-12}$$

$$\text{as } \eta \to -\infty: \ x_{(A)}^{(l)} = x_{(A)0}^{(l)} \tag{9.4.2-13}$$

to find

$$x_{(A)}^{(l)\star} \equiv \frac{x_{(A)}^{(l)} - x_{(A)\text{eq}}^{(l)}}{x_{(A)0}^{(l)} - x_{(A)\text{eq}}^{(l)}}$$

$$= \left[ -\text{erf}\left( \eta \sqrt{\frac{\mathcal{D}^o}{\mathcal{D}_{(AB)}^{(l)}}} \right) + \text{erf}\left( \lambda \sqrt{\frac{\mathcal{D}^o}{\mathcal{D}_{(AB)}^{(l)}}} \right) \right] \left[ 1 + \text{erf}\left( \lambda \sqrt{\frac{\mathcal{D}^o}{\mathcal{D}_{(AB)}^{(l)}}} \right) \right]^{-1} \tag{9.4.2-14}$$

in which

$$\lambda \equiv \frac{h}{\sqrt{4\mathcal{D}^o t}}$$

$$= \text{a constant} \tag{9.4.2-15}$$

and $z_2 = h = h(t)$ denotes the position of the liquid–gas phase interface.

In view of assumption 9.4.2 and Equation (9.4.2-3), the differential mass balances for $A$ and $B$ in the gas phase take the forms (Table 8.5.1-5)

$$\frac{\partial x_{(A)}^{(g)}}{\partial t} + \frac{\partial x_{(A)}^{(g)}}{\partial z_2} v_2^{\diamond(g)}$$

$$= -\frac{1}{c^{(g)}} \frac{\partial J_{(A)2}^{\diamond(g)}}{\partial z_2}$$

$$= \frac{\partial}{\partial z_2} \left( \overline{D}_{11} \frac{\partial x_{(A)}^{(g)}}{\partial z_2} \right) + \frac{\partial}{\partial z_2} \left( \overline{D}_{12} \frac{\partial x_{(B)}^{(g)}}{\partial z_2} \right) \tag{9.4.2-16}$$

and

$$\frac{\partial x_{(B)}^{(g)}}{\partial t} + \frac{\partial x_{(B)}^{(g)}}{\partial z_2} v_2^{\diamond(g)}$$

$$= -\frac{1}{c^{(g)}} \frac{\partial J_{(B)2}^{\diamond(g)}}{\partial z_2}$$

$$= \frac{\partial}{\partial z_2}\left(\overline{D}_{21} \frac{\partial x_{(A)}^{(g)}}{\partial z_2}\right) + \frac{\partial}{\partial z_2}\left(\overline{D}_{22} \frac{\partial x_{(B)}^{(g)}}{\partial z_2}\right) \qquad (9.4.2\text{-}17)$$

From (9.4.2-15),

$$u_2 = \frac{dh}{dt}$$

$$= \lambda \sqrt{\frac{D^o}{t}} \qquad (9.4.2\text{-}18)$$

In view of (9.4.2-8), this permits us to observe that

$$v_2^{\diamond(g)} = \lambda \frac{c^{(g)} - c^{(l)}}{c^{(g)}} \sqrt{\frac{D^o}{t}} \qquad (9.4.2\text{-}19)$$

If, in addition to recognizing (9.4.2-19), we make the change of variable (9.4.2-10), Equations (9.4.2-16) and (9.4.2-17) become

$$\frac{d}{d\eta}\left(\frac{\overline{D}_{11}}{D^o} \frac{dx_{(A)}^{(g)\star}}{d\eta}\right) + \left(\frac{x_{(B)0}^{(g)} - x_{(B)\text{eq}}^{(g)}}{x_{(A)0}^{(g)} - x_{(A)\text{eq}}^{(g)}}\right) \frac{d}{d\eta}\left(\frac{\overline{D}_{12}}{D^o} \frac{dx_{(B)}^{(g)\star}}{d\eta}\right)$$

$$+ 2\left[\eta + \frac{(c^{(l)} - c^{(g)})}{c^{(g)}}\lambda\right] \frac{dx_{(A)}^{(g)\star}}{d\eta} = 0 \qquad (9.4.2\text{-}20)$$

and

$$\left(\frac{x_{(A)0}^{(g)} - x_{(A)\text{eq}}^{(g)}}{x_{(B)0}^{(g)} - x_{(B)\text{eq}}^{(g)}}\right) \frac{d}{d\eta}\left(\frac{\overline{D}_{21}}{D^o} \frac{dx_{(A)}^{(g)\star}}{d\eta}\right) + \frac{d}{d\eta}\left(\frac{\overline{D}_{22}}{D^o} \frac{dx_{(B)}^{(g)\star}}{d\eta}\right)$$

$$+ 2\left[\eta + \frac{(c^{(l)} - c^{(g)})}{c^{(g)}}\lambda\right] \frac{dx_{(B)}^{(g)\star}}{d\eta} = 0 \qquad (9.4.2\text{-}21)$$

in which we have introduced

$$x_{(A)}^{(g)\star} \equiv \frac{x_{(A)}^{(g)} - x_{(A)\text{eq}}^{(g)}}{x_{(A)0}^{(g)} - x_{(A)\text{eq}}^{(g)}}$$

$$\qquad (9.4.2\text{-}22)$$

$$x_{(B)}^{(g)\star} \equiv \frac{x_{(B)}^{(g)} - x_{(B)\text{eq}}^{(g)}}{x_{(B)0}^{(g)} - x_{(B)\text{eq}}^{(g)}}$$

Here $x_{(A)0}^{(g)}$ is the initial mole fraction of species $A$ in the gas phase; $x_{(A)eq}^{(g)}$ is the equilibrium mole fraction of species $A$ at the liquid–gas interface. These equations must satisfy the boundary conditions

$$\text{at } \eta = \lambda : \ x_{(A)}^{(g)\star} = 0$$

$$x_{(B)}^{(g)\star} = 0 \tag{9.4.2-23}$$

and

$$\text{as } \eta \to \infty : \ x_{(A)}^{(g)\star} \to 1$$

$$x_{(B)}^{(g)\star} \to 1 \tag{9.4.2-24}$$

Note that these equations cannot be solved numerically without knowing $\lambda$, $x_{(A)eq}^{(g)}$, and $x_{(B)eq}^{(g)}$.

Recognizing Fick's first law in the liquid, we see that the jump mass balance for species $A$ requires (Section 8.2.1)

$$N_{(A)2}^{(g)} = -c^{(l)}\mathcal{D}_{(AB)}^{(l)} \left.\frac{\partial x_{(A)}^{(l)}}{\partial z_2}\right|_{z_2=h} + u_2\left(c^{(g)}x_{(A)eq}^{(g)} - c^{(l)}x_{(A)eq}^{(l)}\right)$$

$$\tag{9.4.2-25}$$

$$J_{(A)2}^{\diamond(g)} = -c^{(l)}\mathcal{D}_{(AB)}^{(l)} \left.\frac{\partial x_{(A)}^{(l)}}{\partial z_2}\right|_{z_2=h} + c^{(l)}u_2\left(x_{(A)eq}^{(g)} - x_{(A)eq}^{(l)}\right)$$

where we have recognized (9.4.2-8). If we recognize assumption 9.4.2 and make the changes of variable introduced in (9.4.2-10), (9.4.2-19), and (9.4.2-22), this becomes

$$-\frac{\overline{D}_{11}}{D^o}\left(x_{(A)0}^{(g)} - x_{(A)eq}^{(g)}\right)\left.\frac{dx_{(A)}^{(g)\star}}{d\eta}\right|_{\eta=\lambda} - \frac{\overline{D}_{12}}{D^o}\left(x_{(B)0}^{(g)} - x_{(B)eq}^{(g)}\right)\left.\frac{dx_{(B)}^{(g)\star}}{d\eta}\right|_{\eta=\lambda}$$

$$= -\frac{c^{(l)}}{c^{(g)}}\frac{\mathcal{D}_{(AB)}^{(l)}}{D^o}\left(x_{(A)0}^{(l)} - x_{(A)eq}^{(l)}\right)\left.\frac{dx_{(A)}^{(l)\star}}{d\eta}\right|_{\eta=\lambda} + 2\lambda\frac{c^{(l)}}{c^{(g)}}\left(x_{(A)eq}^{(g)} - x_{(A)eq}^{(l)}\right) \tag{9.4.2-26}$$

In a similar fashion, the jump mass balance for species $B$,

$$N_{(B)2}^{(g)} = -c^{(l)}\mathcal{D}_{(AB)}^{(l)} \left.\frac{\partial x_{(B)}^{(l)}}{\partial z_2}\right|_{z_2=h} + u_2\left(c^{(g)}x_{(B)eq}^{(g)} - c^{(l)}x_{(B)eq}^{(l)}\right)$$

$$\tag{9.4.2-27}$$

$$J_{(B)2}^{\diamond} = c^{(l)}\mathcal{D}_{(AB)}^{(l)} \left.\frac{\partial x_{(A)}^{(l)}}{\partial z_2}\right|_{z_2=h} + c^{(l)}u_2\left(x_{(B)eq}^{(g)} - x_{(B)eq}^{(l)}\right)$$

becomes

$$-\frac{\overline{D}_{21}}{D^o}\left(x_{(A)0}^{(g)} - x_{(A)eq}^{(g)}\right)\left.\frac{dx_{(A)}^{(g)\star}}{d\eta}\right|_{\eta=\lambda} - \frac{\overline{D}_{22}}{D^o}\left(x_{(B)0}^{(g)} - x_{(B)eq}^{(g)}\right)\left.\frac{dx_{(B)}^{(g)\star}}{d\eta}\right|_{\eta=\lambda}$$

$$= \frac{c^{(l)}}{c^{(g)}}\frac{\mathcal{D}_{(AB)}^{(l)}}{D^o}\left(x_{(A)0}^{(l)} - x_{(A)eq}^{(l)}\right)\left.\frac{dx_{(A)}^{(l)\star}}{d\eta}\right|_{\eta=\lambda} + 2\lambda\frac{c^{(l)}}{c^{(g)}}\left(x_{(B)eq}^{(g)} - x_{(B)eq}^{(l)}\right) \tag{9.4.2-28}$$

Finally, we must describe the equilibrium at the phase interface. We shall account for the nonideality in the liquid phase by the activity coefficients for respective species. For species $A$,

$$x_{(A)\text{eq}}^{(g)} = \frac{P_{(A)}^{sat}}{P} \gamma_{(A)} x_{(A)\text{eq}}^{(l)} \tag{9.4.2-29}$$

and for species $B$,

$$x_{(B)\text{eq}}^{(g)} = \frac{P_{(B)}^{sat}}{P} \gamma_{(B)} x_{(B)\text{eq}}^{(l)} \tag{9.4.2-30}$$

For the moment, let us assume that we know the liquid diffusion coefficient $\mathcal{D}_{(AB)}^{(l)}$. (Remember that we can define $\mathcal{D}^o$ as $\mathcal{D}_{(AB)}^{(l)}$ or simply 1, whatever we find more convenient.) In this problem there are seven unknowns, $x_{(A)}^{(l)\star}, x_{(A)}^{(g)\star}, x_{(B)}^{(g)\star}, \lambda, x_{(A)\text{eq}}^{(g)}, x_{(B)\text{eq}}^{(g)}$, and $x_{(A)\text{eq}}^{(l)}$, which are determined by solving simultaneously the seven equations (9.4.2-14), (9.4.2-20), (9.4.2-21), (9.4.2-26), and (9.4.2-28) through (9.4.2-30). The two second-order differential equations must be solved consistent with the boundary conditions (9.4.2-23) and (9.4.2-24). [Remember that, in view of Equation (8.4.4-2), the coefficients $\overline{D}_{11}, \overline{D}_{12}, \overline{D}_{21}$, and $\overline{D}_{22}$ are functions of $x_{(A)}^{(g)\star}$ and $x_{(B)}^{(g)\star}$.] But we are assuming here that $\mathcal{D}_{(AB)}^{(l)}$ is unknown. For this reason, we must conduct an experiment that will allow us to determine $\lambda$ and calculate $\mathcal{D}_{(AB)}^{(l)}$.

## Experimental Study

Evaporation of a small amount of a binary liquid mixture was observed in a vertical 70-cm tube. A binary liquid mixture of known composition was introduced into the tube from its bottom with the aid of a valve connected to a large reservoir. As soon as the desired quantity of liquid entered the tube, the valve was closed, and measurements were begun. A video camera in a previously calibrated configuration along with a time-lapse video recorder was used to measure the position of the liquid–gas interface as a function of time. Because of the small resistance to the flow of energy from the surrounding air through the glass tube to the liquid, the liquid temperature remains nearly equal to ambient temperature during evaporation (Lee and Wilke 1954).

Evaporation of a 50 mol% solution of benzene ($A$) in chloroform ($B$) into air ($C$) was studied. Experimental conditions were $T = 24.8°C$, $P = 1.012 \times 10^5$ Pa, $x_{(A)0}^{(l)} = 0.5$, $x_{(A)0}^{(g)} = 0$, and $x_{(B)0}^{(l)} = 0$. We estimated that $\mathcal{D}_{(AC)}^{(g)} = 9.30 \times 10^{-6}$ m$^2$/s [corrected from $9.32 \pm 0.149 \times 10^{-6}$ m$^2$/s at 25°C and 1 atmosphere (Lugg 1968) using Reid et al. (1987, Eq. 11-4.4)] and $\mathcal{D}_{(BC)}^{(g)} = 8.86 \times 10^{-6}$ m$^2$/s [corrected from $8.88 \pm 0.187 \times 10^{-6}$ m$^2$/s at 25°C and 1 atm (Lugg 1968) using Reid et al. (1987, Eq. 11-4.4)]. Although benzene and chloroform cannot exist as a binary vapor at the conditions considered here and their binary diffusion coefficient is not a physical quantity, we have estimated it as $\mathcal{D}_{(BC)}^{(g)} = 3.74 \pm 0.2 \times 10^{-6}$ m$^2$/s (Reid et al. 1987, Eq. 11-4.4); the error has been estimated as 5.4 percent (Reid et al. 1987, p. 590). We estimated $c^{(g)} = 0.0411$ kg mol/m$^3$ (Dean 1979, p. 10-92) and $c^{(l)} = 11.7$ kg mol/m$^3$ (Sanni and Hutchison 1973). Activity coefficients for benzene and chloroform at 25°C were determined using the Margules equation with constants obtained from experimental data (Gmehling, Onken, and Arlt 1980, p. 65).

Figure 9.4.2-2 shows the measured height of the liquid–gas interface as a function of time. The experimental technique used is the same as that described in Section 9.3.1. The dimensionless concentration profiles in the gas and liquid phases are shown as functions of $\eta$ in Figures 9.4.2-3 and 9.4.2-4.

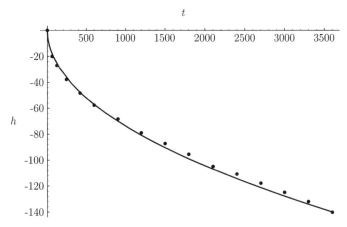

**Figure 9.4.2-2.** Points denote measured positions of the liquid–gas phase interface $h$ ($\mu$m) as a function of $t$ (s) for evaporation of a 50 mol% solution of benzene in chloroform into air at 24.8°C and $P = 1.012 \times 10^5$ Pa. The solid curve is the result of the least-square fit of (9.4.2-15) to these data.

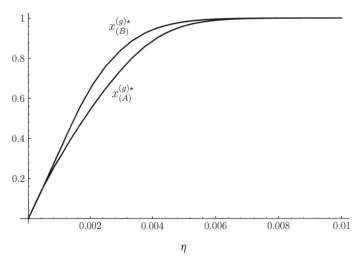

**Figure 9.4.2-3.** The dimensionless mole fractions $x_{(A)}^{(g)\star}$ and $x_{(B)}^{(g)\star}$ as functions of $\eta$ in the gas phase.

### Measurement of Diffusion Coefficient

For simplicity, we have taken $\mathcal{D}^o = 1$. The value of $\lambda$ was determined from a least-square fit of these experimental data using (9.4.2-15), as shown in Figure 9.4.2-2.

Equations (9.4.2-20), (9.4.2-21), (9.4.2-26), and (9.4.2-28) were solved consistent with (9.4.2-23) and (9.4.2-24) for four unknowns: $D_{(AB)}^{(l)}$, $x_{(A)\text{eq}}^{(l)}$, $x_{(A)}^{(g)}$, and $x_{(B)}^{(g)}$ using Mathematica (1993). With assumed values for $D_{(AB)}^{(l)}$ and $x_{(A)\text{eq}}^{(l)}$, Equations (9.4.2-20) and (9.4.2-21) were solved consistent with (9.4.2-23) and (9.4.2-24) in order to determine $x_{(A)}^{(g)}$ and $x_{(B)}^{(g)}$ as functions $\eta$. Because Mathematica (1993) is designed to solve nonlinear ordinary differential equations with all the boundary conditions specified at the same point, we used a shooting

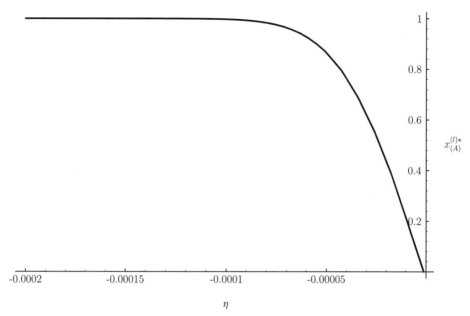

**Figure 9.4.2-4.** The dimensionless mole fraction $x_{(A)}^{(l)\star}$ as a function of $\eta$ in the liquid phase.

method. Equations (9.4.2-26) and (9.4.2-28) were checked. If they were not satisfied, new values of $D_{(AB)}^{(l)}$ and $x_{(A)eq}^{(l)}$ were assumed and the process was repeated.

In this way, we found the binary liquid diffusion coefficient $D_{(AB)}^{(l)}$ at 25°C to be $2.21 \pm 0.048 \times 10^{-9}$ m$^2$/s [corrected from $2.21 \times 10^{-9}$ m$^2$/s at 24.8°C using a popular empirical correlation (Reid et al. 1987, Eq. 11-11.1)]. This can be compared with a previously reported value of $2.46 \times 10^{-9}$ m$^2$/s (Sanni and Hutchison 1973; these authors did not report an error estimate) and an estimated value of $3.79 \times 10^{-9}$ m$^2$/s (Reid et al. 1987, Eq. 11-10.4). [In this last computation, activity coefficients for benzene and chloroform at 25°C were determined using the Margules equation with constants obtained from experimental data (Gmehling et al. 1980, p. 65).]

We attribute the primary error in our result to our estimation of the binary diffusion coefficients for the gas phase, presented in the preceding section. The error that we report in our result has been computed by using first the largest and then the smallest possible diffusion coefficients for the gas phase.

### 9.4.3 Oxidation of Iron

This section is taken from Slattery et al. (1995).

In the oxidation of a metal, there are several steps: gas absorption, surface reaction, and diffusion through one or more layers of metal oxides.

Most prior analyses of high-temperature oxidation of metals are based upon the work of Wagner (1951). Although Wagner considered ionic diffusion through the metal oxide to be the rate-limiting step, he identified the local activity of the oxygen ion with the activity of molecular oxygen at a corresponding partial pressure without explanation. He restricted his theory to simple metal oxides in which the valence of the metal ions has only one value.

Himmel, Mehl, and Birchenall (1953, p. 840) used this theory together with two correction factors to obtain close agreement with experimental data for the high-temperature oxidation of iron. The method by which these correction factors are to be obtained was not clearly explained.

Wagner (1969), Smeltzer (1987), and Coates and Dalvi (1970) have extended this theory to the oxidation of binary alloys. Their result is valid only in the limit of dilute solutions (Section 8.4.6).

In what follows, we develop a new theory for the high-temperature oxidation of iron, in which the rate-limiting step is ternary diffusion of ferric, ferrous, and oxygen ions in the iron oxides that are formed. Like Wagner (1951), we assume that electrical neutrality is maintained at each point within each phase. Unlike Wagner (1951), we will assume that local equilibrium is established at all phase interfaces and that the ions form an ideal solution in oxide phases. Although Wagner (1951) did not directly use the assumption of ideal solutions, in measuring their diffusion coefficients Himmel et al. (1953) followed the analysis of (Steigman, Shockly, and Nix 1939), requiring the unstated assumption of ideal, binary solutions of "iron" and oxygen.

### Problem Statement

In attempting to understand this problem, let us begin with an extreme case: iron exposed to $O_2$ at $1 \times 10^5$ Pa and 1,200°C. From the phase diagram shown in Figure 9.4.3-1, we conclude

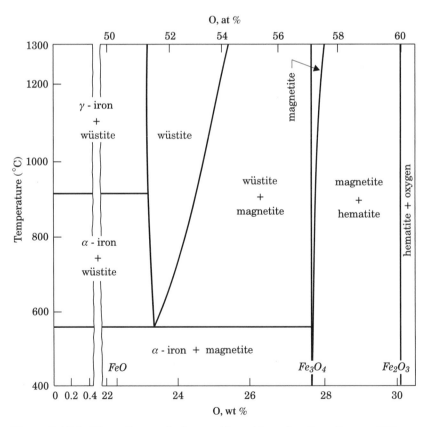

**Figure 9.4.3-1.** Phase diagram for iron and its oxides, taken from Borg and Dienes (1988, p. 115).

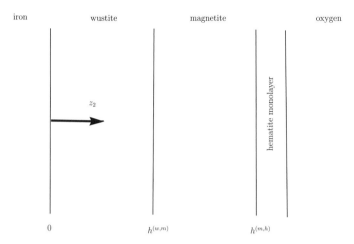

**Figure 9.4.3-2.** Corrosion layer consisting of two nonstoichiometric phases, magnetite and wüstite, covered by a monolayer of hematite.

that, with time, a corrosion layer will develop consisting of two nonstoichiometric phases, magnetite and wüstite, and a monolayer of hematite, as shown in Figure 9.4.3-2.

In analyzing this problem, we will make several assumptions.

1) Equilibrium is established at the three interfaces shown in Figure 9.4.3-2. With this assumption and the stoichiometry of the reactions discussed in assumptions 7 and 8, it will not be necessary to have separate descriptions for the kinetics of the heterogeneous reactions.
2) Neither Fe nor $O_2$ can diffuse through wüstite and magnetite.
3) Wüstite and magnetite are nonstoichiometric, and we will assume that these materials are fully dissociated.
4) The ionic radius of $O^{2-}$ is 1.40 Å, that of $Fe^{2+}$ is 0.76 Å; and that of $Fe^{3+}$ is 0.64 Å (Dean 1979). In a frame of reference such that the iron–wüstite interface is stationary, the ferrous ions $Fe^{2+}$ and ferric ions $Fe^{3+}$ diffuse through a lattice of stationary oxygen ions $O^{2-}$ (Davies, Simnad, and Birchenall 1951, p. 892; Hauffe 1965, p. 285). For this reason, the wüstite and magnetite must be regarded as consisting of three components.
5) Within the wüstite and magnetite phases, $c_{(O^{2-})}$ is a constant, because we assume that the oxygen ions $O^{2-}$ are stationary with respect to the moving boundary (see above). Looking ahead to the jump mass balance for $O^{2-}$ at the wüstite–magnetite interface (9.4.3-42), we see that $c_{(O^{2-})}$ takes the same value in both phases.
6) The oxidation-reduction reaction at the iron–wüstite interface,

$$Fe + 2Fe^{3+} \rightarrow 3Fe^{2+}$$

results in no generation of free electrons.
7) The oxidation-reduction reaction at the hematite–oxygen interface,

$$4Fe^{2+} + O_2 \rightarrow 4Fe^{3+} + 2O^{2-}$$

results in no generation of free electrons. Because we are assuming that the hematite is a monolayer and because equilibrium is established at the magnetite–hematite interface (assumption 1), this reaction can be regarded as taking place at this latter interface.

8) Within the wüstite and magnetite phases, we recognize that we may have the reaction

$$Fe^{3+} + e^- \rightarrow Fe^{2+}$$

9) The $Fe^{2+}$ and $Fe^{3+}$ move in such a way as to preserve local electrical neutrality:

$$2c_{(Fe^{2+})} + 3c_{(Fe^{3+})} - 2c_{(O^{2-})} = 0 \tag{9.4.3-1}$$

10) The oxides form on a flat sheet of iron.
11) In the one-dimensional problem to be considered here, there is no current flow, and, in view of assumptions 7 and 8, there is no free electron flow:

$$2N_{(Fe^{2+})2} + 3N_{(Fe^{3+})2} = 0 \tag{9.4.3-2}$$

Here we have recognized both that $c_{(O^{2-})}$ is independent of position within an oxide and that the oxygen ions $O^{2-}$ are stationary with respect to the moving boundary.

12) Binary diffusion coefficients are taken to be constants.
13) Both wüstite and magnetite are assumed to form ideal solutions of the $Fe^{3+}$, $Fe^{2+}$, and $O^{2-}$ ions. This will allow us to use the Stefan–Maxwell equations developed in and Exercise 8.4.4-1.

We will work in a frame of reference in which the iron–wüstite interface is stationary. In view of the requirement

$$x_{(Fe^{2+})} + x_{(Fe^{3+})} + x_{(O^{2-})} = 1 \tag{9.4.3-3}$$

assumption 10

$$2x_{(Fe^{2+})} + 3x_{(Fe^{3+})} - 2x_{(O^{2-})} = 0 \tag{9.4.3-4}$$

and assumption 11, it is necessary that we seek a solution only for

$$x_{(Fe^{2+})} = x_{(Fe^{2+})}(t, z_2) \tag{9.4.3-5}$$

within the wüstite and magnetite phases.

From Figure 9.4.3-1, we find that, at $1,200°C$ in equilibrium with iron, wüstite has the composition $x_{(O^{2-})} = 0.513$. Recognizing (9.4.3-3) and (9.4.3-4), we have two equations to solve simultaneously for $x_{(Fe^{2+})}$ and $x_{(Fe^{3+})}$ to obtain

$$\text{at } z_2 = 0: \quad x^{(w)}_{(Fe^{2+})} = 0.433 \tag{9.4.3-6}$$

From (9.4.3-1), we have

$$c^{(w)} = \frac{1}{3}c^{(w)}_{(Fe^{2+})} + \frac{5}{3}c^{(w)}_{(O^{2-})} \tag{9.4.3-7}$$

which allows us to write

$$x^{(w)}_{(Fe^{2+})} = \frac{3c^{(w)}_{(Fe^{2+})}}{c^{(w)}_{(Fe^{2+})} + 5c^{(w)}_{(O^{2-})}} \tag{9.4.3-8}$$

or

$$c_{(Fe^{2+})}^{(w)} = \frac{5 x_{(Fe^{2+})}^{(w)} c_{(O^{2-})}^{(w)}}{3 - x_{(Fe^{2+})}^{(w)}} \tag{9.4.3-9}$$

We can now rewrite (9.4.3-6) as

$$\text{at } z_2 = 0: \quad c_{(Fe^{2+})}^{(w)} = c_{(Fe^{2+})eq,\,0}^{(w)}$$

$$= 0.843 \times c_{(O^{2-})}^{(w)}. \tag{9.4.3-10}$$

Arguing in a similar manner, we find

$$\text{at } z_2 = h^{(w,m)}: \quad x_{(O^{2-})}^{(w)} = 0.539$$

$$x_{(Fe^{2+})}^{(w)} = 0.303$$

$$c_{(Fe^{2+})}^{(w)} = c_{(Fe^{2+})eq,\,a}^{(w)}$$

$$= 0.562 \times c_{(O^{2-})}^{(w)} \tag{9.4.3-11}$$

$$\text{at } z_2 = h^{(w,m)}: \quad x_{(O^{2-})}^{(m)} = 0.574$$

$$x_{(Fe^{2+})}^{(m)} = 0.125$$

$$c_{(Fe^{2+})}^{(m)} = c_{(Fe^{2+})eq,\,b}^{(m)}$$

$$= 0.217 \times c_{(O^{2-})}^{(m)} \tag{9.4.3-12}$$

$$\text{at } z_2 = h^{(m,h)}: \quad x_{(O^{2-})}^{(m)} = 0.577$$

$$x_{(Fe^{2+})}^{(m)} = 0.112$$

$$c_{(Fe^{2+})}^{(m)} = c_{(Fe^{2+})eq,\,c}^{(m)}$$

$$= 0.194 \times c_{(O^{2-})}^{(m)} \tag{9.4.3-13}$$

$$\text{at } z_2 = h^{(m,h)}: \quad c_{(Fe^{2+})}^{(h)} = 0 \tag{9.4.3-14}$$

Here, $h^{(w,m)}$ denotes the position of the wüstite–magnetite interface shown in Figure 9.4.3-2, $h^{(m,h)}$ the position of the magnetite–hematite interface, $(w)$ a quantity associated with the wüstite phase, $(m)$ a quantity associated with the magnetite phase, and $(h)$ a quantity associated with the hematite phase. It is helpful to begin by looking at the concentration distributions in each phase separately. Consider first the wüstite.

With assumptions 4, 10, 12, and 14, the Stefan–Maxwell equations (Exercise 8.4.4-1) require that

$$
\frac{\partial x_{(Fe^{2+})}^{(w)}}{\partial z_2} = \frac{1}{c^{(w)}\mathcal{D}_{(Fe^{2+},O^{2-})}^{(w)}}\left(-x_{(O^{2-})}^{(w)}N_{(Fe^{2+})2}^{(w)}\right)
$$

$$
+ \frac{1}{c^{(w)}\mathcal{D}_{(Fe^{2+},Fe^{3+})}^{(w)}}\left(x_{(Fe^{2+})}^{(w)}N_{(Fe^{3+})2}^{(w)} - x_{(Fe^{3+})}^{(w)}N_{(Fe^{2+})2}^{(w)}\right)
$$

$$
= \frac{1}{c^{(w)}\mathcal{D}_{(Fe^{2+},O^{2-})}^{(w)}}\left(-x_{(O^{2-})}^{(w)}N_{(Fe^{2+})2}^{(w)}\right)
$$

$$
+ \frac{1}{c^{(w)}\mathcal{D}_{(Fe^{2+},Fe^{3+})}^{(w)}}\left(-\frac{2}{3}x_{(Fe^{2+})}^{(w)} - x_{(Fe^{3+})}^{(w)}\right)N_{(Fe^{2+})2}^{(w)}
$$

$$
= -x_{(O^{2-})}^{(w)}N_{(Fe^{2+})2}^{(w)}\left(\frac{1}{c^{(w)}\mathcal{D}_{(Fe^{2+},O^{2-})}^{(w)}} + \frac{2}{3}\frac{1}{c^{(w)}\mathcal{D}_{(Fe^{2+},Fe^{3+})}^{(w)}}\right) \qquad (9.4.3\text{-}15)
$$

or

$$
N_{(Fe^{2+})2}^{(w)} = -\frac{c^{(w)2}}{c_{(O^{2-})}^{(w)}}\mathcal{D}_{(Fe^{2+})}^{(w)}\frac{\partial x_{(Fe^{2+})}^{(w)}}{\partial z_2} \qquad (9.4.3\text{-}16)
$$

where

$$
\mathcal{D}_{(Fe^{2+})}^{(w)} \equiv \left(\frac{1}{\mathcal{D}_{(Fe^{2+},O^{2-})}^{(w)}} + \frac{2}{3}\frac{1}{\mathcal{D}_{(Fe^{2+},Fe^{3+})}^{(w)}}\right)^{-1} \qquad (9.4.3\text{-}17)
$$

In view of assumption 9, the differential mass balances for $Fe^{2+}$ and for $Fe^{3+}$ (Table 8.5.1-5) require

$$
\frac{\partial c_{(Fe^{2+})}^{(w)}}{\partial t} + \frac{\partial N_{(Fe^{2+})2}^{(w)}}{\partial z_2} = \frac{r_{(Fe^{2+})}^{(w)}}{M_{(Fe^{2+})}} \qquad (9.4.3\text{-}18)
$$

and

$$
\frac{\partial c_{(Fe^{3+})}^{(w)}}{\partial t} + \frac{\partial N_{(Fe^{3+})2}^{(w)}}{\partial z_2} = \frac{r_{(Fe^{3+})}^{(w)}}{M_{(Fe^{3+})}}
$$

$$
= -\frac{r_{(Fe^{2+})}^{(w)}}{M_{(Fe^{2+})}} \qquad (9.4.3\text{-}19)
$$

In writing this last expression, we have employed assumption 9. Adding (9.4.3-18) and (9.4.3-19) and taking advantage of assumptions 6, 10, and 12, we find

$$
\frac{\partial c_{(Fe^{2+})}^{(w)}}{\partial t} + \frac{\partial N_{(Fe^{2+})2}^{(w)}}{\partial z_2} = 0 \qquad (9.4.3\text{-}20)
$$

As the result of assumptions 4 and 6, the differential mass balance for $O^{2-}$ is satisfied identically. At the end of this analysis, we could use (9.4.3-18) and (9.4.3-19) to compute $r^{(w)}_{(Fe^{2+})}$ and $r^{(w)}_{(Fe^{3+})}$, if they are desired.

Let us look for a solution by first transforming (9.4.3-20) into an ordinary differential equation. In terms of a new independent variable

$$\eta \equiv \frac{z_2}{\sqrt{10\mathcal{D}^{(w)}_{(Fe^{2+})}t/3}} \tag{9.4.3-21}$$

and using assumptions 6 and 10 as well as (9.4.3-16), Equation (9.4.3-20) may be expressed as (remember that $c^{(w)}$ is not a constant)

$$\frac{d^2 c^{(w)}_{(Fe^{2+})}}{d\eta^2} + \eta \frac{dc^{(w)}_{(Fe^{2+})}}{d\eta} = 0 \tag{9.4.3-22}$$

From (9.4.3-10) and (9.4.3-11), the corresponding boundary conditions are

$$\text{at } \eta = 0: \quad c^{(w)}_{(Fe^{2+})} = c^{(w)}_{(Fe^{2+})eq,\,0} \tag{9.4.3-23}$$

and

$$\text{at } \eta = \frac{h^{(w,m)}}{\sqrt{10\mathcal{D}^{(w)}_{(Fe^{2+})}t/3}}: \quad c^{(w)}_{(Fe^{2+})} = c^{(w)}_{(Fe^{2+})eq,\,a} \tag{9.4.3-24}$$

This last expression implies that

$$\frac{h^{(w,m)}}{\sqrt{10\mathcal{D}^{(w)}_{(Fe^{2+})}t/3}} = \lambda^{(w,m)}$$
$$= \text{a constant} \tag{9.4.3-25}$$

Equation (9.4.3-25) describes the thickness $h^{(w,m)}$ of the layer of wüstite as a function of time. Alternatively, it implies that the speed of displacement of the wüstite–magnetite interface

$$u^{(w,m)}_2 = \frac{dh^{(w,m)}}{dt}$$
$$= \lambda^{(w,m)}\sqrt{\frac{5\mathcal{D}^{(w)}_{(Fe^{2+})}}{6t}} \tag{9.4.3-26}$$

Equation (9.4.3-22) can be integrated consistent with (9.4.3-23) to find

$$c^{(w)}_{(Fe^{2+})} - c^{(w)}_{(Fe^{2+})eq,\,0} = C_1 \operatorname{erf}(\eta) \tag{9.4.3-27}$$

where $C_1$ is a constant of integration. The boundary condition (9.4.3-24) requires in view of (9.4.3-25)

$$C_1 = \frac{c^{(w)}_{(Fe^{2+})eq,\,a} - c^{(w)}_{(Fe^{2+})eq,\,0}}{\operatorname{erf}(\lambda^{(w,m)})} \tag{9.4.3-28}$$

We can immediately write down the similar results for magnetite

$$\frac{d^2 c^{(m)}_{(Fe^{2+})}}{d\mu^2} + \mu \frac{dc^{(m)}_{(Fe^{2+})}}{d\mu} = 0 \tag{9.4.3-29}$$

where

$$\mu \equiv \frac{z_2}{\sqrt{10\mathcal{D}^{(m)}_{(Fe^{2+})}t/3}} \qquad (9.4.3\text{-}30)$$

and, by analogy with (9.4.3-7), we have used

$$c^{(m)} = \frac{1}{3}c^{(m)}_{(Fe^{2+})} + \frac{5}{3}c^{(m)}_{(O^{2-})} \qquad (9.4.3\text{-}31)$$

Equation (9.4.3-29) is to be solved consistent with the boundary conditions (9.4.3-12) and (9.4.3-13) or

$$\text{at } \mu = \frac{h^{(w,m)}}{\sqrt{10\mathcal{D}^{(m)}_{(Fe^{2+})}t/3}} : \quad c^{(m)}_{(Fe^{2+})} = c^{(m)}_{(Fe^{2+})\text{eq},\,b} \qquad (9.4.3\text{-}32)$$

and

$$\text{at } \mu = \frac{h^{(m,h)}}{\sqrt{10\mathcal{D}^{(m)}_{(Fe^{2+})}t/3}} : \quad c^{(m)}_{(Fe^{2+})} = c^{(m)}_{(Fe^{2+})\text{eq},\,c} \qquad (9.4.3\text{-}33)$$

We conclude that

$$\frac{h^{(m,h)}}{\sqrt{10\mathcal{D}^{(m)}_{(Fe^{2+})}t/3}} = \lambda^{(m,h)}$$

$$= \text{a constant} \qquad (9.4.3\text{-}34)$$

describes the thickness $h^{(m,h)}$ of the layer of magnetite as a function of time. This is turn implies that the speed of displacement of the magnetite–hematite interface is

$$u_2^{(m,h)} = \frac{dh^{(m,h)}}{dt}$$

$$= \lambda^{(m,h)}\sqrt{\frac{5\mathcal{D}^{(m)}_{(Fe^{2+})}}{6t}} \qquad (9.4.3\text{-}35)$$

Equation (9.4.3-29) can be integrated consistent with (9.4.3-32) through (9.4.3-34) to find

$$c^{(m)}_{(Fe^{2+})} - c^{(m)}_{(Fe^{2+})\text{eq},\,b} = C_2\left[\text{erf}(\mu) - \text{erf}\left(\sqrt{\frac{\mathcal{D}^{(w)}_{(Fe^{2+})}}{\mathcal{D}^{(m)}_{(Fe^{2+})}}}\lambda^{(w,m)}\right)\right] \qquad (9.4.3\text{-}36)$$

and

$$C_2 = \left(c^{(m)}_{(Fe^{2+})\text{eq},\,c} - c^{(m)}_{(Fe^{2+})\text{eq},\,b}\right)\left[\text{erf}\left(\lambda^{(m,h)}\right) - \text{erf}\left(\sqrt{\frac{\mathcal{D}^{(w)}_{(Fe^{2+})}}{\mathcal{D}^{(m)}_{(Fe^{2+})}}}\lambda^{(w,m)}\right)\right]^{-1} \qquad (9.4.3\text{-}37)$$

Here, $C_2$ is a constant of integration.

At the iron–wüstite interface, the jump mass balance for $Fe^{2+}$ (8.2.1-4) is

$$N^{(w)}_{(Fe^{2+})2} = \frac{r^{(i,w)}_{(Fe^{2+})}}{M_{(Fe^{2+})}} \qquad (9.4.3\text{-}38)$$

in which $r^{(i,w)}_{(Fe^{2+})}$ is the rate of production of $Fe^{2+}$ in the iron–wüstite interface. The jump mass balance for $Fe^{3+}$ as well as assumption 7 requires

$$
\begin{aligned}
N^{(w)}_{(Fe^{3+})2} &= \frac{r^{(i,w)}_{(Fe^{3+})}}{M_{(Fe^{3+})}} \\[2mm]
&= -\frac{2}{3} \frac{r^{(i,w)}_{(Fe^{2+})}}{M_{(Fe^{2+})}}
\end{aligned}
\tag{9.4.3-39}
$$

Eliminating $r^{(i,w)}_{(Fe^{2+})}$ between (9.4.3-38) and (9.4.3-39), we have (9.4.3-2), which we require to be satisfied. By assumption 4, the jump mass balance for $O^{2-}$ is satisfied identically. In summary, the jump mass balances at the iron–wüstite interface allow us to compute $r^{(i,w)}_{(Fe^{2+})}$, $r^{(i,w)}_{(Fe^{3+})}$, and $r^{(i,w)}_{(O^{2-})}$ if desired. They will not be required in the analysis that follows.

From the jump mass balance for $Fe^{2+}$ at the wüstite–magnetite interface (8.2.1-4) together with (9.4.3-7), (9.4.3-16), (9.4.3-26), and (9.4.3-31), we find

$$
\text{at } z_2 = h^{(w,m)}: \ N^{(w)}_{(Fe^{2+})2} - N^{(m)}_{(Fe^{2+})2} - \left( c^{(w)}_{(Fe^{2+})} - c^{(m)}_{(Fe^{2+})} \right) u^{(w,m)}_2
$$

$$
= -\frac{c^{(w)2}}{c^{(w)}_{(O^{2-})}} \mathcal{D}^{(w)}_{(Fe^{2+})} \frac{\partial x^{(w)}_{(Fe^{2+})}}{\partial z_2} + \frac{c^{(m)2}}{c^{(m)}_{(O^{2-})}} \mathcal{D}^{(m)}_{(Fe^{2+})} \frac{\partial x^{(m)}_{(Fe^{2+})}}{\partial z_2}
$$

$$
- \left( c^{(w)}_{(Fe^{2+})eq,\, a} - c^{(m)}_{(Fe^{2+})eq,\, b} \right) \lambda^{(w,m)} \sqrt{\frac{5\mathcal{D}^{(w)}_{(Fe^{2+})}}{6t}}
$$

$$
= -\frac{5}{3} \mathcal{D}^{(w)}_{(Fe^{2+})} \frac{\partial c^{(w)}_{(Fe^{2+})}}{\partial z_2} + \frac{5}{3} \mathcal{D}^{(m)}_{(Fe^{2+})} \frac{\partial c^{(m)}_{(Fe^{2+})}}{\partial z_2}
$$

$$
- \left( c^{(w)}_{(Fe^{2+})eq,\, a} - c^{(m)}_{(Fe^{2+})eq,\, b} \right) \lambda^{(w,m)} \sqrt{\frac{5\mathcal{D}^{(w)}_{(Fe^{2+})}}{6t}}
$$

$$
= 0
\tag{9.4.3-40}
$$

or

$$
\text{at } \eta = \lambda^{(w,m)}: \ -\frac{dc^{(w)}_{(Fe^{2+})}}{d\eta} + \sqrt{\frac{\mathcal{D}^{(m)}_{(Fe^{2+})}}{\mathcal{D}^{(w)}_{(Fe^{2+})}}} \frac{dc^{(m)}_{(Fe^{2+})}}{d\mu} - \left( c^{(w)}_{(Fe^{2+})eq,\, a} - c^{(m)}_{(Fe^{2+})eq,\, b} \right) \lambda^{(w,m)}
$$

$$
= -C_1 \frac{2}{\sqrt{\pi}} \exp\left( -\lambda^{(w,m)2} \right) + C_2 \frac{2}{\sqrt{\pi}} \sqrt{\frac{\mathcal{D}^{(m)}_{(Fe^{2+})}}{\mathcal{D}^{(w)}_{(Fe^{2+})}}} \exp\left( -\frac{\mathcal{D}^{(w)}_{(Fe^{2+})}}{\mathcal{D}^{(m)}_{(Fe^{2+})}} \lambda^{(w,m)2} \right)
$$

$$
- \left( c^{(w)}_{(Fe^{2+})eq,\, a} - c^{(m)}_{(Fe^{2+})eq,\, b} \right) \lambda^{(w,m)}
$$

$$
= 0
\tag{9.4.3-41}
$$

In view of assumption 4, the jump mass balance for $O^{(2-)}$ requires

$$c^{(w)}_{(O^{2-})} = c^{(m)}_{(O^{2-})} \tag{9.4.3-42}$$

as indicated in assumption 6. With this result as well as assumptions 10 and 12, the jump mass balance for $Fe^{3+}$ at the wüstite–magnetite interface also reduces to (9.4.3-41).

At the magnetite–hematite interface, the jump mass balance for $Fe^{2+}$ is, by assumption 8,

$$N^{(m)}_{(Fe^{2+})2} - c^{(m)}_{(Fe^{2+})} u_2 = -\frac{r^{(m,h)}_{(Fe^{2+})}}{M_{(Fe^{2+})}} \tag{9.4.3-43}$$

in which $r^{(m,h)}_{(Fe^{2+})}$ is the rate of production of $Fe^{2+}$ in the magnetite–hematite interface. Recognizing that the $Fe^{3+}$ in the hematite moves with the speed of displacement of the interface, the jump mass balance for $Fe^{3+}$, as well as assumption 8, requires

$$N^{(m)}_{(Fe^{3+})2} - c^{(m)}_{(Fe^{3+})} u_2 = -\frac{r^{(m,h)}_{(Fe^{3+})}}{M_{(Fe^{3+})}}$$

$$= \frac{r^{(m,h)}_{(Fe^{2+})}}{M_{(Fe^{2+})}} \tag{9.4.3-44}$$

Adding these equations together, multiplying by 3, and employing assumption 12, we have

$$\text{at } z_2 = h^{(m,h)}: \quad N^{(m)}_{(Fe^{2+})2} - 3\left(c^{(m)}_{(Fe^{2+})} + c^{(m)}_{(Fe^{2+})}\right)u^{(m,h)}_2$$

$$= N^{(m)}_{(Fe^{2+})2} - 3\left(1 + \frac{x^{(m)}_{(Fe^{3+})}}{x^{(m)}_{(Fe^{2+})}}\right)c^{(m)}_{(Fe^{2+})}u^{(m,h)}_2$$

$$= -\frac{c^{(m)2}}{c^{(m)}_{(O^{2-})}}\mathcal{D}^{(m)}_{(Fe^{2+})}\frac{\partial x^{(m)}_{(Fe^{2+})}}{\partial z_2} - 3\left(1 + \frac{x^{(m)}_{(Fe^{3+})}}{x^{(m)}_{(Fe^{2+})}}\right)c^{(m)}_{(Fe^{2+})eq,\,c}\lambda^{(m,h)}\sqrt{\frac{5\mathcal{D}^{(m)}_{(Fe^{2+})}}{6t}}$$

$$= -\frac{5}{3}\mathcal{D}^{(m)}_{(Fe^{2+})}\frac{\partial c^{(m)}_{(Fe^{2+})}}{\partial z_2} - 3\left(1 + \frac{x^{(m)}_{(Fe^{3+})}}{x^{(m)}_{(Fe^{2+})}}\right)c^{(m)}_{(Fe^{2+})eq,\,c}\lambda^{(m,h)}\sqrt{\frac{5\mathcal{D}^{(m)}_{(Fe^{2+})}}{6t}}$$

$$= 0 \tag{9.4.3-45}$$

or

$$\text{at } \mu = \lambda^{(m,h)}: \quad \frac{dc^{(m)}_{(Fe^{2+})}}{d\mu} + 3\left(1 + \frac{x^{(m)}_{(Fe^{3+})}}{x^{(m)}_{(Fe^{2+})}}\right)c^{(m)}_{(Fe^{2+})eq,\,c}\lambda^{(m,h)}$$

$$= C_2\frac{2}{\sqrt{\pi}}\exp\left(-\lambda^{(m,h)2}\right) + 3\left(1 + \frac{x^{(m)}_{(Fe^{3+})}}{x^{(m)}_{(Fe^{2+})}}\right)c^{(m)}_{(Fe^{2+})eq,\,c}\lambda^{(m,h)}$$

$$= 0 \tag{9.4.3-46}$$

Finally, the jump mass balance for $O^{2-}$ at the magnetite–hematite interface (or, literally, the sum of the jump mass balances at this interface and at the hematite–oxygen interface) requires

$$c_{(O^{2-})}^{(m)} u_2 = \frac{r_{(O^{2-})}^{(m,h)}}{M_{(O^{2-})}} \tag{9.4.3-47}$$

Equations (9.4.3-43), (9.4.3-44), and (9.4.3-47) can be used to compute $r_{(Fe^{2+})}^{(m,h)}$, $r_{(Fe^{3+})}^{(m,h)}$, and $r_{(O^{2-})}^{(m,h)}$ if desired.

We can estimate that at $1,200°C$ (Chen and Peterson 1975)[5]

$$\mathcal{D}_{(Fe^{2+})}^{(w)} = 3.80 \times 10^{-11} \, \text{m}^2/\text{s} \tag{9.4.3-48}$$

and (Himmel et al. 1953)

$$\mathcal{D}_{(Fe^{2+})}^{(m)} = 3.59 \times 10^{-12} \, \text{m}^2/\text{s} \tag{9.4.3-49}$$

and (Touloukian 1966, p. 481)

$$\rho^{(w)} = 5.36 \times 10^3 \, \text{kg/m}^3 \tag{9.4.3-50}$$

If we assume that the thermal expansion coefficient is the same for magnetite and wüstite, given the densities at room temperature (Weast 1982, p. B-109) we find

$$\rho^{(m)} = 4.82 \times 10^3 \, \text{kg/m}^3 \tag{9.4.3-51}$$

In view of assumption 6, we can compute (at the iron–wüstite phase interface)

$$\begin{aligned} c_{(O^{2-})}^{(w)} &= c_{(O^{2-})}^{(m)} \\ &= x_{(O^{2-})}^{(w)} \frac{\rho^{(w)}}{M^{(w)}} \\ &= 0.513 \frac{5.36 \times 10^3}{35.5} \\ &= 77.8 \, \text{kg mol/m}^3 \end{aligned} \tag{9.4.3-52}$$

Under these circumstances, we can solve (9.4.3-28), (9.4.3-37), (9.4.3-40), and (9.4.3-45) to get

$$\lambda^{(w,m)} = 0.600$$

$$\lambda^{(m,h)} = 2.04$$

$$C_1 = -36.1$$

$$C_2 = -1.84 \times 10^3 \tag{9.4.3-53}$$

---

[5] We will show in a subsequent manuscript that, when we analyze the experiments of Himmel et al. (1953) and of Chen and Peterson (1975) using our theory, we find that their "self-diffusion coefficients for iron" can be interpreted as our $\mathcal{D}_{(Fe^{2+})}^{(i)}$ in the limit $\mathcal{D}_{(Fe^{2+})}^{(i)} = \mathcal{D}_{(Fe^{3+})}^{(i)}$, where $i = w$ or $m$.

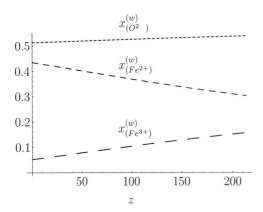

**Figure 9.4.3-3.** Mole fractions of the three ions ($O^{2-}$, $Fe^{2+}$, and $Fe^{3+}$) in the wüstite phase as functions of $z_2$ ($\mu$m) at 1,000 s.

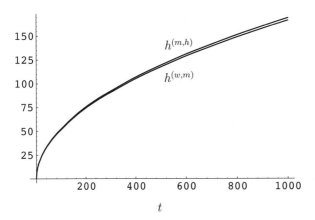

**Figure 9.4.3-4.** The position ($\mu$m) $h^{(w,m)}$ of the wüstite–magnetite interface (upper curve) and the position ($\mu$m) $h^{(m,h)}$ of the magnetite–hematite interface (lower curve) as functions of time $t$ (s) for iron exposed to $O_2$ at $1 \times 10^5$ Pa and 1,200°C.

By way of illustrating these results, Figure 9.4.3-3 shows the mole fractions of the three ions in the wüstite phase at a particular time, 1,000 s. Equations (9.4.3-25) and (9.4.3-33) permit us to plot $h^{(w,m)}$ and $h^{(m,s)}$ as functions of time $t$ in Figure 9.4.3-4.

Davies et al. (1951) observe experimentally that, for iron exposed to $O_2$ at $1 \times 10^5$ Pa and 1,200°C,

$$\frac{d}{dt}\left[\rho^{(w)}h^{(w,m)} + \rho^{(m)}\left(h^{(m,s)} - h^{(w,m)}\right)\right] = \frac{K_{\exp}}{\sqrt{t}}$$

$$= \frac{0.0241}{\sqrt{t}}\ \frac{\text{kg}}{\text{m}^2\text{s}} \qquad (9.4.3\text{-}54)$$

**Table 9.4.3-1.** Comparison of the predictions of (9.4.3-55) with the experimental observations of Davies et al. (1951)

| $T$ (°C) | $K_{exp}$ (kg m$^{-2}$ s$^{-1/2}$) | $K$ (kg m$^{-2}$ s$^{-1/2}$) |
|---|---|---|
| 700 | 0.00077 | 0.000633 |
| 800 | 0.002388 | 0.00254 |
| 900 | 0.005035 | 0.00532 |
| 1,035 | 0.0116 | 0.0135 |
| 1,090 | 0.0142 | 0.0171 |
| 1,200 | 0.024 | 0.026 |

We estimate that

$$\frac{d}{dt}\left[\rho^{(w)}h^{(w,m)} + \rho^{(m)}\left(h^{(m,s)} - h^{(w,m)}\right)\right]$$

$$= \left(\rho^{(w)} - \rho^{(m)}\right)\lambda^{(w,m)}\sqrt{\frac{5\mathcal{D}_{Fe^{2+}}^{(w)}}{6t}} + \rho^{(m)}\lambda^{(m,s)}\sqrt{\frac{5\mathcal{D}_{Fe^{2+}}^{(m)}}{6t}}$$

$$= \frac{K}{\sqrt{t}}$$

$$= \frac{0.026}{\sqrt{t}}\frac{kg}{m^2 s} \tag{9.4.3-55}$$

Further comparisons between their measurements and our predictions for a broad range of temperatures are shown in Table 9.4.3-1. Much of the difference between theory and experiment may be attributable to our rough estimates for the physical properties, particularly (9.4.3-48).

### Discussion

It is important to remember that the Stefan–Maxwell equations were derived for dilute gases (Bird et al. 1960, p. 570). They become empiricisms when they are extended to describe ion diffusion in solids. The coefficient $\mathcal{D}_{(Fe^{2+},Fe^{3+})}$ makes this particularly obvious. As a binary diffusion coefficient, $\mathcal{D}_{(Fe^{2+},Fe^{3+})}$ makes no sense. (A binary solution of two cations is impossible.) There is no contradiction when it is viewed simply as an empirical coefficient.

### Conclusion

The comparison between the calculated and observed values for the rate constant $K$ shown in Table 9.4.3-1 was obtained without the use of adjustable parameters as needed by Himmel et al. (1953, p. 840). In contrast with prior theories (Wagner 1951, 1969; Smeltzer 1987; Coates and Dalvi 1970), neither do we assume dilute solutions in treating this problem of ternary diffusion, nor do we require any thermodynamic data other than the phase diagram.

## 9.5 Boundary-Layer Theory

As we developed boundary-layer theory in Sections 3.5.1 and 6.7.1, we argued for $N_{Re} \gg 1$ that, outside the immediate neighborhood of a flat plate, fluid could be considered to be nonviscous and nonconducting and that, within the immediate neighborhood of a flat plate, a portion of the viscous terms in the differential momentum balance and a portion of the conduction terms in the differential energy balance could be neglected.

Our intention here is to extend the boundary-layer concept to the mass balance for an individual species. In this discussion, we will accept without further argument for $N_{Re} \gg 1$ that, outside the immediate neighborhood of a flat plate, one can develop an argument similar to that given in the introduction to Section 6.6 to neglect diffusion. We will focus our attention on the boundary layer within the immediate neighborhood of the plate.

As discussed in Section 9.2, there are many situations in which there is a close analogy between energy transfer problems and mass transfer problems. In developing boundary-layer theory here, we will focus on those situations in which this analogy breaks down. In particular, we will concern ourselves with processes where diffusion-induced convection cannot be neglected, where there are homogeneous chemical reactions, or where there are heterogeneous chemical reactions.

### 9.5.1  Plane Flow Past a Flat Plate

Let us begin by considering in some detail the same class of flows that we used to introduce boundary-layer theory in Section 3.5.1: plane flow past a flat plate. With reference to Figure 6.7.1-1,

$$\text{at } z_1 = 0: \quad T = T_\infty$$
$$\omega_{(A)} = \omega_{(A)\infty} \tag{9.5.1-1}$$

It is important that we work in terms of mass fractions, since it is the mass-averaged velocity that appears in the differential momentum balance and the differential energy balance. For the moment, we will say no more about the external flow and the conditions at the plate. For simplicity, we limit ourselves to a two-component, incompressible Newtonian fluid with constant physical properties, independent of temperature and concentration. To better illustrate the argument, we will neglect pressure diffusion, forced diffusion, thermal diffusion, and the Dufour effect [see (8.4.3-2)]. In addition, we will assume that there are no homogeneous chemical reactions. The development that follows can be expanded to include these effects with little difficulty.

Following the examples of Sections 3.5.1 and 6.7.1, we will find it convenient to work in terms of the following dimensionless variables:

$$v_i^\star \equiv \frac{v_i}{v_0}, \qquad\qquad T^\star \equiv \frac{T - T_\infty}{T_0 - T_\infty}$$

$$\overline{H}_{(A)}^\star \equiv \frac{1}{\hat{c}(T_0 - T_\infty)}\overline{H}_{(A)}, \; z_i^\star \equiv \frac{z_i}{L} \tag{9.5.1-2}$$

$$t^\star \equiv \frac{t}{t_0}$$

Here, $v_0$ is a magnitude of the velocity characteristic of the plane nonviscous flow outside the boundary layer, $T_0$ is characteristic of the temperature distribution on the plate, $L$ is the length of the plate, and $t_0$ is a characteristic time. The quantities $v_0$ and $T_0$ will be defined in the context of a particular problem, as in the next section.

With the assumption that, for this plane flow,

$$T^\star = T^\star\left(z_1^\star, z_2^\star, t^\star\right)$$
$$\omega_{(A)} = \omega_{(A)}\left(z_1^\star, z_2^\star, t^\star\right) \tag{9.5.1-3}$$

let us develop the boundary-layer equations for momentum, energy, and mass transfer.

## Momentum Transfer

Because viscosity and density are assumed to be constants, independent of temperature and concentration, the development given in Section 3.5.1 still applies. But within the context of specific problems, we must be alert to the effects of diffusion-induced convection and heterogeneous chemical reactions in specifying boundary conditions.

## Energy Transfer

The development of the differential energy balance for a multicomponent boundary layer is similar to that given in Section 6.7.1. Recognizing that the fluid has been assumed to be incompressible, we find that the differential energy balance of Table 8.5.2-1 reduces for this system to

$$\rho\hat{c}\frac{d_{(v)}T}{dt} = -\text{div}\,\epsilon - \sum_{C=A}^{B} \nabla\overline{H}_{(C)}\cdot\mathbf{j}_{(C)} + \text{tr}(\mathbf{S}\cdot\nabla\mathbf{v}) \tag{9.5.1-4}$$

In view of (8.4.3-2) (neglecting the Dufour effect) and Fick's first law from Table 8.5.1-7, the dimensionless form of (9.5.1-4) reduces for this plane flow to

$$\frac{1}{N_{St}}\frac{\partial T^\star}{\partial t^\star} + \frac{\partial T^\star}{\partial z_1^\star}v_1^\star + \frac{\partial T^\star}{\partial z_2^\star}v_2^\star = \frac{1}{N_{Pr}N_{Re}}\left(\frac{\partial^2 T^\star}{\partial z_1^{\star 2}} + \frac{\partial^2 T^\star}{\partial z_2^{\star 2}}\right)$$

$$+ \frac{1}{N_{Sc}N_{Re}}\sum_{C=A}^{B}\left(\frac{\partial\overline{H}_{(C)}^\star}{\partial z_1^\star}\frac{\partial\omega_{(C)}}{\partial z_1^\star} + \frac{\partial\overline{H}_{(C)}^\star}{\partial z_2^\star}\frac{\partial\omega_{(C)}}{\partial z_2^\star}\right)$$

$$+ \frac{2N_{Br}}{N_{Pr}N_{Re}}\left[\left(\frac{\partial v_1^\star}{\partial z_1^\star}\right)^2 + \left(\frac{\partial v_2^\star}{\partial z_2^\star}\right)^2\right] + \frac{N_{Br}}{N_{Pr}N_{Re}}\left(\frac{\partial v_1^\star}{\partial z_2^\star} + \frac{\partial v_2^\star}{\partial z_1^\star}\right)^2 \tag{9.5.1-5}$$

or, in terms of

$$z_2^{\star\star} \equiv \sqrt{N_{Re}}z_2^\star$$
$$v_2^{\star\star} \equiv \sqrt{N_{Re}}v_2^\star \tag{9.5.1-6}$$

Equation (9.5.1-5) becomes

$$\frac{1}{N_{St}}\frac{\partial T^\star}{\partial t^\star} + \frac{\partial T^\star}{\partial z_1^\star}v_1^\star + \frac{\partial T^\star}{\partial z_2^{\star\star}}v_2^{\star\star} = \frac{1}{N_{Pr}}\left(\frac{1}{N_{Re}}\frac{\partial^2 T^\star}{\partial z_1^{\star 2}} + \frac{\partial^2 T^\star}{\partial z_2^{\star\star 2}}\right)$$

$$+ \frac{1}{N_{Sc}}\sum_{C=A}^{B}\left(\frac{1}{N_{Re}}\frac{\partial\overline{H}_{(C)}^\star}{\partial z_1^\star}\frac{\partial\omega_{(C)}}{\partial z_1^\star} + \frac{\partial\overline{H}_{(C)}^\star}{\partial z_2^{\star\star}}\frac{\partial\omega_{(C)}}{\partial z_2^{\star\star}}\right)$$

$$+ \frac{2N_{Br}}{N_{Pr}N_{Re}}\left[\left(\frac{\partial v_1^\star}{\partial z_1^\star}\right)^2 + \left(\frac{\partial v_2^{\star\star}}{\partial z_2^{\star\star}}\right)^2\right] + \frac{N_{Br}}{N_{Pr}}\left(\frac{\partial v_1^\star}{\partial z_2^{\star\star}} + \frac{1}{N_{Re}}\frac{\partial v_2^{\star\star}}{\partial z_1^\star}\right)^2 \tag{9.5.1-7}$$

Here

$$N_{Pr} \equiv \frac{\hat{c}\mu}{k}, \qquad N_{Sc} \equiv \frac{\mu}{\rho \mathcal{D}_{(AB)}}$$

$$N_{Br} \equiv S \frac{\mu v_0^2}{k(T_0 - T_\infty)} \tag{9.5.1-8}$$

Equation (9.5.1-7) suggests that, for $N_{Re} \gg 1$, a fixed value of $N_{Pr}$, and arbitrary values of $N_{St}$ and $N_{Br}$, the dimensionless differential energy balance may be simplified to

$$\frac{1}{N_{St}} \frac{\partial T^\star}{\partial t^\star} + \frac{\partial T^\star}{\partial z_1^\star} v_1^\star + \frac{\partial T^\star}{\partial z_2^{\star\star}} v_2^{\star\star}$$

$$= \frac{1}{N_{Pr}} \frac{\partial^2 T^\star}{\partial z_2^{\star\star 2}} + \frac{1}{N_{Sc}} \sum_{C=A}^{B} \left( \frac{\partial \overline{H}_{(C)}^\star}{\partial z_2^{\star\star}} \frac{\partial \omega_{(C)}}{\partial z_2^{\star\star}} \right) + \frac{N_{Br}}{N_{Pr}} \left( \frac{\partial v_1^\star}{\partial z_2^{\star\star}} \right)^2 \tag{9.5.1-9}$$

Note that in the limit

$$N_{Sc} \gg N_{Pr} \tag{9.5.1-10}$$

Equation (9.5.1-9) simplifies to

$$\frac{1}{N_{St}} \frac{\partial T^\star}{\partial t^\star} + \frac{\partial T^\star}{\partial z_1^\star} v_1^\star + \frac{\partial T^\star}{\partial z_2^{\star\star}} v_2^{\star\star} = \frac{1}{N_{Pr}} \frac{\partial^2 T^\star}{\partial z_2^{\star\star 2}} + \frac{N_{Br}}{N_{Pr}} \left( \frac{\partial v_1^\star}{\partial z_2^{\star\star}} \right)^2 \tag{9.5.1-11}$$

Finally, the development leading to (6.7.1-12) again applies:

$$\text{for } z_2^{\star\star} \to \infty : \quad T^\star \to \tilde{T}^\star \tag{9.5.1-12}$$

where $\tilde{T}^\star$ is the dimensionless temperature distribution for the nonviscous, nonconducting, nondiffusing flow evaluated at the boundary.

## Mass Transfer

The development of the differential mass balance for species $A$ in a boundary layer also is similar to that presented in Section 6.7.1. Beginning with the differential mass balance for species $A$ presented in Table 8.5.1-8,

$$\rho \left( \frac{\partial \omega_{(A)}}{\partial t} + \nabla \omega_{(A)} \cdot \mathbf{v} \right) = \rho \mathcal{D}_{(AB)} \, \text{div} \nabla \omega_{(A)} \tag{9.5.1-13}$$

we find that for our plane flow

$$\frac{1}{N_{St}} \frac{\partial \omega_{(A)}}{\partial t^\star} + \frac{\partial \omega_{(A)}}{\partial z_1^\star} v_1^\star + \frac{\partial \omega_{(A)}}{\partial z_2^\star} v_2^\star = \frac{1}{N_{Sc} N_{Re}} \left( \frac{\partial^2 \omega_{(A)}}{\partial z_1^{\star 2}} + \frac{\partial^2 \omega_{(A)}}{\partial z_2^{\star 2}} \right) \tag{9.5.1-14}$$

or, in terms of $z_2^{\star\star}$ and $v_2^{\star\star}$,

$$\frac{1}{N_{St}} \frac{\partial \omega_{(A)}}{\partial t^\star} + \frac{\partial \omega_{(A)}}{\partial z_1^\star} v_1^\star + \frac{\partial \omega_{(A)}}{\partial z_2^{\star\star}} v_2^{\star\star} = \frac{1}{N_{Sc}} \left( \frac{1}{N_{Re}} \frac{\partial^2 \omega_{(A)}}{\partial z_1^{\star 2}} + \frac{\partial^2 \omega_{(A)}}{\partial z_2^{\star\star 2}} \right) \tag{9.5.1-15}$$

This suggests that, for $N_{Re} \gg 1$, a fixed value of $N_{Sc}$, and an arbitrary value of $N_{St}$, the dimensionless differential energy balance may be simplified to

$$\frac{1}{N_{St}} \frac{\partial \omega_{(A)}}{\partial t^\star} + \frac{\partial \omega_{(A)}}{\partial z_1^\star} v_1^\star + \frac{\partial \omega_{(A)}}{\partial z_2^{\star\star}} v_2^{\star\star} = \frac{1}{N_{Sc}} \frac{\partial^2 \omega_{(A)}}{\partial z_2^{\star\star 2}} \tag{9.5.1-16}$$

Outside the boundary-layer region, diffusion can be neglected with respect to convection in the differential mass balance for species $A$. Let

$$\omega_{(A)}^{(e)} = \omega_{(A)}^{(e)} \left( z_1{}^\star, z_2{}^\star \right)$$

$$= \omega_{(A)}^{(e)} \left( z_1{}^\star, \frac{z_2{}^{\star\star}}{\sqrt{N_{Re}}} \right) \tag{9.5.1-17}$$

where we denote $\omega_{(A)}$ for the nonviscous, nonconducting, nondiffusing external flow. Within a region where both the boundary-layer solution and the external nonviscous, nonconducting, nondiffusing flow are valid

$$\tilde{\omega}_{(A)} \equiv \lim N_{Re} \gg 1 \text{ for } z_1^\star, z_2^{\star\star} \text{ fixed} : \omega_{(A)}^{(e)} \left( z_1{}^\star, \frac{z_2{}^{\star\star}}{\sqrt{N_{Re}}} \right) = \omega_{(A)}^{(e)\star}(z_1{}^\star, 0) \tag{9.5.1-18}$$

For $z_2{}^{\star\star} \gg 1$, we require that $\omega_{(A)}$ from the boundary-layer solution must approach asymptotically the corresponding temperature from the nonviscous, nonconducting, nondiffusing flow:

$$\text{for } z_2{}^{\star\star} \to \infty : \quad \omega_{(A)} \to \tilde{\omega}_{(A)} \tag{9.5.1-19}$$

### 9.5.2 Flow Past Curved Walls and Bodies of Revolution

Following the discussion in Sections 3.5.3 and 6.7.3, we find that the differential mass balance in the boundary layer on a curved wall will almost always have the same form as we found in the preceding section for flow past a flat plate. It is necessary only to work in terms of a slightly different coordinate system and to observe a mild restriction on the curvature of the wall.

In a similar manner, Sections 3.5.6 and 6.7.5 suggest that the form of the differential mass balance in the boundary layer on a body of revolution is similar to that found in Section 9.5.1. But don't forget that there is a problem with the overall differential mass balance as developed in Section 3.5.6. For this reason it may be better to use all of the differential mass balances for all of the species and to avoid using the overall differential mass balance.

## 9.6 Forced Convection in Dilute Solutions

Convection in mass transfer differs fundamentally from convection in energy transfer. In both energy and mass transfer, we can have both forced convection and natural convection, which results from density gradients in the fluid. But in mass transfer, we observe an additional effect: diffusion-induced convection. The motions of the individual species are sufficient in general to require the mass-averaged or molar-averaged velocity distributions to differ from zero.

Our discussion in Sections 9.3.1 through 9.3.4 suggests that diffusion-induced convection can be neglected in sufficiently dilute solutions. More generally, diffusion-induced convection can be neglected with respect to forced convection in dilute solutions.

In Section 9.2, we saw that mass transfer problems take the same mathematical form as energy transfer problems, if we are able to make the following assumptions:

1) The system is isothermal; all viscous dissipation and radiation can be neglected (to ensure that the differential and jump energy balances are satisfied).
2) The system consists of a single phase, so that the jump mass balance for species $A$ need not be considered.
3) The phase is incompressible (to ensure that the overall differential mass balances have the same forms).
4) The system has only two components, or the multicomponent solution is sufficiently dilute that diffusion can be regarded as binary (Section 8.4.6).
5) There are no homogeneous or heterogeneous chemical reactions.
6) The solution is sufficiently dilute that diffusion-induced convection can be neglected.
7) Effects attributable to thermal, pressure, and forced diffusion can be neglected.
8) All physical properties are constants.

Note that mass transfer problems take the same form as energy transfer problems under the conditions noted above, only when it is the mass-averaged velocity $\mathbf{v}$ that appears in the differential mass balance. If one chooses to work in terms of mole fractions $x_{(A)}$ rather than mass fractions $\omega_{(A)}$, the system must be so dilute that $\mathbf{v} \approx \mathbf{v}^\circ$.

Under these conditions, there are no new physical or mathematical issues to be explored. For this reason, we will focus here on homogeneous and heterogeneous chemical reactions in dilute solutions with forced convection. We will stop after just two examples.

### 9.6.1 Unsteady-State Diffusion with a First-Order Homogeneous Reaction

At time $t = 0$, a gas of pure species $A$ is brought into contact with a liquid $B$. Component $A$ diffuses into the liquid phase, where it undergoes an irreversible first-order reaction $A + B \rightarrow 2C$. Let us determine the rate at which species $A$ is absorbed by the liquid phase. For the time of observation, it may be assumed that species $A$ and $C$ are never present in the liquid solution in more than trace amounts.

To somewhat simplify the analysis, let us take the liquid–gas phase interface to be the plane $z_2 = 0$, and let us say that the liquid phase occupies the half-space $z_2 > 0$. The initial condition is that

$$\text{at } t = 0 \text{ for all } z_2 > 0 : \quad x_{(A)} = 0 \tag{9.6.1-1}$$

Since the liquid and gas phases are assumed to be in equilibrium at the phase interface, we require

$$\text{at } z_2 = 0 \text{ for all } t > 0 : \quad x_{(A)} = x_{(A)\text{eq}} \tag{9.6.1-2}$$

where $x_{(A)\text{eq}}$ is presumed to be known a priori. To recognize that the liquid must be supported by an impermeable container, we specify that

$$\text{as } z_2 \rightarrow \infty \text{ for all } t : \quad \mathbf{v}^\circ \rightarrow 0 \tag{9.6.1-3}$$

Because we are dealing with a dilute liquid solution, it seems reasonable to assume both that the solution is ideal and that the density $\rho$ is a constant. But for this dilute solution

$$c \approx \frac{\rho}{M_{(B)}} \tag{9.6.1-4}$$

which suggests that we may assume that the molar density $c$ is nearly a constant as well.

We will seek a solution of the form

$$v_1^\diamond = v_3^\diamond$$
$$= 0$$
$$v_2^\diamond = v_2^\diamond(t, z_2)$$
$$x_{(A)} = x_{(A)}(t, z_2)$$

From the overall differential mass balance of Table 8.5.1-10, we find

$$\frac{\partial v_2^\diamond}{\partial z_2} = 0 \tag{9.6.1-5}$$

This implies

$$v_2^\diamond = v_2^\diamond(t) \tag{9.6.1-6}$$

To be consistent with boundary condition (9.6.1-3), we must require

$$\text{everywhere}: \ v_2^\diamond = 0 \tag{9.6.1-7}$$

It should become clear to you that we have specified a very specialized problem in that (9.6.1-7) requires that the number of moles of components $B$ and $C$ leaving the liquid through the phase interface must be exactly equal to the number of moles of $A$ entering the liquid. We limited ourselves to this physical situation when we said both that the phase interface must be fixed in space at the plane $z_2 = 0$ and that the liquid must be bounded by an impermeable wall as $z_2 \to \infty$ in (9.6.1-3).

Since we are concerned with the concentration distribution of the trace quantity $A$ in an ideal ternary solution, we may use (8.4.6-1) to describe the mass flux vector:

$$\mathbf{N}_{(A)} = c_{(A)}\mathbf{v}^\diamond - c\mathcal{D}_{(Am)}\nabla x_{(A)} \tag{9.6.1-8}$$

We will further simplify the problem by taking $\mathcal{D}_{(Am)}$ to be a constant. In view of (9.6.1-5) and (9.6.1-6) through (9.6.1-8), the differential mass balance for species $A$ from Table 8.5.1-5 specifies

$$\frac{\partial x_{(A)}}{\partial t} = \mathcal{D}_{(Am)}\frac{\partial^2 x_{(A)}}{\partial z_2^2} + \frac{r_{(A)}}{cM_{(A)}} \tag{9.6.1-9}$$

Since this is a first-order, irreversible, homogeneous reaction in a dilute solution, we assume

$$\frac{r_{(A)}}{M_{(A)}} = -k_1''' c_{(A)} \tag{9.6.1-10}$$

The required concentration distribution for species $A$ is, consequently, a solution to

$$\frac{\partial x_{(A)}}{\partial t} = \mathcal{D}_{(Am)}\frac{\partial^2 x_{(A)}}{\partial^2 z_2} - k_1''' x_{(A)} \tag{9.6.1-11}$$

that satisfies both (9.6.1-1) and (9.6.1-2).

Let us begin by taking the Laplace transform of (9.6.1-11):

$$sg = \mathcal{D}_{(Am)} \frac{\partial^2 g}{\partial^2 z_2} - k_1''' g \qquad (9.6.1\text{-}12)$$

Here we define

$$g = g(s, z_2) \equiv \mathcal{L}\left(x_{(A)}(t, z_2)\right) \qquad (9.6.1\text{-}13)$$

It is readily seen that one solution to (9.6.1-12) is of the form

$$g \equiv A \exp(\sqrt{K} z_2) + B \exp(-\sqrt{K} z_2) \qquad (9.6.1\text{-}14)$$

where

$$K \equiv \frac{s + k_1'''}{\mathcal{D}_{(Am)}} \qquad (9.6.1\text{-}15)$$

and the constants $A$ and $B$ are as yet unspecified. Since we must require that

$$\text{as } z_2 \to \infty : \ g \text{ must be finite} \qquad (9.6.1\text{-}16)$$

we have

$$A = 0 \qquad (9.6.1\text{-}17)$$

In terms of the transformed variable $g$, Equation (9.6.1-2) says

$$\text{at } z_2 = 0 : \ g = \frac{1}{s} x_{(A)\text{eq}} \qquad (9.6.1\text{-}18)$$

Consequently,

$$B = \frac{1}{s} x_{(A)\text{eq}} \qquad (9.6.1\text{-}19)$$

In summary,

$$g = \frac{1}{s} x_{(A)\text{eq}} \exp(-\sqrt{K} z_2) \qquad (9.6.1\text{-}20)$$

Taking the inverse Laplace transform of this, we have

$$
\begin{aligned}
x_{(A)} &= \mathcal{L}^{-1}(g) \\
&= x_{(A)\text{eq}} \int_0^t \frac{z_2}{2\sqrt{\pi \mathcal{D}_{(Am)} u^3}} \exp\left(-k_1''' u - \frac{z_2^2}{4\mathcal{D}_{(Am)} u}\right) du
\end{aligned}
\qquad (9.6.1\text{-}21)
$$

or

$$\frac{x_{(A)}}{x_{(A)\text{eq}}} = \frac{2}{\sqrt{\pi}} \int_{z_2/\sqrt{4\mathcal{D}_{(Am)}t}}^{\infty} \exp\left(-\lambda^2 - \frac{k_1''' z_2^2}{4\mathcal{D}_{(Am)} \lambda^2}\right) d\lambda \qquad (9.6.1\text{-}22)$$

Noting that (Churchill 1958, p. 140)

$$\frac{4}{\sqrt{\pi}} \int_r^{\infty} \exp\left(-\lambda^2 - \frac{a^2}{\lambda^2}\right) d\lambda = e^{2a} \operatorname{erfc}\left(r + \frac{a}{r}\right) + e^{-2a} \operatorname{erfc}\left(r - \frac{a}{r}\right) \qquad (9.6.1\text{-}23)$$

we may write (9.6.1-22) in the more useful form (Danckwerts 1950)

$$\frac{2x_{(A)}}{x_{(A)eq}} = \exp\left(z_2\sqrt{\frac{k_1'''}{\mathcal{D}_{(Am)}}}\right)\text{erfc}\left(\zeta + \sqrt{k_1'''t}\right)$$

$$+ \exp\left(-z_2\sqrt{\frac{k_1'''}{\mathcal{D}_{(Am)}}}\right)\text{erfc}\left(\zeta - \sqrt{k_1'''t}\right) \tag{9.6.1-24}$$

Here we have introduced the complementary error function

$$\text{erfc}(x) \equiv 1 - \text{erf}x$$

$$= \frac{2}{\sqrt{\pi}}\int_x^\infty e^{-\lambda^2}d\lambda \tag{9.6.1-25}$$

and we have defined

$$\zeta \equiv \frac{z_2}{\sqrt{4\mathcal{D}_{(Am)}t}} \tag{9.6.1-26}$$

We set out to determine the rate at which species $A$ is absorbed by the liquid phase. This is the same as asking for the flux of species $A$ through the liquid–gas phase interface:

$$N_{(A)2}\big|_{z_2=0} = -c\mathcal{D}_{(Am)}\frac{\partial x_{(A)}}{\partial z_2}\bigg|_{z_2=0}$$

$$= cx_{(A)eq}\sqrt{k_1'''\mathcal{D}_{(Am)}}\left[\text{erf}\sqrt{k_1'''t} + \frac{\exp\left(-k_1'''t\right)}{\sqrt{\pi k_1'''t}}\right] \tag{9.6.1-27}$$

The total amount of $A$ adsorbed per unit area of interface between time $t = 0$ and time $t = t_0$ is consequently

$$\int_0^{t_0} N_{(A)}\big|_{z_2=0}\,dt = cx_{(A)eq}\sqrt{k_1'''\mathcal{D}_{(Am)}}\left[\left(t_0 + \frac{1}{2k_1'''}\right)\text{erf}\sqrt{k_1'''t_0}\right.$$

$$\left. + \sqrt{\frac{t_0}{\pi k_1'''}}\exp(-k_1'''t_0)\right] \tag{9.6.1-28}$$

We are often interested in the limit

$$\text{as } k_1'''t_0 \to \infty: \quad \int_0^{t_0} N_{(A)}\big|_{z_2=0}\,dt$$

$$\to cx_{(A)eq}\sqrt{k_1'''\mathcal{D}_{(Am)}}\left(t_0 + \frac{1}{2k_1'''}\right) \tag{9.6.1-29}$$

**Exercise 9.6.1-1** Fill in the details in going from (9.6.1-20) to (9.6.1-24).

*Hint:* Use the convolution theorem.

**Exercise 9.6.1-2** Derive (9.6.1-28), starting with (9.6.1-24).

**Exercise 9.6.1-3** Repeat the problem discussed in this section, assuming that the liquid has a finite depth $L$. The plane $z_2 = 0$ represents the gas–liquid phase interface; the plane $z_2 = L$ is a wall that is impermeable to all three species.

i) Begin by introducing as dimensionless variables

$$t^\star \equiv k_1''' t, \quad x_{(A)}^\star \equiv \frac{x_{(A)}}{x_{(A)\text{eq}}}$$

$$y \equiv \alpha - z_2 \sqrt{\frac{k_1'''}{\mathcal{D}_{(Am)}}}$$

where

$$\alpha \equiv L \sqrt{\frac{k_1'''}{\mathcal{D}_{(Am)}}}$$

ii) Take the Laplace transform with respect to $t^\star$ to find

$$g(s) = \frac{\cosh(y\sqrt{s+1})}{s \, \cosh(\alpha\sqrt{s+1})}$$

iii) Take the inverse transform to learn that

$$x_{(A)}^\star = 4\pi \sum_{n=1}^{\infty} \left[ \frac{(-1)^n (2n-1)}{(2n-1)^2 \pi^2 + 4\alpha^2} \right]$$

$$\times \left[ \exp\left( -\left[ \frac{(2n-1)^2 \pi^2 + 4\alpha^2}{4\alpha^2} \right] t^\star \right) - 1 \right] \cos\left( \frac{(2n-1)\pi y}{2\alpha} \right)$$

**9.6.2**  Gas Absorption in a Falling Film with Chemical Reaction

An incompressible Newtonian fluid of nearly pure species $B$ flows down an inclined plane as shown in Figure 3.2.5-4 (see Exercise 3.2.5-5). Species $A$ is transferred from the surrounding gas stream to the liquid where it undergoes an irreversible first-order homogeneous reaction. Let us assume that there is no mass transfer from the gas stream to the falling film for $z_1 < 0$:

$$\text{at } z_2 = \delta \text{ for } z_1 < 0 : \quad \mathbf{n}_{(A)} \cdot \boldsymbol{\xi} = 0 \tag{9.6.2-1}$$

Here $\boldsymbol{\xi}$ is the unit normal to the phase interface pointed from the liquid to the gas. Outside the immediate neighborhood of the liquid film, the gas stream has a uniform concentration. If the liquid were in equilibrium with this gas stream, its concentration would be $\rho_{(A)\text{eq}}$. To simplify the problem, we will assume that for $z_1 > 0$ the concentration of the liquid at the phase interface is $\rho_{(A)\text{eq}}$:

$$\text{at } z_2 = \delta \text{ for } z_1 > 0 : \quad \rho_{(A)} = \rho_{(A)\text{eq}} \tag{9.6.2-2}$$

Very far upstream, the liquid is pure species $B$:

$$\text{as } z_1 \to -\infty \text{ for } 0 \leq z_2 \leq \delta : \quad \rho_{(A)} \to 0 \tag{9.6.2-3}$$

We wish to determine the concentration distribution in the boundary layer near the entrance of the adsorption section for $N_{Pe,m} \gg 1$.

In the limit of a dilute solution, diffusion-induced convection may be neglected with respect to forced convection and the velocity distribution in the fluid is the same as that found in Exercise 3.2.5-5:

$$v_1 = v_{1,\text{max}}\left[2\frac{z_2}{\delta} - \left(\frac{z_2}{\delta}\right)^2\right] \tag{9.6.2-4}$$

Let us assume that

$$\rho_{(A)} = \rho_{(A)}(z_1, z_2) \tag{9.6.2-5}$$

The differential mass balance for species $A$ from Table 8.5.1-8 requires

$$v_1\frac{\partial \rho_{(A)}}{\partial z_1} = \mathcal{D}^0_{(AB)}\left(\frac{\partial^2 \rho_{(A)}}{\partial z_1^2} + \frac{\partial^2 \rho_{(A)}}{\partial z_2^2}\right) - k_1''' \rho_{(A)} \tag{9.6.2-6}$$

Determining a solution to this equation that is consistent with boundary conditions (9.6.2-1) through (9.6.2-3) will be very difficult. This suggests that we restrict our attention to the entrance of the adsorption region as we did in Exercise 6.7.6-5.

If we introduce as dimensionless variables

$$\omega_{(A)}^{\star\star} \equiv \frac{\rho_{(A)}}{\rho_{(A)\text{eq}}}$$

$$z_1^{\star} \equiv \frac{z_1}{\delta} \tag{9.6.2-7}$$

$$s^{\star} \equiv 1 - \frac{z_2}{\delta} \tag{9.6.2-8}$$

Equation (9.6.2-6) becomes

$$(1 - s^{\star 2})\frac{\partial \omega_{(A)}^{\star\star}}{\partial z_1^{\star}} = \frac{1}{N_{Pe,m}}\left(\frac{\partial^2 \omega_{(A)}^{\star\star}}{\partial z_1^{\star 2}} + \frac{\partial^2 \omega_{(A)}^{\star\star}}{\partial z_2^{\star 2}}\right) - \frac{N_{Da}}{N_{Pe,m}}\omega_{(A)}^{\star\star} \tag{9.6.2-9}$$

where

$$N_{Pe,m} \equiv \frac{\delta v_{1,\text{max}}}{\mathcal{D}^0_{(AB)}}, \quad N_{Da} \equiv \frac{k_1''' \delta^2}{\mathcal{D}^0_{(AB)}} \tag{9.6.2-10}$$

Since we are primarily interested in the entrance region to the absorption section, our discussion in Exercise 6.7.6-5 suggests that we introduce as an *expanded* variable

$$s^{\star\star} \equiv (N_{Pe,m})^{1/2}s^{\star} \tag{9.6.2-11}$$

In terms of this expanded variable, (9.6.2-9) becomes

$$\left(1 - \frac{s^{\star\star 2}}{N_{Pe,m}}\right)\frac{\partial \omega_{(A)}^{\star\star}}{\partial z_1^{\star}} = \frac{1}{N_{Pe,m}}\frac{\partial^2 \omega_{(A)}^{\star\star}}{\partial z_1^{\star 2}} + \frac{\partial^2 \omega_{(A)}^{\star\star}}{\partial s^{\star\star 2}} - \frac{N_{Da}}{N_{Pe,m}}\omega_{(A)}^{\star\star} \tag{9.6.2-12}$$

In the limit $N_{Pe,m} \gg 1$, this last expression simplifies to

$$\frac{\partial \omega_{(A)}^{\star\star}}{\partial z_1^{\star}} = \frac{\partial^2 \omega_{(A)}^{\star\star}}{\partial s^{\star\star 2}} - \frac{N_{Da}}{N_{Pe,m}}\omega_{(A)}^{\star\star} \tag{9.6.2-13}$$

Since we are neglecting axial diffusion in (9.6.2-13), it seems reasonable to replace boundary conditions (9.6.2-1) and (9.6.2-3) with

$$\text{at } z_1^{\star} = 0 : \quad \omega_{(A)}^{\star\star} = 0 \tag{9.6.2-14}$$

In terms of our dimensionless variables, (9.6.2-2) may be expressed as

$$\text{at } s^{\star\star} = 0: \quad \omega_{(A)}^{\star\star} = 1 \tag{9.6.2-15}$$

Our problem is reduced to finding a solution to (9.6.2-13) that is consistent with boundary conditions (9.6.2-14) and (9.6.2-15).

Equations (9.6.2-13) to (9.6.2-15) belong to the class of problems discussed in Exercise 9.6.2-8. Consequently, the solution of interest here can be determined from Section 3.2.4 as

$$\omega_{(A)}^{\star\star} = \exp\left(-\frac{N_{Da}z_1^{\star}}{N_{Pe,m}}\right)\left[1 - \mathrm{erf}\left(\frac{s^{\star\star}}{\sqrt{4z_1^{\star}}}\right)\right]$$

$$+ \frac{N_{Da}}{N_{Pe,m}} \int_0^{z_1^{\star}} \exp\left(-\frac{N_{Da}z_1^{\star}}{N_{Pe,m}}\right)\left[1 - \mathrm{erf}\left(\frac{s^{\star\star}}{\sqrt{4z_1^{\star}}}\right)\right] dz_1^{\star} \tag{9.6.2-16}$$

Finally, it is interesting to compute the rate at which mass of species $A$ is absorbed per unit width of a film of length $L$:

$$\mathcal{W}_{(A)} \equiv \int_0^L - n_{(A)2}\big|_{z_2=\delta} \, dz_1 \tag{9.6.2-17}$$

It follows from (9.6.2-16) that (Bird et al. 1960, p. 553)

$$\frac{\mathcal{W}_{(A)}}{\rho_{(A)\mathrm{eq}}v_{1,\max}}\sqrt{\frac{k_1'''}{\mathcal{D}_{(AB)}^0}} = \left(\frac{1}{2}+u\right)\mathrm{erf}\sqrt{u} + \sqrt{\frac{u}{\pi}}\,\exp(-u) \tag{9.6.2-18}$$

where, for the sake of convenience, we have introduced

$$u \equiv \frac{N_{Da}L}{N_{Pe,m}\delta} \tag{9.6.2-19}$$

In the limit where there is no chemical reaction, $u \to 0$ and (9.6.2-18) requires

$$\frac{\mathcal{W}_{(A)}}{\rho_{(A)\mathrm{eq}}\sqrt{\mathcal{D}_{(AB)}^0 v_{1,\max}L}} = \sqrt{\frac{4}{\pi}} \tag{9.6.2-20}$$

**Exercise 9.6.2-1** *A general solution for unsteady-state diffusion with a first-order homogeneous reaction* Let us assume that the differential mass balance for species $A$ in a system may be shown to take the form

$$\frac{\partial \omega_{(A)}}{\partial t} + \nabla\omega_{(A)} \cdot \mathbf{v} = \mathcal{D}_{(Am)}\mathrm{div}\,\nabla\omega_{(A)} + k_1'''\omega_{(A)} \tag{9.6.2-21}$$

where $\mathbf{v}$ is known to be independent of time. This equation is to be solved for $\omega_{(A)}$ subject to the conditions that

$$\text{at } t = 0: \quad \omega_{(A)} = 0$$

and

$$\text{at some surfaces}: \quad \omega_{(A)} = \omega_{(A)s}$$

We wish to show that (Danckwerts 1951; Crank 1956, p. 124; Lightfoot 1964)

$$\omega_{(A)} = f \exp\left(k_1''' t\right) - k_1''' \int_0^t f \exp\left(k_1''' \tau\right) d\tau$$

$$= \int_0^t \frac{\partial f}{\partial \tau} \exp\left(k_1''' \tau\right) d\tau \tag{9.6.2-22}$$

Here, $f$ is a solution to the same problem with $k_1''' = 0$.

i) Begin by introducing as dimensionless variables

$$t^\star \equiv k_1''' t, \qquad z_i^\star \equiv z_i \sqrt{\frac{k_1'''}{\mathcal{D}_{(Am)}}}$$

$$\mathbf{v}^\star \equiv \frac{\mathbf{v}}{\sqrt{k_1''' \mathcal{D}_{(Am)}}}$$

ii) Take the Laplace transform of both problems (with and without reaction).
iii) Assume a solution to the original problem of the form

$$\mathcal{L}\left(\omega_{(A)}\right) = a \mathcal{L}(f)$$

where

$$a = a(s)$$

iv) Invert this expression for $\mathcal{L}\left(\omega_{(A)}\right)$ to obtain the desired result.

**Exercise 9.6.2-2** *More on a general solution for unsteady-state diffusion with a first-order homogeneous reaction* (Danckwerts 1951; Crank 1956, p. 124) Repeat Exercise 9.6.2-1 assuming that boundary condition (9.6.1-23) is replaced by

$$\text{at some surfaces :} \quad \nabla \omega_{(A)} \cdot \mathbf{n} = K \left( \omega_{(A)\infty} - \omega_{(A)} \right) \tag{9.6.2-23}$$

Determine that the solution has the same form as (9.6.2-22).

**Exercise 9.6.2-3** *Still more on a general solution for unsteady-state diffusion with a first-order homogeneous reaction* Let us assume that a solution to (9.6.2-21) is to be found consistent with the conditions that

$$\text{at } t = 0 : \quad \omega_{(A)} = \omega_{(A)0}$$

$$\text{at some surfaces :} \quad \omega_{(A)} = 0$$

and

$$\text{at other surfaces :} \quad \nabla \omega_{(A)} \cdot \mathbf{n} = -K \omega_{(A)}$$

Use the approach suggested in Exercise 9.6.2-1 to determine that the solution has the form (Bird et al. 1960, p. 621)

$$\omega_{(A)} = f \exp(k_1''' t)$$

where $f$ is a solution to the same problem with $k_1''' = 0$.

**Exercise 9.6.2-4** *Still more on a general solution for unsteady-state diffusion* **(Metz, personal communication, 1974)** We seek a solution to (9.6.2-21) that is consistent with the conditions that

at $t = 0$ : $\omega_{(A)} = \omega_{(A)0}$

at some surfaces : $\omega_{(A)} = \omega_{(A)s}$

Begin by making the additional change of variable

$$\omega_{(A)}{}^{\star} \equiv \omega_{(A)} - \omega_{(A)s}$$

Use the approach suggested in Exercise 9.6.2-1 to determine that

$$\omega_{(A)} - \omega_{(A)s} = f \exp\left(k_1'''t\right) - \frac{\omega_{(A)s}k_1'''}{\omega_{(A)s} - \omega_{(A)0}} \int_0^t f \exp\left(k_1'''\tau\right) d\tau$$

where $f$ is a solution to the same problem with $k_1''' = 0$.

**Exercise 9.6.2-5** *Still more on a general solution for unsteady-state diffusion* **(Metz, personal communication, 1974)** We seek a solution to (9.6.2-21) that is consistent with the conditions that

at $t = 0$ : $\omega_{(A)} = \omega_{(A)0}$

at some surfaces : $\nabla\omega_{(A)} \cdot \mathbf{n} = K\left(\omega_{(A)\infty} - \omega_{(A)}\right)$

Begin by making the additional change of variable

$$\omega_{(A)}{}^{\star} \equiv \omega_{(A)} - \omega_{(A)\infty}$$

Proceed as in Exercise 9.6.2-4 to conclude that

$$\omega_{(A)} - \omega_{(A)\infty} = f \exp(k_1'''t) - \frac{\omega_{(A)\infty}k_1'''}{\omega_{(A)\infty} - \omega_{(A)0}} \int_0^t f \exp(k_1'''\tau) d\tau$$

where $f$ is a solution to the same problem with $k_1''' = 0$.

**Exercise 9.6.2-6** *Still more on a general solution for unsteady-state diffusion* **(Metz, personal communication, 1974)** We seek a solution to (9.6.2-21) that is consistent with the conditions that

at $t = 0$ : $\omega_{(A)} = \omega_{(A)0}$

at some surfaces : $\omega_{(A)} = \omega_{(A)s}$

at other surfaces : $\nabla\omega_{(A)} \cdot \mathbf{n} = 0$

Conclude that the solution of Exercise 9.6.2-4 again applies.

**Exercise 9.6.2-7** *More on a general solution for unsteady-state diffusion with a first-order homogeneous reaction* We seek a solution to (9.6.2-21) consistent with the conditions that

at $t = 0$ : $\omega_{(A)} = \omega_{(A)0}$ $\qquad\qquad\qquad\qquad\qquad\qquad\qquad$ (9.6.2-24)

at surfaces I : $\omega_{(A)} = \omega_{(A)s}$ $\qquad\qquad\qquad\qquad\qquad\qquad$ (9.6.2-25)

and

at surfaces II : $\nabla\omega_{(A)} \cdot \mathbf{n} = K\left(\omega_{(A)\infty} - \omega_{(A)}\right)$ $\qquad\qquad\qquad$ (9.6.2-26)

Begin by assuming

$$\omega_{(A)} = \omega_{(A)1} + \omega_{(A)2}$$

The function $\omega_{(A)1}$ satisfies (9.6.2-21) as well as (9.6.2-25) and (9.6.2-26), and

at $t = 0$ : $\omega_{(A)1} = 0$

The function $\omega_{(A)2}$ is a solution to (9.6.2-21) consistent with (9.6.2-24),

at surfaces I : $\omega_{(A)2} = 0$

and

at surfaces II : $\nabla \omega_{(A)2} \cdot \mathbf{n} = -K \omega_{(A)2}$

Conclude that a solution to (9.6.2-21) that satisfies (9.6.2-24) through (9.6.2-26) is (Lightfoot 1964; corrected by C. Y. Lin and J. D. Chen in 1977).

$$\omega_{(A)} = f \exp(k_1'''t) - k_1''' \int_0^t f \exp(k_1'''t)\, dt + g \exp(k_1'''t)$$

Here, $f$ is a solution to the system of equations describing $\omega_{(A)1}$ with $k_1''' = 0$; $g$ is a solution to the system of equations describing $\omega_{(A)2}$ with $k_1''' = 0$.

**Exercise 9.6.2-8** *Critical size of an autocatalytic system* (Bird et al. 1960, p. 623) Acetylene gas is thermodynamically unstable. It tends to decompose:

$$H_2C_2(\text{gas}) \rightarrow H_2(\text{gas}) + 2C(\text{solid})$$

One of the steps in this reaction appears to involve a free radical. Since free radicals are effectively neutralized by contact with an iron surface, their concentration is essentially zero at such a surface. This suggests that acetylene gas can be safely stored in steel cylinders of sufficiently small diameter. If the cylinder is too large, the formation of even a small concentration of free radicals is likely to cause a rapidly increasing rate of decomposition according to the overall reaction described above. Since this reaction is exothermic, an explosion may result.

The problem can be readily corrected by filling the cylinder with an iron wool to create a porous medium of iron. Let us determine the critical pore diameter of this iron wool, assuming that the decomposition may be described as a first-order homogeneous reaction.

For an ideal-gas mixture at constant temperature and pressure in a cylindrical pore, use Exercise 9.6.2-3 to determine

$$x_{(A)} = \sum_{n=1}^{\infty} A_n \exp\left([1 - \lambda_n^2 K]t^\star\right) J_0\left(\lambda_n r^\star\right)$$

where

$$J_0(\lambda_n) = 0, \quad n = 1, 2, \ldots$$

$$A_n = \frac{2 \int_0^1 r^\star J_0(\lambda_n r^\star) x_{(A)0}(r^\star)\, dr^\star}{[J_1(\lambda_n)]^2}, \quad n = 1, 2, \ldots$$

and

$$K \equiv \frac{\mathcal{D}_{(Am)}}{R^2 k_1'''}$$

Argue that the acetylene gas can be safely stored, provided

$$R < \lambda_1 \sqrt{\frac{D_{(Am)}}{k_1'''}}$$

$$= 2.40 \sqrt{\frac{D_{(Am)}}{k_1'''}}$$

Here $\lambda_1$ is the first and smallest zero of $J_0(x)$ (Irving and Mullineux 1959, p. 130).

**Exercise 9.6.2-9**    Repeat Exercise 9.6.1-3 using Exercise 9.6.2-1.

**Exercise 9.6.2-10** *Steady-state diffusion in a sphere*    Species $A$ diffuses into a solid sphere of radius $R$, where it is consumed by an irreversible first-order reaction. We will assume that $A$ is never present in more than trace amounts. With the assumption that

$$\text{at } r = R : \; \omega_{(A)} = \omega_{(A)\text{eq}}$$

determine that

$$\frac{r\omega_{(A)}}{\omega_{(A)\text{eq}}} = \frac{\sinh\left(\sqrt{N_{Da}}\,r/R\right)}{\sinh\left(\sqrt{N_{Da}}\right)}$$

where we have introduced the Damköhler number

$$N_{Da} \equiv \frac{k_1''' R^2}{D_{Am}}$$

Conclude that the rate at which $A$ is consumed is

$$\mathcal{W}_{(A)} = 4\pi R \rho \omega_{(A)\text{eq}} D_{Am} \left[1 - \sqrt{N_{Da}} \coth\left(\sqrt{N_{Da}}\right)\right]$$

*Hint:*    Introduce the transformation

$$f \equiv r\omega_{(A)}$$

**Exercise 9.6.2-11** *More on gas absorption in a falling film with chemical reaction*    Let us repeat the problem discussed in this section, attempting to describe the boundary condition at the gas–liquid phase interface more realistically. Rather than saying that the phase interface is in equilibrium with the gas very far away from it, let us describe the mass transfer by means of Newton's "law" of mass transfer (Section 9.2.1):

$$\text{at } z_2 = \delta \text{ for } z_1 > 0 : \; j_{(A)2} = k_{(A)\omega}\left(\omega_{(A)} - \omega_{(A)\text{eq}}\right)$$

*Answer:*

$$\omega_{(A)}^{\star\star} \equiv \frac{\omega_{(A)}}{\omega_{(A)\text{eq}}}$$

$$= F \exp\left(-\frac{N_{Da}}{N_{Pe,m}}z_1^\star\right) + \frac{N_{Da}}{N_{Pe,m}} \int_0^{z_1^\star} F \exp\left(-\frac{N_{Da}}{N_{Pe,m}}z_1^\star\right) dz_1^\star$$

$$F \equiv 1 - \text{erf}\left(\frac{s^{\star\star}}{\sqrt{4z_1^\star}}\right) - \exp\left(\frac{s^{\star\star}}{B} + \frac{z_1^\star}{B^2}\right)\left[1 - \text{erf}\left(\frac{s^{\star\star}}{\sqrt{4z_1^\star}} + \frac{\sqrt{z_1^\star}}{B}\right)\right]$$

*Hint:*    See Exercise 6.7.6-6.

# 10

# Integral Averaging in Mass Transfer

THIS IS OUR CONCLUSION to integral averaging techniques begun in Chapter 4 and continued in Chapter 7. As I mentioned in introducing Chapter 7, the ideas presented here are best understood in the context of Chapter 4. It is in Chapter 4 that I try to spend a little extra time in discussing the motivation for some of the developments. It is also there that some of the key steps common to all the derivations are explained in detail.

## 10.1 Time Averaging

By *turbulent mass transfer*, I mean that at least one of the phases involved in the mass-transfer process is in turbulent flow. For a discussion of the basic concepts and terminology, please refer to Section 4.1.

In the next few sections we shall be concerned with the time average of the differential mass balance for an individual species $A$. Our approach here will be very similar to that taken in Section 7.1, where we discussed turbulent energy transfer.

### 10.1.1 The Time-Averaged Differential Mass Balance for Species A

As in our previous discussions of turbulence (Sections 4.1 and 7.1), we will for simplicity limit ourselves to incompressible fluids. For this limiting case, the differential mass balance of Table 8.5.1-5 becomes

$$\rho \left[ \frac{\partial \omega_{(A)}}{\partial t} + \text{div}\left(\omega_{(A)}\mathbf{v}\right) \right] + \text{div } \mathbf{j}_{(A)} - r_{(A)} = 0 \tag{10.1.1-1}$$

Using the definition introduced in Section 4.1.1, let us take the time average of this equation:

$$\frac{1}{\Delta t} \int_t^{t+\Delta t} \left[ \rho \frac{\partial \omega_{(A)}}{\partial t'} + \text{div}\left(\rho \omega_{(A)}\mathbf{v} + \mathbf{j}_{(A)}\right) - r_{(A)} \right] dt' = 0 \tag{10.1.1-2}$$

The time-averaging operation commutes with partial differentiation with respect to time (see Section 4.1.1) and with the divergence operation:

$$\rho \frac{\partial \overline{\omega_{(A)}}}{\partial t} + \text{div}\left(\rho \overline{\omega_{(A)}} \overline{\mathbf{v}} + \overline{\mathbf{j}_{(A)}}\right) - \overline{r_{(A)}} = 0 \tag{10.1.1-3}$$

It is more common to write this result as

$$\rho \left(\frac{\partial \overline{\omega_{(A)}}}{\partial t} + \nabla \overline{\omega_{(A)}} \cdot \overline{\mathbf{v}}\right) + \text{div}\left(\overline{\mathbf{j}_{(A)}} + \mathbf{j}_{(A)}^{(t)}\right) = \overline{r_{(A)}} \tag{10.1.1-4}$$

where we have introduced the *turbulent mass flux*

$$\mathbf{j}_{(A)}^{(t)} \equiv \rho \left(\overline{\omega_{(A)} \mathbf{v}} - \overline{\omega_{(A)}}\,\overline{\mathbf{v}}\right) \tag{10.1.1-5}$$

When we limit ourselves to binary diffusion and when we recognize that Fick's first law is an appropriate expression for the mass flux, (10.1.1-4) takes the form

$$\rho \left(\frac{\partial \overline{\omega_{(A)}}}{\partial t} + \nabla \overline{\omega_{(A)}} \cdot \overline{\mathbf{v}}\right) = \text{div}\left(\rho \mathcal{D}_{(AB)} \nabla \overline{\omega_{(A)}} - \mathbf{j}_{(A)}^{(t)}\right) + \overline{r_{(A)}} \tag{10.1.1-6}$$

Let us in particular assume that we have an $n$th-order homogeneous reaction

$$r_{(A)} = k_n''' \rho_{(A)}{}^n \tag{10.1.1-7}$$

so that

$$\overline{r_{(A)}} = k_n''' \overline{\rho_{(A)}}{}^n + k_n''' \left(\overline{\rho_{(A)}{}^n} - \overline{\rho_{(A)}}{}^n\right) \tag{10.1.1-8}$$

Notice that, for a first-order reaction,

$$\overline{r_{(A)}} = k_1''' \overline{\rho_{(A)}} \tag{10.1.1-9}$$

The rate of production of the mass of species $A$ per unit volume is not an explicit function of the concentration fluctuations. In contrast, $\overline{r_{(A)}}$ is explicitly dependent upon the concentration fluctuations for higher-order reactions.

The problem posed here by $\mathbf{j}_{(A)}^{(t)}$ is very similar to those encountered in Sections 4.1.1 and 7.1.1. Just as there we had to stop and propose empirical data correlations for the Reynolds stress tensor $\mathbf{S}^{(t)}$ and the turbulent energy flux vector $\mathbf{q}^{(t)}$, we must here stop and formulate empirical representations for the turbulent mass flux $\mathbf{j}_{(A)}^{(t)}$.

Exercise 10.1.1-1 *Turbulent diffusion in dilute electrolytes*   Quite often it is convenient to arrange the computations for a mass-transfer problem a little differently from the arrangement suggested in the text.

i) Determine that an alternative expression for the time-averaged equation of continuity for species $A$ is

$$\frac{\partial \overline{c_{(A)}}}{\partial t} + \text{div}\,\overline{\mathbf{n}_{(A)}} = \frac{\overline{r_{(A)}}}{M_{(A)}}$$

ii) For sufficiently dilute solutions of an electrolyte, find that

$$\overline{\mathbf{N}_{(A)}} = -c\mathcal{D}_{(Am)}\left(\nabla\overline{x_{(A)}} + \frac{\overline{x_{(A)}\epsilon_{(A)}}}{kT}\nabla\Phi\right) + \overline{c_{(A)}}\,\overline{\mathbf{v}} + \frac{1}{M_{(A)}}\mathbf{j}_{(A)}^{(t)}$$

$$= -c\mathcal{D}_{(Am)}\left(\nabla\overline{x_{(A)}} + \frac{\overline{x_{(A)}\epsilon_{(A)}}}{kT}\nabla\Phi\right) + \overline{x_{(A)}}\sum_{B=1}^{N}\overline{\mathbf{N}_{(B)}} + \frac{1}{M_{(A)}}\mathbf{j}_{(A)}^{(t)}$$

Here we have neglected the effects of pressure and thermal diffusion.

## 10.1.2 Empirical Correlations for the Turbulent Mass Flux

Our discussion of empirical data correlations for the turbulent mass flux $\mathbf{j}_{(A)}^{(t)}$ will be relatively brief, inasmuch as it is essentially a duplication of Section 7.1.2.

Our approach is based upon three principles.

1) For changes of frame such that

$$\overline{\mathbf{Q}} \doteq \mathbf{Q} \tag{10.1.2-1}$$

we may use the result of Section 4.1.2 to find that $\mathbf{j}_{(A)}^{(t)}$ is frame indifferent:

$$\mathbf{j}_{(A)}^{(t)*} \equiv \rho\left(\overline{\omega_{(A)}^{*}\mathbf{v}^{*}} - \overline{\omega_{(A)}^{*}}\,\overline{\mathbf{v}^{*}}\right)$$

$$= \rho\left[\overline{\omega_{(A)}^{*}\left(\mathbf{v}^{*} - \overline{\mathbf{v}^{*}}\right)}\right]$$

$$= \rho\mathbf{Q}\cdot\left[\overline{\omega_{(A)}\left(\mathbf{v} - \overline{\mathbf{v}}\right)}\right]$$

$$= \mathbf{Q}\cdot\mathbf{j}_{(A)}^{(t)} \tag{10.1.2-2}$$

Here, $\mathbf{Q}$ is a (possibly) time-dependent, orthogonal, second-order tensor. To obtain (10.1.2-2), we have made use of the fact that a velocity difference is frame indifferent (see Exercise 1.2.2-1).

2) We shall assume that the principle of frame indifference discussed in Section 2.3.1 applies to any empirical correlations developed for $\mathbf{j}_{(A)}^{(t)}$, so long as the changes of frame considered satisfy (10.1.2-1).

3) The Buckingham–Pi theorem (Brand 1957) will be used to further limit the form of any expression for $\mathbf{j}_{(A)}^{(t)}$.

## Example 1: Prandtl's Mixing-Length Theory

Example 1 in Section 7.1.2 suggests that, for the fully developed flow regime in wall turbulence, we assume that the turbulent mass flux be regarded as a function of the density of the fluid, the distance $l$ from the wall, $\overline{\mathbf{D}}$, and $\nabla\overline{\omega_{(A)}}$:

$$\mathbf{j}_{(A)}^{(t)} = \mathbf{j}_{(A)}^{(t)}\left(\rho, l, \overline{\mathbf{D}}, \nabla\overline{\omega_{(A)}}\right) \tag{10.1.2-3}$$

Because we are limiting ourselves to the fully developed flow regime, the diffusivity and viscosity are not included as independent variables. The implications of the principle of frame indifference and of the Buckingham–Pi theorem are spelled out in Section 7.1.2.

A special case of (10.1.2-3) that is consistent with the principle of material frame indifference and the Buckingham–Pi theorem is

$$\mathbf{j}_{(A)}^{(t)} = -\mathcal{D}^\star \rho l^2 \sqrt{2\operatorname{tr}\left(\overline{\mathbf{D}} \cdot \overline{\mathbf{D}}\right)} \nabla \overline{\omega_{(A)}} \tag{10.1.2-4}$$

where $\mathcal{D}^\star$ is a dimensionless constant. Equation (10.1.2-4) should be viewed as the tensorial form of *Prandtl's mixing-length theory* for mass transfer. It is probably worth emphasizing that we should not expect the Prandtl mixing-length theory to be appropriate to the laminar sublayer or buffer zone.

### Example 2: Deissler's Expression for the Region near the Wall

In view of our discussion of Example 2 in Section 7.1.2, we are motivated to propose for the laminar sublayer and the buffer zone

$$\mathbf{j}_{(A)}^{(t)} = \mathbf{j}_{(A)}^{(t)}\left(\rho, \mu, l, \overline{\mathbf{v}} - \mathbf{v}^{(s)}, \nabla \overline{\omega_{(A)}}\right) \tag{10.1.2-5}$$

Remember that $\mathbf{v}^{(s)}$ indicates the velocity of the bounding wall. Deissler (1955) has proposed on empirical grounds that

$$\mathbf{j}_{(A)}^{(t)} = -n^2 \rho l |\overline{\mathbf{v}} - \mathbf{v}^{(s)}| \left[1 - \exp(-n^2 N)\right] \nabla \overline{\omega_{(A)}} \tag{10.1.2-6}$$

with the definition

$$N \equiv \frac{\rho l |\overline{\mathbf{v}} - \mathbf{v}^{(s)}|}{\mu} \tag{10.1.2-7}$$

The $n$ appearing here is meant to be the same as that used in (4.1.3-21) and evaluated in Section 4.1.4. Of course, (10.1.2-6) satisfies the principle of frame indifference, and it is consistent with the Buckingham–Pi theorem (Brand 1957).

### Example 3: Eddy Diffusivity in Free Turbulence

Very far away from any wall in a region of free turbulence, it is common to say that

$$\mathbf{j}_{(A)}^{(t)} = -\rho \mathcal{D}_{(AB)}^{(t)} \nabla \overline{\omega_{(A)}} \tag{10.1.2-8}$$

The scalar $\mathcal{D}_{(AB)}^{(t)}$ is normally assumed to be independent of position. It is known as the *eddy diffusivity*.

In the next section, we will look at a technique that has been used to measure $\mathcal{D}_{(AB)}^{(t)}$.

### 10.1.3 Turbulent Diffusion from a Point Source in a Moving Stream

The following material is taken from Wilson (1904) and Bird et al. (1960, p. 552).

In a region far removed from any bounding walls or surfaces, a fluid of pure species $B$ moves in a steady-state, turbulent flow with a uniform and constant speed $v_0$. With respect to the cylindrical coordinate system $(r, \theta, z)$ shown in Figure 10.1.3-1, the fluid moves in the $z$ direction. Species $A$ is continuously injected into the stream at the origin of this coordinate system. The rate of injection is $W_{(A)}$ (mass per unit time), which can be considered to be so small that the mass-averaged speed of the stream does not deviate appreciably from $v_0$. As species $A$ moves downstream from the point of injection, it diffuses in both the axial and radial directions. We wish to determine the concentration distribution of $A$ in the stream.

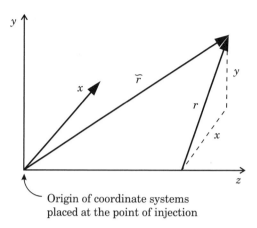

Origin of coordinate systems
placed at the point of injection

**Figure 10.1.3-1.** Coordinate systems used to
describe turbulent diffusion from a point source
in a constant-velocity stream.

Since the region of flow under consideration is very far away from any bounding walls, it
seems reasonable to assume that the flow is in free turbulence and that the turbulent mass flux
vector may be expressed in terms of a constant eddy diffusivity as described in Example 3
of Section 10.1.2. According to the assumptions above, we are justified in assuming that
there is only one nonzero component of the time-averaged velocity vector in the cylindrical
coordinate system indicated:

$$\bar{v}_r = \bar{v}_\theta$$
$$= 0$$
$$\bar{v}_z = v_0$$
$$= \text{a constant}$$
$$(10.1.3\text{-}1)$$

For the problem described, the time-averaged differential mass balance for species $A$, derived
in Section 10.1.1, reduces to

$$v_0 \left( \frac{\partial \overline{\omega_{(A)}}}{\partial z} \right)_r = \left( \mathcal{D}_{(AB)} + \mathcal{D}^{(t)}_{(AB)} \right) \text{div} \, \nabla \overline{\omega_{(A)}} \qquad (10.1.3\text{-}2)$$

Our intuition suggests and, as we shall see later, experimental evidence confirms that $\mathcal{D}_{(AB)} \ll \mathcal{D}^{(t)}_{(AB)}$.

Since the fluid very far downstream from the point of injection is pure species $B$, it seems
reasonable to employ as one boundary condition that in the spherical coordinate system
$(r, \theta)$ suggested in Figure 10.1.3-1

$$\text{as } \tilde{r} \to \infty : \; \overline{\omega_{(A)}} \to 0 \qquad (10.1.3\text{-}3)$$

We must also make a statement about the mass flow rate of species $A$ at the point of
injection. For any constant value of $r$, we can say that

$$W_{(A)} = \int_0^{2\pi} \int_0^{\pi} \overline{n_{(A)\tilde{r}}} \, \tilde{r}^2 \, \sin\theta \, d\theta \, d\varphi$$

$$= 2\pi \int_0^{\pi} \left[ \overline{\rho_{(A)} \, v_{\tilde{r}}} - \rho \left( \mathcal{D}_{(AB)} + \mathcal{D}^{(t)}_{(AB)} \right) \frac{\partial \overline{\omega_{(A)}}}{\partial \tilde{r}} \right] \tilde{r}^2 \, \sin\theta \, d\theta \qquad (10.1.3\text{-}4)$$

If for the moment we assume that

$$\text{as } \tilde{r} \to 0: \quad \rho_{(A)}\tilde{r} \to C_1 = \text{a constant} \tag{10.1.3-5}$$

it follows that

$$\text{as } \tilde{r} \to 0: \quad \int_0^\pi \overline{\rho_{(A)} \, v_{\tilde{r}}} \, \tilde{r}^2 \, \sin\theta \, d\theta$$

$$= \int_0^\pi \overline{\rho_{(A)}} v_0 \tilde{r}^2 \, \sin\theta \, \cos\theta \, d\theta$$

$$= \overline{\rho_{(A)}} v_0 \tilde{r}^2 \int_0^\pi \sin\theta \, \cos\theta \, d\theta$$

$$= 0 \tag{10.1.3-6}$$

and

$$\text{as } \tilde{r} \to 0: \quad W_{(A)}$$

$$= -2\pi \int_0^\pi \rho \left( \mathcal{D}_{(AB)} + \mathcal{D}_{(AB)}^{(t)} \right) \frac{\partial \overline{\omega_{(A)}}}{\partial \tilde{r}} \tilde{r}^2 \, \sin\theta \, d\theta$$

$$= -2\pi \rho \left( \mathcal{D}_{(AB)} + \mathcal{D}_{(AB)}^{(t)} \right) \frac{\partial \overline{\omega_{(A)}}}{\partial \tilde{r}} \tilde{r}^2 \int_0^\pi \sin\theta \, d\theta$$

$$= -4\pi \rho \left( \mathcal{D}_{(AB)} + \mathcal{D}_{(AB)}^{(t)} \right) \frac{\partial \overline{\omega_{(A)}}}{\partial \tilde{r}} \tilde{r}^2 \tag{10.1.3-7}$$

In arriving at (10.1.3-7), we have made use of (10.1.3-5) to reason that

$$\text{as } \tilde{r} \to 0: \quad \tilde{r}^2 \frac{\partial \omega_{(A)}}{\partial \tilde{r}} = -\frac{C_1}{\rho} \tag{10.1.3-8}$$

In a moment we shall return to check (10.1.3-5).

Our next step is to find a solution to (10.1.3-2) that is consistent with boundary conditions (10.1.3-3) and (10.1.3-7). This is a little awkward, since (10.1.3-2) is stated in terms of cylindrical coordinates, whereas boundary conditions (10.1.3-3) and (10.1.3-7) are more naturally given in terms of spherical coordinates. Wilson (1904) at this point made the clever suggestion that we look for a solution in the form of

$$\overline{\omega_{(A)}} = e^{-\alpha z} \varphi(\tilde{r}) \tag{10.1.3-9}$$

Employing this assumed form for the solution, we can calculate that

$$\text{div} \left( \nabla \overline{\omega_{(A)}} \right) = 2\nabla\varphi \cdot \nabla(e^{-\alpha z}) + \varphi \, \text{div} \, \nabla(e^{-\alpha z}) + e^{-\alpha z} \, \text{div} \, \nabla\varphi$$

$$= -2\alpha \, e^{-\alpha z} \left( \frac{\partial \varphi}{\partial z} \right)_r + \alpha^2 \varphi \, e^{-\alpha z} + e^{-\alpha z} \, \text{div} \, \nabla\varphi \tag{10.1.3-10}$$

and

$$\left( \frac{\partial \overline{\omega_{(A)}}}{\partial z} \right)_r = -\alpha \, e^{-\alpha z} \varphi + e^{-\alpha z} \left( \frac{\partial \varphi}{\partial z} \right)_r \tag{10.1.3-11}$$

As a result, (10.1.3-2) becomes

$$\varphi \left( -\frac{\alpha v_0}{\mathcal{D}_{(AB)} + \mathcal{D}_{(AB)}^{(t)}} - \alpha^2 \right) + \left( \frac{\partial \varphi}{\partial z} \right)_r \left( \frac{v_0}{\mathcal{D}_{(AB)} + \mathcal{D}_{(AB)}^{(t)}} + 2\alpha \right) = \text{div} \, \nabla\varphi \tag{10.1.3-12}$$

With the definition

$$\alpha \equiv \frac{v_0}{2\left(\mathcal{D}_{(AB)} + \mathcal{D}_{(AB)}^{(t)}\right)}$$ (10.1.3-13)

Equation (10.1.3-12) further reduces to

$$\alpha^2 \varphi = \frac{2}{\tilde{r}} \frac{\partial \varphi}{\partial \tilde{r}} + \frac{\partial^2 \varphi}{\partial \tilde{r}^2}$$ (10.1.3-14)

The standard change of variable

$$Y \equiv \tilde{r} \varphi$$ (10.1.3-15)

can be used to express (10.1.3-14) as

$$\alpha^2 Y = \frac{\partial^2 Y}{\partial \tilde{r}^2}$$ (10.1.3-16)

solutions to which have the form

$$\begin{aligned} Y &\equiv \tilde{r} \varphi \\ &= A \, \exp(\alpha \tilde{r}) + B \, \exp(-\alpha \tilde{r}) \end{aligned}$$ (10.1.3-17)

For boundary condition (10.1.3-3) to be satisfied, we must require

$$B = 0$$ (10.1.3-18)

(remember that $\alpha$ is negative).

From (10.1.3-9), (10.1.3-15), (10.1.3-17), and (10.1.3-18), we find

$$\overline{\omega_{(A)}} = \frac{A}{\tilde{r}} \exp[\alpha(\tilde{r} - z)]$$ (10.1.3-19)

Finally, boundary condition (10.1.3-7) demands

$$\text{as } \tilde{r} \to 0: \ W_{(A)} \to 4\pi\rho \left(\mathcal{D}_{(AB)} + \mathcal{D}_{(AB)}^{(t)}\right) A$$ (10.1.3-20)

or

$$A \equiv \frac{W_{(A)}}{4\pi\rho \left(\mathcal{D}_{(AB)} + \mathcal{D}_{(AB)}^{(t)}\right)}$$ (10.1.3-21)

In summary, the mass-fraction distribution for species $A$ in the free-turbulence flow described is represented by (10.1.3-13), (10.1.3-19), and (10.1.3-21). We see further that we were justified in assuming (10.1.3-5).

From an experimental point of view, the useful result here is

$$\begin{aligned} \frac{d \ln \left(\tilde{r} \, \overline{\omega_{(A)}}\right)}{d(\tilde{r} - z)} &= \alpha \\ &\equiv -\frac{v_0}{2\left(\mathcal{D}_{(AB)} + \mathcal{D}_{(AB)}^{(t)}\right)} \end{aligned}$$ (10.1.3-22)

If the slope on the left can be evaluated from experimental data, this expression may be used to calculate $\mathcal{D}_{(AB)} + \mathcal{D}_{(AB)}^{(t)}$. Towle and Sherwood (1939) have done this for $CO_2$ injected into a stream of air to conclude that $\mathcal{D}_{(CO_2, air)}^{(t)} \approx 23 \, cm^2/s$, which is several orders of magnitude larger than $\mathcal{D}_{(CO_2, air)}^{0}$.

For a further discussion of this experimental technique, see Sherwood and Pigford (1952, p. 42).

## 10.2   Area Averaging

In what follows, we extend to mass transfer the concept of area averaging introduced in Sections 4.2 and 7.2. The essential point is that sometimes it is advantageous to average the differential mass balance over a cross section normal to the macroscopic mass transfer.

Keep in mind that, whenever one of the integral averaging techniques is used, some information is lost. We are always called upon to compensate for this loss of information by making an approximation or by applying an empirical data correlation. You will notice that the approximation employed in Section 10.2.1 is a little different from those used in Sections 4.2.1 and 7.2.1. Because of the somewhat ad hoc nature of area averaging, I cannot give specific recommendations for the types of approximations to be employed that will be applicable in each and every problem you may encounter. Hopefully, having been warned an approximation will be necessary, you will find the example problems in Sections 4.2.1, 7.2.1, and 10.2.1 sufficient stimuli for your imagination.

I think you will gain the maximum benefit from the next section by reading it in the context of Sections 4.2 and 7.2.

### 10.2.1  Longitudinal Dispersion

At time $t = 0$, we find that for $z > 0$ a very long tube is filled with a pure solvent $\rho_{(A)} = 0$; for $z < 0$, the solvent has a uniform concentration of dissolved material $\rho_{(A)} = \rho_{(A)0}$. For $t > 0$, the fluid is forced to move in the $z$ direction through the tube with a constant volume flow rate. We wish to determine the concentration in the tube as a function of time and position.

In this analysis we will assume that the physical properties of the liquid are constants. It follows that the velocity distribution is independent of composition; its specific form is dictated by the constitutive equation chosen for the extra-stress tensor $\mathbf{S}$. For this analysis, it will not be necessary to choose a particular constitutive equation or to be any more explicit about the velocity distribution.

The implication is that there are no chemical reaction. The equation of continuity for species $A$ requires

$$\frac{\partial \rho_{(A)}}{\partial t} + \frac{1}{r} \frac{\partial \left( r n_{(A)r} \right)}{\partial r} + \frac{1}{r} \frac{\partial n_{(A)\theta}}{\partial \theta} + \frac{\partial n_{(A)z}}{\partial z} = 0 \tag{10.2.1-1}$$

If we use this as a basis for our analysis, we will be faced with solving a partial differential equation.

Let us assume that we are primarily interested in the area-averaged composition

$$\overline{\rho_{(A)}} \equiv \frac{1}{\pi R^2} \int_0^{2\pi} \int_0^R \rho_{(A)} \, r \, dr \, d\theta \tag{10.2.1-2}$$

as a function of time and axial position. This suggests that we take the area average of (10.2.1-1):

$$\frac{\partial \overline{\rho_{(A)}}}{\partial t} + \frac{1}{\pi R^2} \int_0^{2\pi} \int_0^R \left[ \frac{\partial \left( r n_{(A)r} \right)}{\partial r} + \frac{\partial n_{(A)\theta}}{\partial \theta} \right] dr \, d\theta + \frac{\partial \overline{n}_{(A)z}}{\partial z} = 0 \qquad (10.2.1\text{-}3)$$

The second and third terms on the left can be integrated to find

$$\frac{1}{\pi R^2} \int_0^{2\pi} \int_0^R \frac{\partial \left( r n_{(A)r} \right)}{\partial r} dr \, d\theta = 0 \qquad (10.2.1\text{-}4)$$

and

$$\frac{1}{\pi R^2} \int_0^R \int_0^{2\pi} \frac{\partial n_{(A)\theta}}{\partial \theta} d\theta \, dr = 0 \qquad (10.2.1\text{-}5)$$

As a result, (10.2.1-3) assumes the simpler form

$$\frac{\partial \overline{\rho_{(A)}}}{\partial t} + \frac{\partial \overline{n}_{(A)z}}{\partial z} = 0 \qquad (10.2.1\text{-}6)$$

We can express the second term on the left of this equation in terms of composition by using the area average of the $z$ component of Fick's first law:

$$\overline{n_{(A)z}} = \overline{\rho_{(A)} v_z} - \mathcal{D}_{(AB)} \frac{\partial \overline{\rho_{(A)}}}{\partial z} \qquad (10.2.1\text{-}7)$$

Unfortunately, we do not achieve in this way a differential equation for $\overline{\rho_{(A)}}$.

An approximation appears to be in order. Equation (10.2.1-7) suggests that the simplest approach is to say

$$\overline{n_{(A)z}} \doteq \overline{\rho_{(A)}} \, \overline{v}_z - \mathcal{K} \frac{\partial \overline{\rho_{(A)}}}{\partial z} \qquad (10.2.1\text{-}8)$$

To compensate for the fact that the area average of a product is generally not equal to the product of the area averages, we replace the diffusion coefficient by an empirical dispersion coefficient $\mathcal{K}$, which we will assume here to be a constant. Recognizing that

$$\overline{v}_z = \text{a constant} \qquad (10.2.1\text{-}9)$$

we see that (10.2.1-8) enables us to say from (10.2.1-6)

$$\frac{\partial \overline{\rho_{(A)}}}{\partial t} + \overline{v}_z \frac{\partial \overline{\rho_{(A)}}}{\partial z} - \mathcal{K} \frac{\partial^2 \overline{\rho_{(A)}}}{\partial z^2} = 0 \qquad (10.2.1\text{-}10)$$

We must solve this equation consistent with the initial conditions

$$\text{at } t = 0, \text{ for } z > 0 : \ \overline{\rho_{(A)}} = 0 \qquad (10.2.1\text{-}11)$$

and

$$\text{at } t = 0, \text{ for } z < 0 : \ \overline{\rho_{(A)}} = \rho_{(A)0} \qquad (10.2.1\text{-}12)$$

If we think of $\overline{\rho_{(A)}}$ as a function of $t$ and a new independent variable

$$\zeta \equiv z - \overline{v}_z t \qquad (10.2.1\text{-}13)$$

Equation (10.2.1-10) reduces to

$$\left(\frac{\partial \overline{\rho_{(A)}}}{\partial t}\right)_{\zeta} - \mathcal{K}\frac{\partial^2 \overline{\rho_{(A)}}}{\partial \zeta^2} = 0 \tag{10.2.1-14}$$

Our experience in Section 9.2.1 suggests that we think of

$$\overline{\rho_{(A)}^{\star}} \equiv \frac{\overline{\rho_{(A)}}}{\rho_{(A)0}} \tag{10.2.1-15}$$

as a function of a single independent variable

$$\eta \equiv \frac{\zeta}{\sqrt{4\mathcal{K}t}} \tag{10.2.1-16}$$

since this allows us to express (10.2.1-14) as an ordinary differential equation:

$$\frac{d^2 \overline{\rho_{(A)}^{\star}}}{d\eta^2} + 2\eta\frac{d\overline{\rho_{(A)}^{\star}}}{d\eta} = 0 \tag{10.2.1-17}$$

From (10.2.1-11) and (10.2.1-12), the corresponding boundary conditions are

$$\text{as } \eta \to \infty : \ \overline{\rho_{(A)}^{\star}} \to 0 \tag{10.2.1-18}$$

and

$$\text{as } \eta \to -\infty : \ \overline{\rho_{(A)}^{\star}} \to 1 \tag{10.2.1-19}$$

Integrating (10.2.1-17) once we find

$$\frac{d\overline{\rho_{(A)}^{\star}}}{d\eta} = C_1 \exp(-\eta^2) \tag{10.2.1-20}$$

A second integration consistent with boundary condition (10.2.1-18) yields

$$\overline{\rho_{(A)}^{\star}} = -C_1 \int_{\eta}^{\infty} \exp(-\eta^2)\, d\eta$$

$$= -\frac{C_1\sqrt{\pi}}{2}(1 - \operatorname{erf}\eta) \tag{10.2.1-21}$$

To satisfy boundary condition (10.2.1-19) we must set

$$C_1 \equiv -\frac{1}{\sqrt{\pi}} \tag{10.2.1-22}$$

Our final result for the area-averaged composition in the tube is

$$\overline{\rho_{(A)}^{\star}} = \frac{1}{2}(1 - \operatorname{erf}\eta) \tag{10.2.1-23}$$

One aspect of this solution for which we may have some intuitive feeling is the length $L$ of the transition zone in which $\overline{\rho_{(A)}}$ changes from $0.9\rho_{(A)0}$ to $0.1\rho_{(A)0}$. From (10.2.1-23), we can calculate

$$0.8 = \frac{1}{2}(\operatorname{erf}\eta_{0.1} - \operatorname{erf}\eta_{0.9})$$

$$= \operatorname{erf}\eta_{0.1}$$

$$= \operatorname{erf}\left(\frac{L}{4\sqrt{\mathcal{K}t}}\right) \tag{10.2.1-24}$$

We conclude that

$$L = 3.62\sqrt{\mathcal{K}t} \tag{10.2.1-25}$$

Taylor (1953) has also analyzed longitudinal dispersion resulting from the introduction of a concentrated mass of solute in the cross section $z = 0$ at time $t = 0$.

A very interesting theoretical analysis of the dependence of the dispersion coefficient $\mathcal{K}$ upon the diffusion coefficient $\mathcal{D}_{(AB)}$ has been given by Taylor (1953, 1954b,a) and later more carefully by Aris (1956), who concluded that

$$\mathcal{K} = \overline{\mathcal{D}_{(AB)}} + \chi \frac{\overline{v_z}^2 R^2}{\overline{\mathcal{D}_{(AB)}}} \tag{10.2.1-26}$$

Here $\overline{\mathcal{D}_{(AB)}}$ is the area-averaged diffusion coefficient and $\chi$ is a factor that depends upon the shape of the cross section of the tube as well as the variation in the velocity and diffusion coefficient profiles. If the diffusion coefficient is taken to be a constant, the velocity profile parabolic, and the tube cross section circular, they have found

$$\chi = \frac{1}{48} \tag{10.2.1-27}$$

A further refinement has been offered by Gill and Sankarasubramanian (1970).

Exercise 10.2.1-1 *A catalytic tubular reactor*   The open tube shown in Figure 10.2.1-1 is a very simple reactor. For $0 < z < L$, the wall of the tube is a catalyst for the reaction $A \rightarrow B$. You may assume that the physical properties of the liquid are constants and that the catalytic reaction can be described as first order:

$$\text{at } r = R : \quad n_{(A)r} = r_{(A)}{}^{(\sigma)}$$

$$= k_1'' \rho_{(A)}$$

If the liquid very far upstream has a uniform mass density $\rho_{(A)0}$, what is the composition of $A$ in the product downstream?

Use the same approach taken in the text to determine the average composition of the liquid downstream from the reactor.

i) The tubular reactor together with its connecting upstream and downstream sections is illustrated in Figure 10.2.1-1. Conclude that, for the region upstream from the reactor,

$$\frac{\partial \overline{\rho_{(A)}^{\star}}}{\partial z^{\star}} = \frac{1}{N_{Pe}} \frac{\partial^2 \overline{\rho_{(A)}^{\star}}}{\partial z^{\star 2}} \tag{10.2.1-28}$$

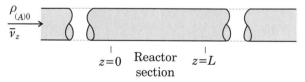

**Figure 10.2.1-1.** For $0 < z < L$, the wall of a very long open tube is a catalyst for the reaction $A \rightarrow B$.

for the reactor itself,

$$\frac{\partial \overline{\rho_{(A)}^{\star}}}{\partial z^{\star}} = \frac{1}{N_{Pe}} \frac{\partial^2 \overline{\rho_{(A)}^{\star}}}{\partial z^{\star 2}} - \frac{N_{Da}}{N_{Pe}} \overline{\rho_{(A)}^{\star}} \tag{10.2.1-29}$$

and for the region downstream of the reactor,

$$\frac{\partial \overline{\rho_{(A)}^{\star}}}{\partial z^{\star}} = \frac{1}{N_{Pe}} \frac{\partial^2 \overline{\rho_{(A)}^{\star}}}{\partial z^{\star 2}} \tag{10.2.1-30}$$

Here

$$N_{Pe} \equiv \frac{L \bar{v}_z}{\mathcal{K}}$$

$$N_{Da} \equiv \frac{2 k_1'' L^2}{R \mathcal{K}}$$

$$\overline{\rho_{(A)}^{\star}} \equiv \frac{\overline{\rho_{(A)}}}{\rho_{(A)0}}$$

$$z^{\star} \equiv \frac{z}{L}$$

What assumptions have been made?

ii) What are the boundary conditions that must be satisfied by this system of equations at the entrance to the reactor, at the exit from the reactor, very far upstream, and very far downstream?

iii) Solve (10.2.1-28) through (10.2.1-30) individually and evaluate the six constants of integration to find the following concentration distribution:

for $z^{\star} \le 0$:

$$\frac{\overline{\rho_{(A)}^{\star}} - 1}{\overline{\rho_{(A)}^{\star}}(0) - 1} = \exp(N_{Pe} z^{\star})$$

where

$$\overline{\rho_{(A)}^{\star}}(0) \equiv g_0 \left[ (1+a) \exp\left(\frac{a N_{Pe}}{2}\right) - (1-a) \exp\left(\frac{-a N_{Pe}}{2}\right) \right]$$

$$g_0 \equiv 2 \left[ (1+a)^2 \exp\left(\frac{a N_{Pe}}{2}\right) - (1-a)^2 \exp\left(\frac{-a N_{Pe}}{2}\right) \right]^{-1}$$

$$a \equiv \left[ 1 + \frac{4 N_{Da}}{N_{Pe}^2} \right]^{1/2}$$

for $0 \le z^{\star} \le 1$:

$$\overline{\rho_{(A)}^{\star}} = g_0 \exp\left(\frac{N_{Pe} z^{\star}}{2}\right) \left\{ (1+a) \exp\left(\frac{a N_{Pe}}{2}[1 - z^{\star}]\right) \right.$$

$$\left. - (1-a) \exp\left(-\frac{a N_{Pe}}{2}[1 - z^{\star}]\right) \right\}$$

for $z^{\star} \ge 1$:

$$\overline{\rho_{(A)}^{\star}} = 2 g_0 a \exp\left(\frac{N_{Pe}}{2}\right)$$

The analysis suggested here is essentially that given by Wehner and Wilhelm (1956) for the fixed-bed catalytic reactor (see Exercise 10.3.8-4).

## 10.3  Local Volume Averaging

We are commonly concerned with chemical reactions in beds of porous pellets impregnated with a catalyst. We now realize that, if we were to pump a waste stream down a disposal well and into a layer of porous rock, we would have to give serious consideration to the possibility that freshwater supplies for surrounding communities might be contaminated. When significant portions of a river's flow are diverted to distant localities (perhaps by a system of aqueducts), saltwater may begin to encroach upon the river's delta region, destroying its previous fertility. Can freshwater be pumped down selected wells in the delta in order to limit the concentration of salt in the soil? In each of these processes one of the controlling features is mass transfer in porous media.

A common approach to mass transfer in porous media has been to view the porous solid and whatever gases and liquids it contains as a continuum and to employ simply the usual differential equation of continuity discussed in Section 8.2.1. In other words, one treats mass transfer in a porous medium as diffusion in a single phase. But there is a fundamental difference. In the case of the bed of porous catalyst pellets, we are concerned with two distinct phases: diffusion takes place in the gas phase, whereas a chemical reaction proceeds at the gas–solid phase boundary. In contrast, intermolecular forces control the rate at which helium moves through Pyrex glass and the rate at which trichloromethane diffuses through a polymer.

Our successful discussions of momentum transfer in Section 4.3 and of the energy transfer in Section 7.3 suggest that we take the same point of view here in studying mass transfer. This means that we should begin by developing the local volume average of the differential equation of continuity for species $A$.

For simplicity, we shall restrict this discussion to a single fluid flowing through a stationary, rigid, porous medium.

### 10.3.1  Local Volume Average of the Differential Mass Balance for Species $A$

We can begin as we did in Section 4.3.1, where we began to develop the local volume average of the differential mass balance. Let us think of a particular point $\mathbf{z}$ in the porous medium and let us integrate the differential mass balance for species $A$ over $R^{(f)}$, the region of space occupied by the fluid within $S$ associated with $\mathbf{z}$:

$$\frac{1}{V} \int_{R^{(f)}} \left( \frac{\partial \rho_{(A)}}{\partial t} + \operatorname{div} \mathbf{n}_{(A)} - r_{(A)} \right) dV = 0 \qquad (10.3.1\text{-}1)$$

We use here one form of the differential mass balance for species $A$ from Table 8.5.1-5. The operations of volume integration and differentiation with respect to time may be interchanged in the first term on the left:

$$\frac{1}{V} \int_{R^{(f)}} \frac{\partial \rho_{(A)}}{\partial t} \, dV = \frac{\partial \overline{\rho_{(A)}}^{(f)}}{\partial t} \qquad (10.3.1\text{-}2)$$

The theorem of Section 4.3.2 can be used to express the second term on the left of (10.3.1-1) as

$$\frac{1}{V} \int_{R^{(f)}} \text{div } \mathbf{n}_{(A)} \, dV = \text{div } \overline{\mathbf{n}_{(A)}}^{(f)} - r''_{(A)} \tag{10.3.1-3}$$

where we have introduced $r''_{(A)}$ as the rate at which species $A$ is produced by a catalytic chemical reaction at the fluid–solid phase interface (see Section 8.2.1):

$$\begin{aligned} r''_{(A)} &\equiv -\frac{1}{V} \int_{S_w} \mathbf{n}_{(A)} \cdot n \, dA \\ &= \frac{1}{V} \int_{S_w} r^{(\sigma)}_{(A)} \, dA \end{aligned} \tag{10.3.1-4}$$

Equations (10.3.1-2) and (10.3.1-3) allow us to express (10.3.1-1) as

$$\frac{\partial \overline{\rho_{(A)}}^{(f)}}{\partial t} + \text{div } \overline{\mathbf{n}_{(A)}}^{(f)} = r''_{(A)} + \overline{r_{(A)}}^{(f)} \tag{10.3.1-5}$$

This is one convenient form for the local volume-averaged differential mass balance for species $A$.

If we had started instead with another form of the differential mass balance for species $A$ from Table 8.5.1-5, we would have found by an entirely analogous train of thought still another form for the local volume-averaged differential mass balance for species $A$:

$$\frac{\partial \overline{c_{(A)}}^{(f)}}{\partial t} + \text{div } \overline{\mathbf{n}_{(A)}}^{(f)} = \frac{r''_{(A)}}{M_{(A)}} + \frac{\overline{r_{(A)}}^{(f)}}{M_{(A)}} \tag{10.3.1-6}$$

This result can, of course, also be obtained by dividing (10.3.1-5) by the molecular weight of species $A$.

In the next sections, I discuss the forms that I might expect empirical correlations for $\overline{\mathbf{n}_{(A)}}^{(f)}$ and $\overline{\mathbf{n}_{(A)}}^{(f)}$ to assume.

## 10.3.2 When Fick's First Law Applies

Let $l_0$ represent a characteristic pore diameter of the structure, and let $\lambda$ be the molecular mean free path (Hirschfelder et al. 1954, p. 10). When the Knudsen number

$$\begin{aligned} N_{Kn} &\equiv \frac{l_0}{\lambda} \\ &> 10 \end{aligned} \tag{10.3.2-1}$$

Fick's first law can be used to describe binary diffusion within a gas in a porous medium [Scott (1962); it can also be used to describe diffusion in multicomponent solutions for the three limiting cases discussed in Section 8.4.6]. Current practice is to always use Fick's first law when talking about binary diffusion in liquids.

We assume that the diffusion coefficient $\mathcal{D}_{(AB)}$ is a constant. We can take the local volume average of Fick's first law from Table 8.5.1-7 to find

$$\overline{\mathbf{n}_{(A)}}^{(f)} = \overline{\rho_{(A)} \mathbf{v}}^{(f)} - \mathcal{D}_{(AB)} \overline{\rho \nabla \omega_{(A)}}^{(f)} \tag{10.3.2-2}$$

The theorem of Section 4.3.2 allows us to say

$$\overline{\nabla \omega_{(A)}}^{(f)} = \nabla \overline{\omega_{(A)}}^{(f)} + \frac{1}{V} \int_{S_w} \omega_{(A)} \mathbf{n} \, dA \tag{10.3.2-3}$$

This allows us to write (10.3.2-2) as

$$\overline{\mathbf{n}_{(A)}}^{(f)} = \langle \rho_{(A)} \rangle^{(f)} \overline{\mathbf{v}}^{(f)} - \langle \rho \rangle^{(f)} \mathcal{D}_{(AB)} \nabla \overline{\omega_{(A)}}^{(f)} - \boldsymbol{\delta}_{(A)} \tag{10.3.2-4}$$

where we define the *mass density tortuosity vector* for species $A$,

$$\boldsymbol{\delta}_{(A)} \equiv \langle \rho_{(A)} \rangle^{(f)} \overline{\mathbf{v}}^{(f)} - \overline{\rho_{(A)} \mathbf{v}}^{(f)} + \mathcal{D}_{(AB)} \overline{\rho \nabla \omega_{(A)}}^{(f)}$$

$$- \langle \rho \rangle^{(f)} \mathcal{D}_{(AB)} \overline{\nabla \omega_{(A)}}^{(f)} + \frac{\langle \rho \rangle^{(f)} \mathcal{D}_{(AB)}}{V} \int_{S_w} \omega_{(A)} \mathbf{n} \, dA \tag{10.3.2-5}$$

The local volume average of the differential mass balance for species $A$ in the form of (10.3.1-5) may be expressed as

$$\frac{\partial \overline{\rho_{(A)}}^{(f)}}{\partial t} + \text{div} \left( \langle \rho_{(A)} \rangle^{(f)} \overline{\mathbf{v}}^{(f)} \right) = -\text{div} \, \mathbf{j}_{(A)}^{(e)} + r_{(A)}'' + \overline{r_{(A)}}^{(f)} \tag{10.3.2-6}$$

where the *effective* mass flux with respect to $\overline{\mathbf{v}}^{(f)}$ is

$$\mathbf{j}_{(A)}^{(e)} \equiv -\langle \rho \rangle^{(f)} \mathcal{D}_{(AB)} \nabla \overline{\omega_{(A)}}^{(f)} - \boldsymbol{\delta}_{(A)} \tag{10.3.2-7}$$

Very similar results can be obtained if we assume the diffusion coefficient $\mathcal{D}_{(AB)}$ is a constant and take the local volume average of another form of Fick's first law from Table 8.5.1-7:

$$\overline{\mathbf{N}_{(A)}}^{(f)} = \langle c_{(A)} \rangle^{(f)} \overline{\mathbf{v}^\diamond}^{(f)} - \langle c \rangle^{(f)} \mathcal{D}_{(AB)} \nabla \overline{x_{(A)}}^{(f)} - \boldsymbol{\Delta}_{(A)} \tag{10.3.2-8}$$

$$\boldsymbol{\Delta}_{(A)} \equiv \langle c_{(A)} \rangle^{(f)} \overline{\mathbf{v}^\diamond}^{(f)} - \overline{c_{(A)} \mathbf{v}^\diamond}^{(f)} + \mathcal{D}_{(AB)} \overline{c \nabla x_{(A)}}^{(f)}$$

$$- \langle c \rangle^{(f)} \mathcal{D}_{(AB)} \overline{\nabla x_{(A)}}^{(f)} + \frac{\langle c \rangle^{(f)} \mathcal{D}_{(AB)}}{V} \int_{S_w} x_{(A)} \mathbf{n} \, dA \tag{10.3.2-9}$$

We will refer to $\boldsymbol{\Delta}_{(A)}$ as the molar-density tortuosity vector for species $A$. In these terms, the local volume average of the differential mass balance for species $A$ in the form of (10.3.1-6) becomes

$$\frac{\partial \overline{c_{(A)}}^{(f)}}{\partial t} + \text{div} \left( \langle c_{(A)} \rangle^{(f)} \overline{\mathbf{v}^\diamond}^{(f)} \right)$$

$$= -\text{div} \, \mathbf{J}_{(A)}^{\diamond(e)} + \frac{r_{(A)}''}{M_{(A)}} + \frac{\overline{r_{(A)}}^{(f)}}{M_{(A)}} \tag{10.3.2-10}$$

where

$$\mathbf{J}_{(A)}^{\diamond(e)} \equiv -\langle c \rangle^{(f)} \mathcal{D}_{(AB)} \nabla \overline{x_{(A)}}^{(f)} - \boldsymbol{\Delta}_{(A)} \tag{10.3.2-11}$$

should be thought of as an effective molar flux vector with respect to $\overline{\mathbf{v}^\diamond}^{(f)}$.

One point worth emphasizing is that

$$\mathbf{j}_{(A)}^{(e)} \neq \frac{M_{(A)} M_{(B)}}{M} \mathbf{J}_{(A)}^{\diamond(e)} \tag{10.3.2-12}$$

The physical meaning of the mass-density tortuosity vector $\delta_{(A)}$ is clarified by noting that, if $\rho_{(A)}$ and $\rho$ are independent of position,

$$\delta_{(A)} = \frac{\rho_{(A)}\mathcal{D}_{(AB)}}{V} \int_{S_w} \mathbf{n}\, dA$$
$$= 0 \tag{10.3.2-13}$$

We have used here the theorem of Section 4.3.2 applied to a constant. In the same way, if $c_{(A)}$ and $c$ are independent of position,

$$\Delta_{(A)} = 0 \tag{10.3.2-14}$$

Because of these simplifications, we shall direct our attention to structures of uniform porosity in the sections that immediately follow.

## 10.3.3 Empirical Correlations for Tortuosity Vectors

In this section, we give three examples of how experimental data can be used to prepare correlations for $\delta_{(A)}$ (or $\Delta_{(A)}$), introduced in Section 10.3.2. Four points form the foundation for this discussion.

1) The tortuosity vector $\delta_{(A)}$ (or $\Delta_{(A)}$) is frame indifferent. For example,

$$\delta_{(A)}^* \equiv \langle \rho_{(A)}^* \rangle^{(f)}\overline{\mathbf{v}}^{(f)} - \overline{\rho_{(A)}^*\mathbf{v}^*}^{(f)} + \mathcal{D}_{(AB)}\overline{\rho^*\nabla\omega_{(A)}^*}^{(f)} - \langle \rho^* \rangle^{(f)}\mathcal{D}_{(AB)}\overline{\nabla\omega_{(A)}^*}^{(f)}$$
$$+ \frac{\langle \rho^* \rangle^{(f)}\mathcal{D}_{(AB)}}{V} \int_{S_w} \omega_{(A)}^*\mathbf{n}^*\, dA$$
$$= \overline{\rho_{(A)}^*\left(\Psi^{-1}\overline{\mathbf{v}^*}^{(f)} - \mathbf{v}^*\right)}^{(f)} + \mathcal{D}_{(AB)}\overline{\rho^*\nabla\omega_{(A)}^*}^{(f)} - \langle \rho^* \rangle^{(f)}\mathcal{D}_{(AB)}\overline{\nabla\omega_{(A)}^*}^{(f)}$$
$$+ \frac{\langle \rho^* \rangle^{(f)}\mathcal{D}_{(AB)}}{V} \int_{S_w} \omega_{(A)}^*\mathbf{n}^*\, dA$$
$$= \overline{\rho_{(A)}\mathbf{Q}\cdot\left(\Psi^{-1}\overline{\mathbf{v}} - \mathbf{v}\right)}^{(f)} + \mathcal{D}_{(AB)}\overline{\rho\mathbf{Q}\cdot\nabla\omega_{(A)}}^{(f)} - \langle \rho \rangle^{(f)}\mathcal{D}_{(AB)}\mathbf{Q}\cdot\overline{\nabla\omega_{(A)}}^{(f)}$$
$$+ \frac{\langle \rho \rangle^{(f)}\mathcal{D}_{(AB)}}{V} \int_{S_w} \omega_{(A)}\mathbf{Q}\cdot\mathbf{n}\, dA$$
$$= \mathbf{Q}\cdot\delta_{(A)} \tag{10.3.3-1}$$

In the second line we observe that the superficial volume average of a superficial volume average is simply the superficial volume average (see Exercise 4.3.7-1); in the third line, we employ the frame indifference of the mass density for species $A$ and the frame indifference of a velocity difference. Here, $\mathbf{Q}$ is a (possibly) time-dependent, orthogonal, second-order tensor. (The molar-density tortuosity vector $\Delta_{(A)}$ can be proved to be frame indifferent in exactly the same manner.)
2) We assume that the principle of frame indifference introduced in Section 2.3.1 applies to any empirical correlation developed for $\delta_{(A)}$ (or $\Delta_{(A)}$).
3) The Buckingham–Pi theorem (Brand 1957) serves to further restrict the form of any expression for $\delta_{(A)}$ (or $\Delta_{(A)}$).

4) The averaging surface $S$ is so large that $\delta_{(A)}$ (or $\mathbf{\Delta}_{(A)}$) may be assumed not to be explicit functions of position in the porous structure, though they very well may be implicit functions of position as a result of their dependence upon other variables.

### Example 1: Nonoriented Porous Solids When Convection Can Be Neglected

We argued in Section 9.6 that diffusion-induced convection may be neglected in the limit of dilute solutions. In this limit with no forced convection, we may neglect the first four terms on the right of (10.3.2-5):

$$\delta_{(A)} = \frac{\langle\rho\rangle^{(f)}\mathcal{D}_{(AB)}}{V}\int_{S_w}\omega_{(A)}\mathbf{n}\,dA \tag{10.3.3-2}$$

For geometrically similar, nonoriented porous media, $\delta_{(A)}$ may be a function of the particle diameter $l_0$, the diffusion coefficient $\mathcal{D}_{(AB)}$, the porosity $\Psi$, as well as some measures of the local concentration distribution such as $\overline{\rho_{(A)}}^{(f)}$ and $\langle\rho\rangle^{(f)}\nabla\overline{\omega_{(A)}}^{(f)}$:

$$\delta_{(A)} = \delta_{(A)}\left(l_0, \mathcal{D}_{(AB)}, \Psi, \overline{\rho_{(A)}}^{(f)}, \langle\rho\rangle^{(f)}\nabla\overline{\omega_{(A)}}^{(f)}\right) \tag{10.3.3-3}$$

For the moment, let us fix our attention on the dependence of $\delta_{(A)}$ upon the vector $\langle\rho\rangle^{(f)}\nabla\overline{\omega_{(A)}}^{(f)}$:

$$\delta_{(A)} = \hat{\delta}_{(A)}\left(\langle\rho\rangle^{(f)}\nabla\overline{\omega_{(A)}}^{(f)}\right) \tag{10.3.3-4}$$

By the principle of frame indifference, the functional relationship between these two variables should be the same in every frame of reference. This means that

$$\begin{aligned}
\delta_{(A)}^* &\equiv \mathbf{Q}\cdot\delta_{(A)} \\
&= \mathbf{Q}\cdot\hat{\delta}\left(\langle\rho\rangle^{(f)}\nabla\overline{\omega_{(A)}}^{(f)}\right) \\
&= \hat{\delta}_{(A)}\left(\mathbf{Q}\cdot\langle\rho\rangle^{(f)}\nabla\overline{\omega_{(A)}}^{(f)}\right)
\end{aligned} \tag{10.3.3-5}$$

or $\hat{\delta}_{(A)}$ is an isotropic function (Truesdell and Noll 1965, p. 22):

$$\hat{\delta}_{(A)}\left(\langle\rho\rangle^{(f)}\nabla\overline{\omega_{(A)}}^{(f)}\right) = \mathbf{Q}^T\cdot\hat{\delta}_{(A)}\left(\mathbf{Q}\cdot\langle\rho\rangle^{(f)}\nabla\overline{\omega_{(A)}}^{(f)}\right) \tag{10.3.3-6}$$

By a representation theorem for a vector-valued isotropic function of one vector (Truesdell and Noll 1965, p. 35), we may write

$$\begin{aligned}
\delta_{(A)} &= \hat{\delta}_{(A)}\left(\langle\rho\rangle^{(f)}\nabla\overline{\omega_{(A)}}^{(f)}\right) \\
&= D_{(A)}\langle\rho\rangle^{(f)}\nabla\overline{\omega_{(A)}}^{(f)}
\end{aligned} \tag{10.3.3-7}$$

where

$$D_{(A)} = \hat{D}_{(A)}\left(\langle\rho\rangle^{(f)}|\nabla\overline{\omega_{(A)}}^{(f)}|\right) \tag{10.3.3-8}$$

Comparing (10.3.3-7) and (10.3.3-8) with (10.3.3-3), we see

$$D_{(A)} = D_{(A)}\left(l_0, \mathcal{D}_{(AB)}, \Psi, \overline{\rho_{(A)}}^{(f)}, \langle\rho\rangle^{(f)}|\nabla\overline{\omega_{(A)}}^{(f)}|\right) \tag{10.3.3-9}$$

An application of the Buckingham–Pi theorem (Brand 1957) allows us to conclude that

$$D_{(A)} = \mathcal{D}_{(AB)}D_{(A)}^\star \tag{10.3.3-10}$$

Here

$$D^{\star}_{(A)} = D^{\star}_{(A)} \left( \Psi, \frac{l_0 \langle \rho \rangle^{(f)} |\nabla \overline{\omega_{(A)}}^{(f)}|}{\overline{\rho_{(A)}}^{(f)}} \right) \tag{10.3.3-11}$$

In summary, Equations (10.3.3-7), (10.3.3-10), and (10.3.3-11) represent probably the simplest form that empirical correlations for the mass-density tortuosity vector $\delta_{(A)}$ can take in a nonoriented porous medium.

### Example 2: Nonoriented Porous Solid Filled with a Flowing Fluid

For geometrically similar nonoriented porous media under conditions such that convection is not negligible, $\delta_{(A)}$ may be thought of as a function of the local particle diameter $l_0$, the diffusion coefficient $\mathcal{D}_{(AB)}$, the porosity $\Psi$, the local volume-averaged velocity of the fluid with respect to the local volume-averaged velocity of the solid $\overline{\mathbf{v}}^{(f)} - \overline{\mathbf{v}}^{(s)}$, as well as some measures of the local mass-density distribution such as $\overline{\rho_{(A)}}^{(f)}$ and $\langle \rho \rangle^{(f)} \nabla \overline{\omega_{(A)}}^{(f)}$:

$$\delta_{(A)} = \delta_{(A)} \left( l_0, \mathcal{D}_{(AB)}, \Psi, \overline{\mathbf{v}}^{(f)} - \overline{\mathbf{v}}^{(s)}, \overline{\rho_{(A)}}^{(f)}, \langle \rho \rangle^{(f)} \nabla \overline{\omega_{(A)}}^{(f)} \right) \tag{10.3.3-12}$$

Let us begin by examining the dependence of $\delta_{(A)}$ upon the two vectors:

$$\delta_{(A)} = \hat{\delta}_{(A)} \left( \overline{\mathbf{v}}^{(f)} - \overline{\mathbf{v}}^{(s)}, \langle \rho \rangle^{(f)} \nabla \overline{\omega_{(A)}}^{(f)} \right) \tag{10.3.3-13}$$

By the principle of frame indifference, the functional relationship between these two variables should be the same in every frame of reference. This means that

$$\begin{aligned} \delta^{\star}_{(A)} &= \mathbf{Q} \cdot \delta_{(A)} \\ &= \mathbf{Q} \cdot \hat{\delta}_{(A)} \left( \overline{\mathbf{v}}^{(f)} - \overline{\mathbf{v}}^{(s)}, \langle \rho \rangle^{(f)} \nabla \overline{\omega_{(A)}} \right) \\ &= \hat{\delta}_{(A)} \left( \mathbf{Q} \cdot \left( \overline{v}^{(f)} - \overline{\mathbf{v}}^{(s)} \right), \mathbf{Q} \cdot \langle \rho \rangle^{(f)} \nabla \overline{\omega_{(A)}}^{(f)} \right) \end{aligned} \tag{10.3.3-14}$$

or $\delta_{(A)}$ is an isotropic function (Truesdell and Noll 1965, p. 22):

$$\begin{aligned} &\hat{\delta}_{(A)} \left( \overline{\mathbf{v}}^{(f)} - \overline{\mathbf{v}}^{(s)}, \langle \rho \rangle^{(f)} \nabla \overline{\omega_{(A)}}^{(f)} \right) \\ &= \mathbf{Q}^{T} \cdot \hat{\delta}_{(A)} \left( \mathbf{Q} \cdot \left[ \overline{\mathbf{v}}^{(f)} - \overline{\mathbf{v}}^{(s)} \right], \mathbf{Q} \cdot \langle \rho \rangle^{(f)} \nabla \overline{\omega_{(A)}}^{(f)} \right) \end{aligned} \tag{10.3.3-15}$$

By an argument similar to that given in Section 7.3.2 (Example 2), we conclude that

$$\delta_{(A)} = D_{(A1)} \langle \rho \rangle^{(f)} \nabla \overline{\omega_{(A)}}^{(f)} - D_{(A2)} \left( \overline{\mathbf{v}}^{(f)} - \overline{\mathbf{v}}^{(s)} \right) \tag{10.3.3-16}$$

where

$$\begin{aligned} D_{(Ai)} = D_{(Ai)} \Big( |\overline{\mathbf{v}}^{(f)} - \overline{\mathbf{v}}^{(s)}|, \langle \rho \rangle^{(f)} |\nabla \overline{\omega_{(A)}}^{(f)}|, \\ \langle \rho \rangle^{(f)} \left( \overline{\mathbf{v}} - \overline{\mathbf{v}}^{(s)} \right) \cdot \nabla \overline{\omega_{(A)}}^{(f)}, \overline{\rho_{(A)}}^{(f)}, l_0, \mathcal{D}_{(AB)}, \Psi \Big) \end{aligned} \tag{10.3.3-17}$$

An application of the Buckingham–Pi theorem (Brand 1957) shows that

$$D_{(A1)} = \mathcal{D}_{(AB)} D^{\star}_{(A1)} \tag{10.3.3-18}$$

and

$$D_{(A2)} = l_0 \langle \rho \rangle^{(f)} |\nabla \overline{\omega_{(A)}}^{(f)}| D^{\star}_{(A2)} \tag{10.3.3-19}$$

Here

$$D^\star_{(Ai)} = D^\star_{(Ai)} \left( N_{Pe}, \frac{\left[\overline{\mathbf{v}}^{(f)} - \overline{\mathbf{v}}^{(s)}\right] \cdot \nabla\overline{\omega_{(A)}}^{(f)}}{|\overline{\mathbf{v}}^{(f)} - \overline{\mathbf{v}}^{(s)}| \, |\nabla\overline{\omega_{(A)}}^{(f)}|}, \ \frac{l_0 \langle\rho\rangle^{(f)} |\nabla\overline{\omega_{(A)}}^{(f)}|}{\overline{\rho_{(A)}}^{(f)}}, \ \Psi \right) \tag{10.3.3-20}$$

and

$$N_{Pe} \equiv \frac{l_0 |\overline{\mathbf{v}}^{(f)} - \overline{\mathbf{v}}^{(s)}|}{\mathcal{D}_{(AB)}} \tag{10.3.3-21}$$

Since $D^\star_{(A2)} = 0$ for $|\nabla\overline{\omega_{(A)}}^{(f)}| = 0$, we can conclude as expected that $\delta_{(A)} = 0$ in this limit.

In summary, (10.3.3-16) and (10.3.3-18) through (10.3.3-20) represent possibly the simplest form that empirical correlations for the mass-density tortuosity vector $\delta_{(A)}$ can take when a fluid flows through a nonoriented porous medium.

For more on this class of empiricisms as well as the traditional description of dispersion (Nikolaevskii 1959; Scheidegger 1961; de Josselin de Jong and Bossen 1961; Bear 1961, 1972; Peaceman 1966), see Chang and Slattery (1988).

### Example 3: Oriented Porous Solids When Convection Can Be Neglected

When convection can be neglected, we saw in Example 1 that (10.3.2-4) reduces to (10.3.3-2). But one should not expect (10.3.3-7), (10.3.3-10), and (10.3.3-11) to describe the mass-density tortuosity vector for a porous structure in which particle diameter $l$ is a function of position. For such a structure, (10.3.3-3) must be altered to include a dependence upon additional vector and possibly tensor quantities. For example, one might postulate a dependence of $\delta_{(A)}$ upon the local gradient in particle diameter as well as upon $\langle\rho\rangle^{(f)} \nabla\overline{\omega_{(A)}}^{(f)}$:

$$\delta_{(A)} = \delta_{(A)} \left( l, \mathcal{D}_{(AB)}, \Psi, \overline{\rho_{(A)}}^{(f)}, \langle\rho\rangle^{(f)}\nabla\overline{\omega_{(A)}}^{(f)}, \nabla l \right) \tag{10.3.3-22}$$

Following essentially the same argument given in Example 2, above, the principle of material frame indifference and the Buckingham–Pi theorem require

$$\delta_{(A)} = E_{(A1)}\langle\rho\rangle^{(f)}\nabla\overline{\omega_{(A)}}^{(f)} + E_{(A2)}\nabla l \tag{10.3.3-23}$$

where

$$E_{(A1)} = \mathcal{D}_{(AB)}E^\star_{(A1)} \tag{10.3.3-24}$$

$$E_{(A2)} = \mathcal{D}_{(AB)}\langle\rho\rangle^{(f)}|\nabla\overline{\omega_{(A)}}^{(f)}|E^\star_{(A2)} \tag{10.3.3-25}$$

and

$$E^\star_{(Ai)} = E^\star_{(Ai)} \left( \frac{\nabla l \cdot \nabla\overline{\omega_{(A)}}^{(f)}}{|\nabla l| \, |\nabla\overline{\omega_{(A)}}^{(f)}|}, \ |\nabla l|, \ \frac{l\langle\rho\rangle^{(f)}|\nabla\overline{\omega_{(A)}}^{(f)}|}{\overline{\rho_{(A)}}^{(f)}}, \ \Psi \right) \tag{10.3.3-26}$$

We expect that $E^\star_{(A2)} = 0$ for $|\nabla\overline{\omega_{(A)}}^{(f)}| = 0$, with the result that $\delta_{(A)} = \mathbf{0}$ in this limit.

Equations (10.3.3-23) through (10.3.3-26) represent possibly the simplest form that empirical correlations for the mass-density tortuosity vector $\delta_{(A)}$ can take in an oriented porous medium, assuming that the orientation of the structure can be attributed to the local gradient of particle diameter.

### 10.3.4 Summary of Results for a Liquid or Dense Gas in a Nonoriented, Uniform-Porosity Structure

I would like to summarize here the results for the case with which the literature has been primarily concerned until now: a liquid or dense gas in a nonoriented, uniform-porosity structure.

In Section 10.3.2, we found that the local volume average of the differential mass balance for species $A$ requires that

$$\frac{\partial \overline{\rho_{(A)}}^{(f)}}{\partial t} + \text{div}\left(\langle \rho_{(A)} \rangle^{(f)} \overline{\mathbf{v}}^{(f)}\right) = -\text{div}\, \mathbf{j}_{(A)}^{(e)} + r''_{(A)} + \overline{r_{(A)}}^{(f)} \tag{10.3.4-1}$$

and

$$\mathbf{j}_{(A)}^{(e)} \equiv -\langle \rho \rangle^{(f)} \mathcal{D}_{(AB)} \nabla \overline{\omega_{(A)}}^{(f)} - \boldsymbol{\delta}_{(A)} \tag{10.3.4-2}$$

should be thought of as the effective mass flux with respect to the volume-averaged, mass-averaged velocity $\overline{\mathbf{v}}^{(f)}$. In arriving at this result, we have assumed only that Fick's first law is applicable. In this way, we have limited the discussion to liquids and gases that are so dense that the molecular mean free path is small compared with the average pore diameter of the structure.

In Section 10.3.3 (Example 2), we suggest that, for a nonoriented porous solid filled with a flowing fluid, the mass density tortuosity vector $\boldsymbol{\delta}_{(A)}$ might be represented by (10.3.3-16) and (10.3.3-18) through (10.3.3-20). In these terms, the effective mass flux can be expressed as

$$\mathbf{j}_{(A)}^{(e)} = -\langle \rho \rangle^{(f)} \mathcal{D}_{(AB)} \left(1 + D_{(A1)}^{\star}\right) \nabla \overline{\omega_{(A)}}^{(f)} + l_0 \langle \rho \rangle^{(f)} |\nabla \overline{\omega_{(A)}}^{(f)}| D_{(A2)}^{\star} \overline{\mathbf{v}}^{(f)} \tag{10.3.4-3}$$

in which

$$D_{(Ai)}^{\star} = D_{(Ai)}^{\star}\left(N_{Pe}, \frac{\overline{\mathbf{v}}^{(f)} \cdot \nabla \overline{\omega_{(A)}}^{(f)}}{|\overline{\mathbf{v}}^{(f)}| \, |\nabla \overline{\omega_{(A)}}^{(f)}|}, \frac{l_0 \langle \rho \rangle^{(f)} |\nabla \overline{\omega_{(A)}}^{(f)}|}{\overline{\rho_{(A)}}^{(f)}}, \Psi\right) \tag{10.3.4-4}$$

and

$$N_{Pe} \equiv \frac{l_0 |\overline{\mathbf{v}}^{(f)}|}{\mathcal{D}_{(AB)}} \tag{10.3.4-5}$$

In arriving at this expression, we have assumed that the porous medium is stationary.

Sometimes it is more convenient to think of the effective mass flux in terms of an effective diffusivity tensor $\mathbf{D}_{(AB)}^{(e)}$:

$$\mathbf{j}_{(A)}^{(e)} = -\langle \rho \rangle^{(f)} \mathbf{D}_{(AB)}^{(e)} \cdot \nabla \overline{\omega_{(A)}}^{(f)} \tag{10.3.4-6}$$

Here

$$\mathbf{D}_{(AB)}^{(e)} \equiv \mathcal{D}_{(AB)} \left(1 + D_{(A1)}^{\star}\right) \mathbf{I} - \frac{l_0 |\nabla \overline{\omega_{(A)}}^{(f)}| D_{(A2)}^{\star}}{\overline{\mathbf{v}}^{(f)} \cdot \nabla \overline{\omega_{(A)}}^{(f)}} \overline{\mathbf{v}}^{(f)} \, \overline{\mathbf{v}}^{(f)} \tag{10.3.4-7}$$

We shall often find it more convenient to work in molar terms. Returning to Section 10.3.2, we found there that the local volume average of the differential mass balance could also be expressed as

$$\frac{\partial \overline{c_{(A)}}^{(f)}}{\partial t} + \text{div}\left(\langle c_{(A)} \rangle^{(f)} \overline{\mathbf{v}^\diamond}^{(f)}\right) = -\text{div}\, \mathbf{J}_{(A)}^{\diamond(e)} + \frac{r''_{(A)}}{M_{(A)}} + \frac{\overline{r_{(A)}}^{(f)}}{M_{(A)}} \tag{10.3.4-8}$$

in which

$$\mathbf{J}_{(A)}^{\diamond(e)} \equiv -\langle c \rangle^{(f)} \mathcal{D}_{(AB)} \nabla \overline{x_{(A)}}^{(f)} - \boldsymbol{\Delta}_{(A)} \tag{10.3.4-9}$$

should be thought of as the effective molar flux of species $A$ with respect to the volume-averaged, molar-averaged velocity $\overline{\mathbf{v}}^{\diamond(f)}$. If we visualize repeating for $\boldsymbol{\Delta}_{(A)}$ the type of analysis given in Section 10.3.3 (Example 2) for a nonoriented porous solid filled with a flowing fluid, we would find by analogy with (10.3.4-3) through (10.3.4-5) that

$$\mathbf{J}_{(A)}^{\diamond(e)} = -\langle c \rangle^{(f)} \mathcal{D}_{(AB)} \left( 1 + D_{(A1)}^{\diamond} \right) \nabla \overline{x_{(A)}}^{(f)} + l_0 \langle c \rangle^{(f)} |\nabla \overline{x_{(A)}}^{(f)}| D_{(A2)}^{\diamond} \overline{\mathbf{v}}^{\diamond(f)} \tag{10.3.4-10}$$

in which

$$D_{(Ai)}^{\diamond} = D_{(Ai)}^{\diamond} \left( N_{Pe}^{\diamond}, \frac{\overline{\mathbf{v}}^{\diamond(f)} \cdot \nabla \overline{x_{(A)}}^{(f)}}{|\overline{\mathbf{v}}^{\diamond(f)}| \, |\nabla \overline{x_{(A)}}^{(f)}|}, \frac{l_0 \langle c \rangle^{(f)} |\nabla \overline{x_{(A)}}^{(f)}|}{\overline{c_{(A)}}^{(f)}}, \Psi \right) \tag{10.3.4-11}$$

and

$$N_{Pe}^{\diamond} \equiv \frac{l_0 |\overline{\mathbf{v}}^{\diamond(f)}|}{\mathcal{D}_{(AB)}} \tag{10.3.4-12}$$

These results can, of course, also be written in terms of an effective diffusivity tensor $\mathbf{D}_{(AB)}^{\diamond(e)}$:

$$\mathbf{J}_{(A)}^{\diamond(e)} = -\langle c \rangle^{(f)} \mathbf{D}_{(AB)}^{\diamond(e)} \cdot \nabla \overline{x_{(A)}}^{(f)} \tag{10.3.4-13}$$

$$\mathbf{D}_{(AB)}^{\diamond(e)} \equiv \mathcal{D}_{(AB)} \left( 1 + D_{(A1)}^{\diamond} \right) \mathbf{I} - \frac{l_0 |\nabla \overline{x_{(A)}}^{(f)}| D_{(A2)}^{\diamond}}{\overline{\mathbf{v}}^{\diamond(f)} \cdot \nabla \overline{x_{(A)}}^{(f)}} \overline{\mathbf{v}}^{\diamond(f)} \overline{\mathbf{v}}^{\diamond(f)} \tag{10.3.4-14}$$

For the sake of simplicity, in what follows we take

$$D_{(Ai)}^{\star} = D_{(Ai)}^{\star}(\Psi) \tag{10.3.4-15}$$

and

$$D_{(Ai)}^{\diamond} = D_{(Ai)}^{\diamond}(\Psi) \tag{10.3.4-16}$$

This allows us to express (10.3.4-3) and (10.3.4-10) as

$$\mathbf{j}_{(A)}^{(e)} = -\langle \rho \rangle^{(f)} A_{(A)} \nabla \overline{\omega_{(A)}}^{(f)} + B_{(A)} \langle \rho \rangle^{(f)} |\nabla \overline{\omega_{(A)}}^{(f)}| \overline{\mathbf{v}}^{(f)} \tag{10.3.4-17}$$

and

$$\mathbf{J}_{(A)}^{\diamond(e)} = -\langle c \rangle^{(f)} A_{(A)}^{\diamond} \nabla \overline{x_{(A)}}^{(f)} + B_{(A)}^{\diamond} \langle c \rangle^{(f)} |\nabla \overline{x_{(A)}}^{(f)}| \overline{\mathbf{v}}^{\diamond(f)} \tag{10.3.4-18}$$

where $A_{(A)}$, $B_{(A)}$, $A_{(A)}^{\diamond}$, and $B_{(A)}^{\diamond}$ are functions of only $\Psi$.

## 10.3.5 When Fick's First Law Does Not Apply

Let $l_0$ represent a characteristic pore diameter of the structure, $\lambda$ the molecular mean free path (Hirschfelder et al. 1954, p. 10), and $N_{Kn} \equiv l_0/\lambda$ the Knudsen number. When

$$0.1 < N_{Kn} < 10 \tag{10.3.5-1}$$

Fick's first law cannot be used to describe binary diffusion in a porous medium (Scott 1962). This means that we must go back to Section 10.3.1 and prepare empirical data correlations for $\overline{\mathbf{n}_{(A)}}^{(f)}$ in (10.3.1-5). The only difficulty is that $\overline{\mathbf{n}_{(A)}}^{(f)}$ is not a frame-indifferent vector.

This suggests that we rewrite (10.3.1-5) in terms of the effective mass flux vector with respect to $\overline{\mathbf{v}}^{(f)}$:

$$\frac{\partial \overline{\rho_{(A)}}^{(f)}}{\partial t} + \operatorname{div}\left(\langle \rho_{(A)} \rangle^{(f)} \overline{\mathbf{v}}^{(f)}\right) = -\operatorname{div} \mathbf{j}_{(A)}^{(e)} + r_{(A)}'' + \overline{r_{(A)}}^{(f)} \tag{10.3.5-2}$$

$$\mathbf{j}_{(A)}^{(e)} \equiv \overline{\mathbf{j}_{(A)}}^{(f)} + \overline{\rho_{(A)}\mathbf{v}}^{(f)} - \langle \rho_{(A)} \rangle^{(f)} \overline{\mathbf{v}}^{(f)} \tag{10.3.5-3}$$

The same reasoning we used in Section 10.3.3 to prepare empirical correlations for $\boldsymbol{\delta}_{(A)}$ may be used here to formulate empirical correlations for $\mathbf{j}_{(A)}^{(e)}$. For example, for a nonoriented porous solid filled with a flowing fluid (see Section 10.3.3, Example 2) our initial guess might be that

$$\mathbf{j}_{(A)}^{(e)} = -\langle \rho \rangle^{(f)} \mathcal{D}_{(AB)} D_{(A1)}^{\star} \nabla \overline{\omega_{(A)}}^{(f)} + D_{(A2)}^{\star} \overline{\rho_{(A)}}^{(f)} \overline{\mathbf{v}}^{(f)} \tag{10.3.5-4}$$

in which

$$D_{(Ai)}^{\star} = D_{(Ai)}^{\star}\left(N_{Pe}, \frac{\overline{\mathbf{v}}^{(f)} \cdot \nabla \overline{\omega_{(A)}}^{(f)}}{|\overline{\mathbf{v}}^{(f)}| \, |\nabla \overline{\omega_{(A)}}^{(f)}|}, \frac{l_0 \langle \rho \rangle^{(f)} |\nabla \overline{\omega_{(A)}}^{(f)}|}{\overline{\rho_{(A)}}^{(f)}}, \Psi\right) \tag{10.3.5-5}$$

and

$$N_{Pe} \equiv \frac{l_0 |\overline{\mathbf{v}}^{(f)}|}{\mathcal{D}_{(AB)}} \tag{10.3.5-6}$$

This also could be thought of in terms of an effective diffusivity tensor $\mathbf{D}_{(AB)}^{(e)}$:

$$\mathbf{j}_{(A)}^{(e)} = -\langle \rho \rangle^{(f)} \mathbf{D}_{(AB)}^{(e)} \cdot \nabla \overline{\omega_{(A)}}^{(f)} \tag{10.3.5-7}$$

$$\mathbf{D}_{(AB)}^{(e)} \equiv \mathcal{D}_{(AB)} D_{(A1)}^{\star} \mathbf{I} - \frac{D_{(A2)}^{\star} \overline{\rho_{(A)}}^{(f)}}{\langle \rho \rangle^{(f)} \left(\overline{\mathbf{v}}^{(f)} \cdot \nabla \overline{\omega_{(A)}}^{(f)}\right)} \overline{\mathbf{v}}^{(f)} \, \overline{\mathbf{v}}^{(f)} \tag{10.3.5-8}$$

If we prefer to think in molar terms, we can introduce the effective molar flux vector with respect to $\overline{\mathbf{v}}^{\diamond(f)}$ in (10.3.1-6):

$$\frac{\partial \overline{c_{(A)}}^{(f)}}{\partial t} + \operatorname{div}\left(\langle c_{(A)} \rangle^{(f)} \overline{\mathbf{v}}^{\diamond(f)}\right) = -\operatorname{div} \mathbf{J}_{(A)}^{\diamond(e)} + \frac{r_{(A)}''}{M_{(A)}} + \frac{\overline{r_{(A)}}^{(f)}}{M_{(A)}} \tag{10.3.5-9}$$

$$\mathbf{J}_{(A)}^{\diamond(e)} \equiv \overline{\mathbf{J}_{(A)}^{\diamond}}^{(f)} + \overline{c_{(A)}\mathbf{v}^{\diamond}}^{(f)} - \langle c_{(A)} \rangle^{(f)} \overline{\mathbf{v}}^{\diamond(f)} \tag{10.3.5-10}$$

Again by analogy with Example 2 in Section 10.3.3, we would hypothesize that, for a nonoriented porous solid filled with a flowing fluid,

$$\mathbf{J}_{(A)}^{\star(e)} = -\langle c \rangle^{(f)} \mathcal{D}_{(AB)} D_{(Ai)}^{\diamond} \nabla \overline{x_{(A)}}^{(f)} + D_{(A2)}^{\diamond} \overline{c_{(A)}}^{(f)} \overline{\mathbf{v}}^{\diamond(f)} \tag{10.3.5-11}$$

$$D_{(Ai)}^{\diamond} = D_{(Ai)}^{\diamond}\left(N_{Pe}^{\diamond}, \frac{\overline{\mathbf{v}}^{\diamond(f)} \cdot \nabla \overline{x_{(A)}}^{(f)}}{|\overline{\mathbf{v}}^{\diamond(f)}| \, |\nabla \overline{x_{(A)}}^{(f)}|}, \frac{l_0 \langle c \rangle^{(f)} |\nabla \overline{x_{(A)}}^{(f)}|}{\overline{c_{(A)}}^{(f)}}, \Psi\right) \tag{10.3.5-12}$$

and

$$N_{Pe}^{\diamond} \equiv \frac{l_0|\overline{\mathbf{v}^{\diamond}}|}{\mathcal{D}_{(AB)}} \tag{10.3.5-13}$$

In terms of an effective diffusivity tensor $\mathbf{D}_{(AB)}^{\diamond(e)}$, we can say

$$\mathbf{J}_{(A)}^{\diamond(e)} = -\langle c \rangle^{(f)} \mathbf{D}_{(AB)}^{\diamond(e)} \cdot \nabla \overline{x_{(A)}}^{(f)} \tag{10.3.5-14}$$

where

$$\mathbf{D}_{(AB)}^{\diamond(e)} \equiv \mathcal{D}_{(AB)} D_{(A1)}^{\diamond} \mathbf{I} - \frac{D_{(A2)}^{\diamond} \overline{c_{(A)}}^{(f)}}{\langle c \rangle^{(f)} \left( \overline{\mathbf{v}}^{\diamond(f)} \cdot \nabla \overline{x_{(A)}}^{(f)} \right)} \overline{\mathbf{v}}^{\diamond(f)} \overline{\mathbf{v}}^{\diamond(f)} \tag{10.3.5-15}$$

Evans, Watson, and Mason (1961) visualize binary diffusion in a porous medium as being described by a ternary diffusion problem, the third species being the stationary porous structure. They refer to this as their "dusty" gas model. If we interpret their variables as being local volume averages, their result can be viewed as a special case of (10.3.5-11) with

$$D_{(A1)}^{\diamond} \equiv \frac{\Psi}{q} \left[ 1 + \frac{\overline{c}^{(s)}}{\overline{c}^{(f)}} \left( \frac{M_{(A)} + M_{(B)}}{M_{(B)}} \right)^{1/2} k_1 \right]^{-1} \tag{10.3.5-16}$$

and

$$D_{(A2)}^{\diamond} + 1 \equiv \left[ 1 + \frac{\overline{c}^{(s)}}{\overline{c}^f} \left( \frac{M_{(A)} + M_{(B)}}{M_{(B)}} \right)^{1/2} k_2 \right]^{-1} \tag{10.3.5-17}$$

Here $\overline{c}^{(s)}$ is the local volume-averaged molar density of solids; $q, k_1$, and $k_2$ are constants characteristic of the porous structure. A slight dependence of $k_1$ and $k_2$ upon the properties of the gas mixture is possible.

When $N_{Kn} < 10$, our local volume-averaged differential momentum balance is no longer applicable. The final form (Darcy's law or its equivalent; see Sections 4.3.4 through 4.3.6) depends upon a constitutive equation for the stress tensor that is not applicable when molecular collisions with the walls of the porous structure become as important as intermolecular collisions. Arguments based upon values of the pressure gradient deduced from Darcy's law are almost certainly not valid.

## 10.3.6 Knudsen Diffusion

When the Knudsen number $N_{Kn} < 0.1$, mass transfer in a porous structure is referred to as Knudsen diffusion (Scott 1962). If we think for the moment in terms of a molecular model, in Knudsen diffusion, collisions between the gas molecules and the walls of the porous structure are more important than collisions between two or more molecules. This suggests that, in a continuum description of Knudsen diffusion, the movement of each species should be independent of all other species present in the gas.

This goal of independence of movement of the various species present will be furthered if $\overline{\rho}^{(f)}$ and $\overline{\mathbf{v}}^{(f)}$ do not appear in the final form of the equation of continuity for any species $A$. Reasoning as we did in Sections 10.3.3 and 10.3.5, we can propose an empirical data correlation for $\mathbf{j}_{(A)}^{(e)}$ that satisfies these conditions.

Let us begin by postulating that

$$\mathbf{j}_{(A)}^{(e)} = \mathbf{j}_A^{(e)}\left(\overline{\rho_{(A)}}^{(f)}, \nabla\overline{\rho_{(A)}}^{(f)}, \overline{\mathbf{v}}^{(f)} - \overline{\mathbf{v}}^{(s)}, l_0, \Psi, R, T, M_{(A)}\right) \qquad (10.3.6\text{-}1)$$

where $R$ is the gas-law constant, $T$ is the temperature, and $M_{(A)}$ is the molecular weight for species $A$. The principle of frame indifference and the Buckingham–Pi theorem (Brand 1957) require for a stationary porous structure

$$\mathbf{j}_{(A)}^{(e)} = -\left(\frac{RT}{M_{(A)}}\right)^{1/2} l_0 D_{(A1)}^\star \nabla\overline{\rho_{(A)}}^{(f)} + D_{(A2)}^\star \overline{\rho_{(A)}}^{(f)} \overline{\mathbf{v}}^{(f)} \qquad (10.3.6\text{-}2)$$

Here

$$D_{(Ai)}^\star = D_{(Ai)}^\star \left(\frac{l_0|\nabla\overline{\rho_{(A)}}^{(f)}|}{\overline{\rho_{(A)}}^{(f)}}, \left(\frac{RT}{M_{(A)}}\right)^{1/2}\frac{1}{|\mathbf{v}|}, \Psi\right) \qquad (10.3.6\text{-}3)$$

In order that $\overline{\mathbf{v}}^{(f)}$ drop out of the final form of the differential mass balance, we take

$$D_{(A2)}^\star = -1 \qquad (10.3.6\text{-}4)$$

and

$$D_{(A1)}^\star = D_{(A1)}^\star \left(\frac{l_0|\nabla\overline{\rho_{(A)}}^{(f)}|}{\overline{\rho_{(A)}}^{(f)}}, \Psi\right) \qquad (10.3.6\text{-}5)$$

In terms of the differential mass balance for species $A$ in the form of (10.3.1-5),

$$\frac{\partial\overline{\rho_{(A)}}^{(f)}}{\partial t} + \operatorname{div}\overline{\mathbf{n}_{(A)}}^{(f)} = r_{(A)}'' + \overline{r_{(A)}}^{(f)} \qquad (10.3.6\text{-}6)$$

Equations (10.3.6-2) and (10.3.6-4) imply

$$\overline{\mathbf{n}_{(A)}}^{(f)} = -D_{(A)Kn}\nabla\overline{\rho_{(A)}}^{(f)} \qquad (10.3.6\text{-}7)$$

where

$$D_{(A)Kn} \equiv \left(\frac{RT}{M_{(A)}}\right)^{1/2} l_0 D_{(A1)}^\star \qquad (10.3.6\text{-}8)$$

is known as the *Knudsen diffusion coefficient*. By comparison, the Knudsen diffusion coefficient is usually said to have the form (Pollard and Present 1948; Carman 1956, p. 78; Evans, Watson, and Mason 1961; Satterfield and Sherwood 1963, p. 17)

$$D_{(A)Kn} = \frac{4}{3}\left(\frac{8RT}{\pi M_{(A)}}\right)^{1/2} l_0 K^\star \qquad (10.3.6\text{-}9)$$

in which the dimensionless coefficient $K^\star$ is characteristic of the porous medium.

Equations (10.3.6-6) and (10.3.6-7) are easily interpreted in molar terms as

$$\frac{\partial\overline{c_{(A)}}^{(f)}}{\partial t} + \operatorname{div}\overline{\mathbf{N}_{(A)}}^{(f)} = \frac{r_{(A)}''}{M_{(A)}} + \frac{\overline{r_{(A)}}^{(f)}}{M_{(A)}} \qquad (10.3.6\text{-}10)$$

and

$$\overline{\mathbf{N}_{(A)}}^{(f)} = -D_{(A)Kn}\nabla\overline{c_{(A)}}^{(f)} \qquad (10.3.6\text{-}11)$$

As I mentioned in concluding the preceding section, for $N_{Kn} < 10$, Darcy's law or its equivalent is no longer applicable. This means that Darcy's law cannot be used to make a statement about the pressure gradient in Knudsen diffusion.

### 10.3.7 The Local Volume Average of the Overall Differential Mass Balance

We have already seen in Chapter 9 that the overall differential mass balance is often very useful in solving mass-transfer problems. This motivates us to look at the local volume average of the overall differential mass balance.

We could directly take the local volume average of the overall differential mass balance in the two forms shown in Table 8.5.1-10. It is easier and completely equivalent to sum (10.3.1-5) and (10.3.1-6) over all species to conclude

$$\frac{\partial \overline{\rho}^{(f)}}{\partial t} + \operatorname{div}\left(\overline{\rho \mathbf{v}}^{(f)}\right) = 0 \tag{10.3.7-1}$$

$$\frac{\partial \overline{c}^{(f)}}{\partial t} + \operatorname{div}\left(\overline{c \mathbf{v}^{\diamond}}^{(f)}\right) = \sum_{A=1}^{N} \left( \frac{r''_{(A)}}{M_{(A)}} + \frac{\overline{r_{(A)}}^{(f)}}{M_{(A)}} \right) \tag{10.3.7-2}$$

We will hereafter refer to these equations as the local volume averages of the overall differential mass balance.

For an incompressible fluid, (10.3.7-1) simplifies considerably to

$$\operatorname{div} \overline{\mathbf{v}}^{(f)} = 0 \tag{10.3.7-3}$$

Incompressible fluids form one of the simplest classes of mass-transfer problems in porous media.

If we can assume that the molar density $c$ is a constant (an ideal gas at constant temperature and pressure), (10.3.7-2) reduces to

$$c \operatorname{div} \overline{\mathbf{v}^{\diamond}}^{(f)} = \sum_{A=1}^{N} \left( \frac{r''_{(A)}}{M_{(A)}} + \frac{\overline{r_{(A)}}^{(f)}}{M_{(A)}} \right) \tag{10.3.7-4}$$

If $c$ is a constant and if the number of moles produced by chemical reactions is exactly equal to the number of moles consumed in these reactions,

$$\sum_{A=1}^{N} \left( \frac{r''_{(A)}}{M_{(A)}} + \frac{\overline{r_{(A)}}^{(f)}}{M_{(A)}} \right) = 0 \tag{10.3.7-5}$$

Equation (10.3.7-2) becomes

$$\operatorname{div} \overline{\mathbf{v}^{\diamond}}^{(f)} = 0 \tag{10.3.7-6}$$

From a mathematical point of view, this class of mass-transfer problems is just as simple as those for incompressible fluids.

### 10.3.8 The Effectiveness Factor for Spherical Catalyst Particles

A catalytic reaction ($A \rightarrow B$) takes place in the gas phase in either a fixed-bed or fluidized reactor. We shall assume that the catalyst is uniformly distributed throughout each of the porous spherical particles of radius $R$ with which the reactor is filled. We wish to focus our attention here upon one of these porous spherical catalyst particles.

We can anticipate that more of the chemical reaction takes place on the catalyst surface in the immediate vicinity of the surface of the sphere than on the catalyst surface distributed around the center of the sphere. This seems obvious when we look at the comparable diffusion paths. What I would like to do here is to examine the overall effectiveness of the catalyst surface in a porous spherical particle. Let us begin by asking about the rate at which species $A$ is consumed by a first-order chemical reaction in the particle:

$$\frac{r''_{(A)}}{M_{(A)}} = -k''_1 a \langle c_{(A)} \rangle^{(f)} \tag{10.3.8-1}$$

Here, $a$ denotes the available catalytic surface area per unit volume.

Since we are dealing with a catalytic reaction,

$$\sum_{C=1}^{N} \frac{\overline{r_{(C)}}^{(f)}}{M_{(C)}} = 0 \tag{10.3.8-2}$$

We can further say that

$$\sum_{C=1}^{N} \frac{r_{(C)}}{M_{(C)}} = 0 \tag{10.3.8-3}$$

since one mole of $A$ is consumed for every mole of $B$ produced. Because we are dealing with a gas, we will idealize the problem to the extent of assuming that the overall molar density $c$ is a constant. Consequently, the local volume average of the overall equation of continuity reduces to

$$\operatorname{div} \overline{\mathbf{v}^\diamond}^{(f)} = 0 \tag{10.3.8-4}$$

It seems reasonable to begin this problem by assuming in spherical coordinates that

$$\overline{v_r^\diamond}^{(f)} = \overline{v_r^\diamond}^{(f)}(r)$$
$$\overline{v_\theta^\diamond}^{(f)} = \overline{v_\varphi^\diamond}^{(f)}$$
$$= 0 \tag{10.3.8-5}$$

and

$$\overline{c_{(A)}}^{(f)} = \overline{c_{(A)}}^{(f)}(r) \tag{10.3.8-6}$$

In view of (10.3.8-5), (10.3.8-4) requires

$$\frac{d}{dr}\left(r^2 \overline{v_r^\diamond}^{(f)}\right) = 0 \tag{10.3.8-7}$$

or

$$\overline{v_r^\diamond}^{(f)} = 0 \tag{10.3.8-8}$$

since we must require $\overline{v_r^\diamond}^{(f)}$ to be finite at the center of the sphere.

Let us assume that the gas in this porous catalyst particle is so dense that Fick's first law applies. For simplicity, we shall assume that $\mathbf{J}_{(A)}^{\diamond(e)}$ can be represented by (10.3.4-18). In view of Equations (10.3.8-5), (10.3.8-6), and (10.3.8-8), there is only one nonzero component of this vector:

$$J_{(A)r}^{\diamond(e)} = -A_{(A)}^{\diamond} \frac{\partial \overline{c_{(A)}}^{(f)}}{\partial r}$$

$$J_{(A)\theta}^{\diamond(e)} = J_{(A)\varphi}^{\diamond(e)} \qquad\qquad (10.3.8\text{-}9)$$

$$= 0$$

Recognizing (10.3.8-1) and (10.3.8-9), we can express the local volume average of the differential mass balance for species $A$ in the form of (10.3.2-10) as

$$\frac{1}{r^2}\frac{d}{dr}\left(r^2 \frac{d\langle c_{(A)}\rangle^{(f)}}{dr}\right) = \frac{k_1'' a}{A_{(A)}^{\diamond}\Psi}\langle c_{(A)}\rangle^{(f)} \qquad\qquad (10.3.8\text{-}10)$$

This differential equation is to be solved consistent with the boundary condition

$$\text{at } r = R + \epsilon : \ \langle c_{(A)}\rangle^{(f)} = c_{(A)0} \qquad\qquad (10.3.8\text{-}11)$$

Here, $\epsilon$ is the diameter of the averaging surface $S$. We shall generally be willing to say that $\epsilon \ll R$.

It is convenient to introduce as dimensionless variables

$$c^{\star} \equiv \frac{\langle c_{(A)}\rangle^{(f)}}{c_{(A)0}}$$

$$r^{\star} \equiv \frac{r}{R+\epsilon} \qquad\qquad (10.3.8\text{-}12)$$

This allows us to write (10.3.8-10) and (10.3.8-11) as

$$\frac{1}{r^{\star 2}}\frac{d}{dr^{\star}}\left(r^{\star 2}\frac{dc^{\star}}{dr^{\star}}\right) = 9\Lambda^2 c^{\star} \qquad\qquad (10.3.8\text{-}13)$$

and

$$\text{at } r^{\star} = 1 : \ c^{\star} = 1 \qquad\qquad (10.3.8\text{-}14)$$

For convenience in comparing the results to be obtained here with those for other particle shapes, we have defined (Aris 1957)

$$\Lambda \equiv \left(\frac{k_1'' a}{A_{(A)}^{\diamond}\Psi}\right)^{1/2}\frac{V_{\mathrm{p}}}{A_{\mathrm{p}}} \qquad\qquad (10.3.8\text{-}15)$$

where $V_{\mathrm{p}}$ and $A_{\mathrm{p}}$ are the volume and area of the bounding surface of the catalyst particle. For a spherical catalyst particle such as we have here,

$$\Lambda \equiv \left(\frac{k_1'' a}{A_{(A)}^{\diamond}\Psi}\right)^{1/2}\frac{R+\epsilon}{3} \qquad\qquad (10.3.8\text{-}16)$$

If we introduce as a change variable

$$u \equiv r^{\star}c^{\star} \qquad\qquad (10.3.8\text{-}17)$$

Equation (10.3.8-13) becomes

$$\frac{d^2u}{dr^{*2}} = 9\Lambda^2 u \tag{10.3.8-18}$$

This can be solved consistent with the conditions that

$$\text{at } r^* = 1 : \quad u = 1 \tag{10.3.8-19}$$

and

$$\text{at } r^* = 0 : \quad u = 0 \tag{10.3.8-20}$$

to find

$$c^* \equiv \frac{\langle c_{(A)} \rangle^{(f)}}{c_{(A)0}}$$

$$= \frac{1}{r^*} \frac{\sinh(3\Lambda r^*)}{\sinh(3\Lambda)} \tag{10.3.8-21}$$

Given this concentration distribution with the catalyst particle, we can calculate the rate at which moles of species $A$ are consumed by chemical reaction:

$$\mathcal{W}_{(A)} \equiv - \int_0^{2\pi} \int_0^\pi N_{(A)r}\big|_{r^*=1} (R+\epsilon)^2 \sin\theta \, d\theta \, d\varphi$$

$$= -4\pi (R+\epsilon)^2 \overline{N_{(A)r}}^{(f)}\bigg|_{r^*=1}$$

$$= -4\pi (R+\epsilon)^2 J_{(A)r}^{\diamond(e)}\bigg|_{r^*=1}$$

$$= 4\pi (R+\epsilon) A_{(A)}^\diamond \Psi c_{(A)0} \frac{dc^*}{dr^*}\bigg|_{r^*=1}$$

$$= 4\pi (R+\epsilon) A_{(A)}^\diamond \Psi c_{(A)0}[3\Lambda \coth(3\Lambda) - 1] \tag{10.3.8-22}$$

In the first line, we have taken advantage of our discussion of integrals of volume-averaged variables in Section 4.3.7. If all the catalytic surface were exposed to fresh fluid, the molar rate of consumption of species $A$ would be

$$\mathcal{W}_{(A)0} = \frac{4}{3}\pi R^3 a k_1'' c_{(A)0} \tag{10.3.8-23}$$

The *effectiveness factor* $\eta$ is defined as

$$\eta \equiv \frac{\mathcal{W}_{(A)}}{\mathcal{W}_{(A)0}} \tag{10.3.8-24}$$

From (10.3.8-22) and (10.3.8-23), it is apparent that the effectiveness factor for spherical catalyst particles is

$$\eta = \left(\frac{R+\epsilon}{R}\right)^3 \frac{1}{3\Lambda^2}[3\Lambda \coth(3\Lambda) - 1] \tag{10.3.8-25}$$

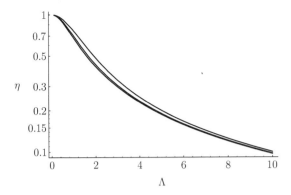

**Figure 10.3.8-1.** Effectiveness factors for porous solid catalysts. Top curve, flat plates (sealed edges); middle curve, cylinders (sealed ends); bottom curve, spheres.

or

$$\eta = \frac{1}{3\Lambda^2}[3\Lambda \coth(3\Lambda) - 1] \tag{10.3.8-26}$$

since we are generally willing to assume

$$\frac{R + \epsilon}{R} \doteq 1 \tag{10.3.8-27}$$

Figure 10.3.8-1 compares (10.3.8-26) for spheres with the analogous expressions for flat plates (Exercise 10.3.8-1) and cylinders (Exercise 10.3.8-2). From a practical point of view, we are fortunate that the effectiveness factor is nearly independent of the macroscopic particle shape.

**Exercise 10.3.8-1** *The effectiveness factor for a flat plate*   Repeat the discussion in the text for a first-order catalytic reaction $A \to B$ taking place in a flat plate (with sealed edges) of thickness $2b$. Conclude that the effectiveness factor is

$$\eta = \frac{1}{\Lambda} \tanh \Lambda$$

where

$$\Lambda^2 = \frac{k_1'' a(b + \epsilon)^2}{A_{(A)}^\circ \Psi}$$

**Exercise 10.3.8-2** *The effectiveness factor for cylinders*   Repeat the analysis in the text for a first-order catalytic reaction $A \to B$ that takes place in a cylindrical catalyst particle (with sealed ends). Determine that the effectiveness factor is given by

$$\eta = \frac{1}{\Lambda} \frac{I_1(2\Lambda)}{I_0(2\Lambda)}$$

where

$$\Lambda^2 = \frac{k_1'' a(R + \epsilon)^2}{4 A_{(A)}^\circ \Psi}$$

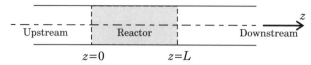

**Figure 10.3.8-2.** Tubular reactor with connecting upstream and downstream sections.

By $I_n(x)$, we mean the modified Bessel function of the first kind (Irving and Mullineux 1959, p. 143).

**Exercise 10.3.8-3** *More on the effectiveness factor for spheres*    Again consider the problem described in the text, but this time assume that the reaction is zero order:

$$\text{for } c_{(A)} > 0 : \quad \frac{r''_{(A)}}{M_{(A)}} = -k''_0 a$$

What is the effectiveness factor?

**Exercise 10.3.8-4** *First-order catalytic reactor*    A catalytic reaction $A \rightarrow B$ is carried out by passing a liquid through a tubular reactor of length $L$ that is packed with catalyst pellets. We wish to determine the volume-averaged mass density of species $A$ as a function of position in the reactor, assuming that species $A$ is consumed by a first-order chemical reaction

$$r''_{(A)} = -k''_1 a \langle \rho_{(A)} \rangle^{(f)}$$

and assuming that the mass density of species $A$ has a uniform value $\rho_{(A)0}$ very far upstream from the entrance to the reactor. Neglect any effects attributable to the development of the velocity profile at the entrance to the reactor.

i) Wehner and Wilhelm (1956) suggest that a tubular reactor should be analyzed together with its connecting upstream and downstream sections, as illustrated in Figure 10.3.8-2. Conclude that, for the open tube upstream from the reactor,

$$\frac{\partial \rho^{\star}_{(A)}}{\partial z^{\star}} = \frac{1}{N_{PeU}} \frac{\partial^2 \rho^{\star}_{(A)}}{\partial z^{\star 2}} \tag{10.3.8-28}$$

for the reactor itself,

$$\frac{\partial \langle \rho^{\star}_{(A)} \rangle^{(f)}}{\partial z^{\star}} = \frac{1}{N_{Pe}} \frac{\partial^2 \langle \rho^{\star}_{(A)} \rangle^{(f)}}{\partial z^{\star 2}} - \frac{N_{Da}}{N_{Pe}} \langle \rho^{\star}_{(A)} \rangle^{(f)} \tag{10.3.8-29}$$

and for the open tube downstream from the reactor,

$$\frac{\partial \rho^{\star}_{(A)}}{\partial z^{\star}} = \frac{1}{N_{PeD}} \frac{\partial^2 \rho^{\star}_{(A)}}{\partial z^{\star 2}} \tag{10.3.8-30}$$

Here

$$N_{PeU} \equiv \frac{L\overline{v_z}^{(f)}}{\mathcal{D}_{(AB)U}}$$

$$N_{Pe} \equiv \frac{\Psi^{-1} L \overline{v_z}^{(f)}}{A_{(A)} + B_{(A)} \overline{v_z}^{(f)}}$$

$$N_{Pe\,D} \equiv \frac{L\bar{v}_z^{(f)}}{\mathcal{D}_{(AB)D}}$$

$$N_{Da} \equiv \frac{\Psi^{-1}k_1''aL^2}{A_{(A)} + B_{(A)}\bar{v}_z^{(f)}}$$

$$\rho_{(A)}^\star \equiv \frac{\rho_{(A)}}{\rho_{(A)0}}$$

$$z^\star \equiv \frac{z}{L}$$

What assumptions have been made in the upstream and downstream sections?

ii) What are the boundary conditions that must be satisfied by this system of equations at the entrance to the reactor, at the exit from the reactor, very far upstream, and very far downstream?

iii) Solve (10.3.8-28) through (10.3.8-30) individually and evaluate the six constants of integration to find the following concentration distribution [Wehner and Wilhelm (1956); see also Exercise 10.2.2-1]:

$$\text{for } z^\star \leq 0 : \quad \frac{1 - \rho_{(A)}^\star}{1 - \rho_{(A)}^\star(0)} = \exp(N_{Pe\,U}z^\star)$$

$$\text{for } 0 \leq z^\star \leq 1 : \quad \langle\rho_{(A)}^\star\rangle^{(f)} = g_0 \exp\left(\frac{N_{Pe}z^\star}{2}\right)$$

$$\times \left\{(1+b)\exp\left(\frac{bN_{Pe}}{2}[1-z^\star]\right) - (1-b)\exp\left(\frac{bN_{Pe}}{2}[z^\star - 1]\right)\right\}$$

$$\text{for } z^\star \geq s1 : \quad \rho_{(A)}^\star = 2bg_0 \exp\left(\frac{N_{Pe}}{2}\right)$$

Here

$$\rho_{(A)}^\star(0) \equiv g_0\left[(1+b)\exp\left(\frac{bN_{Pe}}{2}\right) - (1-b)\exp\left(-\frac{bN_{Pe}}{2}\right)\right]$$

$$g_0 \equiv 2\left[(1+b)^2\exp\left(\frac{bN_{Pe}}{2}\right) - (1-b)^2\exp\left(-\frac{bN_{Pe}}{2}\right)\right]^{-1}$$

$$b \equiv \left[1 + \frac{4N_{Da}}{N_{Pe}^2}\right]^{1/2}$$

## 10.4 Still More on Integral Balances

In the sections that follow, we have two purposes. First, we have one integral balance left to discuss: the integral mass balance for an individual species in a multicomponent mixture. Second, and just as important, we must extend our previous discussions of integral balances to multicomponent systems.

By *multicomponent systems*, I mean systems in which concentration is a function of time or position. If a system consists of more than one species, but concentration is independent of both time and position, the previously developed integral balances apply without change.

The sections that follow are closely related to Sections 4.4 and 7.4. It might be helpful for the reader to review these sections or at least to reread the introductions to these sections, which discuss the place of integral balances in engineering.

There is one point concerning the notation about which the reader should exercise a degree of caution. The entrance and exit surfaces $S_{(\text{ent ex})}$ are to be interpreted in the broadest possible sense to include both

1) surfaces that are unobstructed for flow and across which the individual species are carried primarily by convection and
2) phase interfaces (liquid–liquid, liquid–solid, . . .) across which the individual species are carried primarily by diffusion. We shall refer to these last as the *diffusion surfaces* $S_{(\text{diff})}$.

## 10.4.1 The Integral Mass Balance for Species $A$

Just as in Section 4.4.1 where we developed a mass balance for a system consisting of a single species, we are in the position to develop a mass balance for each individual species present in a multicomponent system.

Let us take the same approach that we have used in developing integral balances for single-component systems. The differential mass balance for species $A$ from Table 8.5.1-5 may be integrated over the system to obtain

$$\int_{R^{(s)}} \left( \frac{\partial \rho_{(A)}}{\partial t} + \text{div } \mathbf{n}_{(A)} - r_{(A)} \right) dV = 0 \tag{10.4.1-1}$$

The first integral on the left can be evaluated using the generalized transport theorem of Section 1.3.2:

$$\frac{d}{dt} \int_{R^{(s)}} \rho_{(A)} \, dV = \int_{R^{(s)}} \frac{\partial \rho_{(A)}}{\partial t} \, dV + \int_{S^{(s)}} \rho_{(A)} \left( \mathbf{v}^{(s)} \cdot \mathbf{n} \right) dA \tag{10.4.1-2}$$

Green's transformation may be used to express the second term as

$$\int_{R^{(s)}} \text{div } \mathbf{n}_{(A)} \, dV = \int_{S^{(s)}} \rho_{(A)} \mathbf{v}_{(A)} \cdot \mathbf{n} \, dA \tag{10.4.1-3}$$

Equations (10.4.1-2) and (10.4.1-3) allow us to express (10.4.1-1) as

$$\frac{d}{dt} \int_{R^{(s)}} \rho_{(A)} \, dV = \int_{S^{(s)}} \rho_{(A)} \left( \mathbf{v}_{(A)} - \mathbf{v}^{(s)} \right) \cdot (-\mathbf{n}) \, dA + \int_{R^{(s)}} r_{(A)} \, dV \tag{10.4.1-4}$$

or

$$\frac{d}{dt} \int_{R^{(s)}} \rho_{(A)} \, dV = \int_{S_{(\text{ent ex})}} \rho_{(A)} \left( \mathbf{v}_{(A)} - \mathbf{v}^{(s)} \right) \cdot (-\mathbf{n}) \, dA + \int_{R^{(s)}} r_{(A)} \, dV \tag{10.4.1-5}$$

Equation (10.4.1-5) is a *general form of the integral mass balance for species $A$* appropriate to single-phase systems.

We will generally find it convenient to account for the effects of diffusion explicitly and write (10.4.1-5) as

$$\frac{d}{dt} \int_{R^{(s)}} \rho_{(A)}\, dV = \int_{S_{(\text{ent ex})}} \rho_{(A)}\left(\mathbf{v} - \mathbf{v}^{(s)}\right) \cdot (-\mathbf{n})\, dA$$

$$+ \int_{S_{(\text{ent ex})}} \mathbf{j}_{(A)} \cdot (-\mathbf{n})\, dA + \int_{R^{(s)}} r_{(A)}\, dV$$

$$= \int_{S_{(\text{ent ex})}} \rho_{(A)}\left(\mathbf{v} - \mathbf{v}^{(s)}\right) \cdot (-\mathbf{n})\, dA + \mathcal{J}_{(A)} + \int_{R^{(s)}} r_{(A)}\, dV \qquad (10.4.1\text{-}6)$$

In words, (10.4.1-6) says that the time rate of change of the mass of species $A$ in the system is equal to the net rate at which the mass of species $A$ is brought into the system by convection, the net rate at which the mass of species $A$ diffuses into the system (relative to the mass-averaged velocity):

$$\mathcal{J}_{(A)} \equiv \int_{S_{(\text{diff})}} \mathbf{j}_{(A)} \cdot (-\mathbf{n})\, dA$$

$$\doteq \int_{S_{(\text{ent ex})}} \mathbf{j}_{(A)} \cdot (-\mathbf{n})\, dA \qquad (10.4.1\text{-}7)$$

and the rate at which the mass of species $A$ is produced in the system by homogeneous chemical reactions. Notice that $\mathcal{J}_{(A)}$ includes the rate of production of species $A$ at the surfaces within or bounding the system either by catalytic reactions or desorption. The surfaces $S_{(\text{diff})}$ generally represent a subset of $S_{(\text{ent ex})}$, since we will almost always be willing to neglect diffusion with respect to convection on those portions of $S_{(\text{ent ex})}$ unobstructed to flow. We will refer to (10.4.1-6) as the *integral mass balance for species A* appropriate to a single-phase system.

As we pointed out in Section 4.4.1, we are more commonly concerned with multiphase systems. Using the approach and notation of Section 4.4.1 and assuming only that we may neglect diffusion with respect to convection on those portions of $S_{(\text{ent ex})}$ unobstructed for flow, we find that the *integral mass balance for species A* appropriate to a multiphase system is

$$\frac{d}{dt} \int_{R^{(s)}} \rho_{(A)}\, dV = \int_{(\text{ent ex})} \rho_{(A)}\left(\mathbf{v} - \mathbf{v}^{(s)}\right) \cdot (-\mathbf{n})\, dA + \mathcal{J}_{(A)} + \int_{R^{(s)}} r_{(A)}\, dV$$

$$+ \int_{\Sigma} \left[\rho_{(A)}\left(\mathbf{v}_{(A)} - \mathbf{u}\right) \cdot \boldsymbol{\xi}\right] dA \qquad (10.4.1\text{-}8)$$

Given the jump mass balance of Section 8.2.1, Equation (10.4.1-8) reduces to (10.4.1-6), and (10.4.1-6) applies equally well to single-phase and multiphase systems.

There are three common types of problems in which the integral mass balance for species $A$ is applied: The rate of diffusion $\mathcal{J}_{(A)}$ may be neglected, it may be the unknown and thus to be determined, or it may be known from previous experimental data. In this last case, one employs an empirical correlation of data for $\mathcal{J}_{(A)}$. In Section 10.4.2, we discuss the form that these empirical correlations should take.

**Exercise 10.4.1-1** *The integral mass balance for species A appropriate to turbulent flows* I recommend following the discussion in Section 4.4.2 in developing the integral mass balance for species $A$ appropriate to turbulent flows.

i) Show that, for single-phase or multiphase systems that do not involve fluid–fluid phase interfaces, we can repeat the derivation of Section 10.4.1 to find

$$\frac{d}{dt} \int_{R^{(s)}} \overline{\rho_{(A)}} \, dV = \int_{S_{(\text{ent ex})}} \left( \overline{\rho_{(A)} \mathbf{v}} - \overline{\rho_{(A)}} \mathbf{v}^{(s)} \right) \cdot (-\mathbf{n}) \, dA + \mathcal{J}_{(A)} + \int_{R^{(s)}} \overline{r_{(A)}} \, dV$$
$$+ \int_{\Sigma} \left[ \left( \overline{\rho_{(A)} \mathbf{v}_{(A)}} - \overline{\rho_{(A)}} \mathbf{u} \right) \cdot \boldsymbol{\xi} \right] dA$$

The time-averaged jump mass balance of Exercise 10.4.1-2 simplifies this to

$$\frac{d}{dt} \int_{R^{(s)}} \overline{\rho_{(A)}} \, dV = \int_{S_{(\text{ent ex})}} \left( \overline{\rho_{(A)} \mathbf{v}} - \overline{\rho_{(A)}} \mathbf{v}^{(s)} \right) \cdot (-\mathbf{n}) \, dA + \mathcal{J}_{(A)} + \int_{R^{(s)}} \overline{r_{(A)}} \, dV$$

Note that, in arriving at these results, we have again neglected diffusion of species $A$ with respect to convection on those portions of $S_{(\text{ent ex})}$ unobstructed for flow.

ii) For single-phase or multiphase systems that include one or more fluid–fluid interfaces, I recommend time averaging the integral mass balance of Section 10.4.1.

**Exercise 10.4.1-2** *Time-averaged jump mass balance for species A*    Determine that the time-averaged jump mass balance for species $A$ applicable to solid–fluid phase interfaces that bound turbulent flows is identical to the balance found in Section 8.2.1:

$$\left[ \left( \overline{\rho_{(A)} \mathbf{v}_{(A)}} - \overline{\rho_{(A)}}^{(f)} \mathbf{u} \right) \cdot \boldsymbol{\xi} \right] = \left[ \overline{\rho_{(A)}} \left( \mathbf{v}_{(A)} - \mathbf{u} \right) \cdot \boldsymbol{\xi} \right]$$
$$= 0$$

## 10.4.2 Empirical Correlations for $\mathcal{J}_{(A)}$

Empirical data correlations for $\mathcal{J}_{(A)}$ ($\overline{\mathcal{J}_{(A)}}$ when dealing with turbulent flows) are prepared in much the same way as our empirical correlations for $Q$, discussed in Section 7.4.2. There are three principal thoughts to be kept in mind.

1) The rate of diffusion of species $A$ from the permeable or catalytic surfaces of the system is frame indifferent:

$$\mathcal{J}_{(A)}^* \equiv \int_{S_{(\text{diff})}} \mathbf{j}_{(A)}^* \cdot (-\mathbf{n}^*) \, dA$$
$$= \int_{S_{(\text{diff})}} \mathbf{j}_{(A)} \cdot (-\mathbf{n}) \, dA$$
$$= \mathcal{J}_{(A)} \tag{10.4.2-1}$$

2) We assume that the principle of frame indifference, introduced in Section 2.3.1, applies to any empirical correlation developed for $\mathcal{J}_{(A)}$.

3) The form of any expression for $\mathcal{J}_{(A)}$ must satisfy the Buckingham–Pi theorem (Brand 1957).

We illustrate the approach in terms of a specific situation.

## Example: Forced Convection in Plane Flow of a Binary Fluid
## Past a Cylindrical Body

An infinitely long cylindrical body is submerged in a large mass of a binary Newtonian fluid. We assume that the surface of the body is in equilibrium with the fluid at the surface and that the mass fraction of species $A$ at the surface is a constant $\omega_{(A)0}$. Outside the immediate neighborhood of the body, the mass fraction of species $A$ has a nearly uniform value $\omega_{(A)\infty}$. In a frame of reference that is fixed with respect to the earth, the cylindrical body translates without rotation at a constant velocity $\mathbf{v}_0$; the fluid at a very large distance from the body moves with a uniform velocity $\mathbf{v}_\infty$. The vectors $\mathbf{v}_0$ and $\mathbf{v}_\infty$ are normal to the axis of the cylinder, so that we may expect that the fluid moves in a plane flow. One unit vector $\boldsymbol{\alpha}$ is sufficient to describe the orientation of the cylinder with respect to $\mathbf{v}_0$ and $\mathbf{v}_\infty$.

It seems reasonable to assume that $\mathcal{J}_{(A)}$ should be a function of

$$\Delta\omega_{(A)} \equiv \omega_{(A)0} - \omega_{(A)\infty} \tag{10.4.2-2}$$

a characteristic fluid density $\rho$, a characteristic fluid viscosity $\mu$, a characteristic diffusion coefficient $\mathcal{D}_{(AB)}$, a length $L$ that is characteristic of the cylinder's cross section, $\mathbf{v}_\infty - \mathbf{v}_0$, and $\boldsymbol{\alpha}$:

$$\mathcal{J}_{(A)} = f\left(\rho, \mu, \mathcal{D}_{(AB)}, L, \mathbf{v}_\infty - \mathbf{v}_0, \boldsymbol{\alpha}, \Delta\omega_{(A)}\right) \tag{10.4.2-3}$$

We recognize that density, viscosity, and the diffusion coefficient may be dependent upon position as the result of their functional dependence upon composition. In referring to $\rho$, $\mu$, and $\mathcal{D}_{(AB)}$ as characteristic of the fluid, we mean that they are to be evaluated at some average or representative composition. Dependence upon

$$\Delta\omega_{(B)} \equiv \omega_{(B)0} - \omega_{(B)\infty} \tag{10.4.2-4}$$

is not included, since

$$\Delta\omega_{(B)} = -\Delta\omega_{(A)} \tag{10.4.2-5}$$

The same argument that we used in discussing Example 1 of Section 7.4.2 may be repeated here to show that the principle of frame indifference and the Buckingham–Pi theorem require that this be of a form[1]

$$N_{Nu(A)} = N_{Nu(A)}\left(N_{Re}, N_{Sc}, \Delta\omega_{(A)}, \frac{\mathbf{v}_\infty - \mathbf{v}_0}{|\mathbf{v}_\infty - \mathbf{v}_0|} \cdot \boldsymbol{\alpha}\right) \tag{10.4.2-6}$$

where the Nusselt, Reynolds, and Schmidt numbers are defined as

$$N_{Nu(A)} \equiv \frac{\mathcal{J}_{(A)}}{\rho\mathcal{D}_{(AB)}L\,\Delta\omega_{(A)}}$$

$$N_{Re} \equiv \frac{L\rho|\mathbf{v}_\infty - \mathbf{v}_0|}{\mu} \tag{10.4.2-7}$$

$$N_{Sc} \equiv \frac{\mu}{\rho\mathcal{D}_{(AB)}}$$

---

[1]  We have anticipated our definition of the mass-transfer coefficient in (10.4.2-8) by our definition of the Nusselt number. The Buckingham–Pi theorem suggests $\mathcal{J}_{(A)}/\rho\mathcal{D}_{(AB)}L$ as a dimensionless group.

We follow Bird et al. (1960, p. 640) in defining a mass-transfer coefficient $k_{(A)\omega}$ as

$$k_{(A)\omega} \equiv \frac{\mathcal{J}_{(A)}}{A\,\Delta\omega_{(A)}} \tag{10.4.2-8}$$

where $A$ is proportional to $L^2$ and denotes the area available for mass transfer.[2] The Nusselt number for species $A$ is usually expressed in terms of this mass-transfer coefficient:

$$N_{Nu(A)} = \frac{k_{(A)\omega}L}{\rho\mathcal{D}_{(AB)}} \tag{10.4.2-9}$$

One computes the rate of diffusion of species $A$ across the permeable surfaces of the system as

$$\mathcal{J}_{(A)} = k_{(A)\omega}A\,\Delta\omega_{(A)} \tag{10.4.2-10}$$

estimating the mass-transfer coefficient $k_{(A)\omega}$ from an empirical data correlation of the form of (10.4.2-6).

---

[2] Equation (10.4.2-8) can easily be rewritten as

$$k_{(A)\omega} = \frac{\mathcal{N}_{(A)} - \omega_{(A)0}\left(\mathcal{N}_{(A)} + \mathcal{N}_{(B)}\right)}{A\,\Delta\omega_{(A)}} \tag{10.4.2-8a}$$

where

$$\mathcal{N}_{(A)} \equiv \int_{S_{(\mathrm{diff})}} \mathbf{n}_{(A)} \cdot (-\mathbf{n})\, dA \tag{10.4.2-8b}$$

and $\omega_{(A)0}$ is the mass fraction of species $A$ at $S_{(\mathrm{diff})}$, assumed to be a constant.

The mass-transfer coefficient defined by (10.4.2-8) differs from that widely used in the literature prior to 1960. The traditional definition is suggested by writing the integral mass balance of Section 10.4.1 as

$$\frac{d}{dt}\int_{R^{(s)}} \rho_{(A)}\,dV = \int_{S_{(\mathrm{ent\ ex})} - S_{(\mathrm{diff})}} \rho_{(A)}\left(\mathbf{v} - \mathbf{v}^{(s)}\right) \cdot (-\mathbf{n})\, dA$$
$$+ \mathcal{W}_{(A)} + \int_{R^{(s)}} r_{(A)}\,dV \tag{10.4.2-8c}$$

where

$$\mathcal{W}_{(A)} \equiv \int_{S_{(\mathrm{diff})}} \rho_{(A)}\left(\mathbf{v}_{(A)} - \mathbf{v}^{(s)}\right) \cdot (-\mathbf{n})\, dA \tag{10.4.2-8d}$$

Equation (10.4.2-8c) again incorporates the assumption that diffusion can be neglected with respect to convection on those portions of $S_{(\mathrm{ent\ ex})}$ unobstructed for flow: $S_{(\mathrm{ent\ ex})} - S_{(\mathrm{diff})}$. The traditional mass-transfer coefficient $K_{(A)\rho}$ is defined as

$$K_{(A)\rho} \equiv \frac{\mathcal{W}_{(A)}}{A\,\Delta\rho_{(A)}}$$

The advantage of working in terms of $\mathcal{W}_{(A)}$ and the traditional mass-transfer coefficient $K_{(A)\rho}$ is that the contribution of convection on $S_{(\mathrm{diff})}$ is automatically taken into account. The disadvantage is that $K_{(A)\rho}$ shows a more complicated dependence upon concentration and mass-transfer rates than does $k_{(A)\omega}$ (Bird et al. 1960, p. 640). In our opinion, this loss outweighs the possible gain in computational ease.

Most empirical correlations for $N_{Nu(A)}$ are not as general as (10.4.2-6) suggests. Most studies are for a single orientation of a body (or a set of bodies such as an array of particles) with respect to the fluid stream. Further, for sufficiently small rates of mass transfer, diffusion-induced convection is not important, and $\Delta\omega_{(A)}$ is so small that its influence can be neglected. Under these conditions, (10.4.2-6) assumes a simpler form (Bird et al. 1960, p. 647):

$$N_{Nu(A)} = N_{Nu(A)}(N_{Re}, N_{Sc}) \tag{10.4.2-11}$$

When chemical reactions and diffusion-induced convection can be neglected (as well as a few other things), there is a strict analogy between energy and mass transfer (see Section 9.2). Since there are more data for energy transfer available in the literature, it is often convenient to identify (10.4.2-11) with the analogous relation in energy transfer.

When diffusion-induced convection is important (larger rates of mass transfer), the dependence of $N_{Nu(A)}$ upon $\Delta\omega_{(A)}$ in (10.4.2-6) cannot be neglected. Because of a shortage of experimental data, the recommended approach at the present time is to derive a simple correction to be applied to empirical correlations of the form of (10.4.2-11) that are restricted to small rates of mass transfer. See Section 9.2.1 as well as the excellent discussion given by Bird et al. (1960, p. 658).

### 10.4.3 The Integral Overall Balances

The derivation of the integral mass balance for a single-component system given in Section 4.4.1 may be repeated almost line for line for a multicomponent system. The only change necessary is that the differential mass balance of Section 1.3.3 must be replaced by the overall differential mass balance of Section 8.3.1. Two forms of the *integral overall mass balance* are found, corresponding to the two forms of the overall differential mass balance given in Table 8.5.1-10:

$$\frac{d}{dt}\int_{R^{(s)}} \rho \, dV = \int_{S_{(ent\ ex)}} \rho\left(\mathbf{v} - \mathbf{v}^{(s)}\right)\cdot(-\mathbf{n})\,dA \tag{10.4.3-1}$$

and

$$\frac{d}{dt}\int_{R^{(s)}} c \, dV = \int_{S_{(ent\ ex)}} c\left(\mathbf{v}^\star - \mathbf{v}^{(s)}\right)\cdot(-\mathbf{n})\,dA + \int_{R^{(s)}} \sum_{A=1}^{N} \frac{r_{(A)}}{M_{(A)}} dV \tag{10.4.3-2}$$

The only assumption made in deriving these results is that the jump overall mass balances of Section 8.3.1 and Exercise 8.3.1-1 are assumed to apply at the phase interface.

The overall momentum, mechanical energy, and moment-of-momentum balances take exactly the same form as those derived for single-component systems in Section 4.4. In the derivations, it is necessary only to replace the differential momentum balance for single-component materials derived in Section 2.2.3 with the overall differential momentum balance in Section 8.3.2. This means that we must interpret $\mathbf{v}$ as the mass-averaged velocity vector and $\mathbf{f}$ as the mass-averaged external force vector.

However, it is necessary to modify the further discussion of the mechanical energy balance in Section 7.4.3. All the results of Tables 7.4.6-1 through 7.4.3-3 are equally applicable to single-component and multicomponent systems, with the exceptions of those for isothermal and isentropic systems. Results comparable to those given there for isothermal and isentropic systems can be prepared, but they are not presented because of their complexity.

The derivation of the integral energy balance given in Section 7.4.1 for single-component systems can be repeated here for multicomponent systems, replacing only the differential energy balance of Section 5.1.3 with the overall differential energy balance of Section 8.3.4. Because of the form of the caloric equation of state for a multicomponent material, not all the results of Tables 7.4.2-1 through 7.4.2-3 carry over immediately to multicomponent systems. In fact, some of the comparable results for multicomponent systems are sufficiently complex to be of marginal usefulness and are not given. For this reason, I thought it might be helpful to restate in Tables 10.4.3-1 to 10.4.3-3 those forms of the integral overall energy balance that are more likely to be useful.

It is necessary only to substitute the differential entropy inequalities of Section 8.3.5 for those of Section 5.2.3 to obtain the following two forms of the integral overall entropy inequality for multiphase systems:

$$\frac{d}{dt} \int_{R^{(s)}} \rho \hat{S} \, dV \geq \int_{S_{(\text{ent ex})}} \rho \hat{S} \left( \mathbf{v} - \mathbf{v}^{(s)} \right) \cdot (-\mathbf{n}) \, dA$$

$$+ \int_{S^{(s)}} \frac{1}{T} \left( \mathbf{q} - \sum_{A=1}^{N} \mu_{(A)} \mathbf{j}_{(A)} \right) \cdot (-\mathbf{n}) \, dA + \int_{R^{(s)}} \rho \frac{Q}{T} \, dV \qquad (10.4.3\text{-}3)$$

and

$$\int_{R^{(s)}} \left\{ -\frac{1}{T^2} \mathbf{q} \cdot \nabla T + \frac{1}{T} \text{tr} \left( \mathbf{S} \cdot \nabla \mathbf{v} \right) - \frac{1}{T} \sum_{A=1}^{N} \mathbf{j}_{(A)} \cdot \left[ T \nabla \left( \frac{\mu_{(A)}}{T} \right) - \mathbf{f}_{(A)} \right] \right.$$

$$\left. - \sum_{A=1}^{N} \frac{1}{T} \mu_{(A)} r_{(A)} \right\} dV \geq 0 \qquad (10.4.3\text{-}4)$$

In arriving at (10.4.3-3), we have assumed that the jump entropy inequality of Section 8.3.5 is valid for all phase interfaces involved. Of the two forms, (10.4.3-3) is probably the more important. Often we will be willing to neglect $\sum_{A=1}^{N} \mu_{(A)} \mathbf{j}_{(A)}$ with respect to $\mathbf{q}$ in the second term on the right, in which case (10.4.3-3) takes on the same form as the result for single-component systems in Section 7.4.4. (The reader is again reminded of the caution issued in the introduction: $S_{(\text{ent ex})}$ includes phase interfaces across which the individual species are carried primarily by diffusion.)

### Exercise 10.4.3-1 *Some additional forms of the entropy inequality*

i) Let us consider a system bounded by fixed, impermeable walls; there are no entrances and exits. The system may consist of any number of phases. Temperature is assumed to be independent of both time and position. Determine that

$$\frac{d}{dt} \int_{R^{(s)}} \rho \left( \hat{A} + \frac{1}{2} v^2 + \varphi \right) dV - \int_{R^{(s)}} \sum_{B=1}^{N} \mathbf{j}_{(B)} \cdot \mathbf{f}_{(B)} \, dV \leq 0$$

We assume here that

$$\mathbf{f} \equiv \sum_{B=1}^{N} \omega_{(B)} \mathbf{f}_{(B)}$$

$$= -\nabla \varphi$$

where $\varphi$ is not an explicit function of time. Discuss under what conditions Helmholtz free energy is minimized at equilibrium for a system of the type described.

**Table 10.4.3-1.** General forms of the integral energy balance applicable to a single-phase system

$$\frac{d}{dt} \int_{R^{(s)}} \rho \left( \hat{U} + \frac{1}{2} v^2 + \varphi + \frac{p_0}{\rho} \right) dV$$

$$= \int_{S_{(\text{ent ex})}} \rho \left( \hat{H} + \frac{1}{2} v^2 + \varphi \right) \left( \mathbf{v} - \mathbf{v}^{(s)} \right) \cdot (-\mathbf{n}) \, dA$$

$$+ \mathcal{Q} - \mathcal{W} + \int_{R^{(s)}} \left( \sum_{A=1}^{N} \mathbf{j}_{(A)} \cdot \mathbf{f}_{(A)} + \rho Q \right) dV$$

$$+ \int_{S_{(\text{ent ex})}} \left[ -(P - p_0) \left( \mathbf{v}^{(s)} \cdot \mathbf{n} \right) + \mathbf{v} \cdot (\mathbf{S} \cdot \mathbf{n}) \right] dA^a$$

$$\frac{d}{dt} \int_{R^{(s)}} \rho \left( \hat{U} + \frac{1}{2} v^2 + \frac{p_0}{\rho} \right) dV$$

$$= \int_{S_{(\text{ent ex})}} \rho \left( \hat{H} + \frac{1}{2} v^2 \right) \left( \mathbf{v} - \mathbf{v}^{(s)} \right) \cdot (-\mathbf{n}) \, dA$$

$$+ \mathcal{Q} - \mathcal{W} + \int_{R^{(s)}} \left( \sum_{A=1}^{N} \mathbf{n}_{(A)} \cdot \mathbf{f}_{(A)} + \rho Q \right) dV$$

$$+ \int_{S_{(\text{ent ex})}} \left[ -(P - p_0) \left( \mathbf{v}^{(s)} \cdot \mathbf{n} \right) + \mathbf{v} \cdot (\mathbf{S} \cdot \mathbf{n}) \right] dA$$

$$\frac{d}{dt} \int_{R^{(s)}} \rho \left( \hat{U} + \frac{p_0}{\rho} \right) dV$$

$$= \int_{S_{(\text{ent ex})}} \rho \left( \hat{U} + \frac{p_0}{\rho} \right) \left( \mathbf{v} - \mathbf{v}^{(s)} \right) \cdot (-\mathbf{n}) \, dA + \mathcal{Q}$$

$$+ \int_{R^{(s)}} \left[ -(P - p_0) \operatorname{div} \mathbf{v} + \operatorname{tr}(\mathbf{S} \cdot \nabla \mathbf{v}) + \sum_{A=1}^{N} \mathbf{n}_{(A)} \cdot \mathbf{f}_{(A)} + \rho Q \right] dV$$

*For an incompressible fluid:*

$$\frac{d}{dt} \int_{R^{(s)}} \rho \hat{U} \, dV = \int_{S_{(\text{ent ex})}} \rho \hat{U} \left( \mathbf{v} - \mathbf{v}^{(s)} \right) \cdot (-\mathbf{n}) \, dA + \mathcal{Q}$$

$$+ \int_{R^{(s)}} \left[ \operatorname{tr}(\mathbf{S} \cdot \nabla \mathbf{v}) + \sum_{A=1}^{N} \mathbf{j}_{(A)} \cdot \mathbf{f}_{(A)} + \rho Q \right] dV$$

*For an isobaric fluid:*

$$\frac{d}{dt} \int_{R^{(s)}} \rho \hat{H} \, dV = \int_{S_{(\text{ent ex})}} \rho \hat{H} \left( \mathbf{v} - \mathbf{v}^{(s)} \right) \cdot (-\mathbf{n}) \, dA + \mathcal{Q}$$

$$+ \int_{R^{(s)}} \left[ \operatorname{tr}(\mathbf{S} \cdot \nabla \mathbf{v}) + \sum_{A=1}^{N} \mathbf{j}_{(A)} \cdot \mathbf{f}_{(A)} + \rho Q \right] dV$$

[a] We assume $\mathbf{f} = \sum_{A=1}^{N} \omega_{(A)} \mathbf{f}_{(A)} = -\nabla \varphi$, where $\varphi$ is not an explicit function of time.

**Table 10.4.3-2.** General forms of the integral energy balance applicable to a multiphase system where the jump energy balance (5.1.3-9) applies

$$\frac{d}{dt} \int_{R^{(s)}} \rho \left( \hat{U} + \frac{1}{2}v^2 + \varphi + \frac{p_0}{\rho} \right) dV$$

$$= \int_{S_{(\text{ent ex})}} \rho \left( \hat{H} + \frac{1}{2}v^2 + \varphi \right) \left( \mathbf{v} - \mathbf{v}^{(s)} \right) \cdot (-\mathbf{n}) \, dA$$

$$+ \mathcal{Q} - \mathcal{W} + \int_{R^{(s)}} \left( \sum_{A=1}^{N} \mathbf{j}_{(A)} \cdot \mathbf{f}_{(A)} + \rho Q \right) dV$$

$$+ \int_{S_{(\text{ent ex})}} \left[ -(P - p_0) \left( \mathbf{v}^{(s)} \cdot \mathbf{n} \right) + \mathbf{v} \cdot (\mathbf{S} \cdot \mathbf{n}) \right] dA + \int_{\Sigma} \left[ \rho \varphi (\mathbf{v} - \mathbf{u}) \cdot \boldsymbol{\xi} \right] dA^a$$

$$\frac{d}{dt} \int_{R^{(s)}} \rho \left( \hat{U} + \frac{1}{2}v^2 + \frac{p_0}{\rho} \right) dV$$

$$= \int_{S_{(\text{ent ex})}} \rho \left( \hat{H} + \frac{1}{2}v^2 \right) \left( \mathbf{v} - \mathbf{v}^{(s)} \right) \cdot (-\mathbf{n}) \, dA$$

$$+ \mathcal{Q} - \mathcal{W} + \int_{R^{(s)}} \left( \sum_{A=1}^{N} \mathbf{n}_{(A)} \cdot \mathbf{f}_{(A)} + \rho Q \right) dV$$

$$+ \int_{S_{(\text{ent ex})}} \left[ -(P - p_0) \left( \mathbf{v}^{(s)} \cdot \mathbf{n} \right) + \mathbf{v} \cdot (\mathbf{S} \cdot \mathbf{n}) \right] dA$$

$$\frac{d}{dt} \int_{R^{(s)}} \rho \left( \hat{U} + \frac{p_0}{\rho} \right) dV$$

$$= \int_{S_{(\text{ent ex})}} \rho \left( \hat{U} + \frac{p_0}{\rho} \right) \left( \mathbf{v} - \mathbf{v}^{(s)} \right) \cdot (-\mathbf{n}) \, dA + \mathcal{Q}$$

$$+ \int_{R^{(s)}} \left[ -(P - p_0) \operatorname{div} \mathbf{v} + \operatorname{tr} (\mathbf{S} \cdot \nabla \mathbf{v}) + \sum_{A=1}^{N} \mathbf{j}_{(A)} \cdot \mathbf{f}_{(A)} + \rho Q \right] dV$$

$$+ \int_{\Sigma} \left[ \rho \left( \hat{U} + \frac{p_0}{\rho} \right) (\mathbf{v} - \mathbf{u}) \cdot \boldsymbol{\xi} + \mathbf{q} \cdot \boldsymbol{\xi} \right] dA$$

*For incompressible fluids:*

$$\frac{d}{dt} \int_{R^{(s)}} \rho \hat{U} \, dV = \int_{S_{(\text{ent ex})}} \rho \hat{U} \left( \mathbf{v} - \mathbf{v}^{(s)} \right) \cdot (-\mathbf{n}) \, dA + \mathcal{Q}$$

$$+ \int_{R^{(s)}} \left[ \operatorname{tr}(\mathbf{S} \cdot \nabla \mathbf{v}) + \sum_{A=1}^{N} \mathbf{j}_{(A)} \cdot \mathbf{f}_{(A)} + \rho Q \right] dV$$

$$+ \int_{\Sigma} \left[ \rho \hat{U} (\mathbf{v} - \mathbf{u}) \cdot \boldsymbol{\xi} + \mathbf{q} \cdot \boldsymbol{\xi} \right] dA$$

*For an isobaric system:*

$$\frac{d}{dt} \int_{R^{(s)}} \rho \hat{H} \, dV = \int_{S_{(ent\ ex)}} \rho \hat{H} \left(\mathbf{v} - \mathbf{v}^{(s)}\right) \cdot (-\mathbf{n}) \, dA + \mathcal{Q}$$

$$+ \int_{R^{(s)}} \left[ \text{tr}(\mathbf{S} \cdot \nabla \mathbf{v}) + \sum_{A=1}^{N} \mathbf{j}_{(A)} \cdot \mathbf{f}_{(A)} + \rho Q \right] dV$$

$$+ \int_{\Sigma} \left[ \rho \hat{H}(\mathbf{v} - \mathbf{u}) \cdot \boldsymbol{\xi} + \mathbf{q} \cdot \boldsymbol{\xi} \right] dA$$

[a] We assume $\mathbf{f} = \sum_{A=1}^{N} \omega_{(A)} \mathbf{f}_{(A)} = -\nabla \varphi$, where $\varphi$ is not an explicit function of time.

**Table 10.4.3-3.** Restricted forms of the integral energy balance applicable to a multiphase system. These forms are applicable in the context of assumptions 1 through 6 in the text.

$$\frac{d}{dt} \int_{R^{(s)}} \rho \left( \hat{U} + \frac{1}{2} v^2 + \varphi + \frac{p_0}{\rho} \right) dV$$

$$= \int_{S_{(ent\ ex)}} \rho \left( \hat{H} + \frac{1}{2} v^2 + \varphi \right) (-\mathbf{v} \cdot \mathbf{n}) \, dA + \mathcal{Q} - \mathcal{W} + \int_{R^{(s)}} \sum_{A=1}^{N} \mathbf{j}_{(A)} \cdot \mathbf{f}_{(A)} \, dV^a$$

$$\frac{d}{dt} \int_{R^{(s)}} \rho \left( \hat{U} + \frac{1}{2} v^2 + \frac{p_0}{\rho} \right) dV$$

$$= \int_{S_{(ent\ ex)}} \rho \left( \hat{H} + \frac{1}{2} v^2 \right) (-\mathbf{v} \cdot \mathbf{n}) \, dA + \mathcal{Q} - \mathcal{W} + \int_{R^{(s)}} \sum_{A=1}^{N} \mathbf{j}_{(A)} \cdot \mathbf{f}_{(A)} \, dV$$

$$\frac{d}{dt} \int_{R^{(s)}} \rho \left( \hat{U} + \frac{p_0}{\rho} \right) dV$$

$$= \int_{S_{(ent\ ex)}} \rho \left( \hat{U} + \frac{p_0}{\rho} \right) (-\mathbf{v} \cdot \mathbf{n}) \, dA + \mathcal{Q}$$

$$+ \int_{R^{(s)}} \left[ -(P - p_0) \, \text{div} \, \mathbf{v} + \text{tr}(\mathbf{S} \cdot \nabla \mathbf{v}) + \sum_{A=1}^{N} \mathbf{j}_{(A)} \cdot \mathbf{f}_{(A)} \right] dV$$

*For incompressible fluids:*

$$\frac{d}{dt} \int_{R^{(s)}} \rho \hat{U} \, dV = \int_{S_{(ent\ ex)}} \rho \hat{U}(-\mathbf{v} \cdot \mathbf{n}) \, dA + \mathcal{Q} + \int_{R^{(s)}} \left[ \text{tr}(\mathbf{S} \cdot \nabla \mathbf{v}) + \sum_{A=1}^{N} \mathbf{j}_{(A)} \cdot \mathbf{f}_{(A)} \right] dV$$

*For an isobaric system:*

$$\frac{d}{dt} \int_{R^{(s)}} \rho \hat{H} \, dV = \int_{S_{(ent\ ex)}} \rho \hat{H}(-\mathbf{v} \cdot \mathbf{n}) \, dA + \mathcal{Q} + \int_{R^{(s)}} \left[ \text{tr}(\mathbf{S} \cdot \nabla \mathbf{v}) + \sum_{A=1}^{N} \mathbf{j}_{(A)} \cdot \mathbf{f}_{(A)} \right] dV$$

[a] We assume $\mathbf{f} = \sum_{A=1}^{N} \omega_{(A)} \mathbf{f}_{(A)} = -\nabla \varphi$, where $\varphi$ is not an explicit function of time.

ii) Consider the same system as above but require in addition that pressure be independent of time and position. Determine that

$$\frac{d}{dt} \int_{R^{(s)}} \rho \left( \hat{G} + \frac{1}{2}v^2 + \varphi \right) dV - \int_{R^{(s)}} \sum_{B=1}^{N} \mathbf{j}_{(B)} \cdot \mathbf{f}_{(B)} \, dV \leq 0$$

Discuss under what conditions Gibbs free energy is minimized at equilibrium for a system of the type described.

## 10.4.4 Example

This example is taken from Bird et al. (1960, p. 707).

A fluid stream containing a waste material $A$ at concentration $\rho_{(A)0}$ is to be discharged into a river at a constant volume rate of flow $Q$. Material $A$ is unstable and decomposes at a rate proportional to its mass density:

$$-r_{(A)} = k_1 \rho_{(A)} \tag{10.4.4-1}$$

To reduce pollution, the stream is to pass through a holding tank of volume $V$ before it is discharged into the river. At time $t = 0$, the fluid begins to flow into the empty tank, which may be considered to have a perfect stirrer. No liquid leaves the tank until it is filled. We wish to develop an expression for the mass density of $A$ in the tank and in the effluent from the tank as a function of time.

This problem should be considered in two parts. First, we must determine the mass density of $A$ in the tank as a function of time during the filling process. The mass density of $A$ in the tank at the moment the tank becomes filled forms the boundary condition for the second portion of the problem: the mass density of $A$ in the tank and in the effluent stream as a function of time.

Figure 10.4.4-1 schematically describes the situation during the filling process. Let us choose our system to be the fluid in the tank. For this system, which has one entrance and no exit, the integral mass balance of Section 10.4.1 requires

$$\frac{d}{dt} \int_{R^{(s)}} \rho_{(A)} \, dV = \int_{S_{(\text{ent})}} \rho_{(A)0} \mathbf{v} \cdot (-\mathbf{n}) \, dA - k_1 \int_{R^{(s)}} \rho_{(A)} \, dV \tag{10.4.4-2}$$

or

$$\frac{d\mathcal{M}_{(A)}}{dt} = \rho_{(A)0}Q - k_1 \mathcal{M}_{(A)} \tag{10.4.4-3}$$

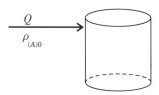

**Figure 10.4.4-1.** Waste tank during filling.

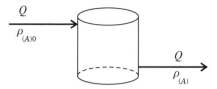

**Figure 10.4.4-2.** Waste tank after filling.

where we denote the mass of species $A$ in the system by

$$\mathcal{M}_{(A)} \equiv \int_{R^{(s)}} \rho_{(A)} \, dV \tag{10.4.4-4}$$

Equation (10.4.4-3) is easily integrated to find

$$\mathcal{M}_{(A)} = \frac{\rho_{(A)0} Q}{k_1} [1 - \exp(-k_1 t)] \tag{10.4.4-5}$$

This means that, when the tank is filled,

$$\text{at } t = \frac{V}{Q} : \quad \rho_{(A)} = \rho_{(A)f} \equiv \frac{\rho_{(A)0} Q}{V k_1} \left[ 1 - \exp\left( -\frac{k_1 V}{Q} \right) \right] \tag{10.4.4-6}$$

Once the waste tank is filled, the discharge line is opened as shown in Figure 10.4.4-2. Our system is still the fluid in the tank, but now we have both an entrance and an exit. The integral mass balance for species $A$ requires

$$\frac{d\mathcal{M}_{(A)}}{dt} = \rho_{(A)0} Q - \frac{\mathcal{M}_{(A)}}{V} Q - k_1 \mathcal{M}_{(A)} \tag{10.4.4-7}$$

This can be integrated using (10.4.4-6) as the boundary condition to find

$$\frac{\rho_{(A)} - \rho_{(A)\infty}}{\rho_{(A)f} - \rho_{(A)\infty}} = \exp\left( -\left[ \frac{Q}{V} + k_1 \right] \left[ t - \frac{V}{Q} \right] \right) \tag{10.4.4-8}$$

Here $\rho_{(A)\infty}$ is the steady-state mass density of species $A$ in the waste tank:

$$\text{as } t \to \infty : \quad \rho_{(A)} \to \rho_{(A)\infty} \equiv \frac{\rho_{(A)0} Q}{Q + k_1 V} \tag{10.4.4-9}$$

**Exercise 10.4.4-1** *Irreversible first-order reaction in a continuous reactor* (Bird et al. 1960, p. 707)   A solution of species $A$ at mass density $\rho_{(A)0}$ initially fills a well-stirred reactor of volume $V$. At time $t = 0$, an identical solution of $A$ is introduced at a constant volume rate of flow $Q$. At the same time, a constant stream of dissolved catalyst is introduced, causing $A$ to disappear according to the expression

$$-r_{(A)} = k_1 \rho_{(A)}$$

where the constant $k_1$ may be assumed to be independent of composition and time. Determine that the mass density of species $A$ in the reactor at any time is given by

$$\frac{\rho_{(A)} - \rho_{(A)\infty}}{\rho_{(A)\infty}} = \frac{k_1 V}{Q} \exp\left[ -\left( \frac{Q}{V} + k_1 \right) t \right]$$

in which

$$\text{as } t \to \infty: \ \rho_{(A)} \to \rho_{(A)\infty} \equiv \frac{\rho_{(A)0}Q}{Q + k_1 V}$$

**Exercise 10.4.4-2** *Irreversible second-order reaction in a continuous reactor* (Bird et al. 1960, p. 708) Repeat Exercise 10.4.4-1 assuming that species $A$ disappears according to the expression

$$-r_{(A)} = k_2 \rho_{(A)}{}^2$$

*Answer:*

$$\rho_{(A)} = -\frac{B}{2k_2}$$

$$+ \left\{ \frac{k_2 V}{BV - Q} + \left( \frac{2k_2}{B + 2k_2\rho_{(A)0}} - \frac{k_2 V}{BV - Q} \right) \exp\left( -\left[ B - \frac{Q}{V} \right]t \right) \right\}^{-1}$$

in which

$$B \equiv \frac{Q}{V}\left( 1 + \sqrt{1 + \frac{4k_2 V \rho_{(A)0}}{Q}} \right)$$

*Hint:* The differential equation to be solved can be put into the form of a Bernoulli equation with the change of variable

$$u \equiv \rho_{(A)} + \frac{Q}{2k_2 V}\left( 1 + \sqrt{1 + \frac{4k_2 V \rho_{(A)0}}{Q}} \right)$$

The resulting Bernoulli equation can in turn be integrated by making another change of variable

$$v \equiv u^{-1}$$

**Exercise 10.4.4-3** *Start-up of a chemical reactor* (Bird et al. 1960, pp. 701 and 708)   Species $B$ is to be formed by a reversible reaction from a raw material $A$ in a chemical reactor of volume $V$ equipped with a perfect stirrer. Unfortunately, $B$ undergoes an irreversible first-order decomposition to a third species $C$. All reactions may be considered to be first order. We use the notation

$$\begin{array}{cc} k_{1B} & k_{1C} \\ A \rightleftharpoons B & \longrightarrow C \\ k_{1A} & \end{array}$$

At time $t = 0$, a solution of $A$ at mass density $\rho_{(A)0}$ that is free of species $B$ is introduced into the initially empty reactor at a constant volume rate of flow $Q$.

i) Determine that during the filling period the mass of species $B$ in the reactor is the following function of time

$$\mathcal{M}_{(B)} = \frac{\rho_{(A)0}Q}{k_{1C}}\left[ 1 + \frac{s_-}{(s_+ - s_-)}\exp(s_+ t) - \frac{s_+}{(s_+ - s_-)}\exp(s_- t) \right]$$

where

$$2s_\pm \equiv -(k_{1A} + k_{1B} + k_{1C}) \pm \sqrt{(k_{1A} + k_{1B} + k_{1C})^2 - 4k_{1B}k_{1C}}$$

*Hint:* Take Laplace transforms of the integral mass balances for species $A$ and $B$.

ii) Prove that $s_+$ and $s_-$ are always real and negative.

*Hint:* Start by showing that

$$(k_{1A} + k_{1B} + k_{1C})^2 - 4k_{1B}k_{1C} = (k_{1A} - k_{1B} + k_{1C})^2 + 4k_{1A}k_{1B}$$

**Exercise 10.4.4-4** *Continuous-flow stirred-tank reactors*   Two successive first-order irreversible reactions $(A \rightarrow B \rightarrow C)$ are to be carried out in a series of continuous-flow, stirred-tank reactors. Derive an expression from which we may find the number of reactors required to give a maximum concentration of $B$ in the product. All reactors are at the same temperature and have the same holding time.

Assuming that $k_1 = k_2$ and that the initial concentrations of $B$ and $C$ in the feed are zero, what is this number when $k_1 = k_2 = 0.1 \text{ h}^{-1}$ and the holding time per tank is 1 h?

# Tensor Analysis

TENSOR ANALYSIS is the language in terms of which continuum mechanisms can be presented in the simplest and most physically meaningful fashion. For this reason, I suggest that those readers who are not already familiar with this subject should read at least a portion of this appendix before starting with the main text.

The degree to which tensor analysis must be mastered depends upon your aims. We have written this appendix with three types of people in mind.

Many first-year graduate students in engineering are anxious to get to interesting applications as quickly as possible. We suggest that they read only those sections marked with double asterisks. They should also understand those exercises marked with double asterisks. Not all of this need be done before embarking on Chapter 1. Sometimes it is helpful to alternate between Chapter 1 and this appendix.

Those students who are somewhat more curious about the foundations of continuum mechanics will want to read the unmarked sections as well as those marked with two asterisks. The unmarked sections not only allow you to be more critical in your reading but are required for a complete understanding of the transport theorem in Section 1.3.2.

The complete Appendix A is recommended for anyone who wishes to do serious research in any of the subareas of continuum mechanics. The single-asterisked sections are required to derive the forms of various results in curvilinear coordinate systems. Without these sections, the curvilinear forms presented in the tables at the end of Chapter 2 cannot be derived; the basis for the discussion of boundary layers on curved walls in Sections 3.5.3 and 3.5.6 cannot be checked; and one is handicapped in working with new descriptions of material behavior or with out-of-the-ordinary coordinate systems.

For those readers who are not sure into which category they fall, we suggest that you begin with the double-asterisked sections. As your interest in the subject grows, it is easy to turn back and read a little more.

## A.1  Spatial Vectors

### A.1.1  Definition**

We visualize that the real world occupies the space $E$ studied in elementary geometry. To each point of $E$ there corresponds a place in the universe.

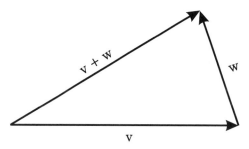

**Figure A.1.1-1.** The parallelogram rule.

Corresponding to each pair $(a, b)$ of the points of $E$ taken in order, there is a directed line segment denoted by $\vec{ab}$. Each directed line segment $\vec{ab}$ is characterized by a length $|ab|$ and direction (with the exception of the zero vector, which has zero length and an arbitrary direction).

Let us define the set of *spatial* vectors to be composed of the set of all directed line segments, with the understanding that two directed line segments that differ only by a parallel displacement represent the same element of the set. We define three operations for this set: addition, multiplication by real scalars, and inner product.

**Addition**    The sum $\mathbf{v} + \mathbf{w}$ of two spatial vectors $\mathbf{v}$ and $\mathbf{w}$ is defined by the familiar parallelogram rule as indicated in Figure A.1.1-1. This sum satisfies the following rules:

$(A_1)$   $\mathbf{v} + \mathbf{w} = \mathbf{w} + \mathbf{v}$
$(A_2)$   $\mathbf{u} + (\mathbf{v} + \mathbf{w}) = (\mathbf{u} + \mathbf{v}) + \mathbf{w}$
$(A_3)$   $\mathbf{v} + \mathbf{0} = \mathbf{v}$. Here $\mathbf{0}$ is the zero spatial vector, which should be regarded as having zero length and arbitrary direction.
$(A_4)$   Given any spatial vector $\mathbf{v}$, there is another spatial vector denoted by $-\mathbf{v}$ such that $\mathbf{v} + (-\mathbf{v}) = \mathbf{0}$.

**Scalar multiplication**    Let $\alpha$ be a real number (scalar) and $\mathbf{v}$ be a spatial vector. We define the spatial vector $\alpha\mathbf{v}$ to have a length $|\alpha|\,|\mathbf{v}|$; the direction of $\alpha\mathbf{v}$ is defined to be the same as that of $\mathbf{v}$ if $\alpha > 0$ and the opposite direction if $\alpha < 0$. Scalar multiplication must satisfy the following rules:

$(M_1)$   $\alpha(\beta\mathbf{v}) = (\alpha\beta)\mathbf{v}$.
$(M_2)$   $1\mathbf{v} = \mathbf{v}$
$(M_3)$   $\alpha(\mathbf{v} + \mathbf{w}) = \alpha\mathbf{v} + \alpha\mathbf{w}$
$(M_4)$   $(\alpha + \beta)\mathbf{v} = \alpha\mathbf{v} + \beta\mathbf{v}$

**Inner product**    The inner product $\mathbf{v} \cdot \mathbf{w}$ of two spatial vectors $\mathbf{v}$ and $\mathbf{w}$ is a real number obtained by multiplying the length of $\mathbf{v}$, the length of $\mathbf{w}$, and the cosine of the angle between the directions $\mathbf{v}$ and $\mathbf{w}$. Alternatively, the inner product of two spatial vectors must satisfy:

$(I_1)$   $\mathbf{v} \cdot \mathbf{w} = \mathbf{w} \cdot \mathbf{v}$
$(I_2)$   $\mathbf{u} \cdot (\mathbf{v} + \mathbf{w}) = \mathbf{u} \cdot \mathbf{v} + \mathbf{u} \cdot \mathbf{w}$
$(I_3)$   $\alpha(\mathbf{v} \cdot \mathbf{w}) = (\alpha\mathbf{v}) \cdot \mathbf{w}$
$(I_4)$   $\mathbf{v} \cdot \mathbf{v} \geq 0$; $\mathbf{v} \cdot \mathbf{v} = 0$, if and only if $\mathbf{v} = \mathbf{0}$

More generally, if for any set of objects we define *addition* and *scalar multiplication* in such a way that the rules $(A_1)$ to $(A_4)$ and $(M_1)$ to $(M_4)$ hold, we define the set to be a *vector space*, and we refer to the elements of the set as *vectors*. If for any vector space an *inner product* that satisfies rules $(I_1)$ to $(I_4)$ is introduced, we refer to the vector space as an *inner product space*. By definition, the set of spatial vectors is an inner product space.

For spatial vectors, we adopt the following abbreviations:

$$\mathbf{v} - \mathbf{w} \equiv \mathbf{v} + (-\mathbf{w}) \tag{A.1.1-1}$$

$$v \equiv |\mathbf{v}| \equiv \sqrt{\mathbf{v} \cdot \mathbf{v}} \tag{A.1.1-2}$$

The nonnegative number $v$ (or $|\mathbf{v}|$) is the *magnitude* or length of the vector $\mathbf{v}$.

**Exercise A.1.1-1** ** Consider the set of all real numbers. If we understand $x + y$ and $\alpha x$ to be the ordinary numerical addition and multiplication, prove that this set constitutes a vector space.

**Exercise A.1.1-2** ** Consider the set $R^n$ of all $n$-tuples of real numbers. If $x = (\xi_1, \ldots, \xi_n)$ and $y = (\eta_1, \ldots, \eta_n)$ are elements of $R^n$, we define

$$x + y = (\xi_1 + \eta_1, \ldots, \xi_n + \eta_n)$$

$$\alpha x = (\alpha \xi_1, \ldots, \alpha \xi_n)$$

$$0 = (0, \ldots, 0)$$

$$-x = (-\xi_1, \ldots, -\xi_n)$$

$$x \cdot y = (\xi_1 \eta_1 + \xi_2 \eta_2 + \cdots + \xi_n \eta_n)$$

Prove that $R^n$ is an inner product space.

## A.1.2  Position Vectors

Any point $z$ in $E$ may be located with respect to another point $O$ by means of the spatial vector $\mathbf{z} \equiv \overrightarrow{Oz}$. It is common practice to refer to $\mathbf{z}$ as the *position vector* of the point $z$ with respect to the *origin O*.

A particular point in $E$ having been designated as the origin $O$, the set of all position vectors, which locate points in $E$ with respect to the origin $O$, is identical to the set of all spatial vectors.

When speaking in general, it rarely makes a difference whether one speaks about *the point z* or *the point whose position vector relative to the origin O is* $\mathbf{z}$. However, in computations we always express locations in terms of position vectors, because we are able to sense only relative locations.

## A.1.3  Spatial Vector Fields

Temperature, concentration, and pressure are examples of real, numerically valued functions of position. We refer to any real numerically valued function of position as a *real scalar field*.

When we think of water flowing through a pipe or in a river, we recognize that the velocity of the water is a function of position. At the wall of the pipe, the velocity of the water is zero; at the center, it is a maximum. The velocity of the water in the pipe is an example of spatial vector-valued function of position. We shall term any spatial vector-valued function a *spatial vector field*.

As another example, consider the *position vector field* $\mathbf{p}(z)$. It maps every point $z$ of $E$ into the corresponding position vector $\mathbf{z}$ measured with respect to a previously chosen origin $O$:

$$\mathbf{z} = \mathbf{p}(z) \tag{A.1.3-1}$$

With the following definitions for addition, scalar multiplication, and inner product, the set of all spatial vector fields becomes an inner product space.

**Addition**    If $\mathbf{v}$ and $\mathbf{w}$ are two spatial vector fields, we define the spatial vector field $\mathbf{v} + \mathbf{w}$ such that at every point $z$ of $E$

$$(\mathbf{v} + \mathbf{w})(z) \equiv \mathbf{v}(z) + \mathbf{w}(z)$$

The addition on the right is that defined for spatial vectors. It is to be understood here that $\mathbf{v} + \mathbf{w}$, $\mathbf{v}$, and $\mathbf{w}$ indicate functions; $(\mathbf{v} + \mathbf{w})(z)$, $\mathbf{v}(z)$, and $\mathbf{w}(z)$ denote the values of these functions at the point $z$.

**Scalar multiplication**    If $\alpha$ is a real scalar field (a real numerically valued function of position; see Halmos (1958, p. 1)) and $\mathbf{v}$ is a spatial vector field, we define the spatial vector field $\alpha\mathbf{v}$ such that at every point $z$, $(\alpha\mathbf{v})(z) \equiv \alpha(z)\mathbf{v}(z)$. The scalar multiplication on the right is that defined for spatial vectors.

**Inner product**    If $\mathbf{v}$ and $\mathbf{w}$ are two spatial vector fields, we define the real scalar field $\mathbf{v} \cdot \mathbf{w}$ such that at every point $z$, $(\mathbf{v} \cdot \mathbf{w})(z) \equiv \mathbf{v}(z) \cdot \mathbf{w}(z)$. The inner product indicated on the right is that defined for spatial vectors.

In the text, we have occasion to discuss many fields besides those already mentioned: stress fields, energy flux fields, mass flux fields, enthalpy fields, .... In developing basic concepts, we are generally concerned with real scalar fields and spatial vector fields. It is in the final results of applications that we usually become concerned with the values of real scalar fields and integrals of real scalar fields (the average temperature in a tank or the average concentration in an exit stream) and the values of spatial vector fields and integrals of spatial vector fields (the force acting on a body or the torque exerted upon on surface).

It is common practice in the literature to refer to real scalar fields and spatial vector fields inexactly as *scalars* and *vectors*. Writers depend upon the context to clarify whether they are talking about the functions (the spatial vector fields) or their values (spatial vectors).

## A.1.4  Basis

Let $\alpha_1, \alpha_2$, and $\alpha_3$ be scalars. We define the set of spatial vectors $\mathbf{e}_1, \mathbf{e}_2, \mathbf{e}_3$ to be *linearly independent*, if

$$\alpha_1\mathbf{e}_1 + \alpha_2\mathbf{e}_2 + \alpha_3\mathbf{e}_3 = \sum_{i=1}^{3} \alpha_i\mathbf{e}_i = \mathbf{0} \tag{A.1.4-1}$$

can hold only when the numbers $\alpha_1$, $\alpha_2$, and $\alpha_3$ are all zero. Geometrically, three vectors are linearly independent if they are not all parallel to one plane.

A *basis* for a vector space $M$ is defined to be a set $\chi$ of linearly independent vectors such that every vector in $M$ is a linear combination of elements of $\chi$.

For example, $(\mathbf{e}_1, \mathbf{e}_2, \mathbf{e}_3)$ are said to form a basis for the set of all spatial vectors if $(\mathbf{e}_1, \mathbf{e}_2, \mathbf{e}_3)$ are linearly independent as in (A.1.4-1) and if every spatial vector $\mathbf{v}$ can be written as a linear combination of them:

$$\mathbf{v} = v_1\mathbf{e}_1 + v_2\mathbf{e}_2 + v_3\mathbf{e}_3$$

$$= \sum_{i=1}^{3} v_i\mathbf{e}_i \tag{A.1.4-2}$$

The numbers $(v_1, v_2, v_3)$ are referred to as the *components* of the vector $\mathbf{v}$ with respect to the basis $(\mathbf{e}_1, \mathbf{e}_2, \mathbf{e}_3)$.

The *dimension* of a finite-dimensional vector space $M$ is defined to be the number of elements in a basis of $M$.

We accept without proof here that the number of elements in any basis of a finite-dimensional vector space is the same as that in any other basis (Halmos 1958, p. 13). It follows as a corollary that a set of $n$ vectors in any $n$-dimensional vector space $M$ is a basis if and only if it is linearly independent or, alternatively, if and only if every vector in $M$ is a linear combination of elements of the set (Halmos 1958, p. 14).

The space of spatial vectors is by definition *three dimensional*. It therefore follows that the space of spatial vector fields must be three dimensional as well.

**Exercise A.1.4-1**   Let

$$\mathbf{a} = \sum_{i=1}^{3} a_i\mathbf{e}_i$$

and

$$\mathbf{b} = \sum_{j=1}^{3} b_j\mathbf{e}_j$$

be two spatial vectors, where the $a_i$ ($i = 1, 2, 3$) and $b_j$ ($j = 1, 2, 3$) are real numbers (scalars). Let $\alpha$ be a real number. Express the spatial vectors $\mathbf{a} + \mathbf{b}$ and $\alpha\mathbf{a}$ as linear combinations of the $\mathbf{e}_i$ ($i = 1, 2, 3$).

**Exercise A.1.4-2**   Prove that a set of $n$ vectors in an $n$-dimensional vector space $M$ is a basis if and only if it is linearly independent, or, alternatively, if and only if every vector in $M$ is a linear combination of elements of the set.

*Hint:*   You may accept without proof that the number of elements in any basis of a finite-dimensional vector space is the same as in any other basis.

**Exercise A.1.4-3**   Let the set $(\mathbf{m}_1, \mathbf{m}_2, \mathbf{m}_3)$ form a basis for the space of spatial vectors. Any spatial vector $\mathbf{v}$ may be expressed as

$$\mathbf{v} = \sum_{i=1}^{3} v_i\mathbf{m}_i$$

Prove that the components of $\mathbf{v}$ with respect to this basis are unique.

**A.1.5**  Basis for the Spatial Vector Fields

A basis $(\mathbf{m}_1, \mathbf{m}_2, \mathbf{m}_3)$ for the space of spatial vector fields is said to be *Cartesian* (McConnell 1957, p. 39) if the basis fields are of unit length (at every point $z$ of $E$ the corresponding spatial vectors are of unit length):

$$\mathbf{m}_1(z) \cdot \mathbf{m}_1(z) = 1$$
$$\mathbf{m}_2(z) \cdot \mathbf{m}_2(z) = 1 \qquad\qquad\qquad\qquad (A.1.5\text{-}1)$$
$$\mathbf{m}_3(z) \cdot \mathbf{m}_3(z) = 1$$

A basis is said to be *orthogonal* if the basis elements are orthogonal to one another (at every point $z$ of $E$ the corresponding spatial vectors are orthogonal to one another):

$$\mathbf{m}_i(z) \cdot \mathbf{m}_j(z) = 0 \quad \text{for } i \neq j \qquad\qquad\qquad (A.1.5\text{-}2)$$

In what follows, as well as in the body of the text, we usually will use an *orthogonal Cartesian* basis (*orthonormal basis*).

A *rectangular Cartesian* basis is the most familiar to us all. Besides being orthonormal, the basis fields have the property that for every two points $x$ and $y$ in $E$,

$$\mathbf{m}_i(x) = \mathbf{m}_i(y) \quad \text{for } i = 1, 2, 3 \qquad\qquad\qquad (A.1.5\text{-}3)$$

This means that both the length and the direction of the basis fields are independent of position in $E$. We shall reserve the symbols $\mathbf{e}_1, \mathbf{e}_2, \mathbf{e}_3$ for such a basis.

Every spatial vector field $\mathbf{u}$ may be written as a linear combination of rectangular Cartesian basis fields $\mathbf{e}_1, \mathbf{e}_2, \mathbf{e}_3$:

$$\mathbf{u} = u_1\mathbf{e}_1 + u_2\mathbf{e}_2 + u_3\mathbf{e}_3 = \sum_{i=1}^{3} u_i\mathbf{e}_i \qquad\qquad\qquad (A.1.5\text{-}4)$$

The quantities $u_1, u_2, u_3$ are known as the *rectangular Cartesian components* of $\mathbf{u}$; in general, they are functions of position in $E$.

A special case is the position vector field defined in Section A.1.3:

$$\mathbf{p} = z_1\mathbf{e}_1 + z_2\mathbf{e}_2 + z_3\mathbf{e}_3 = \sum_{i=1}^{3} z_i\mathbf{e}_i \qquad\qquad\qquad (A.1.5\text{-}5)$$

The rectangular Cartesian components $(z_1, z_2, z_3)$ of the position vector field $\mathbf{p}$ are called the *rectangular Cartesian coordinates* with respect to the previously chosen origin $O$. They are naturally functions of position $z$ in $E$:

$$z_i = z_i(z) \quad \text{for } i = 1, 2, 3 \qquad\qquad\qquad (A.1.5\text{-}6)$$

For this reason we will often find it convenient to think of $\mathbf{p}$ as being a function of the rectangular Cartesian coordinates:

$$\mathbf{z} = \mathbf{p}(z_1, z_2, z_3) \qquad\qquad\qquad (A.1.5\text{-}7)$$

**Exercise A.1.5-1**    If we define

$$\frac{\partial \mathbf{p}}{\partial z_1} \equiv \lim_{\Delta z_1 \to 0} \frac{1}{\Delta z_1}[\mathbf{p}(z_1 + \Delta z_1, z_2, z_3) - \mathbf{p}(z_1, z_2, z_3)]$$

determine that

$$\frac{\partial \mathbf{p}}{\partial z_1} = \mathbf{e}_1$$

With similar definitions for $\partial \mathbf{p}/\partial z_2$ and $\partial \mathbf{p}/\partial z_3$, determine that

$$\frac{\partial \mathbf{p}}{\partial z_2} = \mathbf{e}_2$$

and

$$\frac{\partial \mathbf{p}}{\partial z_3} = \mathbf{e}_3$$

**Exercise A.I.5-2**

    i) Let $(z_1, z_2, z_3)$ and $(\bar{z}_1, \bar{z}_2, \bar{z}_3)$ denote two rectangular Cartesian coordinate systems. If $z_i = z_i(\bar{z}_1, \bar{z}_2, \bar{z}_3)$ for $i = 1, 2, 3$, prove that

$$\frac{\partial \mathbf{p}}{\partial \bar{z}_i} = \sum_{m=1}^{3} \frac{\partial z_m}{\partial \bar{z}_i} \frac{\partial \mathbf{p}}{\partial z_m}$$

    ii) Prove that

$$\bar{\mathbf{e}}_i = \sum_{m=1}^{3} \frac{\partial z_m}{\partial \bar{z}_i} \mathbf{e}_m$$

**Exercise A.I.5-3**    Let $\mathbf{u}$ be a spatial vector field. If the $u_i$ ($i = 1, 2, 3$) are the components of $\mathbf{u}$ with respect to a set of rectangular Cartesian basis fields $(\mathbf{e}_1, \mathbf{e}_2, \mathbf{e}_3)$ and the $\bar{u}_m (m = 1, 2, 3)$ are the components of $\mathbf{u}$ with respect to another set of rectangular Cartesian basis fields $(\bar{\mathbf{e}}_1, \bar{\mathbf{e}}_2, \bar{\mathbf{e}}_3)$, prove that

$$u_i = \sum_{m=1}^{3} \frac{\partial z_i}{\partial \bar{z}_m} \bar{u}_m$$

## A.I.6  Basis for the Spatial Vectors

Any basis $(\mathbf{m}_1, \mathbf{m}_2, \mathbf{m}_3)$ for the spatial vector fields may be used to generate an infinite number of bases for the space of spatial vectors. The values of these functions at any point $z$ of $E$, the spatial vectors

$$\mathbf{m}_i = \mathbf{m}_i(z) \quad \text{for } i = 1, 2, 3 \tag{A.1.6-1}$$

may be used as a basis for the spatial vectors. The basis will depend upon the particular point $z$ chosen, in the sense that the magnitudes and directions of the $\mathbf{m}_i$ may vary with position. The cylindrical and spherical coordinate systems provide good examples of such bases. (Notice that in writing (A.1.6-1) we use the same notation both for the function and for its values.)

Of particular interest is any rectangular Cartesian basis $(\mathbf{e}_1, \mathbf{e}_2, \mathbf{e}_3)$ for the spatial vector fields. The magnitude and direction of the values of these functions are independent of position in $E$. We will often find it convenient to use the values of these functions, the spatial vectors, as a basis for the spatial vectors. For example, we will often express a particular position vector $\mathbf{z}$ and point difference $\mathbf{a} \equiv \overrightarrow{xy}$ in terms of their rectangular Cartesian components:

$$\mathbf{z} = \mathbf{p}(z)$$

$$= \sum_{i=1}^{3} z_i(z)\mathbf{e}_i \tag{A.1.6-2}$$

$$\mathbf{a} \equiv \overrightarrow{xy}$$

$$= \sum_{i=1}^{3} a_i \mathbf{e}_i \tag{A.1.6-3}$$

## A.1.7 Summation Convention

In writing a spatial vector field in terms of its rectangular Cartesian components, notice that the summation is over a repeated index $i$:

$$\mathbf{u} = \sum_{i=1}^{3} u_i \mathbf{e}_i \tag{A.1.7-1}$$

This suggests that we adopt a simpler notation in which we understand that a summation from 1 to 3 is to be performed over every index that appears twice within a single term. This is known as the *summation convention*. It allows (A.1.7-1) to be written as

$$\mathbf{u} = u_i \mathbf{e}_i \tag{A.1.7-2}$$

With this convention, we can write the inner product of two spatial vector fields as

$$\begin{aligned} \mathbf{v} \cdot \mathbf{w} &= (v_i \mathbf{e}_i) \cdot (w_j \mathbf{e}_j) \\ &= v_i w_j (\mathbf{e}_i \cdot \mathbf{e}_j) \\ &= v_i w_j \delta_{ij} \\ &= v_i w_i \end{aligned} \tag{A.1.7-3}$$

In going from (A.1.7-3)$_1$ to (A.1.7-3)$_2$, rules ($I_1$) to ($I_3$) for the inner product have been employed. In proceeding from (A.1.7-3)$_2$ to (A.1.7-3)$_3$, we have recognized that, for a rectangular Cartesian coordinate system,

$$(\mathbf{e}_i \cdot \mathbf{e}_j) = \delta_{ij} \tag{A.1.7-4}$$

where the *Kronecker delta* is defined by

$$\delta_{ij} \equiv \begin{cases} 1 & \text{if } i = j \\ 0 & \text{if } i \neq j \end{cases} \tag{A.1.7-5}$$

We wish to emphasize that the summation convention is not defined for an index that appears more than twice in a single term. If this happens, there are several possibilities:

1) In writing a relation such as

$$\begin{aligned} (\mathbf{v} \cdot \mathbf{w})(\mathbf{q} \cdot \mathbf{n}) &= (v_i w_i)(q_j n_j) \\ &= v_i w_i q_j n_j \end{aligned} \tag{A.1.7-6}$$

we must be careful not to confuse the summation in the expression for $(\mathbf{v} \cdot \mathbf{w})$ with that for $(\mathbf{q} \cdot \mathbf{n})$. Observe that, besides being undefined, $v_i w_i q_i n_i$ is confusing. It might mean

$$v_i w_i q_i n_i \stackrel{?}{=} \sum_{i=1}^{3} (v_i w_i q_i n_i)$$

$$\stackrel{?}{=} \sum_{i=1}^{3} (v_i w_i) \sum_{j=1}^{3} (q_j n_j)$$

$$\stackrel{?}{=} \sum_{i=1}^{3} (v_i q_i) \sum_{j=1}^{3} (w_j n_j)$$

$$\stackrel{?}{=} \sum_{i=1}^{3} (v_i n_i) \sum_{j=1}^{3} (w_j q_j)$$

2) Sometimes a summation is intended over an index that appears more than twice in a single term. The summation sign should be used explicitly in such a case:

$$\mathbf{u} = \sum_{i=1}^{3} \sqrt{g_{ii}} u^i \mathbf{g}_{\langle i \rangle}$$

3) Occasionally an index may appear twice or more in a single term, although no summation is intended. For clarity, make a note to this effect next to the equation:

$$u_{\langle i \rangle} = \frac{u_i}{\sqrt{g_{ii}}}, \quad \text{no summation on } i$$

## A.2 Determinant

### A.2.1 Definitions

Define $e_{ijk}$ and $e^{ijk}$ to have only three distinct values:

0, when any two of the indices are equal;
+1, when $ijk$ is an even permutation of 123;
−1, when $ijk$ is an odd permutation of 123.

The quantities $e_{ijk}$ and $e^{ijk}$ are said to be completely skew symmetric in the indices $ijk$; that is, interchanging any two of these indices changes the sign of the quantity. For the moment we shall have no occasion to use $e^{ijk}$, but we shall return to make use of it later.

Let us introduce the notation $\det(a_{ij})$ for the determinant that has as its typical entry $a_{ij}$:

$$\det(a_{ij}) = \begin{vmatrix} a_{11} & a_{12} & a_{13} \\ a_{21} & a_{22} & a_{23} \\ a_{31} & a_{32} & a_{33} \end{vmatrix} \tag{A.2.1-1}$$

When we expand the determinant $\det(a_{ij})$ by rows, we find we may write

$$\det(a_{ij}) = e_{ijk} a_{1i} a_{2j} a_{3k} \tag{A.2.1-2}$$

Similarly, when we expand the determinant by columns, we have

$$\det(a_{ij}) = e_{ijk}a_{i1}a_{j2}a_{k3} \tag{A.2.1-3}$$

Equations (A.2.1-2) and (A.2.1-3) suggest that we consider the quantity $e_{ijk}a_{im}a_{jn}a_{kp}$. This quantity is completely skew symmetric in the indices $mnp$. As a proof, we have, by relabeling indices,

$$e_{ijk}a_{im}a_{jn}a_{kp} = e_{jik}a_{jm}a_{in}a_{kp}$$
$$= -e_{ijk}a_{in}a_{jm}a_{kp} \tag{A.2.1-4}$$

In the same way, we show that interchanging any two of the indices $mnp$ alters the sign. In view of (A.2.1-3), this suggests that we may write, for an expansion by columns,

$$e_{ijk}a_{im}a_{jn}a_{kp} = \det(a_{rs})e_{mnp} \tag{A.2.1-5}$$

The same type of argument may be used to infer from (A.2.1-2) that, for an expansion by rows,

$$e_{ijk}a_{mi}a_{nj}a_{pk} = \det(a_{rs})e_{mnp} \tag{A.2.1-6}$$

As an example of the use of this notation, consider the product of two determinants:

$$\det(a_{rs})\det(b_{xy}) = \det(a_{rs})e_{ijk}b_{i1}b_{j2}b_{k3}$$
$$= e_{mnp}a_{mi}a_{nj}a_{pk}b_{i1}b_{j2}b_{k3}$$
$$= e_{mnp}(a_{mi}b_{i1})(a_{nj}b_{j2})(a_{pk}b_{k3})$$
$$= \det(a_{us}b_{sv}) \tag{A.2.1-7}$$

Let us introduce a further concept, the *cofactor*. Starting with (A.2.1-5), write

$$e_{rnp}e_{ijk}a_{im}a_{jn}a_{kp} = \det(a_{st})e_{rnp}e_{mnp} \tag{A.2.1-8}$$

In Section A.1.7, we introduced the Kronecker delta $\delta_{rm}$, defined as

$$\delta_{rm} \equiv \begin{cases} 1 \text{ if } r = m \\ 0 \text{ if } r \neq m \end{cases} \tag{A.2.1-9}$$

An equivalent expression, which is sometimes useful, is

$$\delta_{rm} = \frac{1}{2}e_{rnp}e_{mnp} \tag{A.2.1-10}$$

We consequently may rearrange (A.2.1-8) to read

$$\left(\frac{1}{2}e_{rnp}e_{ijk}a_{jn}a_{kp}\right)a_{im} = \det(a_{st})\delta_{rm} \tag{A.2.1-11}$$

or

$$A_{ri}a_{im} = \det(a_{st})\delta_{rm} \tag{A.2.1-12}$$

where we define

$$A_{ri} = \frac{1}{2}e_{rnp}e_{ijk}a_{jn}a_{kp} \qquad\qquad\qquad (A.2.1\text{-}13)$$

The quantity $A_{ri}$ is called the *cofactor* of the element $a_{ir}$ in the determinant $\det(a_{st})$. When the determinant is expanded in full, it is obvious that any element such as $a_{ir}$ appears once in each of a certain number of terms of the expansion; the coefficient of $a_{ir}$ in this expansion is just $A_{ri}$.

For a further discussion of determinants, see McConnell (1957).

**Exercise A.2.1-1**   Starting with (A.2.1-6), prove that

$$a_{mr}A_{ri} = \det(a_{st})\delta_{mi}$$

where $A_{ri}$ is given by (A.2.1-13).

**Exercise A.2.1-2**

i) Show that $\mathbf{e}_{ijk}\mathbf{e}_{mnk}$ takes the values:
   +1, when $i$, $j$ and $m$, $n$ are the same permutation of the same two numbers;
   −1, when $i$, $j$ and $m$, $n$ are opposite permutation of the same two numbers;
   0, otherwise.

ii) Demonstrate that

$$e_{ijk}e_{mnk} = \delta_{im}\delta_{jn} - \delta_{in}\delta_{jm}$$

**Exercise A.2.1-3**

i) Show (A.2.1-10).

ii) Demonstrate that

$$e_{mnp}e_{mnp} = 2\delta_{mm} = 6$$

The results here are to be observed or demonstrated rather than proved.

**Exercise A.2.1-4**   If any two rows or columns are identical, prove that the determinant is zero.

**Exercise A.2.1-5**   In each of the following examples, we start with an equation and proceed to derive another equation. Indicate whether each step in the derivation is valid (can be derived from the previous step) and give reasons.

*Example A*   Given: $b_{ijk}b_{mnk} = \delta_{im}\delta_{jn} - \delta_{in}\delta_{jm}$.
   Step 1: $b_{ijk}b_{mnk}\delta_{jn} = \delta_{im}\delta_{jn}\delta_{jn} - \delta_{in}\delta_{jm}\delta_{jn}$.
   Step 2: $b_{ink}b_{mnk} = 2\delta_{im}$.

*Example B* Given: $b_{ink}b_{mnk} = 2\delta_{im}$.

Step 1: $b_{ijk}b_{mnk}\delta_{jn} = \delta_{im}\delta_{jn}\delta_{jn} - \delta_{in}\delta_{jm}\delta_{jn}$.

Step 2: $b_{ijk}b_{mnk} = \delta_{im}\delta_{jn} - \delta_{in}\delta_{jm}$.

---

## A.3    Gradient of Scalar Field

### A.3.1    Definition**

The gradient of a scalar field $\alpha$ is a spatial vector field denoted by $\nabla\alpha$. The gradient is specified by defining its inner product with an arbitrary spatial vector at all points $z$ in $E$:

$$\nabla\alpha(\mathbf{z}) \cdot \mathbf{a} \equiv \lim_{s \to 0} \frac{\alpha(\mathbf{z} + s\mathbf{a}) - \alpha(\mathbf{z})}{s} \tag{A.3.1-1}$$

The spatial vector $\mathbf{a}$ should be interpreted as the directed line segment or point difference $\mathbf{a} \equiv \overrightarrow{zy}$, where $y$ is an arbitrary point in $E$. In writing (A.3.1-1), we have assumed that an origin $O$ has been specified and we have interpreted $\alpha$ as a function of the position vector $\mathbf{z}$ measured with respect to this origin rather than as a function of the point $z$ itself.

Equation (A.3.1-1) may be rearranged into a more easily applied expression in the following manner.

$$\nabla\alpha(\mathbf{z}) \cdot \mathbf{a} = \lim_{s \to 0} \frac{1}{s} \{\alpha\left([z_1 + sa_1]\mathbf{e}_1 + [z_2 + sa_2]\mathbf{e}_2 + [z_3 + sa_3]\mathbf{e}_3\right)$$

$$- \alpha\left(z_1\mathbf{e}_1 + [z_2 + sa_2]\mathbf{e}_2 + [z_3 + sa_3]\mathbf{e}_3\right)\}$$

$$+ \lim_{s \to 0} \frac{1}{s} \{\alpha\left(z_1\mathbf{e}_1 + [z_2 + sa_2]\mathbf{e}_2 + [z_3 + sa_3]\mathbf{e}_3\right)$$

$$- \alpha\left(z_1\mathbf{e}_1 + z_2\mathbf{e}_2 + [z_3 + sa_3]\mathbf{e}_3\right)\}$$

$$+ \lim_{s \to 0} \frac{1}{s} \{\alpha\left(z_1\mathbf{e}_1 + z_2\mathbf{e}_2 + [z_3 + sa_3]\mathbf{e}_3\right) - \alpha\left(z_1\mathbf{e}_1 + z_2\mathbf{e}_2 + z_3\mathbf{e}_3\right)\}$$

$$= a_1 \lim_{a_1 s \to 0} \frac{1}{a_1 s} \{\alpha\left([z_1 + sa_1]\mathbf{e}_1 + [z_2 + sa_2]\mathbf{e}_2 + [z_3 + sa_3]\mathbf{e}_3\right)$$

$$- \alpha\left(z_1\mathbf{e}_1 + [z_2 + sa_2]\mathbf{e}_2 + [z_3 + sa_3]\mathbf{e}_3\right)\}$$

$$+ a_2 \lim_{a_2 s \to 0} \frac{1}{a_2 s} \{\alpha\left(z_1\mathbf{e}_1 + [z_2 + sa_2]\mathbf{e}_2\right.$$

$$+ [z_3 + sa_3]\mathbf{e}_3\left) - \alpha\left(z_1\mathbf{e}_1 + z_2\mathbf{e}_2 + [z_3 + sa_3]\mathbf{e}_3\right)\right\}$$

$$+ a_3 \lim_{a_3 s \to 0} \frac{1}{a_3 s} \{\alpha\left(z_1\mathbf{e}_1 + z_2\mathbf{e}_2 + [z_3 + sa_3]\mathbf{e}_3\right) - \alpha\left(z_1\mathbf{e}_1 + z_2\mathbf{e}_2 + z_3\mathbf{e}_3\right)\}$$

$$= a_1 \frac{\partial\alpha}{\partial z_1}(\mathbf{z}) + a_2 \frac{\partial\alpha}{\partial z_2}(\mathbf{z}) + a_3 \frac{\partial\alpha}{\partial z_3}(\mathbf{z})$$

$$= a_i \frac{\partial\alpha}{\partial z_i}(\mathbf{z}) \tag{A.3.1-2}$$

In arriving at this result, we have used the definition of the partial derivative in the form

$$\frac{\partial \alpha}{\partial z_2}(\mathbf{z}) \equiv \lim_{\Delta z_2 \to 0} \frac{1}{\Delta z_2} \{ \alpha \, (z_1 \mathbf{e}_1 + [z_2 + \Delta z_2] \mathbf{e}_2 + z_3 \mathbf{e}_3)$$
$$- \alpha \, (z_1 \mathbf{e}_1 + z_2 \mathbf{e}_2 + z_3 \mathbf{e}_3) \} \tag{A.3.1-3}$$

Since **a** is an arbitrary spatial vector, take $\mathbf{a} = \mathbf{e}_i$:

$$\nabla \alpha \cdot \mathbf{e}_i = \frac{\partial \alpha}{\partial z_i} \tag{A.3.1-4}$$

We conclude that

$$\nabla \alpha = \frac{\partial \alpha}{\partial z_i} \mathbf{e}_i \tag{A.3.1-5}$$

**Exercise A.3.1-1**\*\*  Prove that

$$\nabla z_i = \mathbf{e}_i$$

**Exercise A.3.1-2**

i) Let $(z_1, z_2, z_3)$ and $(\bar{z}_1, \bar{z}_2, \bar{z}_3)$ denote two rectangular Cartesian coordinate systems such that

$$z_i = z_i(\bar{z}_1, \bar{z}_2, \bar{z}_3) \quad \text{for } i = 1, 2, 3$$

Prove that

$$\nabla z_i = \frac{\partial z_i}{\partial \bar{z}_m} \nabla \bar{z}_m$$

ii) Show that for these two coordinate systems

$$\mathbf{e}_i = \frac{\partial z_i}{\partial \bar{z}_m} \bar{\mathbf{e}}_m$$

Compare this result with that of Exercise A.1.5-2.

**Exercise A.3.1-3**  Let **u** be some spatial vector field. If the $u_i$ ($i = 1, 2, 3$) are the components of **u** with respect to a set of rectangular Cartesian basis fields $(\mathbf{e}_1, \mathbf{e}_2, \mathbf{e}_3)$ and the $\bar{u}_m$ ($m = 1, 2, 3$) are the components of **u** with respect to another set of rectangular Cartesian basis fields $(\bar{\mathbf{e}}_1, \bar{\mathbf{e}}_2, \bar{\mathbf{e}}_3)$, prove that

$$u_i = \frac{\partial \bar{z}_m}{\partial z_i} \bar{u}_m$$

Compare this result with that of Exercise A.1.5-3.

**Exercise A.3.1-4**  Let $(z_1, z_2, z_3)$ and $(\bar{z}_1, \bar{z}_2, \bar{z}_3)$ denote two rectangular Cartesian coordinate systems such that

$$z_i = z_i(\bar{z}_1, \bar{z}_2, \bar{z}_3) \quad \text{for } i = 1, 2, 3$$

Prove that

$$\frac{\partial \bar{z}_i}{\partial z_j} = \frac{\partial z_j}{\partial \bar{z}_i}$$

**Exercise A.3.1-5** *Normal to a surface*   Let

$$\varphi(\mathbf{z}) = \text{a constant}$$

be the equation of a surface in euclidean point space. We assume here that an origin $O$ has been specified and that $\varphi$ can be interpreted as a function of the position vector $\mathbf{z}$ measured with respect to this origin rather than the point $z$ itself. Prove that $\nabla\varphi(\mathbf{z})$ is orthogonal to the surface.

*Hint:*   Take an arbitrary curve on the surface and let $s$ be a parameter measured along this curve. Consider the implications of $d\varphi/ds = 0$.

---

## A.4   Curvilinear Coordinates

### A.4.1   Natural Basis**

Consider some curve in space and let $t$ be a parameter measured along this curve. Let $\mathbf{p}$ be a position vector-valued function of $t$ along this curve. We define

$$\frac{d\mathbf{p}}{dt}(t) \equiv \lim_{\Delta t \to 0} \frac{\mathbf{p}(t + \Delta t) - \mathbf{p}(t)}{\Delta t} \tag{A.4.1-1}$$

Figure A.4.1-1 suggests that $(d\mathbf{p}/dt)(t)$ is a tangent vector to the curve at the point $t$.

In Section A.1.3 we introduced the position vector field $\mathbf{p}$, which in Section A.1.5 we expressed in terms of its rectangular Cartesian coordinates:

$$\mathbf{p} = z_i \mathbf{e}_i \tag{A.4.1-2}$$

Let us assume that each $z_i$ $(i = 1, 2, 3)$ may be regarded as a function of three parameters $(x^1, x^2, x^3)$, called *curvilinear coordinates*:

$$z_i = z_i(x^1, x^2, x^3) \quad \text{for } i = 1, 2, 3 \tag{A.4.1-3}$$

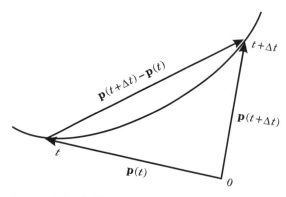

**Figure A.4.1-1.** The points $t$ and $t + \delta t$ on a curve in space.

[Here we use the common notation-preserving device of employing the same symbol for both the function $z_i$ and its value $z_i (x^1, x^2, x^3)$.] For fixed values of $x^1, x^2, x^3$, these equations define surfaces called *curvilinear surfaces*. The curve of intersection of any two curvilinear surfaces defines a curvilinear coordinate line.

The spatial vector field $\mathbf{g}_k$ ($k = 1, 2, 3$) is defined as

$$\mathbf{g}_k \equiv \frac{\partial \mathbf{p}}{\partial x^k} = \frac{\partial(z_j \mathbf{e}_j)}{\partial x^k} = \frac{\partial z_j}{\partial x^k} \mathbf{e}_j \tag{A.4.1-4}$$

At any point $z$, the spatial vector $\mathbf{g}_k(z)$ is tangent to the $x^k$-coordinate curve. Note that, in general, the magnitude and direction of $\mathbf{g}_k(z)$ vary with position $z$ in $E$.

May the three spatial vector fields $(\mathbf{g}_1, \mathbf{g}_2, \mathbf{g}_3)$ be regarded as a new basis for the spatial vector fields? It follows from our discussion in Section A.1.4 that they may, if we can demonstrate that every spatial vector field can be written as a linear combination of them. Since every spatial vector field can be written as a linear combination of the rectangular Cartesian basis fields $(\mathbf{e}_1, \mathbf{e}_2, \mathbf{e}_3)$, all that it is necessary to show is that all the $\mathbf{e}_i$ may be expressed as linear combination of the $\mathbf{g}_k$. The system of linear equations

$$\frac{\partial z_j}{\partial x^k} \mathbf{e}_j = \mathbf{g}_k \tag{A.4.1-5}$$

can be solved for the $\mathbf{e}_j$ so long as the determinant

$$\det \left( \frac{\partial z_j}{\partial x^k} \right) \neq 0 \tag{A.4.1-6}$$

When this condition is satisfied, since

$$\frac{\partial z_m}{\partial x^k} \frac{\partial x^k}{\partial z_n} = \delta_{mn} \tag{A.4.1-7}$$

we may write from (A.4.1-5)

$$\frac{\partial x^k}{\partial z_m} \frac{\partial z_j}{\partial x^k} \mathbf{e}_j = \delta_{mj} \mathbf{e}_j$$

$$= \mathbf{e}_m$$

$$= \frac{\partial x^k}{\partial z_m} \mathbf{g}_k \tag{A.4.1-8}$$

We demonstrate in this way that the $\mathbf{g}_k$ ($k = 1, 2, 3$) may be regarded as a set of basis fields for the spatial vector fields.

We will refer to any set of parameters $(x^1, x^2, x^3)$ that satisfies (A.4.1-6) as a *curvilinear coordinate system*. We refer to the set $(\mathbf{g}_1, \mathbf{g}_2, \mathbf{g}_3)$ as the *natural basis* for this curvilinear coordinate system.

The natural basis fields are orthogonal if

$$\mathbf{g}_i \cdot \mathbf{g}_j = 0 \quad \text{for } i \neq j \tag{A.4.1-9}$$

When the natural basis fields are orthogonal to one another, we say that they form an *orthogonal coordinate system*. Geometrically, it is clear that the cylindrical and spherical coordinate systems of Exercises A.4.1-4 and A.4.1-5 are orthogonal. One may check whether any coordinate system is orthogonal by examining

$$
\begin{aligned}
g_{ij} \equiv \mathbf{g}_i \cdot \mathbf{g}_j &= \frac{\partial \mathbf{p}}{\partial x^i} \cdot \frac{\partial \mathbf{p}}{\partial x^j} \\
&= \frac{\partial z_m}{\partial x^i} \mathbf{e}_m \cdot \frac{\partial z_n}{\partial x^j} \mathbf{e}_n \\
&= \frac{\partial z_m}{\partial x^i} \frac{\partial z_n}{\partial x^j} \delta_{mn} \\
&= \frac{\partial z_m}{\partial x^i} \frac{\partial z_m}{\partial x^j}
\end{aligned}
\tag{A.4.1-10}
$$

In applications, it is usually more convenient to work in terms of an orthogonal Cartesian (orthonormal) basis (see Section A.1.5). The natural basis fields defined by (A.4.1-9) may be normalized to form an orthonormal basis $(\mathbf{g}_{\langle 1 \rangle}, \mathbf{g}_{\langle 2 \rangle}, \mathbf{g}_{\langle 3 \rangle})$:

$$
\mathbf{g}_{\langle i \rangle} \equiv \frac{\mathbf{g}_i}{\sqrt{\mathbf{g}_i \cdot \mathbf{g}_i}} = \frac{\mathbf{g}_i}{\sqrt{g_{ii}}}, \quad \text{no summation on } i
\tag{A.4.1-11}
$$

This basis is referred to as the *physical basis* for the coordinate system. In this text we will not discuss the normalized natural basis except for the case of orthogonal coordinate systems.

Any spatial vector $\mathbf{u}$ may be expressed as a linear combination of the three physical basis vector fields associated with an orthogonal coordinate system:

$$
\mathbf{u} = u_{\langle i \rangle} \mathbf{g}_{\langle i \rangle}
\tag{A.4.1-12}
$$

The three coefficients $(u_{\langle 1 \rangle}, u_{\langle 2 \rangle}, u_{\langle 3 \rangle})$ are referred to as the *physical components* of $\mathbf{u}$ with respect to this particular coordinate system.

In applications, we are almost always concerned with orthogonal coordinate systems and physical components of spatial vector fields. We visualize spatial vectors and spatial vector fields that have some physical interpretation (such as velocity or force) in terms of their physical components. Since we will most readily formulate boundary conditions to differential equations in terms of physical components, we will find it most natural to formulate specific problems to be solved in terms of physical components. Many engineering texts deal with physical components exclusively, never mentioning the covariant and contravariant components discussed in Section A.4.3.

**Exercise A.4.1-1**   Discuss why, in a rectangular Cartesian coordinate system, the basis vector fields $\mathbf{e}_k$ $(k = 1, 2, 3)$ are what we term here the natural basis vectors.

**Exercise A.4.1-2**   If the $\mathbf{g}_i$ $(i = 1, 2, 3)$ are the natural basis vector fields associated with one curvilinear coordinate system $(x^1, x^2, x^3)$ and if the $\overline{\mathbf{g}}_i$ $(i = 1, 2, 3)$ are the natural basis vector fields associated with another curvilinear coordinate system $(\overline{x}^1, \overline{x}^2, \overline{x}^3)$, prove that

$$
\overline{\mathbf{g}}_i = \frac{\partial x^j}{\partial \overline{x}^i} \mathbf{g}_j
$$

**Exercise A.4.1-3**   Prove that for any rectangular Cartesian coordinate system

$$
g_{ij} = \delta_{ij}
$$

Here $\delta_{ij}$ is the Kronecker delta.

**Exercise A.4.1-4**\*\* *Cylindrical coordinates*   Given a cylindrical coordinate system

$$z_1 = x^1 \cos x^2$$
$$= r \cos \theta$$
$$z_2 = x^1 \sin x^2$$
$$= r \sin \theta$$
$$z_3 = x^3$$
$$= z$$

prove that

$$g_{11} = g_{33}$$
$$= 1$$
$$g_{22} = (x^1)^2$$
$$= r^2$$
$$g_{ij} = 0 \quad \text{if } i \neq j$$

and

$$g \equiv \det(g_{ij})$$
$$= r^2$$

**Exercise A.4.1-5**\*\* *Spherical coordinates*   Given a spherical coordinate system

$$z_1 = x^1 \sin x^2 \cos x^3$$
$$= r \sin \theta \cos \varphi$$
$$z_2 = x^1 \sin x^2 \sin x^3$$
$$= r \sin \theta \sin \varphi$$
$$z_3 = x^1 \cos x^2$$
$$= r \cos \theta$$

prove that

$$g_{11} = 1$$
$$g_{22} = \left(x^1\right)^2$$
$$= r^2$$
$$g_{33} = \left(x^1 \sin x^2\right)^2$$
$$= (r \sin \theta)^2$$
$$g_{ij} = 0 \quad \text{if } i \neq j$$

and

$$g \equiv \det\left(g_{ij}\right)$$
$$= r^4 \sin^2 \theta$$

**Exercise A.4.1-6** *Parabolic coordinates*    In the paraboloidal coordinate system

$$z_1 = x^1 x^2 \cos x^3$$
$$z_2 = x^1 x^2 \sin x^3$$
$$z_3 = \frac{1}{2} \left[ \left( x^1 \right)^2 - \left( x^2 \right)^2 \right]$$

the $x^1$ surfaces and $x^2$ surfaces are paraboloids of revolution and the $x^3$ surfaces are planes through the $z^3$ axis. Prove that

$$g_{11} = g_{22}$$
$$= \left( x^1 \right)^2 + (x^2)^2$$
$$g_{33} = \left( x^1 x^2 \right)^2$$

and

$$g_{ij} = 0 \quad \text{if } i \neq j$$

**Exercise A.4.1-7**    Prove that $g_{ij}$ is symmetric in the indices $i$ and $j$.

**Exercise A.4.1-8\*\*** *Changes of coordinates*    We restrict ourselves to an orthogonal curvilinear coordinate system $x^i$ ($i = 1, 2, 3$). We denote $\mathbf{v} = v_i \mathbf{e}_i$.

i) Determine that

$$\mathbf{g}_{(i)} = \frac{1}{\sqrt{g_{ii}}} \sum_{j=1}^{3} \frac{\partial z_j}{\partial x^i} \mathbf{e}_j, \quad \text{no summation on } i$$

ii) Starting with the result of (i), find that

$$v_j = \sum_{i=1}^{3} \frac{1}{\sqrt{g_{ii}}} \frac{\partial z_j}{\partial x^i} v_{(i)}$$

iii) For cylindrical coordinates as defined in Exercise A.4.1-4, show that

$$v_1 = v_r \cos \theta - v_\theta \sin \theta$$
$$v_2 = v_r \sin \theta + v_\theta \cos \theta$$
$$v_3 = v_z$$

Here we introduce the common notation (see Section 2.4.1)

$$v_r \equiv v_{(1)}, \quad v_\theta \equiv v_{(2)}, \quad v_z \equiv v_{(3)}$$

iv) For spherical coordinates as defined in Exercise A.4.1-5, prove that

$$v_1 = v_r \sin \theta \cos \varphi + v_\theta \cos \theta \cos \varphi - v_\varphi \sin \varphi$$
$$v_2 = v_r \sin \theta \sin \varphi + v_\theta \cos \theta \sin \varphi + v_\varphi \cos \varphi$$
$$v_3 = v_r \cos \theta - v_\theta \sin \theta$$

We define here (see Section 2.4.1)

$$v_r \equiv v_{(1)}, \quad v_\theta \equiv v_{(2)}, \quad v_\varphi \equiv v_{(3)}$$

**Exercise A.4.1-9**\*\* *More on changes of coordinates* We again restrict ourselves to an orthogonal curvilinear coordinate system $x^i$ ($i = 1, 2, 3$).

i) Prove that

$$\mathbf{e}_j = \sum_{i=1}^{3} \sqrt{g_{ii}} \frac{\partial x^i}{\partial z_j} \mathbf{g}_{(i)}$$

ii) Starting with the result of (i), find that

$$v_{(i)} = \sqrt{g_{ii}} \sum_{j=1}^{3} \frac{\partial x^i}{\partial z_j} v_j, \quad \text{no summation on } i$$

iii) For cylindrical coordinates as defined in Exercise A.4.1-4, prove that (see Section 2.4.1)

$$\begin{aligned}
v_r &\equiv v_{(1)} \\
&= v_1 \cos\theta + v_2 \sin\theta \\
v_\theta &\equiv v_{(2)} \\
&= -v_1 \sin\theta + v_2 \cos\theta \\
v_z &\equiv v_{(3)} \\
&= v_3
\end{aligned}$$

iv) For spherical coordinates as defined in Exercise A.4.1-5, prove that (see Section 2.4.1)

$$\begin{aligned}
v_r &\equiv v_{(1)} \\
&= v_1 \sin\theta \cos\varphi + v_2 \sin\theta \sin\varphi + v_3 \cos\theta \\
v_\theta &\equiv v_{(2)} \\
&= v_1 \cos\theta \cos\varphi + v_2 \cos\theta \sin\varphi - v_3 \sin\theta \\
v_\varphi &\equiv v_{(3)} \\
&= -v_1 \sin\varphi + v_2 \cos\varphi
\end{aligned}$$

**Exercise A.4.1-10** *More about position vector field*

i) For cylindrical coordinates as defined in Exercise A.4.1-4, determine that the physical components of the position vector field

$$\mathbf{p} = p_{(i)} \mathbf{g}_{(i)}$$

are

$$\begin{aligned}
p_{(1)} &= r \\
p_{(2)} &= 0 \\
p_{(3)} &= z
\end{aligned}$$

ii) For spherical coordinates as defined in Exercise A.4.1-5, determine that

$$\begin{aligned}
p_{(1)} &= r \\
p_{(2)} &= 0 \\
p_{(3)} &= 0
\end{aligned}$$

**Exercise A.4.1-11** *Ellipsoidal coordinates*    In the ellipsoidal coordinate system

$$z_1 = \left[ \frac{(x^1 - a)(x^2 - a)(x^3 - a)}{(b - a)(c - a)} \right]^{\frac{1}{2}}$$

$$z_2 = \left[ \frac{(x^1 - b)(x^2 - b)(x^3 - b)}{(c - b)(a - b)} \right]^{\frac{1}{2}}$$

$$z_3 = \left[ \frac{(x^1 - c)(x^2 - c)(x^3 - c)}{(a - c)(b - c)} \right]^{\frac{1}{2}}$$

Here $a$, $b$, and $c$ are constants such that $a > b > c > 0$. The $x^1$ surfaces are ellipsoids, the $x^2$ surfaces are hyperboloids of one sheet, the $x^3$ surfaces are hyperboloids of two sheets, and all the quadrics belong to the family of confocals

$$\frac{(z_1)^2}{y - a} + \frac{(z_2)^2}{y - b} + \frac{(z_3)^2}{y - c} = 1$$

Prove that

$$g_{11} = \frac{(x^3 - x^1)(x^2 - x^1)}{4(x^1 - a)(x^1 - b)(x^1 - c)}$$

$$g_{22} = \frac{(x^1 - x^2)(x^3 - x^2)}{4(x^2 - a)(x^2 - b)(x^2 - c)}$$

$$g_{33} = \frac{(x^1 - x^3)(x^2 - x^3)}{4(x^3 - a)(x^3 - b)(x^3 - c)}$$

and

$$g_{ij} = 0 \quad \text{if } i \neq j$$

## A.4.2   Dual Basis*

Another interesting set of spatial vector fields associated with a curvilinear coordinate system are the *dual vector fields* $\mathbf{g}^i$ ($i = 1, 2, 3$), defined as the gradients of the curvilinear coordinates:

$$\mathbf{g}^i \equiv \nabla x^i \tag{A.4.2-1}$$

It is reasonable to ask whether the dual vector fields may also be regarded as a basis for the space of spatial vector fields. Before answering this question, let us examine some of the properties of these fields.

The dual vector fields, like all other spatial vector fields, may be expressed as linear combinations of the natural basis:

$$\mathbf{g}^i = g^{ji} \mathbf{g}_j \tag{A.4.2-2}$$

The coefficients $g^{ji}$ may be regarded as being defined by this equation. Let us consider the scalar product of one of the dual vector fields with one of the natural basis fields:

$$\mathbf{g}^i \cdot \mathbf{g}_j = \nabla x^i \cdot \frac{\partial \mathbf{p}}{\partial x^j}$$

$$= \frac{\partial x^i}{\partial z_m} \mathbf{e}_m \cdot \frac{\partial z_n}{\partial x^j} \mathbf{e}_n$$

$$= \frac{\partial x^i}{\partial z_m} \frac{\partial z_n}{\partial x^j} \delta_{mn}$$

$$= \frac{\partial x^i}{\partial z_m} \frac{\partial z_m}{\partial x^j}$$

$$= \delta_j^i \tag{A.4.2-3}$$

Here, $\delta_j^i$ is another form of the Kronecker delta (Sections A.1.7 and A.2.1); the index $i$ is used in the superscript position only to preserve for the reader's eye the relative positions of $i$ and $j$ in the preceding lines. With a minimum of artificiality in the notation, indices associated with the curvilinear coordinates will maintain their relative position (superscript or subscript) in every term of an equation hereafter. We will encourage this symmetry with appropriate choices of notation where necessary, since we will find that it aids our memory and serves as a quick check for certain types of errors. We shall elaborate on this point shortly.

Referring to (A.4.2-2) and Section A.4.1, we see that another way of expressing (A.4.2-3) is to write

$$g^{ki} \mathbf{g}_k \cdot \mathbf{g}_j = g_{jk} g^{ki}$$

$$= \delta_j^i \tag{A.4.2-4}$$

This should remind the reader of the discussion of cofactors in Section A.2.1. The discussion of determinants given there remains valid whether we use superscripts, subscripts, or any appropriate mixture. With this thought in mind, we may recognize $g^{im}$ as the cofactor of the $g_{mi}$ in $\det(g_{rs})$ divided by $\det(g_{rs})$:

$$g^{im} = \frac{1}{2 \det(g_{rs})} e^{ijk} e^{mnp} g_{nj} g_{pk} \tag{A.4.2-5}$$

Notice that in writing (A.4.2-5) we used $e^{ijk}$, which was defined in Section A.2.1 but never used. Our excuse for using it here is that we wish to preserve the symmetry of indices. Notice also how understood summations on indices associated with curvilinear coordinate systems occur between one superscript and one subscript. There will be more on this later.

In arriving at (A.4.2-5), we divided by $\det(g_{rs})$, assuming that $\det(g_{rs})$ cannot be zero. Let us prove this. Starting with the definition of $g_{mn}$, we have

$$g_{mn} \equiv \mathbf{g}_m \cdot \mathbf{g}_n$$

$$= \frac{\partial \mathbf{p}}{\partial x^m} \cdot \frac{\partial \mathbf{p}}{\partial x^n}$$

$$= \frac{\partial z_i}{\partial x^m} \mathbf{e}_i \cdot \frac{\partial z_j}{\partial x^n} \mathbf{e}_j$$

$$= \frac{\partial z_i}{\partial x^m} \frac{\partial z_j}{\partial x^n} \delta_{ij}$$

$$= \frac{\partial z_i}{\partial x^m} \frac{\partial z_i}{\partial x^n} \tag{A.4.2-6}$$

This means that

$$\det(g_{mn}) = \det\left(\frac{\partial z_i}{\partial x^m}\frac{\partial z_i}{\partial x^n}\right) \tag{A.4.2-7}$$

Our discussion of the product of two determinants in Section A.2.1 allows us to write this as

$$\det(g_{mn}) = \left[\det\left(\frac{\partial z_i}{\partial x^m}\right)\right]^2 \tag{A.4.2-8}$$

But the restriction placed upon the definition of a curvilinear coordinate system in Section A.4.1 allows us to conclude that $\det(g_{mn}) \neq 0$.

From (A.4.2-2) and (A.4.2-3), we have that

$$\begin{aligned}
\mathbf{g}^i \cdot \mathbf{g}^j &= g^{ki}\mathbf{g}_k \cdot \mathbf{g}^j \\
&= g^{ki}\delta_k{}^j \\
&= g^{ji}
\end{aligned} \tag{A.4.2-9}$$

Since the scalar product is symmetric, $g^{ij}$ is symmetric in its indices.

Let us prove that the three dual vector fields $\mathbf{g}^i$ ($i = 1, 2, 3$) form another basis for the spatial vector fields. From our discussion in Section A.1.4, it is necessary to show only that every spatial vector field can be written as a linear combination of them. Since we have already shown that every spatial vector field can be written as a linear combination of the natural basis, all that we must demonstrate is that each of the natural basis fields can be expressed as a linear combination of the dual vector fields. Multiplying (A.4.2-2) by $g_{ik}$, summing on $i$, and employing (A.4.2-4)$_2$, we obtain

$$\begin{aligned}
g_{ik}\mathbf{g}^i &= g^{ji}g_{ik}\mathbf{g}_j \\
&= \delta_k{}^j\mathbf{g}_j \\
&= \mathbf{g}_k
\end{aligned} \tag{A.4.2-10}$$

which completes the proof.

**Exercise A.4.2-1**   Prove that in a rectangular Cartesian coordinate system the natural basis vectors $\mathbf{e}_k$ and the dual basis vectors $\mathbf{e}^k$ are identical.

**Exercise A.4.2-2**   If the $\mathbf{g}^i$ ($i = 1, 2, 3$) are the dual basis vector fields associated with one curvilinear coordinate system ($x^1, x^2, x^3$) and if the $\bar{\mathbf{g}}^i$ ($i = 1, 2, 3$) are the dual basis vector fields associated with another curvilinear coordinate system ($\bar{x}^1, \bar{x}^2, \bar{x}^3$), prove that

$$\bar{\mathbf{g}}^i = \frac{\partial \bar{x}^i}{\partial x^j}\mathbf{g}^j$$

**Exercise A.4.2-3**   Prove that in orthogonal coordinate systems the dual basis fields are orthogonal to one another.

**Exercise A.4.2-4**  If the curvilinear coordinates are orthogonal, prove that

$$\det(g_{mn}) = g_{11}g_{22}g_{33}$$

$$g^{11} = \frac{1}{g_{11}}$$

$$g^{22} = \frac{1}{g_{22}}$$

$$g^{33} = \frac{1}{g_{33}}$$

and

$$g_{mn} = g^{mn}$$
$$= 0 \quad \text{if } m \neq n$$

## A.4.3  Covariant and Contravariant Components*

Given a curvilinear coordinate system, we may express every spatial vector field as a linear combination of the natural basis

$$\mathbf{u} = u^i \mathbf{g}_i \qquad\qquad\qquad\qquad\qquad (A.4.3\text{-}1)$$

or as a linear combination of the dual basis

$$\mathbf{u} = u_i \mathbf{g}^i \qquad\qquad\qquad\qquad\qquad (A.4.3\text{-}2)$$

The $u^i$ and $u_i$ are the *contravariant* and *covariant* components of the spatial vector field $\mathbf{u}$.

It is because we concern ourselves with these two sets of bases in dealing with each curvilinear coordinate system that we choose to introduce superscripts as well as subscripts in our notation. We will notice hereafter that, when the summation convention is employed with covariant and contravariant components, one of the repeated indices will be a superscript and one will be a subscript. This is the result of our arbitrary choice of notation in (A.4.3-1) and (A.4.3-2); it is here that the summation between superscripts and subscripts is introduced.

The notation has at least one helpful feature that should be kept in mind. We will see that any equation involving components will have a certain symmetry with respect to indices not involved in summations. For example, if the index $j$ is not repeated and if it occurs as a superscript in one term of the equation, it will occur as a superscript in all terms of the equation.

Why did we not use superscripts as well as subscripts when discussing rectangular Cartesian coordinate systems? In Exercise A.4.2-1, one learns that the natural and dual basis vectors are identical in orthogonal Cartesian coordinate systems. Consequently, it is pointless to distinguish between covariant and contravariant components of vectors in rectangular Cartesian coordinates, and the need for superscripts in addition to subscripts disappears.

Since for any spatial vector field

$$\mathbf{u} = u^i \mathbf{g}_i$$
$$= u^i g_{ki} \mathbf{g}^k$$
$$= u_k \mathbf{g}^k \qquad\qquad\qquad\qquad\qquad (A.4.3\text{-}3)$$

we may write

$$(u^i g_{ki} - u_k)\mathbf{g}^k = 0 \tag{A.4.3-4}$$

The dual basis vector fields are linearly independent and therefore (A.4.3-4) implies that

$$u^i g_{ki} - u_k = 0 \tag{A.4.3-5}$$

or

$$u_k = g_{ki} u^i \tag{A.4.3-6}$$

In the same way,

$$\begin{aligned}
\mathbf{u} &= u_i \mathbf{g}^i \\
&= u_i g^{ji} \mathbf{g}_j \\
&= u^j \mathbf{g}_j
\end{aligned} \tag{A.4.3-7}$$

so that we may identify

$$u^j = g^{ji} u_i \tag{A.4.3-8}$$

We find in this way that the $g_{ij}$ and the $g^{ij}$ may be used to *raise and lower* indices.

Let us determine the relation between the physical components

$$\left( u_{\langle 1 \rangle}, u_{\langle 2 \rangle}, u_{\langle 3 \rangle} \right)$$

of a spatial vector field $\mathbf{u}$ and its contravariant components. From Section A.4.1 and (A.4.3-1), we may write

$$\begin{aligned}
\mathbf{u} &= u^i \mathbf{g}_i \\
&= u^1 \sqrt{g_{11}} \mathbf{g}_{\langle 1 \rangle} + u^2 \sqrt{g_{22}} \mathbf{g}_{\langle 2 \rangle} + u^3 \sqrt{g_{33}} \mathbf{g}_{\langle 3 \rangle} \\
&= u_{\langle i \rangle} \mathbf{g}_{\langle i \rangle}
\end{aligned} \tag{A.4.3-9}$$

We conclude that

$$u_{\langle i \rangle} = \sqrt{g_{ii}} u^i, \quad \text{no summation on } i \tag{A.4.3-10}$$

A similar relation may be obtained for the physical components in terms of the covariant components. From the definition of the physical basis fields in Section A.4.1 and the relation between the dual basis fields and natural basis fields in Section A.4.2, we have

$$\begin{aligned}
\mathbf{g}_{\langle i \rangle} &= \frac{1}{\sqrt{g_{ii}}} g_{ii} \mathbf{g}^i, \quad \text{no summation on } i \\
&= \sqrt{g_{ii}} \mathbf{g}^i, \quad \text{no summation on } i
\end{aligned} \tag{A.4.3-11}$$

In arriving at this result, we have taken advantage of the restriction to orthogonal coordinate systems when discussing physical basis fields. From (A.4.3-2),

$$\begin{aligned}
\mathbf{u} &= u_i \mathbf{g}^i \\
&= \frac{u_1}{\sqrt{g_{11}}} \mathbf{g}_{\langle 1 \rangle} + \frac{u_2}{\sqrt{g_{22}}} \mathbf{g}_{\langle 2 \rangle} + \frac{u_3}{\sqrt{g_{33}}} \mathbf{g}_{\langle 3 \rangle} \\
&= u_{\langle i \rangle} \mathbf{g}_{\langle i \rangle}
\end{aligned} \tag{A.4.3-12}$$

We have consequently that

$$u_{\langle i \rangle} = \frac{u_i}{\sqrt{g_{ii}}}, \quad \text{no summation on } i \tag{A.4.3-13}$$

**Exercise A.4.3-1**   Show that in rectangular Cartesian coordinates:

i) The natural basis fields and the physical basis fields are equivalent.
ii) It is unnecessary to distinguish among covariant, contravariant, and physical components of spatial vector fields in rectangular Cartesian coordinates. We may write all indices as subscripts, employing the summation convention over repeated subscripts so long as we restrict ourselves to *rectangular Cartesian coordinates*.

**Exercise A.4.3-2**

i) Let **u** be some spatial vector field. If the $u^i$ ($i = 1, 2, 3$) are the contravariant components of **u** with respect to one curvilinear coordinate system $(x^1, x^2, x^3)$ and if the $\overline{u}^i$ ($i = 1, 2, 3$) are the contravariant components of **u** with respect to another curvilinear coordinate system $(\overline{x}^1, \overline{x}^2, \overline{x}^3)$, show that

$$u^i = \frac{\partial x^i}{\partial \overline{x}^j} \overline{u}^j$$

ii) Similarly, show that

$$u_i = \frac{\partial \overline{x}^j}{\partial x^i} \overline{u}_j$$

**Exercise A.4.3-3**   *Angle between surfaces*   Show that the angle between two surfaces, $\varphi(x^1, x^2, x^3) =$ a constant and $\psi(x^1, x^2, x^3) =$ a constant, is given by

$$\cos \theta = \frac{g^{mn}(\partial \varphi / \partial x^m)(\partial \psi / \partial x^n)}{[g^{rs}(\partial \varphi / \partial x^r)(\partial \varphi / \partial x^s)g^{uv}(\partial \psi / \partial x^u)(\partial \psi / \partial x^v)]^{1/2}}$$

*Hint:*   See Exercise A.3.1-5.

**Exercise A.4.3-4**   *Angle between coordinate surfaces*   Deduce that the angle $\varphi_{12}$ between the coordinate surfaces $x^1 =$ a constant and $x^2 =$ a constant is given by

$$\cos \varphi_{12} = \frac{g^{12}}{\sqrt{g^{11} g^{22}}}$$

**Exercise A.4.3-5**   Establish that if two surfaces, $\varphi(x^1, x^2, x^3) =$ a constant and $\psi(x^1, x^2, x^3) =$ a constant, cut one another orthogonally,

$$g^{mn} \frac{\partial \varphi}{\partial x^m} \frac{\partial \psi}{\partial x^n} = 0$$

**Exercise A.4.3-6** *More about gradient of scalar field*   Let $\varphi$ be a scalar field. Starting with the expression for $\nabla \varphi$ with respect to a rectangular Cartesian coordinate basis, show that

$$\nabla \varphi = \frac{\partial \varphi}{\partial x^i} \mathbf{g}^i$$

If we restrict ourselves to an orthogonal coordinate system, we have

$$\nabla \varphi = \sum_{i=1}^{3} \frac{1}{\sqrt{g_{ii}}} \frac{\partial \varphi}{\partial x^i} \mathbf{g}_{\langle i \rangle}$$

## A.5    Second-Order Tensors

A second-order tensor field $\mathbf{T}$ is a transformation (or mapping or rule) that assigns to each given spatial vector field $\mathbf{v}$ another spatial vector field $\mathbf{T} \cdot \mathbf{v}$ such that the rules

$$\mathbf{T} \cdot (\mathbf{v} + \mathbf{w}) = \mathbf{T} \cdot \mathbf{v} + \mathbf{T} \cdot \mathbf{w}$$
$$\mathbf{T} \cdot (\alpha \mathbf{v}) = \alpha (\mathbf{T} \cdot \mathbf{v}) \tag{A.5.0-1}$$

hold. By $\alpha$ we mean here a real scalar field. (Note that we are using the dot notation in a different manner here from that used in Section A.1.1 when we discussed the inner product. Our choice of notation is suggestive, however, as will shortly become evident.)

We define the sum $\mathbf{T} + \mathbf{S}$ of two second-order tensor fields $\mathbf{T}$ and $\mathbf{S}$ to be a transformation such that, for every spatial vector field $\mathbf{v}$,

$$(\mathbf{T} + \mathbf{S}) \cdot \mathbf{v} = \mathbf{T} \cdot \mathbf{v} + \mathbf{S} \cdot \mathbf{v} \tag{A.5.0-2}$$

The product $\alpha \mathbf{T}$ of a second-order tensor field $\mathbf{T}$ with a real scalar field $\alpha$ is a transformation such that, for every spatial vector field $\mathbf{v}$,

$$(\alpha \mathbf{T}) \cdot \mathbf{v} = \alpha (\mathbf{T} \cdot \mathbf{v}) \tag{A.5.0-3}$$

The transformations $\mathbf{T} + \mathbf{S}$ and $\alpha \mathbf{T}$ may be easily shown to obey the rules for second-order tensor fields. If we define the zero second-order tensor field $\mathbf{0}$ by the requirement that

$$\mathbf{0} \cdot \mathbf{v} = \mathbf{0} \tag{A.5.0-4}$$

for all spatial vector fields $\mathbf{v}$, we see that the rules $(A_1)$ to $(A_4)$ and $(M_1)$ to $(M_4)$ of Section A.1.1 are satisfied and that the set of all second-order tensor fields constitutes a vector space.

If two spatial vector fields $\mathbf{a}$ and $\mathbf{b}$ are given, we can define a second-order tensor field $\mathbf{ab}$ by the requirement that it transform every vector field $\mathbf{v}$ into another vector field $(\mathbf{ab} \cdot \mathbf{v})$ according to the rule

$$(\mathbf{ab}) \cdot \mathbf{v} \equiv \mathbf{a}(\mathbf{b} \cdot \mathbf{v}) \tag{A.5.0-5}$$

This tensor field $\mathbf{ab}$ is called the *tensor product* or *dyadic product* of the spatial vector fields $\mathbf{a}$ and $\mathbf{b}$. [Another common notation for the tensor product is $\mathbf{a} \otimes \mathbf{b}$ (Halmos 1958, p. 40; Lichnerowicz 1962, p. 29).]

In this text, we use boldface capital letters for second-order tensor fields and boldface lowercase letters for spatial vector fields.

### A.5.1 Components of Second-Order Tensor Fields

If $\mathbf{T}$ is a second-order tensor field and the $\mathbf{e}_j$ ($j = 1, 2, 3$) form a rectangular Cartesian basis for the space of spatial vector fields, we may write

$$\mathbf{T} \cdot \mathbf{e}_j = T_{ij}\mathbf{e}_i \tag{A.5.1-1}$$

The *matrix* $[T_{ij}]$ (the array of the components) of the second-order tensor field $\mathbf{T}$,

$$[T_{ij}] \equiv \begin{bmatrix} T_{11} & T_{12} & T_{13} \\ T_{21} & T_{22} & T_{23} \\ T_{31} & T_{32} & T_{33} \end{bmatrix} \tag{A.5.1-2}$$

tells how the basis fields $\mathbf{e}_j$ are transformed by $\mathbf{T}$.

Let $\mathbf{v}$ be any spatial vector field. Equation (A.5.1-1) allows us to develop an expression for the vector field $\mathbf{T} \cdot \mathbf{v}$ in terms of the rectangular Cartesian components of $\mathbf{v}$:

$$\begin{aligned} \mathbf{T} \cdot \mathbf{v} &= \mathbf{T} \cdot (v_j\mathbf{e}_j) \\ &= v_j\mathbf{T} \cdot \mathbf{e}_j \\ &= v_j T_{ij}\mathbf{e}_i \end{aligned} \tag{A.5.1-3}$$

In this way, for each set of basis fields $(\mathbf{e}_1, \mathbf{e}_2, \mathbf{e}_3)$, we may associate a matrix $[T_{ij}]$ with any second-order tensor field $\mathbf{T}$. This association or correspondence is one-to-one (that is, for the same set of basis fields, the matrices of two different second-order tensor fields are different) (Halmos 1958, p. 67). To prove this, observe that the matrix $[T_{ij}]$ of a second-order tensor field $\mathbf{T}$ completely determines $\mathbf{T}$ [by (A.5.1-3), $\mathbf{T} \cdot \mathbf{v}$ is determined for every $\mathbf{v}$].

Given any second-order tensor field $\mathbf{T}$, which transforms the rectangular Cartesian basis fields according to (A.5.1-1), define a new second-order tensor

$$\mathbf{T}^* \equiv T_{ij}\mathbf{e}_i\mathbf{e}_j \tag{A.5.1-4}$$

which is the sum of nine tensor products (see the introduction to Section A.5). However, we have

$$\begin{aligned} \mathbf{T}^* \cdot \mathbf{e}_j &= (T_{ik}\mathbf{e}_i\mathbf{e}_k) \cdot \mathbf{e}_j \\ &= T_{ik}\mathbf{e}_i\delta_{kj} \\ &= T_{ij}\mathbf{e}_i \end{aligned} \tag{A.5.1-5}$$

indicating that the same matrix $[T_{ij}]$ corresponds to both $\mathbf{T}$ and $\mathbf{T}^*$. Since we showed above that, with the choice of a particular set of rectangular Cartesian basis fields $\mathbf{e}_i$ ($i = 1, 2, 3$), there is a one-to-one correspondence between $(3 \times 3)$ matrices and second-order tensor fields, we conclude that

$$\begin{aligned} \mathbf{T} &= \mathbf{T}^* \\ &= T_{ij}\mathbf{e}_i\mathbf{e}_j \end{aligned} \tag{A.5.1-6}$$

The nine coefficients $T_{ij}$ ($i, j = 1, 2, 3$) are referred to as the *rectangular Cartesian components* of $\mathbf{T}$. We will find this representation for second-order tensor fields in terms of a sum of tensor products of basis fields to be a very useful one.

The *identity tensor field* **I** is a specific example of a second-order tensor field. It transforms every spatial vector field into itself:

$$\begin{aligned}
\mathbf{I} \cdot \mathbf{e}_j &= I_{ij}\mathbf{e}_i \\
&= \mathbf{e}_j \\
&= \delta_{ij}\mathbf{e}_i
\end{aligned} \tag{A.5.1-7}$$

Here $\delta_{ij}$ is the Kronecker delta defined in Section A.1.7. From (A.5.1-7), we have

$$(I_{ij} - \delta_{ij})\mathbf{e}_i = \mathbf{0} \tag{A.5.1-8}$$

But since the rectangular Cartesian basis fields are linearly independent (Section A.1.4), we conclude that

$$I_{ij} = \delta_{ij} \tag{A.5.1-9}$$

Let us pause before pursuing these ideas further to say something about the notation we have chosen to use here. When we write $\mathbf{T} \cdot \mathbf{e}_i$, the dot is to remind us that $\mathbf{T}$ operates on the quantity that follows. It has a completely different significance from the dot in $\mathbf{e}_i \cdot \mathbf{e}_j$, where the dot indicates a scalar product. Although this is a disadvantage to the notation chosen, when any equation is read in context, there is little excuse for confusion. Having been told that $\mathbf{T}$ is a second-order tensor field and that $\mathbf{e}_i$ is a spatial vector field, you will have no occasion to interpret $\mathbf{T} \cdot \mathbf{e}_i$ as the scalar product of two spatial vector fields. The advantage of the notation is that it is suggestive of the operation to be carried out when $\mathbf{T}$ is written as a sum of tensor products:

$$\begin{aligned}
\mathbf{T} \cdot \mathbf{e}_i &= (T_{jk}\mathbf{e}_j\mathbf{e}_k) \cdot \mathbf{e}_i \\
&= T_{jk}\mathbf{e}_j(\mathbf{e}_k \cdot \mathbf{e}_i)
\end{aligned} \tag{A.5.1-10}$$

The dot in $\mathbf{T} \cdot \mathbf{e}_i$ reminds you that, when $\mathbf{T}$ is written as the sum of tensor products, the transformation is accomplished by taking the *scalar product* between the second spatial vector of the tensor product and the spatial vector to be transformed, $\mathbf{e}_i$. The notation adopted here is more common in engineering and applied science texts, where considerable emphasis is placed upon working out problems in specific coordinate systems. Mathematicians adopt a slightly different notation when treating subjects where the introduction of coordinate systems is either avoided or is of secondary importance. If one understands any one system of notation, there is little difficulty in adapting to another.

Let us return to (A.5.1-6) and observe that the set of nine tensor products $\mathbf{e}_i\mathbf{e}_j$ ($i, j = 1, 2, 3$) forms a basis (Section A.1.4) for the vector space of second-order tensor fields (introduction to Section A.5). Certainly, every element of the set of second-order tensor fields is expressible as a linear combination of the $\mathbf{e}_i\mathbf{e}_j$. We must show that the $\mathbf{e}_i\mathbf{e}_j$ are linearly independent. If

$$\begin{aligned}
\mathbf{A} &= A_{ij}\mathbf{e}_i\mathbf{e}_j \\
&= \mathbf{0}
\end{aligned} \tag{A.5.1-11}$$

then

$$\begin{aligned}
\mathbf{A} \cdot \mathbf{e}_k &= (A_{ij}\mathbf{e}_i\mathbf{e}_j) \cdot \mathbf{e}_k \\
&= A_{ij}\mathbf{e}_i\delta_{jk} \\
&= \mathbf{0}
\end{aligned} \tag{A.5.1-12}$$

or

$$A_{ik}\mathbf{e}_i = \mathbf{0} \tag{A.5.1-13}$$

Since the rectangular Cartesian basis fields are linearly independent, this implies that

$$A_{mk} = 0 \tag{A.5.1-14}$$

We conclude that the nine tensor products of the form $\mathbf{e}_i\mathbf{e}_j$ are linearly independent and form a basis for the vector space of second-order tensors. [As a by-product, we find that the vector space of second-order tensors is nine dimensional (Section A.1.4).]

In physical applications it is often convenient to introduce an orthogonal curvilinear coordinate system. If the $\mathbf{g}_{\langle i \rangle}$ ($i = 1, 2, 3$) are the associated physical basis fields (Section A.4.1), we may write by analogy with (A.5.1-1)

$$\mathbf{T} \cdot \mathbf{g}_{\langle i \rangle} = T_{\langle ji \rangle}\mathbf{g}_{\langle j \rangle} \tag{A.5.1-15}$$

By the same argument that led us to (A.5.1-6), we may write

$$\mathbf{T} = T_{\langle ij \rangle}\mathbf{g}_{\langle i \rangle}\mathbf{g}_{\langle j \rangle} \tag{A.5.1-16}$$

where the nine coefficients $T_{\langle ij \rangle}$ ($i, j = 1, 2, 3$) are referred to as the *physical components* of $\mathbf{T}$. [The set of nine tensor products $\mathbf{g}_{\langle i \rangle}\mathbf{g}_{\langle j \rangle}$ ($i, j = 1, 2, 3$) forms another basis for the vector space of second-order tensor fields.]

### Exercise A.5.1-1

i) For an orthogonal coordinate system, prove that (A.5.1-16) holds for every second-order tensor field $\mathbf{T}$.

ii) Prove that the nine tensor products $\mathbf{g}_{\langle i \rangle}\mathbf{g}_{\langle j \rangle}$ ($i, j = 1, 2, 3$) form a basis for the vector space of second-order tensor fields.

### Exercise A.5.1-2

*i) Given any curvilinear system, prove that every second-order tensor field $\mathbf{T}$ may be written as

$$\mathbf{T} = T^{ij}\mathbf{g}_i\mathbf{g}_j$$

The nine coefficients $T^{ij}$ ($i, j = 1, 2, 3$) are referred to as the *contravariant components* of $\mathbf{T}$.

ii) Prove that the nine tensor products $\mathbf{g}_i\mathbf{g}_j$ ($i, j = 1, 2, 3$) form a basis for the space of second-order tensor fields.

### Exercise A.5.1-3

i) Given any curvilinear coordinate system, prove that every second-order tensor field $\mathbf{T}$ may be written as

$$\mathbf{T} = T_{ij}\mathbf{g}^i\mathbf{g}^j$$

The nine coefficients $T_{ij}$ ($i, j = 1, 2, 3$) are referred to as the *covariant components* of $\mathbf{T}$.

ii) Prove that the nine tensor products $\mathbf{g}^i\mathbf{g}^j$ ($i, j = 1, 2, 3$) form a basis for the space of second-order tensor fields.

### Exercise A.5.1-4

*i) Given any curvilinear coordinate system, prove that every second-order tensor field **T** may be written as

$$\mathbf{T} = T_{.i}^{j}\mathbf{g}^{i}\mathbf{g}_{j}$$

The nine coefficients $T_{.i}^{j}$ ($i, j = 1, 2, 3$) are referred to as the *mixed* components of **T** *covariant* in $i$ and *contravariant* in $j$.

ii) Prove that the nine tensor products $\mathbf{g}^{i}\mathbf{g}_{j}$ ($i, j = 1, 2, 3$) form a basis for the space of second-order tensor fields.

### Exercise A.5.1-5

*i) Given any curvilinear coordinate system, prove that every second-order tensor field **T** may be written as

$$\mathbf{T} = T_{.j}^{i}\mathbf{g}_{i}\mathbf{g}^{j}$$

The nine coefficients $T_{.j}^{i}$ ($i, j = 1, 2, 3$) are referred to as the *mixed* components of **T**, *contravariant* in $i$ and *covariant* in $j$.

ii) Prove that the nine tensor products $\mathbf{g}_{i}\mathbf{g}^{j}$ ($i, j = 1, 2, 3$) form a basis for the space of second-order tensor fields.

### Exercise A.5.1-6*    Show that

$$T_{\langle ij\rangle} = \frac{T_{ij}}{\sqrt{g_{ii}g_{jj}}}, \quad \text{no summation on } i \text{ and } j$$

$$T_{\langle ij\rangle} = T_{.j}^{i}\frac{\sqrt{g_{ii}}}{\sqrt{g_{jj}}}, \quad \text{no summation on } i \text{ and } j$$

and

$$T_{\langle ij\rangle} = T_{i.}^{j}\frac{\sqrt{g_{jj}}}{\sqrt{g_{ii}}}, \quad \text{no summation on } i \text{ and } j$$

### Exercise A.5.1-7    Show that the identity tensor has the following equivalent forms:

$$\mathbf{I} = \delta_{i}^{\ j}\mathbf{g}_{j}\mathbf{g}^{i}$$
$$= \delta_{j}^{\ i}\mathbf{g}^{j}\mathbf{g}_{i}$$
$$= g^{ij}\mathbf{g}_{i}\mathbf{g}_{j}$$
$$= g_{ij}\mathbf{g}^{i}\mathbf{g}^{j}$$

[The $g_{ij}$ and $g^{ij}$ are usually referred to as the covariant metric tensor (components) and contravariant metric tensor (components). These names will not be used here, since they are not consistent with the form of presentation chosen.]

### Exercise A.5.1-8*    Show that it is unnecessary to distinguish among covariant, contravariant, and physical components of second-order tensor fields in *rectangular Cartesian coordinate systems*. We may therefore write all indices as subscripts, employing the summation convention

over repeated subscripts *so long as we restrict ourselves to rectangular Cartesian coordinate systems.*

### Exercise A.5.1-9

*i) Let **T** be some second-order tensor field. If the $T^{ij}$ ($i, j = 1, 2, 3$) are the contravariant components of **T** with respect to one curvilinear coordinate system $(x^1, x^2, x^3)$ and if the $\overline{T}^{ij}$ ($i, j = 1, 2, 3$) are the contravariant components of **T** with respect to another curvilinear coordinate system $(\overline{x}^1, \overline{x}^2, \overline{x}^3)$, show that

$$T^{ij} = \frac{\partial x^i}{\partial \overline{x}^m} \frac{\partial x^j}{\partial \overline{x}^n} \overline{T}^{mn}$$

ii) Similarly, show that

$$T_{ij} = \frac{\partial \overline{x}^m}{\partial x^i} \frac{\partial \overline{x}^n}{\partial x^j} \overline{T}_{mn}$$

$$T_i^{\cdot j} = \frac{\partial \overline{x}^m}{\partial x^i} \frac{\partial x^j}{\partial \overline{x}^n} \overline{T}_m^{\cdot n}$$

and

$$T_{\cdot j}^i = \frac{\partial x^i}{\partial \overline{x}^m} \frac{\partial \overline{x}^n}{\partial x^j} \overline{T}_{\cdot n}^m$$

### Exercise A.5.1-10

*i) If the $T_{ij}$ ($i, j = 1, 2, 3$) are the *covariant* components of a second-order tensor field **T** and the $T^{ij}$ ($i, j = 1, 2, 3$) are the *contravariant* components, show that

$$T_{ij} = g_{im} g_{in} T^{mn}$$

ii) Similarly, show that

$$T_{i.}^{\ j} = g_{im} T^{mj}$$

$$T_{.j}^i = g^{im} T_{mj}$$

and

$$T^{ij} = g^{im} g^{jn} T_{mn}$$

We find here that the $g_{ij}$ and the $g^{ij}$ may be used to raise and lower indices. (Compare with the relations between covariant and contravariant components of spatial vector fields found in Section A.4.3.) We use the dot in writing $T_{.j}^i$ and $T_{i.}^{\ j}$ to remind ourselves which index has been raised or lowered; this is unnecessary when dealing with symmetric second-order tensors (Section A.5.2).

### Exercise A.5.1-11
Let **T** be some second-order tensor field. If the $T_{ij}$ are the components of **T** with respect to some rectangular Cartesian coordinate system $(z_1, z_2, z_3)$ and the $\overline{T}_{mn}$ ($m, n = 1, 2, 3$) are the components of **T** with respect to another rectangular Cartesian coordinate system $(\overline{z}_1, \overline{z}_2, \overline{z}_3)$, show that

$$T_{ij} = \frac{\partial z_i}{\partial \overline{z}_m} \frac{\partial z_j}{\partial \overline{z}_n} \overline{T}_{mn}$$

and

$$T_{ij} = \frac{\partial \bar{z}_m}{\partial z_i} \frac{\partial \bar{z}_n}{\partial z_j} \bar{T}_{mn}$$

### Exercise A.5.1-12

    i) Let $\mathbf{T} = T_{ij}\mathbf{e}_i\mathbf{e}_j$ and $\mathbf{S} = S_{ij}\mathbf{e}_i\mathbf{e}_j$ be two second-order tensor fields, where the $T_{ij}$ and $S_{ij}$ $(i, j = 1, 2, 3)$ are real scalar fields. Let $\alpha$ be a real scalar field. Express the second-order tensor fields $\mathbf{T} + \mathbf{S}$ and $\alpha\mathbf{T}$ as linear combinations of the $\mathbf{e}_i\mathbf{e}_j$.

    ii) Express the second-order tensor fields $\mathbf{T} + \mathbf{S}$ and $\alpha\mathbf{T}$ as linear combinations of the $\mathbf{g}_i\mathbf{g}^j$.

## A.5.2 Transpose of a Second-Order Tensor Field**

Let $\mathbf{T}$ be any second-order tensor field. We define $\mathbf{T}^T$, the *transpose* of $\mathbf{T}$, to be that second-order tensor field such that, if $\mathbf{u}$ and $\mathbf{v}$ are any two spatial vector fields,

$$(\mathbf{T} \cdot \mathbf{u}) \cdot \mathbf{v} = \mathbf{u} \cdot (\mathbf{T}^T \cdot \mathbf{v}) \tag{A.5.2-1}$$

To determine the relation of $\mathbf{T}^T$ to $\mathbf{T}$, let

$$\mathbf{u} = \mathbf{e}_i$$
$$\mathbf{v} = \mathbf{e}_j$$
$$\mathbf{T} = T_{mn}\mathbf{e}_m\mathbf{e}_n \tag{A.5.2-2}$$
$$\mathbf{T}^T = T_{rs}^T\mathbf{e}_r\mathbf{e}_s$$

Here the $\mathbf{e}_i$ $(i = 1, 2, 3)$ represent a set of rectangular Cartesian basis fields. From (A.5.2-1), we can compute

$$(\mathbf{T} \cdot \mathbf{e}_i) \cdot \mathbf{e}_j = \mathbf{e}_i \cdot (\mathbf{T}^T \cdot \mathbf{e}_j)$$
$$(T_{mn}\mathbf{e}_m\mathbf{e}_n \cdot \mathbf{e}_i) \cdot \mathbf{e}_j = \mathbf{e}_i \cdot (T_{rs}^T\mathbf{e}_r\mathbf{e}_s \cdot \mathbf{e}_j) \tag{A.5.2-3}$$
$$T_{ji} = T_{ij}^T$$

If we represent $\mathbf{T}$ as indicated in (A.5.2-2)$_3$, we may represent its transpose by

$$\mathbf{T}^T = T_{nm}\mathbf{e}_m\mathbf{e}_n \tag{A.5.2-4}$$

Since for any spatial vector field $\mathbf{w}$

$$\mathbf{T} \cdot \mathbf{w} = T_{ij}w_j\mathbf{e}_i$$
$$= T_{ji}^T w_j\mathbf{e}_i$$
$$= w_j T_{ji}^T\mathbf{e}_i \tag{A.5.2-5}$$

we are prompted to introduce the definition

$$\mathbf{w} \cdot \mathbf{T}^T \equiv \mathbf{T} \cdot \mathbf{w} \tag{A.5.2-6}$$

One may think of this operation as being carried out in the following manner:

$$\mathbf{w} \cdot \mathbf{T}^T = (w_k\mathbf{e}_k) \cdot (T^T{}_{ji}\mathbf{e}_j\mathbf{e}_i)$$
$$= w_j T_{ji}^T\mathbf{e}_i \tag{A.5.2-7}$$

A second-order tensor field $\mathbf{T}$ is said to be *symmetric* if it is equal to its transpose:

$$\mathbf{T} = \mathbf{T}^T \tag{A.5.2-8}$$

In terms of their rectangular Cartesian components, we have

$$
\begin{aligned}
T_{ij} &= T_{ij}^T \\
&= T_{ji}
\end{aligned} \tag{A.5.2-9}
$$

A second-order tensor field $\mathbf{T}$ is said to be *skew symmetric* if

$$\mathbf{T} = -\mathbf{T}^T \tag{A.5.2-10}$$

The relation between the rectangular Cartesian components of these two tensor fields is

$$
\begin{aligned}
T_{ij} &= -T_{ij}^T \\
&= -T_{ji}
\end{aligned} \tag{A.5.2-11}
$$

An *orthogonal* tensor field or transformation of the space of spatial vector fields is one that preserves lengths and angles. If $\mathbf{u}$ and $\mathbf{v}$ are any two spatial vector fields and $\mathbf{Q}$ is an orthogonal tensor field, we require

$$(\mathbf{Q} \cdot \mathbf{u}) \cdot (\mathbf{Q} \cdot \mathbf{v}) = \mathbf{u} \cdot \mathbf{v} \tag{A.5.2-12}$$

But this means that

$$
\begin{aligned}
\mathbf{u} \cdot \left[ \mathbf{Q}^T \cdot (\mathbf{Q} \cdot \mathbf{v}) \right] &= \mathbf{u} \cdot \left[ (\mathbf{Q}^T \cdot \mathbf{Q}) \cdot \mathbf{v} \right] \\
&= \mathbf{u} \cdot \mathbf{v}
\end{aligned} \tag{A.5.2-13}
$$

or

$$\mathbf{u} \cdot \left[ (\mathbf{Q}^T \cdot \mathbf{Q}) \cdot \mathbf{v} - \mathbf{v} \right] = 0 \tag{A.5.2-14}$$

Since $\mathbf{u}$ and $\mathbf{v}$ are arbitrary spatial vector fields, we conclude that

$$(\mathbf{Q}^T \cdot \mathbf{Q}) \cdot \mathbf{v} = \mathbf{v} \tag{A.5.2-15}$$

and

$$\mathbf{Q}^T \cdot \mathbf{Q} = \mathbf{I} \tag{A.5.2-16}$$

It can be shown further that (see Exercise A.5.2-3)

$$\mathbf{Q} \cdot \mathbf{Q}^T = \mathbf{Q}^T \cdot \mathbf{Q} = \mathbf{I} \tag{A.5.2-17}$$

Here we introduce the notation $\mathbf{A} \cdot \mathbf{B}$, where $\mathbf{A}$ and $\mathbf{B}$ are any two second-order tensor fields. If $\mathbf{v}$ is any spatial vector field, the spatial vector field $\mathbf{A} \cdot (\mathbf{B} \cdot \mathbf{v})$ is obtained by first applying the transformation $\mathbf{B}$ to the spatial vector field $\mathbf{v}$ and then applying the transformation $\mathbf{A}$ to the result. We may think of $\mathbf{A} \cdot (\mathbf{B} \cdot \mathbf{v})$ as being obtained from $\mathbf{v}$ as the consequence of one transformation $\mathbf{A} \cdot \mathbf{B}$:

$$\mathbf{A} \cdot (\mathbf{B} \cdot \mathbf{v}) = (\mathbf{A} \cdot \mathbf{B}) \cdot \mathbf{v} \tag{A.5.2-18}$$

where

$$
\begin{aligned}
\mathbf{A} \cdot \mathbf{B} &\equiv (A_{ij}\mathbf{e}_i\mathbf{e}_j) \cdot (B_{km}\mathbf{e}_k\mathbf{e}_m) \\
&= A_{ij}B_{jm}\mathbf{e}_i\mathbf{e}_m
\end{aligned} \tag{A.5.2-19}
$$

[This observation gives us another view of second-order tensor fields. Any second-order tensor field **A** is a transformation that assigns to any other second-order tensor field **B** another second-order tensor field **A** · **B** defined by (A.5.2-19). However, not all transformations of the space of second-order tensor fields into itself are of this form (Hoffman and Kunze 1961, p. 69).]

Let us determine the transpose of **A** · **B**, where **A** and **B** are second-order tensor fields. If **u** and **v** are any two spatial vector fields,

$$[(\mathbf{A} \cdot \mathbf{B}) \cdot \mathbf{u}] \cdot \mathbf{v} = (\mathbf{B} \cdot \mathbf{u}) \cdot (\mathbf{A}^T \cdot \mathbf{v})$$
$$= \mathbf{u} \cdot [(\mathbf{B}^T \cdot \mathbf{A}^T) \cdot \mathbf{v}] \qquad (A.5.2\text{-}20)$$

We conclude that

$$(\mathbf{A} \cdot \mathbf{B})^T = \mathbf{B}^T \cdot \mathbf{A}^T \qquad (A.5.2\text{-}21)$$

**Exercise A.5.2-1**   If **T** is a second-order tensor field,

*i) Show that, with respect to any curvilinear coordinate system,

$$T_{ij} = T^T_{ji}$$

and

$$T^{ij} = T^{Tji}$$

ii) Show that, with respect to any orthogonal curvilinear coordinate system,

$$T_{\langle ij \rangle} = T^T_{\langle ji \rangle}$$

**Exercise A.5.2-2**   If **A** and **B** are any two second-order tensor fields,

*i) Show that, with respect to any curvilinear coordinate system,

$$\mathbf{A} \cdot \mathbf{B} = A_{ij} B^{jk} \mathbf{g}^i \mathbf{g}_k$$
$$= A^{ij} B_{jk} \mathbf{g}_i \mathbf{g}^k$$
$$= A^{j}_{i.} B_{jk} \mathbf{g}^i \mathbf{g}^k$$
$$= A^{ij} B^{k}_{j.} \mathbf{g}_i \mathbf{g}_k$$

ii) Show that, with respect to any orthogonal curvilinear coordinate system,

$$\mathbf{A} \cdot \mathbf{B} = A_{\langle ij \rangle} B_{\langle jk \rangle} \mathbf{g}_{\langle i \rangle} \mathbf{g}_{\langle k \rangle}$$

**Exercise A.5.2-3**   Starting with (A.5.2-16), prove (A.5.2-17).

**Exercise A.5.2-4** *Isotropic second-order tensors*   Let **A** be a second-order tensor field and **Q** an orthogonal tensor field. If

$$\mathbf{Q} \cdot \mathbf{A} \cdot \mathbf{Q}^T = \mathbf{A}$$

we refer to **A** as an *isotropic* second-order tensor. Prove that

$$\mathbf{A} = a\mathbf{I}$$

where $a$ is a scalar field.

*Hint:* Let [**Q**] denote the matrix (array) of the components of **Q** with respect to an appropriate basis.

i) Let

$$[\mathbf{Q}] = \begin{bmatrix} 1 & 0 & 0 \\ 0 & 1 & 0 \\ 0 & 0 & -1 \end{bmatrix}$$

to conclude that $A_{13} = A_{31} = A_{23} = A_{32} = 0$.

ii) Let

$$[\mathbf{Q}] = \begin{bmatrix} 1 & 0 & 0 \\ 0 & -1 & 0 \\ 0 & 0 & 1 \end{bmatrix}$$

to conclude that $A_{12} = A_{21} = A_{23} = A_{32} = 0$.

iii) Let

$$[\mathbf{Q}] = \begin{bmatrix} 0 & 1 & 0 \\ 1 & 0 & 0 \\ 0 & 0 & 1 \end{bmatrix}$$

to conclude that $A_{11} = A_{22}$.

iv) Let

$$[\mathbf{Q}] = \begin{bmatrix} 1 & 0 & 0 \\ 0 & 0 & 1 \\ 0 & 1 & 0 \end{bmatrix}$$

to conclude that $A_{22} = A_{33}$.

**Exercise A.5.2-5**  In Exercises A.4.1-8 and A.4.1-9, we studied the relations between the rectangular Cartesian components of a spatial vector field **v** and the physical components of this vector field with respect to an orthogonal curvilinear coordinate system. Let us consider these relationships from a different point of view.

i) Consider the transformations **A** and **B** such that

$$\mathbf{g}_{\langle j \rangle} = \mathbf{A} \cdot \mathbf{e}_j = A_{(ij)} \mathbf{e}_i$$

and

$$\mathbf{e}_j = \mathbf{B} \cdot \mathbf{g}_{\langle j \rangle} = B_{(ij)} \mathbf{g}_{\langle i \rangle}$$

Prove that **A** and **B** are orthogonal transformations and

$$\mathbf{B} = \mathbf{A}^T$$

$$\mathbf{A} = \mathbf{B}^T$$

Note that **A** and **B** are *not* second-order tensors. They are transformations relating two sets of spatial vector fields that happen to be bases for the space of all spatial vector fields. The parentheses around the subscripts of the coefficients $A_{(ij)}$ and $B_{(ij)}$ are used to remind the reader that they are not components of second-order tensors. The definitions for an orthogonal transformation, transpose of a transformation, ... are the same as for second-order tensors. The summation convention will continue to be used.

ii) Prove that for any vector field

$$\mathbf{v} = v_i \mathbf{e}_i = v_{(j)} \mathbf{g}_{(j)}$$

we have

$$v_i = A_{(ij)} v_{(j)} = B_{(ji)} v_{(j)}$$

and

$$v_{(i)} = B_{(ij)} v_j = A_{(ji)} v_j$$

iii) Starting with the results of Exercise A.4.1-8(iii) and (iv), immediately write down the results of Exercise A.4.1-9(iii) and (iv).

iv) Starting with the results of Exercise A.4.1-9(iii) and (iv), immediately write down the results of Exercise A.4.1-8(iii) and (iv).

## A.5.3  Inverse of a Second-Order Tensor Field

We say that a second-order tensor field $\mathbf{A}$ is *invertible* when the following conditions are satisfied:

1. If $\mathbf{u}_1$ and $\mathbf{u}_2$ are spatial vector fields such that $\mathbf{A} \cdot \mathbf{u}_1 = \mathbf{A} \cdot \mathbf{u}_2$, then $\mathbf{u}_1 = \mathbf{u}_2$.
2. There corresponds to every spatial vector field $\mathbf{v}$ at least one spatial vector field $\mathbf{u}$ such that $\mathbf{A} \cdot \mathbf{u} = \mathbf{v}$.

If $\mathbf{A}$ is invertible, we define a second-order tensor field $\mathbf{A}^{-1}$, called the *inverse* of $\mathbf{A}$, as follows: If $\mathbf{v}_1$ is any spatial vector field, by property 2 we may find a spatial vector field $\mathbf{u}_1$ for which $\mathbf{A} \cdot \mathbf{u}_1 = \mathbf{v}_1$. Say that $\mathbf{u}_1$ is not uniquely determined, such that $\mathbf{v}_1 = \mathbf{A} \cdot \mathbf{u}_1 = \mathbf{A} \cdot \mathbf{u}_2$. By property 1, $\mathbf{u}_1 = \mathbf{u}_2$ and we have a contradiction. The spatial vector field $\mathbf{u}_1$ is uniquely determined. We define $\mathbf{A}^{-1} \cdot \mathbf{v}_1$ to be $\mathbf{u}_1$.

To prove that $\mathbf{A}^{-1}$ satisfies the linearity rules for a second-order tensor field (A.5.0-1), we may evaluate $\mathbf{A}^{-1} \cdot (\alpha_1 \mathbf{v}_1 + \alpha_2 \mathbf{v}_2)$, where $\alpha_1$ and $\alpha_2$ are real scalar fields. If $\mathbf{A} \cdot \mathbf{u}_1 = \mathbf{v}_1$ and $\mathbf{A} \cdot \mathbf{u}_2 = \mathbf{v}_2$, we have

$$\mathbf{A} \cdot (\alpha_1 \mathbf{u}_1 + \alpha_2 \mathbf{u}_2) = \alpha_1 \mathbf{A} \cdot \mathbf{u}_1 + \alpha_2 \mathbf{A} \cdot \mathbf{u}_2$$
$$= \alpha_1 \mathbf{v}_1 + \alpha_2 \mathbf{v}_2 \qquad\qquad (A.5.3\text{-}1)$$

This means that

$$\mathbf{A}^{-1} \cdot (\alpha_1 \mathbf{v}_1 + \alpha_2 \mathbf{v}_2) = \alpha_1 \mathbf{u}_1 + \alpha_2 \mathbf{u}_2$$
$$= \alpha_1 \mathbf{A}^{-1} \cdot \mathbf{v}_1 + \alpha_2 \mathbf{A}^{-1} \cdot \mathbf{v}_2 \qquad\qquad (A.5.3\text{-}2)$$

It follows immediately from the definition that, for any invertible transformation $\mathbf{A}$,

$$\mathbf{A}^{-1} \cdot \mathbf{A} = \mathbf{A} \cdot \mathbf{A}^{-1}$$
$$= \mathbf{I} \qquad\qquad (A.5.3\text{-}3)$$

If **A**, **B**, and **C** are second-order tensor fields such that

$$\mathbf{A} \cdot \mathbf{B} = \mathbf{C} \cdot \mathbf{A}$$
$$= \mathbf{I} \tag{A.5.3-4}$$

let us show that **A** is invertible and $\mathbf{A}^{-1} = \mathbf{B} = \mathbf{C}$. If $\mathbf{A} \cdot \mathbf{u}_1 = \mathbf{A} \cdot \mathbf{u}_2$, we have from (A.5.3-4)

$$(\mathbf{C} \cdot \mathbf{A}) \cdot \mathbf{u}_1 = (\mathbf{C} \cdot \mathbf{A}) \cdot \mathbf{u}_2$$
$$\mathbf{u}_1 = \mathbf{u}_2 \tag{A.5.3-5}$$

This fulfills property 1 of an invertible transformation. The second property is also satisfied. If **v** is any spatial vector field and if $\mathbf{u} = \mathbf{B} \cdot \mathbf{v}$, by (A.5.3-4)

$$\mathbf{A} \cdot \mathbf{u} = (\mathbf{A} \cdot \mathbf{B}) \cdot \mathbf{v}$$
$$= \mathbf{v} \tag{A.5.3-6}$$

Now that we have proved **A** to be invertible, from (A.5.3-4)

$$\mathbf{A}^{-1} \cdot \mathbf{A} \cdot \mathbf{B} = \mathbf{C} \cdot \mathbf{A} \cdot \mathbf{A}^{-1}$$
$$= \mathbf{A}^{-1}$$
$$\mathbf{B} = \mathbf{C} \tag{A.5.3-7}$$
$$= \mathbf{A}^{-1}$$

In this way, we have shown that (A.5.3-3) is valid for some second-order tensor field $\mathbf{A}^{-1}$ if and only if **A** is invertible.

As a trivial example of an invertible second-order tensor field, we have the identity transformation, for which $\mathbf{I}^{-1} = \mathbf{I}$. Neither the zero tensor field **0** nor a tensor product **ab** is invertible.

For any orthogonal transformation **Q**, we have that

$$\mathbf{Q}^{-1} = \mathbf{Q}^T \tag{A.5.3-8}$$

This discussion is based upon that given by Halmos (1958, p. 62).

### Exercise A.5.3-1

i) The second-order tensor fields **A** and **B** are invertible. Show that $(\mathbf{A} \cdot \mathbf{B})^{-1} = \mathbf{B}^{-1} \cdot \mathbf{A}^{-1}$.
ii) Show that, if $\alpha \neq 0$ and **A** is invertible, $(\alpha \mathbf{A})^{-1} = (1/\alpha)\mathbf{A}^{-1}$.
iii) Show that, if **A** is invertible, $\mathbf{A}^{-1}$ is invertible and $(\mathbf{A}^{-1})^{-1} = \mathbf{A}$.

### Exercise A.5.3-2

i) If **A** is invertible and (A.5.3-3) holds, show that $\det(A_{ij}) \neq 0$, where the $A_{ij}$ denote the rectangular Cartesian components of **A**.
ii) Beginning with an equation of the same general form as (A.2.1-12), show that, if $\det(A_{ij}) \neq 0$, **A** must be invertible.

## A.5.4 Trace of a Second-Order Tensor Field**

Let **a** and **b** be spatial vector fields, let $\alpha$ be a scalar field, and let **S** and **T** be second-order tensor fields. An operation "tr" that assigns to each second-order tensor **T** a number tr(**T**) is called a *trace* if it obeys the following rules:

$$\text{tr}(S + T) = \text{tr}(S) + \text{tr}(T)$$
$$\text{tr}(\alpha T) = \alpha \, \text{tr}(T)$$
$$\text{tr}(ab) = a \cdot b$$

With respect to a rectangular Cartesian coordinate system, the trace of any second-order tensor $T = T_{ij}e_i e_j$ may be written as

$$
\begin{aligned}
\text{tr}(T) &= T_{ij} \, \text{tr}(e_i e_j) \\
&= T_{ij}(e_i \cdot e_j) \\
&= T_{ij}\delta_{ij} \\
&= T_{ii}
\end{aligned}
\tag{A.5.4-1}
$$

The trace of a second-order tensor may be thought of as the sum of the diagonal components in the matrix $[T_{ij}]$.

### Exercise A.5.4-1

i) Determine that the trace of a second-order tensor product does not depend upon the coordinate system being used.

ii) Show that the trace of a second-order tensor field does not depend upon the coordinate system being used.

**Exercise A.5.4-2**  The second-order tensor $T$ may be expressed with respect to two different rectangular Cartesian coordinate systems as

$$
\begin{aligned}
T &= T_{ij}e_i e_j \\
&= \overline{T}_{mn}\overline{e}_m \overline{e}_n
\end{aligned}
$$

Without using the definition of the trace or the results of Exercise A.5.4-1, prove that

$$T_{ii} = \overline{T}_{mn}$$

### Exercise A.5.4-3

*i) Show that, with respect to any curvilinear coordinate system,

$$
\begin{aligned}
\text{tr}(T) &= T^{\;j}_{.i} \\
&= T^{\;\;j}_{i.} \\
&= g_{ij}T^{ij} \\
&= g^{ij}T_{ij}
\end{aligned}
$$

*ii) Show that, with respect to any orthogonal curvilinear coordinate system,

$$\text{tr}(T) = T_{\langle ii \rangle}$$

**Exercise A.5.4-4**

*i) If **A** and **B** are second-order tensor fields, show that, with respect to any curvilinear coordinate system,

$$\text{tr}\,(\mathbf{A} \cdot \mathbf{B}) = A^{ij}B_{ji}$$
$$= A_{ij}B^{ji}$$
$$= A^{j}_{i.}B^{i}_{j.}$$
$$= A^{i}_{.j}B^{j}_{.i}$$

ii) Show that, with respect to any orthogonal curvilinear coordinate system,

$$\text{tr}\,(\mathbf{A} \cdot \mathbf{B}) = A_{\langle ij\rangle}B_{\langle ji\rangle}$$

**Exercise A.5.4-5**

i) Let **A** and **B** be second-order tensor fields. We define

$$(\mathbf{A}, \mathbf{B}) \equiv \text{tr}\,(\mathbf{A} \cdot \mathbf{B}^{T})$$

Show that $(\mathbf{A}, \mathbf{B})$ satisfies the requirements for an inner product in the vector space of second-order tensor fields.

ii) We define the *length* of a second-order tensor field as

$$\|A\| \equiv \sqrt{(\mathbf{A}, \mathbf{A})}$$

If $(\mathbf{A}, \mathbf{B}) = 0$, we say that **A** and **B** are orthogonal to each other.

Consider an orthogonal curvilinear coordinate system. Show that the set of nine second-order tensor products $(\mathbf{g}_{m}\mathbf{g}_{n})$ $(m, n = 1, 2, 3)$ is an *orthonormal* basis for the space of second-order tensor fields (with respect to the inner product and length as defined above). This justifies labeling the components $T_{\langle mn\rangle}$ of the second-order tensor field **T** as *physical* components.

---

## A.6 Gradient of Vector Field

### A.6.1 Definition**

The gradient of a spatial vector field **v** is a second-order tensor field denoted by $\nabla\mathbf{v}$. The gradient is specified by defining how it transforms an arbitrary spatial vector at all points $z$ in $E$:

$$\nabla\mathbf{v}(\mathbf{z}) \cdot \mathbf{a} = \lim_{s \to 0} \frac{1}{s}[\mathbf{v}(\mathbf{z} + s\mathbf{a}) - \mathbf{v}(\mathbf{z})] \tag{A.6.1-1}$$

The spatial vector **a** should be interpreted as the directed line segment or point difference $\mathbf{a} \equiv \overrightarrow{zy}$, where $y$ is an arbitrary point in $E$. In writing (A.6.1-1), we have assumed that an origin $O$ has been specified and we have interpreted **v** as a function of the position vector **z** measured with respect to this origin rather than as a function of the point $z$ itself.

By analogy with our discussion of the gradient of a scalar field in Section A.3.1, we have that

$$(\nabla \mathbf{v}) \cdot \mathbf{a} = \frac{\partial \mathbf{v}}{\partial z_j} a_j \tag{A.6.1-2}$$

For the particular case $\mathbf{a} = \mathbf{e}_j$,

$$(\nabla \mathbf{v}) \cdot \mathbf{e}_j = \frac{\partial \mathbf{v}}{\partial z_j} = \frac{\partial v_i}{\partial z_j} \mathbf{e}_i \tag{A.6.1-3}$$

In reaching this result, we have noted that the magnitudes and directions of the rectangular Cartesian basis fields are independent of position. Comparing (A.6.1-3) with (A.5.1-1) and (A.5.1-6), we conclude that[1]

$$\nabla \mathbf{v} = \frac{\partial v_i}{\partial z_j} \mathbf{e}_i \mathbf{e}_j \tag{A.6.1-4}$$

The trace (Section A.5.4) of the gradient of a spatial vector field $\mathbf{v}$ is a familiar operation:

$$\text{tr}(\nabla \mathbf{v}) = \frac{\partial v_i}{\partial z_j} \text{tr}(\mathbf{e}_i \mathbf{e}_j)$$

$$= \frac{\partial v_i}{\partial z_i} \tag{A.6.1-5}$$

It is more common to refer to this operation as the *divergence* of the spatial vector field $\mathbf{v}$. Several symbols for this operation are common:

$$\text{div } \mathbf{v} \equiv \nabla \cdot \mathbf{v}$$

$$\equiv \text{tr}(\nabla \mathbf{v})$$

$$= \frac{\partial v_i}{\partial z_i} \tag{A.6.1-6}$$

### A.6.2  Covariant Differentiation*

In Section A.6.1, we arrived at the components of the gradient of a spatial vector field $\mathbf{v}$ with respect to a rectangular Cartesian coordinate system. Here we derive an expression for the mixed components of $\nabla \mathbf{v}$ with respect to any curvilinear coordinate system (see Section A.5.1).

In Section A.6.1 we showed that

$$\nabla \mathbf{v} \cdot \mathbf{a} = \frac{\partial \mathbf{v}}{\partial z_i} a_i \tag{A.6.2-1}$$

---

[1] Although we believe this to be the most common meaning for the symbol $\nabla \mathbf{v}$, some authors define (Morse and Feshbach 1953, p. 65; Bird et al. 1960, p. 723)

$$\nabla \mathbf{v} \equiv \frac{\partial v_i}{\partial z_j} \mathbf{e}_j \mathbf{e}_i$$

Where we would write $(\nabla \mathbf{v}) \cdot \mathbf{w}$, they say instead $\mathbf{w} \cdot (\nabla \mathbf{v})$. As long as either definition is used consistently, there is no difference in any derived result.

In terms of curvilinear coordinates, we may express this operation as

$$\nabla \mathbf{v} \cdot \mathbf{a} = \frac{\partial \mathbf{v}}{\partial x^j} \frac{\partial x^j}{\partial z_i} a_i$$

$$= \frac{\partial \mathbf{v}}{\partial x^j} \overline{a}^j \tag{A.6.2-2}$$

where the $\overline{a}^j$ ($j = 1, 2, 3$) represent the contravariant curvilinear components of the spatial vector $\mathbf{a}$ (see Section A.4.3). Let us examine the quantity

$$\frac{\partial \mathbf{v}}{\partial x^j} = \frac{\partial}{\partial x^j}(v^i \mathbf{g}_i)$$

$$= \frac{\partial v^i}{\partial x^j} \mathbf{g}_i + v^i \frac{\partial \mathbf{g}_i}{\partial x^j} \tag{A.6.2-3}$$

Unlike the basis fields $\mathbf{e}_j$ of rectangular Cartesian coordinates, the natural basis fields of curvilinear coordinates are functions of position. In differentiation, they cannot be treated as constants:

$$\frac{\partial \mathbf{g}_i}{\partial x^j} = \frac{\partial}{\partial x^j}\left(\frac{\partial \mathbf{p}}{\partial x^i}\right)$$

$$= \frac{\partial^2}{\partial x^j \, \partial x^i}(z_k \mathbf{e}_k)$$

$$= \frac{\partial^2 z_k}{\partial z^j \, \partial x^i} \mathbf{e}_k \tag{A.6.2-4}$$

We saw in Section A.4.1 that

$$\mathbf{e}_k = \frac{\partial x^m}{\partial z_k} \mathbf{g}_m \tag{A.6.2-5}$$

This allows us to write

$$\frac{\partial \mathbf{g}_i}{\partial x^j} = \frac{\partial^2 z_k}{\partial x^j \, \partial x^i} \frac{\partial x^m}{\partial z_k} \mathbf{g}_m$$

$$= \left\{ \begin{matrix} m \\ j \quad i \end{matrix} \right\} \mathbf{g}_m \tag{A.6.2-6}$$

where we define

$$\left\{ \begin{matrix} m \\ j \quad i \end{matrix} \right\} \equiv \frac{\partial^2 z_k}{\partial x^j \, \partial x^i} \frac{\partial x^m}{\partial z_k} \tag{A.6.2-7}$$

These symbols are known as the *Christoffel symbols of the second kind*.
   *Christoffel symbols of the first kind* are defined by

$$[ji, p] \equiv g_{pm} \left\{ \begin{matrix} m \\ j \quad i \end{matrix} \right\} \tag{A.6.2-8}$$

From Section A.4.1, we have that

$$g_{pm} = \frac{\partial z_n}{\partial x^p} \frac{\partial z_n}{\partial x^m} \tag{A.6.2-9}$$

and we express (A.6.2-8) as

$$[ji, p] = \frac{\partial z_n}{\partial x^p} \frac{\partial z_n}{\partial x^m} \frac{\partial^2 z_k}{\partial x^j \partial x^i} \frac{\partial x^m}{\partial z_k}$$

$$= \frac{\partial^2 z_k}{\partial x^j \partial x^i} \frac{\partial z_k}{\partial x^p} \tag{A.6.2-10}$$

Equation (A.6.2-8) also allows us to write

$$\left\{ \begin{matrix} r \\ j \quad i \end{matrix} \right\} = g^{rp}[ji, p] \tag{A.6.2-11}$$

Although (A.6.2-7) is sufficient to define the Christoffel symbols of the second kind, it is rarely used in practice. Equation (A.6.2-9) may be differentiated to obtain

$$\frac{\partial g_{ij}}{\partial x^k} = \frac{\partial^2 z_m}{\partial x^k \partial x^i} \frac{\partial z_m}{\partial x^j} + \frac{\partial z_m}{\partial x^i} \frac{\partial^2 z_m}{\partial x^k \partial x^j} \tag{A.6.2-12}$$

Two similar expressions may be obtained by rotating the indices $i$, $j$, and $k$:

$$\frac{\partial g_{jk}}{\partial x^i} = \frac{\partial^2 z_m}{\partial x^i \partial x^j} \frac{\partial z_m}{\partial x^k} + \frac{\partial z_m}{\partial x^j} \frac{\partial^2 z_m}{\partial x^i \partial x^k} \tag{A.6.2-13}$$

and

$$\frac{\partial g_{ki}}{\partial x^j} = \frac{\partial^2 z_m}{\partial x^j \partial x^k} \frac{\partial z_m}{\partial x^i} + \frac{\partial z_m}{\partial x^k} \frac{\partial^2 z_m}{\partial x^j \partial x^i} \tag{A.6.2-14}$$

Adding (A.6.2-13) and (A.6.2-14) and subtracting (A.6.2-12), we get

$$\frac{\partial g_{kj}}{\partial x^i} + \frac{\partial g_{ik}}{\partial x^j} - \frac{\partial g_{ij}}{\partial x^k} = 2\frac{\partial^2 z_m}{\partial x^i \partial x^j} \frac{\partial z_m}{\partial x^k}$$

$$= 2[ij, k] \tag{A.6.2-15}$$

From (A.6.2-11) and (A.6.2-15), we have another expression for the Christoffel symbols of the second kind, which is usually found to be more convenient to use in practice than (A.6.2-7):

$$\left\{ \begin{matrix} m \\ j \quad i \end{matrix} \right\} = g^{mk}[ij, k]$$

$$= \frac{g^{mk}}{2} \left( \frac{\partial g_{kj}}{\partial x^i} + \frac{\partial g_{ik}}{\partial x^j} - \frac{\partial g_{ij}}{\partial x^k} \right) \tag{A.6.2-16}$$

Let us return to (A.6.2-3) and write, with the help of (A.6.2-6),

$$\frac{\partial \mathbf{v}}{\partial x^j} = \frac{\partial v^i}{\partial x^j}\mathbf{g}_i + v^i \left\{ \begin{matrix} m \\ j \quad i \end{matrix} \right\} \mathbf{g}_m$$

$$= \left[ \frac{\partial v^i}{\partial x^j} + \left\{ \begin{matrix} i \\ j \quad m \end{matrix} \right\} v^m \right] \mathbf{g}_i$$

$$= v^i{}_{,j}\mathbf{g}_i \tag{A.6.2-17}$$

Here we define

$$v^i{}_{,j} \equiv \frac{\partial v^i}{\partial x^j} + \left\{ \begin{matrix} i \\ j \quad m \end{matrix} \right\} v^m \tag{A.6.2-18}$$

The quantity $v^i{}_{,j}$ is referred to as the $j$th *covariant derivative* of the $i$th contravariant component of the spatial vector field $\mathbf{v}$. This allows us to write (A.6.2-2) as

$$(\nabla \mathbf{v}) \cdot \mathbf{a} = v^i{}_{,j} a^j \mathbf{g}_i \qquad (\text{A.6.2-19})$$

with the understanding that the $a^j$ represent the contravariant components of the spatial vector $\mathbf{a}$ in the curvilinear coordinate system under consideration. If we follow the practice, introduced in Section A.5.1, of expressing second-order tensor fields as sums of tensor products of basis fields, we may represent $\nabla \mathbf{v}$ as

$$\nabla \mathbf{v} = v^i{}_{,j} \mathbf{g}_i \mathbf{g}^j \qquad (\text{A.6.2-20})$$

The nine quantities of the form $v^i{}_{,j}$ ($i, j = 1, 2, 3$) represent the mixed components of the second-order tensor field $\nabla \mathbf{v}$.

In (A.6.2-3), we expressed $\mathbf{v}$ as a linear combination of the natural basis vectors. How are these expressions altered when we express $\mathbf{v}$ as a linear combination of the dual basis vectors? We have in this case

$$
\begin{aligned}
\frac{\partial \mathbf{v}}{\partial x^j} &= \frac{\partial}{\partial x^j}(v_i \mathbf{g}^i) \\
&= \frac{\partial v_i}{\partial x^j} \mathbf{g}^i + v_i \frac{\partial \mathbf{g}^i}{\partial x^j}
\end{aligned}
\qquad (\text{A.6.2-21})
$$

Our major problem is to obtain an expression for $\partial \mathbf{g}^i / \partial x^j$.

From Section A.4.2,

$$\mathbf{g}_i \cdot \mathbf{g}^j = \delta_i{}^j \qquad (\text{A.6.2-22})$$

Taking the derivative of this expression with respect to $x^k$, we obtain

$$\frac{\partial \mathbf{g}_i}{\partial x^k} \cdot \mathbf{g}^j + \mathbf{g}_i \cdot \frac{\partial \mathbf{g}^j}{\partial x^k} = 0 \qquad (\text{A.6.2-23})$$

or

$$
\begin{aligned}
\mathbf{g}_i \cdot \frac{\partial \mathbf{g}^j}{\partial x^k} &= -\frac{\partial \mathbf{g}_i}{\partial x^k} \cdot \mathbf{g}^j \\
&= -\left\{ \begin{matrix} m \\ k \quad i \end{matrix} \right\} \mathbf{g}_m \cdot \mathbf{g}^j \\
&= -\left\{ \begin{matrix} j \\ k \quad i \end{matrix} \right\}
\end{aligned}
\qquad (\text{A.6.2-24})
$$

The spatial vector $\partial \mathbf{g}^j / \partial x^k$, like any other element of the vector space of spatial vectors, may be written as a linear combination of the dual basis vectors:

$$\frac{\partial \mathbf{g}^j}{\partial x^k} = A_{kt}{}^j \mathbf{g}^t \qquad (\text{A.6.2-25})$$

where the coefficients $A_{kt}{}^j$ are yet to be determined. The scalar product of this equation with $\mathbf{g}_i$ yields from (A.6.2-24)

$$
\begin{aligned}
\mathbf{g}_i \cdot \frac{\partial \mathbf{g}^j}{\partial x^k} &= A_{ki}{}^j \\
&= -\left\{ \begin{matrix} j \\ k \quad i \end{matrix} \right\}
\end{aligned}
\qquad (\text{A.6.2-26})
$$

This allows us to write

$$\frac{\partial \mathbf{g}^j}{\partial x^k} = - \begin{Bmatrix} j \\ k \quad i \end{Bmatrix} \mathbf{g}^i \tag{A.6.2-27}$$

Returning to (A.6.2-21), we may write

$$\frac{\partial \mathbf{v}}{\partial x^j} = \frac{\partial v_i}{\partial x^j} \mathbf{g}^i - v_i \begin{Bmatrix} i \\ j \quad k \end{Bmatrix} \mathbf{g}^k$$

$$= \left[ \frac{\partial v_i}{\partial x^j} - \begin{Bmatrix} k \\ j \quad i \end{Bmatrix} v_k \right] \mathbf{g}^i$$

$$= v_{i,j} \mathbf{g}^i \tag{A.6.2-28}$$

where we define the symbol $v_{i,j}$ as

$$v_{i,j} \equiv \frac{\partial v_i}{\partial x^j} - \begin{Bmatrix} k \\ j \quad i \end{Bmatrix} v_k \tag{A.6.2-29}$$

The quantity $v_{i,j}$ is referred to as the $j$th *covariant derivative* of the $i$th covariant component of the spatial vector field **v**.

From (A.6.2-2) and (A.6.2-28), we obtain

$$(\nabla \mathbf{v}) \cdot \mathbf{a} = v_{i,j} a^j \mathbf{g}^i \tag{A.6.2-30}$$

again with the understanding that the $a^j$ represent the contravariant components of the spatial vector **a** in whatever curvilinear coordinate system is under consideration. In terms of our discussion in Section A.5.1, we conclude that

$$\nabla \mathbf{v} = v_{i,j} \mathbf{g}^i \mathbf{g}^j \tag{A.6.2-31}$$

The $v_{i,j}$ represent the covariant components of $\nabla \mathbf{v}$.

Similar to (A.6.2-20), equation (A.6.2-31) can be written as

$$\nabla \mathbf{v} = v_{i,j} g^{ik} \mathbf{g}_k \mathbf{g}^j$$

$$= v_{k,j} g^{ki} \mathbf{g}_i \mathbf{g}^j \tag{A.6.2-32}$$

Equation (A.6.2-20) in turn may be written as

$$\nabla \mathbf{v} = (g^{ki} v_k)_{,j} \mathbf{g}_i \mathbf{g}^j \tag{A.6.2-33}$$

We conclude that

$$(g^{ki} v_k)_{,j} = g^{ki} v_{k,j} \tag{A.6.2-34}$$

(Remember here that the nine tensor products of the form $\mathbf{g}_i \mathbf{g}^j$ were shown to be linearly independent in Section A.5.1.) This means that the $g^{ki}$ may be treated as constants with respect to covariant differentiation.

**Exercise A.6.2-1** Show that

$$(g_{in}v^n)_{,j} = g_{in}v^n{}_{,j}$$

implying that the $g_{in}$ may be treated as constants with respect to covariant differentiation.

**Exercise A.6.2-2** Starting with

$$(g_{in}v^n)_{,j} = \frac{\partial(g_{in}v^n)}{\partial x^j} - \left\{ \begin{matrix} k \\ j \quad i \end{matrix} \right\} g_{kn}v^n$$

rework Exercise A.6.2-1.

**Exercise A.6.2-3** *Rectangular Cartesian coordinates* Show that, in rectangular Cartesian coordinates,

$$\left\{ \begin{matrix} i \\ j \quad k \end{matrix} \right\} = 0$$

and covariant differentiation reduces to ordinary partial differentiation.

**Exercise A.6.2-4** *Cylindrical coordinates* Show that, in cylindrical coordinates, where

$$z_1 = x^1 \cos x^2$$
$$= r \cos \theta$$
$$z_2 = x^1 \sin x^2$$
$$= r \sin \theta$$
$$z_3 = x^3$$
$$= z$$

the only nonzero Christoffel symbols of the second kind are

$$\left\{ \begin{matrix} 1 \\ 2 \quad 2 \end{matrix} \right\} = -r$$

and

$$\left\{ \begin{matrix} 2 \\ 1 \quad 2 \end{matrix} \right\} = \left\{ \begin{matrix} 2 \\ 2 \quad 1 \end{matrix} \right\}$$
$$= \frac{1}{r}$$

**Exercise A.6.2-5** *Spherical coordinates* Show that, in spherical coordinates, where

$$z_1 = x^1 \sin x^2 \cos x^3$$
$$= r \sin \theta \cos \varphi$$
$$z_2 = x^1 \sin x^2 \sin x^3$$
$$= r \sin \theta \sin \varphi$$
$$z_3 = x^1 \cos x^2$$
$$= r \cos \theta$$

the only nonzero Christoffel symbols of the second kind are

$$\left\{ \begin{matrix} & 1 & \\ 2 & & 2 \end{matrix} \right\} = -r$$

$$\left\{ \begin{matrix} & 1 & \\ 3 & & 3 \end{matrix} \right\} = -r \, \sin^2\theta$$

$$\left\{ \begin{matrix} & 2 & \\ 1 & & 2 \end{matrix} \right\} = \left\{ \begin{matrix} & 2 & \\ 2 & & 1 \end{matrix} \right\}$$

$$= \left\{ \begin{matrix} & 3 & \\ 1 & & 3 \end{matrix} \right\}$$

$$= \left\{ \begin{matrix} & 3 & \\ 1 & & 3 \end{matrix} \right\}$$

$$= \frac{1}{r}$$

$$\left\{ \begin{matrix} & 2 & \\ 3 & & 3 \end{matrix} \right\} = -\sin\theta \, \cos\theta$$

and

$$\left\{ \begin{matrix} & 3 & \\ 2 & & 3 \end{matrix} \right\} = \left\{ \begin{matrix} & 3 & \\ 3 & & 2 \end{matrix} \right\}$$

$$= \cot\theta$$

## A.7   Third-Order Tensors

### A.7.1   Definition

Our discussion of third-order tensor fields closely parallels the treatment of second-order tensor fields in the introduction to Section A.5.

A third-order tensor field $\beta$ is a transformation (or mapping or rule) that assigns to each given spatial vector field $\mathbf{v}$ a second-order tensor field $\beta \cdot \mathbf{v}$ such that the rules

$$\beta \cdot (\mathbf{v} + \mathbf{w}) = \beta \cdot \mathbf{v} + \beta \cdot \mathbf{w}$$
$$\beta \cdot (a\mathbf{v}) = a(\beta \cdot \mathbf{v})$$

(A.7.1-1)

hold. The quantity $a$ denotes a real scalar field. (Note that we are using the dot notation in a different manner here than we used it in Section A.1.1 when we discussed the inner product. Our choice of notation is suggestive in the same way that it was in our treatment of second-order tensor fields, as will be clear shortly.)

We define the sum $\alpha + \beta$ of two third-order tensor fields $\alpha$ and $\beta$ to be a transformation such that, for every spatial vector field $\mathbf{v}$,

$$(\alpha + \beta) \cdot \mathbf{v} = \alpha \cdot \mathbf{v} + \beta \cdot \mathbf{v}$$

(A.7.1-2)

The product $a\beta$ of a third-order tensor field $\beta$ with a real scalar field $a$ is a transformation such that, for every spatial vector field $\mathbf{v}$,

$$(a\beta) \cdot \mathbf{v} = a(\beta \cdot \mathbf{v}) \tag{A.7.1-3}$$

The transformations $\alpha + \beta$ and $a\beta$ may be easily shown to obey the rules for a third-order tensor field. If we define the zero third-order tensor $\mathbf{0}$ by the requirement that

$$\mathbf{0} \cdot \mathbf{v} = \mathbf{0}$$

for all spatial vector fields $\mathbf{v}$, rules $(A_1)$ to $(A_4)$ and $(M_1)$ to $(M_4)$ of Section A.1.1 are satisfied, and the set of all third-order tensor fields constitutes a vector space.

If three spatial vector fields $\mathbf{a}$, $\mathbf{b}$, and $\mathbf{c}$ are given, we can define a third-order tensor field $\mathbf{abc}$ by the requirement that

$$(\mathbf{abc}) \cdot \mathbf{v} = \mathbf{ab}(\mathbf{c} \cdot \mathbf{v}) \tag{A.7.1-4}$$

holds for all spatial vector fields $\mathbf{v}$. This tensor field $\mathbf{abc}$ is called the *tensor product* of the spatial vector fields $\mathbf{a}$, $\mathbf{b}$, and $\mathbf{c}$.

## A.7.2 Components of Third-Order Tensor Fields

If $\beta$ is a third-order tensor field and the $\mathbf{e}_k$ ($k = 1, 2, 3$) form a rectangular Cartesian basis for the space of spatial vector fields, we may write

$$\beta \cdot \mathbf{e}_k = \beta_{ijk} \mathbf{e}_i \mathbf{e}_j \tag{A.7.2-1}$$

The set of 27 coefficients $\beta_{ijk}$ ($i, j, k = 1, 2, 3$) will hereafter be referred to as the coefficient matrix $[\beta_{ijk}]$, reminiscent of the nomenclature used for second-order tensor fields. The coefficient matrix $[\beta_{ijk}]$ tells us how the basis fields $\mathbf{e}_k$ are transformed by $\beta$. If $\mathbf{v}$ is any spatial vector field,

$$\begin{aligned}
\beta \cdot \mathbf{v} &= \beta \cdot (v_k \mathbf{e}_k) \\
&= v_k \beta \cdot \mathbf{e}_k \\
&= v_k \beta_{ijk} \mathbf{e}_i \mathbf{e}_j
\end{aligned} \tag{A.7.2-2}$$

In this way, for each set of basis fields $(\mathbf{e}_1, \mathbf{e}_2, \mathbf{e}_3)$, we may associate a set of 27 coefficients with any third-order tensor field $\beta$. This association or correspondence is one-to-one. [For the same set of basis fields, two different third-order tensor fields will have two different coefficient matrices. To prove this, observe that the coefficient matrix $[\beta_{ijk}]$ of a third-order tensor field $\beta$ completely determines $\beta$; by (A.7.2-2), $\beta \cdot \mathbf{v}$ is determined for every $\mathbf{v}$.]

Using the arguments applied in the discussion of second-order tensor fields (see Section A.5.1), we may show that every third-order tensor field $\beta$ may be written as a sum of 27 tensor products:

$$\beta = \beta_{ijk} \mathbf{e}_i \mathbf{e}_j \mathbf{e}_k \tag{A.7.2-3}$$

The coefficients $\beta_{ijk}$ are the same as those introduced in (A.7.2-1). They are referred to as the rectangular Cartesian components of $\beta$.

Continuing in the same fashion, we can show that the 27 tensor products $\mathbf{e}_i\mathbf{e}_j\mathbf{e}_k$ ($i, j, k = 1, 2, 3$) are linearly independent. Since every third-order tensor field $\beta$ can be expressed as a linear combination of the $\mathbf{e}_i\mathbf{e}_j\mathbf{e}_k$, we conclude that the 27 tensor products $\mathbf{e}_i\mathbf{e}_j\mathbf{e}_k$ form a set of basis fields for the space of third-order tensor fields. Consequently, the vector space of third-order tensor field is 27 dimensional.

In physical applications, we often find it convenient to speak in terms of an orthogonal curvilinear coordinate system. If the $\mathbf{g}_{(k)}$ ($k = 1, 2, 3$) are the associated physical basis fields (Section A.4.1), we may write by analogy with (A.7.2-1)

$$\beta \cdot \mathbf{g}_{(k)} = \beta_{(ijk)}\mathbf{g}_{(i)}\mathbf{g}_{(j)} \tag{A.7.2-4}$$

By the same argument that leads us to (A.7.2-3), we may consequently write

$$\beta = \beta_{(ijk)}\mathbf{g}_{(i)}\mathbf{g}_{(j)}\mathbf{g}_{(k)} \tag{A.7.2-5}$$

where the 27 coefficients $\beta_{(ijk)}$ ($i, j, k = 1, 2, 3$) are referred to as the physical components of $\beta$. The set of 27 tensor products $\mathbf{g}_{(i)}\mathbf{g}_{(j)}\mathbf{g}_{(k)}$ ($i, j, k = 1, 2, 3$) forms another basis for the vector space of third-order tensor fields.

**Exercise A.7.2-1**    Prove that every third-order tensor field $\beta$ may be written as a sum of 27 tensor products (A.7.2-3).

**Exercise A.7.2-2**    Prove that the 27 tensor products $\mathbf{e}_i\mathbf{e}_j\mathbf{e}_k$ are linearly independent and that these tensor products form a basis for the space of third-order tensor fields.

**Exercise A.7.2-3**

i) For an orthogonal curvilinear coordinate system, prove that (A.7.2-5) holds for every third-order tensor field $\beta$.

ii) Prove that the 27 tensor products $\mathbf{g}_{(i)}\mathbf{g}_{(j)}\mathbf{g}_{(k)}$ ($i, j, k = 1, 2, 3$) form a basis for the vector space of third-order tensor fields.

**Exercise A.7.2-4**    Let $\beta$ be some third-order tensor field. If the $\beta_{ijk}$ are the components of $\beta$ with respect to one rectangular Cartesian coordinate system ($z_1, z_2, z_3$) and the $\overline{\beta}_{ijk}$ ($i, j, k = 1, 2, 3$) are the components with respect to another rectangular Cartesian coordinate system ($\overline{z}_1, \overline{z}_2, \overline{z}_3$), prove that

$$\beta_{ijk} = \frac{\partial \overline{z}_m}{\partial z_i}\frac{\partial \overline{z}_n}{\partial z_j}\frac{\partial \overline{z}_p}{\partial z_k}\overline{\beta}_{mnp}$$

and

$$\beta_{ijk} = \frac{\partial z_i}{\partial \overline{z}_m}\frac{\partial z_j}{\partial \overline{z}_n}\frac{\partial z_k}{\partial \overline{z}_p}\overline{\beta}_{mnp}$$

**Exercise A.7.2-5**

*i) Given any curvilinear coordinate system, prove that every third-order tensor field $\beta$ may be written as

$$\beta = \beta_{ijk}\mathbf{g}^i\mathbf{g}^j\mathbf{g}^k$$

The 27 coefficients $\beta_{ijk}$ ($i, j, k = 1, 2, 3$) are referred to as the *covariant components* of $\beta$.

ii) Prove that the 27 tensor products $\mathbf{g}^i\mathbf{g}^j\mathbf{g}^k$ ($i, j, k = 1, 2, 3$) form a basis for the space of third-order tensor fields.

**Exercise A.7.2-6**

*i) Given any curvilinear coordinate system, prove that every third-order tensor field $\beta$ may be written as

$$\beta = \beta^{ijk}\mathbf{g}_i\mathbf{g}_j\mathbf{g}_k$$

The 27 coefficients $\beta^{ijk}$ ($i, j, k = 1, 2, 3$) are referred to as the *contravariant components* of $\beta$.

ii) Prove that the 27 tensor products $\mathbf{g}_i\mathbf{g}_j\mathbf{g}_k$ ($i, j, k = 1, 2, 3$) form a basis for the space of third-order tensor fields.

**Exercise A.7.2-7\*** Prove that

$$\beta_{\langle ijk \rangle} = \frac{\beta_{ijk}}{\sqrt{g_{ii}g_{jj}g_{kk}}} \quad \text{no summation on } i, j, k$$

and

$$\beta_{\langle ijk \rangle} = \sqrt{g_{ii}g_{jj}g_{kk}}\,\beta^{ijk} \quad \text{no summation on } i, j, k$$

**Exercise A.7.2-8\*** Prove that it is unnecessary to distinguish among covariant, contravariant, and physical components of third-order tensor fields *in rectangular Cartesian coordinate systems*. We may write all indices as subscripts, employing the summation convention over repeated subscripts *so long as we restrict ourselves to rectangular Cartesian coordinate systems*.

**Exercise A.7.2-9**

*i) Let $\beta$ be some third-order tensor field. If the $\beta^{ijk}$ ($i, j, k = 1, 2, 3$) are the contravariant components of $\beta$ with respect to one curvilinear coordinate system $(x^1, x^2, x^3)$ and if the $\overline{\beta}^{ijk}$ ($i, j, k = 1, 2, 3$) are the contravariant components of $\beta$ with respect to another curvilinear coordinate system $(\overline{x}^1, \overline{x}^2, \overline{x}^3)$, prove that

$$\beta^{ijk} = \frac{\partial x^i}{\partial \overline{x}^m}\frac{\partial x^j}{\partial \overline{x}^n}\frac{\partial x^k}{\partial \overline{x}^p}\overline{\beta}^{mnp}$$

ii) Similarly, prove that

$$\beta_{ijk} = \frac{\partial \overline{x}^m}{\partial x^i}\frac{\partial \overline{x}^n}{\partial x^j}\frac{\partial \overline{x}^p}{\partial x^k}\overline{\beta}_{mnp}$$

**Exercise A.7.2-10\*** If the $\beta_{ijk}$ ($i, j, k = 1, 2, 3$) are the covariant components of a third-order tensor field $\beta$ and the $\beta^{mnp}$ ($m, n, p = 1, 2, 3$) are the contravariant components, prove that

$$\beta_{ijk} = g_{im}g_{jn}g_{kp}\beta^{mnp}$$

**Exercise A.7.2-11** We will have occasion to use a particular third-order tensor field defined by its components with respect to a rectangular Cartesian coordinate system $(z_1, z_2, z_3)$:

$$\epsilon \equiv e_{ijk}\mathbf{e}_i\mathbf{e}_j\mathbf{e}_k$$

The quantity $e_{ijk}$ is defined in Section A.2.1.

i) Prove that with respect to any other rectangular Cartesian coordinate system $(\bar{z}_1, \bar{z}_2, \bar{z}_3)$, we have

$$\epsilon = \left[ \det \left( \frac{\partial z_m}{\partial \bar{z}_n} \right) \right] e_{ijk} \bar{\mathbf{e}}_i \bar{\mathbf{e}}_j \bar{\mathbf{e}}_k$$

ii) Starting with an expression for the components of the identity tensor, prove that

$$\left[ \det \left( \frac{\partial z_m}{\partial \bar{z}_n} \right) \right]^2 = 1$$

In view of the discussion in Section A.9.2 and the result of Exercise 1.2.1-2, it is necessary that the rectangular Cartesian coordinate system used here in defining $\epsilon$ be right-handed.

**Exercise A.7.2-12**

i) The third-order tensor field $\epsilon$ is defined in Exercise A.7.2-11. Prove that for an arbitrary curvilinear coordinate system we may write

$$\epsilon = \epsilon^{ijk} \mathbf{g}_i \mathbf{g}_j \mathbf{g}_k$$

where

$$\epsilon^{ijk} \equiv \frac{1}{\sqrt{g}} e^{ijk}$$

*ii) Prove that we may also write

$$\epsilon = \epsilon_{ijk} \mathbf{g}^i \mathbf{g}^j \mathbf{g}^k$$

where

$$\epsilon_{ijk} \equiv \sqrt{g} e_{ijk}$$

iii) For an orthogonal curvilinear coordinate system, prove that

$$\epsilon = e^{ijk} \mathbf{g}_{\langle i \rangle} \mathbf{g}_{\langle j \rangle} \mathbf{g}_{\langle k \rangle}$$
$$= e_{ijk} \mathbf{g}_{\langle i \rangle} \mathbf{g}_{\langle j \rangle} \mathbf{g}_{\langle k \rangle}$$

**A.7.3** Another View of Third-Order Tensor Fields

If $\beta$ is a third-order tensor field and if $\mathbf{u}$ and $\mathbf{v}$ are spatial vector fields, then $\beta \cdot \mathbf{u}$ is a second-order tensor field and $(\beta \cdot \mathbf{u}) \cdot \mathbf{v}$ is a spatial vector field. For convenience, let us introduce the notation

$$\beta : \mathbf{uv} \equiv (\beta \cdot \mathbf{u}) \cdot \mathbf{v} \tag{A.7.3-1}$$

In particular,

$$\beta : \mathbf{e}_i \mathbf{e}_j = (\beta \cdot \mathbf{e}_i) \cdot \mathbf{e}_j$$
$$= (\beta_{mni} \mathbf{e}_m \mathbf{e}_n) \cdot \mathbf{e}_j$$
$$= \beta_{mji} \mathbf{e}_m \tag{A.7.3-2}$$

This suggests that we may use $\beta$ to define a transformation (or mapping or rule) that assigns to every tensor product **uv** of spatial vector fields a spatial vector field $\beta : \mathbf{uv}$ such that the rules

$$\beta : (\mathbf{ab} + \mathbf{uv}) = \beta : \mathbf{ab} + \beta : \mathbf{uv}$$
$$\beta : (a\mathbf{ab}) = a(\beta : \mathbf{ab}) \tag{A.7.3-3}$$

hold. The quantity $a$ denotes a real scalar field.

Since every second-order tensor field **T** may be written as a linear combination of tensor products (see Section A.5.1), we have

$$\begin{aligned} \beta : \mathbf{T} &= \beta : (T_{ij}\mathbf{e}_i\mathbf{e}_j) \\ &= T_{ij}\beta : \mathbf{e}_i\mathbf{e}_j \\ &= T_{ij}\beta_{mji}\mathbf{e}_m \end{aligned} \tag{A.7.3-4}$$

It follows immediately from the rules given in (A.7.3-3) that, for any two second-order tensor fields **S** and **T** and for any scalar field $a$, we have the rules

$$\beta : (\mathbf{S} + \mathbf{T}) = \beta : \mathbf{S} + \beta : \mathbf{T}$$
$$\beta : (a\mathbf{T}) = a(\beta : \mathbf{T}) \tag{A.7.3-5}$$

If $\epsilon$ is the third-order tensor defined in Exercise A.7.2-11 and **B** is any skew-symmetric second-order tensor,

$$\mathbf{b} \equiv \epsilon : \mathbf{B} \tag{A.7.3-6}$$

is known as the corresponding *axial vector*. The vorticity vector, defined by (3.4.0-8), is the axial vector corresponding to the second-order vorticity tensor (2.3.2-4).

---

## A.8 Gradient of Second-Order Tensor Field

### A.8.1 Definition

The gradient of a second-order tensor field **T** is a third-order tensor field denoted by $\nabla\mathbf{T}$. The gradient is specified by defining how it transforms an arbitrary spatial vector at all points $z$ in $E$:

$$\nabla\mathbf{T}(\mathbf{z}) \cdot \mathbf{a} = \lim_{s \to 0} \frac{1}{s}[\mathbf{T}(\mathbf{z} + s\mathbf{a}) - \mathbf{T}(\mathbf{z})] \tag{A.8.1-1}$$

The spatial vector **a** should be interpreted as the directed line segment or point difference $\mathbf{a} \equiv \vec{zy}$, where $y$ is an arbitrary point in $E$. In writing (A.8.1-1), we have assumed that an origin $O$ has been specified and we have interpreted **T** as a function of the position vector **z** measured with respect to this origin rather than as a function of the point $z$ itself.

By analogy with our discussions of the gradient of a scalar field in Section A.3.1 and of the gradient of a spatial vector field in Section A.6.1, we have

$$(\nabla \mathbf{T}) \cdot \mathbf{a} = \frac{\partial \mathbf{T}}{\partial z_j} a_j \tag{A.8.1-2}$$

For the particular case $\mathbf{a} = \mathbf{e}_k$,

$$(\nabla \mathbf{T}) \cdot \mathbf{e}_k = \frac{\partial \mathbf{T}}{\partial z_k}$$

$$= \frac{\partial T_{ij}}{\partial z_k} \mathbf{e}_i \mathbf{e}_j \tag{A.8.1-3}$$

In reaching this result, we have noted that the magnitudes and directions of the rectangular Cartesian basis fields are independent of position. Comparing (A.8.1-3) with (A.7.2-1) and (A.7.2-3), we conclude that[2]

$$\nabla \mathbf{T} = \frac{\partial T_{ij}}{\partial z_k} \mathbf{e}_i \mathbf{e}_j \mathbf{e}_k \tag{A.8.1-4}$$

A common operation is the *divergence* of a second-order tensor field $\mathbf{T}$:

$$\operatorname{div} \mathbf{T} \equiv \nabla \cdot \mathbf{T}$$

$$\equiv \frac{\partial T_{ij}}{\partial z_j} \mathbf{e}_i \tag{A.8.1-5}$$

### A.8.2  More on Covariant Differentiation*

In Section A.8.1 we arrive at the components of the gradient of a second-order tensor field $\mathbf{T}$ with respect to a rectangular Cartesian coordinate system. Here we derive expressions for the mixed components of $\nabla \mathbf{T}$ with respect to any curvilinear coordinate system.

In Section A.8.1, we showed that

$$(\nabla \mathbf{T}) \cdot \mathbf{a} = \frac{\partial \mathbf{T}}{\partial z_m} a_m \tag{A.8.2-1}$$

In terms of curvilinear coordinates, we may express this operation as

$$(\nabla \mathbf{T}) \cdot \mathbf{a} = \frac{\partial \mathbf{T}}{\partial x^k} \frac{\partial x^k}{\partial z_m} a_m$$

$$= \frac{\partial \mathbf{T}}{\partial x^k} \overline{a}^k \tag{A.8.2-2}$$

where the $\overline{a}^k$ ($k = 1, 2, 3$) represent the contravariant curvilinear components of the position vector $\mathbf{a}$.

---

[2] Although we believe that this is the most common meaning for the symbol $\nabla \mathbf{T}$, some authors define (Bird et al. 1960, pp. 723 and 730)

$$\nabla \mathbf{T} \equiv \frac{\partial T_{ij}}{\partial z_k} \mathbf{e}_k \mathbf{e}_i \mathbf{e}_j$$

Where we would write $(\nabla \mathbf{T}) \cdot \mathbf{v}$, they say instead $\mathbf{v} \cdot (\nabla \mathbf{T})$. As long as either definition is used consistently, there is no difference in any derived result.

On the basis of our discussion in Section A.6.2, we may express

$$\frac{\partial \mathbf{T}}{\partial x^k} = \frac{\partial}{\partial x^k}(T^{ij}\mathbf{g}_i\mathbf{g}_j)$$

$$= \frac{\partial T^{ij}}{\partial x^k}\mathbf{g}_i\mathbf{g}_j + T^{ij}\frac{\partial \mathbf{g}_i}{\partial x^k}\mathbf{g}_j + T^{ij}\mathbf{g}_i\frac{\partial \mathbf{g}_j}{\partial x^k}$$

$$= \frac{\partial T^{ij}}{\partial x^k}\mathbf{g}_i\mathbf{g}_j + T^{ij}\left\{\begin{matrix} r \\ k\;\;i \end{matrix}\right\}\mathbf{g}_r\mathbf{g}_j + T^{ij}\mathbf{g}_i\left\{\begin{matrix} r \\ k\;\;j \end{matrix}\right\}\mathbf{g}_r$$

$$= \left[\frac{\partial T^{ij}}{\partial x^k} + \left\{\begin{matrix} i \\ k\;\;r \end{matrix}\right\}T^{rj} + \left\{\begin{matrix} j \\ k\;\;r \end{matrix}\right\}T^{ir}\right]\mathbf{g}_i\mathbf{g}_j$$

$$= T^{ij}{}_{,k}\mathbf{g}_i\mathbf{g}_j \tag{A.8.2-3}$$

Here we define the symbol $T^{ij}{}_{,k}$ as

$$T^{ij}{}_{,k} \equiv \frac{\partial T^{ij}}{\partial x^k} + \left\{\begin{matrix} i \\ k\;\;r \end{matrix}\right\}T^{rj} + \left\{\begin{matrix} j \\ k\;\;r \end{matrix}\right\}T^{ir} \tag{A.8.2-4}$$

The quantity $T^{ij}{}_{,k}$ is referred to as the $k$th *covariant derivative* of the $ij$ contravariant component of the second-order tensor field $\mathbf{T}$.

This allows us to write (A.8.2-2) as

$$(\nabla \mathbf{T}) \cdot \mathbf{a} = T^{ij}{}_{,k}\overline{a}^k\mathbf{g}_i\mathbf{g}_j \tag{A.8.2-5}$$

If we follow the practice introduced in Section A.7.2 of expressing third-order tensor fields as sums of tensor products of basis fields, we may represent $\nabla \mathbf{T}$ as

$$\nabla \mathbf{T} = T^{ij}{}_{,k}\mathbf{g}_i\mathbf{g}_j\mathbf{g}^k \tag{A.8.2-6}$$

The $T^{ij}_{,k}$ $(i, j, k = 1, 2, 3)$ are consequently the *mixed* components of $\nabla \mathbf{T}$.

**Exercise A.8.2-1**  Let $\mathbf{T}$ be any scalar second-order tensor field. For any curvilinear coordinate system, prove the following:

i)   $\nabla \mathbf{T} = T_{ij,k}\mathbf{g}^i\mathbf{g}^j\mathbf{g}^k$
   where

$$T_{ij,k} \equiv \frac{\partial T_{ij}}{\partial x^k} - \left\{\begin{matrix} r \\ k\;\;i \end{matrix}\right\}T_{rj} - \left\{\begin{matrix} r \\ k\;\;j \end{matrix}\right\}T_{ir}$$

ii)   $\nabla \mathbf{T} = T^i{}_{.j,k}\mathbf{g}_i\mathbf{g}^j\mathbf{g}^k$
   where

$$T^i{}_{.j,k} \equiv \frac{\partial T^i{}_{.j}}{\partial x^k} + \left\{\begin{matrix} i \\ k\;\;r \end{matrix}\right\}T^r{}_{.j} - \left\{\begin{matrix} r \\ k\;\;j \end{matrix}\right\}T^i{}_{.r}$$

iii)   $\nabla \mathbf{T} = T_{i.,k}{}^j\mathbf{g}^i\mathbf{g}_j\mathbf{g}^k$
   where

$$T_{i.,k}{}^j \equiv \frac{\partial T_{i.}{}^j}{\partial x^k} - \left\{\begin{matrix} r \\ k\;\;i \end{matrix}\right\}T_{r.}{}^j + \left\{\begin{matrix} j \\ k\;\;r \end{matrix}\right\}T_i{}^r$$

## Exercise A.8.2-2

   i) Prove that

$$\nabla \mathbf{I} = 0$$

   ii) Conclude that

$$g_{ij,k} = 0$$

   and

$$g^{ij}{}_{,k} = 0$$

**Exercise A.8.2-3**   We may define a fourth-order tensor field $\Theta$ to be transformation (or mapping or rule) that assigns to each spatial vector field $\mathbf{v}$ a third-order tensor field $\beta$ such that the rules

$$\Theta \cdot (\mathbf{v} + \mathbf{w}) = \Theta \cdot \mathbf{v} + \Theta \cdot \mathbf{w}$$

$$\Theta \cdot (a\mathbf{v}) = a(\Theta \cdot \mathbf{v})$$

hold. Here $a$ is a real scalar field.

The gradient of a third-order tensor field $\beta$ is the fourth-order tensor field denoted by $\nabla\beta$. If $\mathbf{a} = a_i \mathbf{e}_i$ indicates the directed line segment or point difference $\mathbf{a} \equiv \overrightarrow{zy}$, where $y$ is an arbitrary point in $E$, then $\nabla\beta(\mathbf{z})$ is the linear transformation that assigns to $\mathbf{a}$ the third-order tensor field given by the following rule:

$$\nabla\beta(\mathbf{z}) \cdot \mathbf{a} = \lim_{s \to 0} \frac{1}{s} [\beta(\mathbf{z} + s\mathbf{a}) - \beta(\mathbf{z})]$$

We conclude by an argument analogous to the one used in discussing the gradient of a second-order tensor field:

$$\nabla\beta = \beta^{ijk}{}_{,m} \mathbf{g}_i \mathbf{g}_j \mathbf{g}_k \mathbf{g}^m$$
$$= \beta_{ijk,m} \mathbf{g}^i \mathbf{g}^j \mathbf{g}^k \mathbf{g}^m$$

where

$$\beta^{ijk}{}_{,m} \equiv \frac{\partial \beta^{ijk}}{\partial x^m} + \left\{ \begin{matrix} i \\ m \quad r \end{matrix} \right\} \beta^{rjk} + \left\{ \begin{matrix} j \\ m \quad r \end{matrix} \right\} \beta^{irk} + \left\{ \begin{matrix} k \\ m \quad r \end{matrix} \right\} \beta^{ijr}$$

and

$$\beta_{ijk,m} \equiv \frac{\partial \beta_{ijk}}{\partial x^m} - \left\{ \begin{matrix} r \\ m \quad i \end{matrix} \right\} \beta_{rjk} - \left\{ \begin{matrix} r \\ m \quad j \end{matrix} \right\} \beta_{irk} - \left\{ \begin{matrix} r \\ m \quad k \end{matrix} \right\} \beta_{ijr}$$

   i) Prove that (see Exercises A.7.2-11 and A.7.2-12)

$$\nabla\epsilon = 0$$

   ii) Conclude that

$$\epsilon^{ijk}{}_{,m} = 0$$

   and

$$\epsilon_{ijk,m} = 0$$

## Exercise A.8.2-4

i) By writing out in full that

$$\epsilon_{rst,p} = 0$$

and putting $r = 1, s = 2, t = 3$, prove that (McConnell 1957, p. 155)

$$\frac{\partial \log \sqrt{g}}{\partial x^p} = \left\{ \begin{matrix} m \\ m \quad p \end{matrix} \right\}$$

ii) Writing out

$$g^{rs}{}_{,t} = 0$$

in full, deduce from (i) that (McConnell 1957, p. 155)

$$\frac{1}{\sqrt{g}} \frac{\partial \left( \sqrt{g} g^{rs} \right)}{\partial x^s} + \left\{ \begin{matrix} r \\ m \quad n \end{matrix} \right\} g^{mn} = 0$$

iii) Using the result of (i), prove that (McConnell 1957, p. 155)

$$\text{div } \mathbf{v} = v^r{}_{,r} = \frac{1}{\sqrt{g}} \frac{\partial \left( \sqrt{g} v^r \right)}{\partial x^r}$$

---

## A.9  Vector Product and Curl

### A.9.1  Definitions**

Let $\mathbf{a}$ and $\mathbf{b}$ be two spatial vector fields. Students are often advised to remember the *vector product* $(\mathbf{a} \wedge \mathbf{b})$ with respect to a set of rectangular Cartesian coordinates basis fields in the form of a determinant:

$$(\mathbf{a} \wedge \mathbf{b}) = \begin{vmatrix} \mathbf{e}_1 & \mathbf{e}_2 & \mathbf{e}_3 \\ a_1 & a_2 & a_3 \\ b_1 & b_2 & b_3 \end{vmatrix}$$

$$= (a_2 b_3 - a_3 b_2)\mathbf{e}_1 + (a_3 b_1 - a_1 b_3)\mathbf{e}_2 + (a_1 b_2 - a_2 b_1)\mathbf{e}_3 \qquad (A.9.1\text{-}1)$$

With respect to the physical basis fields of an orthogonal curvilinear coordinate system, the vector product takes a similar form:

$$(\mathbf{a} \wedge \mathbf{b}) = \begin{vmatrix} \mathbf{g}_{(1)} & \mathbf{g}_{(2)} & \mathbf{g}_{(3)} \\ a_{(1)} & a_{(2)} & a_{(3)} \\ b_{(1)} & b_{(2)} & b_{(3)} \end{vmatrix}$$

$$= \left( a_{(2)} b_{(3)} - a_{(3)} b_{(2)} \right) \mathbf{g}_{(1)} + \left( a_{(3)} b_{(1)} - a_{(1)} b_{(3)} \right) \mathbf{g}_{(2)}$$

$$+ \left( a_{(1)} b_{(2)} - a_{(2)} b_{(1)} \right) \mathbf{g}_{(3)} \qquad (A.9.1\text{-}2)$$

The direction of $(\mathbf{a} \wedge \mathbf{b})$ is found by the

**Right-hand rule**   When the index finger points in the direction **a** and the middle finger points in the direction **b**, the thumb points in the direction **a** ∧ **b**.

A similar mnemonic device is often suggested for the components of the *curl* of a vector field **v** with respect to a set of rectangular Cartesian basis fields:

$$
\text{curl } \mathbf{v} = \begin{vmatrix} \mathbf{e}_1 & \mathbf{e}_2 & \mathbf{e}_3 \\ \partial/\partial z_1 & \partial/\partial z_2 & \partial/\partial z_3 \\ v_1 & v_2 & v_3 \end{vmatrix}
$$

$$
= \left( \frac{\partial v_3}{\partial z_2} - \frac{\partial v_2}{\partial z_3} \right) \mathbf{e}_1 + \left( \frac{\partial v_1}{\partial z_3} - \frac{\partial v_3}{\partial z_1} \right) \mathbf{e}_2 + \left( \frac{\partial v_2}{\partial z_1} - \frac{\partial v_1}{\partial z_2} \right) \mathbf{e}_3 \tag{A.9.1-3}
$$

We find it convenient to adopt a more compact notation. Instead of (A.9.1-1) and (A.9.1-2), we write for the vector product

$$
(\mathbf{a} \wedge \mathbf{b}) = e_{ijk} a_j b_k \mathbf{e}_i \tag{A.9.1-4}
$$

and

$$
(\mathbf{a} \wedge \mathbf{b}) = e_{ijk} a_{\langle j \rangle} b_{\langle k \rangle} \mathbf{g}_{\langle i \rangle} \tag{A.9.1-5}
$$

where $e_{ijk}$ is defined in Section A.2.1. Rather than (A.9.1-3), we prefer to write for the curl of a vector field **v**

$$
\text{curl } \mathbf{v} = e_{ijk} \frac{\partial v_k}{\partial z_j} \mathbf{e}_i \tag{A.9.1-6}
$$

Equations (A.9.1-4) through (A.9.1-6) are suggested by our presentation of determinants in Section A.2.1.

**Exercise A.9.1-1\*\***   Show that the curl of the gradient of a scalar field $\alpha$ is identically zero:

$$
\text{curl} (\nabla \alpha) = 0
$$

**Exercise A.9.1-2\*\***   Show that, for any spatial vector field **v**,

$$
\text{div curl } \mathbf{v} = 0
$$

**Exercise A.9.1-3\*\***   Show that, for any spatial vector field **v**,

$$
\text{curl curl } \mathbf{v} = \nabla (\text{div } \mathbf{v}) - \text{div} (\nabla \mathbf{v})
$$

**Exercise A.9.1-4\*\***   Three spatial vectors **a**, **b**, **c** may be viewed as forming three edges of a parallelepiped. If **a**, **b**, **c** have the same orientation as the first two fingers and thumb on the right hand, show that $(\mathbf{a} \wedge \mathbf{b}) \cdot \mathbf{c}$ determines the volume of the corresponding parallelepiped. This is known as the *right-hand rule*.

## A.9.2   More on the Vector Product and Curl

Our discussion in Section A.9.1 suggests that we take as our formal definition for the vector product of any two spatial vector fields **a** and **b**

$$
(\mathbf{a} \wedge \mathbf{b}) \equiv \boldsymbol{\epsilon} : (\mathbf{ba})
$$
$$
= e_{ijk} a_j b_k \mathbf{e}_i \tag{A.9.2-1}
$$

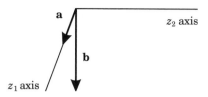

**Figure A.9.2-1.** The vector **b** lies in the coordinate plane $z_3 = 0$.

where (see Exercises A.7.2-11 and A.7.2-12)

$$\epsilon = e_{ijk}\mathbf{e}_i\mathbf{e}_j\mathbf{e}_k \tag{A.9.2-2}$$

The $\mathbf{e}_m$ ($m = 1, 2, 3$) are to be interpreted as any convenient set of rectangular Cartesian basis fields.

To find the geometrical interpretation of the spatial vector field ($\mathbf{a} \wedge \mathbf{b}$), let us fix our attention on a particular point $z$ in Figure A.9.2-1. For brevity, we adopt the inexact notation $\mathbf{a} = \mathbf{a}(z)$ and $\mathbf{b} = \mathbf{b}(z)$, and we introduce a particular rectangular Cartesian coordinate system, the origin of which coincides with the point $z$. We have chosen the $z_1$ axis along $\mathbf{a}$ and the $z_2$ axis perpendicular to $\mathbf{a}$, but in the plane of $\mathbf{a}$ and $\mathbf{b}$. In this special coordinate system, we have

$$\begin{aligned} a_i &= (a, 0, 0) \\ b_i &= (b \cos\theta, \, b \sin\theta, \, 0) \end{aligned} \tag{A.9.2-3}$$

in which $a$ and $b$ denote the magnitudes of **a** and **b**. The components of $\mathbf{c} \equiv (\mathbf{a} \wedge \mathbf{b})$ with respect to this coordinate system are

$$c_i = (0, 0, ab \sin\theta) \tag{A.9.2-4}$$

We conclude that ($\mathbf{a} \wedge \mathbf{b}$) lies along the perpendicular to the plane of **a** and **b** and that its magnitude is $ab \sin\theta$. We must still decide on its direction along this line.

We observe from (A.9.2-1) that

$$c^2 = e_{ijk}c_i a_j b_k \tag{A.9.2-5}$$

is a positive scalar. This suggests that we examine the sign of

$$e_{ijk}d_i a_j b_k \tag{A.9.2-6}$$

where **d** is any spatial vector. If we take the same rectangular Cartesian coordinate system indicated in Figure A.9.2-1, we see that

$$e_{ijk}d_i a_j b_k = d_3 ab \sin\theta \tag{A.9.2-7}$$

The quantity (A.9.2-6) will vanish, only if $d_3 = 0$ or only if one vector becomes coplanar with the other two. If we continuously deform the triad (**a**, **b**, **d**) in such a way that it never becomes coplanar, we see that (A.9.2-6) varies continuously, but it always retains the same sign. Let us deform it continuously, until **a** coincides with the positive $z_1$ axis, **b** with the positive $z_2$ axis, and **d** with the $z_3$ axis. The quantity (A.9.2-6) must be positive, if **d** coincides with the positive $z_3$ axis, or negative, if **d** coincides with the negative $z_3$ axis.

Returning to (A.9.2-5), we see that $\mathbf{a}$, $\mathbf{b}$, and $\mathbf{c} \equiv (\mathbf{a} \wedge \mathbf{b})$ must have the same orientation with respect to one another as the coordinate system axes. In *right-handed* coordinate systems, the basis fields $\mathbf{e}_1$, $\mathbf{e}_2$, $\mathbf{e}_3$ have the same orientation as the index finger, middle finger, and thumb on the right hand. We clarify the definition (A.9.2-2) of $\epsilon$ by requiring that the rectangular Cartesian coordinate system be a right-handed one. Consequently, the direction of $(\mathbf{a} \wedge \mathbf{b})$ is found by the right-hand rule given in Section A.9.1.

We will generally find it convenient to limit ourselves to right-handed coordinate systems. Left-handed coordinate systems are discussed in Exercise A.9.2-4.

If the tensor product $\mathbf{ba}$ in (A.9.2-1) is replaced by the gradient of a spatial vector field $\mathbf{v}$, we have the definition of the curl of $\mathbf{v}$:

$$\text{curl } \mathbf{v} \equiv \epsilon : (\nabla \mathbf{v}) \tag{A.9.2-8}$$

With respect to a rectangular Cartesian coordinate system, we have

$$\text{curl } \mathbf{v} = e_{ijk} \frac{\partial v_k}{\partial z_j} \mathbf{e}_i \tag{A.9.2-9}$$

**Exercise A.9.2-1**

*i) Let $\mathbf{a}$ and $\mathbf{b}$ be any two spatial vector fields. Show that, with respect to any curvilinear coordinate system, we may write

$$(\mathbf{a} \wedge \mathbf{b}) = \epsilon^{ijk} a_j b_k \mathbf{g}_i$$

and

$$(\mathbf{a} \wedge \mathbf{b}) = \epsilon_{ijk} a^j b^k \mathbf{g}^i$$

ii) Show that, with respect to an orthogonal curvilinear coordinate system,

$$(\mathbf{a} \wedge \mathbf{b}) = e_{ijk} a_{\langle j \rangle} b_{\langle k \rangle} \mathbf{g}_{\langle i \rangle}$$

**Exercise A.9.2-2*** Let $\mathbf{v}$ be any spatial vector field. Show that with respect to any curvilinear coordinate system we may write

i) $\text{curl } \mathbf{v} = \epsilon^{ijk} v_{k,j} \mathbf{g}_i$ and
ii) $\text{curl } \mathbf{v} = \epsilon_{ijk} v^k{}_{,m} g^{mj} \mathbf{g}^i$

**Exercise A.9.2-3**   If $\mathbf{v}$ is a spatial vector field, show that

$$\text{curl } \mathbf{v} = \text{div} (\epsilon \cdot \mathbf{v})$$

where $\epsilon \cdot \mathbf{v}$ is a second-order tensor field.

**Exercise A.9.2-4**

i) Let the $\bar{\mathbf{e}}_j$ ($j = 1, 2, 3$) be a set of left-handed rectangular Cartesian basis fields. Use the right-hand rule and Exercise A.7.2-11 to show that

$$\bar{\mathbf{e}}_1 \cdot (\bar{\mathbf{e}}_2 \wedge \bar{\mathbf{e}}_3) = \det \left( \frac{\partial z_i}{\partial \bar{z}_j} \right)$$
$$= -1$$

Here the coordinates $z_i$ ($i = 1, 2, 3$) refer to some right-handed rectangular Cartesian basis $\mathbf{e}_i$ ($i = 1, 2, 3$); the coordinates $\bar{z}_j$ ($j = 1, 2, 3$) refer to the left-handed basis $\bar{\mathbf{e}}_j$ ($j = 1, 2, 3$).

From this result and Exercise A.7.2-11, we conclude that, when left-handed rectangular Cartesian coordinate systems are employed, we must write

$$\epsilon = -e_{ijk}\mathbf{e}_i\mathbf{e}_j\mathbf{e}_k$$

ii) Show that, when dealing with left-handed curvilinear coordinate systems, we must take the negative square root in Exercise A.7.2-11 and write

$$\epsilon^{ijk} = -\frac{1}{\sqrt{g}}e^{ijk}$$

and

$$\epsilon_{ijk} = -\sqrt{g}e_{ijk}$$

## A.10  Determinant of Second-Order Tensor

### A.10.1 Definition

Let the $\mathbf{g}_i$ ($i = 1, 2, 3$) be a set of basis fields for an arbitrary curvilinear coordinate system. At any point $z$ of $E$, the basis fields may be thought of as three edges of a parallelepiped as shown in Figure A.10.1-1. We wish to introduce the magnitude of the determinant of a second-order tensor field $\mathbf{T}$ at the point $z$ as the ratio of the volume of the parallelepiped spanned by $(\mathbf{T} \cdot \mathbf{g}_1, \mathbf{T} \cdot \mathbf{g}_2, \mathbf{T} \cdot \mathbf{g}_3)$ to the volume of the parallelepiped spanned by $(\mathbf{g}_1, \mathbf{g}_2, \mathbf{g}_3)$. We choose the sign of the determinant of $\mathbf{T}$ to be positive, if $(\mathbf{T} \cdot \mathbf{g}_1, \mathbf{T} \cdot \mathbf{g}_2, \mathbf{T} \cdot \mathbf{g}_3)$ and $(\mathbf{g}_1, \mathbf{g}_2, \mathbf{g}_3)$ have the same orientation (as the first two fingers and thumb on either the left or right hand); it is negative if they have the opposite orientation.

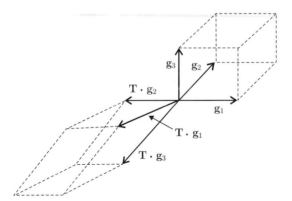

**Figure A.10.1-1.** Parallelepipeds spanned by $(\mathbf{g}_1, \mathbf{g}_2, \mathbf{g}_3)$ and by the transformations of these basis vectors.

This, together with Exercise A.9.1-4, suggests that we take as our formal definition of the scalar field

$$\det \mathbf{T} \equiv \frac{[(\mathbf{T} \cdot \mathbf{g}_1) \wedge (\mathbf{T} \cdot \mathbf{g}_2)] \cdot (\mathbf{T} \cdot \mathbf{g}_3)}{(\mathbf{g}_1 \wedge \mathbf{g}_2) \cdot \mathbf{g}_3} \tag{A.10.1-1}$$

To better appreciate the relationship of $\det \mathbf{T}$ to the concept of the determinant introduced in Section A.2.1, let us apply this definition to a set of rectangular Cartesian basis fields:

$$\begin{aligned}
\det \mathbf{T} &= \frac{[(\mathbf{T} \cdot \mathbf{e}_1) \wedge (\mathbf{T} \cdot \mathbf{e}_2)] \cdot (\mathbf{T} \cdot \mathbf{e}_3)}{(\mathbf{e}_1 \wedge \mathbf{e}_2) \cdot \mathbf{e}_3} \\
&= \frac{T_{j1} T_{k2} T_{i3} (\mathbf{e}_j \wedge \mathbf{e}_k) \cdot \mathbf{e}_i}{(\mathbf{e}_1 \wedge \mathbf{e}_2) \cdot \mathbf{e}_3} \\
&= T_{j1} T_{k2} T_{i3} \frac{e_{ijk}}{e_{312}} \\
&= e_{ijk} T_{i1} T_{j2} T_{k3} \\
&= \det(T_{mn}) \tag{A.10.1-2}
\end{aligned}$$

When $\mathbf{T}$ is expressed in terms of its physical components with respect to an orthogonal curvilinear coordinate system, the results of Section A.2.1 are again directly applicable (Exercise A.10.1-2). But in general, minor modifications must be made (Exercise A.10.1-3). It is easy to show (see Exercise A.10.1-4)

$$\det(\mathbf{S} \cdot \mathbf{T}) = (\det \mathbf{S})(\det \mathbf{T}) \tag{A.10.1-3}$$

$$\det(\mathbf{T}^{-1}) = \frac{1}{\det \mathbf{T}} \tag{A.10.1-4}$$

and

$$\det(\mathbf{T}^T) = \det \mathbf{T} \tag{A.10.1-5}$$

This discussion is based upon that given by Coleman et al. (1966, p. 102).

**Exercise A.10.1-1**    In terms of the components of $\mathbf{T}$ with respect to a rectangular Cartesian coordinate system, deduce that

$$e_{ijk} T_{im} T_{jn} T_{kp} = \det \mathbf{T} e_{mnp}$$

**Exercise A.10.1-2**    In terms of the physical components $\mathbf{T}$ with respect to an orthogonal curvilinear coordinate system, deduce that

$$e_{ijk} T_{\langle im \rangle} T_{\langle jn \rangle} T_{\langle kp \rangle} = \det \mathbf{T} e_{mnp}$$

**Exercise A.10.1-3\***    For an arbitrary curvilinear coordinate system,

$$\begin{aligned}
T &= T^{ij} \mathbf{g}_i \mathbf{g}_j \\
&= T_{ij} \mathbf{g}^i \mathbf{g}^j \\
&= T_i{}^k \mathbf{g}^i \mathbf{g}_k
\end{aligned}$$

prove that

$$\epsilon_{ijk} T^{im} T^{jn} T^{kp} = \det \mathbf{T} \epsilon^{mnp}$$

$$\epsilon^{ijk} T_{im} T_{jn} T_{kp} = \det \mathbf{T} \epsilon_{mnp}$$

and

$$\epsilon^{ijk} T_i{}^m T_j{}^n T_k{}^p = \det \mathbf{T} \epsilon^{mnp}$$

**Exercise A.10.1-4**   Prove (A.10.1-3) through (A.10.1-5).

**Exercise A.10.1-5**   Prove the following rule for differentiation of determinants:

$$\frac{d_{(m)}(\det \mathbf{T})}{dt} = (\det \mathbf{T}) \left[ \text{tr} \left( \mathbf{T}^{-1} \cdot \frac{d_{(m)} \mathbf{T}}{dt} \right) \right]$$

*Hint:*   Begin with

$$\det T_{rs} = \frac{1}{6} e_{ijk} e_{mnp} T_{im} T_{jn} T_{kp}$$

**Exercise A.10.1-6**

i) If $\mathbf{A}$ is invertible and (A.5.3-3) holds, show that $\det \mathbf{A} \neq 0$.
ii) Beginning with an equation of the same general form as (A.2.1-12) of Section A.2.1, show that, if $\det \mathbf{A} \neq 0$, $\mathbf{A}$ must be invertible.

**Exercise A.10.1-7** *Orthogonal tensors*   Prove that

i) if $\mathbf{Q}$ is an orthogonal second-order tensor,

$$\det \mathbf{Q} = \pm 1$$

ii) If $\det \mathbf{Q} = -1$ and if $(\mathbf{e}_1, \mathbf{e}_2, \mathbf{e}_3)$ is a right-handed triad,

$$\{ \mathbf{Q} \cdot \mathbf{e}_1, \mathbf{Q} \cdot \mathbf{e}_2, \mathbf{Q} \cdot \mathbf{e}_3 \}$$

is left-handed. In the context of changes of frame in Section 1.2.1, $\mathbf{Q}$ may be thought of as both a rotation and a reflection with $\det \mathbf{Q} = -1$.

## A.11   Integration

### A.11.1 Spatial Vector Fields**

A volume, surface, or line integration is an addition of quantities associated with different points in space. When we integrate a spatial vector field (a spatial vector-valued function of position), we add spatial *vectors* associated with different points in space.

Let $\mathbf{v}$ be some spatial vector field and let $z$ and $y$ denote two points in space. By the parallelogram rule for the addition of spatial vectors (Section A.1.2), we may write

$$\mathbf{v}(z) + \mathbf{v}(y) = v_i(z) \mathbf{e}_i + v_i(y) \mathbf{e}_i$$
$$= [v_i(z) + v_i(y)] \mathbf{e}_i \tag{A.11.1-1}$$

In applying the parallelogram rule here, we take advantage of the fact that the direction and magnitude of the rectangular Cartesian basis fields are independent of position in space.

Equation (A.11.1-1) suggests how we should proceed with the integration of a spatial vector field. Let us consider an integration of $\mathbf{v}$ over some region $R$ (this might be a curve, surface, or volume):

$$\int_R \mathbf{v} \, dR = \int_R v_i \mathbf{e}_i \, dR = \left( \int_R v_i \, dR \right) \mathbf{e}_i \tag{A.11.1-2}$$

Since the magnitude and direction of the rectangular Cartesian basis fields are independent of position in space, they may be treated as constants with respect to integration. By (A.11.1-2) we have transformed the problem of integration of a spatial vector field to the familiar problem of integration of three scalar fields.

**Exercise A.11.1-1\*** Let $\mathbf{v} = v_i \mathbf{e}_i$ for some rectangular Cartesian coordinate system $(z_1, z_2, z_3)$ and $\mathbf{v} = \overline{v}_j \overline{\mathbf{g}}^j$ for some curvilinear coordinate system $(\overline{x}^1, \overline{x}^2, \overline{x}^3)$.

i) Show that

$$\int_R \mathbf{v} \, dR = \int_R v_i \, dR \mathbf{e}_i$$
$$\neq \int_R \overline{v}_j \, dR \, \overline{\mathbf{g}}^j$$

ii) Show that

$$\int_R \mathbf{v} \, dR = \int_R \frac{\partial \overline{x}^j}{\partial z_i} \overline{v}_j \, dR \, \mathbf{e}_i$$

Sometimes it is most convenient to express the integration of a spatial vector field in terms of its covariant or contravariant components with respect to some curvilinear coordinate system, even though the integration is carried out with respect to some rectangular Cartesian coordinate system.

**Exercise A.11.1-2\*\*** *Second-order tensors* Extend the discussion of this section to second-order tensor fields.

## A.11.2 Green's Transformation

Our objective here is to develop *Green's transformation*, a special case of which is the divergence theorem or Gauss's theorem.

Let $\varphi$ be any scalar, vector, or tensor. If $\mathbf{n}$ is understood to be the outwardly directed unit normal to the closed surface $S_m$ shown in Figure A.11.2-1, we may approximate

$$\int_{S_m} \varphi \mathbf{n} \, dA \approx [\varphi(z_1 + \Delta z_1, z_2, z_3) - \varphi(z_1, z_2, z_3)] \, \Delta z_2 \Delta z_3 \mathbf{e}_1$$
$$+ [\varphi(z_1, z_2 + \Delta z_2, z_3) - \varphi(z_1, z_2, z_3)] \, \Delta z_1 \Delta z_3 \mathbf{e}_2$$
$$+ [\varphi(z_1, z_2, z_3 + \Delta z_3) - \varphi(z_1, z_2, z_3)] \, \Delta z_1 \Delta z_2 \mathbf{e}_3 \tag{A.11.2-1}$$

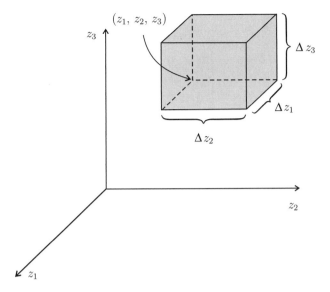

**Figure A.11.2-1.** The element of volume $\Delta V_m \equiv \Delta z_1 \Delta z_2 \Delta z_3$; the closed bounding surface of $\Delta V_m$ is $S_m$.

Dividing through by $\Delta V_m = \Delta z_1 \Delta z_2 \Delta z_3$, we see that in the limit as

$$\Delta z_1 \to 0,\ \Delta z_2 \to 0,\ \Delta z_3 \to 0: \quad \frac{1}{\Delta V_m} \int_{S_m} \varphi \mathbf{n}\, dA = \nabla \varphi \qquad (A.11.2\text{-}2)$$

Now consider a region of space that has a volume $V$ and a closed bounding surface $S$. Let $\mathbf{n}$ be the outwardly directed unit normal spatial vector field to the surface $S$. Referring to Figure A.11.2-2, we may define a volume integral to be obtained as the result of a limiting process in which we visualize the region $R$ to be approximated by $K$ parallelepipeds:

$$\int_R \nabla \varphi\, dV = \text{limit max } \Delta V_m \to 0: \quad \sum_{m=1}^{K} (\nabla \varphi)_m \Delta V_m \qquad (A.11.2\text{-}3)$$

By $(\nabla \varphi)_m$ we mean the value of $\nabla \varphi$ evaluated at some point in the $m$th parallelepiped. From (A.11.2-2) and (A.11.2-3), we conclude that

$$\int_R \nabla \varphi\, dV = \text{limit max } \Delta V_m \to 0: \quad \sum_{m=1}^{K} \int_{S_m} \varphi \mathbf{n}\, dA$$

$$= \int_S \varphi \mathbf{n}\, dA \qquad (A.11.2\text{-}4)$$

In arriving at this result, we have noted in Figure A.11.2-2 that the contributions to $S_m$ and $S_{m+1}$ cancel on that portion of their boundary that they share in common. We will refer to (A.11.2-4) as *Green's transformation*.

If $\mathbf{v}$ is a spatial vector field, Green's transformation requires

$$\int_R \nabla \mathbf{v}\, dV = \int_S \mathbf{v} \mathbf{n}\, dA \qquad (A.11.2\text{-}5)$$

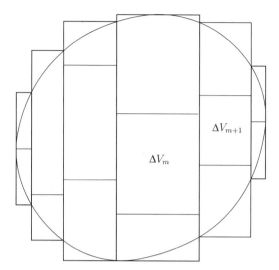

**Figure A.11.2-2.** The approximation of the region $R$ by $K$ parallelepipeds.

A special case is the familiar *divergence theorem* or Gauss's theorem:

$$\int_R \operatorname{div} \mathbf{v} \, dV = \int_S \mathbf{v} \cdot \mathbf{n} \, dA \tag{A.11.2-6}$$

Let $\mathbf{T}$ be any second-order tensor field. Green's transformation says that

$$\int_R \nabla \mathbf{T} \, dV = \int_S \mathbf{T} \mathbf{n} \, dA \tag{A.11.2-7}$$

It is a special case of this last equation that we will have several occasions to use:

$$\int_R \operatorname{div} \mathbf{T} \, dV = \int_S \mathbf{T} \cdot \mathbf{n} \, dA \tag{A.11.2-8}$$

For an alternative view of Green's transformation and for a further application, see Exercises A.11.2-1 and A.11.2-2.

This discussion is based upon that given by Ericksen (1960, p. 815).

### Exercise A.11.2-1

    i) Assuming $\varphi$ is a scalar field, write the $i$ component of (A.11.2-4) in rectangular Cartesian coordinates.

    ii) Given a spatial vector field $\varphi$, write the $ij$ component of (A.11.2-4) in rectangular Cartesian coordinates.

    iii) Assuming $\varphi$ is a second-order tensor field, write the $ijk$ component of (A.11.2-4) in rectangular Cartesian coordinates.

    iv) Use the result of (ii) to derive (A.11.2-6).

    v) Use the result of (iii) to derive (A.11.2-8).

**Exercise A.11.2-2**  Let **v** be a spatial vector field. Show as another case of Green's transformation that

$$\int_R \text{curl } \mathbf{v} \, dV = \int_S (\mathbf{n} \wedge \mathbf{v}) \, dA$$

*Hint:*  Use Exercise A.11.2-1 (iii).

## A.11.3 Change of Variable in Volume Integrations**

Let $(x^1, x^2, x^3)$ denote one system of coordinates and $(\overline{x}^1, \overline{x}^2, \overline{x}^3)$ another, such that

$$x^1 = x^1(\overline{x}^1, \overline{x}^2, \overline{x}^3)$$

$$x^2 = x^2(\overline{x}^1, \overline{x}^2, \overline{x}^3)$$

$$x^3 = x^3(\overline{x}^1, \overline{x}^2, \overline{x}^3)$$

We know that

$$\int_R F(x^1, x^2, x^3) \, dx^1 dx^2 dx^3$$

$$= \int_R F\left(x^1(\overline{x}^1, \overline{x}^2, \overline{x}^3), \ldots\right) \left| \det\left(\frac{\partial x^i}{\partial \overline{x}^j}\right) \right| d\overline{x}^1 d\overline{x}^2 d\overline{x}^3 \tag{A.11.3-1}$$

where $R$ is the region over which the integrations are to be performed.

If one coordinate system in (A.11.3-1) is rectangular Cartesian, we have (see Exercise A.11.3-1)

$$\int_R F(z_1, z_2, z_3) \, dz_1 dz_2 dz_3$$

$$= \int_R F(z_1(x^1, x^2, x^3), \ldots) \sqrt{g} \, dx^1 dx^2 dx^3 \tag{A.11.3-2}$$

Here $g \equiv \det(g_{ij})$.

**Exercise A.11.3-1**  Prove that

$$\left| \det\left(\frac{\partial z_i}{\partial x^j}\right) \right| = \sqrt{g}$$

and that (A.11.3-2) holds.

**Exercise A.11.3-2**  What form does (A.11.3-2) take for the case where the $(x^1, x^2, x^3)$ denote

i) cylindrical coordinates?
ii) spherical coordinates?

**Exercise A.11.3-3**

i) Consider some curve in space and let $t$ be a parameter along this curve. We know from Section A.4.1 that $(d\mathbf{p}/dt)$ is tangent to the curve at the point $t$. Prove that

$$\left(\frac{d\mathbf{p}}{dt} \cdot \frac{d\mathbf{p}}{dt}\right)(dt)^2 = \left(\frac{dz_1}{dt}dt\right)^2 + \left(\frac{dz_2}{dt}dt\right)^2 + \left(\frac{dz_3}{dt}dt\right)^2$$

We define the length of a differential segment of a curve to be the differential of arc length

$$(ds)^2 = \left(\frac{dp}{dt} \cdot \frac{dp}{dt}\right)(dt)^2$$

This means that

$$\frac{d\mathbf{p}}{ds} \cdot \frac{d\mathbf{p}}{ds} = 1$$

and

$$(ds)^2 = (dz_1)^2 + (dz_2)^2 + (dz_3)^2$$

ii) In any coordinate system, three coordinate curves intersect at each point in space. Choose a particular point and denote the unit tangent vectors to the three coordinate curves at this point by $d\mathbf{p}/ds_{(1)}, d\mathbf{p}/ds_{(2)}$, and $d\mathbf{p}/ds_{(3)}$. Here $s_{(i)}$ denotes arc length measured along the $x^i$ coordinate curve.

Let us obtain an expression for the differential volume $dV$ of the parallelepiped formed from $d\mathbf{p}/ds_{(1)}, d\mathbf{p}/ds_{(2)}$, and $d\mathbf{p}/ds_{(3)}$ with sides of length $ds_{(1)}, ds_{(2)}$, and $ds_{(3)}$. Starting with the definition

$$dV \equiv \left(\frac{d\mathbf{p}}{ds_{(1)}} \wedge \frac{d\mathbf{p}}{ds_{(2)}}\right) \cdot \frac{d\mathbf{p}}{ds_{(3)}} \, ds_{(1)} ds_{(2)} ds_{(3)}$$

show that

$$dV = \sqrt{g} \, dx^1 dx^2 dx^3$$

In this way, we may suggest the result of (A.11.3-1).

# More on the Transport Theorem

THE DERIVATION OF THE TRANSPORT THEOREM in Section 1.3.2 may be unsatisfactory for the reader who has chosen to read only the double-asterisked sections of Appendix A. One may understandably object to using Exercise A.10.1-5, which one has not proved, in order to arrive at (1.3.2-4). In what follows, an alternative derivation of the transport theorem is suggested, in which an obvious but additional statement is required about the time rate of change of the volume of the system.

Starting with (1.3.2-1):

$$\frac{d}{dt} \int_{R_{(m)}} \Psi \, dV = \int_{R_{(m)}} \left( \frac{d_{(m)} \Psi}{dt} + \frac{\Psi}{J} \frac{d_{(m)} J}{dt} \right) dV \tag{B.0.3-1}$$

let us set

$$\Psi = 1 \tag{B.0.3-2}$$

to obtain

$$\frac{dV_{(m)}}{dt} = \int_{R_{(m)}} \frac{1}{J} \frac{d_{(m)} J}{dt} \, dV \tag{B.0.3-3}$$

This last equation gives us an expression for the time rate of change of the volume $V_{(m)}$ associated with the material body.

Intuitively, we can say that the rate at which the volume of the body increases can be related to the net rate at which the bounding surface of the body moves in an outward direction:

$$\frac{dV_{(m)}}{dt} = \int_{S_{(m)}} \mathbf{v} \cdot \mathbf{n} \, dA \tag{B.0.3-4}$$

An application of Green's transformation (Section A.11.2) allows us to express this last equation as

$$\frac{dV_{(m)}}{dt} = \int_{R_{(m)}} \operatorname{div} \mathbf{v} \, dV \tag{B.0.3-5}$$

By eliminating the time rate of change of the volume of the body between (B.0.3-3) and (B.0.3-5), we find

$$
\int_{R_{(m)}} \left( \frac{1}{J} \frac{d_{(m)} J}{dt} - \text{div } \mathbf{v} \right) dV = 0 \tag{B.0.3-6}
$$

But this statement is true for any body or for any portion of a body (since a portion of a body is itself a body). We conclude that the integrand in (B.0.3-6) must be zero:

$$
\frac{1}{J} \frac{d_{(m)} J}{dt} = \text{div } \mathbf{v} \tag{B.0.3-7}
$$

Equation (B.0.3-7) is just what we need to put (B.0.3-1) in the form of the transport theorem (1.3.2-9):

$$
\frac{d}{dt} \int_{R_{(m)}} \Psi \, dV = \int_{R_{(m)}} \left( \frac{d_{(m)} \Psi}{dt} + \Psi \, \text{div } \mathbf{v} \right) dV \tag{B.0.3-8}
$$

From here the alternative forms of the transport theorem (1.3.2-10) and (1.3.2-11), as well as the generalized transport theorem (1.3.2-12), follow as described in Section 1.3.2.

The advantage of this discussion over that given in Section 1.3.2 is that we have been able to avoid the use of Exercise A.10.1-5 in arriving at (B.0.3-7). The disadvantage is that we have adopted an intuitively motivated step (B.0.3-4) in the course of our proof.

# Bibliography

Acrivos, A., M. J. Shah, and E. E. Petersen (1960). Momentum and heat transfer in laminar boundary-layer flows of non-Newtonian fluids past external surfaces. *AIChE J.*, **6**, 312–317.

Acrivos, A., M. J. Shah, and E. E. Petersen (1965). On the solution of the two-dimensional boundary-layer flow equations for a non-Newtonian power law fluid. *Chem. Eng. Sci.*, **20**, 101–105.

Acrivos, A. and T. D. Taylor (1964). The Stokes flow past an arbitrary particle, the slightly deformed sphere. *Chem. Eng. Sci.*, **19**, 445–451.

Adivarahan, P., D. Kunii, and J. M. Smith (1962). Heat transfer in porous rocks through which single-phase fluids are flowing. *Soc. Pet. Eng. J.*, **2**, 290–296.

Anderson, T. B. and R. Jackson (1967). A fluid mechanical description of fluidized beds. *Ind. Eng. Chem. Fundam.*, **6**, 527–538.

Aris, R. (1956). On the dispersion of a solute in a fluid flowing through a tube. *Proc. R. Soc. London, Ser. A*, **235**, 67–77.

Aris, R. (1957). On shape factors for irregular particles – I. The steady state problem. Diffusion and reactions. *Chem. Eng. Sci.*, **6**, 262–268.

Arnold, J. H. (1944). Studies in diffusion: III. Unsteady-state vaporization and adsorption. *Trans. Am. Inst. Chem. Eng.*, **40**, 361–378.

Arnold, K. R., and H. L. Toor (1967). Unsteady diffusion in ternary gas mixtures. *AIChE J.*, **13**, 909–914.

Ashare, E., R. B. Bird, and J. A. Lescarboura (1965). Falling cylinder viscometer for non-Newtonian fluids. *AIChE J.*, **11**, 910–916.

Bachmat, Y. (1972). Spatial macroscopization of processes in heterogeneous systems. *Isr. J. Technol.*, **10**, 391–398.

Barrer, R. M. (1951). *Diffusion in and Through Solids*. Cambridge, UK: Cambridge University Press.

Batchelor, G. K. (1959). *The Theory of Homogeneous Turbulence*. London: Cambridge University Press.

Batchelor, G. K. (1967). *An Introduction to Fluid Mechanics*. Cambridge, UK: Cambridge University Press.

Bear, J. (1961). On the tensor form of dispersion in porous media. *J. Geophys. Res.*, **66**, 1185.

Bear, J. (1972). *Dynamics of Fluids in Porous Media*. New York: American Elsevier.

Bedingfield, C. H., Jr., and T. B. Drew (1950). Analogy between heat transfer and mass transfer. A psychometric study. *Ind. Eng. Chem.*, **42**, 1164–1173.

Beebe, N. H. F. (1998, March). authidx: An author editor indexing package. *Tugboat*, **19**, 12–19.

Bird, R. B. (1957). The equations of change and the macroscopic mass, momentum, and energy balances. *Chem. Eng. Sci.*, **6**, 123–131.

Bird, R. B. (1965a). The equations of change and the macroscopic balances. *Chem. Eng. Prog. Symp. Ser.*, **61**(58), 1–15.

Bird, R. B. (1965b). Experimental tests of generalized Newtonian models containing a zero-shear viscosity and a characteristic time. *Can. J. Chem. Eng.*, **43**, 161–168.

Bird, R. B. (1993). The basic concepts in transport phenomena. *Chem. Eng. Educ.*, Spring, 102–109.

Bird, R. B., R. C. Armstrong, and O. Hassager (1977). *Dynamics of Polymeric Liquids, Volume 1: Fluid Mechanics* (1st ed.). New York: Wiley.

Bird, R. B., R. C. Armstrong, and O. Hassager (1987). *Dynamics of Polymeric Liquids, Volume 1: Fluid Mechanics* (2nd ed.). New York: Wiley.

Bird, R. B. and C. F. Curtiss (1959). Tangential Newtonian flow in annuli – I. Unsteady state velocity profiles. *Chem. Eng. Sci.*, **11**, 108–113.

Bird, R. B., O. Hassager, R. C. Armstrong, and C. F. Curtiss (1977). *Dynamics of Polymeric Liquids, Volume 2: Kinetic Theory*. New York: Wiley.

Bird, R. B., W. E. Stewart, and E. N. Lightfoot (1960). *Transport Phenomena*. New York: Wiley.

Birkhoff, G. (1955). *Hydrodynamics a Study in Logic, Fact and Similitude*. New York: Dover.

Blasius, H. (1908). Grenzschichten in flüssigkeiten mit kleiner reibung. *Z. Math. Phys.*, **56**, 1–37. English translation, National Advisory Committee for Aeronautics, Tech. Memo. 1256.

Bohr, N. (1909). Determination of the surface tension of water by the method of jet vibrations. *Philos. Trans. R. Soc. London, Ser. A*, **209**, 281–317.

Borg, R. J., and G. J. Dienes (1988). *An Introduction to Solid State Diffusion*. New York: Academic.

Boussinesq, J. (1877). *Mém. prés. par Div. Savants Acad. Sci. Paris*, **23**, 46.

Bowen, R. M. (1967). Toward a thermodynamics and mechanics of mixtures. *Arch. Rational Mech. Anal.*, **24**, 370–403.

Brand, L. (1957). The Pi theorem of dimensional analysis. *Arch. Rational Mech. Anal.*, **1**, 35–45.

Brinkman, H. C. (1949a). A calculation of the viscous force exerted by a flowing fluid on a dense swarm of particles. *Appl. Sci. Res.*, **A1**, 27–34.

Brinkman, H. C. (1949b). On the permeability of media consisting of closely packed porous particles. *Appl. Sci. Res.*, **A1**, 81–86.

Brown, G. M. (1960). Heat or mass transfer in a fluid in laminar flow in a circular or flat conduit. *AIChE J.*, **6**, 179–183.

Brown, H. (1940). On the temperature assignments of experimental thermal diffusion coefficients. *Phys. Rev.*, **58**, 661–662.

Burkhalter, J. E., and E. L. Koschmieder (1974). Steady supercritical taylor vortices after sudden starts. *Phys. Fluids*, **17**, 1929–1935.

Carman, P. C. (1956). *Flow of Gases Through Porous Media*. New York: Academic.

Carslaw, H. S., and J. C. Jaeger (1959). *Conduction of Heat in Solids* (2nd ed.). London: Oxford University Press.

Carty, R., and J. T. Schrodt (1975). Concentration profiles in ternary gaseous diffusion. *Ind. Eng. Chem. Fundam.*, **14**, 276–278.

Chang, S. H., and J. C. Slattery (1988). A new description of dispersion. *Transport in Porous Media*, **3**, 515–527.

Chen, W. K., and N. L. Peterson (1975). Effect of the deviation from stoichiometry on cation self-diffusion and isotope effect in wüstite, $Fe_{1-x}O$. *J. Phys. Chem. Solids*, **36**, 1097–1103.

Chu, J. C., R. L. Getty, L. F. Brennecke, and R. Paul (1950). *Distillation Equilibrium Data*. New York: Reinhold.

Churchill, R. V. (1958). *Operational Mathematics* (2nd ed.). New York: McGraw-Hill.

Churchill, R. V. (1960). *Complex Variables and Applications* (2nd ed.). New York: McGraw-Hill.

Coates, D. E., and A. D. Dalvi (1970). An extension of the Wagner theory of alloy oxidation and sulfidation. *Oxid. Met.*, **2**, 331–347.

Colburn, A. P. (1933). A method of correlating forced convection heat transfer data and a comparison with fluid friction. *Trans. Am. Inst. Chem. Eng.*, **29**, 174–210.

Colburn, A. P., and T. B. Drew (1937). The condensation of mixed vapors. *Trans. Am. Inst. Chem. Eng.*, **33**, 197–215.

Cole, J. D. (1968). *Perturbation Methods in Applied Mathematics*. Waltham, MA: Blaisdell.

Coleman, B. D. (1962). Kinematical concepts with applications in the mechanics and thermodynamics of incompressible viscoelastic fluids. *Arch. Rational Mech. Anal.*, **9**, 273–300.

Coleman, B. D. (1964). Thermodynamics of materials with memory. *Arch. Rational Mech. Anal.*, **17**, 1–46.

Coleman, B. D., H. Markovitz, and W. Noll (1966). *Viscometric Flows of Non-Newtonian Fluids*. New York: Springer-Verlag.

Coleman, B. D., and W. Noll (1959). On certain steady flows of general fluids. *Arch. Rational Mech. Anal.*, **3**, 289–303.

Coleman, B. D., and W. Noll (1961). Recent results in the continuum theory of viscoelastic fluids. *Ann. N.Y. Acad. Sci.*, **89**, 672–714.

Coleman, B. D., and W. Noll (1962). Steady extension of incompressible simple fluids. *Phys. Fluids*, **5**, 840–843.

Condiff, D. W. (1969). On symmetric multicomponent diffusion coefficients. *J. Chem. Phys.*, **51**, 4209–4212.

Corrsin, S. (1963). Turbulence: experimental methods. In S. Flügge and C. Truesdell, (eds.), *Handbuch der Physik*, Vol. 8/2. Berlin: Springer-Verlag.

Crank, J. (1956). *The Mathematics of Diffusion*. London: Oxford University Press.

Curtiss, C. F. (1956). Kinetic theory of nonspherical molecules. *J. Chem. Phys.*, **24**, 225–241.

Curtiss, C. F. (1968). Symmetric gaseous diffusion coefficients. *J. Chem. Phys.*, **49**, 2917–2919.

Curtiss, C. F., and R. B. Bird (1998). Multicomponent diffusion. (To be submitted for publication.)

Curtiss, C. F., and J. O. Hirschfelder (1949). Transport properties of multicomponent gas mixtures. *J. Chem. Phys.*, **17**, 550–555.

Cussler, E. L., Jr., and E. N. Lightfoot, Jr. (1963a). Multicomponent diffusion in restricted systems. *AIChE J.*, **9**, 702–703.

Cussler, E. L., Jr., and E. N. Lightfoot, Jr. (1963b). Multicomponent diffusion in semi-infinite systems. *AIChE J.*, **9**, 783–785.

Dahler, J. S., and L. E. Scriven (1961). Angular momentum of continua. *Nature*, **192**, 36–37.

Danckwerts, P. V. (1950). Absorption by simultaneous diffusion and chemical reaction. *J. Chem. Soc. Faraday Trans.*, **46**, 300–304.

Danckwerts, P. V. (1951). Absorption by simultaneous diffusion and chemical reaction into particles of various shapes and into falling drops. *J. Chem. Soc. Faraday Trans.*, **47**, 1014–1023.

Darcy, H. P. (1856). *Les Fontaines Publiques de la Ville de Dijon*. Paris: Dalmont.

Daubert, T. E., and R. P. Danner (1989). *Physical and Thermodynamic Properties of Pure Chemicals: Data Compilation*. New York: Hemisphere.

Davies, M. H., M. T. Simnad, and C. E. Birchenall (1951). On the mechanism and kinetics of the scaling of iron. *J. Met.*, Oct. 889–896.

de Josselin de Jong, G., and M. J. Bossen (1961). Discussion of paper by Jacob Bear, "On the tensor form of dispersion in porous media." *J. Geophys. Res.*, **66**, 3623.

Deal, B. E., and A. S. Grove (1965). General relationship for the thermal oxidation of silicon. *J. Appl. Phys.*, **36**(12), 3770–3778.

Dean, J. A. (1979). *Lange's Handbook of Chemistry*. (12th ed.). New York: McGraw-Hill.

Defay, R., I. Prigogine, A. Bellemans, and D. H. Everett (1966). *Surface Tension and Adsorption*. New York: Wiley.

DeGroff, H. M. (1956a). Comments on viscous heating. *J. Aeronaut. Sci.*, **23**, 978–979.

DeGroff, H. M. (1956b). On viscous heating. *J. Aeronaut. Sci.*, **23**, 395–396.

Deissler, R. G. (1950). Analytical and experimental investigation of adiabatic turbulent flow in smooth tubes. Technical Note 2138, NACA.

Deissler, R. G. (1955). Analysis of turbulent heat transfer, mass transfer, and friction in smooth tubes at high Prandtl and Schmidt numbers. Technical Report 1210, NACA.

Delhaye, J. M. (1977a). Instantaneous space-averaged equations. In S. Kakaç and F. Mayinger (eds.), *Two-Phase Flows and Heat Transfer*, Vol. 1, pp. 81–90. Washington, DC: Hemisphere.

Delhaye, J. M. (1977b). Space/time and time/space-averaged equations. In S. Kakaç and F. Mayinger (eds.), *Two-Phase Flows and Heat Transfer*, Vol. 1, pp. 101–114. Washington, DC: Hemisphere.

Delhaye, J. M. (1981a). Composite-averaged equations. In J. M. Delhaye, M. Giot, and M. L. Riethmuller, (eds.), *Thermohydraulics of Two-Phase Systems for Industrial Design and Nuclear Engineering*, pp. 181–186. New York: Hemisphere.

Delhaye, J. M. (1981b). Instantaneous space-averaged equations. In J. M. Delhaye, M. Giot, and M. L. Riethmuller, (eds.), *Thermohydraulics of Two-Phase Systems for Industrial Design and Nuclear Engineering*, pp. 159–170. New York: Hemisphere.

Denbigh, K. G. (1963). *The Principles of Chemical Equilibrium*. London: Cambridge University Press.

Denn, M. M., C. J. S. Petrie, and P. Avenas (1975). Mechanics of steady spinning of a viscoelastic liquid. *AIChE J.*, **21**, 791–799.

Dhawan, S. (1953). Direct measurements of skin friction. Technical Report 1121, NACA.

Drew, D. A. (1971). Averaged field equations for two-phase media. *Stud. Appl. Math.*, **L**(2), 133–139.

Eckert, E. R., and J. R. M. Drake (1959). *Heat and Mass Transfer* (2nd ed.). New York: McGraw-Hill.

Edwards, D. A., H. Brenner, and D. T. Wasan (1991). *Interfacial Transport Processes and Rheology*. Boston: Butterworth-Heinemann.

Ericksen, J. L. (1960). Tensor fields. In S. Flügge (ed.), *Handbuch der Physik*, Vol. 3/1. Berlin: Springer-Verlag.

Evans, R. B., G. M. Watson, and E. A. Mason (1961). Gaseous diffusion in porous media at uniform pressure. *J. Chem. Phys.*, **35**, 2076–2083.

Fisher, R. J., and M. M. Denn (1976). A theory of isothermal melt spinning and draw resonance. *AIChE J.*, **22**, 236–246.

Fredrickson, A. G. (1964). *Principles and Applications of Rheology*. Englewood Cliffs, NJ: Prentice-Hall.

Gagon, D. K., and M. M. Denn (1981). Computer simulation of steady polymer melt spinning. *Polym. Eng. Sci.*, **21**, 844–853.

George, H. H. (1982). Model of steady-state melt spinning at intermediate take-up speeds. *Polym. Eng. Sci.*, **22**, 292–299.

Gibbs, J. W. (1928). *Collected Works*, Vol. 1. New York: Longmans.

Gill, W. N., and R. Sankarasubramanian (1970). Exact analysis of unsteady convective diffusion. *Proc. R. Soc. London, Ser. A.*, **316**, 341–350.

Gin, R. F., and A. B. Metzner (1965). Normal stresses in polymeric solutions. *Proceedings of the Fourth International Congress on Rheology*, **2**, 583–601.

Gmehling, J., U. Onken, and W. Arlt (1980). *Vapor-Liquid Equilibrium Data Collection, Aromatic Hydrocarbons*, Vol. 1/7. Frankfurt/Main: DECHEMA.

Goldstein, S. (1938). *Modern Developments in Fluid Dynamics*. London: Oxford University Press.

Graetz, L. (1883). Ueber die Wärmeleitungsfähigkeit von Flüssigkeiten. *Ann. Phys. Chem.*, **18**, 79–94.

Graetz, L. (1885). Ueber die Wärmeleitungsfähigkeit von Flüssigkeiten. *Ann. Phys. Chem.*, **25**, 337–357.

Gray, W. G. (1975). A derivation of the equations for multi-phase transport. *Chem. Eng. Sci.*, **30**, 229–233.

Gray, W. G., and K. O'Neil (1976). On the general equations for flow in porous media and their reduction to Darcy's law. *Water Resour. Res.*, **12**, 148.

Grew, K. E., and T. L. Ibbs (1952). *Thermal Diffusion in Gases*. New York: Cambridge University Press.

Grossetti, E. (1958). Born–Lertes effect on the dipolar rotation of liquids. *Nuovo Cimento*, **10**, 193–199.

Grossetti, E. (1959). Dipolar rotation effect in liquids. *Nuovo Cimento*, **13**, 350–353.

Gurtin, M. E. (1971). On the thermodynamics of chemically reacting fluid mixtures. *Arch. Rational Mech. Anal.*, **43**, 198–212.

Gurtin, M. E., and A. S. Vargas (1971). On the classical theory of reacting fluid mixtures. *Arch. Rational Mech. Anal.*, **43**, 179–197.

Gutfinger, C., and R. Shinnar (1964). Velocity distributions in two-dimensional laminar liquid-into-liquid jets in power-law fluids. *AIChE J.*, **10**, 631–639.

Hajiloo, A., T. R. Ramamohan, and J. C. Slattery (1987). Effect of interfacial viscosities on the stability of a liquid thread. *J. Colloid Interface Sci.*, **117**, 384–393

Halmos, P. R. (1958). *Finite-Dimensional Vector Spaces*. Princeton, NJ: Van Nostrand.

Happel, J., and H. Brenner (1965). *Low Reynolds Number Hydrodynamics*. Englewood Cliffs, NJ: Prentice-Hall.

Hauffe, K. (1965). *Oxidation of Metals*. New York: Plenum Press.

Hein, P. (1914). Untersuchungen über den kritischen Zustand. *Z. Phys. Chem. (Munich)*, **86**, 385–410.

Hermes, R. A., and A. G. Fredrickson (1967). Flow of viscoelastic fluids past a flat plate. *AIChE J.*, **13**, 253–259.

Herzog, R. O. and K. Weissenberg (1928). *Kelloid Z.*, **46**, 277. See also *Physics Today*, **21** (8), 13 (1968).

Hiemenz, K. (1911). Die grenzschicht an einem in den gleichförmigen flüssigkeitsstrom eingetauchten geraden kreiszylinder. *Dinglers Polytechnisches J.*, **326**, 321–324.

Himmel, L., R. F. Mehl, and C. E. Birchenall (1953). Self-diffusion of iron oxides and the Wagner theory of oxidation. *J. Met.*, June, 827–843.

Hinze, J. O. (1959). *Turbulence*. New York: McGraw-Hill.

Hirschfelder, J. O., C. F. Curtiss, and R. B. Bird (1954). *Molecular Theory of Gases and Liquids*. New York: Wiley. Corrected with notes added 1964.

Hoffman, K., and R. Kunze (1961). *Linear Algebra*. Englewood Cliffs, NJ: Prentice-Hall.

Howarth, L. (1934). On the calculation of steady flow in the boundary layer near the surface of a cylinder in a stream. Technical Report 1632, Aeron. Res. Committee (Great Britain).

Howarth, L. (1938). On the solution of the laminar boundary layer equations. *Proc. R. Soc. London, Ser. A*, **164**, 547–579.

Hsu, H. W., and R. B. Bird (1960). Multicomponent diffusion problems. *AIChE J.*, **6**, 516–524.

Huang, J. H., and J. M. Smith (1963). Heat transfer in porous media with known pore structure (Alundum). *J. Chem. Eng. Data*, **8**, 437–439.

Illingworth, C. R. (1950). Some solutions of the equations of flow of a viscous compressible fluid. *Proc. Cambridge Philos. Soc.*, **46**, 469–478.

Irving, J., and N. Mullineux (1959). *Mathematics in Physics and Engineering*. New York: Academic.

Jahnke, E., and F. Emde (1945). *Tables of Functions*. New York: Dover.

Jakob, M. (1949). *Heat Transfer*, Vol. 1. New York: Wiley.

Jakob, M. (1957). *Heat Transfer*, Vol. 2. New York: Wiley.

Jamieson, D. T. (1975). *Liquid Thermal Conductivity: A Data Survey to 1973*. Edinburgh: Edinburgh Press.

Jiang, T. S., M. H. Kim, V. J. Kremesec, Jr., and J. C. Slattery (1987). The local volume-averaged equations of motion for a suspension of non-neutrally buoyant spheres. *Chem. Eng. Commun.*, **50**, 1–30.

Johnson, M. W. (1977). A note on the functional form of the equations of continuum thermodynamics. *J. Non-Newtonian Fluid Mech.*, **3**, 297–305.

Kaplan, W. (1952). *Advanced Calculus*. Cambridge, MA: Addison-Wesley.

Karim, S. M., and L. Rosenhead (1952). The second coefficient of viscosity of liquids and gases. *Rev. Mod. Phys.*, **24**, 108–116.

Kase, S., and T. Matsuo (1965). Studies on melt spinning. I fundamental equations on dynamics of melt spinning. *J. Polym. Sci., Part A: Polym. Chem.*, **3**, 2541–2554.

Kays, W. M. (1955). Numerical solutions for laminar-flow heat transfer in circular tubes. *Trans. Am. Soc. Mech. Eng.*, **77**, 1265–1272.

Kays, W. M., and A. L. London (1964). *Compact Heat Exchangers* (2nd ed.). New York: McGraw-Hill.

Kellogg, O. D. (1929). *Foundations of Potential Theory*. New York: Ungar.

Kittredge, C. P., and D. S. Rowley (1957). Resistance coefficients for laminar and turbulent flow through one-half-inch valves and fittings. *Trans. ASME*, **79**, 1759–1766.

Koschmieder, E. L. (1979). Turbulent Taylor vortex field. *J. Fluid Mech.*, **93**, 515–527.

Krieger, I. M., and H. Elrod (1953). Direct determination of the flow curves of non-Newtonian fluids. II. Shearing rate in the concentric cylinder viscometer. *J. Appl. Phys.*, **24**, 134–136.

Krieger, I. M., and S. H. Maron (1952). Direct determination of the flow curves of non-Newtonian fluids. *J. Appl. Phys.*, **23**, 147–149, 1412.

Krieger, I. M., and S. H. Maron (1954). Direct determination of the flow curves of non-Newtonian fluids. III. Standardized treatment of viscometric data. *J. Appl. Phys.*, **25**, 72–75.

Kunii, D., and J. M. Smith (1960). Heat transfer characteristics of porous rocks. *AIChE J.*, **6**, 71–78.

Kunii, D., and J. M. Smith (1961a). Heat transfer characteristics of porous rocks: Thermal conductivities of unconsolidated particles with flowing fluids. *AIChE J.*, **7**, 29–34.

Kunii, D., and J. M. Smith (1961b). Thermal conductivities of porous rocks filled with stagnant fluid. *Soc. Pet. Eng. J.*, **1**, 37–42.

Lamb, H. (1945). *Hydrodynamics* (6th ed.). New York: Dover.

Landau, L. D., and E. M. Lifshitz (1987). *Fluid Mechanics* (2nd ed.), Volume 6, *Course of Theoretical Physics*. Elmsford, NY: Pergamon Press.

Lapple, C. E. (1949). Velocity head simplifies flow computation. *Chem. Eng.*, **56**(5), 96–104.

Laufer, J. (1953). The structure of turbulence in fully developed pipe flow. Technical Note 2954, NACA.

Lee, C. Y., and C. R. Wilke (1954). Measurement of vapor diffusion coefficient. *Ind. Eng. Chem.*, **46**, 2381–2387.

Leigh, D. C. (1968). *Nonlinear Continuum Mechanics*. New York: McGraw-Hill.

Lertes, P. (1921a). Der Dipolrotationseffekt bei dielektrischen Flüssigkeiten. *Z. Phys.*, **6**, 56–68.

Lertes, P. (1921b). Der Dipolrotationseffekt bei dielektrischen Flüssigkeiten. *Phys. Z.*, **22**, 621–623.

Lertes, P. (1921c). Untersuchungen über Rotationen von dielektrischen Flüssigkeiten im elektrostatischen Drehfeld. *Z. Phys.*, **4**, 315–336.

Levich, V. G. (1962). *Physicochemical Hydrodynamics*. Englewood Cliffs, NJ: Prentice-Hall.

Lewis, G. N., M. Randall, K. S. Pitzer, and L. Brewer (1961). *Thermodynamics* (2nd ed.). New York: McGraw-Hill.

Lichnerowicz, A. (1962). *Elements of Tensor Calculus*. New York: Wiley.

Lie, L. N., R. R. Razouk, and B. E. Deal (1982). High pressure oxidation of silicon in dry oxygen. *J. Electrochem. Soc.*, **129**(12), 2828–2834.

Liepmann, H. W., and S. Dhawan (1951). Direct measurements of local skin friction in low-speed and high-speed flow. In *Proc. First US Nat. Congr. Appl. Mech.*, pp. 869, as quoted by Schlichting (1979, p. 149).

Lightfoot, E. N. (1964). Unsteady diffusion with first-order reaction. *AIChE J.*, **10**, 278, 282–284.

Lin, C. C., and W. H. Reid (1963). Turbulent flow, theoretical aspects. In S. Flügge and C. Truesdell (eds.), *Handbuch der Physik*, Vol. 8/2, Berlin: Springer-Verlag.

Lin, C. Y., and J. C. Slattery (1982a). Thinning of a liquid film as a small drop or bubble approaches a fluid-fluid interface. *AIChE J.*, **28**, 786–792.

Lin, C. Y., and J. C. Slattery (1982b). Three-dimensional, randomized, network model for two-phase flow through porous media. *AIChE J.*, **28**, 311.

Livingston, P. M., and C. F. Curtiss (1959). Kinetic theory of nonspherical molecules. IV. Angular momentum transport coefficient. *J. Chem. Phys.*, **31**, 1643–1645.

Lodge, A. S. (1964). *Elastic Liquids*. New York: Academic.

Lugg, G. A. (1968). Diffusion coefficients of some organic and other vapors in air. *Anal. Chem.*, **40**, 1072–1077.

Maass, O. (1938). Changes in the liquid state in the critical temperature region. *Chem. Rev.*, **23**, 17–28.

Marle, C. M. (1967). Écoulements monophasique en milieu poreux. *Rev. Inst. Français du Pétrole*, **22**, 1471–1509.

Marsh, K. N. (ed.) (1994). *TRC Thermodynamic Tables–Hydrocarbons*. College Station, TX: Thermodynamics Research Center, Texas A&M University.

Masamune, S., and J. M. Smith (1963a). Thermal conductivity of beds of spherical particles. *Ind. Eng. Chem. Fundam.*, **2**, 136–143.

Masamune, S., and J. M. Smith (1963b). Thermal conductivity of porous catalyst pellets. *J. Chem. Eng. Data*, **8**, 54–58.

Mathematica (1993). Version 2.2. Champaign, IL: Wolfram Research, Inc.

Mathematica (1996). Version 3.0. Champaign, IL: Wolfram Research, Inc.

Matovich, M. A., and J. R. A. Pearson (1969). Spinning a molten threadline. *Ind. Eng. Chem. Fundam.*, **8**, 512–520.

Matsuhisa, S., and R. B. Bird (1965). Analytical and numerical solutions for laminar flow of the non-Newtonian Ellis fluid. *AIChE J.*, **11**, 588–595.

McConnell, A. J. (1957). *Applications of Tensor Analysis*. New York: Dover.

McIntosh, R. L., J. R. Dacey, and O. Maass (1939). Pressure, volume, temperature relations of ethylene in the critical region. II. *Can. J. Res., Sect. B*, **17**, 241–250.

Merk, H. J. (1959). The macroscopic equations for simultaneous heat and mass transfer in iotropic, continuous and closed systems. *Appl. Sci. Res. A*, **8**, 73–99.

Mhetar, V., and J. C. Slattery (1997). The Stefan problem of a binary liquid mixture. *Chem. Eng. Sci.*, **52**, 1237–1242.

Milne-Thomson, L. M. (1955). *Theoretical Hydrodynamics* (3rd ed.). New York: Macmillan.

Mischke, R. A., and J. M. Smith (1962). Thermal conductivity of alumina catalyst pellets. *Ind. Eng. Chem. Fundam.*, **1**, 288–292.

Morgan, A. J. A. (1957). On the Couette flow of compressible, viscous, heat-conducting, perfect gas. *J. Aeronaut. Sci.*, **24**, 315–316.

Morse, P. M., and H. Feshbach (1953). *Methods of Theoretical Physics, Part 1*. New York: McGraw-Hill.

Müller, I. (1968). A thermodynamic theory of mixtures of fluids. *Arch. Rational Mech. Anal.*, **28**, 1–39.

Müller, W. (1936). Zum Problem der Anlaufströmung einer Flüssigkeit im geraden Rohr mit Kreisring- und Kreisquerschnitt. *Z. Angew. Math. Mech.*, **16**, 227–238.

Newman, J. (1969). Preprint 18646. Berkeley, CA: Lawrence Radiation Laboratory, University of California.

Newman, J. S. (1973). *Electrochemical Systems*. Englewood Cliffs, NJ: Prentice-Hall.

Nikolaevskii, V. N. (1959). Convective diffusion in porous media. *J. Appl. Math. Mech.*, **23**, 1492.

Nikuradse, J. (1942). Laminare Reibungsschichten an der längsangeströmten Platte. Technical report, Zentrale f. wiss. Berichtswesen, Berlin, as reported by Schlichting (1979, p. 149).

Noll, W. (1958). A mathematical theory of the mechanical behavior of continuous media. *Arch. Rational Mech. Anal.*, **2**, 197–226.

Noll, W. (1962). Motions with constant stretch history. *Arch. Rational Mech. Anal.*, **11**, 97–105.

Norton, F. J. (1961). Permeation of gaseous oxygen through vitreous silica. *Nature*, **191**, 701.

Oldroyd, J. G. (1950). On the formulation of rheological equations of state. *Proc. R. Soc. London, Ser. A*, **200**, 523–541.

Oldroyd, J. G. (1965). Some steady flows of the general elastico-viscous liquid. *Proc. R. Soc. London, Ser. A*, **283**, 115–133.

Palmer, H. B. (1952). Ph. D. thesis, University of Wisconsin, as reported by Hirschfelder *et al.* (1954).

Patel, J. G., M. G. Hedge, and J. C. Slattery (1972). Further discussion of two-phase flow in porous media. *AIChE J.*, **18**, 1062–1063.

Paton, J. B., P. H. Squires, W. H. Darnell, F. M. Cash, and J. F. Carley (1959). In E. C. Burnhardt (ed.), *Processing of Thermoplastic Materials*. New York: Reinhold.

Pawlowski, J. (1953). Bestimmung des reibungsgesetzes der nichtnewtonschen Flüssigkeiten aus den Viskositätsmessungen mit Hilfe eines Rotationsviskosimeters. *Kolloid-Z.*, **130**, 129–131.

Peaceman, D. (1966). Improved treatment of dispersion in numerical calculation of multidimensional miscible displacement. *Soc. Pet. Eng. J.*, **6**, 213.

Peng, K. Y., L. C. Wang, and J. C. Slattery (1996). A new theory of silicon oxidation. *J. Vac. Sci. Technol., B*, **14**, 3316–3320.

Phan-Thien, N. (1978). A nonlinear network viscoelastic model. *J. Rheol. (N. Y.)*, **22**, 259–283.

Philippoff, W. (1935). Zur Theorie der Strukturviskosität. I. *Kolloid-Z.*, **71**, 1–16.

Pollard, W. G., and R. D. Present (1948). On gaseous self-diffusion in long capillary tubes. *Phys. Rev.*, **73**, 762–774.

Prager, W. (1945). Strain hardening under combined stresses. *J. Appl. Phys.*, **16**, 837–843.

Prandtl, L. (1925). Bericht über Untersuchungen zur ausgebildeten Turbulenz. *Z. Angew. Math. Mech.*, **5**, 136–139.

Prata, A. T., and E. M. Sparrow (1985). Diffusion driven nonisothermal evaporation. *J. Heat Transfer*, **107**, 239–242.

Prausnitz, J. M. (1969). *Molecular Thermodynamics of Fluid-Phase Equilibria*. Englewood Cliffs, NJ: Prentice-Hall.

Proudman, I., and J. R. Pearson (1957). Expansions at small Reynolds numbers for the flow past a sphere and a circular cylinder. *J. Fluid Mech.*, **2**, 237–262.

Quintard, M., and S. Whitaker (1994). Transport in ordered and disordered porous media, I: The cellular average and the use of weighting functions. *Transp. Porous Media*, **14**, 163–177.

Rabinowitsch, B. (1929). Über die Viskosität und Elastizität von Solen. *Z. Phys. Chem. A*, **145**, 1–26.

Ranz, W. E., and W. R. Marshall, Jr. (1952a) Evaporation from drops, Part I. *Chem. Eng. Prog.*, **48**(3), 141–146.

Ranz, W. E., and W. R. Marshall, Jr. (1952b). Evaporation from drops, Part II. *Chem. Eng. Prog.*, **48**(4), 173–180.

Rayleigh, L. (1878). On the capillary phenomena of jets. *Proc. R. Soc. London*, **29**, 71–97.

Reid, R. C., J. M. Prausnitz, and B. E. Poling (1987). *The Properties of Gases and Liquids* (4th ed.). New York: McGraw-Hill.

Reiner, M. (1945). A mathematical theory of dilatancy. *Am. J. Math.*, **67**, 350–362.

Reiner, M. (1960). *Deformation, Strain and Flow* (2nd ed.). New York: Interscience.

Richardson, J. F. (1959). The evaporation of two-component liquid mixtures. *Chem. Eng. Sci.*, **10**, 234–242.

Robinson, E., W. A. Wright, and G. W. Bennett (1932). Constant evaporating systems. *J. Phys. Chem.*, **36**, 658–663.

Rouse, H., and S. Ince (1957). *History of Hydraulics*. State University of Iowa: Iowa Institute of Hydraulic Research.

Rubinstein, L. I. (1971). The Stefan problem. In *Translations of Mathematical Monographs*, Vol. 27. Providence, RI: American Mathematical Society.

Rumscheidt, F. D., and S. G. Mason (1962). Break-up of stationary liquid threads. *J. Colloid Sci.*, **17**, 260–269.

Rutland, D. F., and G. J. Jameson (1971). A non-linear effect in the capillary instability of liquid jets. *J. Fluid Mech.*, **46**, 267–271.

Sabersky, R. H., A. J. Acosta, and E. G. Hauptmann (1989). *Fluid Flow* (3rd ed.). New York: Macmillan.

Sakiadis, B. C. (1961a). Boundary-layer behavior on continuous solid surfaces, I. The boundary layer on a continuous flat surface. *AIChE J.*, **7**, 26–28.

Sakiadis, B. C. (1961b). Boundary-layer behavior on continuous solid surfaces, II. Boundary-layer equations for two-dimensional and axisymmetric flow. *AIChE J.*, **7**, 221–225.

Sampson, R. A. (1891). On Stokes's current function. *Phil. Trans. R. Soc. London, Ser. A*, **182**, 449–518.

Sanni, A. S., and P. Hutchison (1973). Diffusivities and densities for binary liquid mixtures. *J. Chem. Eng. Data*, **18**, 317–322.

Satterfield, C. N., and T. K. Sherwood (1963). *The Role of Diffusion in Catalysis*. Reading, MA: Addison-Wesley.

Scheidegger, A. E. (1961). General theory of dispersion in porous media. *J. Geophys. Res.*, **66**, 3273.

Scheidegger, A. E. (1963). Hydrodynamics in porous media. In S. Flügge and C. Truesdell, (eds.), *Handbuch der Physik*, Vol. 8/2. Berlin: Springer-Verlag.

Schlichting, H. (1979). *Boundary-Layer Theory* (7th ed.). New York: McGraw-Hill.

Schmidt, E., and W. Beckmann (1930). Das temperatur- und geschwindigkeitsfeld von einer Wärme abgebenden, senkrechten platte bei natürlicher konvektion. *Forsch. Ing.-Wes.*, **1**, 391–404.

Schowalter, W. R. (1960). The application of boundary-layer theory to power-law pseudoplastic fluids: similar solutions. *AIChE J.*, **6**, 24–28.

Scott, D. S. (1962). Gas diffusion with chemical reaction in porous solids. *Can. J. Chem. Eng.*, **40**, 173–177.

Serrin, J. (1959). Mathematical principles of classical fluid mechanics. In S. Flügge and C. Truesdell (eds.), *Handbuch der Physik*, Vol. 8/3, Berlin: Springer-Verlag.

Sha, W. T., and J. C. Slattery (1980). Local volume-time average equations of motion for dispersed, turbulent, multiphase flows. Technical Report NUREG/CR-1491: ANL-80-51, Argonne National Laboratory, Argonne, IL.

Shertzer, C. R., and A. B. Metzner (1965). Measurement of normal stresses in viscoelastic materials at high shear rates. *In Proceedings of the Fourth International Congress on Rheology*, **2**, 603–618.

Sherwood, T. K., and R. L. Pigford (1952). *Adsorption and Extraction* (2nd ed.). New York: McGraw-Hill.

Shimizu, I., N. Okui, and T. Kikutani (1985). Simulation of dynamics and structure formation in high-speed melt spinning. In A. Ziabicki and H. Kawai (eds.), *High-Speed Fiber Spinning*, pp. 173–201. New York: John Wiley.

Siegel, R., E. M. Sparrow, and T. M. Hallman (1958). Steady laminar heat transfer in a circular tube with prescribed wall heat flux. *Appl. Sci. Res., Sect. A*, **7**, 386–392.

Sisko, A. W. (1958). The flow of lubricating greases. *Ind. Eng. Chem.*, **50**, 1789–1792.

Slattery, J. C. (1959). *Non-Newtonian Flow about a Sphere*. Ph. D. thesis, University of Wisconsin, Madison.

Slattery, J. C. (1961). Analysis of the cone-plate viscometer. *J. Colloid Sci.*, **16**, 431–437.

Slattery, J. C. (1962). The flow external to a non-Newtonian boundary layer. *Chem. Eng. Sci.*, **17**, 689–691.

Slattery, J. C. (1964). Unsteady relative extension of incompressible simple fluids. *Phys. Fluids*, **7**, 1913–1914.

Slattery, J. C. (1965). Scale-up for viscoelastic fluids. *AIChE J.*, **11**, 831–834.

Slattery, J. C. (1966). Spinning of a Noll simple fluid. *AIChE J.*, **12**, 456–460.

Slattery, J. C. (1967). Flow of viscoelastic fluids through porous media. *AIChE J.*, **13**, 1066–1071.

Slattery, J. C. (1968a). Dimensional considerations in viscoelastic flows. *AIChE J.*, **14**, 516–518; See also *ibid.*, **15**, 950 (1969).

Slattery, J. C. (1968b). Multiphase viscoelastic flow through porous media. *AIChE J.*, **14**, 50–56.

Slattery, J. C. (1969). Single-phase flow through porous media. *AIChE J.*, **15**, 866–872.

Slattery, J. C. (1970). Two-phase flow through porous media. *AIChE J.*, **16**, 345–352.

Slattery, J. C. (1974). A fundamental approach to mass transfer in porous media. In A. Dybbs (ed.), *Workshop on Heat and Mass Transfer in Porous Media*. Cleveland, OH: Case Western Reserve University, Department of Fluid, Thermal and Aerospace Sciences.

Slattery, J. C. (1981). *Momentum, Energy, and Mass Transfer in Continua* (2nd ed.). Malabar, FL Krieger; (1st ed.), New York: McGraw-Hill, 1972.

Slattery, J. C. (1990). *Interfacial Transport Phenomena*. New York: Springer-Verlag.

Slattery, J. C., and R. A. Gaggioli (1962). The macroscopic angular momentum balance. *Chem. Eng. Sci.*, **17**, 893–895.

Slattery, J. C., and C. Y. Lin (1978). Fick's first law compared with the Stefan–Maxwell equation in describing constant evaporating mixtures. *Chem. Eng. Commun.*, **2**, 245–247.

Slattery, J. C., and V. Mhetar (1996). Unsteady-state evaporation and the measurement of a binary diffusion coefficient. *Chem. Eng. Sci.*, **52**, 1511–1515.

Slattery, J. C., K. Y. Peng, A. M. Gadalla, and N. Gadalla (1995). Analysis of iron oxidation at high temperatures. *Ind. Eng. Chem. Res.*, **34**, 3405–3410.

Slattery, J. C., and R. L. Robinson (1996). Effects of induced convection upon the rate of crystallization. *Chem. Eng. Sci.*, **51**, 1357–1363.

Smeltzer, W. W. (1987). Diffusional growth of multiphase scales and subscales on binary alloys. *Mater. Sci. and Eng.*, **87**, 35–43.

Smith, G. F. (1965). On isotropic integrity bases. *Arch. Rational Mech. Anal.*, **18**, 282–292.

Smith, J. M., and H. C. Van Ness (1959). *Introduction to Chemical Engineering Thermodynamics* (2nd ed.). New York: McGraw-Hill.

Smith, J. M., and H. C. Van Ness (1987). *Introduction to Chemical Engineering Thermodynamics* (4th ed.). New York: McGraw-Hill.

Smith, V. G., W. A. Tiller, and J. W. Rutter (1955). A mathematical analysis of solute redistribution during solidification. *Can. J. Phys.*, **33**, 723–745.

Spearot, J. A., and A. B. Metzner (1972). Isothermal spinning of molten polyethylenes. *Trans. Soc. Rheology*, **16**, 495–518.

Spencer, A. J., and R. S. Rivlin (1959). Further results in the theory of matrix polynomials. *Arch. Rational Mech. Anal.*, **4**, 214–230.

Stefan, J. (1889). Über einige probleme der theorie der wärmeleitung. *S.-B. Wien-Akad. Mat. Natur.*, **98**, 473–484.

Steigman, J., J. Shockly, and F. C. Nix (1939). The self-diffusion of copper. *Phys. Rev.*, **56**, 13–21.

Stevenson, W. H. (1968). On a transformation of the equation describing steady diffusion through a stagnant gas. *AIChE J.*, **14**, 350–351.

Szymanski, P. (1932). Quelques solutions exactes des équations de l'hydrodynamique du fluide visqueux dans le cas d'un tube cylindrique. *J. Math. Pure Appl.*, **11**(9), 67–107.

Taylor, G. (1953). Dispersion of soluble matter in solvent flowing slowly through a tube. *Proc. R. Soc. London, Ser. A.*, **219**, 186–203.

Taylor, G. (1954a). Conditions under which dispersion of a solute in a stream of solvent can be used to measure molecular diffusion. *Proc. R. Soc. London, Ser. A.*, **225**, 473–477.

Taylor, G. (1954b). The dispersion of matter in turbulent flow through a pipe. *Proc. R. Soc. London, Ser. A.*, **223**, 446–468.

Taylor, G. I. (1932). The transport of vorticity and heat through fluids in turbulent motion. *Proc. R. Soc. London, Ser. A*, **135**, 685–702.

Taylor, T. D., and A. Acrivos (1964). On the deformation and drag of a falling viscous drop at low Reynolds number. *J. Fluid Mech.*, **18**, 466–476.

Tiller, W. A. (1991). *The Science of Crystallization: Macroscopic Phenomena and Defect Generation*. Cambridge: Cambridge University Press.

Tomotika, S. (1935). On the instability of a cylindrical thread of a viscous liquid surrounded by another viscous fluid. *Proc. R. Soc., London, Ser. A.*, **150**, 322–337.

Toor, H. L. (1964a). Solution of the linearized equations of multicomponent mass transfer. *AIChE J.*, **10**, 448–455.

Toor, H. L. (1964b). Solution of the linearized equations of multicomponent mass transfer, II. Matrix methods. *AIChE J.*, **10**, 460–465.

Toor, H. L., C. V. Seshadri, and K. R. Arnold (1965). Diffusion and mass transfer in multicomponent mixtures of ideal gases. *AIChE J.*, **11**, 746–747.

Touloukian, Y. S. (1966). *Recommended Values of the Thermophysical Properties of Eight Alloys, Major Constituents and Their Oxides*. Lafayette, IN: Thermophysical Properties Research Center, Purdue University.

Towle, W. L., and T. K. Sherwood (1939). Eddy diffusion. *Ind. Eng. Chem.*, **31**, 457–462.

Townsend, A. A. (1956). *The Structure of Turbulent Shear Flow*. London: Cambridge University Press.

Truesdell, C. (1952). The mechanical foundations of elasticity and fluid dynamics. *Arch. Rational Mech. Anal.*, **1**, 125–300; Corrected reprint, *International Science Review Series*, Vol. 8/1. New York: Gordon and Breach, 1966.

Truesdell, C. (1962). Mechanical basis of diffusion. *J. Chem. Phys.*, **37**, 2336–2344.

Truesdell, C. (1964). The natural time of a viscoelastic fluid: Its significance and measurement. *Phys. Fluids*, **7**, 1134–1142.

Truesdell, C. (1966a). *The Elements of Continuum Mechanics*. New York: Springer-Verlag.

Truesdell, C. (1966b). *Six Lectures on Modern Natural Philosophy*. New York: Springer-Verlag.

Truesdell, C. (1969). *Rational Thermodynamics*. New York: McGraw-Hill.

Truesdell, C. (1977). *A First Course in Rational Continuum Mechanics*. New York: Academic.

Truesdell, C., and W. Noll (1965). The non-linear field theories of mechanics. In S. Flügge, (ed.), *Handbuch der Physik*, Vol. 3/3. Berlin: Springer-Verlag.

Truesdell, C., and R. A. Toupin (1960). The classical field theories. In S. Flügge, (ed.), *Handbuch der Physik*, Vol. 3/1. Berlin: Springer-Verlag.

Turian, R. M., and R. B. Bird (1963). Viscous heating in the cone-and-plate viscometer, II. Newtonian fluids with temperature-dependent viscosity and thermal conductivity. *Chem. Eng. Sci.*, **18**, 689–696.

Van Dyke, M. (1982). *An Album of Fluid Motion*. Stanford, CA: The Parabolic Press.

Vaughn, M. W., and J. C. Slattery (1995). Effects of viscous normal stresses in thin draining films. *Ind. Eng. Chem. Res.*, **34**, 3185–3186.

Wagner, C. (1951). Diffusion and high temperature oxidation of metal. In *Atom Movement*, pp. 153–171. American Society for Metals.

Wagner, C. (1969). The distribution of cations in metal oxide and metal sulphide solid solutions formed during the oxidation of alloys. *Corrosion Science*, **9**, 91–109.

Wallis, G. B. (1969). *One-Dimensional Two-Phase Flow*. New York: McGraw-Hill.

Washburn, E. W. (1928). *International Critical Tables of Numerical Data, Physics, Chemistry and Technology*, Vol. 3. New York: McGraw-Hill.

Washburn, E. W. (1929). *International Critical Tables of Numerical Data, Physics, Chemistry and Technology*, Vol. 5. New York: McGraw-Hill.

Wasserman, M. L., and J. C. Slattery (1964). The surface profile of a rotating liquid. *Proc. Phys. Soc. London*, **84**, 795–802.

Weast, R. C. (1982). *CRC Handbook of Chemistry and Physics* (63rd ed.). Boca Raton, FL: CRC Press.

Weber, C. (1931). Zum zerfall eines flüssigkeitsstrahles. *Z. Angew. Math. Mech.*, **11**, 136–154.

Wehner, J. F., and R. H. Wilhelm (1956). Boundary conditions of flow reactor. *Chem. Eng. Sci.*, **6**, 89–93.

Werlé, H. (1974). Le tunnel hydrodynamique au service de la recherche Aérospatiale. Technical Report 156, ONERA, France.

Whitaker, S. (1966). The equations of motion in porous media. *Chem. Eng. Sci.*, **21**, 291–300.

Whitaker, S. (1967a). Diffusion and dispersion in porous media. *AIChE J.*, **13**, 420–427.

Whitaker, S. (1967b). Velocity profile in the Stefan diffusion tube. *Ind. Eng. Chem. Fundam.*, **6**, 476.

Whitaker, S. (1968). *Introduction to Fluid Mechanics*. Englewood Cliffs, NJ: Prentice-Hall.

Whitaker, S. (1969). Advances in theory of fluid motion in porous media. *Ind. Eng. Chem.*, **61**(12), 14–28.

Whitaker, S. (1973). The transport equations for multiphase systems. *Chem. Eng. Sci.*, **28**, 139.

White, J. L., and A. B. Metzner (1965). Constitutive equations for viscoelastic fluids with application to rapid external flows. *AIChE J.*, **11**, 324–330.

Wilcox, W. R. (1969). Validity of the stagnant film approximation for mass transfer in crystal growth and dissolution. *Mat. Res. Bull.*, **4**, 265–274.

Wilke, C. R. (1950). Diffusional properties of multicomponent gases. *Chem. Eng. Progr.*, **46**, 95–104.

Willhite, G. P., J. S. Dranoff, and J. M. Smith (1963). Heat transfer perpendicular to fluid flow in porous rocks. *Soc. Pet. Eng. J.*, **3**, 185–188.

Willhite, G. P., D. Kunii, and J. M. Smith (1962). Heat transfer in beds of fine particles (heat transfer perpendicular to flow). *AIChE J.*, **8**, 340–345.

Wilson, H. A. (1904). *Cambridge Philos. Soc.*, **12**, 406.

Winkler, C. A., and O. Maass (1933). Density discontinuities at the critical temperature. *Can. J. Phys.*, **9**, 613–629.

Wright, W. A. (1933). A method for a more complete examination of binary liquid mixtures. *J. Phys. Chem.*, **37**, 233–243.

Yau, J., and C. Tien (1963). Simultaneous development of velocity and temperature profiles for laminar flow of a non-Newtonian fluid in the entrance region of flat ducts. *Can. J. Chem. Eng.*, **41**, 139–145.

Zeichner, G. R. (1973). Spinnability of viscoelastic fluids. Master's thesis, Univ. of Delaware, as quoted by Fisher and Denn (1976).

Ziabicki, A. (1976). *Fundamentals of Fibre Formation*. New York: Wiley.

# Author/Editor Index

# Index